Non-Petroleum Automotive Transportation

Carl Arthur MacCarley

Non-Petroleum Automotive Transportation

Springer

Carl Arthur MacCarley
Department of Electrical Engineering
California Polytechnic State University
San Luis Obispo, CA, USA

ISBN 978-3-031-85027-1 ISBN 978-3-031-85028-8 (eBook)
https://doi.org/10.1007/978-3-031-85028-8

This Springer imprint is published by the registered company Springer Nature Switzerland AG
The registered company address is: Gewerbestrasse 11, 6330 Cham, Switzerland

Preface

Why This Book?

Automotive transportation is vital to our way of life. It has evolved through multiple technologies and energy sources. The story probably begins in the *Golden Age of Electric Vehicles* between 1890 and 1910, when horse power still ruled, and there were more electric vehicles operating on US and European roads than combustion engine vehicles of any type—internal combustion engines (ICEs) or steam engines (continuous combustion engines or CCEs). Deployed almost entirely in urban areas including New York, Baltimore, Paris, Stuttgart, and London, battery-electric vehicles were the fascination of affluent owners and taxi operators: quiet, less odiferous, and easy to operate requiring no hand crank starting. Electric power was sourced almost exclusively from coal-burning power plants, some dedicated solely for charging electric vehicles. Energy storage was almost exclusively via lead-acid batteries (with the exception of Edison nickel-iron cells), as they would continue to be for the next 110 years.

Competing vehicle technologies used many different liquid hydrocarbon fuels, usually purchased at local drug or general stores. These included minimally refined petroleum derivatives, vegetable oils, whale oil, benzine (as gasoline is still called in much of Europe), toluene, acetone, methanol (wood alcohol), and ethanol (grain alcohol). Standardized gasoline, as we know it, would not become common for another 20 years. But with the incredibly high energy density and low cost of petroleum fuels, introduction of the low-cost Ford Model T in 1908 [1], and subsequent technical innovations such as the Kettering ignition and electric starter, ICE vehicles eclipsed battery-electric vehicles almost completely by 1920. Short-lived electric car manufacturers were replaced by start-up ICE vehicle manufacturers. Paved roads replaced dirt streets and paths, and transcontinental highways and autobahns were built, all to enable the unprecedented privilege of high-speed personal transportation using gasoline or diesel fuel. Electric cars were relegated to special applications, such as indoor-operated or underground mining vehicles, although electric rail continued to grow in popularity, as it remains today since it does not require energy storage onboard the vehicle.

Regarding pollution … All horseless vehicular innovations, electric or ICE, were touted as major improvements over the emissions of horses that were responsible for serious urban sanitation problems. Coal was king, burned by locomotives and ships and for domestic heating, electricity generation, and town gas production for lighting and cooking. The connection between air pollution and health was soon recognized but was accepted as a cost of progress. This was the signature of the industrial revolution in England and Western Europe and later in the USA.

Over the next 100 years, periods of interest in alternatives to gasoline and diesel fuel were frequent, usually followed by public epiphanies about automotive air pollution or "gas shortages" due to the global politics of oil. Until the 1970s, world oil reserves were assumed to be limitless, an assumption that was challenged in 1956 following the publication of Standard Oil Co. engineer Edwin Hubbert's "Hubbert Curve" for peak oil, which predicted that a geopolitical crisis would occur not when reserves were depleted, but when oil extraction could not keep pace with demand. Shortages of World Wars I and II, the Suez Canal conflict of 1956, and the Israel-Egypt Six-Day War in 1967 brought home the importance of free-flowing oil to Western Europe, but had little effect on North America due to its large indigenous oil reserves. Following the oil shortages of 1973 and 1979, attitudes in the USA changed dramatically due to growing dependence on inexpensive oil imported from the Persian Gulf controlled by the Organization of the Petroleum Exporting Countries (OPEC) cartel. The myth that every local gas station would always have (inexpensive) fuel for your car was shattered for these brief periods, although, by the 1980s, quickly forgotten.

Starting in the mid-1970s worldwide, e.g., exemplified by the 1972 publication of "Limits of Growth" [2], the addiction to earth-sourced petroleum was recognized as a primary driver of not only world economics and politics, but also increasingly dire climate changes. By 2010, over one billion [3] petroleum-powered vehicles were operating daily worldwide. Petroleum and its energy and environmental sustainability impacts were now the topic of daily media headlines. An explosion of (mis)information with various degrees of credibility began in the 1990s via the ubiquitous Internet, a medium that gives voice to anyone, lacking the vetting process of formal publication that had previously assured some degree of credibility since the time of Gutenberg.

At the time of writing (2024), we have more motive energy options available than ever before in history. In addition to electric propulsion, a wide range of alternative combustion fuels, most notably alcohols, vegetable oils, and gaseous fuels, have been considered and supported at times by government and private funding. Unfortunately, also appearing were "amazing" engine accessories whose claims often violate the laws of thermodynamics. Greenwashing remains a powerful marketing tool with a gullible public and politicians. Within this noise were a few pracical "transition fuels" and related technologies that could potentially bridge the existing fuel distribution infrastructure with a tolerable degree of modification.

A renaissance of electric propulsion began in the late 1990s with popular major-manufacturer vehicles such as the GM EV-1 introduced in 1996 (lead-acid and nickel metal hydride or NiMh batteries), the Honda Insight Hybrid (NiCad and NiMh batteries) in 1998, and the Toyota Prius Hybrid in 2001 (NiMh batteries). But the technical development that enabled this revolution was lithium-based batteries having energy storage densities 3–5 times greater than previous lead-acid, NiCad, or NiMh batteries. After decades of battery research, this was the turning point. The EV revolution then gained momentum from attractive high-power electric vehicles: the AC Propulsion T-Zero circa 1999, the 2008 Tesla Roadster, and, most significantly, the 2012 Nissan Leaf. As of 2023, over a 100 EV and plug-hybrid models have been subsequently introduced by major and start-up manufacturers worldwide, and the number is growing.

The apparent consensus of car manufacturers and government entities worldwide is that the future is electric vehicles with energy stored in lithium batteries or hydrogen for fuel cells. Almost all major world manufacturers had pledged to sunset all ICE automobile production in favor of battery-electrics, mostly by 2030. To support these vehicles, large investments are done in subsidized public and captive (e.g., Tesla-only) charging infrastructure. Despite their significantly higher purchase costs and lingering concerns about range and charging time, the advantages of electric cars and trucks were well established, appealing to a public that wants to help alleviate climate change or improve energy independence, as long as it is not too inconvenient. The rush to electric automobiles has largely ignored the impacts of this transition on electric power generation, distribution, and storage. The immediate retort to such concerns is "solar and wind energy," despite the well-known misalignment between time of production and time of demand. Proposed energy storage solutions have been plagued by a lack of practical data and sheer magnitude of the problem—the huge amount of energy that must be stored and released daily either locally or in the grid to accommodate intermittent solar and wind generators. With the lowest number of operating nuclear power plants in the USA since the 1960s, and 10+ year approval/design/build time for new nuclear facilities, even that unpopular but carbon-free generation option is off the table until well after the transition to electric transportation energy has played out.

Meanwhile, the broad hopes assigned to vehicle electrification have almost completely narrowed commercial and government interest to battery-electric or fuel-cell electric vehicles, displacing renewable fuel alternatives, some that have comparatively lower overall carbon footprints than EVs charged from the current US and world mix of energy sources for grid electricity. Objective direct comparisons of the full range of motive energy options have been rare in both scientific literature and popular media.

It is my objective here to examine in one place a full range of options for ground transportation, and present the technologies, advantages, and limitations of each. These include battery and fuel-cell electric vehicles, alternative combustion fuels, and the status quo—gasoline and diesel fuels.

Energy and environmental issues have become one of the largest areas of public debate, driven by desperation to address environmental and energy sustainability challenges in an information ecology ranging from voluminous academic and government research papers to rants on public Internet forums. This is a sociotechnical area with countless experts and public voices, often reaching different conclusions. Technical and policy changes occur almost daily. It is not as simple as science vs. opinion. Differing answers to the same technical questions rely heavily on the assumptions and data sources. In the material herein, I will do my best to state my assumptions and sources so that my results can be checked by others.

I will attempt to identify greenwashing[1] regardless of the source or objective, since this phenomenon has become so widespread as to sabotage solutions that are actually scientifically sound, practical, and not encumbered by profit motives or career/political ambitions.

This is not a report commissioned by any entity, but it relies heavily on data and methods published by many others. Among the many insightful contributors to knowledge in this area, I wish to give particular recognition to Dr. David MacKay, author of "Sustainability Without the Hot Air" [4], a courageous and inspirational work by a pragmatic environmentalist, released to the public domain upon his passing in 2008. His work preceded the recent electric vehicle renaissance. This book extends his critical approach to an era in which petroleum is no longer synonymous with cars, but focuses on technologies that could provide the most effective alternatives. My treatment will follow Dr. MacKay's lead with a first-person perspective involving simplifications as appropriate for ease of understanding by readers knowledgeable in their own fields but not wellversed about automobiles or their energy sources.

San Luis Obispo, CA, USA — Carl Arthur MacCarley
2024

References

1. https://www.history.com/this-day-in-history/ford-motor-company-takes-its-first-order. Accessed 12 Dec 2022.
2. D.H. Meadows, D.L. Meadows, J. Randers, W. Behrens III, *Limits to Growth* (Club of Rome, 1972)
3. https://www.greencarreports.com/news/1093560_1-2-billion-vehicles-on-worlds-roads-now-2-billion-by-2035-report. Accessed 1 Mar 2023.
4. D.J.C. MacKay, Sustainable energy without the hot air, https://www.withouthotair.com/download.html. Accessed June 2022.

[1] Greenwashing: "The act or practice of making a product, policy, activity, etc. appear to be more environmentally friendly or less environmentally damaging than it really is," Merriam Webster dictionary of American English. Accessed 2023.

Contents

About the Author

Carl Arthur MacCarley has been researching alternative fuels, electric automobiles, and related applications since the late 1970s. He has been serving as a Professor of Electrical Engineering at California Polytechnic State University, San Luis Obispo, since 1987 (Department Chair 2000–2003 and 2010–2013, emeritus since 2022), and continues to teach as an adjunct at the Munich University of Applied Sciences. His academic specialization is automotive electronics and control systems. Between 1996 and 2008, he served as Principal Engineer for Loragen Corporation, a government and commercial automotive research contractor. He continues to represent auto manufacturers and suppliers as a legal expert witness.

Prof. MacCarley's academic and corporate research has been funded by the US Department of Energy, the California Department of Transportation, Daimler Benz, Fuji-Isuzu, and other automotive manufacturers. He is an associate editor for the SAE and a reviewer for IEEE publications. His extensive practical technical skills in these areas have helped to enhance his teaching and research. He has converted over 30 vehicles to alternative fuel or electric propulsion and developed innovations in electronic fuel injection and alternative drivetrains. He has been commuting daily using self-constructed methanol-powered and electric vehicles since 1996.

Introduction

1

Objective and Audience

The purpose of this book is to pull together in one place information about alternatives to the continued use of petroleum to fuel automobiles and, to a lesser degree, rail, air, and marine transportation. Much of the content in this book can be found dispersed across a large number of other sources that I have cited as references herein, and most are accessible online. My objectives are to discuss all practical automotive fuel/energy options in the same conversation and context, using common units and evaluation criteria, spanning the technical, environmental, economic, and social topics of common relevance to all. To compare and contrast dissimilar alternatives to gasoline and diesel fuel for vehicle propulsion. To introduce at a sufficient but not overwhelming level the most important engineering, environmental, and economic analysis tools and resources needed to see beyond the media hype, and misleading popular solutions that actually benefit only specific entities or political objectives rather than the greater good of humanity. To fill in the history of non-petroleum energy options to enable learning from the mistakes and successes of the past rather than repeatedly rediscovering "innovations" that could easily be found with an adequate literature search. To identify opportunities and areas of need for scientific and sociotechnical advancement in support of energy, environmental, economic, and political sustainability. And, when appropriate, to highlight the human drama, colorful personalities, and backstories behind major developments (Fig. 1.1).

The intended audience is college-level (or experience-equivalent) readers from all fields of study who are interested in alternative fuels and energy sources for automobiles but may have a limited prior background in automotive technology or the engineering of sustainable systems. The only necessary preparation is a basic high school or college chemistry course that includes chemical reactions and the properties of common elements. A basic understanding of energy concepts and terminology is also assumed, for example, the difference between power (e.g., kilowatts) and energy (kilowatt-hours). Everything else needed is covered or reviewed

C. A. MacCarley, *Non-Petroleum Automotive Transportation*,
https://doi.org/10.1007/978-3-031-85028-8_1

Fig. 1.1 Hypothetical image of a "green" car generated by 123rf.com in response to the author's prompt

herein. Detailed calculations are relegated to the appendices to allow verification but avoid diluting the main storyline with detailed math.

The scope of this book is very broad, and there are inevitably some topics not included. I have attempted to cover what I believe to be the most important and often most contentious issues, with facts both in support of and against each technical or policy argument. For the most controversial topics such as hydrogen, ethanol, methanol, biodiesel, e-fuels, carbon credits, EV marketing, and the changing world economic order, I venture an optional *Commentary* section at the end of the chapter in which I present my conclusions rather than leaving the reader to make sense of the mass of material in an overall context. I emphasize that the *Commentary* contains my opinions rather than the referenced facts that make up the chapter.

The Transportation Energy Quagmire

For over 120 years, petroleum has been an incredible gift from the earth to humankind that has radically advanced standards of living, but at the cost of irreversible damage to our humble planet. We have extracted and burned it at an unprecedented rate. Within a century (an infinitesimally short period in geological time), we have managed to extract over half[1] of the earth's accessible hydrocarbon resources, which were formed over a period of 300 million years by the decomposition of organic materials under pressure and temperature in the earth's crust. This gift is an energy-dense liquid fuel or gas for combustion that has had no serious rivals due to its very

[1] Based upon known or projected extractable hydrocarbon resources, from multiple references such as Hannah Ritchie and Pablo Rosado [1].

low cost, ease of extraction, minimal processing requirements, and convenient distribution. With the exception of nuclear fuel, gasoline and diesel fuel represent the most compact portable energy forms commonly available, allowing the storage of huge quantities of energy in small spaces and the transport of this energy rapidly and reasonably safely. It is hard to compete with these attributes. The automobiles and other transportation vehicles that utilize this resource have played a major role in defining the technical progress, cultures, economies, and conflicts of the world over the past century. This is especially true in newer countries such as the USA, with cities whose growth occurred largely during the *Age of Oil*. Petroleum has enabled countless products that we now take for granted, including plastics, agricultural chemicals, pharmaceuticals, construction materials, fabrics, advanced electronics, tools, and toys, and helped to set the standards of sustenance and status that now define our wants and needs. Petroleum could legitimately be called the lifeblood of the twentieth and early twenty-first centuries. Yet we continue to burn it as fuel because it is so easily extracted from the earth's crust and is such a compact and convenient form of portable energy.

We are now aware of the finiteness of this resource, as well as the environmental consequences of its use. While ideal combustion in air can theoretically generate only water vapor and carbon dioxide, actual combustion produces unintended toxins and carcinogens which remain a health concern. But for the entirety of the twentieth century, little if any attention was given to the climate-related impacts of the non-toxic exhaust products, water and carbon dioxide. It turns out that these non-toxic emissions may have a greater long-term effect on life on earth. These are *Greenhouse Gases*[2] which change the atmospheric chemistry of the planet, driving an irreversible trend toward increased average air temperatures and more frequent extreme weather events.

The damaging effects of greenhouse gases have driven climate change at a rate that is astounding considering the geological timescale required for the formation of petroleum.

Climate change is already well addressed in every form of media or publication. The scope of this book is limited to technical alternatives to petroleum for its dominant and most problematic application: internal combustion engines in road vehicles. The large majority are powered by gasoline or diesel fuel, both blends of various hydrocarbon liquids refined from crude oil. The automobile has played a major role in defining our current civilization, despite growing concern about both its limited supply and the climatic consequences of its continued use. Today (2023), with an estimated 3 billion petroleum-fueled automobiles operating on the planet and global infrastructure in place to support its continued use, we have become addicted to this short-lived gift/curse (Fig. 1.2).

[2] Greenhouse gases are any gases that in the atmosphere limit the reflection of infrared energy from the surface of the earth to black space, resulting in increased surface heat retention leading to global warming.

Fig. 1.2 Commuter traffic near San Francisco, California, USA, 2017. Photo from https://www.thedetroitbureau.com/wp-content/uploads/2017/08/Traffic-snarl.jpg. Public domain

Greenhouse Gases (GHGs) and Their Global Warming Potential (GWP)

The combustion, respiration, decay, or various other reactions involving combining hydrocarbons with atmospheric oxygen all produce carbon dioxide, the primary GHG of concern. Since transportation, especially automobiles, burns huge amounts of hydrocarbon fuel, it is a major contributor to the climate change dilemma.

But CO_2 is not the only GHG, or even the worst. Water vapor is a natural GHG, especially in the form of clouds in the troposphere. But it is a finite resource on the surface of the earth, and it enters and leaves the atmosphere continually in the form of precipitation or evaporation. The GHGs of concern related to climate change are *anthropogenic*, that is, the result of human activities, often involving the extraction and use of fossil fuels from the crust of the earth. Hydrocarbons extracted from the earth are not intrinsically harmful; it is their dominant position as combustion fuels that have made them so. But they are also the most important substances related to almost all of the major advances of civilization.

The anthropogenic GHGs that are the drivers of climate change are each assigned an empirical factor, the *Global Warming Potential* (GWP). GWP is a metric of the effectiveness of each at retaining solar heat in the troposphere that otherwise would be radiated into black space to maintain the thermal balance of the planet. GWP is normalized to the effect of carbon dioxide, which is given GWP = 1. Numbers greater than one produce warming effects greater than carbon dioxide by that multiplicative factor. Table 1.1 lists the six anthropogenic GHGs of greatest concern, two of which are strongly related to automobiles.

GHG Emissions Other than Carbon Dioxide [3]

- **Methane** (CH_4) is estimated to have a GWP of 28 over 100 years (EPA), but as much as GWP = 85 for a shorter horizon (period of effect assessment). CH_4 emitted today lasts in the atmosphere about a decade. But CH_4 also absorbs much

Table 1.1 Anthropogenic greenhouse gases and their global warming potentials (GWP)

Greenhouse gas	GWP	From cars?	Notes
Carbon dioxide (CO_2)	1	Yes	The reference for GWP
Methane (CH_4)	25–72	Yes	Depends on analysis period, 100–10 years
Nitrous oxide (N_2O)	298	No	Almost entirely from agriculture
Synthetic refrigerants (xFCs)	1430 for R134	No	Some xFCs have GWPs >10,000
Hydrogen	12–49	Not yet	Depends on lifetime, 100–10 years
Sulfur hexafluoride (SF6)	24,000	No	Grid power switching arc suppression

Data from US EPA [2]

more solar energy than CO_2. The net effect of the shorter lifetime and higher energy absorption is reflected in its GWP. CH_4 GWP also accounts for some indirect effects, such as the fact that CH_4 is a precursor to ozone, and ozone is itself a GHG.

- **Oxides of nitrogen (NO and NO_2)** from automobiles have the properties of greenhouse gases, but *are not* considered as such because of their high reactivity and short lifetime in the atmosphere. It is **nitrous oxide** N_2O, an otherwise non-reactive gas, that is a powerful GHG with a GWP 265–298 times that of CO_2 for a 100-year timescale. It originates almost entirely from monocrop agriculture.
- **Chlorofluorocarbons (CFCs)**, hydrofluorocarbons (HFCs), hydrochlorofluorocarbons (HCFCs), perfluorocarbons (PFCs), and other synthetic refrigerants and fire-suppression agents are high-GWP gases because for a given amount of mass, they trap substantially more heat compared with CO_2. The GWPs for these gases are in *the thousands or tens of thousands*. Fortunately, their anthropogenic emissions are small if the systems in which they are contained do not leak.
- **Sulfur Hexafluoride** (SF_6) is a synthetic gas invented by Brown Bovari ABB in the mid-1970s. It is the worst of all known greenhouse gases, with a GWP of 24,000. Fortunately, its only significant use is in small quantities for arc suppression in high-voltage electric power switching equipment. SF_6 was revolutionary for this application, allowing safer high-voltage DC transmission. If the switching equipment does not leak, SF_6 is not released into the atmosphere and is therefore not a global warming problem.
- **Hydrogen** occurs naturally in the atmosphere but only in ppm quantities. Because of this, it was only recently recognized as an indirect greenhouse gas. Its impact on global warming could increase dramatically if the widespread use of hydrogen for energy reaches full deployment. Its GWP differs greatly over different periods of observation. It has a much higher GWP during its first several weeks/months in the atmosphere, but over the most common 100 year GWP evaluation period, it disappears almost entirely due to its reactivity. (More about this in the Chap. 9.)

Notes

- N_2O (nitrous oxide) is a GHG, whereas NO and NO_2 (NO_x) are not (at least not directly). N_2O is not produced in significant quantities in HC combustion. It is non-toxic, whereas NO and NO_2 are potentially lethal regulated exhaust emissions.
- CH_4, NO_x, and HFCs are more potent GHGs than CO_2, but there is less of them being released into the atmosphere compared with CO_2, at least at this time. CH_4 emissions have been increasing at a greater pace than CO_2, for reasons to be discussed in the Chaps. 8, 9, and 10.
- HFC/CFCs are synthetic refrigeration fluids, not combustion emissions. They are relatively inert, but if released into the atmosphere, they are powerful contributors to global warming due to their extremely high GWPs.
- E85 (ethanol) and CNG (compressed natural gas) vehicles tested by the EPA for MY 2009 were both found to have *greater* NO_x and methane (respectively) emissions per mile than gasoline or diesel vehicles. For ethanol, this is related to the fuel combustion properties, while for CNG this is mostly due to inconsistent fuel delivery using gaseous fuel carburetors. These emissions are largely eliminated by modern catalytic converters as long as they are not excessive.
- Although the combustion of any hydrocarbon fuel except pure carbon (coal) produces water vapor, its role as an *anthropogenic* greenhouse gas is usually considered inconsequential. As a major component of the atmosphere, water vapor has by far the largest total global warming impact, because there is so much of it. But water is part of the natural water cycle with an approximately constant average atmospheric concentration when observed over multi-year time periods. It is actually critical to the natural greenhouse-related thermal stability of the planet. Anthropogenic water emissions are only a problem if emitted directly into the upper troposphere (say, as aircraft contrails), where they have a much more profound global warming effect.

What Is Meant by Sustainability?

In the common vernacular, "sustainability" has evolved to usually mean the continuation of human, animal, and plant life and the comfortable natural ecosystem of the planet for an indefinite period in the future while maintaining standards of living, cultures, and the natural environments that we have become accustomed to. The American English online reference *Dictionary.com* defines *sustainable* as "pertaining to a system that maintains its own viability by using techniques that allow for continual reuse." Prior to the 1990s, the terms sustainable and sustainability [4] referred exclusively to the ability to keep something going indefinitely, e.g., a sustainable chemical reaction or a sustainable business model. Since then, its usage is often tied to nebulous environmental and/or energy goals, with different meanings to different people. This is not just a problem of linguistics. *What we choose to sustain* guides our objectives and actions. To resolve the ambiguity, in this book I will qualify every mention of sustainability more specifically as:

- **Energy Sustainability**: Long-term viability of an energy source or solution. Can we rely on it indefinitely to meet our needs, especially in view of ever-increasing demand?
- **Environmental Sustainability**: Assuring future survivability and quality of life on earth. Will our current trajectory lead to the continued livability of the planet, or will we have to accept environmental consequences that threaten life as we know it on Earth?
- **Economic Sustainability**: Financial viability compared with other alternatives. Is this the best way to spend limited funds?
- **Political Sustainability**: The durability of a solution in the presence of changes in politics and power structures at all levels, local to geopolitical.

It is certainly possible to continue to subdivide these categories further. For example, energy sustainability could be separated into multiple-use sectors such as residential vs industrial vs transportation. But these four categories encapsulate the major competing factors that must be considered when finding truly "sustainable" solutions.

These four sustainability objectives are almost always in conflict. Energy sustainability almost always comes with environmental impacts that affect environmental sustainability. And ultimately, decisions to take action are controlled by economic and political sustainability considerations. For example, petroleum is certainly a sustainable energy and economic solution (at least for a few more decades) due to low cost and well-established production industries, but its extraction and use are in direct conflict with environmental sustainability objectives. Oil may support economic sustainability for countries rich with these resources, but it is rarely politically sustainable, as it has been at the root of many of the armed conflicts of the past century.

Renewable Energy

The adjective "renewable" has evolved in recent years to describe almost any energy option other than coal, oil or natural gas. But its formal definition is much more narrow. The original distinction was that the energy source had a *regenerative* or *closed* carbon cycle, with the natural environment returning all or most carbon dioxide that we add to the atmosphere from combustion back into fuel or an immediate precursor, resulting in a net-zero carbon balance. For example, if all the CO_2 released into the atmosphere by combustion, cultivation, and processing of a plant-based fuel is absorbed by photosynthesis within a few growing seasons, the fuel qualifies as *renewable* (Fig. 1.3).

But what if only some of the carbon is actually recycled in this way? Or what if the net carbon cycle is positive, but the net energy balance is negative, i.e., more energy is required to make the fuel/energy than its usable energy value? And what if we are talking about wind and solar energy, in which no carbon is generated so no carbon recycling is needed? A slightly more lenient definition has evolved,

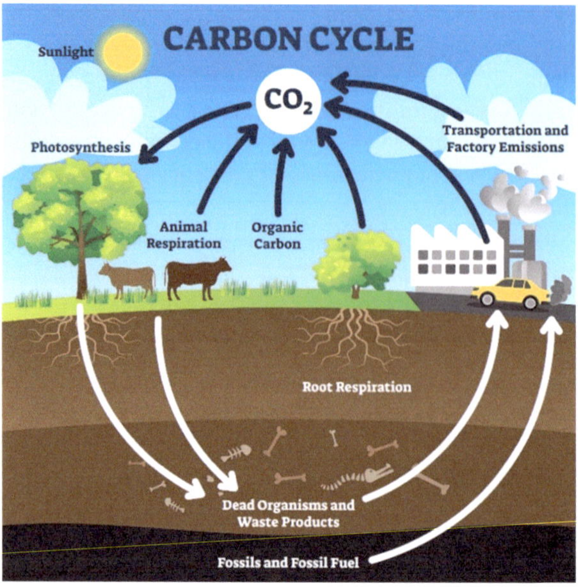

Fig. 1.3 Depiction of a closed carbon cycle in which anthropogenic CO_2 emissions are balanced by organic CO_2 uptake. Graphic from Andrea Bierema Systems Thinking and the Carbon Cycle, openbooks.lib.msu.edu. Unrestricted use

apparently allowing any type of energy that does not involve generation of CO_2 (e.g., solar, wind) or fuels derived from plants, vegetation, biomass, peat, or even municipal refuse incineration to be referred to as *renewable energy,* regardless of the carbon balance or energy utilization efficiency. For example, ethanol produced by fermentation of corn is burned in car engines releasing CO_2 into the atmosphere. If within a few crop cycles, carbon dioxide equal to the amount released from the combustion is assimilated by photosynthesis in the leaves of a future corn plant, it should logically qualify as a renewable fuel. But significant additional energy in the form of fossil fuels and electricity is required to convert the corn into ethanol (e.g., drying, fermentation, distillation), and the CO_2 contributions of these energy sources should be included. While a closed carbon cycle may still apply to just the plant/fuel relationship, the net carbon emissions are then much larger.

Another implied attribute of renewable fuels or energy is the ***time constant*** of the carbon cycle. Technically, *all* fossil fuels fall within the definition of a renewable fuel—they began as plant life that absorbed CO_2 from the atmosphere or water, decayed, were buried by geologic changes, and later were unearthed in solid, liquid, or gaseous form to be burned for energy, returning the carbon to the atmosphere. What's wrong with this definition? The amount of time required for the closed carbon cycle, or its *time constant*. In the case of fossil fuels, it's typically 100–300 million years. The modern terms *renewable fuels* or *renewable energy* carry with them the implication of a carbon cycle usually no longer than about ten years, or no carbon cycle at all.

As mentioned, completely carbon-free energy sources such as solar, wind, geothermal, tidal, or small-scale hydroelectric are also referred to as renewable, because the outcome is the same: zero net carbon or carbon neutral, *even if nothing is being renewed*. But other zero-carbon sources, such as nuclear and large-scale hydropower, are excluded from the renewable class because of the considerations unrelated to their net carbon emissions.

At this time, the following energy sources are generally referred to as *renewable* and have qualified for various forms of government assistance in the USA and the EU:

- Solar (PV or thermal).
- Wind.
- Geothermal.
- Small-scale local hydropower.
- Ethanol.
- Biodiesel.
- Bio-derived aviation fuels.
- Biomethane.
- Hydrogen made by a "green" or, arguably, a "blue" process.
- Municipal refuse incineration that generates electricity.
- Synfuels made from natural carbon reacted with "green" hydrogen.

While the following energy sources are referred to as *non-renewable*:

- Gasoline.
- Diesel fuel.
- Coal.
- Natural gas (methane from drilled wells).
- Nuclear energy.
- Combustion of some forms of biomass without natural origins, e.g., cultivated crop waste.
- Methanol made from natural gas or coal.
- Any synfuels that are derived from fossil fuels.

In the interest of maintaining consistency with its current meaning, I will use the term "renewable" to refer to energy sources that have been given this designation in popular media, even if they do not involve the carbon cycle or anything actually being renewed.

Alternative Fuels and Energy

Since the 1960s, the term "alternative fuels" or "alternative energy" has evolved, generally meaning any form of fuel other than gasoline or diesel, or any source of energy other than those that have been traditionally used to power vehicles, stationary engines, domestic or commercial heating or electric power demand, mechanized production, etc. This definition would include natural gas as a vehicle fuel, but not as a fuel for electric power generation, domestic heating, or production of other fuels or products such as plastics. LPG (propane and butane) would also be considered "alternative" if used as a vehicle fuel, but not as a fuel for home heating and cooking or other traditional uses. Solar, wind, geothermal, or any non-traditional forms of usable energy would also be considered "alternative". In sum, the "alternative" definition just distinguishes previously unused (or rarely used) fuels and energy sources from traditional ones. Note that there is no implication that an alternative fuel or energy source is environmentally sustainable, renewable, or even beneficial. In fact, the alternative vehicle fuels most popular in the 1970s and 1980s were distinctly not zero or even low carbon, and only some could be considered renewable within the modern definition.

In this text, I will use the term "alternative" only to identify fuels or energy sources that have been seriously considered as replacements for traditional fuels or energy sources in their most common applications.

Green Energy and Green Technology

"Green" is possibly the most abused descriptor in the field of alternative energy. Due to its ambiguity and the creativity of marketers, it has been attached to almost everything. "Green" energy is often used synonymously with renewable energy, although these are separate distinctions (Fig. 1.4).

"Green" has often become a red flag, misleading the public as to the environmental sustainability of a product, proposal, or plan. It pervades media articles, press conferences, government funding opportunities, and political agendas. It is the origin of the word "greenwashing" and often directly implies it.

Fig. 1.4 Green agriculture meme. From https://www.publicdomainpictures.net/pictures/300000/velka/erneuerbare-energie.jpg. Public domain

Sustainable Automotive Transportation

The field of energy technologies and their environmental impacts is extremely broad. Automotive (rubber-tired, on and off-road) transportation plays a dominant role in world energy use and accounts for the largest single contributor of anthropogenic greenhouse gas emissions. For 100 years, gasoline and diesel fuel have been impossible to compete with. However, the energy landscape for automobiles is changing faster than ever before in automotive history. Petroleum is not the only motive power option, either directly or indirectly. Only about 25 gaseous, liquid, and solid alternative fuels and their attendant technologies have demonstrated the potential to replace and often improve upon the energy, environmental, and financial sustainability of gasoline and diesel fuel for internal combustion engines. These include renewable fuels and portable fuels or energy carriers that can be derived from zero or low-carbon sources such as solar, wind, geothermal, or nuclear. In only two of these cases, solar PV and wind, are the initial energy from electricity. Generation of electricity most often involves the combustion of natural gas or coal, nuclear fission, or geothermal heat. Since electric vehicles are almost universally charged using utility grid energy, they are not innocent of the consequences of the combustion of fossil fuels.

It is a harsh reality that all practical forms of energy incur some environmental harm. This includes electric and hydrogen fuel cell electric vehicles despite their classification as "zero-emission vehicles" (ZEVs). An optimal solution for one transportation application may not be optimum for another. As reported in Bloomberg, in an August 2023 interview with BMW CEO Oliver Zipse, he took issue with the current European Union's industrial policy focused exclusively on battery electric or hydrogen fuel cell electric vehicles, railing against the trend "to rely on a single drive technology, rather than a breadth of options." [5] This is a popular opinion, but we ignore it at our collective peril.

By far the most popular alternatives at this time are vehicles powered by electricity stored in batteries or generated onboard by hydrogen fuel cells. After over a century of R&D, developments in electrochemistry have only, in the past two decades, resulted in practical batteries capable of storing a sufficient amount of electrical energy to compete with the hundreds of miles of range provided by even a small volume of gasoline or diesel fuel. Hydrogen, itself an excellent combustion fuel, is currently the only practical energy source for the electrochemical fuel cells of fuel cell electric vehicles (FCVs). Like battery electric vehicles, the emission-free conversion of hydrogen and air into electricity and water by the fuel cell qualifies them as zero-emission vehicles (ZEVs),[3] *as long as the emissions of the energy sources required to generate hydrogen are not considered.*

Not since the *Golden Age of Electric Vehicles*, circa 1890–1920, have battery electric vehicles been considered mainstream. And with the new norm comes new

[3] "Zero-Emissions" actually only refers to specific regulated hazardous gases from vehicle exhaust. This designation is currently restricted to electric and hydrogen vehicles, which is dubious since it does not consider the emissions related to the production of the fuel or the electricity source.

challenges: electric generation, distribution, and vehicle recharging infrastructure just as a start.

It is no surprise that energy and environmental issues have always been contentious areas of public debate, driven by desperation to concurrently address environmental and energy sustainability challenges balanced by economic sustainability issues such as jobs and taxpayer costs. Complicating the debate is that most solutions may take a generation or more to prove successful (or not). Alternative automotive energy sources, renewable or otherwise, have been the subject of research, demonstration, adulation, condemnation, hype, and sometimes outright fraud for the entirety of the twentieth and the twenty-first century years that followed. Information is ubiquitous, but knowledge is rare. Thousands, possibly millions of both credible and questionable reports, opinions, and speculations related to future combustion fuels and electric vehicles are spread over a vast number of media forums. In May 2023, a Google® search on keywords "renewable energy" yielded 435 million hits, "alternative fuels" yielded 152 million hits, and "electric vehicles" yielded 593 million hits. Like so many other techno-political topics, the challenge now for people seeking to become better informed is not a lack of information; it is sorting knowledge from noise.

Complicating these challenges further is an information ecology that ranges from jargon-heavy academic research papers to uninformed rants on public forums. It is not as simple as science vs pseudoscience when it is hard to tell them apart. Differing answers to the same technical questions depend on the assumptions and experimental conditions, often unstated, that underlie the analysis. Comparisons of options are between apples and oranges. This book is an attempt to assist the reader in sorting the fruit.

About Greenwashing

Greenwashing[4] is a colloquial term referring to the misleading use of "green" arguments to support the sale of a product or the adoption of a policy. In the USA, the nineteenth-century term for this type of pseudo-science deception was "snake oil" referring to worthless tonics and elixirs sold by slick traveling salesmen. Now the entire Internet is a traveling medicine show. Financial gain or career ambitions very often influence public opinion leading to counterproductive government policies and expenditures of taxpayer funds. Even legitimate "green marketing" and "green causes" frequently devolve into little more than greenwashing that polarizes rational public discussion and obscures better solutions.

[4] Merriam Webster dictionary of American English. Greenwashing: "The act or practice of making a product, policy, activity, etc. appear to be more environmentally friendly or less environmentally damaging than it really is." Accessed 30 August 2023.

The Best We Can Do

The world is late in realizing the extent of the thermal problem created by anthropogenic greenhouse gas emissions. And worldwide, political and financial forces continue to diminish or deny this situation, despite radical climate change well underway.

There are no Holy Grail solutions.[5] When considering environmental and energy technology options and policies, there are always tradeoffs; all solutions have some weaknesses and limitations, and the complexity makes these difficult to resolve, especially in this era of "alternative facts."[6] It takes work, objectivity, and logical skepticism. The situation has now become urgent. We no longer have time or resources to waste on mistakes.

Disclaimer and Acknowledgments

This is not a report commissioned by any institution or sponsor. It relies on the published reports of many others and is at risk of error if either the information sources or my use or transcription of them is in error. I am ultimately responsible for all content, which may or may not reflect the positions of the universities I am affiliated with, the California Polytechnic State University, San Luis Obispo, California, and the Munich University of Applied Science in Munich, Germany.

The inspiration for this book is the late Sir Dr. David MacKay, author of "Sustainability Without the Hot Air" [6] in 2008. Dr. McKay was a passionate environmental physicist who had the courage to call out the serious lack of rigor and herd mentality of popular energy and environmental assumptions and policies. His work in 2008 preceded the electric vehicle renaissance, so a secondary objective of this book is to extend his analytical but readable approach into an era in which petroleum is no longer the only or best option for transportation.

There are many people and institutions to thank for their support and contributed knowledge. Among these are William Van Vorst (UCLA, deceased), Paul Greiling (emeritus, UCLA), Colin Ferguson (Professor, Purdue Univ.), Nicholis Zart (automotive journalist), Greg Dolan, CEO, Methanol Institute (methanol and DME fuel support for 25+ years), Peter Livingston (Professor, Cal Poly BRAE Dept), Joseph Finegold (UCLA Hydrogen Car project, 1972–1978), Leroy Lacey (Methanol Racing teammate, 1981) and many former students in the USA and Germany that are now employed in automotive industry, energy and environmental research, or at government and non-profit agencies in positions that can make a difference.

[5] "Holy Grail" is a Christian reference to an ultimate quest or perfect solution. Derived from Arthurian legend, first appearing in written form in the twelfth century by French poet Chretien de Troyes in "Perceval, the Story of the Grail."

[6] "Alternative facts" is a term created and made popular by Kelly Anne Conway, Counselor to President Donald Trump, defending his false assertion about record-setting attendance at his inauguration in 2017.

References

1. H. Ritchie, P. Rosado, *Fossil Fuels* (Our World in Data, 2017) https://ourworldindata.org/fossil-fuels. Accessed 10 June 2023
2. US Environmental Protection Agency (EPA), Global warming potential. (2015), http://www.epa.gov/oms/climate/documents/420f11041.pdf Accessed 15 March 2018
3. US EPA, Understanding global warming potentials. (2024), https://www.epa.gov/ghgemissions/understanding-global-warming-potentials. Accessed 1 May 2024
4. Merriam Webster Dictionary of American English, https://www.merriam-webster.com/dictionary/sustainability. Accessed 6 June 2023
5. Electric cars are winning out because of consumers, not politicians. (Bloomberg Online, 2023), https://www.bloomberg.com/news/newsletters/2023-09-06/electric-cars-are-winning-out-because-of-buyers-not-politicians?cmpid=BBD090623_hyperdrive&utm_medium=email&utm_source=newsletter&utm_term=230906&utm_campaign=hyperdrive. Accessed 6 Sept 2023
6. D.J.C. MacKay, *Sustainable Energy Without the Hot Air* (Cambridge Press) open-access online at https://www.withouthotair.com/download.html Accessed 30 June 2022

Transportation Energy Realities

2

This chapter summarizes the environmental impacts associated with transportation, especially automobiles and trucks. There are many excellent resources cited as references to provide additional detail on any specific subtopic. To start, we look at the tradeoffs associated with the development and use of any form of energy, then narrow the discussion to the underlying thermodynamic principles, energy quality, energy carriers vs. sources, mobile source emissions, and attempts at regulatory solutions in the USA (Fig. 2.1).

All Energy Sources Have Costs, But Some More than Others

Since the discovery of fire, human civilization has relied on external sources of energy, with demand increasing exponentially since the first industrial revolution in England, 1760–1840.

All sources of energy have costs and consequences, although some more than others. The relative costs and consequences of each are perennial topics of intense debate. At one extreme might be coal, with the lowest cost but the worst direct environmental impacts in production and use. At the other extreme might be solar photovoltaic energy, which has comparatively tiny impacts attributed to the production of solar panels and its land use issues. Petroleum lies somewhere between these extremes but is responsible for the most significant total impact simply because of the sheer volume of its use as a transportation fuel.

Anthropogenic Environmental Impacts

- Toxic air pollutants: HC (hydrocarbons), CO (carbon monoxide), NO_x (nitrogen oxide and dioxide), SO_2 (sulfur dioxide), O_3 (ozone), aldehydes such as formaldehyde, lead, particulates, over 100 identified carcinogens, and radiological materials.

C. A. MacCarley, *Non-Petroleum Automotive Transportation*,
https://doi.org/10.1007/978-3-031-85028-8_2

Fig. 2.1 High-voltage transmission tower. https://www.publicdomainpictures.net/en/view-image.php?image=314422&picture=electricity-energy-co2-sunset. Public domain

- Greenhouse gases: CO_2 (carbon dioxide), CH_4 (methane), N_2O (nitrous oxide), CFCs (chlorofluorocarbons), SF_6 (sulfur hexafluoride), and indirectly H_2 (hydrogen).
- Water pollution: Direct toxins such as lead, dioxins, heavy metals, pesticides, herbicides, biotoxins, loss of dissolved oxygen to algae blooms, biological waste, and chemical process wastes.
- Indirect greenhouse promoters: Disruption of natural CO_2 absorbers, destruction of trees by land clearing, and loss of non-cultivated plants and aquatic plants.
- Wildlife: Destruction of species due to loss of habitat or hunting, pesticides accumulated in the food chain, and illegal capture or killing of endangered species.
- Human health: Direct chemical toxicity—airborne, waterborne, or contact. Carcinogenic particulates leading to respiratory diseases or cancer.

Looking at world energy use from 1850 to 2000 in the United Nations data of Fig. 2.2, there is no indication that this trend will change, barring some natural or human-made cataclysmic event. Radical increases in energy have been associated with industrial revolutions in the UK, Continental Europe, the USA, and most recently in China and Southeast Asia. It is well-established that regardless of country, living standards (as measured by metrics of economic well-being) are closely correlated with energy use.

Figure 2.3 reveals the correlation between individual prosperity measured as per capita GDP and per capita energy consumption in all sectors.

As shown in the Sankey Diagram [1] of Fig. 2.4 from the US Energy Information Agency (EIA) for the USA in 2021, 36% of the total energy used across all sectors comes from petroleum. The largest consumer of oil (90%) is transportation, dominated by automobiles and trucks.

Since the late 1800s, most of the usable energy sources for human progress have involved the combustion of naturally occurring compounds of carbon and hydrogen. Combustion of wood represented an evolutionary milestone by making hunted

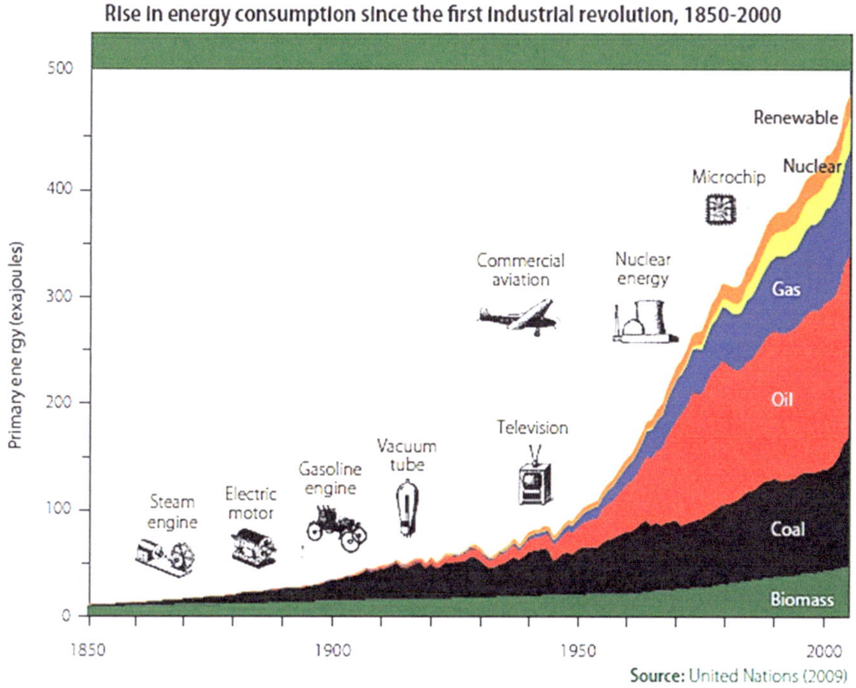

Fig. 2.2 World energy consumption, 1850–2000, United Nations infographic. Public Domain

game edible and providing warmth and light. In the nineteenth century, whale oil fueled lamps for illumination, horses and oxen were fueled by biomass, and coal has fueled almost everything for over a millennia. Since the early twentieth century, natural gas has fueled electric power generation, home heating, and cooking. Most relevant to this discussion is petroleum from the earth, the source of portable energy that made modern transportation possible, at least until now.

Non-combustion energy sources include solar photovoltaic and thermal collection, wind energy due to atmospheric pressure differences, hydropower provided by the earth's solar evaporation/condensation cycle, geothermal provided by the thermal mass of the earth, tidal energy attributed to diurnal ocean levels and monthly changes in the distance of the moon to the earth, and nuclear fission energy fueled by isotopes of uranium or plutonium.

Regardless of the energy source, all involve environmental impacts, usually air or water pollution, harm to natural habitat, or ultimate climate change. Some are worse than others, and often in ways that are difficult to compare (e.g., measurable air pollution from fossil fuels vs risk of accident consequences with nuclear). These are not the only impacts. Economists remind us that large divisions of wealth and political power derive from energy or the lack thereof. The national interests of world superpowers always include and are often dominated by energy resources. Geopolitical impacts have and will continue to lead to wars. The terms *Environmental*

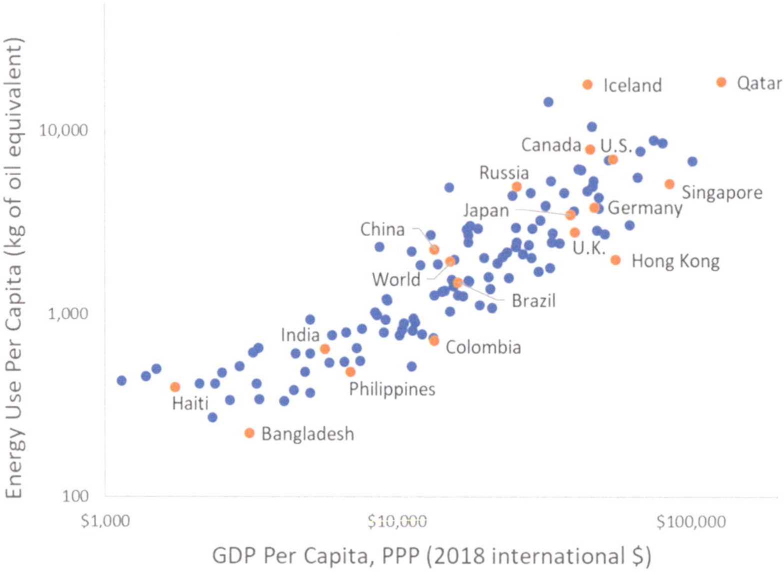

Fig. 2.3 Global energy use per capital vs. GDP per capita. From IEA (International Energy Agency) 2018. "Key World Energy Statistics." Accessed June 2023. Public domain

Justice or *Climate Justice* have evolved to refer to the widely different levels of harmful environmental effects on various groups of people or habitats, segregated by economic class or political situation. Worldwide CO_2 emissions from the combustion of fossil fuels continue to increase (Fig. 2.5).

Focusing on just the USA, the 2022 data of Fig. 2.6 from the US Environmental Protection Agency (EPA) shows anthropogenic greenhouse gas emissions by energy use sector [2]. At 28%, the largest emitting sector is transportation.

California, the "greenest" state in the USA, is even more profoundly impacted by transportation, specifically by the automobile. The largest segments of GHG emissions in the USA (28%) and California (39%) are from transportation, the majority of which is from passenger automobiles. As energy sources for transportation shift from fossil fuels to electric power, the *total* energy use may not change significantly. However, the electrification of transportation will transfer the energy demand to the electric power grid. USA-wide, transportation plus electric generation contribute 53% of total GHG emissions. In California, the total is 63%. As the transportation slice shrinks due to reduced combustion fuels, the electricity use sectors will grow. The takeaway is that for the actual CO_2e emissions attributed to transportation to be reduced, electric generation from other fuels or energy sources must produce much lower GHG emissions over their entire production and use cycles compared with the current fossil fuel situation. This requires a different way of thinking about the challenges than the previous monotonic focus on automotive tailpipe emissions. And the emissions of most significant interest are no longer limited to local or regional impacts—GHGs generated *anywhere* affect the entire planet.

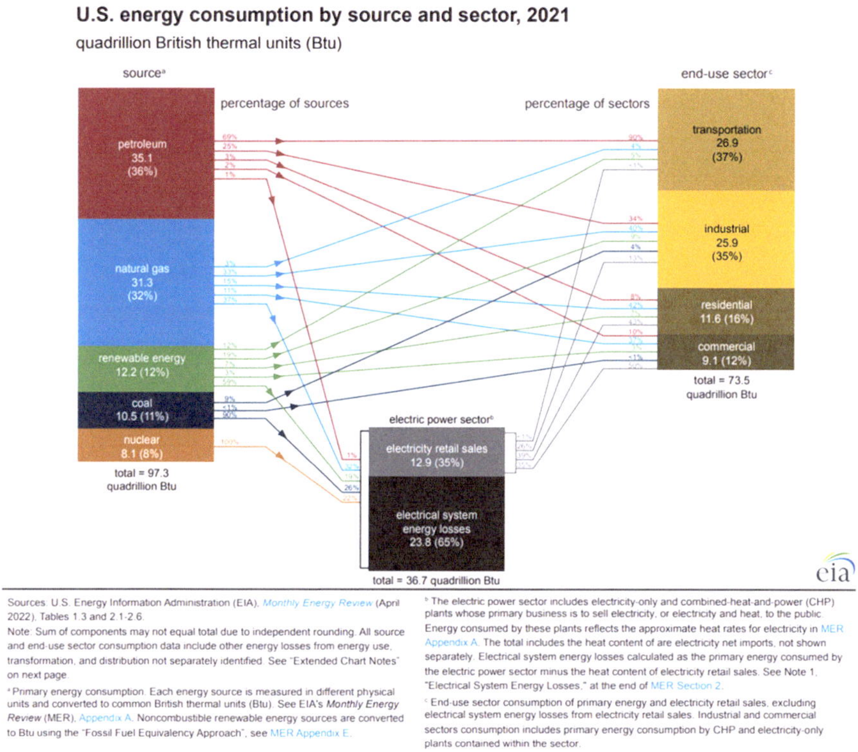

Fig. 2.4 Sankey diagram for energy use by sector in the USA, for 2021. US EIA 2022, Public Domain

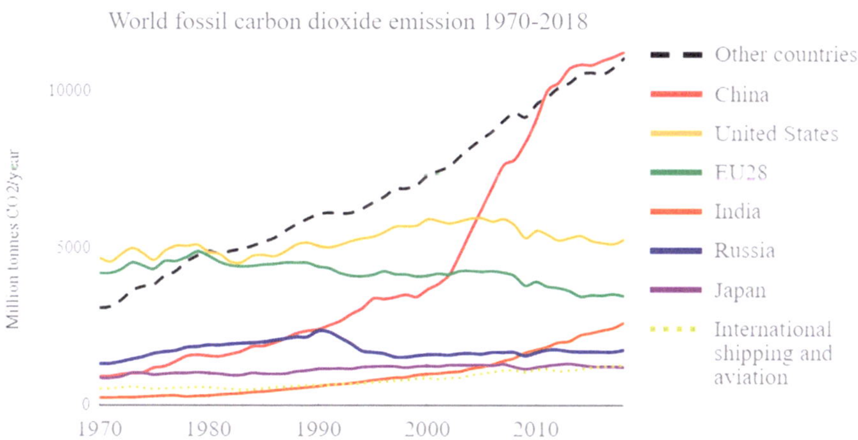

Fig. 2.5 Carbon dioxide emissions 1970–2018, parametric with country. Graphic from International Energy Commission, https://www.iea.org/data-and-statistics/data-product/greenhouse-gas-emissions-from-energy. Public domain

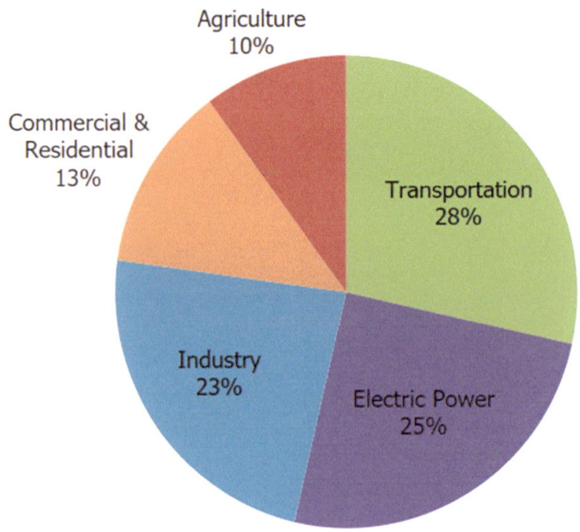

Fig. 2.6 US greenhouse gas emissions by sector, 2022. US EIA. Public Domain

The Carbon Footprint

Environmental harm is subject to multiple interpretations and metrics that are often ill-defined. However, this ambiguity should not deter efforts to assess and quantify consequences for life on earth, especially in times of world economic and political instability when these concerns are often ignored. The *carbon footprint* is a relatively new term used to quantify individual or institutional impacts due to their GHG emissions. Approximately defined, an entity's carbon footprint, either personal or institutional, is an estimate of how much carbon, most in the form of carbon dioxide or methane, the entity is directly or indirectly responsible for releasing into the atmosphere by their activities.

Shown in Fig. 2.7 is a sample of an online calculator provided by the US EPA to estimate the amount of carbon emitted by a typical household. The calculator takes voluntary answers to energy-related questions and generates the graphic below of your household CO_2 compared with the US average. Green represents home energy, including air heating and cooling, water heating, cooking, lighting, and electric appliances and tools.

What may be surprising about this bar graph is that on a per-household basis, the most significant component of the US average household carbon footprint is from transportation, which is predominantly automotive transportation, as shown in the EIA 2021 bar graph of Fig. 2.8, which breaks down transportation into its specific modes. From this data, energy use by light-duty vehicles, freight and commercial light trucks, and buses accounted for 10.66 Mbbl/d[1] or 84% of the entire transportation sector total (12.71 Mbbl/d). And since only a tiny percentage of cars

[1] Mbbl/d means the energy equivalent of a million barrels of oil per day.

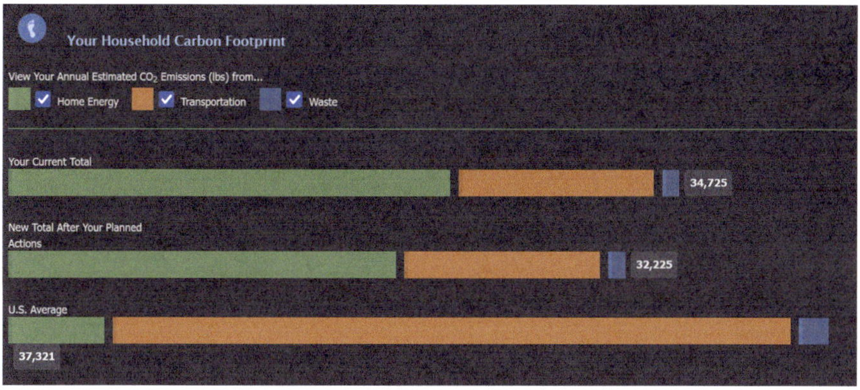

Fig. 2.7 US Environmental Protection Agency. Sample of annual energy use by a US household, https://www3.epa.gov/carbon-footprint-calculator/. Public domain

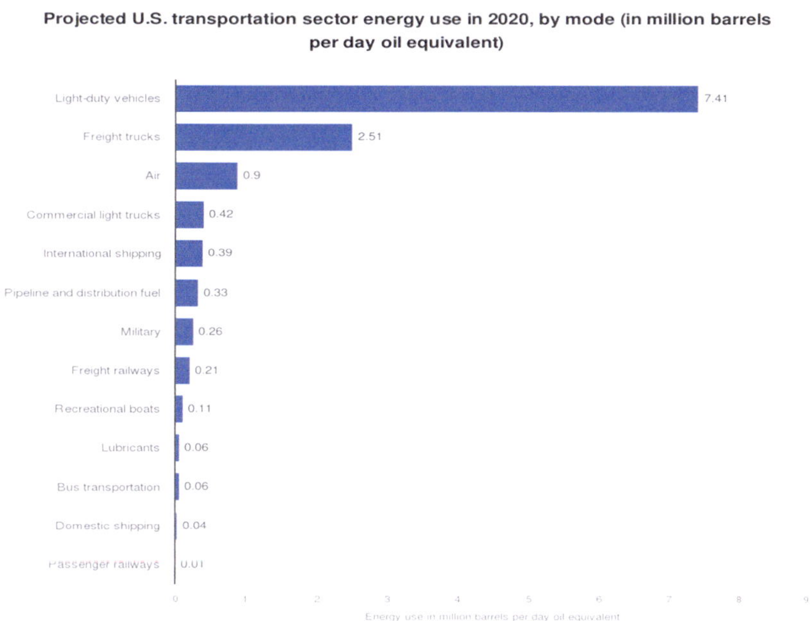

Fig. 2.8 Breakdown of US transportation energy users by mode. Data from US Energy Information Agency and US FHWA, 2021; graphic created by Statistica, university license

are currently electric, this energy is almost entirely in the form of gasoline or diesel fuel.

I can only cover here some of the major milestones in international and US efforts to remediate greenhouse gases related to the transportation use of petroleum fuels.

In 1992, seven (initially) of the industrialized nation members of the United Nations formed the United Nations Framework Convention on Climate Change (UNFCCC). In 2015, 196 member nations signed the *Paris Agreement,* a treaty between participating countries intended to limit global warming to 1.5 °C Celsius above pre-industrial levels. Note that in late 2023, this limit had already been exceeded. This agreement specified the goals and actions each country committed to help achieve this goal, referred to as Nationally Determined Contributions (NDCs). These are reviewed and updated every five years and are self-reported. As of 2024, 198 countries have signed the accord, and 189 have at least declared themselves to be parties to the agreement. NDCs are voluntary, specific to each country, and, in most cases, overly ambitious.

Figure 2.9 is a plot generated by Statistica [3] showing relative progress in 2024 (change from the time of signing accord in 2015) of G20 Forum countries toward meeting their NDCs. Each country sets its own climate goals which vary from 1.5 °C in the UK, to >4.0 °C in Saudi Arabia, India, Indonesia, and Kenya [4]. The exceptional progress shown by some countries in Fig. 2.9 is relative to their chosen temperature goal, and the time of signing the accord (most in 2015) when they may have had comparatively poor environmental records. Four categories of positive action are considered in this data presentation: Per capita GHG emissions, renewable energy deployment, per capita energy use, and a mostly subjective assessment of improvements in climate change policy. In the years leading up to the COP28 (Conference of the Parties) meeting in November/December 2023, there has been much greater international concurrence about the dire state of climate change and more aggressive actions [5]. Progress has been impeded by political turmoil, especially the Russian cutoff of natural gas to eastern Europe as part of the invasion of Ukraine which forced an unexpected reversion to previously mothballed coal fired power plants throughout Europe. According to the Guardian (UK) [6], none of the G20 nations have fully met their intended progress toward their commitments. Most are not even close. Outside of the media coverage of world meetings, few are convinced that the revised 2.0% limit is achievable with the current policy trajectories, despite their official recognition that 1.5% was the "point of no return."

In the USA, the *Energy Policy Act of 2005* created the Renewable Fuel Standard (RFS) that amended the Clean Air Act revision of 1990. It included a schedule for *required* biofuel adoption starting with 15.14 billion liters (fuel unspecified) in 2006 and rose to 28.39 billion liters by 2012. The Act was primarily concerned with energy independence in the wake of the Sept 11, 2001 terrorist attacks in the USA. But for a fuel to qualify as a renewable fuel under the RFS, the EPA must determine that the fuel can achieve a reduction in greenhouse gas (GHG) emissions as compared to a 2005 petroleum baseline. The Act emphasized "drop-in fuels"[2] that could directly replace gasoline or diesel fuel without blending. The term "Fuel Pathways" was inaugurated, referring to the complete "well-to-wheels" process, and the entire administrative journey from demonstration to legal requirement.

Four renewable fuel types were specified [7]:

[2]The term "Drop-in" implied that the fuel is nearly interchangeable with gasoline or diesel fuel, and could be dispensed from an existing gas or diesel fuel pump.

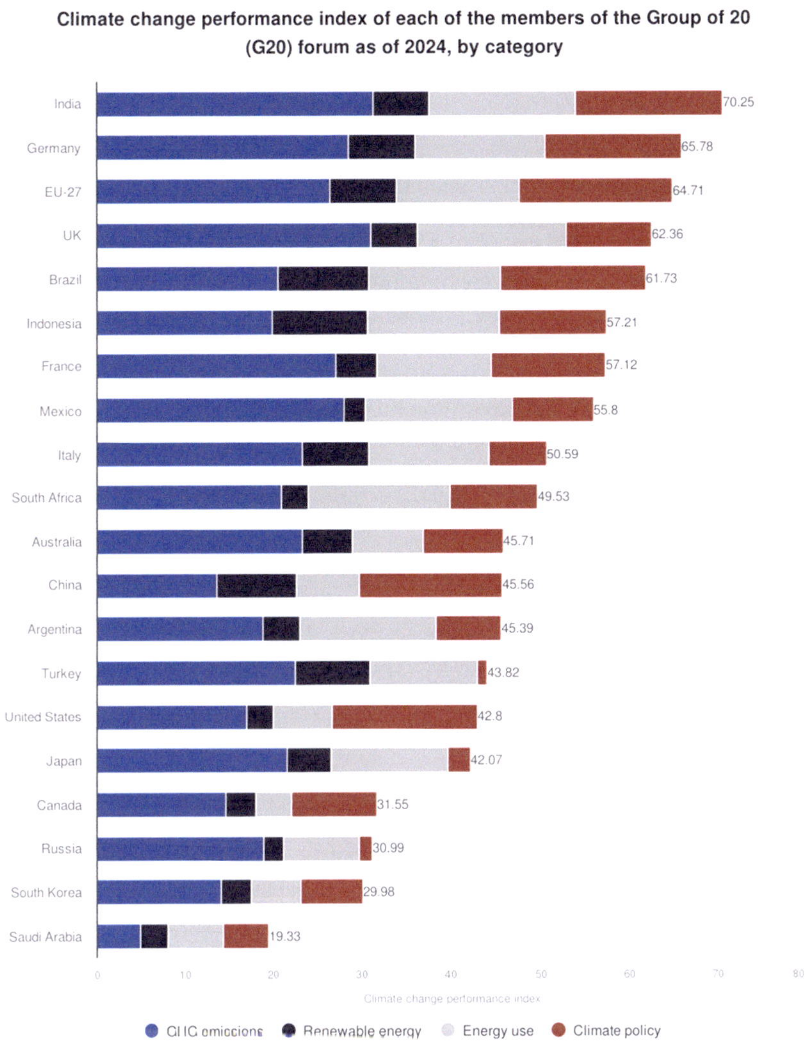

Fig. 2.9 Progress toward Climate Change Goals by G20 signers of the Paris Accord. Source New Climate Institute; Graphic from Statistica "Climate change performance index of each of the members of the Group of 20 (G20) forum as of 2024, by category." https://www-statista-com.calpoly.idm.oclc.org/statistics/1406610/climate-change-performance-index-g20-by-country/. University License

- Biomass-based diesel
- Cellulosic biofuel
- Advanced biofuel
- Total renewable fuel

Practically, these classes restricted "renewable fuels" to ethanol for spark ignition engines or biodiesel for compression ignition engines.

The *Energy Independence and Security Act of 2007* later replaced the RFS with RFS2, a Revised Renewable Fuel Standard. The new Act and the RFS2 included a requirement that the biofuel provides a minimum reduction in CO_2 emissions of 21%, including emissions from the production of the fuel, assuming that ethanol and biodiesel were the only fuels that could meet this requirement. The RFS2 included an expansion of mandated biofuel adoption starting with 34.07 billion liters in 2008, increasing to 136.26 billion liters by 2022 [8].

Energy Carriers vs. Energy Sources

We often refer to electricity as an energy source, e.g., "my car runs on electricity." With the possible exception of Ben Franklin's 1752 alleged attempt to prove the electrical nature of lightning by flying a kite in a storm, electricity is not itself an energy source, merely an ***energy carrier***. Other energy carriers include hydrogen, synthetic fuels such as methanol, kinetic energy stored in flywheels, and chemical reactants. *Sources* of energy include fossil fuels, nuclear fuel, solar, wind, geothermal and hydroelectric.

An energy carrier is a medium or mechanism by which energy generated using some other source can be transferred or stored. It is created from other (original) energy sources, be those fossil fuels, solar radiation, wind, tides, geothermal heat, or nuclear fission. This distinction is critical when evaluating transportation energy options, since it is the original source of the energy that must be used as the starting point in any fair cycle efficiency or environmental impact analyses. This can sometimes be a gray area, so for our purposes the distinction can be stated as a question: Did we collect the energy (e.g., solar irradiation or wind), or did we make it from something that we collected, drilled, or mined?

When analyzing the efficiency of any energy-related process, it is important to relate all comparisons back to some common energy form, and whenever possible, start with the energy source, not the energy carrier which transports the source energy. For example, an electric vehicle is a zero-emission vehicle only if the entirety of the electricity used to charge its battery comes directly from a zero-emitting source such as home solar PV or wind. But the overwhelming majority of electric vehicles draw energy from the utility grid, which gets it energy from many different sources, some renewable but most not. The majority are combustion processes that produce CO_2. Therefore, the actual carbon emissions of a vehicle powered by electricity, hydrogen, or e-fuels must be related back to the energy sources that actually feed the grid, and their composite carbon dioxide emissions

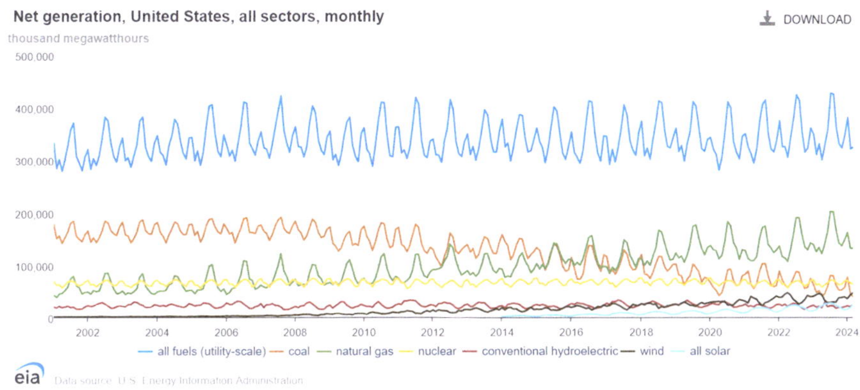

Fig. 2.10 US EIA, https://www3.eia.gov/electricity/data/browser/. Public Domain

normalized to each unit of energy, kg CO_2/kWh. What are the actual energy sources used to generate electric power? The mix of sources varies by country and state, with the annual historical distribution for the USA listed in data source [9] (Fig. 2.10).

The majority (60+% in 2022) of US grid electricity still comes from natural gas or coal. Overall, the mix of all of the actual energy sources including these fuels along with solar, wind, nuclear, and hydroelectric *produces approximately 400g CO_2 per kWh in the USA (2022 EPRI data)* and worse in most other countries. This is the operating carbon footprint of electric vehicles. The arithmetic may actually be worse than the operating energy use figures suggest, since they do not take into account the inefficiencies associated with the generation, transmission, and distribution of electricity, and, more importantly, the carbon emissions from the extraction of minerals, manufacturing, and disposal or recycling of the EV's batteries. Similar tradeoffs exist in some form or another for any motive energy source or carrier, including all of the other viable candidates to be discussed in later chapters.

But it is a start, even if not a perfect solution. The EV transition anticipates the day when electricity, both utility and distributed, is entirely or at least majority renewable; when *zero emissions* actually means *zero*, or at least significantly lower than the present EV carbon emissions from electric generation.

The Tyranny of Thermodynamics

Critical to understanding energy and environmental options are the first and second laws of the four laws of thermodynamics. Rather than involving physical but obscure measures such as *entropy*, here is the simplistic way that I first learned these concepts. This helps to see why, for the numerous schemes proffered as energy or environmentally sustainable, the more energy conversion steps required, the lower the efficiency (Fig. 2.11).

Fig. 2.11 Combustion: the energy conversion process behind most of human progress. https://pixabay.com/photos/flames-fire-burn-hot-shining-heat-2041271/ Free for use

First Law of Thermodynamics: No Free Lunch

You cannot make energy from nothing. There is no such thing as a perpetual motion machine. You need energy to get stuff done. It must come from some source: chemical reactants, electric charge, thermal (heat), EM radiation including light energy, kinetic (energy of motion), potential (including the earlier mentioned sources as well as the energy required to elevate a mass in gravity), nuclear fission or fusion, etc.

Second Law of Thermodynamics: Can't Even Get What You Pay For

Every time the form of energy is changed via some process, the amount of energy coming out of the process *is always less* than the amount of energy that went into the process. You always lose. Where does the missing energy go? Usually to unwanted heat, although mechanical, acoustic, radiation or liquid–gas phase changes can also account for the energy loss. *Maximum energy efficiency* is achieved by improving the efficiency of each energy conversion process and by *minimizing the number of energy conversion steps*, a reality that is very often ignored in simplistic analyses, commercial promotions, and even government agency reports that advocate particular energy or environmental innovations.

This will be discussed in greater detail later, but here is a simple example to illustrate the effect of multiple conversions on overall process efficiency: a simple comparison of a battery electric vehicle with an ICE vehicle. The original energy source for both is natural gas:

Burning natural gas in a utility power plant to generate electric energy has a *thermal efficiency* (kWh of electricity output divided by kWh of fuel energy input) that varies from 40% for a new conventional steam turbine to 60% for the best combined cycle (gas turbine plus steam turbine) power plants. This means that 40–60% (50% avg) of the fuel energy is lost before the electricity is delivered to the loads it is intended to power. There are also other energy losses: electricity transmission and distribution (5.9% US avg), battery charging (15% for an average mix of public Level 2 and Level 3 chargers), lithium battery round-trip efficiency losses (RTE, 15%), power electronics heat (5%), and the electric motor and drivetrain losses (10%). Combined, the energy conversion efficiency due to these combined losses is

$$(1-0.50)(1-0.06)(1-0.15)(1-0.15)(1-0.05)(1-0.10)=29\% \text{ efficiency or } 71\% \text{ energy lost}$$

For comparison, if an otherwise-identical vehicle with a 30% efficient IC engine, including 10% gas compression energy and similar 10% drivetrain losses, was powered directly by the same natural gas that fueled the power plant, it would be 24.7% *efficient*:

$$(0.30)(1-0.10)(1-0.10)=24.3\% \text{ efficiency or } 75.7\% \text{ energy loss}$$

While the EV is more efficient, the direct combustion of the original fuel by the car is not significantly worse. Natural gas was used here as the common energy source because similar vehicles are available that are fueled by either electricity or compressed natural gas. Here is a list of assumptions and notes for these calculations:

1. The comparison does not include any of the amortized energy investments in the manufacturing of an ICV or EV, most notably the battery which is known to be highly energy-intensive and environmentally consequential.
2. Very few ICVs are driving around on natural gas (the 1998–2015 Honda GX [10] was one of the only mass-market natural gas automobiles). Gasoline would have been a more realistic comparison fuel, and its ICE efficiency is slightly lower than natural gas [11]. But the energy at the power plant is made from natural gas, not gasoline. It is also worth noting that a natural gas powered car is more similar to an EV than to a gasoline car, considering the longer time required for refueling, limited range, and inconvenience of owning a non-mainstream vehicle.
3. Data are averages from multiple references or from personal experience working on these fuel technologies. There are higher and lower average efficiency values published by a myriad of information sources, most on the Internet. If, for example, the analysis assumed entirely fast charging of EVs, the EV energy losses would be greater. If the natural gas engine efficiency was lower than 30%, the ICV losses would increase. If a more accurate US average grid energy efficiency (40% in 2023, see Chap. 14) was used instead of 50% for the combined cycle natural gas plant, the EV efficiency would be proportionately lower.
4. A more comprehensive analysis is presented in Chap. 14 *Electric Vehicles*, which includes all factors and uses the lower grid efficiency.

Energy Quality

The notion of *energy quality* is important to a fair analysis of practical energy-related processes. Its formal definition, based on entropy, is not very helpful in practice. An easy way to think about energy quality: *The more things you can do with an energy form, the higher its quality.*

Here is an example: *Heat* produced by incident sunlight is considered a *low quality* energy form, since it can only be used to increase the temperature of something, e.g., heat water or air. The heating of a surface with solar radiation is a high-efficiency process. On the other hand, electricity is generated by a solar photovoltaic panel at a much lower efficiency, but it is a *high-quality* energy form. It can be used to heat something, but it can also directly illuminate a light bulb or LED at night, turn an electric motor via electromagnetism, create or amplify sound, make possible telecommunications and computation, power an induction stove, or power a heat pump that can more efficiently heat water or air. Creating a high-quality form of energy from a lower quality form always incurs thermodynamic losses. This is why it is misleading to compare a Joule or Watt-hour of electric energy with a Joule or Watt-hour of a thermal source or combustion fuel that generates heat. The same energy content in the form of electricity is much more useful and therefore higher quality.

There is always an advantage to using the lowest quality form of energy that can do a given job. For example, electric resistive heating is nearly 100% efficient for converting electricity to heat, but when the efficiency of the electricity generation and distribution is included, its overall fuel-to-heat efficiency and cost is inferior to the direct combustion of the fuel to generate heat. This is an underappreciated flaw of electric propulsion when the electric energy is generated from the utility grid.

Greenhouse Gases and the Thermal Balance of the Earth

While the scientific foundations of climate science were established by several great thinkers of the eighteenth through the twentieth century, the Swedish physicist Svante Arrhenius is credited with creating in 1896 the first model of climate change [12], coupled to the effects of heat-trapping gases (later known as greenhouse gases). While attempting to explain the effect of cloud cover on night temperatures, he identified greenhouse gases in the atmosphere that are thermal absorbers and insulators. They reflect or absorb some of the incident solar infrared radiation. Relevant to global warming, they prevent re-radiation from the surface of the earth. The retention of this additional heat near the surface is the greenhouse effect. It is a natural process that is actually critical to life on this planet. We rely on it to maintain an average surface temperature of about 14.6 °C (58.3 °F).

Oxygen and nitrogen, the primary components of the atmosphere, are transparent to infrared radiation and neither reflect nor block re-radiation. They are not greenhouse gases.

Water vapor and condensate in clouds are by far the largest contributors to the overall greenhouse effect. Their global warming potency is weak, but there is a lot of it. Water vapor in the atmosphere averages 2–3%. On a molar average basis, water vapor is about half as effective as a GHG compared with carbon dioxide. The effect varies highly with the form of the water vapor, e.g., humidity vs. clouds. It occurs naturally and has been in balance in its vapor, liquid, and solid forms for millennia. Water vapor is recycled in the atmosphere by precipitation and evaporation. But the total water on the planet in all forms is relatively invariant.

Carbon dioxide, methane, and ozone are much less prevalent in the atmosphere than water vapor, but they are more potent GHGs compared with water vapor. CO_2 is created by combustion, chemical reactions, organic decay, and animal respiration. CO_2 is variable: it increases with the total carbon available for combustion, removed by plant life. Methane is the natural result of anaerobic (without oxygen) decomposition of organic matter. These occur naturally in parts per million (ppm) concentrations in the atmosphere and remain relatively constant if not supplemented by anthropogenic (man-made) activities.

Carbon from the earth's primordial atmosphere is retained in the crust of the earth from millions of years of organic decay. The result is what we now refer to as fossil fuels: oil, gas, coal, tar, and peat. When released into the atmosphere by the combustion of these substances, the greenhouse effect of the planet is altered. The release of carbon stored underground from organic decomposition and geological processes that took place over 300+ million years is then abruptly, in less than two centuries, released into the atmosphere. The natural greenhouse effect that has stabilized earth temperatures for millennia is irrevocably altered.

As discussed earlier, other anthropogenic GHGs include nitrous oxide, and synthesized gases such as CFC and PFC refrigerant gases and sulfur hexafluoride (SF6) used as a high-voltage insulating gas in electric power distribution. They have extremely high infrared absorption and insulating properties and are therefore powerful greenhouse gases.

The solar-thermal balance of the planet is delicate, and the equilibrium we have enjoyed for millennia is easily upset. Even a small increase in the greenhouse effect due to added GHGs causes more heat to be trapped, leading to a net increase in atmospheric and surface temperature.

While there are some stabilizing mechanisms in the atmosphere, they have relatively small effects that counteract the thermal instability due to the synergy between CO_2 and H_2O in the atmosphere: For example, if temperatures rise due to slightly more CO_2, more water is vaporized and enters the atmosphere. Since water is itself a greenhouse gas, heat entrapment and temperature are further increased. This is a positive feedback cycle that, when combined with weaker stabilizing mechanisms, yields a ***metastable*** system, one that can only be pushed so far off its equilibrium point before reaching the point of run-way climate change, sometimes referred to as the "tipping point."

Also, increased differential heating of the earth's surface and troposphere leads to greater local atmospheric pressure and temperature differentials, which are drivers of extreme weather events [13].

This simplified explanation is confirmed by advanced climate simulation models, such as Argonne National Laboratory's crowd-sourced GREET model in the USA or the EU Euro-Cordex model [14].

Why Are Greenhouse Gases so Closely Tied to Our Energy Use?

The overwhelming scientific consensus is that carbon dioxide (CO_2) and methane (CH_4) are the primary *anthropogenic* (human-made) drivers of global warming. Note, however, that when we use the terms carbon emission, carbon credits, or carbon capture, we are incorrectly implying that carbon in various compounds is the bad actor affecting climate change. But not all greenhouse gases contain carbon, e.g., nitrous oxide (N_2O) from agricultural operations, or sulfur hexafluoride (used in high-voltage power distribution switches) are among the most extreme greenhouse gases, far more potent than CO_2, yet they contain no carbon. It is not the carbon per se that is the problem—it is the infrared absorption and reflection properties of a gas that make them greenhouse gases and lead to atmospheric warming.

While the increase in the average atmospheric temperature is the most commonly cited metric, it has become increasingly obvious in recent years that a radical increase in extreme weather events is a more immediate concern to civilization today. Of course, if sea levels rise sufficiently with the melting of polar ice, the ramifications will be very severe, but arguably, we have adequate warning of this trend. At least enough time to relocate coastal real estate developments that could be underwater in the next 50–100 years. The radical uptick in the number and severity of extreme weather events worldwide (severe storms, droughts, floods) is of much more immediate concern.

Related to petroleum combustion in automobiles, the greenhouse gas of primary concern is carbon dioxide. Nitrous oxide is also a potent greenhouse gas, but it is generated almost entirely by agriculture and industrial processes and is not directly associated with automobiles or transportation, with the possible exception of ammonia combustion [15].

Similarly, methane emissions are a serious contributor to climate change, having a global warming potential (GWP) between 28 and 84 times that of carbon dioxide (depending on the assumptions of the analysis). Methane, the primary component of natural gas, is more associated with agriculture and biomass decomposition than with automotive emissions. But that does not mean it is not a major product of the oil extraction industry. In the USA, as a direct result of improved drilling techniques, approximately 26% of all natural gas is extracted as an unintended by-product of hydrofracturing *for oil*. Methane from extraction sites far from natural gas pipelines or liquefaction facilities has no market value. It is flared or vented. Despite state and federal reporting requirements, there is little reliable data on the amount of natural gas flared or vented. For example, in Texas, with the largest number of oil and gas drilling operations, flared or vented gas is exempt from the state's 7.5% natural gas tax, leaving no incentive to reduce this huge waste of energy. As stated in a 2019 report by the US EIA [16] (Fig. 2.12),

Fig. 2.12 Natural gas flares at undisclosed well site. https://gml.noaa.gov/ccgg/behind_the_scenes/flaring.jpg. Public domain

> The Environmental Defense Fund (EDF) performed a similar analysis of the NOAA data, analyzing flaring rates and volumes in the Permian during 2017. The results indicated that Permian operators alone burned 104 Bcf [billion cubic feet] of natural gas, which equated to 4.4% of all gas produced in Texas in 2017. However, industry only reported 55 Bcf of gas burned. NOAA satellite data shows operators burning almost 8% of produced gas.

A worse scenario is the leakage or intentional *venting* of natural gas into the atmosphere rather than flaring, which avoids detection from satellite imagery. Considering the high global warming potential of methane (the dominant component of natural gas), the release of unburned gas is a much greater contributor to climate change than if the gas was flared. Figure 2.13 is a photograph of the Eagle Ford Shale Play in southern Texas taken from the International Space Station. Each of the "thousand points of light" outside the two cities is a gas flare disposing of natural gas at a hydrofracturing site.

The suspected underreporting of flaring and venting, in addition to countless small leaks from gas compressors and pipelines, has in March 2024 motivated the launch on a multi-satellite SpaceX rocket of a dedicated methane detection satellite, *MethaneSAT* [17], privately funded by the non-profit International Environmental Defense Fund. It was not the first satellite for this purpose, but it included a higher-resolution infrared spectrometer that allowed better localization of the sources of methane by coupling data with precise weather observations. Considering the severity of the problem, it is disappointing that this expensive undertaking had to be underwritten by a private non-profit organization rather than a national space agency such as NASA (USA) or CNSA (China) [18].

GHGs from Automobiles

The combustion of petroleum products in an Internal Combustion Vehicle (ICV) generates several greenhouse gases of variable efficacy. These include carbon dioxide, water vapor, oxides of nitrogen, particulates and trace unburned gases having global warming potential. These GHG emissions add to other natural and

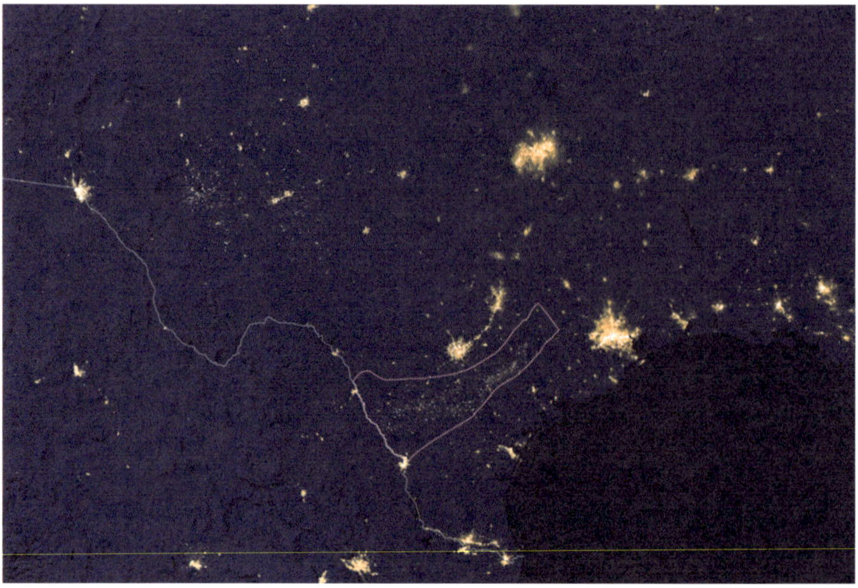

Fig. 2.13 NASA Landsat 5 image 16 February 2016 of the Eagle Ford Shale Play in southern Texas, USA showing natural gas flares. Each flare could have provided enough natural gas to continuously power a fleet of service and delivery trucks. Image source: https://eoimages.gsfc.nasa.gov/images/imagerecords/87000/87725/eagleford_vir_2016046_lrg.png. Public domain

anthropogenic sources of greenhouse gases unrelated to hydrocarbon combustion. How do we assess the relative heat-trapping effects of each greenhouse gas?

Since CO_2 is the anthropogenic GHG of greatest concern for climate change, it is common practice to report all GHGs in terms of their potency compared with CO_2. The metric CO_2e is used for this purpose, equalizing the climate harm that is done by each GHG with an equivalent mass of CO_2. In comparing the relative climate change impact of different combustion fuels, I will follow established practice and report all GHGs as CO_2e.

To allow comparison of relative climate impacts, we apply a multiplier to a particular greenhouse gas to normalize its potency with that of carbon dioxide. This factor is called the *Global Warming Potential* (GWP), and it differs radically between various GHGs. The GWP factor multiplied times the mass of the given GHG gives the metric CO_2e, or *carbon dioxide equivalent*. We use this as a standardized metric for global warming impact by referring to all gases affecting global warming in terms of the equivalent impact of CO_2.

Table 2.1 repeats the list of greenhouse gases and their GWPs from Chap. 1, Table 1.1, but identifies the total impact of each on global warming and the percentage of each that occurs naturally compared with the percentage that is generated by human activities.

Table 2.1 GHGs are segregated by natural vs. anthropogenic origins, and impacts on global warming are reported as CO_2e on a mass basis

Greenhouse gas	Warming effect	% natural	% human	GWP
Water vapor*	95.00%	94.999%	0.001%	0.001–0.0005
Carbon dioxide (CO_2)*	3.62%	3.502%	0.117%	1
Methane (CH_4)*	3.62%	3.502%	0.117%	25–75
Nitrous oxide (N_2O)	0.360%	0.294%	0.066%	293
Hydrogen (H_2)	Trivial	0.1%	99.9%	12-40
Synthetic refrigerants (xFC)	0.950%	0%	0.950%	1600–10,000
Electrical insulator (SF_6)	Undetectable	0%	100%	24,000
Total	100.00%	99.72	0.28%	

Transportation-related GHGs are identified with asterisks. Data from http://www.geocraft.com/WVFossils/greenhouse_data.html

References

1. Monthly energy review, Tables 1.3 and 2.1-2.6. (U.S. Energy Information Administration, 2022), https://flowcharts.llnl.gov/. Accessed May 2023.
2. Sources of greenhouse gas emissions. (US EPA), https://www.epa.gov/ghgemissions/sources-greenhouse-gas-emissions 2022. Accessed May 2023.
3. Climate change performance index of each of the members of the Group of 20 (G20) forum as of 2024, by category. (Statistica), https://www-statista-com.calpoly.idm.oclc.org/statistics/1406610/climate-change-performance-index-g20-by-country/. Accessed 20 March 2024.
4. Which countries are meeting their Paris agreement goals? (Statistica, Climate Action Tracker), https://cdn.statcdn.com/Infographic/images/normal/26102.jpeg
5. COP 28: What was achieved and what happens next? (United Nations Climate Change Report), https://unfccc.int/cop28/5-key-takeaways. Accessed 20 March, 2024.
6. Visualised: how all of G20 is missing climate goals — but some nations are closer than others. (The Guardian), https://www.theguardian.com/environment/2023/dec/07/visualised-how-all-of-g20-is-missing-climate-goals-but-some-nations-are-closer-than-others?CMP=Share_AndroidApp_Other. Accessed 20 March 2024.
7. Overview for renewable fuel standard. (US Environmental Protection Agency), https://www.epa.gov/renewable-fuel-standard-program/overview-renewable-fuel-standard.
8. J. Lewandrowski, J. Rosenfeld, D. Pape, T. Hendrickson, K. Jaglo, K. Moffroid, The greenhouse gas benefits of corn ethanol – assessing recent evidence. Biofuels **11**(3), 361–375 (2020). https://doi.org/10.1080/17597269.2018.1546488
9. Congressional briefing - 2022 sustainable energy in America factbook. (Data Source: US Energy Information Agency (EIA)), https://www.statista.com/statistics/528603/distribution-electricity-net-generation-in-the-us-by-fuel-type/. Accessed 1 June 2023
10. Honda GX review. (Automotorpad). https://automotorpad.com/honda/627620-the-honda-civic-gx-natural-gas-vehicle.html. Accessed 12 Jan 2024.
11. R. Bosch, *Automotive Handbook*, 10th edn. (2020)
12. How 19th century scientists predicted global warming. (JSTOR Daily), https://daily.jstor.org/how-19th-century-scientists-predicted-global-warming/. Accessed 20 April 2024.
13. http://www.skepticalscience.com/water-vapor-greenhouse-gas.htm
14. Regional climate model (EURO-CORDEX) - new high-resolution climate change projections for European impact research. (European Union, Environmental Information Systems, 2019), https://www.eea.europa.eu/data-and-maps/data/external/euro-cordex-new-high-resolution. Accessed 20 April 2024.

15. K.A. Smith, *Nitrous oxide and climate change* (Earthscan, London, 2010) as cited in "What's Your Impact?" https://whatsyourimpact.org/greenhouse-gases/nitrous-oxide-emissions. Accessed Dec 2022.
16. Natural gas flaring and venting: State and federal regulatory overview, trends and impacts. (US Dept of Energy, Office of Oil and Natural Gas, 2019). https://www.energy.gov/sites/prod/files/2019/08/f65/Natural%20Gas%20Flaring%20and%20Venting%20Report.pdf. Accessed 20 March 2024.
17. A new satellite will track climate-warming Pollution. Here's why that's a big deal. (National Public Radio (NPR) Climate, 2024), https://www.npr.org/2024/03/05/1235694992/a-new-satellite-will-track-climate-warming-pollution-heres-why-thats-a-big-deal. Accessed 20 March 2024.
18. Wikipedia, China National Space Administration, https://en.wikipedia.org/wiki/China_National_Space_Administration. Accessed 20 March 2024.

3 Automobiles and the Environment

The automobile has played a major role in defining civilization as we know it today. But in its roughly 130-year existence, automobiles have played a major role in altering the atmospheric composition, most notably due to the carbon dioxide (CO_2) generated by hydrocarbon combustion. Consequently, transportation, including land, sea, and air, has been a major contributor of CO_2 that is now known to be the primary anthropogenic contributor to global warming. Within the entire field of transportation, passenger cars produce the largest share of CO_2 emissions [1] (Fig. 3.1).

It's worthwhile, then, to briefly review how automobiles have helped to lead us to the current threshold of climate decline in such a short time and examine ways by which we might stabilize the damage already done.

Since the 1920s, after the brief golden years of electric vehicles, almost all automobiles and light trucks have burned fossil fuels to produce motive power, in particular, gasoline and diesel fuel, and early approximations of these. The energy conversion devices that turn fuel energy into motive energy have almost entirely been *internal combustion engines* (ICEs) as opposed to *external combustion engines* such as steam engines. Interest in ICEs began with the quest for stationary and mobile mechanical power sources that were lighter and required less attention than the steam engines that had ruled since 1689. Many names appear on patents as early as 1794, when English inventor Robert Street constructed a crude stationary ICE fueled by petroleum.

In 1807, French engineers Nicéphore and Claude Niépce built an engine fueled by coal dust, which they used to power a boat. The first four-wheeled vehicle powered by an ICE was probably the hydrogen-oxygen-powered wagon of Swiss engineer François Isaac de Rivaz, who in 1813 demonstrated a travel distance of 100 m. In 1823, Samuel Brown patented the first practical ICE in the USA, which was used to power a boat and a carriage in 1827–1828. Several similar automobile inventions in England, the USA, Italy, and Belgium were patented or published, but none were commercially successful [2].

C. A. MacCarley, *Non-Petroleum Automotive Transportation*,
https://doi.org/10.1007/978-3-031-85028-8_3

Fig. 3.1 Benz Motorwagen, 1886, Germany, which is considered the first successful automobile, constructed by Rheinische Gasmotorenfabrik Benz & Cie. (Mercedes-Benz) Image from Deutsches Automotive Museum, Munich DE. Public domain

In 1876, German engineers Nicolaus Otto, Gottlieb Daimler, and Wilhelm Maybach patented the first *compressed charge, four-stroke cycle engine,* the configuration used in almost all automobiles today. Their configuration was that of a piston or reciprocating engine, in which a combustible mixture of air and fuel is inducted into a closed cylinder, compressed by the motion of a piston, and then ignited with an electric spark, causing a large increase in pressure that pushes the piston in the other direction to rotate a crankshaft, doing external work, thus the term "reciprocating" or back and forth.

The other fundamental configuration that survived the test of time was the compression ignition (aka diesel) engine, patented in 1892 by German engineer Rudolf Diesel, but not used in production automobiles until the 1930s by Citroen and Peugeot of France, and Mercedes-Benz and Record, both of Germany. Many other IC engine configurations have been conceived since then but failed to capture market success. The one exception was the rotary (non-piston) engine developed in 1954 by German engineer Felix Wankel, which powered passenger cars manufactured from the 1960s through the 1990s by NSU of Germany and Mazda of Japan.

Electric propulsion came on the personal transportation scene with the successful demonstration of the English Ayrton-Perry Electric Tricycle in 1881, soon followed by many battery electric cars that achieved at least some measure of commercial success. By 1900, the automotive world was dominated by electric cars. It is fair to say that the practical history of the automobile as a form of personal transportation started with battery electric vehicles (BEVs). Between 1890 and 1910, more electric vehicles were operating on US and European roads than combustion engine vehicles of any type—ICEs or external combustion steam engines. Deployed almost entirely in urban areas including New York, Baltimore, Paris, Stuttgart, and London, BEVs fascinated affluent owners and taxi operators in terms of being quiet, less odiferous, easy to operate, and requiring no hand-crank starting [3]. Electric power was sourced almost exclusively from coal-burning power plants, most almost solely used for charging electric vehicles. Energy storage was almost exclusively via lead-acid batteries, as it would continue for 110 years.

However, ICE competitors to electric vehicles improved gradually, and even though still smelly and dirty compared with clean electric vehicles, they gained popularity because of their greater range and non-dependence upon charging facilities, which were only available in affluent urban areas. Liquid fuels were usually purchased at the local drug store (chemist) or general store. Fuels were minimally refined petroleum derivatives, vegetable oils, whale oil, benzine (as gasoline is still referred to in much of Europe), toluene, acetone, methanol (wood alcohol), and ethanol (grain alcohol). Standardized gasoline as we know it would not become common for another twenty years. But with the incredibly high energy density and low cost of petroleum fuels, the introduction of the low-cost Ford Model T in 1908 [4] (Fig. 3.2), and subsequent technical innovations such as the Kettering inductive ignition and electric starter, ICE vehicles eclipsed BEVs almost completely after 1920. Short-lived electric car manufacturers were replaced by startup ICE vehicle manufacturers. Paved roads gradually replaced dirt streets and wagon paths, and by the 1950s, limited-access highways and autobahns were built, all to enable the unprecedented convenience of high-speed personal transportation using gasoline or diesel fuel. Electric cars were relegated to special applications, such as indoor or underground vehicles, although electric rail continued to grow in popularity, as it remains today since it does not require energy storage onboard the vehicle.

Fig. 3.2 1912 Ford Model T Roadster. The Model T was the first mass-produced car in the United States (possibly the world) making it affordable for the general population. Photo by Randy von Liski. Flicker. CC0 license. https://www.flickr.com/photos/myoldpostcards/5012726699/

Automobile Air Pollution: Early Perspective

In the context of automobiles, we usually think of *pollution* as the noxious components of ICE automobile engine exhaust. But in the late nineteenth and early twentieth centuries, electric and petroleum-powered automobiles were considered significant improvements over the emissions of horses, responsible for serious urban sanitation and air quality problems. Coal remained king for locomotives, ships, domestic heating, electricity generation, and town gas production for lighting and cooking, but it was an inconvenient fuel for personal transportation. The connection between air pollution and health was eventually recognized, but there was no choice but to accept it as a cost of progress. Autocars were just too desirable. The opaque air that was the signature of the Industrial Revolution in England and Western Europe was matched 100 years later by the automobile smog of major US cities such as Los Angeles.

Over the next 100 years, periods of interest in alternatives to gasoline and diesel fuel were frequent, all following unexpected "gas shortages" due to the global politics of oil or public epiphanies about the consequences of automotive air pollution. Until the 1970s, world oil reserves were assumed by most people to be of no concern to them, an assumption that was challenged in 1956 following the publication of Standard Oil Co. engineer Edwin Hubbert's predictive analysis of oil resources in the USA. The "Hubbert Curve", a logit [5] supply/demand model for petroleum, observed that a geopolitical oil crisis would occur not when reserves were depleted but when oil extraction could not keep pace with growing demand. Gasoline shortages during World Wars I and II, the Suez Canal blockade in 1956, and the Israel-Egypt Six-Day War in 1967 brought home the importance of free-flowing oil to Western Europe but had little effect on North America due to its large indigenous reserves. Following the oil shortages of 1973 and 1979, attitudes in the USA changed dramatically due to growing dependence on inexpensive oil imported from the Persian Gulf, controlled by the newly formed Organization of the Petroleum Exporting Countries (OPEC) cartel. The assumption that gas stations would always have (inexpensive) fuel for your car was shattered during these periods, but it was forgotten as soon as cheap gas became available again in the 1980s.

Starting in the early 1970s, exemplified by the 1972 publication of the book "*Limits to Growth*," our dependency on a finite resource of low-cost petroleum was recognized as an impending conundrum: a primary driver of world prosperity and politics but a serious health hazard and political flashpoint. Automotive emission controls first deployed in 1968 were only marginally effective in reducing the toxic exhaust emissions hydrocarbon (HC), carbon monoxide (CO), and NO_x, referred to as *regulated emissions*. It was not until 1977 that truly effective auto emission controls were deployed, following the development of the 3-way catalytic converter and oxygen-sensing feedback fuel injection, technologies that have been used on all licensed vehicles in the USA since 1991. It took another decade for awareness to grow about an even more insidious problem: the impact of GHGs, especially CO_2, on the climate. CO_2 had always been considered a benign trace gas in the atmosphere. It is nontoxic, the byproduct of natural and anthropogenic (human) activities, industrial processes and automotive engines. A decade passed before another epiphany: methane (CH_4) released into the atmosphere from natural gas extraction and agricultural operations was also a potent GHG. The new environmental concern was GHGs that, even in minimal concentrations, increased the heat retention of the atmosphere.

The engine control technologies that so effectively reduce regulated automotive emissions do nothing to reduce CO_2 emissions. In fact, they slightly increase them since all automotive emission controls seek to convert HC, CO, and NO_x into CO_2, water vapor, and nitrogen. Unlike regulated emissions that are directly toxic and/or carcinogenic, CO_2 is as fundamental to carbon-based life on earth as water. All higher forms of life require or expire it. For gasoline- or diesel-fueled vehicles, CO_2 emissions are exactly proportional to the mass of fuel burned. Another big difference is that the effects of CO_2 are global, unlike the local impacts of regulated emissions. Where and how it enters the atmosphere does not matter. Whether the CO_2 is released from a vehicle tailpipe, the stack of an electric power station or a home furnace makes no difference to its climate change consequences. CO_2 emissions in a city in Southeast Asia eventually have the same impact on a city in Europe as the emissions originating in that city. This distinction is still not fully grasped by much of the public, who are accustomed to being able to escape the pollution of the city by a vacation in a forest. The *global warming* problem caused by anthropogenic (human-made) CO_2 and CH_4 can only be addressed by a major reduction in processes that generate CO_2 of which the combustion of fossil fuels is the leading contributor.

By 2010, over 1 billion [6] petroleum-powered vehicles were operating daily worldwide, and by 2023, that number had increased to 1.47 billion [7]. Petroleum and its energy and environmental sustainability impacts were now the topic of daily media headlines. However, society lacks the collective will to accept and address the predicament, even as global temperatures continue to break records year after year, and radical climate events have exploded in number and severity. The ubiquitous Internet has played a major role in the propagation of misinformation, leading to the paralysis of meaningful collective action and pseudosolutions motivated more by financial or political gain than climate change remediation.

In 2021, in the USA, energy used by powered transportation accounted for 28% of all forms of energy in all uses [8]. In California, considered the "greenest state," 39% of all GHG emissions are attributed to transportation [9] (Fig. 3.3).

Since the dominance of the ICE starting in the early twentieth century, petroleum has been the primary source of energy for transportation in the USA and most Western countries. In 2021, petroleum products accounted for 90% of the total energy use in the US transportation sector [10]. The breakdown by exact vehicle types has not been as well documented as the distribution of the final forms of refined petroleum, illustrated in the EIA pie chart [11] of Fig. 3.4 that shows that gasoline and distillates (e.g., diesel fuel) account for 77% of all petroleum. Hence, they are prime candidates for mitigating anthropogenic global warming.

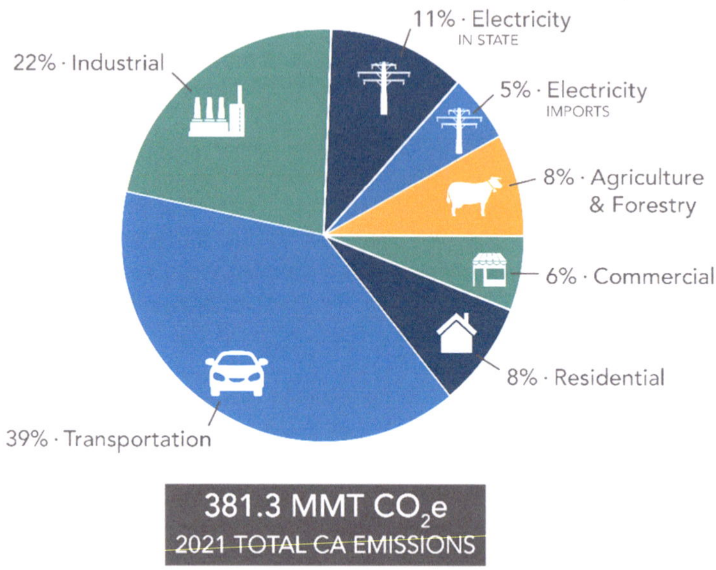

Fig. 3.3 Percent of GHG emissions attributed to each usage sector in California, USA, 2021. Data from California Air Resources Board, 2023. Graphic from https://ww2.arb.ca.gov/ghg-inventory-graphs. Public domain

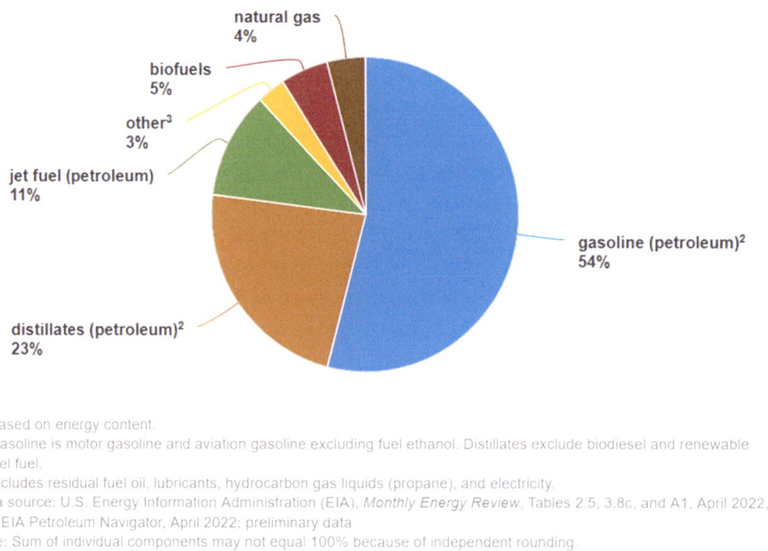

Fig. 3.4 End products of petroleum processing in the USA, 2023. More than ¾ goes to gasoline and diesel fuel. From US Energy Information Agency, “Transportation Fuels” graphic from https://www.energy.gov/energysaver/transportation-fuels. Public domain

Automotive GHG Emissions

GHGs Generated by Combustion Engines

In the USA and much of the world, transportation accounts for the largest fraction of anthropogenic GHGs [12].

GHGs reflect heat (long-wave infrared radiation) back to the earth, causing the gradual atmospheric temperature rise known as global warming. Simplistically stated, a gas molecule can be, but not necessarily is, a good infrared absorber if it is heteronuclear (has two or more different atoms) with an uneven charge distribution between the atoms. This allows the molecule to change its dipole moment, which enables it to absorb and re-emit IR radiation. These types of molecular arrangements are the most potent GHGs [13].

All fossil fuels (*hydrocarbons*) burn to produce GHGs, most significantly, CO_2. We have become accustomed to reporting "carbon intensity" as CO_2e, which is the global warming impact relative to CO_2 for any GHG.

The combustion of gasoline, gasoline blends, diesel, or biodiesel fuel also generates *regulated* pollutants which are directly toxic or carcinogenic:

- HC: HC, including non-methane organic gases (NMOGs), are carcinogenic.
- CO: It is not itself a GHG but reacts with oxygen in the atmosphere to become CO_2.
- NO_x (NO and nitrogen dioxide NO_2) Highly toxic and precursors to ozone.
- HCHO (formaldehyde): It is highly toxic and emitted mostly by alcohol fueled vehicles.
- CH_4: It is a potent GHG. As an automotive emission, it is primarily a concern for natural gas vehicles, and for hydrogen production from natural gas.
- Particulates (carbon and hydrocarbon emissions in solid form): They are responsible for respiratory diseases and much more of a problem in diesel engines.

Regarding NO_x: Combustion of almost anything, regardless of carbon content, produces some NO and NO_2, collectively called NO_x. Any time air is heated to a high temperature, even for a few milliseconds, some of the nitrogen and oxygen molecules react to form NO_x. This is why even the combustion of hydrogen and ammonia, both carbon-free fuels, produces NO_x in quantities varying from near-zero at the lean AFRs (low power, low temperature), to the same or higher than gasoline at stoichiometric ratio (high power, high combustion temperature). NO_x gases are extremely toxic and photoreactive in the atmosphere, serving as precursors to ozone. They are *regulated* automotive emissions subject to legal constraints by the EPA (USA) and EEA (EU), and are of special concern for diesel engines because they operate at higher combustion temperatures due to their higher compression ratios. Although NO and NO_2 have the chemical properties to act as GHGs, they are usually not of global warming concern because of their high reactivity, which makes them short-lived in the atmosphere.

However, the third oxide of nitrogen, N_2O aka "laughing gas," is a very potent and stable GHG, with a much higher global warming potential (GWP) than CO_2 or even CH_4. But it is not a significant automotive combustion product. It is produced almost entirely by agriculture and the agricultural chemical industry [14].

Of all the GHGs, the largest transportation-related contributor to global warming is CO_2, even though it has the lowest GWP, as illustrated in Fig. 3.5. There is a lot of it in the atmosphere.

The EPA and EIA report exhaust GHGs as "CO_2e" (CO_2 equivalent), by multiplying the mass of the GHG by its GWP, which adjusts the physical totals to reflect the relative greenhouse potency equivalent to the mass of CO_2 that would have the same effect [12].

For modern catalytic converter-equipped gasoline vehicles and exhaust-after-treated diesel vehicles, almost the entirety of the GHG effect is attributed to the CO_2 emissions alone. For example, the US EPA's "GHG Score" and resultant "Smart Car" ratings *are based solely on CO_2 emissions* or the CO equivalents of other automotive emissions. Note that this assumption is not valid for older (pre-1990) non-catalyst-equipped vehicles that have much greater regulated emissions.

For gasoline/diesel vehicles, the fuel consumption and official fuel economy ratings, as established in multi-cycle dynamometer testing, are actually measured by monitoring the mass of CO_2 emitted in the exhaust. During testing, a standardized reference gasoline or diesel fuel with known carbon mass concentrations is used. This relies on the fact that except for trace amounts of carbon tied up in the hydrocarbon and carbon monoxide emissions, the CO_2 emissions are simply *proportional to the inverse of the vehicle fuel economy stated in miles/gallon (mpg)*, or, in

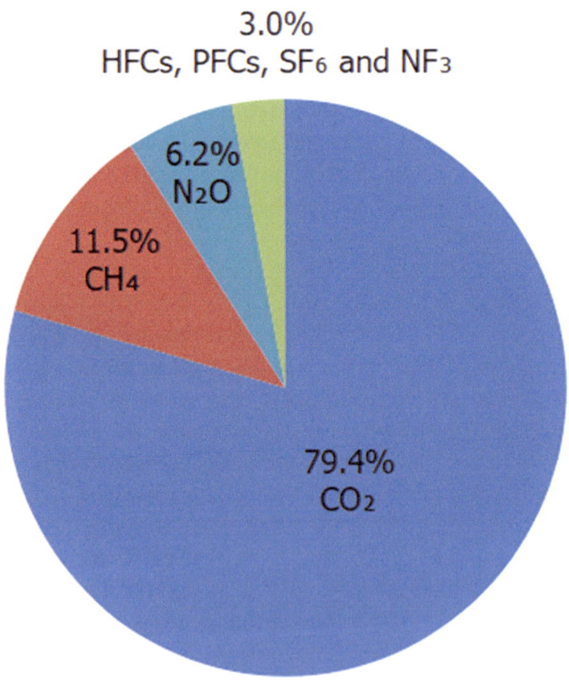

Fig. 3.5 2021 breakdown of global anthropogenic greenhouse gas emissions. EIA data and graphics. Public domain

countries other than the USA, *directly proportional to the vehicle fuel economy stated as liters/kilometer (L/km).*

The US EPA reports [15] the following equivalences for gasoline and diesel fuel:

$$\mathbf{1\,US\,gallon\ of\ gasoline\ without\ ethanol = 8887\,g\,CO_2.}$$

$$\mathbf{1\,US\,gallon\ of\ diesel\ No.2 = 10{,}180\,g\,CO_2.}$$

To be clear, this means that the consumption of each gallon of gasoline in a vehicle contributes about 8.9 kg (19.6 lbs) of CO_2. For perspective, adult human respiration generates 0.7–0.9 kg of CO_2 daily [16].

Different Fuels Produce Different CO_2 Emissions

While it is true that the CO_2 emissions of an automobile are proportional to its fuel (energy) consumption rate, it is not true that all fuels produce the same amount of CO_2 for the same energy content. The exhaust CO_2 can be very different since the carbon mass fraction and the specific combustion energy of the fuel will be different. When measured relative to distance traveled (MPGe, g CO_2/mile), the energy content of a fuel equilibrates to the distance traveled.

This is significant in our quest to reduce the carbon intensity of transportation, since the use of a fuel that produces $x\%$ as much CO_2 as gasoline per unit of energy will produce $x\%$ lower CO_2 per km or mile. This is in addition to the difference in the carbon intensity of the production process of the given fuel compared with gasoline or diesel fuel.

Focusing only on tailpipe CO_2 emissions, we can easily calculate and compare the relative CO_2 emissions in g/mi for any combustion fuel. What we will see is that *there are few alternative fuels, either liquids or gases, that produce CO_2 emissions as high as gasoline or diesel* fuel per unit of distance traveled.

Comparing Combustion Fuels in Terms of Their CO_2 g/mile

What mass of CO_2 gas is produced by the *stoichiometric* combustion of a fuel to generate a given amount of fuel energy (equivalent to a given number of km or miles traveled)?

As discussed in Chapter 4 *Engines and Fuels*, *stoichiometric* ratio means that the engine is burning the exact mass ratio between the air and the fuel such that all available fuel and all available oxygen are used in the combustion.

For the most common automotive fuels, gasoline and diesel fuel, multiple government and private agencies in the USA and worldwide (US EPA, EIA, ASTM, SAE, CARB, and the IEA and EEA in the EU) test and publish official data annually on the combustion energy and carbon content for current regional gasoline blends. Most alternative fuels are pure substances; that is, they are not blends of hydrocarbons like gasoline or diesel. This makes it possible to determine CO_2 emissions from their chemical formulas rather than experimental testing.

CO_2 Generated Per Unit of Combustion Energy

For the combustion of any hydrocarbon fuel in an ICE, exhaust water leaves the cylinder as vapor. No energy is extracted from the condensation of the vapor into liquid water. Therefore, IC engine fuels always use the lower heating value (LHV) or net heat content, not the HHV or gross heat content. Whenever referring to automotive fuels, "Heat Content" or "Fuel Energy" or "Specific Calorific Value" refers to the LHV.

Table 3.1 shows the results of a US DOE online fuel property comparison website, including the LHV of each fuel [17].

Notes:

- BTU = British Thermal Unit, a legacy unit of energy still in use in the USA. 1 BTU = 1.055 kJ = 10^{-3} MJ.
- The LHV for 2023 US gasoline varies from 112 to 116 BTU/gal. Its average is 114,102 BTU/gal, or 42.7 MJ/kg.
- E0 refers to gasoline without ethanol, whereas what is usually sold as "gasoline" or "petrol" at the pump in North America, Japan, and Western Europe is E10 with 10% ethanol.

Are Vehicular CO_2 Emissions Regulated by the US EPA?

Surprisingly, no. But they are incentivized via gasoline use per mile. Based on the GWP weightings above, the EPA created a scoring system for ranking of vehicles according to their GHG emissions, which are inversely proportional to fuel economy (mpg) [18] (Fig. 3.6).

Notes from this chart:

- In 2020, to achieve a GHG rating of 10, a vehicle must get 45 + mpg gasoline (reduced from 2016 when this number was 46 + mpg). Current EPA projections to achieve a GHG rating of 10 in the years 2027 through 2032 are 44.4–54.4 mpg, respectively [19].
- Complete Bin/Tier regulated emissions standards can be found online [20].
- Individual vehicle data from EPA certification tests can be found online [21].
- The current list of SmartWay or SmartWay Elite requirements is online [22].

EPA GHG Classifications for Gasoline Vehicles

The complexity of automotive emission regulations has grown to breathtaking levels. One indirect indicator is the current (2024) number of different vehicle emission classifications:

Zero-emission vehicles (ZEV) are either BEVs or hydrogen fuel cell vehicles (FCVs). Partial zero-emission vehicles (PZEV) are non-plug hybrids. Advanced technology zero-emission vehicles (ATZEV) added plug hybrids under a separate classification.

Among gasoline-powered vehicles, there are the California classifications: Low Emission Vehicle (LEV Options I and II), Transitional Low Emission Vehicle

Table 3.1 Comparison of selected alternative fuel properties

	Gasoline	Biodiesel	Compressed natural gas (CNG)	Ethanol/E100	Methanol	Hydrogen
Chemical structure	C_4H_x to $C_{12}H_x$ + 10% ethanol for E10	Methyl esters of C_{12} to C_{22} fatty acids	CH_4 (majority), C_2H_6, and inert gases	Simple alcohol C_2H_5OH	Simplest alcohol CH_3OH	Diatomic gas, H_2
Source material	Crude oil	Fats and oils from soybeans, waste cooking oil, animal fats, and rapeseed	Underground reserves and renewable biogas	Corn, grains, or agricultural waste (cellulose)	Natural gas, coal, or woody biomass	Natural gas, methanol, and electrolysis of water
Gasoline or diesel gallon equivalent (GGE or DGE)	1 gal = 1.00 GGE (1 GGE = 0.88 DGE)	*B100* 1 gal = 0.93 DGE *B20* 1 gal = 0.99 DGE	1 lb. = 0.18 GGE	1 gal = 0.67 GGE	1 gal = 0.50 GGE	1 lb. = 0.45 GGE 1 kg = 1.0 GGE
Energy comparison	1 gallon of gasoline has nominally 97%–100% of the energy in 1 GGE	1 gallon of B100 has 93% of the energy in 1 DGE 1 gallon of B20 has 99% of the energy in 1 DGE	5.66 lb., or 123.57 ft. [3], of CNG has the same energy as 1 GGE	E85 contains 73%–83% of the energy same vol of gasoline E100 has 67% of energy in 1 GGE	1 gallon of methanol contains 50% of the energy as 1 GGE	2.2 lbs. of H_2 has approximately the same energy as 1 GGE
Energy content (LHV)	112,114–116,090 Btu/gal	*B100* 119,550 Btu/gal *B20* 126,700 Btu/gal	20,160 Btu/lb	76,330 Btu/gal for E100	57,250 Btu/gal	51,585 Btu/lb 33.3 kWh/kg

(continued)

Table 3.1 (continued)

	Gasoline	Biodiesel	Compressed natural gas (CNG)	Ethanol/E100	Methanol	Hydrogen
Physical state	Liquid	Liquid	Compressed gas, typically 25 MPa	Liquid	Liquid	Compressed gas, typically 70 MPa
Cetane number	N/A	45–65	N/A	0–54	N/A	N/A
Octane (LON) number	84–93	N/A	120+	110	112	130+
Autoignition temperature	257 °C	N/A	540 °C	423 °C	481 °C	566–582 °C
Operational and maintenance issues	Fuel degrades quickly due to evaporation of lighter components	Low-sulfur diesel fuel has low lubricity	High-pressure tanks require periodic certification	Ethanol-compatible parts needed	Methanol-compatible parts needed. Toxic if ingested	High-pressure tanks require periodic certification
Energy security impacts	From crude oil, 70% of petroleum consumption	Biodiesel is domestically produced from seed crops. It is renewable to the extent that its energy balance is positive	Natural gas is obtained almost entirely from drilling	Ethanol is produced domestically, potentially renewable	Methanol is domestically produced from natural gas, but can be made from renewable resources	Green hydrogen is produced by electrolysis using electricity. But most hydrogen is produced from natural gas or coal

Table data generated by the US DOE fuel property comparison tool. https://afdc.energy.gov/fuels/properties?fuels=GS,BD,CNG,ETH,ME,HY. Public domain

Rating	MPG (gas)	CO_2 (g/mile)
10	≥ 45	0-200
9	39-44	201-231
8	34-38	232-265
7	30-33	266-301
6	27-29	302-335
5	23-26	336-395
4	20-22	396-456
3	17-19	457-539
2	15-16	540-613
1	≤ 14	≥ 614

	Emission Standards		
	US EPA		California Air Resources Board
Rating	Tier 2	Tier 3	LEV III
10	Bin 1	Bin 0	ZEV
9	--	--	--
8	--	Bin 20	SULEV20
7	Bin 2	Bin 30	SULEV30
6	--	Bin 50	ULEV50
5	Bin 3	Bin 85, Bin 70	ULEV70
4	--	--	--
3	Bin 4	Bin 125, Bin 110	ULEV125
2	--	--	--
1	Bin 5-8	Bin 160	LEV160

Fig. 3.6 EPA Tier 2 bins 1–10. From https://www.epa.gov/energy/ghg-equivalencies-calculator-calculations-and-references. Public domain

Table 3.2 US EPA CO_2 and MPG bins, in Tier 2, 2021

Rating	MPG (gas)	CO_2 (g/mile)
10	≥53	0—169
9	43—52	170—209
8	36—42	210—250
7	31—35	251—291
6	27—30	292—335
5	23—26	336—395
4	20—22	396—456
3	17—19	457—539
2	15—16	540—613
1	≤14	≥614

(TLEV), Ultra-Low Emission Vehicle (ULEV I and II), Super Ultra-Low Emission Vehicle (SULEV20, SULEV30, SULEV50, SULEV70, SULEV125, and SULEV I and II), and Inherently Low Emission Vehicles (ILEV). The ILEV designation applies only to alternative fuel vehicles such as hydrogen or methanol which inherently produce lower regulated emissions, evaporative emissions, and CO_2. Not to be left out of the acronym competition started by the California Air Resources Board, in about 2003, the US EPA introduced Tiers and Bins. They also aggregated many of the classes into a single classification, the National Low Emission Vehicle (NLEV).

This recitation does not even include the diesel-related emission classes. A complete description of each emission classification can be found in ref. [23].

Regarding GHG emissions, Table 3.2 is a list of the EPA classifications for MY 2021 gasoline vehicles for rating "bins" within "Tier 2," which contains most light-duty gasoline vehicles. Since CO_2 is proportional to gasoline consumption, the table is indexed by MPG, ranging from less than 14 mpg for Bin 1 to greater than 53 MPG for Bin 10.

CO_2 Is Incentivized by Corporate Average Fuel Economy (CAFE) Standards

Within a few years of the passage of the US Federal Clean Air Act in 1970, fuel economy targets were established by the newly founded EPA for US auto manufacturers. These targets took the form of Federal CAFE standards, which provided an incentive for US auto manufacturers to produce more fuel-efficient cars and trucks. The concern at the time was energy security, motivated by the fuel shortages of 1973 and 1979 caused by fuel production restrictions implemented by OPEC to increase the value of imported crude oil at a time when the USA had become dependent upon imports from the Middle East. GHG emissions wouldn't become a factor until nearly 30 years later.

The periodically updated CAFE standards are developed by the US EPA and enforced by the National Highway Transportation Safety Administration (NHTSA).

The novel idea was based upon the observation that all of the major manufacturers at the time produced a wide range of vehicles, from subcompact cars through heavy trucks. Separate target fuel economy (mpg) values were established for four different broad classes of vehicles. Within each class, the "fleet average" fuel economy (mpg) was required to be better than the CAFE target as published in the US Federal Register. This allowed some vehicles to fail to meet the standard by producing others that exceeded the standard.

CAFE standards have mutated in various ways over the intervening 50 years, but are still in force today, with variations in other major auto manufacturing countries, in particular, Japan. CAFE standards continue to be perhaps the most important regulatory tool to help reduce automotive CO_2 emissions (Fig. 3.7).

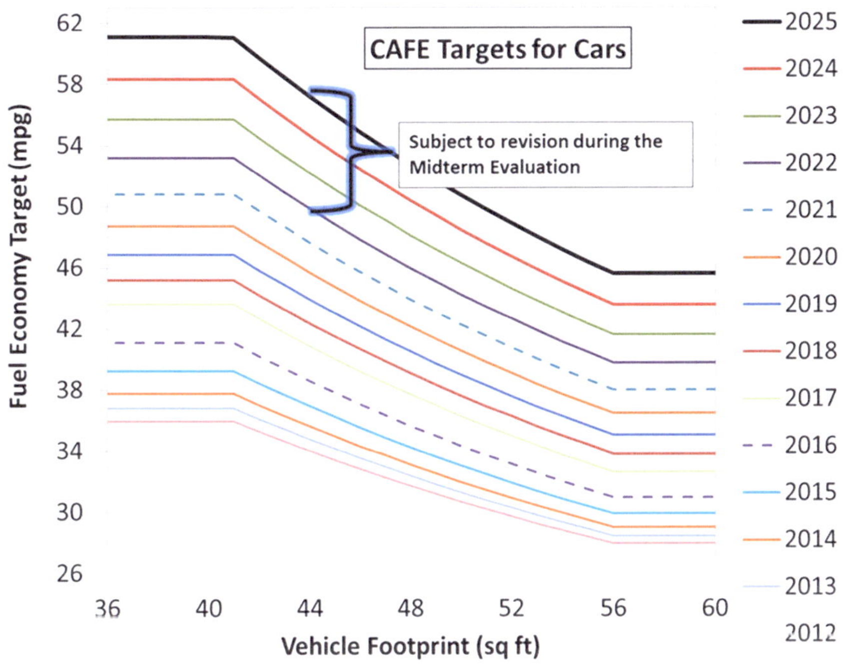

Fig. 3.7 2012–2025 CAFE standards (MPG) by year and vehicle footprint. From U.S. Department of Energy—Fact #941: September 5, 2023. https://www.energy.gov/eere/vehicles/fact-941-september-5-2016-mid-term-evaluation-corporate-average-fuel-economy-standards. Public domain

How to Calculate CO_2 Emissions for Fuels Other Than Gasoline

Find the carbon content of fuel in g carbon/g fuel.

Each carbon atom will become a CO_2 molecule in combustion.

12 g carbon → 44 g CO_2

This gives the g of CO_2 per g of fuel

Determine the car's energy requirement (MJ or kJ) to travel one mile from its MPG rating, using the MPGe and energy content of the fuel.

Divide energy requirement per km or mile by the specific energy (LHV in kJ/g) of the fuel to get g of fuel per km or mile.

Multiply g of fuel per km or mile by CO_2 per g of fuel to get CO_2 per km or mile.

Appendix 2 demonstrates the calculation of CO_2 emissions for any liquid and gaseous fuels. CO_2 emissions are also discussed in Chapter 4 *Engines and Fuels*.

Regulatory Missteps

Early Automotive Emission Controls

Positive intentions and actual results have often been misaligned in regulatory efforts. For example, in the USA, the first EPA automotive emissions regulations that took effect in 1968 were intended to reduce carbon monoxide (CO) and hydrocarbon (HC) emissions, which they marginally did, but at the cost of increased NO_x emissions. This was followed in the early 1970s with crude exhaust gas recirculation methods intended to reduce NO_x emissions, but at the cost of HC, CO, reduced power, and increased fuel consumption. Truly effective control of all regulated emissions from gasoline engines didn't become available until 1979 with the introduction of oxygen-sensing feedback fuel injection control coupled with a new generation of catalytic converters that reduced CO, HC, and NO_x concurrently [24].[1]

Emission Regulations Favor Larger, Less Efficient Vehicles

As of 2025, the average size, mass, and power output of passenger cars and light trucks in the USA have never been larger, even compared with the 1950s and 1960s, years known for chrome-laden land yachts. Auto manufacturers justify this as "what the consumer demands," downplaying the influence of their lifestyle-based advertising or the successful lobbying efforts to manipulate regulations to create false perceptions of environmental concern. IC engines have become remarkably more fuel-efficient due to advanced engine design and control technology, but these advances cannot keep up with the physics of ever-larger and heavier vehicles. As reported in citation [25], Dan Becker, director of the Center for Biological Diversity's *Safe Climate Transport* campaign, referring to current EPA emission regulations, commented:

[1] The Bosch Lambda-Sound system was introduced first on 1979 Volvo cars using the B21 engine.

> The biggest pickup trucks are allowed very gentle treatment. If you create a loophole, that's what they will drive through. Vehicles are getting larger and larger because the larger the vehicle, the weaker the standard.

Possibly the first US regulatory action that led to the continuous skew by US manufacturers toward trucks and heavier vehicles was the little-remembered "Chicken Tax" of 1964, which, via executive order, enacted a 25% protective tariff on the import of pickup trucks to protect domestic production in this high-profit segment of the US vehicle market. It was a meaningless gesture at the time, since 1964 was prior to the first imports of Japanese or German compact trucks. The justification (almost ridiculous) by the "Big Five" US automakers was the popularity of an obscure variant of the popular Volkswagen Type 2 Transporter called the Kombi, which was fitted with a truck bed replacing the rear half of the passenger cabin. (It is now a rare collector car.) This tariff was the start of protectionist tariff policies that have supported the growth of sales of US-manufactured trucks and anything that could be even barely classified as a truck, including minivans and later sport utility vehicles (SUVs). As documented in citation [26].

> The *Chicken tax* is a 25 percent tariff on light trucks (and originally on potato starch, dextrin, and brandy) imposed in 1964 by the USA under President Lyndon B. Johnson in response to tariffs placed by France and West Germany on importation of U.S. chicken.
>
> Eventually, the tariffs on potato starch, dextrin, and brandy were lifted, but since 1964 this form of protectionism has remained in place to give US domestic automakers an advantage over imported competitors. A 2003 Cato Institute study called the tariff 'a policy in search of a rationale' [27].

The trend toward *supersized*[2] vehicles has been growing in the USA and to a lesser extent worldwide, since 1980. Larger vehicles command higher profit margins. Until 2023, the Insurance Institute for Highway Safety (IIHS) considered safety in terms of vehicle occupant protection, not with regard to the safety of the vehicle's targets. A look at the EPA mileage ratings of cars and trucks confirms the obvious: oversized vehicles get undersized fuel economy [28]. While the average weight of automobiles decreased from 1970 to 1980, the trend since 1980 has been monotonically larger, heavier vehicles. The average mass of automobiles and light trucks in the USA increased from about 1450 kg in 1980 to 2200 kg in 2022. For petroleum-fueled vehicles, CO_2 emissions are proportional to fuel consumption: low miles per gallon (mpg) = high CO_2.

In 2012 in the USA, under pressure from major automakers, the US EPA modified regulations establishing less-restrictive emissions and fuel economy standards for larger vehicles, a policy referred to as "attribute-based GHG standards for light-duty vehicles" [29]. While EPA regulations had distinguished between "passenger" and "non-passenger" vehicles since the introduction of CAFE standards in 1975, the

---

[2] Supersized (adjective) and supersize (verb) are words that were first used in 1987 by the McDonalds fast food chain to describe the option for extra large orders of French fries or beverages. It was eventually assimilated into American vernacular and is now applied to anything made unnecessarily big (including cars).

2012 modification established different fuel economy and emission standards based upon the size of the vehicle, determined by the wheelbase multiplied by its track length (width). *Larger cars were allowed higher fuel use and higher regulated emissions.* Industry response (all manufacturers for the US market) was as expected: more and ever-larger trucks and SUVs, which could be dubiously classified as *non-passenger vehicles* as a result of the Energy Independence and Security of 2007 (see Appendix 1). The US automotive vehicle mix changed, reversing the late-1970s trend toward smaller and more efficient cars. The shift toward increasingly massive light trucks and SUVs for passenger use has had a global impact. The SUV designation emerged in the 1990s, which were "supersized" station wagons that fortuitously (for the manufacturer) were allowed to be classified as light trucks, evading the stricter safety requirements of automobiles. Another fabricated class, the *Crossover SUV*, had by 2023 replaced *all* compact and subcompact cars manufactured in the USA. Although some were/are smaller than the station wagons that preceded them, they were treated as a subclass of SUVs, and therefore allowed the light truck classification.

Some of the increased mass can be attributed to increased emissions and safety equipment requirements. But these factors could not explain such a large change.

As reported by Reuters [30],

> An 8 square foot increase in the footprint of a vehicle can allow for 2% to 3% more carbon dioxide emissions, according to industry experts.
>
> SUVs were the second-largest contributor to the increase in global CO_2 emissions since 2010 after the power sector—the International Energy Agency warned in a report in October.
>
> We didn't expect that SUVs would be so prevalent—said Margo Oge, former head of the U.S. Environmental Protection Agency's office of transportation air quality.

In 2024, marketing continues to appeal to everything from luxury image to patriotism to convince buyers that they need larger vehicles, e.g., with "Tough Truck" imagery and the promise of greater luxury in SUVs nearly the size of studio apartments. As explained in the Climate Edition of The New Republic [31] in 2023:

> … between 2012 and 2021 the EPA found that the percentage of new vehicle sales classified as passenger cars and those classified as light trucks has essentially flipped. In 2012, 64 percent of new vehicle sales were classified as passenger vehicles, while 34 percent were classified as light trucks. By 2021, light trucks accounted for 63 percent of sales while passenger vehicles accounted for 37 percent of sales. 'Sedans have largely been replaced with taller vehicles such as truck-like SUVs and crossover utility vehicles (CUVs),' the agency writes. Pickup trucks' share of new cars sales jumped from 10 to 16 percent over the same period. During that time, the overall average footprint of new cars grew by more than 5 percent.
>
> American autos aren't bigger because consumers have suddenly embraced off-roading, the construction trades, or home improvement projects. They're bigger because automakers want to escape regulations. Each manufacturer is required to comply with boutique greenhouse gas emissions standards, which are calculated based on the size and capabilities of the cars in their fleets. Smaller cars are held to different standards than larger cars. So are

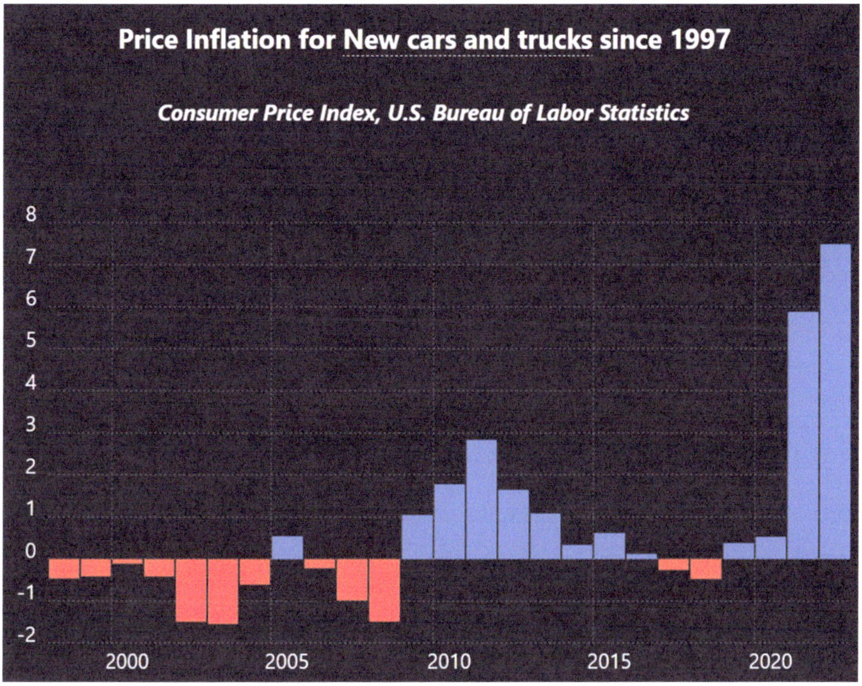

Fig. 3.8 Percent change in US average cost of new cars, 1997–2022, adjusted for inflation. Bureau of Labor Statistics, https://www.in2013dollars.com/New-cars-and-trucks/price-inflation. Public domain

those with specialty features like all-wheel drive or large towing capacities. By changing the makeup of their fleets, in other words, car companies can change the standards to which they're held.

The motivation for manufacturers to shift production to large trucks and SUVs goes beyond regulatory compliance. Struggling to maintain profitability with small cars and compact trucks, automakers were motivated to shift production to larger more expensive trucks and SUVs that provided greater profit margins for themselves and for their franchised dealers. It is actually a double win for manufacturers. Figure 3.8 shows year-to-year cost increases normalized to 2013-constant USD;[3] the retail mean cost of cars across the model range has dramatically increased, especially in the years since, compared with the much lower rate of increase during the prior 20 years [32].

The proliferation of electric vehicles since 2012 has only exacerbated this irony. As discussed in detail in Chap. 14, larger electric SUVs and trucks disguise their increased mass due to the batteries needed by electric vehicles to achieve the range and performance that attract buyers. It is no surprise that the majority of new electric

[3]This is intended to differentiate the trend in the People's Republic of China, in which vehicle costs, especially for battery electric vehicles, have steadily declined over the past 5 years.

vehicles offered for sale in the USA weigh over 2300 kg 5000 lbs. a weight class that, in the (pre-climate awareness) 1980s, contained almost exclusively commercial trucks and vans.[4] In 2023, new electric vehicles weighed an average of 30% more than equivalent gasoline vehicles, with an incremental weight of 1000 lbs. due to the battery [33] and the mass of the car needed to safely carry it. The incremental weight difference of electric trucks is as high as 2900 lbs. It is not surprising that manufacturers now avoid listing vehicle curb weights among their technical specifications on product web pages,[5] especially for electric vehicles.

By restructuring model lines with larger more luxurious versions at minimal incremental manufacturing cost, it is possible to extract higher profits from consumers willing to pay higher Manufacturer's Suggested Retail Price (MSRP). The bigger-is-better strategy also reinvigorates dealers who realize higher profits from higher cost vehicles, and routinely engage in the practice of additional dealer markups to extract even more money from purchases of popular models [34].

In March 2024, MSN reported [35]:

> 80-percent Of New Vehicles Sold In 2022 Were SUVs And Trucks.

GM Authority reported [36]:

> Thanks to strong demand and high profit margins, it's really no secret that full-size trucks and SUVs serve as cash cows for General Motors.

In June 2023, Reuters reported [37]:

> Last year, GM's average per-vehicle earnings before interest and taxes on all of its trucks and SUVs was $10,678.

The underlying reason for this trend? As reported in the New York Times [38].

> G.M.'s Profits From Trucks and S.U.V.s Fuel Its Electric Quest.

Lamenting the lack of a US-made truck that is not grossly oversized, Benjamin Hunting of IH (InsideHook) commented [39],

> The same well-oiled machine that keeps big-boned pickups flying off dealer lots has also invested billions of dollars indoctrinating buyers in the belief that parking anything less than the Titanic in the driveway is tantamount to treason.

The hidden ramifications of the increased weight of electric vehicles will be discussed further in the Chap. 14.

---

[4] A 1988 Ford F350 XLT Dually truck had an unladen weight of 4725 lbs. (2143 kg). Less than a 2022 Tesla S P100D, 4890–5029 lbs (2218–2281 kg) (Ford and Tesla manufacturer data).

[5] For example, Ford specifications for F-150 series trucks, https://media.ford.com/content/dam/fordmedia/North%20America/US/product/2024/f150/2024%20Ford%20F-150%20Tech%20Specs.pdf

Alternative Fuel Options Have Always Been Available

Figure 3.9 is a group finish line photograph of the competitors in the 1970 Clean Air Car Race, a national rally for alternative fuel vehicles from Caltech in Pasadena, California to MIT in Cambridge, Massachusetts. Fuels included CNG, LPG, hydrogen, methanol, ethanol, vegetable oil, and wood gas [40]. Over 50 years later, the world has more automotive energy options than ever before in history. In addition to battery electric propulsion, a wide range of alternative combustion fuels have been considered and sometimes supported by government and private funding. Early university alternative fuel student projects have rarely received media attention, being generally perceived by media and the public as "college students having fun." But these activities were the genesis of many of the ideas that have become options today, and they helped to create public awareness of alternatives to petroleum for transportation.

And just as environmentalists who tracked declining recoverable oil reserves before the 1990s were counting on this scarcity to save the environment, hydraulic fracturing technology changed everything, allowing previously depleted oil fields to be productive again. This innovation is expected to extend the oil Hubbert Curve by between 15 and 50 years.

The Renaissance of electric propulsion began in the late 1990s when a few major manufacturers cautiously introduced BEVs such as the GM EV-1 in 1996, which used in its last production year nickel-metal hydride (NiMh) batteries. However, the development that enabled the EV revolution was lithium-based batteries, having

Fig. 3.9 Competitors in 1970 Clean Air Car Race, a rally for alternative fuel vehicles from Caltech to MIT. Fuel sources included battery electric, CNG, LPG, hydrogen, methanol, ethanol, and wood. Photo from https://alum.mit.edu/sites/default/files/images/Slice_202009_Clean_Air_Car_Race_court.jpg. Public domain

energy storage densities 3–6 times greater than previous lead-acid, NiCad, or NiMH batteries. Government incentives for producing and purchasing electric vehicles have been key to driving this transition. Indeed, it is unlikely that any of the newly launched electric vehicle companies circa 2010 could have survived without some of the most significant public investments in history, not only in the USA but even more so in China and Europe. A new business model emerged based on government-backed loans, grants, tax incentives, carbon offset, and regulatory credits, often without adequate oversight. These will be explained in Chap. 17. No politician would risk an accusation of not doing their part to support energy independence and the emergent notion of sustainability. The traditional coupling between sales revenue and costs became secondary. The current EV market leader, Tesla, only emerged from negative margins in 2020, and by 2022 reported 26.7% operating margin, the largest ever in the auto industry. But the majority came from the sale of $1.6 billion USD in regulatory credits [41]. Most EV *subsidy harvesters* came and went in a few months to years, typically after overly optimistic market predictions faded and investor expectations were not met. Still, other firms such as hydrogen truck maker Nikola, and EV hopefuls Faraday Future and Fisker remained in business for a decade or longer, spending down investor funds and government loans without ever returning a profit, eventually ending in bankruptcy.

The Importance of Finding the Best, Not Just Expedient, Solutions

The current (2023) consensus of government entities and most car manufacturers worldwide is that the future is electric vehicles with energy stored in lithium batteries or delivered via hydrogen fuel cells. Almost all major world manufacturers have pledged to sunset all ICE automobile production in favor of BEVs, some as early as 2030. To support these vehicles, large government investments have been made in subsidized public and captive (e.g., Tesla) charging infrastructure, or in hydrogen fueling stations (in California). Despite their significantly higher purchase costs and lingering concerns about range and charging time, electric and hydrogen cars and trucks appeal to an affluent segment of the public that wants to help alleviate climate change or improve energy independence, as long as it is not too inconvenient. The rush to electric automobiles has occurred so quickly that the impacts of this transition on electric power generation, distribution, and storage have suddenly become urgent public policy issues.

The anticipated windfall of usable renewable electricity from solar and wind may not play out as planned due to the time misalignment and unpredictability between production and demand. In the USA, grid energy storage is a major target of the massive 2022 Federal Inflation Reduction Act (IRA). But the sheer magnitude of the challenge eludes most discussions. In an all-electric future that relies heavily on solar and wind intermittent energy resources (IERs), a nearly unfathomable amount of energy must be stored and released every day, either locally or in the utility grid. Meanwhile, a huge increase in electricity demand is in progress due to the transition of automotive energy from petroleum to electric. This topic will be discussed later in the Chap. 14.

At this time, grid power generation is not appreciably less carbon-intensive than the direct combustion of some alternative fuels. But there are reasons to be hopeful that electric power generation can decarbonize. Just recently, there have been public announcements about the potential of deep geothermal energy, which is beyond the capability of current drilling technology. And geologic hydrogen discoveries have led to nearly religious levels of optimism. And there is the traditionally maligned option of nuclear (fission) energy, a carbon-free but risk-laden option that has been in decline in the USA, Europe, and other Western countries since the 1970s. With the lowest number of operating nuclear power plants in the USA since the 1960s, and 10 + years of approval/design/build time for new nuclear facilities, even if a radical investment in nuclear energy is made right now, the energy contributions will remain off the table until after the transition to EVs is targeted to be complete.

In planning our transportation future, all-electric or hydrogen transportation's merits and commercial advantages are well publicized and enticing. But realities are deeper than the sound bites. BEVs and hydrogen FCVs are not always the optimum solution. Worldwide, despite the meteoric market growth that peaked in late 2023 of BEVs and FCVs, approximately 99% of all automobiles and trucks are still powered by ICEs [42].[6] The year 2035 is the target date for nearly complete electrification in the USA and Europe, but according to the New York Times,

> Automakers are now shifting to electric vehicles, which could make up one-quarter of new sales by 2035, analysts project. But at that point, only 13 percent of vehicles on the road would be electric. Even in 2050, when electric vehicles are projected to make up 60 percent of new sales, the majority of vehicles on the road would still run on gasoline [43].

Recently, enthusiasm for BEVs has become less sanguine now that the early adopters and affluent fashion-forward markets are nearly saturated, leaving market success in the hands and checkbooks of everyday automobile owners. As of late 2023, an abrupt shift in popularity from BEVs to plug hybrids is obvious from sales data for each [44]. There is growing skepticism that perhaps BEVs were over-promoted and over-incentivized. The realities of higher cost, shorter range, refueling inconvenience, and safety concerns matter. The shortcomings of the electric power infrastructure can eventually be overcome, but at a huge cost in the USA alone. As reported by the Wall Street Journal [45]:

> EPRI projects utility companies overall will spend $1.5 trillion to $1.8 trillion on infrastructure and operations by 2030.

Another subtle point is missed when, in virtually all reports, electric vehicles are compared only with the status quo: gasoline and diesel fuel. The eventual demise of fossil fuels is a foregone conclusion. The decisions before us now are how to replace

[6] After a decade of rapid growth, in 2020 the global electric car stock hit the 10 million mark, a 43% increase over 2019, representing a 1% increased share of the US vehicle market.

them with the *best* of all available options for a given application. Comparisons might be more relevant if they admitted all viable alternatives to petroleum, including some of the options discussed in this book that have remarkably favorable energy and environmental footprints, even compared with electric vehicles.

References

1. International Energy Agency data consolidated by Statista, DossierPlus, The potential impact of battery electric vehicles on CO_2 emissions, p. 7 Passenger cars produce the largest share of CO_2 emissions. (2022), https://www.statista.com/statistics/1185535/transport-carbon-dioxide-emissions-breakdown/. Accessed 11 June 2023
2. Wikipedia, History of the internal combustion engine, https://en.wikipedia.org/wiki/History_of_the_internal_combustion_engine
3. E. Wakefield, *History of the Electric Car* (SAE Press, 1994)
4. https://www.history.com/this-day-in-history/ford-motor-company-takes-its-first-order. Accessed 12 Dec 2022
5. What is a Logit Function and Why Use Logistic Regression? https://www.theanalysisfactor.com/what-is-logit-function/
6. https://www.greencarreports.com/news/1093560_1-2-billion-vehicles-on-worlds-roads-now-2-billion-by-2035-report. Accessed 1 March 2023
7. R. Stumpf. Here's about how many cars there are in the world in 2023. (The Drive, 2023), https://www.thedrive.com/guides-and-gear/how-many-cars-are-there-in-the-world. Accessed 10 Dec 2023
8. U.S. Energy Information Administration, Monthly Energy Review, Table 2.1, 2022, preliminary data. https://www.eia.gov/energyexplained/use-of-energy/transportation.php. Accessed May 2023
9. Current California GHG Emission Inventory Data 2000–2021 GHG Inventory (2023), California Air Resources Board, https://ww2.arb.ca.gov/ghg-inventory-data. Accessed 10 March 2024
10. U.S. Energy Information Administration, Monthly Energy Review, 2022, preliminary data as reported in https://www.eia.gov/energyexplained/use-of-energy/transportation.php. Accessed May 2023
11. U.S. Energy Information Administration (EIA), Monthly energy review, tables 2.5, 3.8c, and A1, April 2022, and EIA petroleum navigator, April 2022; preliminary data
12. US EPA, Inventory of U.S. Greenhouse Gas Emissions and Sinks: 1990–2021
13. Earth Science Stack Exchange, https://earthscience.stackexchange.com/questions/18354/why-is-carbon-dioxide-a-greenhouse-gas-whereas-ammonia-is-not. Accessed 11 Jan 2024
14. H. Tian, R. Xu, J.G. Canadell, et al., A comprehensive quantification of global nitrous oxide sources and sinks. Nature **586**, 248–256 (2020) https://doi.org/10.1038/s41586-020-2780-0
15. US EPA Publication, Greenhouse gas emissions from a typical passenger vehicle, p.2. EPA-420-F-18-008 (2018)
16. The Globe Program, Scientist' blog. Carbon dioxide Part 2: Walk, drive a car or ride a bike? (2008), https://globe.gov. Accessed 1 Nov 2024
17. US Dept. of Energy, Fuel properties comparison, https://afdc.energy.gov/fuels/properties. Accessed 15 May 2024
18. US EPA, Greenhouse gases equivalencies calculator - Calculations and references. https://www.epa.gov/energy/ghg-equivalencies-calculator-calculations-and-references. Accessed 14 May 2024
19. US EPA, Greenhouse gas rating. https://www.epa.gov/greenvehicles/greenhouse-gas-rating. Accessed 1 Sept 2024

20. Federal and California light-duty vehicle emissions standards for air pollutants. (US EPA), https://nepis.epa.gov/Exe/ZyPDF.cgi?Dockey=P100O1EZ.pdf. Accessed 1 July 2024
21. US Dept of Energy. fueleconomy.gov, https://www.fueleconomy.gov/feg/Find.do?action=sbsSelect. Accessed 1 July 2024
22. Find a smart-way vehicle. (US EPA) https://www.epa.gov/greenvehicles/find-smartway-vehicle. Accessed 1 July 2024
23. Federal and California light-duty vehicle emissions standards for air pollutants EPA-420-B-19-043. (US EPA, 2019), https://nepis.epa.gov/Exe/ZyPDF.cgi?Dockey=P100XCIV.pdf. Accessed 1 Dec 2024
24. Sensing the adequate mixture – the Bosch Lambda sensor. (Robert Bosch GmbH, 2024), https://www.bosch.com/stories/40-years-of-bosch-lambda-sensor/. Accessed 10 March 2024.
25. K. Aronoff, The truck-size loophole in the EPA's car emissions rule. (The New Republic, 2024), https://newrepublic.com/article/180263/epa-tailpipe-emissions-loophole. Accessed 15 June 2024
26. Chicken tax (Wikipedia, 2024), https://en.wikipedia.org/wiki/Chicken_tax. Accessed 10 Aug 2024
27. Policy studies. Competition in currency: The potential for private money. (Cato Institute, 2012), https://upstatement.cato.org/publications.html?#. Accessed 1 May 2024
28. FuelEconomy.gov—Search by EPA Size Class. (US Dept of Energy) https://www.fueleconomy.gov/feg/byclass/byEPAclassNF.shtml Accessed 15 June 2024
29. Model year 2010 fuel economy guide. US Dept of Energy document DOE/EE-0328. (US Environmental Protection Agency, 2020), Updated 15 Sept 2020. https://www.fueleconomy.gov/feg/pdfs/guides/FEG2010.pdf. Accessed 15 March 2024
30. J. White, B. Rigby, How U.S. emissions rules encourage larger SUVs and trucks. (Reuters Business, 2020), https://www.reuters.com/article/idUSKBN21D1KK/. Accessed 20 March 2024
31. K. Aronoff. Remember that time the EPA killed the sedan? (New Republic, Climate, 2024), https://newrepublic.com/article/180044/epa-small-cars-sedan-suvs. Accessed 23 March 2024
32. New cars and trucks priced at $20,000 in 1997 → $23,280.27 in 2022. (US Bureau of Labor Statistics, as published by Official Data Foundation/Alioth LLC, 2024), https://www.in2013dollars.com/New-cars-and-trucks/price-inflation. Accessed 20 March 2024
33. R. Heaps (Kelly Blue Book, 2024), https://www.kbb.com/car-advice/heaviest-electric-vehicles/. Accessed 1 Aug 2024
34. T. Duffy, The 10 new cars with the craziest dealer markups right now. (Motoring News, 2023). https://www.gearpatrol.com/cars/g41331731/cars-biggest-dealer-markup/. Accessed 10 March 2024
35. O. Hicks, The real reason why SUVs are so popular in the U.S. (MSN republishing article from TopSpeed, 2024) https://www.msn.com/en-us/autos/news/the-real-reason-why-suvs-are-so-popular-in-the-us/ar-AA1ifCrg. Accessed 23 March 2024
36. T. Hawkins, Here's how much profit GM makes on full-size trucks and SUVs. (GM Authority, 2023) https://gmauthority.com/blog/2023/06/heres-how-much-profit-gm-makes-on-full-size-trucks-and-suvs/. Accessed 20 March 2024
37. P. Lienert, J. White. GM could reap billions by building combustion trucks and SUVs through 2035. (Reuters Autos and Transportation, 2023) https://www.reuters.com/business/autos-transportation/gm-could-reap-billions-by-building-combustion-trucks-suvs-through-2035-2023-06-13/. Accessed 20 March 2024
38. N.E. Boudette. G.M.'s profits from trucks and S.U.V.s fuel its electric quest (New York Times, 2021), https://www.nytimes.com/2021/02/10/business/gm-profits-electric-cars.html. Accessed 10 Feb 2021
39. B. Hunting, Americans are obsessed with the wrong trucks. (Inside Hook, 2024), https://www.insidehook.com/lifestyle/americans-obsessed-wrong-trucks-ford-maverick Accessed 3 July 2024

40. How a bunch of college kids convinced Detroit to cut smog in 1970. (Hemmings Motor News, 2014), https://www.hemmings.com/stories/2014/09/24/how-a-bunch-of-college-kids-convinced-detroit-to-cut-smog-in-1970. Accessed 1 June 2023
41. F. Zandt, *How Successful is Tesla* (Statista, 2024)
42. International Energy Agency. "Global EV Outlook 2021: Accelerating ambitions despite the pandemic".(International Energy Agency (IEA), Clean Energy Ministerial, and Electric Vehicles Initiative (EVI), 2021) https://www.iea.org/reports/global-ev-outlook-2021. Accessed 16 May 2021.
43. Electric cars are coming. How long until they rule the road? (New York Times, 2021), https://www.nytimes.com/interactive/2021/03/10/climate/electric-vehicle-fleet-turnover.html
44. P. Valdes-Dapena, How EVs became such a massive disappointment. (CNN, 2023), https://www.cnn.com/2024/02/25/cars/what-happened-with-electric-vehicle-sales-in-2023/index.html Accessed 1 July 2024
45. B. Ziegler, Can the power grid handle a wave of new electric vehicles? (Wall Street Journal Reports: Energy, 2023), https://www.wsj.com/articles/url-us-power-grid-electric-vehicles-ev-charging-11675444994 Accessed 1 July 2024

4 Engines and Fuels

An automobile's engine or electric motor is an energy conversion device, transforming stored fuel or electrochemical energy into mechanical work. For internal combustion vehicles (ICVs), the energy conversion mechanism is the combustion of fuel and air in a cylinder, which increases the pressure in a cylinder, forcing a piston to move. The linear motion of the piston is translated into rotary motion by a crankshaft, which transfers mechanical power through a transmission and drivetrain, ultimately causing the rotation of the drive wheels.

For battery electric vehicles (BEVs) or fuel cell vehicles (FCVs), the electric motor converts electrical energy from a battery into rotary motion to drive the wheels. For a hybrid car, both can provide the motive power, either in series or parallel. In this chapter, the focus is just on combustion fuels and internal combustion engines (Figs 4.1 and 4.2).

Automotive internal combustion engines are usually classified by the type of fuel and how the fuel is ignited. While small utility engines such as those in chainsaws or leaf blowers operate on two strokes, intake/compression, and power/exhaust, all automotive engines operate with the same four "strokes," each a half-turn of the crankshaft so that power is produced only once every two rotations of the crankshaft. This is true for both gasoline and diesel engines, the difference being:

- Spark Ignition (SI) = Gasoline, alcohols, gaseous fuels: Fuel is ignited with a spark plug.
- Compression Ignition (CI) = Diesel, biodiesel, DME: Fuel ignites as it is injected into cylinder.

C. A. MacCarley, *Non-Petroleum Automotive Transportation*,
https://doi.org/10.1007/978-3-031-85028-8_4

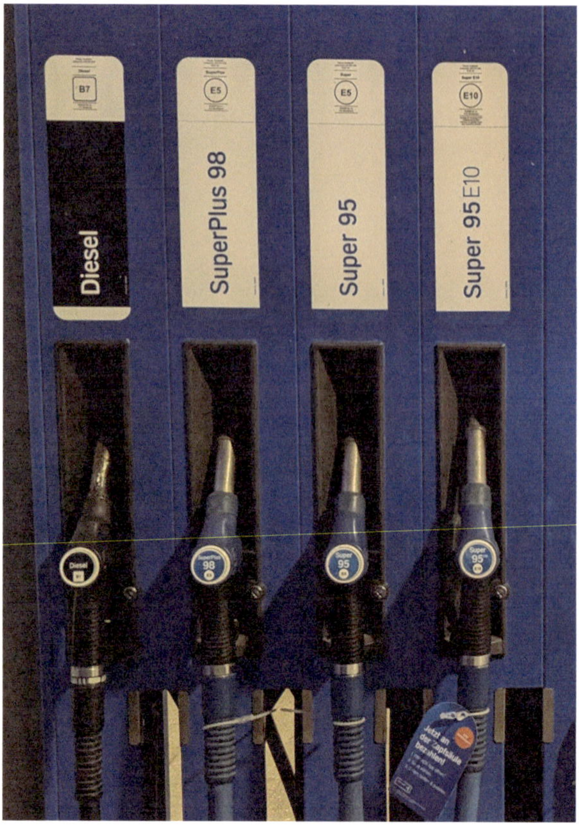

Fig. 4.1 Fuel pump dispensing two grades of E5, as well as E10 and diesel fuel at ARAL petrol station in Munich, Germany. (Photo: author)

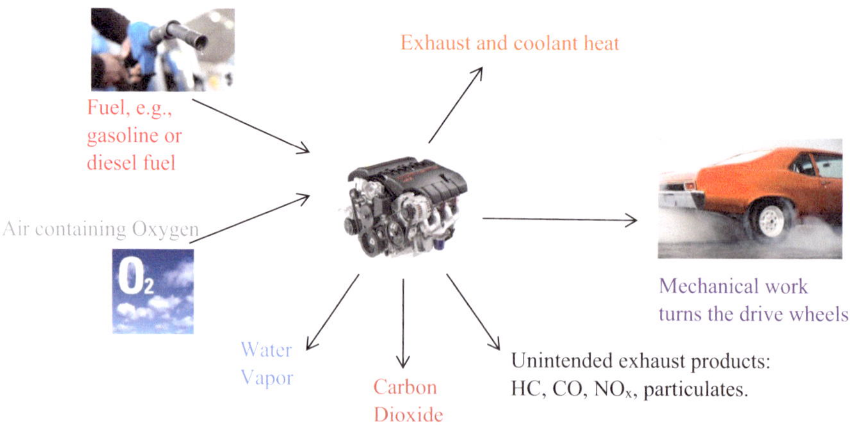

Fig. 4.2 The energy conversion processes of an IC engine. (By author, using PD clip art)

Attributes of a Good Motor Fuel

We are used to assuming that automobile engines can only run on ***gasoline (aka petrol, benzin) or diesel fuel***. But, in fact, *any liquid or gas* that combusts in air is a potential IC engine fuel. But some are better than others (Fig. 4.3). What we want is:

- Maximum Efficiency: Greatest mechanical work per unit of fuel energy. *Minimize waste heat per unit of fuel energy. While the engine mechanics are primarily responsible for thermal efficiency, the fuel properties can also affect this.*
- Maximum Power: Highest rate of output energy generation. *This means we maximize the rate of energy conversion.*
- Minimum regulated pollutants from combustion: Assure that we minimize the generation of output gases other than carbon dioxide and water vapor.
- Minimum greenhouse gas emissions (mostly carbon dioxide).
- It can be extracted inexpensively directly as a natural resource or can be produced from an inexpensive feedstock. Renewable natural resources preferred if long-term costs and environmental consequences are considered.
- The fuel or energy production process is energy efficient and has minimal environmental consequences.
- It can be distributed inexpensively and quickly with minimal infrastructure requirements.
- It can be stored locally safely and with minimum energy loss.
- It can be dispensed into a vehicle safely, quickly, conveniently, and at acceptable cost.
- It can be stored onboard the vehicle safely and at a high energy density (volume and mass density). This determines the travel range of the vehicle.
- Power output of engine can be easily controllable.
- Will not cause excessive wear or otherwise damage the engine.
- It is aligned with the financial and/or political objective of those involved in its extraction, production, cultivation, transportation, promotion, and point of sale.

Fig. 4.3 Image depicting a fictional "green" future based on renewable fuels, generated by Dall-e from author's prompt

As mentioned in the *Introduction*, any fuel other than gasoline or diesel fuel can be referred to as an *alternative fuel*. Some are considered *renewable fuels* if their complete production and use cycle does not produce positive amounts of carbon dioxide, although this definition is often applied to any fuel or energy source that starts with plants or biomass, even if not carbon-free.

How Combustion Fuels Produce Mechanical Power

Combustion is the reaction between fuel and oxygen initiated by a spark. It converts the fuel and oxygen into (mostly) water vapor and carbon dioxide while releasing heat:

$$\text{Fuel} + \text{Oxygen} \rightarrow \text{Carbon dioxide} + \text{Water} + \text{Heat}$$

A large amount of combustion heat is released, and there is a slight increase in the molar volume of the products (1:0 intake mole → 1.23 exhaust moles for gasoline). This increases the pressure in the cylinder, pushing on the piston, which turns a crankshaft, converting the linear motion of the piston into rotational work—the engine's rotational power output.

The Four-Stroke Engine

The four-stroke reciprocating (piston) engine is found in almost every automobile, truck, or motorcycle on the road today. It is not the only mechanical configuration—the two-stroke and Wankel rotary engines are others. Still, it is the best configuration to simultaneously meet the emissions, efficiency, reliability, and cost criteria of modern automobiles.

Two complete rotations of the crankshaft are required for a complete engine cycle. The two rotations are divided into four *stokes*, or specific mechanical processes. During each stroke, the crankshaft rotates 180° (π radians), so a complete cycle takes 720° of rotation (Fig. 4.4).

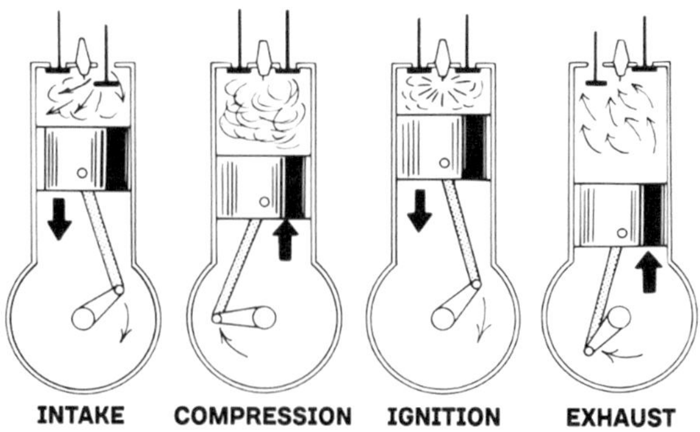

Fig. 4.4 Simplified diagram of a four-stroke IC engine. (https://openclipart.org/detail/295364/4stroke-engine-cycle. Public domain)

Intake: Starting at the top of the cylinder, the piston descends with the intake valve open. Air is drawn into the combustion chamber. Fuel is added to the air by a fuel injector or carburetor.

Compression: The intake valve closes. The piston rises, compressing the air and fuel *mixture* or *charge*.

Power: Shortly before the piston reaches the top of the cylinder, a spark plug ignites the air and fuel. The mixture starts to burn rapidly increasing the pressure in the cylinder. This pushes the piston down, creating a force on the crankshaft that causes it to rotate, producing mechanical work. *This is the only stroke that creates output power*. The others are just for getting the fuel and air into the cylinder, compressing it, and letting the exhaust out.

Exhaust: When the piston reaches the bottom of its stroke, the valve opens, and the pressurized exhaust leaves via the exhaust manifold and pipe. As the piston rises, the remaining exhaust is cleared from the cylinder. The loud exhaust noise occurs at the moment the exhaust valve opens, and the still-pressurized exhaust gases are released from the cylinder.

Repeat: The cycle repeats, starting with the intake stroke.

The Status Quo: Gasoline

Gasoline (aka petrol, benzin, benzine, benzina, gasoline, gasorin, E0) is a mixture of many hydrocarbons spanning carbon numbers from propane (C_3H_8) to dodecane ($C_{12}H_{26}$). The composition is variable batch-to-batch and season-to-season, within standards set by regulatory organizations in each country. All components (except for trace sulfur compounds) are hydrocarbons, meaning their molecules contain only hydrogen and carbon. For this book, there is no need to go into the chemistry of the 100+ hydrocarbons that can be found in gasoline, but their types may be referred to later. Below are sound-bite descriptions of the main components of gasoline (not including added ethanol in E10).

- Alkanes (paraffins): Straight-chain saturated hydrocarbons such as n-hexane or n-octane. 15–40% ('n' is an optional designation meaning *normal*, containing only single carbon bonds)
- Alkenes (olefins): Unsaturated hydrocarbons such as hexene: approximately 10% (Contain at least one carbon double bond that displaced some hydrogen)
- Iso-alkanes: "Branched" alkanes such as 2-2-4 isopentane, aka "iso-octane": 25–40% (Alkanes with one or more hydrogen atoms replaced by CH_3, called methyl groups)
- Cycloalkanes (naphthenes): Saturated cyclic hydrocarbons such as cyclohexane: 10–20% (Alkanes that are arranged in a ring but contain only single bonds between carbon atoms)
- Aromatics: Unsaturated cyclic hydrocarbons such as benzine or toluene: up to 25% (Unsaturated cyclic hydrocarbons have alternating single and double bonds)

Crude oil undergoes extensive processing to isolate the various hydrocarbon components (fractional distillation), catalytic cracking to break down larger compounds into smaller compounds, usually branched alkanes, hydrocracking that adds hydrogen to unsaturated hydrocarbons, reforming to synthesize other hydrocarbons, and catalytic desulfurization and denitrogenation to remove sulfur and nitrogen compounds [1].

The important attributes that matter to us in predicting the carbon dioxide emissions of a fuel is the hydrogen and the carbon content, since when burned, the hydrogen atoms form water vapor, and the carbon atoms form carbon dioxide.

Saturated means that the hydrocarbon contains the maximum number or hydrogen atoms bonded to each carbon atom, e.g., hexane C_6H_{14}. Alkanes are saturated.

Unsaturated means that fewer than the maximum number of hydrogen atoms are present, with carbon–carbon double or triple bond at their bond sites, e.g., hexene C_6H_{12}. Alkenes are unsaturated.

An intuitive metric for differentiating the global warming impacts of various gasoline compositions as well as various alternative fuels is the *hydrogen-to-carbon ratio*. Saturated hydrocarbons have higher H:C ratios than unsaturated hydrocarbons, producing more water vapor and less CO_2 for a given amount of fuel energy.

Since gasoline is a blend of many hydrocarbons, it cannot be represented by a single hydrocarbon molecule in a stoichiometry equation. A surrogate pseudo-molecule with approximately the same molecular composition and specific calorific value (fuel energy) as the US 2018 average for gasoline is $C_{7.5}H_{15.2}$.[1] A simplified representation of the stoichiometric combustion reaction for gasoline and oxygen can be written as

$$C_{7.5}H_{15.2} + 11.3O_2 \rightarrow 7.5CO_2 + 7.6H_2O + \text{heat}$$

Gasoline by itself is a poor SI fuel due to its propensity to auto-ignite at points in the combustion chamber prior to the arrival of the flame front in the late compression and early power stroke. This results in destructive *engine knock* (aka ping) that can be clearly heard outside the engine. Elimination of knock requires the addition of anti-knock additives such as TEL, MTBE, TAME, methanol, or since 2006, ethanol. All motor fuels contain some form of anti-knock additive. The phenomena of knock and anti-knock additives will be discussed later.

[1] For energy content and density purposes, I derived this surrogate hydrocarbon molecule to match the density, hydrogen-carbon ratio and lower heating value of US gasoline as specified in the 2018 CFR. It is not a real substance. But it is usable as a single-molecule equivalent to actual gasoline in all computations in this book.

Gasoline Standards

The seasonal variability of gasoline composition makes it nearly impossible to calculate the exact AFR, energy content, and carbon dioxide emissions per mass of gasoline. These must be experimentally determined by government and professional society standard-setting organizations predominately those listed below. In the USA, specifications for gasoline are enforced by the Environmental Protection Agency (EPA) for all states except California and states that adopt California standards which have always been more rigorous than the federal "49 state" standards [2]. In the EU, standards are enforced by the International Energy Agency (IEA) [3].

API = American Petroleum Institute: http://www.api.org/news-policy-and-issues/news/2015/08/18/api-additional-oil-and-gas-regulations-u
AGI = American Gas Institute (for CNG, LNG, LPG): https://www.aga.org/
ASTM = American Society of Testing and Materials: http://www.astm.org/
CFR = Code of Federal Regulations: https://www.archives.gov/federal-register/cfr/ or
ASME/CFR = Cooperative Fuel Research: https://www.asme.org/about-asme/who-we-are/engineering-history/landmarks/50-cooperative-fuel-research-engine
SAE = Society of Automotive Engineers: https://www.sae.org/
EPA = US Environmental Protection Agency average online: https://www3.epa.gov/

The Importance of Precise Fuel Control

A spark ignition (gasoline) engine is controlled by throttling the intake airflow, not by controlling the fuel flow. The air and fuel must almost always form a mixture that is *stoichiometric* or *ideal*, i.e., an exact match between air and fuel so that all of the air and all of the fuel are consumed in the combustion, leaving no leftover air or fuel. This ensures complete combustion, which is critical to both the engine's efficiency and emissions. The stoichiometric mass ratio of air to fuel is referred to as the ideal ***AFR***, which is a fixed property for any fuel. *Methods for calculating the stoichiometric mass air/fuel ratio for any fuel are presented in Appendix 3.*

Table 4.1 lists the stoichiometric mass AFRs for various SI engine fuels. Gasoline and diesel fuel ideal AFRs were obtained from the cited references. All others are calculated in Appendix 3.

If the intake air/fuel mixture contains excessive fuel (a rich mixture), the unburned fuel is wasted causing poor efficiency and it forms high HC and CO emissions. If there is excessive air (a lean mixture), combustion quality degrades and power output is reduced from optimal. A lean but combustible mixture of gasoline and air can generate increased NO_x emissions due to higher peak combustion temperatures. If excessively lean, misfire (failure to ignite) can occur, allowing the entire unburned fuel/air mixture to flow out in the exhaust, even more polluting than a rich mixture.

Table 4.1 Stoichiometric mass AFRs for selected liquid fuels

Fuel	Stoichiometric mass AFR
Gasoline (2020 US, Regular, E0)	14.8[a]
E10 (gasoline with 10% ethanol)	14.1
Diesel (D2)	14.5[b]
Ethanol (E100)	9.0
E85 (nominally 85% ethanol, 15% gasoline)	9.8
Methanol (M100)	6.4

[a] Mean value for EU gasoline from Robert Bosch GmbH, Automotive Handbook, 10th ed. 2018. The slight increase in air mass to fuel mass compared with the traditional 14.7 AFR indicates a trend toward a slightly higher carbon-to-hydrogen ratio, possibly due to lower-quality crude oil
[b] Reference [4]

There are exceptions to the stoichiometric fuel/air requirement depending on the operational mode of the engine. The job of the electronic fuel control is to dispense exactly the *right* amount of fuel relative to the intake air to meet the operational needs of the engine under all conditions. At full throttle, slightly excess fuel (rich mixture) assures the utilization of all air in the cylinder, even though the excess fuel will be wasted and HC formation increased. During initial engine startup while cold, the fuel is also enriched. During deceleration, fuel is usually shut off completely. A slightly lean mixture is preferred to maximize efficiency and minimize HC and CO emissions during low power and mid-range cruise conditions. In none of these conditions are the deviations from stoichiometric large. The slight excess air assures that all fuel is consumed, which is preferable to a truly stoichiometric mixture that may not be perfectly mixed, leaving some of the fuel unburned.

The Normalized Mass Air/Fuel Ratio (AFR)

The Greek letter ***lambda*** λ is commonly used to express the air-to-fuel mass ratio normalized to the stoichiometric AFR:

$$\lambda = \frac{\text{AFR}}{\text{Stoichiometric AFR}}$$

AFR is the ratio of the air mass to the fuel mass. λ is the actual AFR divided by the stoichiometric AFR [5].

Prior to about 1980, but now making a comeback in research literature, is the normalized fuel-to-air *equivalence ratio* denoted by the Greek letter ***phi*** ϕ which is the numeric inverse of lambda: $\phi = \frac{1}{\lambda}$.

When the fuel and air relationship is *stoichiometric* or *chemically correct*, $\phi = \lambda = 1$. For a rich mixture, $\phi > 1$. For lean mixture, $\phi < 1$.

The job of approximately maintaining $\lambda = 1$ over the entire range of engine operating and environmental conditions is performed by either a carburetor or fuel injection, discussed below.

Carburetors

For almost 100 years, liquid fuel was metered into the intake air stream by a purely mechanical device called a *carburetor*, for example shown in Fig. 4.5.

Referring to Fig. 4.6, carburetors work on the Bernoulli Effect: Air passes through a constricted section of the intake (aka venturi, throat, barrel, or choke), which causes a small pressure drop. Located in this constriction is a small tube called a jet from which liquid fuel is drawn by the slight vacuum at a rate that theoretically should be proportional to the air velocity. The ratio of fuel to air should, ideally, create a mixture with a constant mass AFR at any airflow rate. A carburetor can be calibrated to provide the exactly intended AFR, but only at one flow condition. And even for the calibrated flow rate, the relationship will vary with the ambient absolute air pressure, temperature, humidity, and even the orientation of the carburetor. Not to mention variations in the fuel itself that may require changes in the AFR to maintain a stoichiometric mixture. Over the entire range of engine speeds and throttle positions, the air/fuel mixture will deviate significantly from ideal, causing either excess fuel or air in the resulting mixture, which leads to increased HC, CO, and NO_x emissions.

Possibly the most significant improvement in carburetor design was the variable venturi (aka constant velocity or constant depression) carburetor that adjusts the throat area in response to the downstream vacuum. As the airflow increases, the venturi diameter and the jet orifice diameter increase to maintain the ratio of fuel to

Fig. 4.5 Autolite 750 CFM 4-barrel carburetor from 1971 Ford 5.7L V8 engine. (Photo: author)

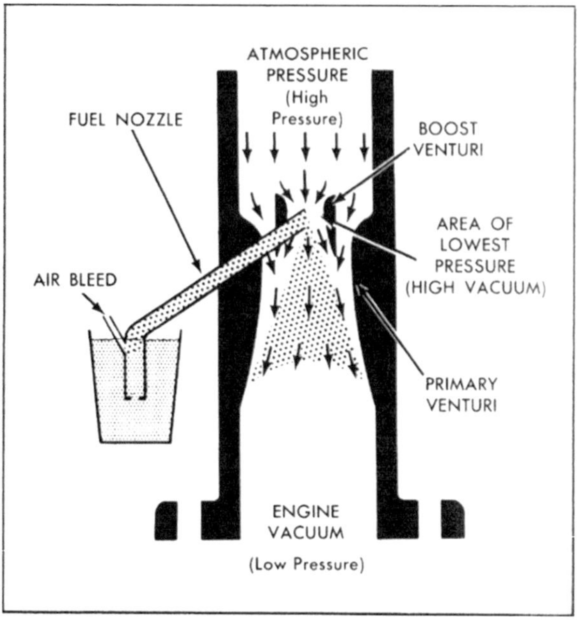

Fig. 4.6 Simplified diagram showing how a carburetor uses the Bernoulli Effect to draw liquid fuel into the intake air stream. (Image by Rusty Petrovic. https://carburetor.ca/tech/articles/CarbBasics/CarbBasics3.html. With permission)

air. The first patent for such a device was (probably) the *SU* carburetor invented by Herbert Skinner in 1904, but not widely used on cars until the 1930s. The legendary Weber Del Oro twin-choke carburetor invented in the 1920s also incorporated variable venturis. Yet despite these improvements, a century of inventions and hundreds of patents for carburetors that incorporated various fuel flow correction remedies, carburetors cannot be made to produce a consistent AFR over the entire airflow range, especially at partial throttle. By the 1950s, the design and setup of carburetors had taken on an air of tribal knowledge and conspiracy theories. Tales of mysterious carburetors that were claimed to give any vehicle over 100 mpg continued until the 1990s, and some still persist today.

By way of example, one of the more innovative (but grossly exaggerated) examples was an invention by Charles Pogue in the early 1930s of a carburetor that prevaporized gasoline using exhaust heat—US Patent 1,750,354 (see Fig. 4.7). While most automobiles enhanced fuel vaporization by providing intake manifold heating via engine coolant or exhaust gas, the Pogue device attempted to fully vaporize all fuel before induction into the cylinders. This could indeed improve the poor fuel atomization and consistency produced by conventional carburetors of the time. It therefore could increase fuel economy in poorly designed engines, but never to the 200 mpg claimed by the inventor and believed by the conspiracy-eager media. The complex fuel-heating device was also a potential safety hazard since it entrained a large volume of easily-ignited vaporized gasoline-air mixture. A scan of the Web

The Mystery of the 200-Miles Per-Gallon Pogue Carburetor

For the first time in 14 years, the inventor of one of the most controversial automotive advances of the past two decades reveals the story behind its strange disappearance

By I. T. Galanoy

One of the few photographs ever taken of inventor Charles Nelson Pogue was snapped in 1936 before he and his machine disappeared from the public eye.

Just a Start

Pogue, who until recently conducted his experiments at Winnipeg claims he can step up the carburetor to give 400 miles per gallon, but Hammell is counting, he says, on 200 miles per gallon, "at least for a start."

Twice Pogue's laboratory at Winnipeg was broken into, and working models of his carburetor stolen. It was understood, however, neither model was complete.

Pogue has reported exhaustive tests made with a popular low-priced car, equipped with the new carburetor, showed an average mileage of 216 miles per gallon of standard gasoline.

Fig. 4.7 Charles Pogue, circa 1935. (Image public domain, as reproduced in Automotive American, 5 June 2024. "Charles Nelson Pogue and the 200 mpg Winnipeg Carburetor" https://automotiveamerican.com/2024/06/05/charles-nelson-pogue-and-the-200mpg-winnipeg-carburetor/. Newspaper article and photo circa 1935, out of copyright)

reveals that some still believe that its disappearance was the result of auto industry, oil industry, or government malfeasance [6].

Electronic Fuel Injection

Electronic fuel injection allowed more precise fuel control using sensors that monitored the intake airflow, manifold absolute pressure, and/or throttle position to determine the correct fuel quantity for a given condition. An electronically determined amount of fuel is sprayed under moderate pressure into each intake port or into the intake manifold to create the desired AFR for any engine speed, load, temperature, and ambient pressure. Mechanical fuel injection had been used for several decades prior to the introduction of electronics, e.g., in high-performance racing and aircraft engines. It provided some improvement in AFR control as well as cylinder-to-cylinder mixture consistency. And for aircraft, it made possible the ability to fly inverted (carbs of the time only worked well if right-side-up). But its ability to maintain a desired AFR over the entire operational range of the engine was only slightly better than the carburetors of the 1950s–1970s. Mechanical fuel injection systems are still used today for methanol or ethanol fueled racing engines in drag racing or tractor pull competitions, since alcohol fuels are much more tolerant of poorly-regulated (rich) AFRs than gasoline.

Electronically controlled fuel injection made its debut in 1957 with the *Bendix ElectroJector* system that was sold as an option on a limited number of Chrysler and American Motors sedans 1957–1959. It was ahead of its time, and the electronic components beyond the ability of mechanics to service it. In 1957, there were no dire emissions or fuel economy fears to justify a system that was much more expensive than a carburetor. Gasoline typically cost $0.29 per gallon in the USA. Only one example of an intact ElectroJector system exists today [7]. In the early 1960s, Bendix sold the patents for electronic fuel injection to Robert Bosch GmbH that redesigned and improved it, releasing the first commercially successful EFI system in 1967, the Bosch D-Jetronic (D stood for *Druck*, the German word for pressure) system that was available as an option on some Volkswagen and Volvo vehicles. By this time, growing automobile air pollution in urban areas provided a strong consumer incentive. Less successful EFI systems appeared from other manufacturers, ranging from the AE-Brico system used on Aston Martin DB-series automobiles in the late 1960s, the GM/Bendix EFI system first used on the 1979 Cadillac Seville, and the Nippon Denso system first used on the 1971 Toyota Crown. The first EFI system on a production motorcycle was engineered by Mitsubishi of Japan and used on the 1980 Kawasaki Z1000-H1. All the early EFI variations except the AE-Brico and Mitsubishi systems were based on the Bosch D-Jetronic system that measured manifold absolute air pressure (i.e., air density) to determine the appropriate fuel quantity. The Bosch D-Jetronic system was followed in the mid-1970s by the L-Jetronic (L stood for *Luft*, the German word for air) system that relied on a manifold airflow sensor to provide the primary information for fuel calculation. Aiming for the low-cost market, Bosch concurrently produced the K-Jetronic or CIS (continuous injection system) system that was actually an airflow-controlled mechanical injection system, although it did employ electronic sensors for altitude and temperature fuel correction. (K stood for *Kontinuierlich*, the German word for continuous.).

The original Bosch D, L, LE, K, and similar systems by other manufacturers operated "open loop" without any form of AFR feedback. This limited their fuel control accuracy to the initial calibration of the system based upon the test conditions of the manufacturer or tuner. But even without this feature, the fuel economy and emissions improvements were consistently superior to carburetors.

Timed electronic injection systems such as those mentioned above (except the K-Jetronic) controlled fuel delivery by modulating the "open time" of the solenoid valves referred to as injectors (Fig. 4.8), typically between 1.0 and 12.0 ms. The amount of fuel delivered by a gasoline fuel injector is approximately proportional to the amount of time the injector is open, called the injector pulse duration. Injectors are designed for a specific mass flow rate at a given differential pressure, with 3 bar (43.5 psi) considered the reference pressure for calibration purposes. The injected fuel quantity is calculated electronically using multiple engine sensors: intake airflow, intake manifold absolute pressure, throttle position, engine coolant temperature, and atmospheric temperature.

Injection is synchronized with the engine or camshaft rotation such that one or two (for the L-Jetronic) injections occurred for each intake stroke. In further development, it was found that injection need not actually occur just before or during each intake stroke, although most modern port injection systems do synchronize injection with a specific rotational angle of the camshaft. The injectors of the D-series Bosch-type systems injected once every intake stroke, synchronized in pairs, using a rotational position sensor located in the ignition distributor. The

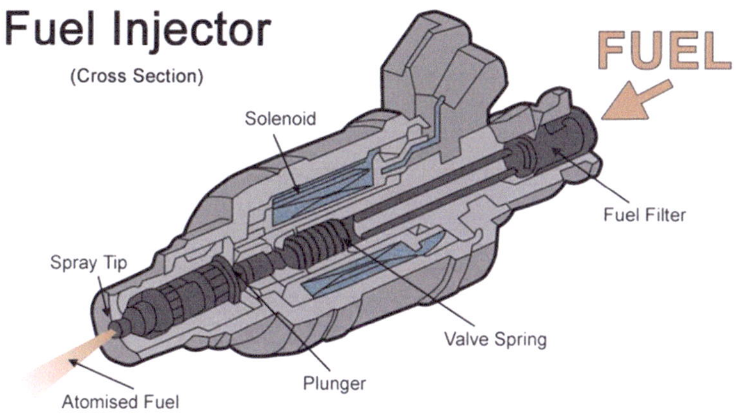

Fig. 4.8 Bosch electrically actuated gasoline fuel injector, first generation, 1967–1995. (Image from Wikipedia, GNU Free Documentation License. https://en.wikipedia.org/wiki/File:Fuelinjector.png)

L-series systems injected all cylinders at the same time, once every crankshaft rotation, regardless of where the injection occurred relative to the intake of each cylinder. This produced two injections for each intake stroke and allowed the use of a crankshaft position sensor that was more immune to ignition system electrical noise. The early Mitsubishi/Kawasaki system modified the L-Jetronic control method by skipping every other crankshaft rotation, providing only a single injection for each intake stroke, still with all cylinders injecting at the same time. For high-RPM motorcycle engines, this reduced the problem of the maximum injection period being too short for adequate fuel control.

Exhaust Oxygen Sensing Feedback Fuel Control

The world-changing improvement in automotive emission control occurred following the development of a durable two-stage (aka three-way because it was effectve with all three regulated pollutant gases) catalytic converter at Engelhard Corp. in the USA [8], and a reliable exhaust oxygen sensor by Robert Bosch [9], both in 1975. Bosch released the Lambda-Sond (λ-sensing) EFI system in 1976 that was offered on MY 1977 Volvo 240/260 sedans (Fig. 4.9).

Feedback fuel control was effective at keeping the AFR within the narrow window in which the three-way catalyst could operate correctly. It adapted to varying fuel composition and atmospheric conditions. The combined fuel control and catalyst technologies resulted in a dramatic reduction in HC, CO, and NO_x. Air pollution regulatory agencies in the USA, Europe, and Japan soon adopted more aggressive emission limits that were within the capability of the new technology but beyond the ability

Fig. 4.9 1977 Volvo 240 with Bosch Lambda Sond system. (Image from Volvo media. With permission)

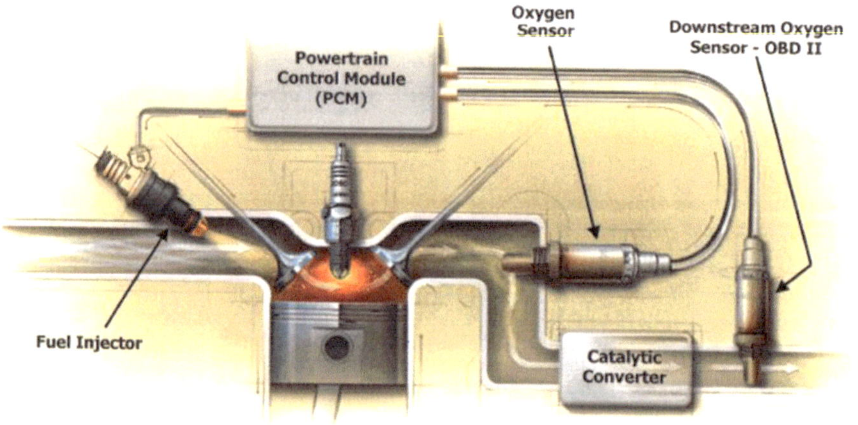

Fig. 4.10 Diagram showing the arrangement of the main components of an oxygen-sensing feedback fuel control and catalytic converter. (Image Copyright 2020 AA1Car.com and CarleySoftware.com. Used with permission)

of carburetors or open-loop EFI systems. By the 2000s, regulated emissions had fallen to levels over 100 times lower than a carbureted engine could achieve (Fig. 4.10).

In the late 1970s and early 1980s, there were various stories about the reluctance of the rest of the auto industry to adopt this innovation, which was more costly than traditional carburetors, and that established Bosch as the dominant world supplier of engine control electronics and components. Objections ranged from increased cost to the consumer, reliability of the electronic components and sensors, and the hazards of increased fuel pressure in the event of a leak. Political compromises allowed a more gradual transition to the stricter emission standards. One of these transition technologies was a classic case of retrograde cost-saving engineering: the modification of a carburetor to continuosly adjust the metering jet size in response to a signal from an oxygen sensor. Between 1978 and the mid-1980, these *feedback carburetors* were proffered as an acceptable alternative to oxygen-sensing EFI (they were not). They had some ability to improve fuel control, but they retained all the inherent limitations of carburetors. By the mid-1980, most manufacturers had relented, and oxygen feedback EFI systems with three-way catalytic converters had become the standard for new cars and light trucks. In 1991 in the USA, new EPA emission regulations made it effectively illegal to sell cars and light trucks that did not adopt electronic fuel injection with these advancements in emission control. One of the only exceptions allowed was the 1991–1994 Subaru Justy, the smallest car sold in the USA at that time.

This new generation of fuel and emission control systems was based on the confluence of four technologies from different engineering and science areas.

- Electronic fuel injection (port, throttle body, or direct cylinder): This facilitated precise control of the fuel quantity delivered to each cylinder, an objective that is unobtainable by carburetors.
- The high-temperature zirconium dioxide exhaust oxygen sensor [10], example shown in Fig. 4.11. A solid-state electrochemical cell that, once it is up to exhaust temperature, produces a voltage in the range of 0–0.7 V, indicative of the absence or presence respectively of at least trace oxygen in the in the exhaust stream. The presence of excess oxygen indicates a lean air/fuel mixture. The absence of any oxygen indicates that the mixture has excess fuel. Early exhaust oxygen sensor had response delays as long as 0.50 seconds (modern sensors about 0.10 s), which motivated the need for the next innovation.

Fig. 4.11 First-generation exhaust oxygen sensor. (Photo from journal4research.org. Public domain)

- Adaptive control methods allowed the electronic fuel controller to compensate for the finite response time of the oxygen sensor, correcting a computer *fuel map* at the engine's operating condition from a fraction of a second in the past, compensating for the response delay of the oxygen sensor. When that condition is visited again, the fuel quantity in the corresponding cell of the map has already been corrected. This creates the ability of the controller to *learn* the engine characteristics as the vehicle is driven, continuously updating the adaptive fuel map. In early versions, the updated map memory would be lost if the vehicle's 12 V battery was disconnected, requiring that after reconnection, the vehicle had to be driven long enough to visit enough operational conditions to reprogram the fuel map. This limitation changed after the development of low-cost nonvolatile solid-state computer memory in the 1990s. Figure 4.12 is an example of an adaption map for the simplest control method: engine speed (N) and throttle position (alpha) are the primary sensor inputs for measuring the intake airflow (thus the designation as an alpha-N system). The oxygen sensing feature is critical in this application to assure that the fuel map is updated not only for changes in intake air flow but for variable fuel composition, as required by *Flexible Fuel* vehicles that use variable blends of alcohol and gasoline.

Map 1 - Fuel - Cylinder 1.2.3.4 - Gear 1.2.3.4.5.6

RPM	% Throttle 0	2	5	10	15	20	40	60	80	100
500	236	84	44	29	19	21	23	12	10	10
750	136	35	10	5	5	5	5	5	5	5
1000	190	18	5	6	5	5	5	5	5	5
1250	190	34	14	11	5	5	35	5	5	5
1500	190	80	64	21	8	5	49	6	5	5
1750	196	54	66	44	6	8	6	5	5	5
2000	202	78	64	57	5	10	7	5	5	5
2250	190	64	26	55	7	12	3	6	6	8
2500	178	72	106	47	38	24	8	5	5	7
2750	178	92	74	60	43	78	18	3	5	7
3000	178	98	62	45	48	62	66	6	6	5
3250	202	104	56	46	31	47	70	50	7	5
3500	202	132	94	56	47	55	62	58	11	3
3750	174	98	80	64	56	54	50	72	27	10
4000	156	114	80	50	60	58	37	43	17	10
4250	136	86	74	46	49	60	55	33	5	5
4500	118	80	64	51	62	62	56	24	5	5
4750	122	90	88	39	68	62	57	17	5	5
5000	92	82	78	39	74	72	12	13	5	0
5250	38	49	60	40	39	66	14	5	0	1
5500	27	41	20	28	17	36	12	0	0	1
5750	12	33	26	12	48	15	8	0	0	0
6000	5	11	21	25	5	12	5	0	1	1
6250	0	6	16	18	6	17	0	0	0	0

Fig. 4.12 Sample adaptive fuel control map for aftermarket DynoJet Power Commander® PC5 EFI controller. Each cell contains a number representative of the fuel correction generated by exhaust oxygen feedback. (Software interface by Dynojet Research, https://www.dynojet.com/. Map data and image by author)

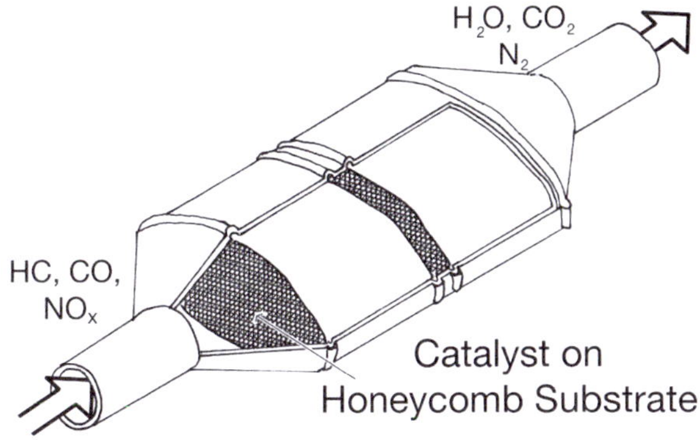

Fig. 4.13 Three-way catalytic converter, showing two catalytic matrices—NO_x *reduction* followed by HC and CO *oxidation*

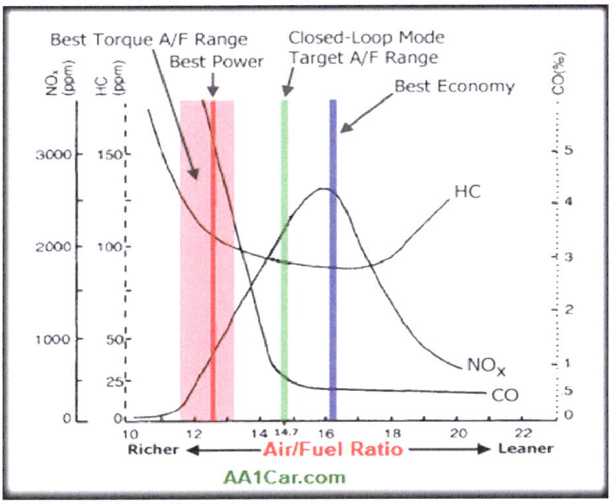

Too much fuel but slightly higher power Just right Not enough fuel but slightly higher efficiency

Fig. 4.14 Efficiency of a three-way catalytic converter for HC, CO and NO_x as a function of AFR. (Original graphic from AA1car.com https://www.aa1car.com/library/converter.htm. With permission)

- A triumph of materials engineering, the three-way catalytic converter shown in Fig. 4.13 is capable of *oxidizing* HC and CO in the exhaust into carbon dioxide and water, and also *reducing* NO_x into elemental nitrogen and benign oxygen compounds. It actually does this in two stages, but got its name "three-way" to emphasize that it handled all three regulated emission classes. The catalytic converter can only perform this amazing feat if the air/fuel ratio is just slightly lean of stoichiometric. This precise AFR control is only possible because of the oxygen-sensing feedback fuel control (above). Thus, the synergy between these four technologies. The effectiveness of the three-way catalytic converter (TWC) for reducing HC, CO, and NO_x as a function of the AFR is shown in Fig. 4.14.

Direct Cylinder Gasoline Injection (DCGI)

The maintenance of a stoichiometric AFR is important when the intake charge in the cylinder must have a *homogeneous composition*, that is, the same AFR everywhere in the cylinder at the time of ignition. There is no choice but to do this if the fuel is inducted along with the air into the cylinder. But it is known that fuel efficiency at partial loads can be improved by charge *stratification* in which a richer mixture is concentrated around the spark plug, while the overall mixture is lean. This can be accomplished if the fuel is injected directly into the cylinder near the spark plug, some before the intake valve closes and some *after* it has closed. The potential efficiency and emissions advantages of charge stratification have been known for decades. The concept was originally tried in the Honda *CVCC* engines in the 1970s and the Mitsubishi MCA-Jet engine circa 1980, which used a small additional intake valve to admit a rich mixture near the spark plug while the overall mixture was lean. But this approach lacked the ability to add fuel after the intake valve closed at end of the intake stroke. Lacking the ability to inject fuel directly into the cylinder, these engines were not as effective as had been hoped. But this became possible with the development in the mid-2000s of high-pressure fuel injection pumps for gasoline that could continuously pressurize a fuel rail to as high as 3000 psi (20 MPa) and electrically actuated fuel injectors that could open at very high fuel pressures and withstand cylinder temperatures.

First introduced in 1954 on the Mercedes 300SL but not widely adopted until after 2000, gasoline direct-cylinder injection (aka GDCI, GDI, DCI, DISI, SIDI) injects fuel at high pressure directly into the cylinder rather than into the intake air stream. Fuel can be either injected while the intake valve is open to form a homogeneous charge, or into a small area around the spark plug after the valve has closed to form a local fuel cloud that is ignitable by the spark plug. The engine operates in dual modes—*stratified charge* at part loads and idle, and *homogeneous charge* for full power.

Feedback AFR control is coordinated with the dual operating modes, with stratified charge operation allowed to produce excess oxygen in the exhaust, but the usual stoichiometric operation at high loads. Overall, GDCI achieves improved efficiency and further reduction in emissions at light loads—a car's operational condition most of the time. As of 2024, most hybrids, performance and luxury cars used DCGI systems (Fig. 4.15).

Effects of Advanced Automotive Emission Controls on Air Quality in the USA

The health benefits of feedback fuel control and the three-way catalytic converter cannot be understated. Figures 4.16 and 4.17 reveal that in the Los Angeles smog basin between 1960 and 2010, while gasoline consumption nearly tripled, there was a 50-fold decrease in atmospheric VOCs: a net reduction of 146:1. (VOCs = Volatile Organic Compounds = carbon compounds in gaseous form at NTP, almost the same as HC emissions for a gasoline engine.)

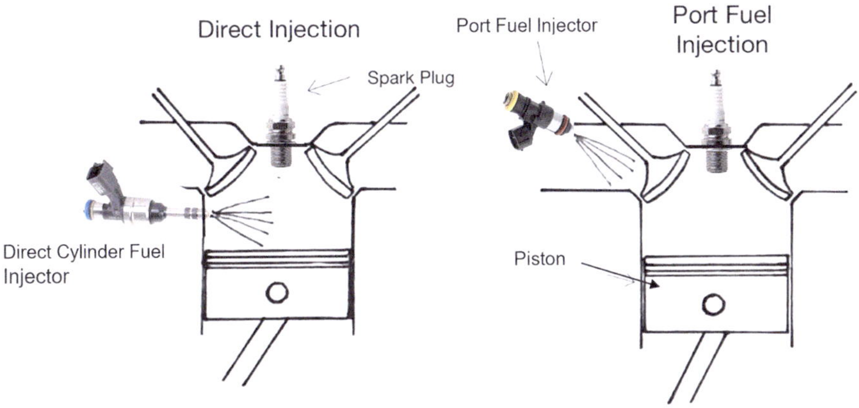

Fig. 4.15 Diagram of cylinder with direct gasoline fuel injection (GDI) left, vs. port fuel injection right. (Graphic by Olivia Hoffsis)

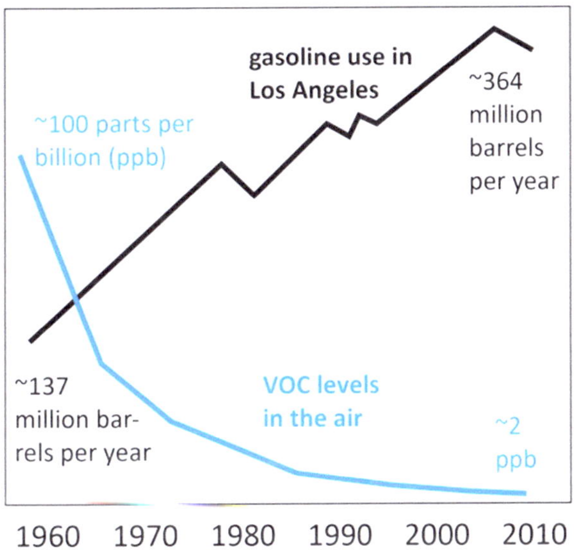

Fig. 4.16 VOC (HC) atmospheric concentrations in Los Angeles, 1960–2010. (From https://www.epa.gov/transportation-air-pollution-and-climate-change/accomplishments-and-success-air-pollution-transportation. Public domain)

Year: 1960
100 ppb/137 Mbbl/year = 0.730 ppb-year/Mbbl gasoline
Year: 2010
2 ppb/364 Mppb = 0.005 ppb-year/Mbbl
where:
ppb = parts per billion (mass fraction) in atmosphere
bbl = barrel of gasoline = 42 gallons
Mbbl = one million barrels

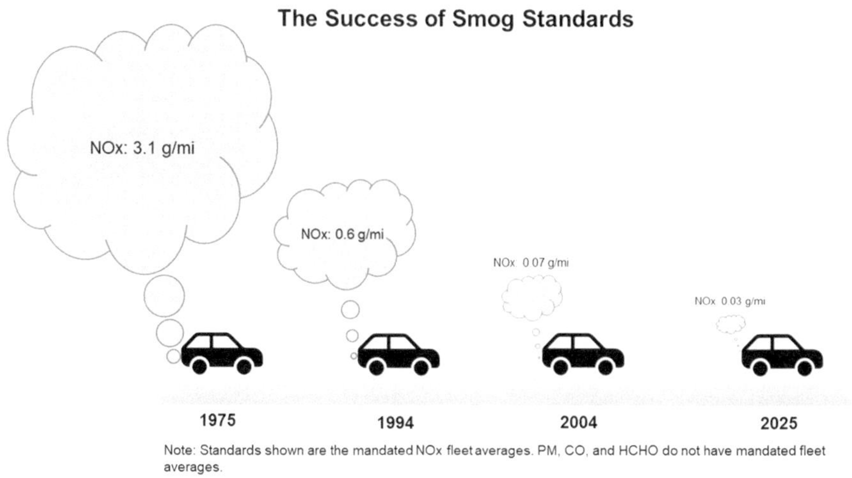

Fig. 4.17 US EPA graphic: Evolution of mandated fleet average standards for NO_x emissions in the USA, 1975–2025. (https://www.epa.gov/greenvehicles/smog-vehicle-emissions. Public domain)

Can an Automobile Emission Control System Reduce CO_2?

No. The CO_2 emissions are unique properties of the fuel, specifically the carbon content. The only way to reduce CO_2 emissions for gasoline or diesel fueled vehicles is to reduce the fuel consumption or use a different fuel with a higher H:C ratio.

Gaseous Fuels (Methane, Natural Gas, LPG, Hydrogen, and Syngas)

Differences Between Gaseous and Liquid Fuels

- Gaseous fuels take up a significant volume of the gaseous intake charge, reducing the amount of air available for combustion. This effectively reduces the displacement of the engine, reducing its power output.
- Liquid fuels, inducted while still in their liquid state, are assumed to take up so little volume compared with the air that the entire intake volume is assumed to be air.
- This does not affect the AFR, just the amount of air and fuel that can be inducted into an engine of a given displacement.
- Gaseous fuels are already vaporized, so cold starting an engine is much easier. And there is no need for fuel enrichment during the warm-up period, assuring cleaner start-up emissions. This is also a great benefit in very cold climates in which liquid fuels may be unable to vaporize to the point that they can be ignited.

- But the vaporization of a liquid fuel reduces the temperature of the air/fuel charge, which increases its density. A denser intake charge increases the amount of fuel and air in the cylinder, increasing the power output of the engine. Gaseous fuels lack this vaporization cooling effect (unless they can be injected as a liquid, which is possible for propane but not methane or hydrogen). The intake cooling effect also reduces the sensitivity to intake backfiring. The lack of evaporative cooling is one of the factors that make methane and hydrogen more sensitive to this phenomena (in addition to their lower ignition energy).
- Gaseous fuels usually have high octane ratings but very low minimum ignition energy requirements. These effects counteract each other to limit the maximum power output of the engine.
- Gaseous fuels usually have higher H:C ratios than liquid fuels. This means lower CO_2 emissions.
- Intake charge formation is more challenging than for liquid fuels. Gaseous fuel injection requires much larger injector flow rates, while traditional propane or natural gas carburetors (gas mixers, as in Fig. 4.18) have difficulty maintaining a constant AFR for all intake flow rates [11]. An electronic sequential gaseous fuel injection system was used on the Honda GX NGV, and components are available from Penske and Delphi [12] (Fig. 4.19 shows the fuel injectors of the natural gas injected engine).
- If aspirated correctly (not over-fueled), all of the gaseous fuels discussed in this book produce lower HC and CO emissions than gasoline, and zero particulates. NO_x is usually similar to gasoline since it is not a property of the fuel, but rather the heating of the air.

Natural Gas and Biomethane Differ Significantly Depending on Location

Natural gas varies widely in its composition between different wells and gas fields. Methane is always the dominant fuel component, but the balance gases include both inert (CO_2, nitrogen, and hydrogen sulfide) and other fuel gases (ethane, carbon monoxide, hydrogen) [13].

Natural gas H (Siberia) (mass %): 98% methane + 1% ethane +1% nitrogen.

Natural gas L (North Sea) (mass %) 83%: CH_4 + 4% C_2H_6 + 1% C_3H_8 + 2% CO_2 + 10% N_2.

Natural gas (US average 2019) = (volume %): 95% methane, 2.5% ethane, 1.6% nitrogen, 0.7% carbon dioxide, 0.2% other non-combustible gases [14].

Producer gas (syngas, wood gas): 7–22% CO + 16–20% H_2 + 2.3% CH_4 + 10–15% CO_2 + 50–55% N_2 [15].

Biomethane is a broad classification including any mixtures of methane and other gases that are formed by biological processes, either naturally or synthetically. Typical components of bio-natural gas (aka biomethane) from various sources are shown in Table 4.2 as volume percentages.

Fig. 4.18 IMPCO gas mixer for operation of engine on propane (LPG). The standard since the late 1960s, still in widespread use in 2023. (Image from IMPCO catalog. https://store.nashfuel.com/products/impco-lpg-propane-carburetor-mixer-ca100-ca100-180. Used with permission)

Fig. 4.19 Electrically actuated natural gas fuel injectors of the type used by Honda on the 1998–2015 Honda Civic GX NGV. Also used for custom CNG and LPG conversions. (Image: eBay. Public domain)

Calculation of the Stoichiometric Mass AFR for Gaseous Fuels

Appendix 3 presents methods for calculation of the stoichiometric *mass* AFR for any gaseous fuel or blend of fuels. Here are some mass AFR results calculated from the compositions of the fuel and air:

Hydrogen: 34.1
Methane: 17.3
Ethane: 16.1
Propane: 15.6
Natural gas (Russia, H): 17.1
Natural gas (US avg, 2020): 16.3

Table 4.2 Composition of various biomethane gases, volume percentages

	Digester gas	Landfill gas	Natural gas (Dutch, 2007)
Methane (vol %)	60–70	35–65	89
Other HC (vol %)	0	0	9.4
Hydrogen (vol %)	0	0–3	0
CO (vol %)	30–40	15–50	0.67
N_2 (vol %)	~0.2	5–40	0.28
O_2 (vol %)	0	0–5	0
H_2S (vol ppm)	0–4000	0–100	2.9
Ammonia (vol ppm)	~100	~5	0
LHV (kWh/m^3_{NTP})	6.5	4.4	11.0

Data from Petersson and Wellinger, 2009 as reported in Ana Martha Coutiño, Teresa Berdugo Vilches. *Barriers and solutions to the successful diffusion of dual-fuel trucks in Europe. An innovation systems approach.* Chalmers University of Technology, Sweden, January 2011

Unlike liquid fuels, the *volume* AFRs for gaseous fuels can be very different due to the density of the gas at a given temperature and pressure compared with a liquid fuel.

Gaseous Fuels Take Up Intake Volume, Reducing Power

Gaseous fuels take up significant volume in the intake air stream. This reduces the amount of air available for combustion in the air/fuel mixture, reducing the power output of the engine as if it had a smaller displacement.

How to find the volume percentage taken up by a fuel in a stoichiometric air/fuel mixture? First have to determine the stoichiometric *volume* AFR (not the usual mass AFR).

Example: Propane C_3H_8

Propane is the main component of LPG (liquid petroleum gas)

Mass AFR = 15.6 (calculated)

Propane molar density = 44 g/mole

Air molar density = 29 g/mole

$$\text{Volume AFR} = \text{Mass AFR} \times \frac{\text{Molar(volume) fuel density}}{\text{Molar(volume) air density}}$$

$$= \frac{15.6 \text{ g air}}{\text{g propane}} \frac{44 \text{ g/mole propane}}{29 \text{ g/mole air}} = 23.67 \frac{\text{air vol}}{\text{fuel vol}}$$

Fraction of intake fuel-air charge occupied by fuel

$$\frac{1 \text{ mole propane}}{23.67 \text{ moles air} + 1 \text{ mole propane}} = 0.0405 = 4.05\%$$

This is effectively a 4.05% reduction in the engine displacement, resulting in a 4.05% reduction in power compared with the fuel being aspirated in a liquid form. If a gaseous fuel is injected directly into the cylinder after the intake valve has closed, it does not displace any of the intake air. High-pressure gaseous direct cylinder injection (GDCI) has been implemented experimentally[2] [16], but are technically challenging and not justified for recovery of the displaced air volume, which is more easily compensated by intake charge pressurization (super/turbo-charging).

How Does this Volume Compare with that of a Liquid Fuel, e.g., Gasoline?

$$\left(\frac{14.7\text{ g air}}{1\text{ g gasoline}}\right)\left(\frac{\dfrac{1\text{ mole air}}{29\text{ g air}}\times\dfrac{1\text{ L air}}{0.0416\text{ mole air}}}{\dfrac{1\text{ mL gasoline}}{0.75\text{ g}}\times\dfrac{1\text{ L}}{1000\text{ mL}}\text{density}}\right)=9.14\times10^{3}\text{ L air/L gasoline}$$
$$=\text{gasoline volume AFR}$$

where

$$n=\frac{PV}{RT}=\frac{(1\text{ atm})(1\text{ L})}{\left(0.08206\dfrac{\text{atm L}}{\text{K mole}}\right)(293\text{ K})}=0.0416\text{ moles air per Liter at NTP}$$

$$\text{Fraction of intake charge volume occupied by liquid gasoline:}$$
$$=\frac{\text{Fuel moles}}{\text{Air moles}+\text{Fuel moles}}=\frac{1\text{ L gasoline}}{9.14\times10^{3}\text{ L air}+1\text{ L gasoline}}=1.09\times10^{-4}=0.0104\%$$

Although some gasoline vaporizes prior to induction, the majority enters the cylinder as an atomized liquid, not as vapor. Since the liquid volume is so small compared with the air volume, *we usually ignore the volume occupied by any liquid fuel* when calculating engine power and CO_2 emissions.

Some Calculated Volume Fractions for Gaseous Fuels

In Table 4.3, the volume fraction occupied by fuel in the intake charge for natural gas or hydrogen is seen to be quite large. The power output of an engine is reduced proportionally with this reduction in intake air.

[2] One of the several injection configurations tested was a direct cylinder hydrogen injection engine that allowed stratified charge formation.

Table 4.3 Mass AFR, volume AFR, and percent of intake volume taken up by fuel in a stoichiometric mixture. The large fraction for gaseous fuels is responsible for a significant power output reduction at full throttle

	Air/fuel mass ratio	Air/fuel vol ratio[a]	Fuel fraction of intake volume[a]
Liquid fuel			
Gasoline (unvaporized)	14.7:1	9090:1	0.011% (trivial)
Ethanol (unvaporized)	9:1	6666:1	0.015% (trivial)
Gaseous fuel			
Natural gas (H)	17.1:1	6.21:1	9.43%
Propane	15.6:1	23.7:1	4.05%
Hydrogen	34:1	2.34:1	30.0%

[a] $\text{Vol AFR} = \text{Mass AFR} \times \frac{\text{Molar(vol) fuel density}}{\text{Molar(vol) air density}}$, $\text{Displaced air volume fraction} = \frac{1}{1 + \text{vol AFR}}$

Calculating CO_2 Emissions of Fuels

The carbon in carbon dioxide comes from the carbon in the fuel. Each carbon atom in the fuel becomes one carbon dioxide molecule in the exhaust.

CO_2 per Gallon or Liter

The CO_2 product resulting from the combustion of gasoline is usually reported as the carbon mass (in grams) for a given mass or volume of fuel.

2022 EPA data for regular E10 gasoline: 2421 g carbon per gallon [17].

Since each carbon atom in the fuel creates one molecule of carbon dioxide when burned, the CO_2 emissions from the combustion of one gallon of E10 gasoline can be found simply by multiplying the carbon mass per gallon by the ratio of the carbon molar density over the carbon dioxide molar density.

$$(2421 \text{ g carbon/gal}) \frac{44 \text{ g } CO_2}{12 \text{ g C}} = 8877 \text{ g } CO_2 \text{ per gallon}$$

For US average diesel No. 2 (aka D2) with a carbon density of 2776 g carbon per gallon

$$(2776 \text{ g carbon/gal}) \frac{44 \text{ g } CO_2}{12 \text{ g C}} = 10{,}180 \text{ g } CO_2 \text{ per gallon}$$

CO_2 per Mile or km

The CO_2 emissions of a fuel per unit of energy are the product of the carbon mass fraction of the fuel times the mass energy density of the fuel. But it is more intuitive to think of the energy in terms of how far the car can travel with that amount of energy, which is specified in the USA as miles per gallon (MPG), or as liters per km everywhere else, for a specific vehicle.

$$\left(\frac{\text{Carbon g/gal}}{\text{MPG}}\right)\left(\frac{44\text{ g CO}_2}{12\text{ g C}}\right) = \text{g CO}_2/\text{mile}$$

For E10 gasoline with 2421 g C/gal, a 44 MPG car would emit

$$\left(\frac{2421\text{ g C/gal}}{44\text{ miles/gal}}\right)\left(\frac{44\text{ g CO}_2}{12\text{ g C}}\right) = 202\text{ g CO}_2/\text{mile}$$

CO_2 emissions are discussed in more detail in the Appendix 2, that shows how to calculate CO_2 emissions for any fuel, in g CO_2/(km or mile).

Here are some calculated carbon dioxide emissions for fuels other than gasoline, using a 44 mpg 2012 Honda Civic Hybrid as the reference vehicle for all fuels, and a 0.38 kWh/mile 2012 electric vehicle, in metric and US units:

Methane	91 g/km	147 g/mi
Natural gas	94 g/km	151 g/mi
2012 Tesla Model S P85:	97 g/km	156 g/mi (0.38 kWh/mile, 390 g CO_2/kWh generation emissions, 7.8% line loss)
Biomethane (sewage gas)	110 g/km	177 g/mi
Methanol	116 g/km	187 g/mi
Ethanol	118 g/km	191 g/mi
E85	119 g/km	192 g/mi
E10	125 g/km	202 g/mi
Gasoline	130 g/km	209 g/mi
Wood gas	190 g/km	306 g/mi

Among the gaseous fuels listed above, methane has the lowest specific CO_2 emissions, while wood gas has the highest. Among liquid fuels, methanol is the cleanest while gasoline is the dirtiest. The effective CO_2 emissions per km or mile of a 2012 Tesla S EV charged from the electric power grid lies between natural gas and methanol.

Engine Knock and the Octane Rating of a Fuel

One of the most important properties of an SI engine fuel is its resistance to knock (aka ping). As mentioned previously, engine knock is the propensity of fuel to auto-ignite prior to the arrival of the combustion flame front in the cylinder. The result is a characteristic knocking sound, and possible damage to the engine (Fig. 4.20). The *octane rating* of a fuel is a measure of its ability to avoid knock. The scale is between two reference fuels. The rating is measured experimentally using a special single-cylinder engine that has a dynamically adjustable compression ratio, called a Cooperative Fuel Research (CFR) engine. Two reference fuels are used. Normal heptane C_7H_{16} is highly prone to knock and therefore assigned an octane rating of **zero**. 2-2-4 trimethyl pentane (popularly but incorrectly referred to as iso-octane) is resistant to knock and assigned an octane rating of **100**. At the time, it was assumed that this would be the highest level of knock resistance needed. Knock ratings above 100 are extrapolated beyond the original linear scale which only spanned 0 to 100. Differences between the extrapolation methods have resulted in different reported values for high-octane fuels such as methanol.

There are two versions of the knock rating test: *Research Octane Number* (RON) and *Motor Octane Number* (MON) which are performed at different engine speeds and knock intensity thresholds. RON is usually higher than MON. In the USA, the *Legal Octane Number* (LON) of a fuel is the numeric average of the RON and MON.

Gasoline without additives has a rather low octane rating, typically between 70 and 90. If used without the addition of some anti-knock compound, it would likely experience knock at higher power levels, even for moderate compression ratios.

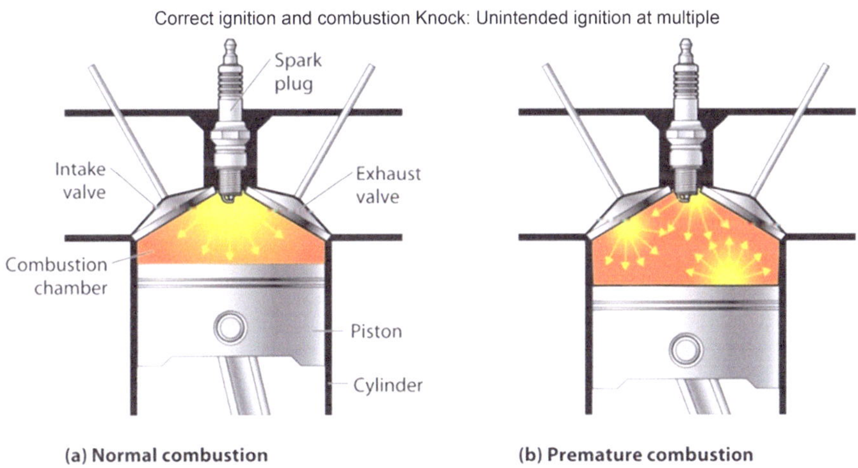

Fig. 4.20 Comparison of proper ignition and combustion with knock. (From B. Averill, General Chemistry: Principles, Patterns, and Applications, 2011. Open Textbook Library, online at https://open.umn.edu/opentextbooks/textbooks/69)

Driven by the need to increase the power output of engines, especially for military piston engine aircraft during WWII, fuel additives called "octane enhancers" were developed to improve the octane rating of gasoline. These are added in small quantities to gasoline at the refinery. In the USA, the following additives have been used:

- TEL (tetraethyl lead) was first synthesized by German chemist Carl Jacob Lowig in 1853, but it was not until 1921 in the USA that its value as an anti-knock additive for gasoline was discovered. It was patented for that purpose by General Motors Corp and subsequently commercialized by the newly formed *Ethyl Corp* having only one product: TEL. It was an exceptionally effective fuel additive in that as little as 1–3 g (0.6–1.8 mL) per gallon of straight-run gasoline could increase its research octane rating by 10–15 points, allowing engine compression ratios to be increased from previously typical 7 or 8:1 to as high as 12:1, which significantly improved both power output and engine efficiency while eliminating engine knock. By the 1950s many automobiles required "ethyl" gasoline, a designation for high-octane gasoline now referred to as *premium* gasoline [18] (Fig. 4.21).

  TEL remained in use for this purpose from 1921 until 1986, 13 years after the US EPA in 1973 recommended that it be phased out due to concerns over the toxic and carcinogenic health effects of the lead aerosols that it created when burned. The final motivation to discontinue use was the introduction of catalytic converters in the 1970s, which were damaged by the lead in the exhaust.

  The historical significance of TEL has probably never been given proper attention. It enabled the high-performance aircraft engines of WWII. Arguably, without TEL, the US car culture and horsepower race of the 1960s may not have evolved as it did. The aftermarket spark plug industry flourished during the TEL decades, since lead deposits formed quickly on the center electrode, leading to misfire and requiring frequent replacement. Engine durability suffered due to lead deposits that formed on exhaust valves and piston heads. Meanwhile, three generations of children grew up in freeway-centric cities such as Los Angeles, California, breathing lead particulates.

  Unfortunately, the replacement anti-knock additives adopted in the 1970s were not much better health-wise. It wasn't until the transition to oxygenated hydrocarbons, first methanol, and later ethanol that the tradeoffs between public health and high-octane gasoline was resolved. Immediately after the phaseout of TEL, several ethers and aromatics were briefly adopted:

- MTBE (methyl tertiary-butyl ether) used 1974–2000. Highly toxic and long-lived if leaked into groundwater from underground gasoline storage tanks at gas stations.
- ETBE (ethyl tertiary-butyl ether) 1974–2000. Rarely used, equivalent to MTBE but more expensive and less toxic.
- BTEX (benzene, toluene, xylene, and ethyl-benzene) 1980–2000. Aromatics are highly carcinogenic and toxic. Only limited use in the USA.
- TAME (tertiary amyl methyl ether) or TAEE (tertiary amyl ethyl ether) used in the EU but rarely in the USA. Similar or worse health impacts than MTBE.

Fig. 4.21 Gas (petrol) pump, circa late 1950s, dispensing high octane gasoline containing tetraethyl lead (Ethyl™). The gasoline cost is shown in the small window at the bottom: $40.9 per gallon. (Image: https://www.rawpixel.com/image/6112536/vintage-texaco-gas-station-usa-jan-2016. Public domain)

- Methanol 1980–1990 in the form of M5 (5% methanol in gasoline by volume, aka *gasohol*). With an extrapolated RON of approximately 130, methanol is a very effective octane additive in gasoline. And compared with the previous octane-enhancing additives, it decomposes rapidly if leaked into the environment and produces no toxins when burned. However, the incompatibility of legacy carbureted fuel system metals and elastomers and the lack of industry advocates led to gasohol being unfairly vilified by mechanics and fuel vendors and eventually phased out in favor of ethanol, which had the same limitations but was more politically favorable.
- Ethanol 1990–present in the form of E10 (10% ethanol by volume). Initially, it was also referred to as gasohol by consumers since the transition from methanol to ethanol occurred largely under the public radar. Ethanol has similar anti-knock properties and material compatibility issues as methanol. It is more expensive and energy-intensive to produce, but corn-fermented ethanol is heavily subsidized (in the USA) at both production and user levels. Since the late 1990s in the

USA, 10% ethanol must, by federal law, be added to gasoline, with a more recent federal 15% requirement being phased in to increase ethanol utilization. It was also adopted throughout much of South America, Europe, and Asia, but made from crops other than corn, e.g., sugarcane in South America.

Ideal and Real Engine Cycles and Processes

The Otto cycle model is an idealization of the operation of a four-stroke SI engine. It facilitates the calculation of many engine parameters by assuming that the intake and exhaust valves open and close instantaneously; combustion occurs instantaneously at TDC; no internal friction and no heat loss occur through the cylinder walls or piston. It is usually of greater theoretical than practical interest, but it provides an understanding of what is happening in each of the four engine strokes and therefore is directly relevant to how engine specifications and fuels affect efficiency, power output, and the cylinder temperatures responsible for NO_x emissions.

Figure 4.22 is an example of a cylinder pressure vs. volume plot for an idealized Otto cycle engine, as it goes through its four strokes: intake, compression, power, and exhaust.

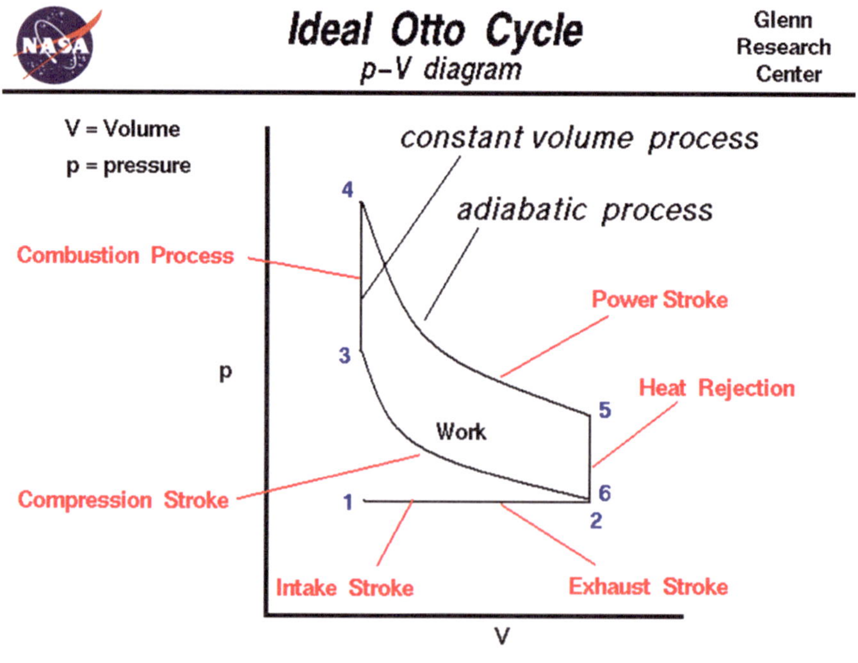

Fig. 4.22 Otto cycle pressure-volume plot for the four stokes of an idealized engine. (From NASA, Ideal Otto Cycle P-V diagram. Public domain. https://www.grc.nasa.gov/WWW/K-12/airplane/otto.html)

Appendix 5 presents the derivation of the Otto cycle model at an appropriate level of complexity for calculation of engine efficiency, power output, and the effect of changes in engine specifications.

Compression Ratio and Thermal Efficiency

One very practical result from the Otto cycle model is a formula for relating the theoretical thermal efficiency of an engine to the compression ratio (CR). The simplified relationship is:

$$\text{Thermal Efficiency}(\%) \approx 100\left(1-\frac{1}{\text{CR}^{0.2}}\right)$$

This relationship is well known by engine designers and performance tuners, since it shows how increasing the CR can significantly improve the ability of the engine to extract useful power from the fuel. Better utilization of the energy available from a given amount of fuel means both improved efficiency (fuel economy) *and* higher power output. In fact, if the octane rating of gasoline was higher, engines in cars would use higher CRs to improve both of these important attributes.

This relationship is not particularly accurate in an absolute sense since it includes several simplifications in the Otto cycle calculations. But it is useful for determining changes in efficiency or power as a result of changes in the compression ratio.

For our example, an engine with a CR = 10 would yield 36.9% theoretical thermal efficiency. If the engine was then modified by "shaving" the cylinder head to increase the compression ratio to CR = 12, the theoretical thermal efficiency would increase to 39.2%, a 7.3% increase. The theoretical efficiency numbers are not accurate, but the percentage increase as a result of increasing the CR is usually a good prediction.

A fuel with a higher octane number allows a higher compression ratio without encountering knock. It may be surprising that all viable alternative fuels have higher octane ratings than gasoline, so they can benefit in thermal efficiency and power output from an increased engine CR, in addition to whatever increase they may gain from the fuel itself, discussed in the next section. For example, methanol has an extrapolated research octane number (RON) of 130 compared with E10 gasoline which has an RON of 91 or gasoline without ethanol that has an RON of 87. It is not uncommon to build racing engines that run on pure methanol with CRs has high as 15:1, not only increasing their power output, but also increasing efficiency which partially compensates for the lower energy density of the liquid fuel.

It should, however, be noted that increasing the compression ratio of an engine running on any fuel will increase NO_x emissions, because it increases the peak combustion temperature. This is true for all hydrocarbons including alcohol fuels. But at least for alcohols and natural gas, any increase in NO_x as a result of increasing the CR will be well within the ability of a stock vehicle's catalytic converter to reduce NO_x to a degree as much or greater than gasoline levels, in terms of grams per km per mile.

Power Output Potential of Various Fuels

It may come as a surprise that different fuels intrinsically provide different amounts of power from an engine of a given displacement. This is because different fuels, liquid or gaseous, pack different amounts of combustion energy into a given volume of a stoichiometric fuel/air intake charge.

Appendix 4 provides details on how to calculate the theoretical power output for any given fuel.

But it is worthwhile to review the method here using examples of one liquid fuel, ethanol, and one gaseous fuel, methane.

A common engine and an operating condition assumption is needed:

- 1.0 L (total of all cylinders) displacement.
- Tested at full throttle at the speed (assume 6000 RPM) at which maximum power is produced.
- 22 °C, 1 atm.
- Same thermal efficiency (28%) for both fuels.

For Both Liquid and Gaseous Fuels

1 L of air is inducted into the engine every two engine revolutions.

Find the number of moles of air in 1 L at 1 atm and 22 °C.

From ideal gas law:

$$n = \frac{PV}{RT} = \frac{(1\ \text{atm})(1\ \text{L})}{\left(0.08206\frac{\text{atm L}}{\text{K mole}}\right)(295\ \text{K})} = 0.0413\ \text{moles/L}$$

Air has a molar density of 29 g/mole of which 6.73 g is oxygen.

$$0.0413\ \text{moles air}\frac{29\ \text{g}}{\text{mole air}} = 1.20\ \text{g air}$$

At 6000 RPM, the rate of airflow into the engine with the throttle wide open is:

$$6000\ \text{RPM}\left(\frac{1\ \text{min}}{60\ \text{s}}\right)\left(\frac{1\ \text{intake}}{2\ \text{revolutions}}\right)\left(\frac{1.0\ \text{L air}}{1\ \text{intake}}\right) = 50\ \text{L/s air inducted by the engine.}$$

Liquid Fuel Example: Ethanol

The same mass of air is inducted every intake stroke: 1.20 g air. Since the stoichiometric mass AFR for ethanol is 9.0:1, the mass of fuel that will burn completely in the amount of air is therefore

$$\frac{1.20 \text{ g air}}{(9.0 \text{ g air / g ethanol})} = 0.133 \text{ g ethanol}$$

The LHV energy of ethanol is 26.8 kJ/g, so 0.133 g ethanol releases 3.56 kJ of energy for every liter of air inducted.

At 6000 RPM, the engine inducts 50 L/s with each liter resulting in 3.56 kJ of energy. The *power* output is therefore:

$$50 \text{ L air / s} \left(\frac{1 \text{ k W}}{1 \frac{\text{kJ}}{\text{s}}} \right) \left(\frac{3.56 \text{ kJ fuel}}{1 \text{ L air}} \right) (0.28 \text{ thermal efficiency}) \left(\frac{1.34 \text{ HP}}{1 \text{k W}} \right) = 66.8 \text{ HP}$$

(4.7% greater power than gasoline).

Gaseous Fuel Example: Methane

The calculations for a gaseous fuel are the same as for a liquid fuel except that some of the intake air is displaced by the volume of the fuel, reducing the amount of air that is inducted. To find out how much air is displaced, we need to know the air/fuel *volume ratio* of methane to air in a stoichiometric mixture. This is easier to calculate than the mass AFR, just by looking at the stoichiometric combustion equation:

$$CH_4 + 2O_2 \rightarrow CO_2 + 2H_2O$$

Since 1 mole of any ideal gas at a given temperature and pressure occupies the same volume, 1 mole (or volume) of methane combines with 2 moles (or volume) of O_2.

Since air is only 21% oxygen, 2 moles of O_2 are contained in 9.52 moles of air. This means that the stoichiometric *air/fuel volume ratio* is 9.52:1.

1 L of intake charge therefore contains

$$\frac{1 \text{ L } CH_4}{1 \text{ L } CH_4 + 9.52 \text{ Lair}} = 0.095 \text{ L methane and}$$

$$\frac{9.52 \text{ L air}}{1 \text{ L } CH_4 + 9.52 \text{ L air}} = 0.905 \text{ L air}$$

Compared with liquid fuels, this amounts to a 9.5% reduction in the intake air volume, and therefore a 9.5% in the fuel volume to be combined with this air, leading to a 9.5% reduction in the output power.

At 6000 RPM the 1.0 L engine inducts 50 L/s of *total intake mixture*, which contains

Intake flow rate fraction that is fuel mass for this volume = mass flow rate of fuel

$$(50 \text{ L air} + \text{fuel/s})\left(0.095 \frac{\text{L methane}}{\text{L of air} + \text{methane}}\right)\left(\frac{16 \text{ g methane/mole}}{24.2 \text{ L methane/mole}}\right)$$
$$= 3.14 \text{ g methane/s}$$

Since the LHV energy of methane is 50 kJ/g and the engine efficiency is the same as the other fuels, 28%, this flow rate of methane produces a power output of

$$(3.14 \text{ g methane/s})\left(\frac{50.0 \text{ kJ}}{\text{g methane}}\right)(0.28 \text{ thermal efficiency})\left(\frac{1 \text{ kW}}{1\frac{\text{kJ}}{\text{s}}}\right)\left(\frac{1.34 \text{ HP}}{1 \text{ kW}}\right)$$
$$= 58.9 \text{ HP}$$

(7.7% less power than gasoline).

Calculated Power Output of Selected Fuels

We can calculate and compare the power output in HP per liter of engine displacement, for any fuel, using the same assumptions as the preceding examples. *Appendix 4 shows the complete calculations for each fuel.* Only the results are listed here, from best (highest power) to worst (lowest power).

Methanol	68.3 HP/L
Ethanol	66.8 HP/L
Gasoline	63.8 HP/L
Methane	58.9 HP/L
Natural gas (H)	58.6 HP/L
Wood gas	35.6 HP/L

It is apparent from this list why methanol and ethanol are preferred racing fuels. The numbers above are actually conservative since the power outputs for alcohols are further increased due to their charge air-cooling effects. Higher octane ratings also allow higher compression ratios without risk of knock, improving power output and efficiency.

Selected Internet Tutorial References for Internal Combustion Engines

There is an ***overwhelmingly large amount of information*** published on the Web for self-instruction on automobiles, trucks, motorcycles, and how an internal combustion engine works. There is also a fair amount of misinformation. Here are some suggestions for a few vetted online resources that may be time-efficient in learning about the workings of internal combustion engines.

As always, the web links could expire at any time, but they worked as of July 1, 2024.

Basics of four-cycle SI engine operation:
http://animagraffs.com/how-a-car-engine-works/ (animated four-stroke engine)

Automobiles and automotive systems:
http://auto.howstuffworks.com/car.htm (general—key automotive components)
http://www.familycar.com/classroom/ (general—"family car" common systems)
Glossary of automotive electronic terms (general—cut through the jargon)

Internal combustion (IC) Engines—types and operation:
http://en.wikipedia.org/wiki/Internal_combustion_engine (general)
http://www.youtube.com/watch?v=_y6PS-2j2Ug (four-cycle engine cycles explained)
http://www.youtube.com/watch?v=QXLsna21FWo&feature=related (animated engine build)
http://www.youtube.com/watch?v=-xRKZPXHXd4&feature=related (Ford Eco-boost engine animation)

Unusual IC engine configurations:
http://www.youtube.com/watch?v=9UxfScCTrTo&feature=related (Wankel or rotary engine)
http://www.youtube.com/watch?v=Z1LrmLHOUpg&feature=related (Swash-plate or axial vector motor)

References

1. American Chemical Society, Gasoline. Chem. Eng. News, **83**(8), 37 (2005), https://pubsapp.acs.org/cen/whatstuff/stuff/8308gasoline.html. Accessed 1 July 2024
2. US Environmental Protection Agency, All U.S. gasoline (except California) average summer gasoline properties, https://www.epa.gov/sites/default/files/2017-02/documents/all-us-gasoline-except-california.pdf. Accessed 1 May 2018
3. International Energy Agency, Technology collaboration programme. Diesel and gasoline, https://www.iea-amf.org/content/fuel_information/diesel_gasoline. Accessed 1 May 2018
4. Robert Bosch GmbH, Mean value for Diesel No. 2, in *Automotive Handbook*, 10th edn., (SAE International, 2018), p. 345
5. HP Academy, Why lambda is BETTER than AFR [QUICK TECH], https://www.hpacademy.com/technical-articles/afr-vs-lambda/. Accessed 20 June 2024
6. B. Mikkelson, Miracle Carburetor: a miraculous car that gets 200 miles to the gallon is sold by mistake then reclaimed by the factory and is never seen again. Snopes (1999), https://www.snopes.com/fact-check/nobodys-fuel/. Accessed 1 May 2024
7. Automotive History Preservation Society, Bendix "Electrojector" fuel injection (1957–1959), https://autohistorypreservationsociety.org/1957-1959-bendix-electrojector-fuel-injection/. Accessed 1 June 2023

8. Catalytic Converter Solutions, The invention of the catalytic converter: a retrospective on the innovators, https://catalyticconvertersolutions.com/the-invention-of-the-catalytic-converter/. Accessed 1 July 2024
9. Robert Bosch GmbH, Sensing the adequate mixture—the Bosch Lambda Sensor, https://www.bosch.com/stories/40-years-of-bosch-lambda-sensor/. Accessed 1 June 2024
10. SST, O_2 SENSORS—Zirconium dioxide (ZrO_2) oxygen sensor operating principle guide, https://www.sstsensing.com/wp-content/uploads/2016/05/AN-0043_rev5_Zirconia-Sensor-Operating-Principle-and-Construction-Guide.pdf. Accessed 1 Oct 2024
11. Impco by eControls, Fuel regulation, https://www.impcotechnologies.com/products/. Accessed 10 May 2024
12. Phinia, holding company for DELPHI®, DELCO REMY® and HARTRIDGE® formed in 2023, Gaseous fuel injectors, http://delphi.com/manufacturers/auto/powertrain/alternative-fuel-systems/multec-cng-injector. Accessed 10 May 2024
13. Robert Bosch GmbH, Properties of gaseous fuels, in *Automotive Handbook*, 10th edn., (SAE International, 2018), p. 346
14. North American Energy Standards Board, Natural gas composition data, in *Natural Gas Specs Sheet*, (North American Energy Standards Board, 2003), https://www.naesb.org/pdf2/wgq_bps100605w2.pdf. Accessed 1 Sept 2020
15. A.K. Rajvanshi, Chapter 4: Biomass gasification, in *Alternative Energy in Agriculture*, vol. 2, (CRC Press, 1986), pp. 83–102
16. C.A. MacCarley, Electronic fuel injection techniques for hydrogen fueled IC engines, MS Thesis, UCLA School of Engineering and Applied Science, 1978
17. US Environmental Protection Agency, Greenhouse gas emissions from a typical passenger vehicle (2023), EPA-420-F-23-014, https://www.epa.gov/greenvehicles/greenhouse-gas-emissions-typical-passenger-vehicle. Accessed 10 May 2024
18. M. Posch, Tetraethyl lead: the solution to one, and cause of many new problems. Hackaday (2024), https://hackaday.com/2020/01/15/tetraethyl-lead-the-solution-to-one-and-cause-of-many-new-problems/. Accessed 14 Nov 2024

Alternatives to Petroleum

5

Overview of Fuels for Automobiles

Gasoline is only one of many fuels that can power combustion engines (Fig 5.1). This chapter is a complete list of fuels that have been used at some time in the past century as energy sources or carriers for internal combustion engines (ICEs). Only a few could be considered practical, except if there are no other alternatives. In the chapters to follow, the practical subset of this list will be discussed in detail, with attention to technical feasibility and each of the four facets of sustainability.

Spark Ignition (SI) Fuels

Usable in modified gasoline engines

- Ethanol and E85 (fermented or cellulosic)
- Methanol and M85 (reformed from natural gas or biomass pyrolysis)
- Propanol, butanol, and higher alcohols
- Natural gas as CNG or LNG
- Biomethane—landfill, sewage, or agricultural effluent
- LPG (propane/butane)
- Hydrogen for combustion (compressed, liquid, or chemical hydrides)
- Petro-synfuels
- e-Fuels (derived from syngas)
- Ammonia
- Coal gas (carbon monoxide)
- Wood gas (low-quality syngas)
- P-Fuel (promoted by Pure Energy Corp, 1999)
- Acetylene

C. A. MacCarley, *Non-Petroleum Automotive Transportation*,
https://doi.org/10.1007/978-3-031-85028-8_5

Fig. 5.1 Gasoline is sooo yesterday. (From https://pixabay.com/photos/gas-station-gas-pump-petrol-fuel-1638021. Public domain)

Compression-Ignition (CI) Fuels

Usable in modified diesel engines

- Vegetable and seed oils
- Recycled or waste vegetable oils
- Biodiesel and renewable diesel
- Mineral/biodiesel blends (B10, B20)
- Bunker oil (heavy low-quality petroleum)
- Dimethyl ether (DME) and DME blends
- Coal dust (for blast injection)

Noncombustion Fuels

Battery electric vehicles (BEVs) and fuel cell vehicles (FCVs)

- Electric vehicles that store energy in batteries (BEVs)
- Battery-supercapacitor vehicles
- Exchangeable battery vehicles (EBVs)
- Roadway-powered electric vehicles
- Liquid or solid material-recharged BEVs
- Hydrogen fuel cell electric vehicles (FCEVs)
- Hybrid electric vehicles (HEVs)

Other Mobile Energy Sources

- Compressed air and hydraulic
- Kinetic (flywheel) energy storage
- Human-electric hybrid (e.g., electric-assisted bicycles)

- Spring or elastomer mechanical power
- Roadway-powered (electric induction)
- Animal power (low-tech default option)

Considerations for Evaluation of Each Fuel Option

- Energy source or energy carrier?
- Renewability of fuel—associated natural cycles and time constants
- Fuel energy density
- Safety issues and relevant regulations
- Relative efficiency and power output
- Onboard storage requirements and vehicle range
- Toxicity (regulated) emissions (NMOG, CO, NO_x, VOCs)
- Carbon dioxide (greenhouse gas) emissions
- Engine operational characteristics and issues due to fuel
- Production, distribution, and refueling infrastructure requirements
- End-of-life issues for mechanical components or batteries
- Full cycle cost "well-to-wheel"
- Market viability
- Political considerations and economic sustainability

Characteristics of Selected Fuels that Could Replace Gasoline - Spark Ignition (SI) Engine fuels

Ethanol (C_2H_5OH and E85)

- An SI liquid fuel.
- Approximately 2/3 the energy density of gasoline.
- Can use existing gasoline fuel distribution system.
- Requires minimal engine modifications, but elastomer and metal compatibility, and cold starting and lubricity issues.
- Currently 100% produced from fermentation of starchy crops, e.g., corn, but massive federal research investments in promise of cellulosic ethanol.
- Government subsidies at multiple levels disguise actual economics.
- A renewable fuel if energy positive; conflicting studies. DOE concludes average energy gain of only 10% over the petroleum used to produce it.
- Current "Green Fuel" program: E85 (85% ethanol) flex-fuel vehicles.
- Over 3000 E85-equipped stations in the USA due to tax incentives.
- Over ten million E85 FF cars on road, but fewer than 5% actually use E85.
- Retail cost about the same as gasoline on a volumetric basis.
- Food vs. fuel issue for fermented (corn) ethanol.

Methanol (CH_3OH) and M85

- An SI liquid fuel
- Approximately 1/2 the volumetric energy density of gasoline
- Requires minimal engine modifications—elastomer and metal compatibility, and cold starting and lubricity worse than ethanol
- A popular racing fuel—improved power potential and high octane
- Can use existing gasoline fuel distribution system
- In the USA, currently 100% produced from natural gas
- Much produced from flare gas from oil extraction or refining
- Can be efficiently produced from almost any source of carbon, from coal to biomass—aka *wood alcohol*
- A renewable fuel if produced from biomass.
- Flex-fuel M85 vehicles considered (politically) a failure during Alternative Fuel Utilization Program of the 1990s
- Currently costs less per energy unit than gasoline in bulk, but no distribution system

Propanol (C_3H_7OH)

- An SI liquid fuel
- Like ethanol, approximately 2/3 the energy density of gasoline
- Can use existing gasoline fuel distribution system
- Can be produced by fermentation of cellulose, but more difficult than ethanol
- Produced thermochemically from natural gas, but not as efficiently as methanol
- Chemical root of drug-store isopropyl alcohol, which is 2-isopropanol
- Not usually considered a viable fuel alternative compared with other alcohols

LPG (Liquid Petroleum Gas, Propane, Butane)

- An SI dual-phase[1] fuel.
- Currently the same or more expensive than gasoline per unit of energy, not including gas compression costs.
- Requires minimal engine modifications—technology in place for many years.
- Reduction in power output compared with gasoline.
- Liquefies at low pressure (under 200 psi). Does not require gas compression to high pressure like methane or hydrogen.
- Slightly reduced mass energy storage density compared with gasoline.
- Readily available anywhere in the USA—a domestic fuel.
- Not a renewable fuel—obtained from wells, the same as liquid petroleum.
- Dual-fuel capability is easily implemented.

[1] Dual-phase fuel means that it is a liquid at room temperature under moderate pressure, but immediately vaporizes at atmospheric pressure.

Natural Gas (CNG, LNG), Methane (CH_4), Biomethane (Biogas)

- An SI gaseous fuel
- Currently the least expensive form of portable fuel per unit of energy, excluding gas compression costs
- Requires minimal engine modifications—technology in place for many years.
- Some lubricity issues
- Theoretically high octane, but knock and backfire potential at high outputs due to low ignition energy
- Reduced power output compared with gasoline in the same engine
- Poor energy storage density—requires high pressure storage. Typical fill: 3600 psi. Vehicle range limited by fuel storage
- Readily available anywhere in the USA, few high-pressure fueling stations
- Not a renewable fuel unless produced from organic waste
- Dual-fuel capability possible

Hydrogen (H_2, as a Combustion Fuel)

- An SI gaseous fuel.
- An energy carrier, not an energy resource.
- A renewable fuel if only if it is made from a renewable energy source, e.g., solar or wind electricity and electrolysis.
- One of only two viable non-carbon-based combustion fuels.
- Zero emissions or GHGs except for NO_x. Considered the cleanest fuel.
- Highest thermal efficiency of any combustion fuel.
- Requires engine modification. Backfire problem, timed fuel injection required.
- Lowest volumetric energy density of any fuel—cannot store much onboard.
- Nontoxic. Minimal actual combustion hazard.
- Many storage options, all questionable, none provide adequate range:
  - Pressurized storage at 35–70 MPa (5,000 to 10,000 psi).
  - Metal hydrides—typical hydrogen density 1.2–5%.
  - Cryogenic liquid—20 K.
  - Onboard reformation of liquid organic hydrogen carriers (LOHCs) such as methanol or ammonia.
  - Chemical hydrides—compounds that can be reacted controllably onboard a vehicle to produce hydrogen.
- Cost depends on energy source and onboard storage method. Usually high.

Characteristics of Fuels that Could Replace Diesel - Compression Ignition (CI) Engine Fuels

Vegetable Oils, Biodiesel, Renewable Diesel Fuel

- All are diesel (CI) engine fuels. liquid fuels.
- Almost all vegetable and seed oils *are already* diesel fuels.

- Biodiesel is vegetable oil that has been processed to make its physical properties more like mineral diesel fuel (Diesel No. 2, or No. 1 for cold climates).
- Biodiesel is made by transesterification of vegetable or animal fats with methanol or ethanol.
- All vegetable oils and biodiesel have naturally low or zero sulfur content.
- Biodiesel has slightly lower energy density than raw vegetable oil or Diesel No. 2 (D2).
- Cetane ratings similar or slightly lower than D2.
- Similar particulates and NO_x emissions compared with D2.
- Approximately the same CO_2 emissions as D2.
- B20 (max 20% blend with D2) is the maximum percentage mixture allowed by diesel manufacturers. Little or no climate benefit.
- Veg/seed oil (e.g., soy or canola/rapeseed) subsidized costs currently similar to D2.
- Despite subsidies, B20 cost is slightly higher than D2.
- Can use existing diesel fuel distribution system with no modifications. A "drop-in" replacement for diesel fuel.
- Food vs. fuel issues. Vegetable and seed oils are edible.

Dimethyl Ether (DME, CH_3OCH_3)

- A CI dual-phase fuel.
- A synthetic fuel made from natural gas, but can be made efficiently from waste biomass.
- Currently produced from natural gas, with a process similar to that for methanol.
- Liquid at room temp under moderate pressure (4.6 bar, 66 psi), similar to butane.
- Higher cetane rating than diesel fuel (D2).
- Wider limits of flammability than D2 help combustion quality.
- Produces nearly zero particulates even if overfueled.
- 90% lower NO_x emissions than D2 or ULSD2.
- Zero sulfur.
- Only about 50% of fuel energy density of D2, requiring larger (pressurized) fuel tank.
- Zero lubricity—requires lubricity additives, typically 10% D2 or veg oil.
- Requires different fuel injection equipment than D2 or biodiesel.
- Could use existing widespread propane fuel distribution system.
- No current use in the USA, but partially deployed as fuel for large trucks in China and Northern Europe.

Vehicles Powered by Electric Motors

Battery Electric Vehicles (BEVs)

- Non-combustion propulsion. No tailpipe.
- Energy source delivered to vehicle is electricity, which is generated from some original energy source.
- Solar, wind, or geothermal are renewable electricity sources.
- Energy is stored in electrochemical battery.
- In use since 1881.
- Currently considered to be the future of automobiles.
- Electricity is available almost everywhere, but dependent upon electric generation, distribution, and public charging facilities.
- Simple mechanical design.
- Heavy vehicle weight due to battery mass. Battery energy density: 35 Wh/kg for lead acid, 150–300 Wh/kg for lithium batteries, vs. 12,200 Wh/kg for gasoline.
- Range limited due to battery energy density.
- Home charging requires 6–8 h. High-rate partial charging in under 30 min.
- Limited charging infrastructure.
- Requires major upgrade to distribution grid, generation, and storage capacity—a problem only recently being addressed.
- Renewability and environmental impacts depend on the source of energy. Zero emissions if electricity is carbon-free, but worse emissions than gasoline if electricity is produced from fossil fuels.

Direct Solar Power

- Only usable for ultra-lightweight vehicles.
- Dependent upon incident sunlight.
- Surface area for photovoltaic cells must be larger than the entire vehicle surface area for power level capable of propulsion for any except the lightest possible vehicle.
- Solar power of vehicle accessories (e.g., air conditioning) or trickle charging of BEVs have been the only practical applications (Fig. 5.2).

Hybrid Electric Vehicles (HEVs)

- Dual propulsion systems—ICE + electric motor.
- Not a renewable energy vehicle.
- Braking energy recovery—the largest benefit.
- Electric propulsion at low speeds.
- Most auto manufacturers offer hybrids.

Fig. 5.2 General Motors Sunraycer, winner of Sun Race Australia, 1987. Solar cells occupied much of the surface area of the vehicle. (Creative Commons CC0)

- Multiple hybrid configurations
  - Series hybrid
  - Parallel hybrid
  - Power-split hybrid
- Plug hybrid allows electric or petroleum refueling.
- Unless it is a plug hybrid, it is just a high fuel economy gasoline vehicle.

Hydrogen FCVs

- Non-combustion propulsion.
- Fuel cell is a continuously refueled battery.
- Hydrogen is currently the only fuel for FCVs.
- Direct methanol cells under development for past 40 years.
- H_2 FCVs are actually BEVs with the batteries charged by a hydrogen-air fuel cell.
- Potentially longer range than BEVs, but onboard fuel storage a major limitation.
- Longer refueling time than other alternative fuels, but faster than battery recharging.
- Operational attributes otherwise identical to BEVs.
- Questionable carbon footprint when the entire fuel cycle is considered.
- Carbon neutrality requires electrolytically generated hydrogen from zero-carbon source.
- Most expensive of all viable fuel options.
- Supported by oil producers as an alternative way to continue the use of fossil fuels.
- Highly publicized and politicized.

Fig. 5.3 Amish horse and buggy on rural road in Pennsylvania, USA. (https://www.peakpx.com/572073/black-wooden-horse-carriage. Creative Commons CC0)

Animal Power

- Not a viable automobile fuel or energy source for transportation on highways, but horse-drawn private and agricultural vehicles remain legal (with some restrictions) on most urban and rural surface streets and roads worldwide (Fig. 5.3).
- A low-tech transportation option if all technology and fuel sources are unavailable.
- Hay or oats power.
- The primary motive power source worldwide for over two millennia.
- A unique pollution problem that was one of the major incentives for shifting to electric or internal combustion vehicles, originally called horseless carriages in 1900.

6 Methanol

The Simplest Synthetic Fuel

Methanol (aka methyl alcohol, wood alcohol, carbinol, CH_3OH) has been used as a fuel source since the seventeenth century. A compact history of methanol can be found in citation [1]. Other than hydrogen, it was the first synthetic fuel. In World Wars I and II, Axis countries used it extensively as a vehicle motor fuel, replacing scarce petroleum [2].

As an IC engine fuel, its salient characteristics are as follows:

- A spark ignition (SI) engine fuel.
- Approximately 1/2 the volumetric or mass energy density of gasoline. Requires nearly twice[1] the fuel tank volume for equivalent energy storage.
- Distribution same as gasoline, but higher cost due to lower energy density (Fig. 6.1).
- Neat (pure, M100) methanol has very poor lubricity, requiring the addition of organic oil additives such as castor or lesquerella [3] oil to prevent excessive cylinder wall and piston ring wear.
- Cold starting not possible at temperatures below about 4 deg C. All-weather use requires startup preheating mechanisms similar to diesel engines, or a volatile fuel additive such as gasoline or DME.
- Highly hygroscopic; will absorb humidity from the air over time. Not a problem for daily automotive use, but a storage issue if vehicle is unused for long periods, or methanol is stored in underground tanks for long periods at refueling facilities.

[1] The lower energy density of methanol is partially offset by its higher energy efficiency. In 40 years of driving on M99 (methanol with 1% castor oil), I have found that it requires on average 1.7 times the fuel tank volume to match the range of a gallon of gasoline (E10). The same result appeared in a report by Mark Looper available online at http://www.altfuels.org/backgrnd/altftype/m85.shtml, 2023. Accessed June 5, 2023.

C. A. MacCarley, *Non-Petroleum Automotive Transportation*,
https://doi.org/10.1007/978-3-031-85028-8_6

Fig. 6.1 Fueling a methanol-converted vehicle with M99 (99% methanol, 1% castor oil). (Photo: author)

- A popular racing fuel. Very high octane allows a high compression ratio which improves efficiency and power. Seven percent greater power output than gasoline, significantly greater than ethanol.
- Incompatible with many metals and elastomers designed exclusively for gasoline compatibility prior to the mid-1980s. Fuel systems designed to accept E85 flex fuel are fully compatible.
- Can use existing fueling infrastructure, requiring only the upgrade of elastomers in fuel dispensing equipment.
- In the USA, currently 100% produced from natural gas due to lowest cost.
- Can also be produced from almost any hydrocarbon, from coal to agricultural biomass. Higher production energy efficiency and lower cost than cellulosic or fermented ethanol.
- Small-scale methanol production from natural gas at remote well sites could provide a safe and environmentally less impactful means to transport and utilize natural gas that is currently flared or vented due to high cost of transport.
- If made from waste biomass, methanol is typically 90+% energy-positive relative to the energy required for production.
- Well-established technology with a large body of experience.
- No “food vs. fuel” concerns.
- M85 flex-fuel vehicles (FFVs) were considered a failure for nontechnical reasons during the DOE Alternative Fuel Utilization Program (AFUP) of the 1990s.
- A popular alternative fuel in areas with limited gasoline availability such as Western China.
- Currently (2024), the cost per unit of energy is lower compared with gasoline in bulk, but there is no distribution system. Almost no M85 flex-fuel cars remain in service from the first FFV program in the late 1990s.
- Carbon dioxide can be reacted with electrolytic hydrogen to synthesize methanol. If the CO_2 is a waste product obtained by carbon capture from a power plant exhaust stack or if it is extracted from the atmosphere, this process forms the basis of carbon capture and utilization or storage (CCUS), a supplemental benefit. But this requires large amounts of electricity and is only viable if excess renewable electricity is available, e.g., geothermal, hydroelectric, nuclear, or isolated solar and wind.

- Fire safety risk is considered roughly equivalent to diesel fuel [4], but opinions vary for different applications. A methanol fire is harder to ignite and easier to extinguish, but its nonluminescent flame and low radiated heat make it more difficult to detect by first responders. The practice of adding 15% gasoline in M85 was partially justified to provide flame luminance.

DIY Methanol Production

Methanol can be made in a number of different ways, including thermochemical and fermentation methods.

Dirty methanol can be made almost trivially from woody biomass, requiring only a crude apparatus and tribal knowledge rather than an understanding of the chemistry. The simple process involves heating biomass in a closed vessel (pyrolysis) to form Biomass Pyrolysis Syngas (BPS) that includes a wide range of organic compounds (H_2, CO, N_2, CO_2), methane, steam, and various other vaporized hydrocarbons and bio-oils. Followed by passing the gas over or through an inexpensive catalyst, usually copper and fractional distillation to separate the methanol and condensed liquid hydrocarbons from the water generated (Fig. 6.2). It is likely that early producers of wood alcohol were not aware of the catalytic properties of the copper plumbing that they traditionally used in their "stills." The just-because-we've-always-done-it-that-way copper construction worked because it was an effective catalyst for the synthesis of methanol from the BPS, which if the biomass was wood, is referred to as *wood gas* [5], generated by heating wood chips in the absence of air. A more detailed discussion of methanol from biomass appears later in this chapter.

Even made by such a crude apparatus, the combustion emissions of the resulting condensed liquid (mostly methanol, water and other liquid hydrocarbons) can be cleaner than the combustion emissions of the original wood feedstock used to make it, and even impure methanol was a much more compact and cleaner form of energy. In the nineteenth century, a single wagonload of methanol contained the energy equivalent of many wagonloads of firewood and was easier to transport. *Wood alcohol* had commercial value as a solvent and a compact fuel for illumination and cooking compared with the only alternatives: coal, wood, or organic oils Alcohol lamps are still used today in laboratories when natural gas for a *Bunsen Burner* is not available.

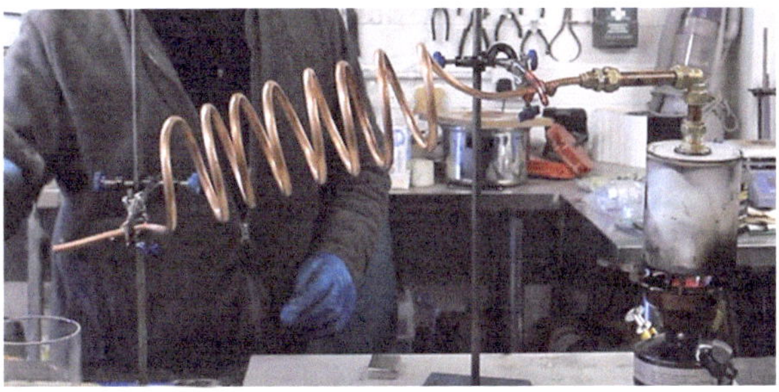

Fig. 6.2 Crude lab apparatus for making methanol from wood chips. (Frame from YouTube video by Robert Murray-Smith; https://www.youtube.com/watch?v=IGKlpicqldE. Used with permission)

One of many DIY guides for making methanol from wood can be found in citation [6]. Instructional write-ups and YouTube videos (beware, some are hazardous) are widely available, but a concise and correct description of the various biomass processes can be found in the 1995 NREL publication *Methanol from Biomass* [7]. A recent summary of the various designs for biomass-to-syngas gasifiers and reformers can be found in the 2022 paper *Reforming processes for syngas production: A mini-review on the current status, challenges, and prospects for biomass conversion to fuels* [8].

Methanol from Natural Gas

Sixty percent of methanol worldwide is made by variations of the (1925, German) Fischer-Tropsch process, which involves catalytic synthesis of syngas produced by reforming natural gas. 35% is produced from coal, 80% of which occurs in China [9]. The production of methanol from coal is considerably more carbon-intensive than from natural gas [10], although this depends on the type of coal and the process technology. Currently, less than 5% of methanol is produced from other sources, mostly forms of biomass [11].

Steam methane reforming (SMR) is the first step for production of methanol from natural gas. Natural gas and steam are mixed in the absence of air and passed over a (typically nickel) catalyst inside stainless steel tubes that are heated to 700–1000 °C [12, 13]. The process requires considerable external energy to form the steam and maintain the reactor temperature. The heat is usually provided by the combustion of the natural gas feedstock, but is partially offset by recycling the heat generated in the exothermic synthesis second step. The composition of the resulting synthesis gas (syngas) can vary with the composition of the natural gas, but assuming for simplicity that it is entirely methane, the resulting syngas will consist of hydrogen and carbon monoxide with some carbon dioxide.

$$CH_4 + H_2O + \text{heat} \rightarrow CO + 3H_2 \left(\text{SMR}\right)$$

With twice the amount of steam required for the SMR reaction, an additional H_2 molecule is generated, and CO_2 replaces the CO.

$$CH_4 + 2H_2O + \text{heat} \rightarrow CO_2 + 4H_2 \left(\text{SMR with excess } H_2O\right)$$

By control of the steam addition, a composite of both reactions yields

$$2CH_4 + 3H_2O + \text{heat} \rightarrow CO + CO_2 + 7H_2 \left(\text{Synthesis Gas and } CO_2\right)$$

The syngas is then converted to methanol by passing through a copper/zinc (typical) catalyst bed at a lower temperature (200–300 °C) but higher pressure (50–100 bar, 725–1450 psig).

Methanol is synthesized by the reaction:

$$CO + 2H_2 \rightarrow CH_3OH + \text{heat}$$

The reaction is exothermic, and the heat produced can be used to partially supplement the heat required by the first step in which the syngas was produced from the natural gas. The net energy efficiency of the process is reported to be 56–57% for large-scale plants.

The CO_2 generated by SMR with excess water can be further reacted with the excess net hydrogen to yield additional methanol and water, up to the limit of the excess H_2 after the second and first reactions.

$$CO_2 + 3H_2 \rightarrow CH_3OH + H_2O$$

The composite reaction is

$CO + CO_2 + 7H_2 \rightarrow 2CH_3OH + 2H_2 + H_2O$ (yields additional methanol and leaves some excess hydrogen)

All of the carbon from the methane feedstock is accounted for in the resulting methanol, and hydrogen and water are left over. The hydrogen gas is easily removed following fractional distillation of the vapor-phase methanol and water products. Hydrogen has other uses, including supplementation of the first step heat requirements for natural gas reforming.

Overall, from input to final output with the water ignored,

$$CH_4 + \text{heat} \rightarrow 2CH_3OH + 2H_2$$

From a mass balance point of view, 32 g of methane yields 64 g of liquid methanol. And unlike hydrogen production by SMR, the process (ideally) produces no CO2 except for that generated in the heat addition.

Overall, the process is net endothermic, requiring external heat, but considerably less than required for SMR hydrogen production from natural gas.

An alternative process called *tri-reforming* is possible which performs reforming and synthesis in a common reactor to directly utilize the heat generated from synthesis to supplement the external heat required for reforming, usually provided by natural gas combustion [14]. It does not appear that the process is yet in large-scale commercial use, reported to be because of challenges with the poisoning of the catalyst [15].

Prior to the final distillation step, the result is "wet methanol" (typically 82% methanol and 18% water with small amounts of ethanol and dissolved hydrocarbon gases from the natural gas feedstock). Water vapor condenses at a higher temperature than methanol vapor, leaving up to 99.85% methanol [2].

As mentioned previously, methanol can also be made using a carbon-free source of hydrogen and carbon dioxide produced by CCS from power generation or industrial process exhaust [16], or by DCCS [17]. The hydrogen is presumably produced using renewable or excess utility electricity by electrolysis. This is not a logical use of electricity if it is itself generated from the usual mix of grid energy sources. But it makes sense if it is produced entirely from renewable sources or is the result of excess generation during mid-day sun hours when solar generation overwhelms the power grid. One exemplary case is Carbon Recycling International (CRI) that has been producing methanol in Iceland under the trade name "Vulcanol" [18] using CO_2 captured from a nearby geothermal power plant and geothermal electric power generation for electrolytic hydrogen production [19]. Iceland is unique with its excess geothermal heat and electricity resources, so this is not necessarily a transferable technology for carbon sequestration. The much-hyped use of carbon extracted from the air by DCCS to make methanol has been studied since at least 2016 [20], but no commercial facilities are currently in operation.

For fuel-grade methanol, the raw methanol/water/impurities need only be distilled sufficiently to remove the water since the impurities are usually combustible and do not appreciably compromise the combustion value or emissions of the fuel. Fuel-grade methanol need not be pure.[2] Ethanol and other HC impurities do not appreciably degrade the fuel value. It is known that a small (<1% volumetric) amount of water absorbed from the atmosphere only minimally reduces the combustion quality of the fuel in an engine. More problematic is that blends of wet methanol with gasoline are prone to phase separation. As a practical test of water contamination, water content above about 2–3% in M85 or methanol with a lubricant additive such as castor oil is indicated by a slightly cloudy appearance, as the gasoline or organic oil starts to separate from the methanol. In my experience, even slightly cloudy M85 can still be used as a fuel for daily commuting, but with some loss of performance and startability. A mixture of as much as 75% water and 25% methanol is actually still flammable, although it cannot be used as fuel [2]. But it is worth noting for fire suppression purposes. A maximum of 1.5% water content is legally allowed in the USA to be sold as M100 fuel [21], although any water content above about 0.1% would be considered unacceptable (wet methanol) by professional racers. The distinction between usable fuel-grade methanol and 99.9% pure racing fuel or 99.999% laboratory-grade methanol is significant since the production cost of slightly impure methanol is significantly lower than high-purity methanol.

Methanol from Biomass

Updating centuries-old methods described previously for the crude synthesis of methanol from wood chips, modern methods for making methanol from biomass still essentially rely on three major steps. These are basically the same as those used for production of methanol from natural gas or coal, except for the low quality of the syngas.

Heating dry carbonaceous biomass in the absence of air produces the syngas (BPS), a mixture of hydrogen and carbon monoxide, carbon dioxide, and a fairly large percentages of impurities characteristic of the type of biomass. For example, wood pyrolysis for small-scale methanol synthesis typically yields a syngas composition:

Nitrogen (N_2): 50–54%
Carbon monoxide (CO): 17–22%
Hydrogen (H_2): 12–20%
Carbon dioxide (CO_2): 9–11%
Methane (CH_4): 2–3%

[2] Based upon personal experience with methanol-fueled vehicles using fuel with up to 2% water, methanol contaminated with even a fraction of a percent water is considered unacceptable as a racing fuel, but is still usable as a regular motor fuel. Legally, neat methanol (M100) in the USA is allowed to contain up to 1.5% water.

For wood waste and other types of biomass with a low hydrogen-to-carbon ratio, supplemental hydrogen or methane[3] may have to be added to assure the 2:1 hydrogen-to-carbon monoxide molar ratio required for consumption of all of the hydrogen and carbon monoxide in the second (synthesis) step:

$$HxCy\,(\text{variable biomass}) + \frac{y}{2}O_2 + \text{heat} \rightarrow y\,CO + \frac{x}{2}H_2\,(+\text{variable } CO_2)$$

The high-temperature syngas at moderate pressure is passed over a catalyst, usually an inexpensive copper-zinc alloy compound, which converts it to a mixture of methanol and water plus other contaminants previously mentioned that we ignore in the equations that follow. This reaction generates heat, which can supplement the heat required by the previous pyrolysis step:

$$CO + 2H_2 \rightarrow CH_3OH + \text{heat}$$

and if additional H_2 is available

$$CO_2 + 3H_2 \rightarrow CH_3OH + H_2O$$

Nitrogen and any unreacted CO_2 pass through and are released into the atmosphere.

The third step is fractional distillation to separate the methanol from the water and other unintended coproducts. As the vaporized product leaves the reactor, it cools, and the water condenses first, followed by ethanol and other trace alcohols, leaving mostly methanol vapor which then condenses. The boiling point of methanol (64 °C) is lower than water (100 °C) or other alcohols such as ethanol (78 °C) at atmospheric pressure. If high-purity methanol is desired, additional fractional distillation stages may be necessary to remove contaminants that has a lower boiling temperature than methanol. For fuel-grade methanol, it is mainly the removal of the water that is necessary, since the other minor fractions are usually combustible hydrocarbons. Efficiency for the overall process can be as high as 40% to 45%, with the majority of the CO_2 emissions coming from the combustion heat required for reformation and distillation.

While not yet considered viable at large scale, biomass-derived methanol seems to have potential for small-scale or local production from woody biomass and agricultural residues. Aside from the minimal cost (at least at the source) of the waste biomass feedstock, the Methanol Institute [22] takes an optimistic view of the environmental sustainability of biomass-derived methanol, additionally noting that CO_2 and methane would be generated anyway by the natural decay of biomass left in a field or plowed under:

> It should be noted that conversion of wood, agricultural and municipal wastes to methanol can be an effective greenhouse mitigation. These wastes generate methane (under anaerobic conditions), which is released to the atmosphere. Methane is a much stronger green-house

[3]As noted in the previous section, reformation of CH_4 yields an excess of hydrogen, which can offset the hydrogen deficiency of biomass.

gas than CO2. Thus, direct conversion of these wastes to fuels and eventually to CO2 through combustion can result in a decreased impact on climate change." And "If all the 'discarded' wastes are converted to methanol, about 10 billion gallons of methanol can be generated per year.

In other words, depending on the biomass quality, the consequences of NOT using biomass to make methanol can be an increase in its GHG impact while discarding its fuel value.

According to a 2022 report published by Energy Central [23],

Production from renewable sources, such as from biomethane, solid biomass, municipal solid waste (or MSW, which contains a considerable fraction of organic waste), and renewable energy, has a low carbon footprint. Most of these pathways achieve 10–40 g CO_2 eq/MJ, and some pathways even have negative emissions (−55 g CO_2 eq/MJ for methanol from biomethane from cow manure) which means effectively that CO_2 is removed from the atmosphere or that the pathway avoids emissions that would have otherwise taken place in other processes.

The ICCT (International Council on Clean Transportation), reporting on the total carbon impacts of the fossil and renewable production paths for methanol as a marine fuel, published the bar chart of Fig. 6.3 with CO_{2e} calculated using a 100-year GWP and the GREET 2020 model [24]. This shows the large difference between the carbon impact of fossil (gray or blue) methanol vs. cellulosic biomethanol. The lowest impact (negative impact) case is e-methanol, but like all other e-fuels that originate from renewable electricity, this path is a questionable use of this high-quality energy form.

The Methanol Institute published the projections of Fig. 6.4 for the worldwide growth of renewable methanol from various waste sources, as well as e-methanol

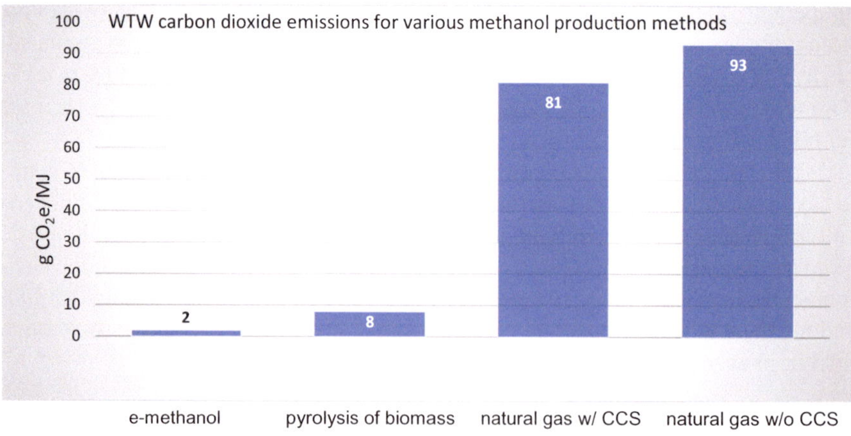

Fig. 6.3 Total CO_{2e} emissions for each of the four production pathways for methanol. e-Methanol assumes carbon-free electricity. CCS recovery percentage is highly variable. (Graphic by author from data published by Abigail Martin, International Council on Clean Transportation, in "A step forward for 'green' methanol and its potential to deliver deep GHG reductions in maritime shipping," 1 September 2021. https://theicct.org/a-step-forward-for-green-methanol-and-its-potential-to-deliver-deep-ghg-reductions-in-maritime-shipping%E2%80%AF/)

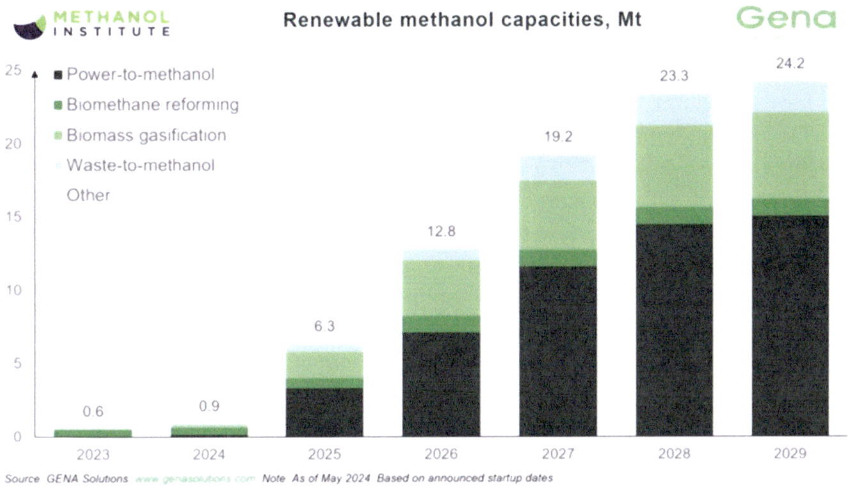

Fig. 6.4 Projected renewable methanol growth 2023–2029, identifying the energy source and process. (From GENA Solutions, published by the Methanol Institute in "Innovation Outlook: Renewable Methanol" 2024, ISBN 978-92-9260-320-5. Public domain)

identified as "Power-to-methanol". Other arguments for and against methanol from biomass can be found in citations [25, 26].

Worldwide, methanol production from biomass is currently only a tiny percentage of production compared with natural gas or coal. This is because the cost and convenience of these feedstocks are far too low to justify the production of renewable methanol. Methanol from biomass is more amenable to small-scale local operations than traditional large-scale commercial operations due to the cost of recovery and transportation of off-site biofeedstocks. This may explain why biomethanol has only gained traction in areas of scarce or expensive gasoline such as Scandinavia, Southern Africa, and especially Western China (which is also aggressively promoting methanol made from indigenous coal).

Methanol Production Summary Points

1. When produced from natural gas (methane), *excess hydrogen is produced* that can potentially be used to provide the supplemental hydrogen needed if low-quality biomass is used concurrently for methanol production. Excess hydrogen may also be used as a carbon-free source of heat for the reformation step, replacing the combustion of natural gas.
2. Most industrial production of methanol from natural gas does not utilize or account for the energy value of the excess hydrogen generated.
3. In current practice, the overall efficiency of natural gas-to-methanol at the medium to large scale is reported across two studies to be between 64% and 72% (mean value of 61%) [27]. For the range of all feedstocks from which

methanol can be made, including coal, wood, and various forms of biomass, one study in 1976 reported efficiencies of 50–65% [28]. The low/high spread seems to be dependent on the degree to which the heat of synthesis is recycled for reformation, or the excess hydrogen (if any) is accounted for.

4. The net CO_2 emissions from the production of methanol from natural gas are reported to average 1.7[4] kg (3.8 lb) of CO_2 per US gallon of methanol product [29] at existing production facilities. If this CO_2 is included along with the 3.91 kg CO_2/gallon combustion emissions of methanol in an engine, the total WTW carbon intensity is 5.61 kg CO_2/gallon of methanol.
5. For comparison, the refining of gasoline from crude oil is reported to generate an average of 2.73[5] kg CO_2/gallon of finished gasoline [30]. Added to the 8.89 kg CO_2/gallon produced by its combustion, this totals 11.62 kg CO_2/gallon of gasoline.
6. But gasoline has more than twice the volumetric energy content of methanol. Equating to equal fuel energy using the volume LHV energy density of 59.0 MJ/gallon methanol and 118.0 MJ/gallon with gasoline, the CO_2 emission per unit of fuel energy of methanol is 95.1 g CO_2/MJ, only slightly less than gasoline at 98.5 g CO_2/MJ.
7. A vehicle using (gray) methanol produced from natural gas would have a WTW carbon footprint only slightly less than gasoline.
8. If produced from biomass and the carbon content of the biomass is considered renewable for carbon accounting purposes, the footprint of methanol can be half that of gasoline.
9. For "blue methanol" that includes some level of CCS to reduce CO_2 emissions, the comparison would strongly favor methanol over gasoline, as will be discussed in the next section.
10. Methanol from cellulosic biomass with 50% CCS would likely produce the lowest net climate impact of all liquid fuels. It is unknown if any such facilities are in operation at this time.

Methanol Toxicity

Methanol, like gasoline, is highly toxic in small quantities. But unlike gasoline, it lacks a strong odor or taste. Methanol mixed with ethanol is hard to detect without test equipment. Consequently, the accidental or nefarious substitution of methanol for beverage ethanol is a well-known cause of death by ingestion [31]. In the USA, this was common during the prohibition years 1919–1933 due to lack of quality control by amateur producers of bootleg ethanol (aka *moonshine*). Methanol poisoning unfortunately still occurs [32], especially in countries that heavily tax or fail

[4] CO_2 emissions of methanol production by SMR + synthesis include emissions from the balance heat source, assumed to be natural gas. No CCS is used.

[5] Published data on gasoline refinery emissions are highly variable, from 0.92 to 3.35 g CO_2/gal. The cited 2.73 g CO_2/gallon from citation [30] is the median of all data samples.

to regulate beverage ethanol. The risk of accidental poisoning from homemade ethanol is always present, since some small fraction of methanol is usually produced in crude stills that lack adequate process controls. This is why, e.g., home brewers of hard liquor usually discard the first 5% of the distillation product, most produced prior to the distillation column or cooling coil being fully up to temperature. The importance of removing methanol from the intended ethanol product is the root of the tradition of distilling vodka multiple times [33], popularly portrayed as a means to improve the taste. This sales pitch may be true if other impurities are present, but the practice of 3–10 distillations originated in the need to be assured against contamination by methanol [34] that is always present to some small degree in fermented ethanol. In the EU, vodka is required by law to not exceed 10 g/100 L methanol content.

For perspective, consider the conclusions of the US National Institute of Health (NIH), regarding ingestion or inhalation injuries related to gasoline [35]:

> [Gasoline] Poisoning accounted for 13% of injuries and 17% of deaths. The primary poisoning injury pattern was ingestion; the primary fatality pattern was inhalation, with about half of those associated with deliberate abuse.

Directly comparable statistics for *fuel* methanol are difficult to isolate, since the overwhelming majority of toxicity injuries are due to accidental ingestion of methanol as a contaminant in beverage ethanol [31]. And deliberate (recreational) inhalation of methanol is unlikely as an injury vector due to the much lower vapor pressure of methanol compared with gasoline. I could find no record of poisonings from the ingestion of M85.

Fire Safety

As discussed above, methanol has a long history as a racing fuel. But its fire characteristics and suppression procedures are different than those for gasoline, with the lack of a visible flame being the greatest concern.

In 1964, a seven-car crash kills drivers Dave MacDonald and Eddie Sachs on the second lap of the Indianapolis 500, in the ensuing gasoline fire. Johnny Rutherford, who was also involved in the crash, survived, because his methanol-fueled car had not ignited. The gasoline fire produced highly luminous flames and thick black smoke that obscured the view of drivers approaching the initial two-car crash. Five vehicles drove directly into the crash, unable to see it. Also a factor: In a gasoline fire, fuel spreads across the tarmac and cannot be extinguished using water, since gasoline floats on top of it. Extinguishing a gasoline fire requires special foam firefighting agents. Some of the drivers caught in the compound wreck were burned to death. By comparison, methanol fires are more difficult to ignite and can be extinguished with water, the flames are nearly invisible, radiant heat is very low, and there is virtually no smoke from the fuel (although materials from the vehicles will impart some smoke and luminosity). Subsequently, methanol was required in all United States Auto Club (USAC) sanctioned races starting the following racing season [36].

Fig. 6.5 Methanol fire while refueling at 1981 Indianapolis 500. Invisible flame was difficult to detect. Still frame from video provided by Indianapolis Fire Department. (Used with permission)

But the invisibility and smoke-free characteristics of a methanol fire create a different set of problems for first responders. One of the most infamous incidents occurred in the 1981 Indianapolis 500, when a methanol fire occurred while refueling and the invisibility of burning methanol (and inexperience of the pit crew) leads to a confused response [37, 38] (Fig. 6.5):

> When Rick Mears pitted on lap 58, fuel began to gush from the refueling hose before it had been connected to the car. Fuel sprayed over the car, Mears and his mechanics, then ignited when it contacted the engine. Methanol burns with a transparent flame and no smoke, and panic gripped the pit as crew members and spectators fled from the invisible fire. Mears, on fire from the waist up, jumped out of his car and ran to the pit wall, where a safety worker, not seeing the fire, tried to remove Mears' helmet. Meanwhile, Mears' fueler, covered in burning fuel, waved his arms frantically to attract the attention of the fire crews already converging on the scene. By this time the safety worker attending to Mears had fled, and Mears, in near panic at being unable to breathe, leaped over the pit wall toward another crewman carrying a fire extinguisher, who dropped the extinguisher and also fled. Mears tried to turn the extinguisher on himself, but at this point his father, Bill Mears, having already pulled Rick's wife Deena to safety, grabbed the extinguisher and put out the fire. His mechanics had also been extinguished, and the pit fire crew arrived to thoroughly douse Mears' car.
>
> Thanks to quick action by Bill Mears and the fact that methanol produces less heat than gasoline, no one was seriously hurt in the incident.

Mr. Mears suffered burns to his face which was exposed through his helmet. Years later, the near tragedy was parodied in the 2006 Will Ferrell comedy film *Talladega Nights: The Ballad of Ricky Bobby* [39].

Over the years, other methanol fires have occurred in organized racing. But methanol and ethanol continue to be considered safer than gasoline to this day. The methanol-only rule stayed in effect until 2005, when USAC, followed by IndyCar and FIA, changed rules to require ethanol blends, a more expensive biofuel fuel that has similar fire characteristics to methanol, but was championed by race sponsors aligned with the US agriculture industry.

Despite these incidents, to this day, methanol remains the preferred fuel for ultra-performance racing such as drag racing, sprint car racing, tractor-pulling competitions, hill climbs, and any type of racing in which high levels of supercharging boost are used. While maximizing power output is the primary objective, the relative safety of methanol compared with gasoline is also a factor.

A practical guide to methanol fuel safety for racers can be found in citation [40].

Methanol as an Octane Additive

Methanol, like all alcohols and ethers, is an oxygenated hydrocarbon, or *oxygenate*. Oxygenates contain one or more oxygen atoms. The presence of the oxygen in alcohols decreases the energy density (by mass and volume), but otherwise imparts characteristics that are advantageous when added to gasoline.

EPA testing in typical (carbureted) vehicles of the 1970s had showed reductions in hydrocarbon (HC) and carbon monoxide (CO) emissions with the addition of oxygenates (as little as 3%) to gasoline. While alcohols do produce lower levels of these emissions on their own, the measured benefits from tests at the time were probably more due to the fact that carburetors were terrible at maintaining a constant air-fuel ratio (AFR), so to prevent "lean spots" in the throttle response, they were usually set overall rich. The additives simply "leaned out" the overall AFR, a trick that some mechanics of that era used to help over-polluting cars pass the newly instituted federal- or state-mandated annual or biennial emissions tests.[6]

But much more important is that oxygenates usually have very high anti-knock properties (octane ratings), so they could replace the previously standard additive tetraethyl lead (TEL), which for decades had been added to premium gasoline to reduce knocking, but released highly toxic lead aerosols in the exhaust.

The 1977 revision of the Clean Air Act and its amendments allowed only "substantially similar gasoline," which required that any oxygenated additive be approved by the US EPA before being allowed to be added to gasoline. As discussed previously, in 1981, the EPA outlawed the use of TEL in all except specialized motor fuels, and approved the blending of methyl tertiary butyl ether (MTBE) to replace TEL. A limit of 11% by volume was originally set, but extended to 15% in 1988 [41]. However, in the early 1990s in the USA, widespread problems were found with gasoline leakage from underground gasoline storage tanks. Due to the extreme toxicity and carcinogenicity of MTBE, this can be a significant public health concern if even small amounts of MTBE entered groundwater [42]. The soon led to the substitution of 5% methanol, and soon after, 10% ethanol to replace MTBE.

As a knock suppression additive, methanol was very effective in as little as a 3% blend. Blends of 3–5% methanol in gasoline were referred to as *gasohol*, soon to become a derogatory term among mechanics and car owners. The methanol additive was unjustly accused of reducing the volumetric mileage (MPG) of the vehicle by consumers that did not understand the difference in the volumetric energy content of alcohols. But it was legitimately responsible for criticisms that it caused problems in the carbureted vehicles of the era, which were never designed for alcohol compatibility. At only a 3% concentration, the negative effects were

[6] Personal experience tuning performance engine in the 1970s–1990s.

actually minimal, but methanol-based gasohol became fodder for anti-environmental complaints among automotive manufacturers and owners. Methanol reacted with the pot metal[7] bodies of carburetors and fuel pumps, eroded the terneplate (lead/tin) coating of fuel tanks, and softened or swelled hoses and gaskets that were engineered only for gasoline compatibility. These problems were eventually corrected by auto manufacturers by inexpensive materials changes, but the reputation endured—which perpetuated arguments in favor of substituting ethanol, E10 rather than M5, although both alcohols had nearly identical material incompatibilities.

The US AFUP, 1989–1999

Methanol was given a formal public trial in the late 1990s in the USA. But it was destined to fail by design of the program. Following successful research demonstrations of methanol vehicles at universities and DOE research facilities, the state of California (USA) requested bids for M85 (85% methanol, 15% gasoline by volume) fueled cars from major manufacturers [43]. In 1981, Ford subcontracted with Bill Stropp Enterprises, a successful race driver and engine builder for Ford Performance, to modify a small number of Ford EXPs for use of M85 or M100, for public relations events and eventual delivery to Los Angeles County. One such event was the 1981 Future Fuels Challenge[8,9,10] an ill-fated transcontinental fuel economy competition for alternative fuel vehicles. The Ford team fielded two methanol EXPs, one piloted by accomplished race driver Lyn St. James.

Ford's modified cars were placed in demonstration service, and some were passed onto local utilities and Verizon Communications. However, these early methanol conversions retained the gasoline carburetor, which was incapable of maintaining a constant AFR or providing adequate fuel/air mixing compared with more precise electronic fuel injection (EFI) systems that were available in 1981. This learning experience further supported the undeserved indictment of methanol as a "bad" fuel, prone to cold starting limitations and high emissions. An additional 100 vehicles (Ford Escorts with EFI) were later delivered, but only four fueling sites remained available, all in California. In 1983, Ford custom-built 582 additional M85 Escort models, delivering

[7] Pot metal is a broadly defined term for various alloys of copper, lead, zinc, tin, aluminum, and (rarely) cadmium, used because of its low molding temperature and its resistance to corrosion by gasoline. Copper, zinc, and aluminum contents make it highly susceptible to corrosion by oxygenated (alcohol) fuel additives. The use of methanol in a gasoline carburetor will ruin it within a few days.

[8] Little information remains accessible, despite extensive media coverage at the time. One online article that recounts this event is Jeff Davis, Mother Earth News (MEN), May 2006. *The MEN Gasifier*, http://gasifiers.bioenergylists.org/node/7. Accessed 20 June 2024. The scoring metric was energy use per vehicle weight. No official winner was declared because the promoter fled the country with the prize money and did not pay the SCCA sanctioning body. But the SCCA reported the unofficial winner as the author in the electronically fuel injected car shown in Fig. 6.10, built under a DOE Appropriate Technology grant.

[9] The trademark record for the Future Fuels Challenge event can be found at https://trademarks.justia.com/733/09/future-fuels-challenge-rally-73309046.html. Accessed 20 June 2024.

[10] A query to the Perplexity AI web site about this event can be found at https://www.perplexity.ai/search/Tell-me-about-Qfig1xjSS2.dJzP6dO3_NA. Accessed 20 June 2024.

Fig. 6.6 Ford advertising postcard for 1996 Ford Taurus M85 Flex Fuel Vehicle. (Image provided by Alden Jewel. Used with permission)

501 of these to the State of California and exporting 81 to New Zealand, Sweden, Norway, the UK, and Canada for trial testing. No published data from these trials could be found [28, 44]. By 1984 (in the USA), gasoline prices fell to record low levels, killing public interest and government investment in alternative fuels.

Nevertheless, from circa 1989 to 1999, the California Air Resources Board (CARB) implemented the AFUP that subsidized the production of M85 FFVs and the M85 fueling infrastructure to support them, entirely in California.[11]

Between 1985 and 1992, Ford and Chrysler delivered 705 M85 FFVs in California and Canada, including the 1.6L Ford Escort, the 3.0L Taurus, and a Crown Victoria police version. These could operate on M85, E85, or gasoline. In 1993, federal legislation was passed to encourage the production of M85 FFVs.

In 1996, a new M85 FFV Ford Taurus was developed (Fig. 6.6) that evolved into the first commercial production of an ethanol (E85) FFV. The momentum of the M85 FFV program declined, and by the end of the 1990s, the emphasis shifted entirely to E85 FFVs, as it remains today. Although ethanol was more expensive to produce, it provided a benefit to the US agriculture industry that was under particular economic stress at the time.

During the years of the federal version of the AFUP, M85 production and distribution was handled on a national scale primarily by two oil companies (Chevron and Mobil Oil) which had negative incentive to make it successful. Typically priced at \$2 USD/gallon[12] at a time when gasoline averaged \$1.25/gallon, it was more than three times as expensive as gasoline for the same energy, i.e., travel distance. And dispensing was usually relegated to a single older fuel pump in poor condition and not methanol-compatible, often located inconveniently. Unused M85 stored in vented underground gasoline storage tanks eventually absorbed atmospheric water, degrading its quality as a fuel and making it prone to

[11] For example, the Ford Taurus, Crown Victoria, and F-series trucks of the late 1990s were available with an M85 FFV option, at the same retail price as non-FFV versions. Most were purchased by government agencies and utility fleets.

[12] Personal experience refueling a methanol-converted vehicle in California, 1996–2003.

phase separation in the underground tank.[13] The public shunned M85, and some FFVs (gasoline/M85) were never run on methanol. Some FFV owners were unaware that their car had that capability, since dealers were known to sell FFV models for the same retail price as non-FFVs, even though the subsidized cost to the dealer was lower for the FFV.[14]

Methanol lacked a strong political advocate, and the cards were stacked against it by both the oil industry entrusted with its production and distribution, and the agriculture industry that advocated strongly for the use of corn ethanol over methanol. By the 2000s, M85 and M5 had been completely replaced by E85 and E10 that, due to its higher volumetric energy density, appeared to the user to provide better mileage.[15] As additives, methanol and ethanol are equally effective at knock reduction. But in the USA, ethanol was heavily promoted as the biofuel that was "best for America."

Methanol has rarely been mentioned since the transition to ethanol, despite its potential as a transition fuel that can be made from a wide range of both renewable and fossil feedstocks [45]. As of 2023 in the USA, methanol had long been forgotten except by a few DIY aficionados.[16] But this has not been the case in other countries, especially China that has embraced methanol as the combustion fuel of the future.

As stated on the Zhejiang Geely Holding Group product website, March 2022 [46]:

> …methanol vehicles allow for a 70% reduction in CO_2 emissions compared to gasoline models. Previous demonstration trials of non-hybrid Geely Emgrand methanol sedans in Iceland achieved an average well to wheel emissions of 46 g of CO_2 per km, even lower than electric vehicles in most countries.
>
> With methanol fuel price at Chinese pumps around 3RMB/L (0.42EUR/L or 1.8USD/Gallon), fuel cost per 100km is roughly 27RMB (3.85 Euros or 4.28USD). In the face of rising global gas prices, methanol fuel offers a viable cleaner affordable alternative.

In a few areas with less-developed transportation infrastructure, homemade methanol remains a popular replacement for scarce or expensive gasoline. In Nigeria, methanol dominates as a replacement for petroleum [47]. And methanol from coal is commonly used as a motor fuel in rural areas of China, which produces

[13] Personal observations, also confirmed in lecture by Greg Dolan, Director of Methanol Institute, 2012.

[14] This assertion is based on personal experience in the mid-1990s in California.

[15] Improved compatibility of ethanol and methanol in fuel systems was almost entirely due to the 1990 implementation of EPA regulations that forced the adoption of alcohol-compatible electronic fuel injection for all vehicles sold in the USA starting in 1991.

[16] The author is such a methanol aficionado—see *My Methanol Motorcycle*, https://digitalcommons.calpoly.edu/cgi/viewcontent.cgi?article=1292&context=eeng_fac

Fig. 6.7 Fleet of Geely Emgrand taxis in Guiyang. (Photo from an article by Hazel Nicole Carreon of Geely, Jan 11, 2024; https://global.geely.com/en/news/2024/geely-e-fuel-solution-carbon-neutral-future. Public domain)

over 60% of methanol worldwide. Chinese government investment and promotion of methanol fuel production are the largest in the world [48, 49]. Fleets of methanol taxis are increasingly common in major Chinese cities, as shown in Fig. 6.7.

Methanol Fuel Characteristics

As a motor fuel, despite its lower specific calorific value (combustion energy density, lower heating value) compared with gasoline, methanol has several desirable attributes.

Anti-knock (Octane) Rating

As mentioned earlier, methanol added to gasoline will significantly increase its anti-knock or octane properties. Methanol on its own (M100) has a very high Research Octane Number (RON), reported as anywhere from 107 [50] to 130 [51] by different sources. The use of methanol as a primary fuel or as a fuel additive allows the use of higher compression ratios in SI engines that would otherwise cause destructive engine knock with gasoline. “Blended Octane” ratings do not necessarily follow the ratio of alcohol to gasoline. Disproportionate knock resistance improvements can occur even for low alcohol/gasoline percentage blends. Gasoline sold as E10, 10% (max) ethanol in regular gasoline, has a typical RON = 91–95 compared to 85 without alcohol. The same is true of M3, a 3% methanol additive. In my practical experience, hand-blending up to 10% methanol into regular grade E10 gasoline improves knock resistance at least to the level of premium grade gasoline (not laboratory verified). Most modern vehicles use oxygen-sensing feedback fuel control that will correct the AFR up to this composition.

Higher Octane Makes Possible Higher Efficiency

The resistance to engine knock of methanol enables the use of higher engine compression ratios without risk of engine damage. As explained in Chap. 4, higher compression ratios produce higher engine efficiency, approximately according to

$$\text{Thermal Efficiency}(\%) \approx 100\left(1 - \frac{1}{\text{CR}^{0.2}}\right)$$

A CR of 10 is typically the highest used in production cars due to the octane rating of pump gasoline. For CR=10, the formula above predicts a theoretical mechanical efficiency of 36.9%. But methanol has no problem with CR=12, which would, according to this simplified formula, provide an efficiency of 39.2%, a 6.2% increase over CR=10.

High Heat of Vaporization

Regarding power and torque, a stoichiometric methanol-air mixture releases 7% more combustion energy than the same volume of a stoichiometric gasoline-air mixture, meaning that even if an engine is crudely converted from gasoline engine to methanol, it will develop at least 7% greater torque and horsepower at full throttle. And this effect is combined with the much greater cooling effect of methanol due to its much higher heat of vaporization and lower vaporization temperature, which produces a cooler and therefore denser fuel-air intake charge. This yields a higher mass of air and fuel in the cylinder simply because of the lower temperature. Competition engines that employ supercharging or turbocharging use methanol rather than gasoline to take advantage of this charge cooling property. In fact, they routinely use grossly over-rich methanol fuel-air ratios just to maximize the cooling effect of its heat of vaporization. Grossly over-rich methanol-air mixtures are possible because of the very wide flammability limits of methanol. This is not the case for gasoline, that would be prone to misfire and poor combustion at this level of overfueling. The cooling property is characteristic of all alcohols including ethanol, but none more so than methanol that has about five times the latent heat of vaporization of gasoline. This cooling property increases the density of the fuel-air mixture, increasing the energy release upon combustion. It is responsible for a power increase compared with gasoline, and combined with supercharging, makes possible the exceptionally high power outputs required in racing.

For a fully emissions-compliant methanol conversion of a gasoline car, a stoichiometric mixture is used except during transients. The combined effects of methanol's intrinsically higher charge energy density, further increased due to vaporization cooling, and the use of an increased CR can be expected to yield a minimum power increase of at least 10% even without super/turbocharging. And since the high octane rating allows high super/turbocharge boost pressures, power outputs on methanol can easily be much higher than gasoline.

Ignition Timing

The combustion flame front propagation speed is the a major determinant of the charge combustion time in an engine. A shorter (faster) burn time allows combustion to be initiated later (with less ignition advance), closer to the Top Dead Center (TDC). Less energy is lost to the early phase of combustion while still in the compression

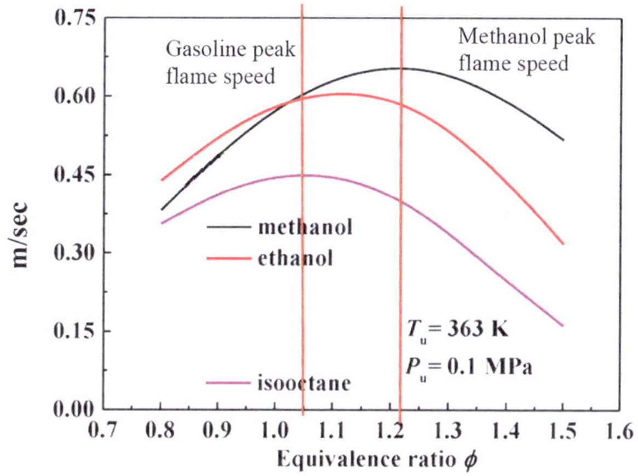

Fig. 6.8 Laboratory-measured laminar flame speeds for stoichiometric combustion of ethanol, methanol, and "isooctane" at 90 °C and 1 bar. Plot curves from citations [52, 53]. Red lines designate equivalence ratios at the peak flame speeds of methanol and isooctane, a standardized surrogate for gasoline

stroke, allowing combustion to occur more completely during the expansion stroke in which it produces mechanical work. The engine operates at higher thermal efficiency, closer to the ideal Otto cycle. The faster the burn time, the higher the possible engine efficiency and power if the timing advance is reduced accordingly.

The flame speed and therefore combustion time are dependent upon the peak compression pressure and temperature, which are primarily determined by the compression ratio. But the most prominent effect on flame speed is the AFR which is controlled by the fuel quantity and intake air density. Laminar flame speeds are usually measured in a flame tube, which is a long Pyrex tube in which the flame front propagation rate can be recorded using high-speed photography. The conditions are very different from those in the combustion chamber at the time of ignition, so that measurements are usually interpreted by comparison with other reference fuels. In the flame speed plots of Fig. 6.8 from citations [52, 53], the conditions were atmospheric pressure and temperature of 90 °C. Comparing stoichiometric fuel/air mixtures of several alcohols with gasoline, alcohol flame speeds exceed those of gasoline at all equivalent ratios Φ, which infers that mean best torque (MBT) ignition timing advance requirement would be less for both alcohols compared with gasoline. Methanol and ethanol have about the same flame speed at a stoichiometric ratio ($\Phi = 1$), but methanol achieves higher peak flame speed for rich mixtures, e.g., $\Phi = 1.22$, the peak flame speed of methanol is approximately 40% faster than that of gasoline.

While many other factors are involved, a stoichiometric homogeneous mixture of methanol and air typically burns about 30% faster than gasoline, and for rich mixtures, up to 60% faster. This trend confirms observations with experimentally determining MBT (mean best torque) ignition advance on a dynamometer for methanol/gasoline dual-fuel engines. The MBT timing of methanol engines intended for legal road use is almost always less advanced than gasoline. Many professional race tuners have observed the opposite: that methanol requires greater ignition advance. This is indeed true for the very rich methanol mixtures, $\Phi > 2$ used in racing, that would not even be combustible for gasoline. [17] But at the near-stoichiometric

[17] Observations based on personal experience with racing methanol engines. There are probably exceptions.

mixture of an emission-controlled engine($\Phi = 1.2$), noting the peak flame speed shown in Fig. 6.8, methanol combustion time is considerably faster than that of gasoline, prescribing less advance for MBT timing.

Cold Starting

There is a negative consequence of the higher latent heat and high vaporization temperatures of alcohols, especially methanol: cold starting. Methanol is a pure substance that has a distinct vaporization temperature rather than the wide band of vaporization temperatures of gasoline due to its broad composition range. In my experience from 30 years of commuting on methanol fuel, starting a neat methanol engine at air temperatures below approximately 4 °C is nearly impossible. Fortunately, the addition of an additive like gasoline or dimethyl ether, or injecting a small amount of a gaseous fuel such as propane or butane into the intake manifold will allow the engine to start, although it will continue to run rough until the engine warms up. Other solutions involve electric manifold preheating, or heating the tips of the fuel injectors to flash-vaporize some of the fuel at the moment of injection.[18] Regardless of the workaround, after about a minute of operation on a rich fuel-air mixture, the increased temperature of the cylinder surfaces improves the fuel vaporization sufficiently to allow the engine to run normally. However, during this cold start period, the excess fuel greatly increases hydrocarbon emissions, and if very rich will lead to the formation of formaldehyde in the exhaust (discussed below). The cold starting challenges of methanol are shared by ethanol, although methanol is more problematic than ethanol. Cold starting is less of a problem in temperate climates, such as Brazil, which has mandated ethanol blends since the late 1970s.

Regulated Exhaust Emissions

With the highest hydrogen-to-carbon ratio of any liquid fuel, methanol combustion generates more water vapor and less carbon dioxide per unit of combustion energy, making it intrinsically lower in CO_2 per mile than gasoline or any other liquid fuel. Referring to GHG calculations presented later in this chapter it can be seen that among all liquid fuels, methanol provides the lowest CO_2 emissions per distance traveled.

For EPA-regulated tailpipe emissions, methanol (at stoichiometric AFR) produces lower HC, CO, and particulate emissions than gasoline or ethanol.[19] I have observed that NO_x emissions measured in EPA CVS testing are typically about the same as gasoline or ethanol, although they are engine-dependent and can exceed gasoline if higher engine compression ratios are used (which would not be possible for gasoline). This is expected because NO_x production is not directly related to the

[18] Heated tip injectors (HTI) were patented by Delphi and are currently in use in Brazil to aid in cold starting E100-fueled engines.

[19] HC and CO emission data from testing from Denver Research Institute/US DOE Multi-fuel Datsun 200SX by Environmental Testing Corp, Aurora, Colorado, June 10, 1981.

composition of the fuel, but rather to the heating of the inert nitrogen in the air during combustion. Higher temperature combustion and longer burn times allow some nitrogen and oxygen in the air to combine to form NO and NO_2, or to a lesser degree, N_2O. This is the reason that compression ignition (diesel) engines, with their very higher compression ratios and high peak combustion temperatures, produce more NO_x than SI engines regardless of the type of the fuel. Fortunately, modern three-way catalytic converters used on gasoline vehicles have been found to be approximately as effective for alcohol fuels, including reduction of NO_x.

As mentioned above, methanol like all alcohol fuels, can produce an additional class of toxic emissions that is not usually found with gasoline: aldehydes, e.g., formaldehyde and acetaldehyde, especially while starting cold with a rich mixtures [54]. This was recognized in the 1980s as an important consideration for M85 and later E85 flex-fuel cars. Aldehydes were subsequently added to the EPA- and CARB (California)-regulated emissions list. Aldehydes are directly toxic and irritating to the eyes and lungs. Anyone that has attended a racing event in which methanol fuel is used (often in a blend with nitromethane) will be familiar with formaldehyde emissions. For regulated emission-controlled cars, this a problem during starting only since after warm-up, the engine runs a stoichiometric fuel-air mixture. But in racing or off-road applications, excessively rich air-fuel mixtures are the norm, and emissions are not a concern. More important is that the unburned excess fuel in the exhaust of a drag racing vehicle allows exhibitions of flaming exhaust to impress crowds.

Phase Separation

Alcohols including methanol and ethanol have both polar and nonpolar chemical properties. What that means in the fuel world is that in pure (neat) form they are infinitely soluble in (nonpolar) gasoline, and they are also infinitely soluble in (polar) water. But in a ternary (three-part) blend of methanol, gasoline and even a small amount of water, the mixture is prone to phase separation—formation of different liquid strata when not thoroughly mixed. Alcohol/gasoline blends between 10% and 90% are susceptible to phase separation due to atmospheric water absorption, causing the formation of a separate layer of gasoline above a layer of alcohol and water in the fuel tank. This stratification becomes more likely at low temperatures. The water content migrates to the alcohol layer, which is why water-absorbing fuel additives such as HEET® that are sold in cold climates to remove water from gasoline are usually just (overpriced) methanol.

Phase separation was a particular problem for the former M85 FFVs, and remains a problem for current E85 FFVs, which are designed to run on any volumetric alcohol-gasoline blend from 15% to 85%. For the 10% maximum blend in E10, this is usually not a problem, but phase separation as well as cold starting problems have justified EPA exceptions to the required use of E10 during the winter in cold climates such as (in North America) North Dakota, Minnesota, Canada and Alaska. E0 (gasoline containing no ethanol) is also sold seasonally at petrol stations in Central and Northern European countries.

Can You Convert a Gasoline Vehicle to Run on Methanol?

Yes. Easily. But is methanol a legal road fuel in the USA? The US Federal Energy Policy Act of 2005 allowed "alcohol" fuels rather than exclusively ethanol. But under EPA and California CARB regulations, any modification of the fuel system, including modification to use alternative fuels, is restricted. Like all other after-market engine performance modifications (a multi-billion dollar industry), this is a gray area, thus the caveat is added "For off-road use only." And just for completeness… state road taxation regulations for methanol are still in place from the AFUP in the 1990s, but since there are no longer any gas stations that dispense properly taxed M85, if methanol or any untaxed alternative fuel is used on public roads, it could be considered a form of tax evasion. A more detailed discussion of US federal and state laws regarding engine modification to alcohol operation can be found in Chap. 7 in the context of ethanol that is subject to the same conversion restrictions.

A vehicle converted after it was manufactured to run on M85 or M100 will be rejected in a California biennial emission test if the engine control module (ECM) firmware has been modified or the engine shows evidence of non-CARB-approved modifications. But nonobvious modifications to allow flexible fueling that do not interfere with manufacturer's gasoline-only operation are a gray area that would not likely be noticed in an inspection. For vehicles that predate OBD-II requirements that took effect in 1996, there should be no problem passing the alternative "STAR-enhanced" (actual tailpipe emission sampling) procedure on alcohol, because the measured CO and HC emissions would almost surely be lower than those on gasoline.

The conversion or original manufacture of a port or manifold fuel-injected gasoline engine/vehicle to methanol operation requires basically the same modifications as those for conversion to operation on ethanol/E85. Detailed information on the engine conversion can be found in Chap. 7, so only the differences will be discussed here.

The material compatibility issues for methanol are the same as for ethanol, but a bit more severe. Fortunately, elastomers and metals that are compatible with ethanol in newer cars are compatible with methanol. Because all cars since the late 1990's are compatible with E10, and many late-model US-manufactured cars are already FFVs, their compatibility with ethanol gives them nearly equivalent compatibility with methanol.

Like ethanol, the main difference between operation on gasoline and either ethanol or methanol is just a different calibration of the injected fuel quantity: for methanol, approximately 47% greater fuel per injection compared with E10 pump gasoline; or for E85, approximately 15–30% greater than E10; the variability is due to the fact that what is sold as E85 can be anything from 51–83% ethanol on a volume basis, not 85% as the name would suggest. This broad definition of E85 (the official flex fuel since 2000) was intended to allow for volumetric blends of ethanol in gasoline to be seasonally adjusted locally from 51% to 83% [55].

All emission-certified vehicles sold in the USA since 1990 (except for the 1991–1993 Subaru Justy, the smallest car sold in the USA) incorporate oxygen-sensing feedback control of the AFR, as discussed previously. However, the range of control authority is typically no more than ±10–20%. This limited engine control authority cannot compensate for the use of either E100, M100, E85, or M85 in an engine calibrated only for gasoline. But it has been my experience for vehicles from the 1990s to the present that small blends of as much as 12.5% (1:7 alcohol-gasoline ratio) are tolerated without any problems, other than slightly rough operation in cold weather until the engine warms up.

Because of the minimal difference between gasoline, ethanol, and methanol operations, it is possible to convert almost any gasoline vehicle quite easily to alcohol operation, either methanol or ethanol. And as previously mentioned, while such a DIY conversion isn't emission legal under federal or California law, if done properly the tailpipe emissions will almost surely be cleaner than the vehicle's EPA-measured gasoline emissions. The fuel substitution will not be in violation of EPA or CARB emission percentage or g/mile limits, but will only violate regulations that prevent "unauthorized" modifications of the fuel system components.[20] The best (and legal) approach is to start with a current E85-capable vehicle, and make only minor modifications to increase the allowable AFR range to accommodate methanol. This may be as simple as firmware recalibration to increase the fuel quantity. This is vehicle-dependent, and access to the internal firmware is impossible without manufacturer-authorized tools or the use of a gray-market remapping "tuner." Any changes to the engine control firmware may be detected in a biennial vehicle inspection since the test procedure checks for illegal changes in the firmware via connection to the OBD-II diagnostic port.

Methanol requires approximately twice the amount of fuel per injection pulse compared with gasoline. The fuel amount can be changed by any one or combination of the following modifications:

1. Replace the fuel injectors with parts having a larger flow coefficient. This is the simplest approach but precludes multi-fuel operation of the engine since each fuel would require injectors with different calibrations. This and method 3 avoid the maximum engine speed limitation discussed below, which could (would) be a problem if methods 1 or 4 are used alone.
2. Increase the fuel rail pressure (injection pressure) without changing the injectors, such that mass fuel delivery is approximately doubled. Since the flow rate of a noncompressible fluid (like gasoline or methanol) increases as the square root of the pressure increase ratio. A 60 psig gasoline rail pressure would have to be increased to 240 psi to accommodate M100, which is impractical.
3. Rescale the injection pulse duration proportional to the 2:1 fuel delivery ratio required for methanol compared with gasoline. This requires a modification of the fuel injection controller's internal "fuel map," or the use of an aftermarket fuel remapping device that simply doubles the duration of each injection pulse (example shown in Fig. 6.9). However, since methanol will require twice the

[20] This engine conversion instruction is intended for off-road applications due the current illegality of making any modifications to engine fuel systems in the USA.

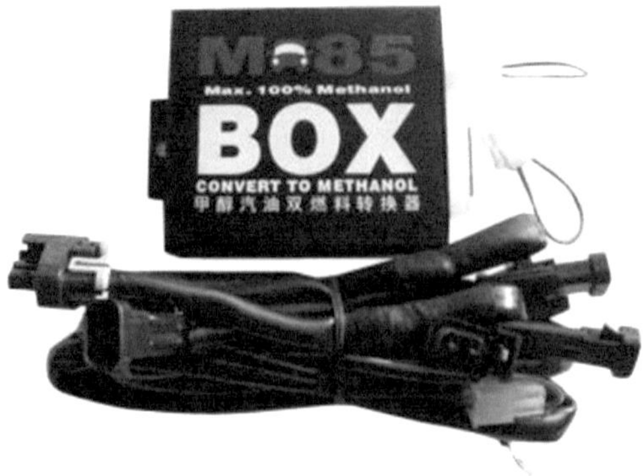

Fig. 6.9 Example of a low-cost aftermarket methanol conversion device that extends the fuel injection pulse duration to double the volumetric fuel delivery under any engine condition. The device works well except at high RPM when the maximum pulse duration may exceed the total allowable time for the injection of fuel in each complete engine cycle, which is two rotations of the engine. (https://www.aliexpress.us/?spm=a1z5k.7633538.0.0.55565a870FymP9&gatewayAdapt=glo2usa. Public domain)

pulse duration, these simple pulse-changing devices usually run into a limitation at the engine's maximum RPM, since the extended pulse duration may exceed the available injection time, which is a minimum at maximum speed.

For example, a manifold- or port-injected engine at maximum power may require up to the entire two rotations of a four-stroke engine for injection at redline RPM. If the highest anticipated engine speed is 6000 RPM, the maximum time available for each fuel injection at that speed would be

$$\frac{1000\,\text{ms}}{1\text{s}}\ \frac{60\text{s}}{1\,\text{min}}\ \frac{1\,\text{min}}{6000\,\text{revs}}\ \frac{2\,\text{revs}}{1\,\text{intake}} = 20\,\text{ms per intake stroke}$$

If the maximum gasoline fuel delivery requires a 12-ms injection pulse, then the maximum injection pulse for methanol would be 24 ms, which is longer than the allowed injection time at 6000 RPM. Consequently, to avoid the maximum speed limitation, the conversion of an engine from gasoline to methanol usually requires a combination of Method 3 and at least one of the methods 1 or 2 e.g., slightly higher flow injectors and/or higher fuel pressure, in addition to rescaled injection pulse duration.

The calculation above assumes that fuel can be injected any time during the two rotations of the crankshaft that make up a complete engine cycle. But fuel and air are only inducted during the (theoretically) 180° intake stroke while the intake valve is open, which is only ¼ of the complete engine cycle. Over the years since the Bosch D-Jetronic system introduced in 1967, it has been found that there is indeed some advantage to injecting during the same radial timing for each cylinder. But these affects have been experimentally found to be minor, at most only slightly affecting emissions because of the variable amount of time for fuel-air mixing. The ubiquitous Bosch L-Jetronic fuel injection systems that dominated the market from 1973 through the mid-1980s simply strobed all the injectors at the same time, twice for every engine intake stroke. This allowed triggering the EFI system using the ignition timing crankshaft position sensor, saving components without any noticeable effect on efficiency, emissions, or power output. And the very popular Bosch K-Jetronic (aka continuous injection system [CIS]), like many mechanical non-pulsed injection systems dating to the early 1960s, injected continuously during the entire 720° engine cycle.

As mentioned in Chap. 4 *Engines and Fuels*, many of the newest automobiles, including almost all hybrids and high-performance cars, use direct cylinder gasoline fuel injection (DCGI or DCI), also known as Direct Injection Spark Ignition (DISI) [56]. With fuel injected at very high pressure (similar to diesel injection) directly into the cylinder of an engine, there is considerably greater flexibility in the timing of the injection pulse or pulses. As discussed in Chap. 4, in a DCI engine, fuel can enter the cylinder even after the intake valve is closed. This allows two modes of combustion: *homogeneous charge formation*, essentially the same as port injection assumed in the discussion above, with an approximately stoichiometric amount of fuel injected during each engine cycle prior to ignition; and alternatively, *stratified charge formation* which involves the injection of a less-than-stoichiometric amount of fuel into the cylinder in close proximity to the spark plug. This locally rich but overall-lean mixture allows net-lean operation which conserves fuel at light loads or idle. While not absolutely necessary, DCI is usually implemented on engines equipped with variable valve timing, allowing engine operation with the *Atkinson Cycle* which is implemented by variably delaying the closure of the intake valve past the end of the intake stroke, in effect reducing the effective engine displacement by allowing partial reverse flow of intake air from the cylinder back into the intake manifold prior to compression. For DCI engines, the requirement for very high fuel rail pressure and the complexity of the injection control puts these engines outside the realm of aftermarket modification, only possible by the manufacturer.

Figures 6.10 and 6.11 show an early multifuel methanol/ethanol/gasoline car commissioned by the US Department of Energy to demonstrate and compare these fuels in actual road operation. Its use of oxygen-sensing feedback fuel injection also made it the lowest emission alcohol-fueled vehicle at that time, and unintentionally the first Flex Fuel vehicle.

Fig. 6.10 Modified 1981 Nissan 200SX, capable of tri-fuel methanol/ethanol/gasoline operation. (Constructed by author in 1981 at the Denver Research Institute in Colorado under a DOE contract to demonstrate and compare multi-fuel operational characteristics. Photo: author)

Fig. 6.11 Engine of 1981 DRI/DOE Multi-fuel Demonstration Vehicle showing dual-fuel injection systems. (Photo: author)

Where to Obtain Methanol, and How Much Does It Cost?

Methanol is a commonly available racing fuel, usually purchased in 5 gallon or 55 gallon drum quantities for off-road use. At the retail level, it is more expensive than gasoline, e.g., Torco Racing Fuels, on June 16, 2024:

Local Pick-up Price before tax [57]:
per 1 gal $6.00 · 5 gal pail $58.00 · 55 gal drum $295.00

This is effectively twice the current cost of gasoline in California, since methanol contains only about half the energy per volume compared with gasoline, but achieves slightly higher efficiency.

The economics vary greatly in countries with higher gasoline costs. From Indian Express (India, September 2024) [58]:

> Methanol gets made from coal and costs only RS 22 per liter as against the prevailing price of about RS 80 per liter for petrol, the Transport Minister said.
>
> Union Road and Highway Minister Nitin Gadkari said on Saturday the government would be soon announcing a policy calling for 15 per cent blending of methanol in petrol to make it cheaper and also reduce pollution.

Since at that time, 1 USD = 83.68 INR, 1 US gallon of methanol cost about $1.05. Even considering the 50% energy content reduction for methanol, the cost per km in India was still about half of the cost of petrol, a major improvement in any country that must import most of its petroleum.

On June 16, 2024, the international commodity market price of methanol was $1.04 USD/gal [59] compared with pretax gasoline at $2.39/gal [60], making methanol approximately 9% less expensive per unit of energy.

Note again that in the USA, all methanol is currently made from natural gas. In China and India, it is almost entirely made from coal. Neither scenario allows methanol to be considered a renewable fuel, and their carbon footprints are tied to the energy sources used to make them. But this situation is due to economics alone. Natural gas and coal are the cheapest fossil fuels, and methanol can be easily and efficiently made

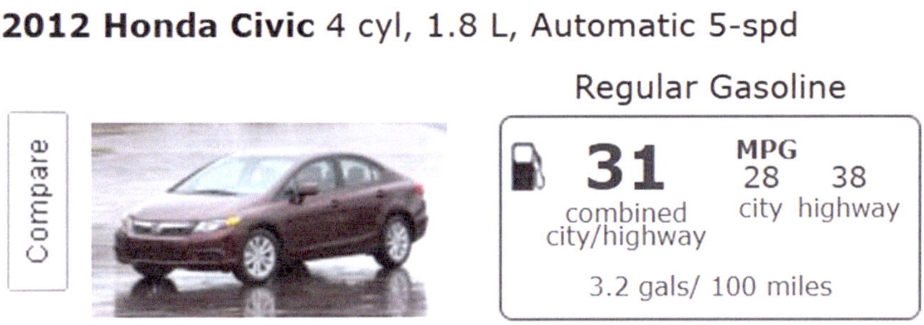

Fig. 6.12 US EPA regulatory data for 2012 Honda Civic 1.8L. From US Dept of Energy "Fueleconomy.gov". (https://www.fueleconomy.gov/feg/Find.do?action=sbsSelect. Public domain)

from them. It is the wide range of renewable energy sources that can be used to make methanol efficiently that create the potential for its use as a transition fuel.

CO_2 Emissions of Methanol Compared with Gasoline

A 2012 Honda Civic-base non-hybrid model is used as the comparison platform for all fuels in the analyses below. It has an EPA combined fuel economy rating of 31 MPG gasoline. This selection was chosen because in 2012, Honda offered many otherwise identical variants of the Civic including one that runs on natural gas and a non-plug hybrid (Fig. 6.12). These will be used later for comparisons of ethanol, natural gas, and electric vehicles.

Gasoline (Regular Grade, Without Ethanol)

2012 Honda Civic LX that gets 31 MPG = 50 km/gal

The carbon dioxide produced by combustion of 1 US gallon of gasoline is 8877 g CO_2/gal gasoline[21]

$$\frac{\left(\frac{8887\,\text{g CO}_2}{\text{gallon}}\right)}{\left(\frac{50\text{ km}}{\text{gallon}}\right)} = 178\,\text{g CO}_2\,/\,\text{km}$$

Methanol (M100)

Same vehicle but operating on methanol.

Assume same energy used per km for all fuels, equivalent to 50 km/gallon gasoline (US).

Gasoline density = 750 g/L

[21] Data from 2020 US Code of Federal Regulations (CFR).

$$\left(\frac{\text{gallon gasoline}\,(\text{US})}{50.0\text{ km}}\right)\left(\frac{3.79\text{L}}{\text{gallon}}\right)\left(\frac{750\text{ g}}{\text{L}}\right)\left(\frac{41.5\text{ kJ}}{\text{g}}\right)=2.36\text{MJ}/\text{km}\quad(\text{for any fuel})$$

Methanol energy density (LHV) = 19.7 kJ/g

$$\frac{2.36\text{MJ}/\text{km}}{19.7\text{MJ}/\text{kg methanol}}=120\text{ g methanol}/\text{km}$$

Stoichiometric combustion:

$$\text{CH}_3\text{OH}+1.5\text{O}_2\rightarrow CO_2+2\text{H}_2\text{O}$$

$$12+3+16+1=32\text{g methanol}\rightarrow 12+32=44\text{g CO}_2$$

$$120\text{g methanol}\cdot\left(\frac{44\text{g } CO_2}{32\text{g } CH_3OH}\right)=165\text{g } CO_2/\text{km}$$

7.3% reduction compared with gasoline.

Power Output of Methanol Compared with Gasoline

Gasoline (Regular Grade, Without Ethanol)

2012 Honda Civic, 31 MPG = 50 km/gal gasoline

Manufacturer specifications [61]: 140 HP at 6500 RPM

For comparison with other fuels, this measured output needs to be matched with the calculated HP to determine the thermal efficiency of the engine.

Air volume flow at 6500 RPM (assuming 100% volumetric efficiency):

$$\left(\frac{1.8\text{L}}{2\text{ revs}}\right)\left(\frac{6500\text{ revs}}{\text{min}}\right)\left(\frac{1\text{ min}}{60\text{s}}\right)=97.5\text{L air}/\text{s}$$

Air mass flow at 6500 RPM at NTP (20 °C, 1 atm)

$$\left(97.5\frac{\text{L air}}{\text{s}}\right)\left(\frac{1.204\text{g air}^*}{\text{L air}}\right)=117\text{g air}/\text{s}\ {}^*\text{from Ideal Gas Law, Appendix}$$

Fuel mass flow at 6500 RPM, for gasoline AFR = 14.7 g air/g fuel [62].

$$\left(\frac{117\text{g air}}{\text{s}}\right)\left(\frac{\text{g gasoline}}{14.7\text{g air}}\right)=7.96\text{g gasoline}/\text{s}$$

Amount of power in this fuel flow:

$$\left(\frac{7.96\text{g gas}}{\text{s}}\right)\left(\frac{41.5\text{ kJ}}{\text{g gas}}\right)\left(\frac{1\,kw\,\text{s}}{1\text{ kJ}}\right)\left(\frac{1.34\,HP_{\text{US}}}{1\,kw}\right)\eta=443\times\eta HP_{\text{US}}\text{ at } NTP$$

Where η is the thermal efficiency of the engine. It is typically about 30% for internal combustion engines of this type. We can find it exactly by equilibrating the equation above with the manufacturer-specified horsepower on gasoline, 140 *HP*:

$$\eta = \frac{140\ \mathrm{HP_{US}}}{443\ \mathrm{HP_{US}}} = 31.6\%$$

Methanol (M100)

Same air volume flow at 6500 RPM (assuming 100% volumetric efficiency): 97.5 L/s

Same air mass flow at 6500 RPM at NTP (20 °C, 1 atm): 117 g/s

Mass AFR for a stoichiometric mixture is 6.45 g air/g methanol (calculated below*)

Fuel mass flow at 6500 RPM:

$$\left(\frac{117\mathrm{g\ air}}{\mathrm{s}}\right)\left(\frac{\mathrm{g\ methanol}}{6.45\mathrm{g\ air}}\right) = 18.1\mathrm{g\ methanol/s}$$

Amount of power in this fuel flow:

$$\left(\frac{18.1\mathrm{g\ methanol}}{\mathrm{s}}\right)\left(\frac{19.7\ \mathrm{kJ}}{\mathrm{g\ methanol}}\right)\left(\frac{1\ \mathrm{kws}}{1\ \mathrm{kJ}}\right)\left(\frac{1.34\ \mathrm{HP_{US}}}{1\ \mathrm{kw}}\right)\eta = 478\times\eta\mathrm{HP_{US}}\ \text{at NTP}$$

We now use the engine thermal efficiency found previously with the engine running on gasoline: 31.6%.

Actual power output including mechanical efficiency:

$$478\times 0.316\mathrm{HP_{US}} = 151\mathrm{HP_{US}} \quad 7.9\ \%\ \text{greater power than gasoline}$$

*Calculation of mass AFR used in calculation above:

Air contains 21% oxygen by molar volume.
Combustion oxygen/fuel molar ratio:

$$CH_3OH + 1.5O_2 \rightarrow CO_2 + 2H_2O \Longrightarrow \text{1 mole of methanol combines with 1.5 moles of oxygen.}$$

Fuel/air mass ratio:

$$\left(\frac{1\ \mathrm{mole\ methanol}}{1.5\ \mathrm{moles}\ O_2}\right)\left(\frac{32\mathrm{g\ methanol}}{1\ \mathrm{mole\ methanol}}\right)\left(\frac{0.21\ \mathrm{mole}\ O_2}{1\ \mathrm{mole\ air}}\right)\left(\frac{\mathrm{mole\ air}}{29\mathrm{g\ air}}\right) = 0.155\mathrm{g\ methanol/g\,air} \Rightarrow \mathrm{AFR} = 6.45$$

In sum, methanol intrinsically provides 7.9% high power while emitting 7.3% less CO_2 per km.

Onboard Methanol Reforming for Hydrogen Vehicles

As will be discussed in greater detail in Chap. 9 *Hydrogen*, methanol is one of a small number of *liquid organic hydrogen carriers* (LOHCs) that can be used as a mobile hydrogen storage medium. Onboard a vehicle, methanol and water can be reformed in a single reactor into hydrogen, carbon dioxide, via essentially the reverse of the process by which methanol is synthesized from syngas produced by SMR [63]. As with all energy conversions, there are energy losses in either direction. And the CO_2 byproduct contributes carbon emissions.

As discussed earlier in the chapter, the production of methanol from natural gas via syngas is less carbon-intensive than the production of hydrogen from natural gas. Ideally, all (or most) of the carbon liberated from the SMR reformer can be used in the subsequent methanol synthesis step, and the waste heat from the exothermic synthesis can partially offset the heat required by the highly endothermic reformer [64].

However, the same cannot be said for the reverse of the synthesis step, reforming methanol and water to produce hydrogen and carbon dioxide as a byproduct. This is exacerbated by the small scale of the onboard process, which is essentially a small refinery.

The quality of the reformed gas may also be a consideration if it is to be used by a hydrogen fuel cell, which is intolerant of CO or CO_2. This is a major difference between hydrogen used by fuel cells and hydrogen used as a combustion fuel. Reformed dirty hydrogen is not a problem for combustion hydrogen vehicles, since the contaminants (CO, CO_2, water, formaldehyde) have only a minor effect on the combustion of hydrogen in the engine.

If the onboard reformer is fueled exclusively by methanol without water, the water gas shift does not occur, reducing the hydrogen yield by 33% but leaving CO rather than CO_2, which itself is a combustion fuel. The CO combusted in the engine will result in the same CO_2 emissions as if the methanol and water were reformed to H_2 and CO_2. This was the approach taken by Finegold et al. at SERI in 1984, in which H_2 and CO from reformed methanol were used successfully to power the IC engine in the Chevrolet Citation [65].

The CO_2 produced by the reformer can be minimized by precise control of the reformer temperature and control of the water necessary to drive unreformed CO to CO_2 plus additional hydrogen. Conventional methods for carbon capture and use have been demonstrated on a lab scale by post-processing the concentrated CO_2 byproduct with electrolytically generated hydrogen to produce additional recyclable methanol, but have not yet been reduced to practice in transportation applications. This is understandable considering the overall complexity and size of the multistep methanol-to-hydrogen fuel system onboard the vehicle. Applied to FCVs, the addition of a methanol reformer, hydrogen gas separator, and methanol fuel tank to the already-complex hydrogen fuel cell and battery electric drive system of an H_2 FCV leads to a bulky system that is probably more amenable to larger scale applications than automobiles. Nevertheless, startup companies such as Element-1 [66],

Blue World Technologies [67], and SerEnergy [68] mentioned in the next section are confident that the use of an onboard methanol reformer with a hydrogen fuel cell can power electric vehicles in a carbon-neutral way, assuming that the methanol was produced renewably [69].

Hydrogen FCVs that rely on onboard reforming of a LOHC such as methanol to generate hydrogen onboard can be portrayed as a *black box*, a vehicle that is fueled by methanol. In a strange twist on EVs and FCVs, the LOHC hydrogen vehicle would simply transfer the CO_2 emissions back to the vehicle's exhaust, rather than leaving them at the power generation or hydrogen SMR facility. The tailpipe CO_2 emissions would cost the "zero-emission vehicle (ZEV)" designation that places hydrogen vehicles (fuel cell or combustion) in the same sought-after regulatory category as electric vehicles.

Methanol Fuel Cells

Methanol is the only liquid hydrocarbon that has been seriously considered as a potential reactant for a room-temperature fuel cell.

In practice, the term *methanol fuel cell* has been used to describe two very different devices: direct methanol fuel cells (DMFCs) and indirect methanol fuel cells (IMFCs). A DMFC is an electrochemical fuel cell that utilizes liquid methanol and air as reactants. An IMFC is the combination of a hydrogen fuel cell with an onboard methanol-to-syngas reformer to produce the hydrogen onboard. But the motor still derives all of the electric power from a hydrogen fuel cell as shown in Fig. 6.13.

Indirect Methanol Fuel Cells (IMFCs)

As discussed above, reforming methanol to produce hydrogen and carbon dioxide using heat and a catalyst is a proven process. This can work in a hydrogen combustion vehicle because the required heat for the endothermic reforming reaction can be provided by the exhaust of an IC engine that burns the hydrogen. The resultant syngas is cooled, and the hydrogen is (ideally) separated from the carbon dioxide, while the carbon dioxide and unintended carbon monoxide are vented into the atmosphere. The otherwise low efficiency of the two energy conversion steps is improved by the recycling of the heat generated by the fuel cell to supplement the heat requirement of the reformer. But if the hydrogen is then used as a fuel cell reactant, the combination is sometimes (misleadingly) referred to as an Indirect Methanol Fuel Cell (IMFC). Onboard a vehicle, the vehicle then can be referred to as a *methanol fuel cell vehicle* (MFCV). A lot of energy conversion processes are involved. When combined with the RTE of the battery and efficiency of the EV motor, the overall efficiency is about the same (30–35%) as the simple combustion of methanol in an

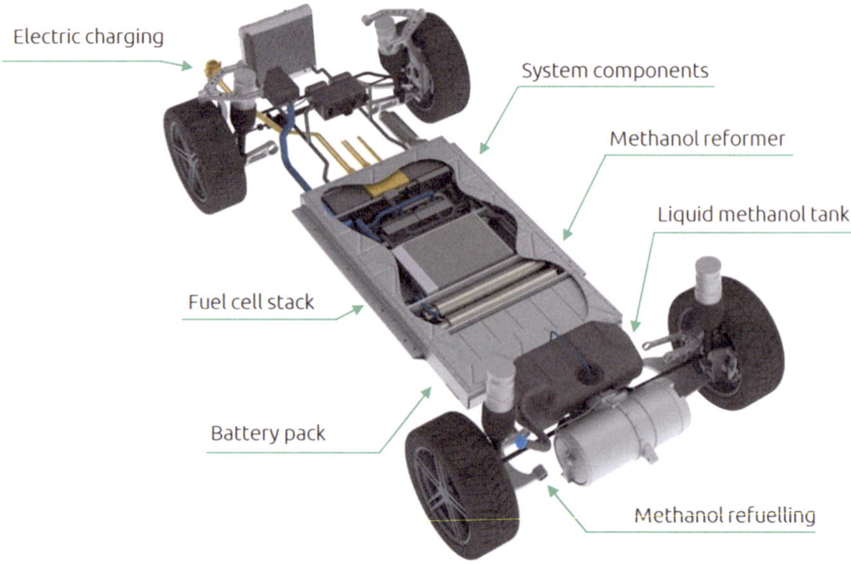

Fig. 6.13 Indirect Methanol Fuel Cell (IMFC) drivetrain under development by Blue World Technologies for Karma Automotive. (https://www.methanol.org/wp-content/uploads/2020/04/Methanol-Fuel-Cells-Blue-World-Technologies-Chongqing-2019.pdf. Slide 8. Accessed 15 June 2024. Used with permission)

ICE. This begs the question, why? One of several recent startup companies producing IMFCs, Blue World Technologies [70], states that the 45% efficiency of their integrated IMFC is greater than that of a DMFC [71], a claim that is reasonable considering that existing DMFCs are reported to have efficiencies no better than ~40% [72]. Blue World Technologies explains that the greater efficiency of their IMFC technology is due to the utilization of byproduct heat, although this is already common practice.

All currently publicized methanol fuel cell manufacturers and MFCVs use IMFCs, not DMFCs [73].

The nonobvious issues with IMFCs for powering EVs can best be illustrated by looking at the specifications published by one of the two[22] recent startup companies that have announced that they will soon be selling methanol FCVs. The Roland Gumpert Nathalie [74] shown in Fig. 6.14 is billed as a "hypercar powered by a methanol fuel cell." It was announced in 2019 at the Geneva Auto Show, with expected delivery by 2022, although not yet in production in 2024. The performance and range specifications of this car are at the level of a Lucid Air or Tesla S P100D [75, 76].

[22] The other company is Karma Motors, as reported in Engadget. *EV maker Karma wants to power electric cars with a methanol fuel cell*. https://www.engadget.com/ev-maker-karma-is-developing-a-methanepowered-hydrogen-drivetrain-112052660.html. Accessed 16 July 2024.

Fig. 6.14 "Gumpert Nathalie Methanol Fuel Cell Hypercar." (From: https://www-rolandgumpert-com.translate.goog/?_x_tr_sl=auto&_x_tr_tl=en&_x_tr_hl=en. Promotional photo, non-copyright)

The Nathalie might more accurately be described as a luxury battery electric supercar with a small but very complex methanol-fueled onboard range extender. It has a 400 kW (536 HP_{US}) electric drivetrain and a large traction battery of sufficient capacity to give it a claimed 0–100 kph (0–62 mph) acceleration time of 2.5 s and a 306 kph (190 mph) top speed. These stellar performance specs and the novelty of its advertised methanol fuel cell power source justify the $455,000 USD announced price.

The "methanol fuel cell" in the Nathalie is the aforementioned IMFC produced by Blue World Technologies [70] with a maximum electrical power output of 15 kW (20 HP_{US}). With this maximum output, the fuel cell by itself could not continuously power the vehicle except possibly at a very slow speed.

The battery capacity is not disclosed by the manufacturer, but according to AutoBlog [77], Gumpert quotes a 178 kWh *combined* energy storage capacity including both the 17.2 gallon methanol tank and the battery capacity. This provides a claimed vehicle range of "510 miles at 75 miles per hour. When driven in Eco mode, the range extends to about 745 miles."

Gumpert clarifies on their website that the 15 kW IMFC provides

> Enough energy to supply the vehicle with its basic energy while driving. When driving in the city, taking breaks and driving at low speeds, the Nathalie Fuel Cell increases the charge of the *buffer* battery.

Using information provided on the Roland Gumpert and Blue World Technologies websites, the "buffer battery" capacity can be estimated by subtracting from the 178-kWh total energy figure the electric energy that can be produced by the IMFC from 17.2 gallons of methanol at the IMFC manufacturer's specified 45% efficiency [78]. This calculation suggests that the lithium battery onboard the Nathalie would have a capacity of 135 kWh, one of the largest batteries used on any electric coupe:

$$
\begin{aligned}
&178\,\text{kWh} - (0.45)(17.2\,\text{gal MeOH})(5.6\,\text{kWh / gal MeOH}) \\
&\quad = 135\,\text{kWh battery capacity}
\end{aligned}
$$

If the total 178 kWh energy storage of the battery and the IMFC output from the 17.2 gallon methanol tank are used to travel the specified 510-mile (821 km) range:

$$\frac{510\ \text{miles}}{178\ \text{kWh stored energy}} = 2.87\,\text{miles}/\text{kWh equivalent to } 96.7\ \text{mpge}$$

$$\left(1\,\text{kWh} = 3.7\,\text{gge}\right)$$

This is just slightly less than the 106 mpge average EPA fuel economy rating of all electric automobiles sold in the USA. However, according to Gumpert, the test condition was not an EPA or EEA test cycle, but "continuous travel at 75 mph," so the actual equivalence to an EPA standard MPGe rating is unclear.

The mass of a state-of-the-art 135-kWh lithium battery would be large even for a full-size electric SUV. Using the industry average of 16 kg/kWh, it could be expected to weigh 2160 kg (4753 lb), heavier than an entire Tesla S P100D that has a 100-kWh battery and similar performance. However, such a large battery would explain and validate its exemplary acceleration numbers.

With the maximum 15 kW output of the IMFC providing the 43.3 kWh of energy available from its 17.2 gallon tank of methanol, *self-charging* the battery from the IMFC would require more than three fills of the methanol tank and 9.0 hours of fuel cell operation. The fact that its IMFC is actually a hydrogen fuel cell with a small methanol-to-hydrogen reformer is not obvious from the advertised specifications. As of 2024, the Gumpert Nathalie is not yet in production [76].

Direct Methanol Fuel Cells (DMFCs)

DMFCs (actual methanol fuel cells) may be considered a subset of PEMFCs (polymer electrolyte membrane fuel cells), in which methanol (CH_3OH) is used as the reactant rather than hydrogen. They are fundamentally different than IMFCs because methanol is used directly rather than being reformed to generate hydrogen that powers a hydrogen fuel cell.

Methanol-air-fuel cells are laboratory-proven and have been in use for small low-power applications as primary (non-rechargeable) batteries for decades. They are particularly well suited to long-life *low current* applications such as telemetry power supplies at remote locations. The promise of a practical DMFC suitable for onboard automotive power generation has been a quest since the 1960s. An online search yields hundreds of research papers and texts. A cross-sectional diagram of a DMFC is shown in Fig. 6.15 [79, 80].

Could a DMFC be used to power an automobile? The limitations most often cited are a lower power density and conversion efficiency compared with hydrogen-air cells, and the carbon dioxide exhaust that is generated from the oxidation methanol. But by comparison, the energy density and handling convenience of methanol are vastly better than hydrogen in any storage form, making them an attractive power option if their limitations could be overcome. A DMFC, even with a larger size than an H_2 fuel cell, would have a lower energy and power density, although the complete energy storage and power system would have a lower volume compared with a compressed hydrogen FCV.

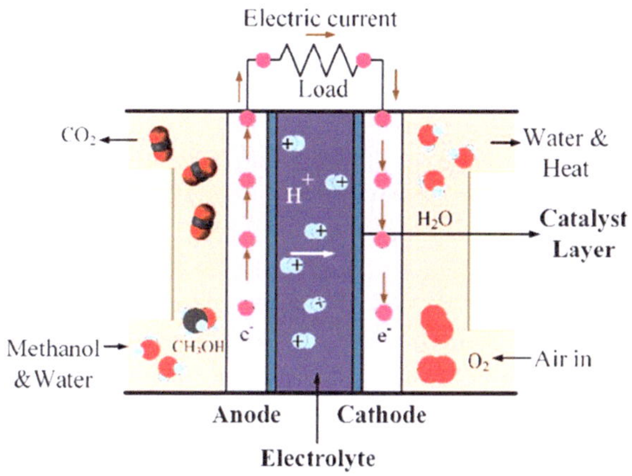

Fig. 6.15 Conceptual diagram of an alkaline electrolyte direct methanol fuel cell. (From Development of a Novel Technological Readiness Assessment Tool for Fuel Cell Technology. July 2020, IEEE Access. https://doi.org/10.1109/ACCESS.2020.3009193. License CC BY 4.0)

The well-to-wheel CO_2 emissions are also a concern. As clear from the stoichiometry equations below, 1 mole (32 g) of methanol will generate 1 mole (44 g) of carbon dioxide in the DMFC, about the same as methanol combustion in an ICE (note that methanol combustion generates the lowest CO_2 emissions of any liquid fuel). So other than the slightly better efficiency of the fuel cell and electric drivetrain compared with an ICE, a DMFC vehicle would produce nearly the same CO_2 tailpipe emissions per distance as a methanol ICV, reducing its atvantages and countermanding its classification as a ZEV.

In a DMFC, the anode draws hydrogen ions directly from the dilute liquid or vaporized methanol, eliminating the need for a fuel reformer. The reactions that occur in the DMFC are as follows:

Anode: $CH_3OH + H_2O \rightarrow CO_2 + 6H + 6e^-$

Cathode: $3/2O_2 + 6H + 6e^- \rightarrow 3H_2O$

Efficiencies of about 40% have been reported at 40–90 °C. Higher efficiencies are possible at higher temperatures. For perspective, the average efficiency of an ICE optimized for methanol is 30–35%.

Practical implementation of a DMFC faces problems with the electrolytic membrane, and the removal of the water exhaust product: Methanol permeates through the electrolytic membrane (the same type of polymer membrane used for hydrogen). And water continuously dilutes the methanol reactant at the anode, and cannot easily be removed without considerable energy investment. [81]. This is acceptable for a batch process, and indeed, methanol-air batteries have been available for specialized applications for several decades. DMFC power generators use methanol cartridges and can operate for months at low current levels before replacing a methanol cartridge (the generator cannot simply be refueled by filling a methanol tank) [82].

Commentary

All factors considered, methanol is probably the most promising among the alternative combustion fuels available for automobiles at this time. It can be produced with higher efficiency than any other synthetic fuels, starting with either fossil fuels or biomass. Its cost is currently the lowest per unit of energy of all ICE fuels except natural gas. However, it is not currently considered a renewable fuel because it is almost entirely made from natural gas or coal, due to the low cost of production from these feedstocks. For fossil-fuel-produced methanol, the overall well-to-wheel CO_2 emissions are not much different than those of gasoline. But since methanol can also be easily produced from a wide range of renewable energy sources, it is a promising candidate for a "future-proof" fuel that can help with the transition from fossil fuels to electric propulsion, still utilizing the existing gasoline distribution infrastructure.

In an ICE, it provides the highest efficiency, highest power output per displacement, and lowest emissions of both CO_2 and regulated pollutants of any available liquid fuel. It is considered safer than gasoline by professional racers. Its poor lubricity requires the addition of some organic lubricant additives, typically castor or lesquerella oil, which have no appreciable effect on its combustion characteristics other than imparting some luminosity to its flame in a fire. It is more toxic than gasoline if ingested, which can happen if accidently mixed with ethanol. It is much less harmful than petroleum in the event of a spill, and it can be diluted by water to the point of nonflammability. It is biodegradable and, in the event of a spill or leak, relatively benign to soil microbial life or aquatic life. Without additives, it requires some provision to aid in engine starting in cold climates—a problem for which simple solutions are available. Methanol contains about half the energy per unit volume of gasoline, requiring twice the fuel tank volume compared with gasoline.

Methanol is also a hydrogen carrier that can be stored compactly on a vehicle and thermally reformed to produce hydrogen, avoiding the bulk and hazards of high pressure or liquid hydrogen storage. But the combination of an onboard methanol reformer and hydrogen fuel cell provides little advantage for efficiency and CO_2 emissions compared with the direct combustion of methanol in an optimized methanol ICE.

Methanol has potential as a fuel cell reactant for DMFCs. Methanol fuel cells are currently limited to efficiencies to well below hydrogen fuel cells. They are currently only suitable for low-power high energy density power sources such as remote telemetry, where they are batch refueled, more like primary batteries than continuous fuel cells. It does not seem likely in the near term, but if improved electrolytic materials are found and cells can be engineered for continuous operation, DMFCs could offer a potential upside similar in significance to the development of advanced lithium batteries.

References

1. G. Bozzano, F. Manenti, Efficient methanol synthesis: perspectives, technologies and optimization strategies. Prog. Energy Combust. Sci. **56**, 71–105 (2016). https://doi.org/10.1016/j.pecs.2016.06.001
2. Methanex Corporation, Vancouver, Canada, Methanol: 19th century wood alcohol emerges as an alternative fuel source for the 21st century, https://www.methanex.com/sites/default/files/news/media-resources/Methanol_19th_Century_Wood_Alcohol_2017.pdf
3. B. Moser, S. Cermak, T. Isbell, Evaluation of castor and lesquerella oil derivatives as additives in biodiesel and ultralow sulfur diesel fuels. Energy Fuels **22**, 1349–1352 (2008)
4. C.A. MacCarley, *Methanol Fuel Safety—A Practical Guide* (California Polytechnic State University, San Luis Obispo Digital Commons, 2013), https://digitalcommons.calpoly.edu/cgi/viewcontent.cgi?article=1345&context=eeng_fac
5. Photo from "Survival Mastery" web site, How to make methanol, https://survival-mastery.com/diy/how-to-make-methanol.html. Accessed 5 June 2023
6. Cut The Wood, Wood alcohol, what it is and how it is made (2023), https://cutthewood.com/guides/wood-facts/wood-alcohol/. Accessed 1 May 2024
7. US Dept of Energy, National renewable Energy Laboratory, *Methanol from Biomass*, NREL/SP-420-5570-Rev.2, DE93010018 (1995), https://www.nrel.gov/docs/legosti/old/5570r2.pdf
8. J.J.B. Caballero, I.N. Zaini, W. Yang, Reforming processes for syngas production: a mini-review on the current status, challenges, and prospects for biomass conversion to fuels. Appl. Energy Combust. Sci. **10**, 100064 (2022)
9. C. Li, H. Bai, Y. Lu, J. Bian, Y. Dong, H. Xu, Life-cycle assessment for coal-based methanol production in China. J. Clean. Prod. **188**, 1004–1017 (2018). https://doi.org/10.1016/j.jclepro.2018.04.051
10. US DOE National Energy Technology Laboratory, Direct liquefaction process, https://www.netl.doe.gov/research/carbon-management/energy-systems/gasification/gasifipedia/direct-liquefaction. Accessed 15 May 2024
11. US DOE National Renewable Energy Laboratory, Methanol from biomass (1995), https://www.nrel.gov/docs/legosti/old/5570r2.pdf. Accessed 1 June 2023
12. US Dept of Energy, National Energy Technology Laboratory, Syngas conversion to methanol, https://www.netl.doe.gov/research/carbon-management/energy-systems/gasification/gasifipedia/methanol. Accessed 20 May 2024
13. Penn State, Alternative fuels from biomass sources: Fischer-Tropsch process to generate liquid fuels: methanol production, https://www.e-education.psu.edu/egee439/node/679
14. C. Song, W. Pan, Tri-reforming of methane: a novel concept for catalytic production of industrially useful synthesis gas with desired H_2/CO ratios. Catal. Today **98**(4), 463–484 (2004)
15. T.P.M. Xuan-Huynh Pham, J.-I.H. Ashik, A.P. Alonso, D. Pla, M. Gómez, D.P. Minh, Review on the catalytic tri-reforming of methane—Part II: Catalyst development. Appl. Catal. A Gen. 623, 118286 (2021)
16. Hydrocarbon Processing, World's largest CO_2-to-methanol plant starts production (2022), https://www.hydrocarbonprocessing.com/news/2022/10/worlds-largest-co2-to-methanol-plant-starts-production
17. Institute for Carbon Removal Law & Policy, Fuel out of thin air: CO_2 capture form air and conversion to methanol (2020), https://research.american.edu/carbonremoval/2020/09/29/fuel-out-of-thin-air-co2-capture-from-air-and-conversion-to-methanol/. Accessed 20 May 2024
18. J. Kumagai, Methanol-fueled cars could drive us toward an Emissionless future. IEEE Spectr. (2018), https://spectrum.ieee.org/methanolfueled-cars-could-drive-us-toward-an-emissionless-future
19. Carbon Recycling International, CO_2 and methanol, https://carbonrecycling.com/co2-methanol. Accessed 25 May 2024
20. L. Zyga, Carbon dioxide captured from air can be directly converted into methanol fuel (2016), https://phys.org/news/2016-01-carbon-dioxide-captured-air-methanol.html. Accessed 15 May 2024

21. Methanol Institute, Aluminum alloy tanker trucks and rail cars, Methanol Safe Handling Technical Bulletin (undated), https://www.methanol.org/wp-content/uploads/2016/06/Materials-Selection-for-Neat-Methanol-Service_FINAL.pdf
22. Improving methanol production efficiency and reducing carbon dioxide emissions, p. 30, www.methanol.org. Accessed 7 Mar 2018
23. C. Rattan, *Carbon Footprint of Methanol, Energy Central* (Methanol Institute, 2022), https://energycentral.com/c/og/carbon-footprint-methanol. Accessed 15 June 2024
24. A. Martin, *A Step Forward for "Green" Methanol and Its Potential to Deliver Deep GHG Reductions in Maritime Shipping* (The International Council on Clean Transportation, 2021), https://theicct.org/a-step-forward-for-green-methanol-and-its-potential-to-deliver-deep-ghg-reductions-in-maritime-shipping%E2%80%AF/. Accessed 15 June 2024
25. A.I. Osman, N. Mehta, A.M. Elgarahy, A. Al-Hinai, A.'a.H. Al-Muhtaseb, D.W. Rooney, Conversion of biomass to biofuels and life cycle assessment: a review. Environ. Chem. Lett. **19**, 4075–4118 (2021). https://doi.org/10.1007/s10311-021-01273-0
26. Kristiina, D. Vogt, Portable methanol production using woody waste: does it play a role in community-based biodiesel production, https://deq.mt.gov/files/Energy/EnergizeMT/EnergySites/Documents/Biodiesel_Production_Educ_Presentations/KVogt_Pablo_NCAT_10_31_07.pdf. Accessed 15 May 2024
27. M. Allard, Issues associated with widespread utilization of methanol. SAE Trans. (2000). https://doi.org/10.4271/2000-01-0005
28. T. Reed, Efficiencies of methanol production from gas, coal, waste or wood. Am. Chem. Soc. **21**, 2 (1976)
29. Improving methanol production efficiency and reducing carbon dioxide emissions, www.methanol.org. Accessed 7 Mar 2018
30. Sustainability (2020), https://innovationorigins.com/en/producing-gasoline-and-diesel-emits-more-co2-than-we-thought/. Accessed 1 June 2024
31. S. Muskopf, Methanol and the chemistry of a poison (2022), https://www.biologycorner.com/2022/09/10/methanol-and-the-chemistry-of-a-poison/
32. One of many documentary reports: Dying for a drink: methanol poisoning in Bali (2016), https://www.flightstofancy.com/methanol-poisoning-bali/
33. Barison Industry (Italy), Methanol: what it is and how it is handled in distillation processes, https://www.barisonindustry.com/en/news/methanol-what-it-is-and-how-it-is-handled-in-distillation-processes. Accessed 15 May 2024
34. How to avoid methanol when distilling, https://diydistilling.com/how-to-avoid-methanol-when-distilling/. Accessed 5 June 2023
35. D.A. Drago, Gasoline-related injuries and fatalities in the United States, 1995–2014. Int J Inj Contr Saf Promot (2018). https://doi.org/10.1080/17457300.2018.1431947
36. Wikipedia, Timeline of alcohol fuel (2024), https://en.wikipedia.org/wiki/Timeline_of_alcohol_fuel. Accessed 30 May 2024
37. Wikipedia, Indianapolis 500 (1981), http://en.wikipedia.org/wiki/1981_Indianapolis_500. Accessed 15 May 2024
38. B. Lohnes, Invisible fire video: a wild methanol fire in the Pits at the 1981 Indy 500 (2019), https://bangshift.com/bangshiftapex/invisible-fire-video-a-wild-methanol-fire-in-the-pits-at-the-1981-indy-500/. Accessed 10 May 2024
39. Video clip of sequence from Talladega nights: the ballad of Ricky Bobby, https://filmschoolrejects.com/10-scenes-we-love-from-talladega-nights-the-ballad-of-ricky-bobby-e7ef1844f622/ or https://www.imdb.com/video/vi713556249/?ref_=tt_vi_i_2. Accessed 1 Sept 2022
40. C.A. MacCarley, Methanol fuel safety—a practical guide (2013), https://digitalcommons.calpoly.edu/cgi/viewcontent.cgi?httpsredir=1&article=1345&context=eeng_fac
41. US Dept of Energy, Energy Information Agency (EIA), MTBE, oxygenates, and motor gasoline (2000), https://www.eia.gov/outlooks/steo/special/pdf/mtbe.pdf. Accessed 11 Nov 2023
42. L. Bromberg, W.K. Cheng, *Methanol as an Alternative Transportation Fuel in the USA: Options for Sustainable and/or Energy-Secure Transportation*, Final Report: UT-Battelle Subcontract Number: 4000096701 (Sloan Automotive Laboratory Massachusetts Institute of

Technology, Cambridge, MA, 2010), https://afdc.energy.gov/files/pdfs/mit_methanol_white_paper.pdf. Accessed 20 May 2024
43. Wikipedia, Flex-fuel vehicle, https://en.wikipedia.org/wiki/Flexible-fuel_vehicle#cite_note-Methanolstory-8. Accessed 10 June 2023
44. R.J. Nichols, The methanol story: a sustainable fuel for the future. J. Sci. Ind. Res. **62**, 97–105 (2003)
45. Yahoo/Finance, Methanol market set for explosive growth, projected to reach US $77.6 billion by 2032—persistence market research (2024), https://finance.yahoo.com/news/methanol-market-set-explosive-growth-170200292.html. Accessed 15 May 2024
46. Zhejiang Geely Holding Group, Geely Auto Group press release (2022), https://zgh.com/media-center/?lang=en. Accessed 12 Dec 2023
47. Methanol fuel technology to employ 40m Nigerians, and Nigeria, Israel to sign pact on methanol fuel technology, Daily Trust (2020), https://dailytrust.com/nigeria-israel-to-sign-pact-on-methanol-fuel-technology/
48. Z. Yangarchive, China is betting big on another gas engine alternative: methanol cars. MIT Technol. Rev. (2022), https://www.technologyreview.com/2022/09/30/1060508/china-betting-methanol-cars/. Accessed 20 May 2024
49. Methanol Institute, China: the leader in methanol transportation, https://www.methanol.org/wp-content/uploads/2020/04/China-Methanol-Fact-Sheet-1.pdf. Accessed 10 June 2023
50. International Energy Agency, Technology Collaboration on Advanced Motor Fuels, Methanol, https://iea-amf.org/content/fuel_information/methanol. Accessed 1 May 2024
51. *Bosch Automotive Handbook*, 10th edn. (2018) Methanol RON=130
52. Q. Li, W. Jin, Z. Huang, Laminar flame characteristics of C1–C5 primary alcohol-isooctane blends at elevated temperature. Energies **9**(7), 511 (2016). https://doi.org/10.3390/en9070511
53. Ö.L. Gulder, Laminar burning velocities of methanol, isooctane, and isooctane/methanol blends. Combust. Sci. Technol. **33**(1–4), 179–192 (1983)
54. S.C. Bhatia, Advanced renewable energy systems, Sec. 21.6, in *Ethanol as an Automotive Fuel*, (WPI, Science Direct, 2014), pp. 523–572
55. US Alternative Fuels Data Center, US Department of Energy, Ethanol fuel basics, https://afdc.energy.gov/fuels/ethanol-fuel-basics. Accessed 1 July 2015
56. *Bosch Automotive Handbook*, 10th edn. (Distributed in North America by SAE International, Warrendale PA), pp. 659–661
57. Torco Racing Fuels, M100 racing fuel online quotation (2024), https://torcoracefuel.net/products/methanol-race-fuel
58. Indian Express, Govt to unveil policy on methanol blending in petrol to cut price and pollution, says Nitin Gadkari (2024), https://indianexpress.com/article/india/govt-to-unveil-policy-on-methanol-blending-in-petrol-to-cut-price-and-pollution-says-nitin-gadkari-4975377/. Accessed 23 Sept 2024
59. Trading Economics, Methanol (2024), https://tradingeconomics.com/commodity/methanol
60. Trading Economics, Gasoline (2024), https://tradingeconomics.com/commodity/gasoline
61. Edmunds, Used 2012 Honda Civic—specifications and features, https://www.edmunds.com/honda/civic/2012/features-specs/. Accessed 1 June 2024
62. *Robert Bosch Automotive Handbook*, 10th edn. (SAE International, 2018), p. 345
63. Shanghai Jiao Tong University, On-board methanol catalytic reforming for hydrogen production—a review. Int. J. Hydrogen Energy **46**(43), 22303–22327 (2021)
64. N. Dummer et al., Methane oxidation to methanol. Chem. Rev. **123**(9), 6359–6411 (2023)
65. J.G. Finegold, G.P. Glinski, G.E. Voecks, *Dissociated Methanol Citation: Final Report* (US DOE Solar Energy Research Institute, 1984) SERI/TR-235-2083
66. R. Schluter, Element 1 Corp (e1), Methanol to H_2 supply systems (MHSS)—fuel cell mobility CaFCP Working Group Meeting (2019), https://www.methanol.org/wp-content/uploads/2020/04/Methanol-to-H2-Fuel-Cell-Mobility.pdf. Accessed 15 June 2024
67. Blue World Technologies, Fueling the future with methanol fuel cells | hydrogen + fuel cells EUROPE, Presentation at H2FC Hannover (2023), https://www.youtube.com/watch?v=ZQezOmL3Owo

68. Ser Energy. Reformed methanol fuel cells—enabling the methanol vision, Presentation at H2FC Hannover (2022), https://www.youtube.com/watch?v=nNuZGRhPd24. Accessed 15 June 2024
69. Blue World Technologies, Superior HT PEM fuel cell technology, https://www.blue.world/products/. Accessed 1 Sept 2024
70. Blue World Technologies, With methanol fuel cells we reduce the cost of going green!, https://www.blue.world/. Accessed 5 Aug 2024.
71. Ser Energy, Clean power, https://www.methanol.org/wp-content/uploads/2020/04/Serenergy-Introduction-Methanol-Institute_July-15th-1-1.pdf. Accessed 15 June 2024
72. K. Scott, L. Xing, Fuel cell engineering, in *Advances in Chemical Engineering*, (2012), https://www.sciencedirect.com/topics/chemistry/direct-methanol-fuel-cell. Accessed 10 June 2024
73. Methanol Institute, Fuel cells, https://www.methanol.org/fuel-cells/. Accessed 15 June 2024
74. R. Gumpert, Fuel cell—permanent energy, https://www-rolandgumpert-com.translate.goog/nathalie/methanol-fuel-cell/?_x_tr_sl=auto&_x_tr_tl=en&_x_tr_hl=en-US. Accessed 15 June 2024
75. R. Gumpert, *Nathalie, 100% Driving Machine And Electric Super Sports Car*, 1st edn., https://www-rolandgumpert-com.translate.goog/?_x_tr_sl=auto&_x_tr_tl=en&_x_tr_hl=en-US. Accessed 15 June 2024
76. G. Natalie, Direct methanol fuel cell, https://www.autoblog.com/2020/03/05/roland-gumpert-nathalie-methanol-fuel-cell-supercar/. Accessed 15 June 2024
77. J. Ramsey, AutoBlog (2020), https://www.autoblog.com/2020/03/05/roland-gumpert-nathalie-methanol-fuel-cell-supercar/. Accessed 20 June 2024.
78. M.F. Jensen. *CCO at Blue World Technologies*, Slide 12, Methanol Vehicles Exhibition, Chongqing (2019), https://www.methanol.org/wp-content/uploads/2020/04/Methanol-Fuel-Cells-Blue-World-Technologies-Chongqing-2019.pdf. Accessed 15 June 2024
79. N.H. Behling, Fuel cells and the challenges ahead, 2.4.1 Principles of operation and characteristics, in *Fuel Cells* (2013)
80. J. Kendall, DMFC diagram, Innovating the methanol fuel cell, SAE Mobilus Online (2021), Also appears on product web page of Green Century, http://www.greencentury.cn/en/index.php?s=/Home/article/lists/cp/1/category/8.html. Accessed 1 Aug 2024
81. C. Hartnig, J. Scholta, Direct methanol fuel cells, in *Materials for Fuel Cells*, (2008), https://www.sciencedirect.com/topics/engineering/direct-methanol-fuel-cell. Accessed 1 June 2023
82. Evergreen Energy Technologies, Calgary, Canada. http://www.evergreen-eti.com/products/methanol-cartridges or https://www.mysmartfuelcell.com/methanol-fuel-cell

Ethanol 7

Summary of Ethanol Fuel Characteristics

- An excellent spark ignition (SI) engine fuel.
- Sold as E85, a blend of 85% ethanol and 15% gasoline (nominal) by volume.
- An octane additive in a 10% blend with gasoline, called E10. Replaces the previous toxic octane additives tetraethyl lead (TEL) and methyl tertiary butyl ether (MTBE).
- Contains two-thirds the combustion energy per volume compared with gasoline. This means approximately 50% more fuel volume is required to match the energy content of gasoline.
- A "nearly drop-in" substitute for gasoline requiring minimal engine modifications.
- Same as beverage alcohol, so it must be denatured (made toxic) immediately after distillation by the addition of a small amount of gasoline or methanol.
- Elastomer and metal compatibility issues for older cars that were equipped with carburetors. No longer an issue since all cars sold in the USA now use materials compatible with ethanol.
- Cold starting remains a problem for ethanol used as the primary blend component. E100 (pure ethanol) requires provisions for manifold or fuel preheating to ensure starting at temperatures below about 5 °C (42 °F).
- Can use existing gasoline fuel distribution with no modifications[1].
- Higher octane than gasoline. Allows higher compression ratio for greater thermal efficiency and higher power. E85 is popular with amateur racers since it can be obtained at regular fuel stations and is less expensive than professional racing fuels.

[1] Mike Kass, Oak Ridge National Laboratory. Compatibility of Fueling Infrastructure Materials in Ethanol Blended Fuels. https://afdc.energy.gov/files/pdfs/ethanol_fueling_compatibility.pdf Accessed 15 June 2024.

C. A. MacCarley, *Non-Petroleum Automotive Transportation*,
https://doi.org/10.1007/978-3-031-85028-8_7

- Currently, 99+% of all ethanol is produced from the fermentation of starchy crops: corn in the USA, sugar cane in South America, or wheat in Europe and Asia.
- Despite decades of research investments, cellulosic (nonfermented) ethanol remains impractical due to its processing energy requirements and/or high cost of required enzymes.
- Ethanol is classified in the USA as a renewable fuel due to its plant origin, although the fossil fuel energy required for its cultivation and production (natural gas, diesel fuel) is equal to 62% of its fuel energy value. (In 2023, the USDA concluded that US production was 38% energy positive.)
- When land diversion is included in the analysis, all independent studies conclude that corn ethanol has a net negative climate impact, a conclusion refuted by advocate organizations.
- All gasoline sold in the USA and most of Western Europe contains 10% ethanol (E10) by volume. Since 2011, 15% ethanol (E15) has been allowed in some states.
- E85 is the USA's designated alternative fuel for flex-fuel vehicles (FFVs).
- Over 3000 E85-equipped stations in the USA, but many are not operational due to low demand or lack of maintenance.
- Government subsidies at multiple levels (farm, refinery, FFV manufacturer, fuel user).
- The retail price of E85 has been regulated at about 80% of the price of E10 gasoline on a volume basis, but the energy cost is slightly higher than 100% on an equivalent energy basis. Energy content per gallon translates to miles per gallon (mpg).
- Over ten million E85 FFVs are on the road in the USA, but very few use E85 due to higher cost or poor availability, despite subsidies (Fig. 7.1).
- Forty percent of the national corn crop and 30 million acres of farmland are dedicated to making fuel ethanol in the USA.
- Food versus fuel issues: Fuel production competes with food production. Federal and State subsidies.

Ethanol goes by many names depending on its application: ethanol, ethyl alcohol, grain alcohol, shellac thinner, 200 proof vodka, Everclear®, moonshine, or simply "alcohol" in a beverage and DUI context. It has only two carbon atoms, making it the second simplest alcohol compared with methanol. It has historically been made by fermentation of almost any plant source of natural sugar or starch (grains, grapes, cane, succulents, and fruit). Any plant that can be used to make fermented fuel ethanol is also a food for human or animal consumption. Corn is the dominant ethanol crop in North America, sugar cane in South America, and wheat in Europe. Any plant from which beverage alcohol can be made is a potential fuel ethanol source. Ethanol is (arguably) the oldest known human intoxicant [1], mentioned in the earliest recorded history. A flammable liquid with density between water and

Fig. 7.1 E85 Flex-Fuel light truck, 2009. (https://commons.wikimedia.org/wiki/File:FlexFuel_GM_badge_64_MIA_12_2008_retouched.jpg. Photo by Mariordo Ortiz, Wikipedia Creative Commons Attribution 3.0 Unported license)

gasoline, it has combustion characteristics that make it a favorable replacement for gasoline, either alone or blended with gasoline.

Ethanol is currently endorsed by the U.S. Department of Energy, Environmental Protection Agency, and Department of Agriculture as the *only* viable substitute for gasoline. Its use as a 10% fuel additive alone generates a demand many times greater than the next most common biofuel, biodiesel [2], which is used exclusively as a partial replacement for diesel fuel. In the early years of the twentieth century, prior to the proliferation and standardization of petroleum-derived gasoline, ethanol was a common fuel for spark ignition (SI) automobile engines. Its current production steps are essentially the same as for beverage ethanol, with the important difference that ethanol intended for fuel use is "denatured" (made non-consumable) immediately following final distillation, usually by the addition of a small percentage of gasoline or methanol so that it cannot be siphoned off for beverage use or subjected to beverage ethanol taxes.

The distillation objectives for fuel ethanol are a bit different than those for beverage alcohol. Beverage ethanol is rarely produced with less than 10% water content (180 proof), but it must contain no other contaminants, especially methanol, which is an unintended byproduct of almost any fermentation. Fuel ethanol, however, is degraded by the presence of water, with little regard for methanol content.

With minimal engine modifications, ethanol can replace gasoline by simple modification of the air-fuel ratio to make it "richer" (AFR=9.0 for neat ethanol vs. 14.8 for gasoline). The most noticeable difference to the vehicle operator following such a conversion has to do with starting the engine in cold weather. Unlike gasoline, a blend of many hydrocarbons with a wide range of vaporization temperatures and pressures, ethanol is a pure substance with fixed vaporization properties. At air temperatures below approximately 5 °C at normal atmospheric air pressure, it does not vaporize adequately to form a flammable mixture, so that starting an engine becomes difficult or impossible, and startup emissions are poor. This is part of the

Fig. 7.2 Mason Jar of homebrewed ethanol, aka *moonshine*. (Photo by Tim Smith, http://timsmithmoonshine.com. 12 June 2021. With permission)

justification for the addition of 17% (explained later in this section) or more gasoline (by volume) in E85.

Ethanol fuel has had a place in American folklore, especially since 1919 following the ratification of the 18th Amendment to the US Constitution, known as the *Prohibition* (repealed by the 21st Amendment in 1933). Folk legend has it that the high-performance cars used by moonshiners (illegal ethanol producers) to transport ethanol from back-country distilleries to urban speakeasies (clandestine pubs) were fueled by the same ethanol they sold as a beverage. This may be true in some cases, but most such accounts were exaggerated because of the inability of crude stills (distillation apparatuses) to adequately desiccate the final product, leaving it with excessive water content for fuel use. Also, gasoline at that time was inexpensive compared with the beverage value of the ethanol, so its use as a fuel would have only made sense if access to gasoline was scarce (which it was in some backwoods areas of southern states) (Fig. 7.2).

Ethanol Is an Excellent Combustion Fuel

If we ignore the many issues associated with its production, ethanol is an excellent alternative motor fuel, with an octane rating and exhaust emissions (regulated and GHG) superior to gasoline. This is true for both straight denatured ethanol E100 and ethanol blended with gasoline as E85, nominally stated to be a mixture of ethanol 85% and gasoline 15% by volume.[2] Almost any gasoline-powered engine can be modified to run on ethanol or ethanol blends.

[2] The 85%/15% volumetric ratio of ethanol to gasoline in E85 is only the nominal upper limit for its ethanol content. In the USA, the exact ratio of fuel ethanol to hydrocarbon may vary under ASTM Specification 5798, allowing ethanol content ranging from 51% to 83%. Ref: Handbook for

US Regulations Pertinent to Use of Ethanol Fuel

Under US Federal law, a DIY conversion of a gasoline vehicle to run on E85 or E100 is considered an illegal modification, even if the resulting emissions are improved compared with gasoline. California, through the California Air Resources Board, has traditionally been allowed to establish its own emissions and certification regulations because they have always been stricter than Federal standards. Rigid rules prohibiting almost any aftermarket engine modification are enforced by periodic (biennial in California) vehicle inspections. Current (2024) regulations applicable to alcohol conversions can be found in citations [3, 4].

In brief, conversions of MY 2004 and later vehicles are only allowed by the original vehicle manufacturer following a certification process equivalent to the original gasoline certification of the car. A distinction is made between 2003 and earlier vehicles if a retrofit kit that had been approved for use on 1993 and earlier vehicles is used. The text of the California regulation [3] is as follows:

> At the option of the retrofit system manufacturer, the standards and test procedures for approval of systems designed to convert 1994 and subsequent through 2003 model year motor vehicles to use alcohol or alcohol/gasoline fuels may be used for approval of systems designed to convert 1993 and earlier model year motor vehicles to use alcohol or alcohol/gasoline fuels in lieu of the 'California Exhaust Emission Standards and Test Procedures for Systems Designed to Convert Motor Vehicles Certified for 1993 and Earlier Model Years to Use Alcohol or Alcohol/Gasoline Fuels.' The standards and test procedures for approval of systems designed to convert 2004 and subsequent model year motor vehicles to use alcohol or alcohol/gasoline fuels in lieu of the original certification fuel system are contained in the 'California Certification and Installation Procedures for Alternative Fuel Retrofit Systems for 2004 and Subsequent Model Year On-Road Motor Vehicles and Engines.'

The last sentence simply refers back to itself, leaving the regulation a bit unclear. An example of an approval letter issued to a well-established modifier of school buses to operate on propane (or possibly also natural gas) can be found at citation [5].

The ability of a vehicle to run on E85 or any combination of ethanol and gasoline incurs very little additional cost to the vehicle manufacturer. *Flex fuel* or *variable fuel* vehicles (FFVs) are capable of adjusting the air-fuel ratio to match the composition of whatever fuel blend is in the tank, which can vary from 10% ethanol to 85% ethanol. From an operational point of view, flex-fuel vehicles exhibit driving characteristics indistinguishable from conventional E10 vehicles, although driver-observed fuel economy measured in miles per gallon (not MPGe) is less than E10, since the volumetric energy content of ethanol is about two-thirds that of gasoline.

Handling, Storing, and Dispensing E85 and Other Ethanol-Gasoline Blends. US Dept of Energy. https://afdc.energy.gov

How to Convert an Engine to Run on Ethanol or E85

What follows is for information only. Aftermarket engine modifications without EPA (US) or CARB (California) approval are allowed only for off-road use.

The main difference between gasoline and alcohol operation is the stoichiometric (chemically correct) air-fuel mass ratio (AFR). Modification for ethanol simply requires increasing the amount of fuel delivered to the engine with each intake stroke by approximately 56% for E100 ethanol compared with E10 or slightly less if the intended fuel is E85, which can legally include 17–50% gasoline. The exact AFR calculations for any particular fuel blend are presented in Appendix 3 *Calculation of Ideal Mass AFR*.

For modern cars and trucks, the conversion task is simplified due to the fact that (with only one exception) every new automobile or light truck sold in the USA since 1991 has been equipped with electronic fuel injection and feedback fuel control that constantly adjusts the injected fuel quantity to maintain a stoichiometric AFR. In EFI systems, the fuel quantity per injection is determined by a combination of the injector nozzle flow coefficient, the fuel pressure, and the amount of time that the injector is opened during each two-revolution cycle of the four-stroke engine.

The modification of the fuel quantity per injection can be done a number of ways. Most obvious is the replacement of the fuel injection valves (injectors) with parts having a larger flow coefficient. There is a healthy aftermarket for high-volume fuel injectors for racing/performance applications, which can also serve as ethanol conversion parts. Another approach is to increase the fuel rail pressure (injection pressure) such that mass fuel delivery is increased by 56% compared with stock E10. Since the flow rate of a incompressible fluid such as gasoline or ethanol increases as the square root of the differential pressure ratio, the fuel rail *pressure* (usually 3–4.5 bar or 43–65 psig for a port-injection system) must be increased by a factor of 2.42 to achieve the desired ethanol fuel delivery. This would mean that a typical 3 bar (43 psi) gasoline rail pressure would have to be increased to 7.3 bar (104 psig). Fuel pumps for port or manifold injection systems usually can handle a maximum of about 80 psi, which is beyond the capability of most aftermarket fuel injection pumps.

The last and most common method is to increase the injector pulse duration to deliver approximately 60% more fuel (by mass) than the original gasoline delivery. In electronic fuel-injected engines, the injection pulse duration typically varies from 1 to 12 ms over the full fuel control range. This is true for either port or manifold injection configurations, and direct cylinder injection (DCI or direct injection spark ignition aka DISI) engines, even though the latter method requires much higher fuel pressure than port injection. The engine control computer also known as the engine control module (ECM) or engine control unit (ECU) determines the injection pulse duration to provide a stoichiometric mixture over most of the engine's speed and load range. Deviations from stoichiometric occur during engine warm-up and full power (slightly rich), or at idle or light load (slightly lean).

An appropriate pulse duration for each engine condition is programmed into the computer's "fuel map," which specifies the fuel pulse duration for any measured air flow rate and engine speed (with correction for air temperature). The map is adaptive based on feedback of oxygen in the exhaust, which is beyond the point of this discussion. DIY kits or piggyback fuel controllers can be purchased that simply rescales the injection pulse by whatever ratio necessary for the fuel composition, letting the oxygen sensing feedback continuously modify the map entries for stoichiometric operation. Any of the aftermarket "tuners"[3] can be used to do this. These devices are intended for ("off-road only") performance tuning but can serve as excellent devices to rescale fuel delivery for ethanol.

Inexpensive E85 "plug-n-play" conversion kits are also available from China, which simply intercept and extend the fuel injection pulse duration by approximately 56% (for E100). Extending the original pulse duration is an effective means to increase the fuel quantity over the entire engine speed range, as long as the resulting maximum fuel pulse duration does not exceed the time required for the 720° rotation of the engine between injections (port injection assumed). This depends on the maximum engine speed and the manufacturer's selection of the pulse duration to fuel quantity scaling. In most cars with low redline RPM limits, this is not a problem. However, in high-performance cars with high maximum engine speeds and large fuel delivery requirements, the lengthening of the pulse will be limited at some engine RPM. Replacing the fuel injectors with higher flow units will be necessary in addition to or in place of the 56% pulse extension.

As mentioned previously, aftermarket conversion of cars that run on alcohol, e.g., E85 or M85, is rare in the USA due to the regulatory environment and the general decline of practical skills for DIY repair or modification of cars in the USA [6]. But in areas of the world that lack a reliable gasoline supply infrastructure, ethanol and methanol engine conversions are quite common. They are also popular among *survivalists*[4] planning for an energy-starved future.

Figure 7.3 shows an inexpensive E85 conversion kits that can be purchased on the internet from China that intercept and extend the fuel injection pulse duration by about 50% for ethanol or 100% for methanol, effective but not legal in the USA under EPA and CARB regulations.

Regarding the compatibility of the materials in the fuel system: tubing or fittings in contact with ethanol may have to be replaced. The components most at risk of ethanol compatibility problems are the fuel pump, fuel injectors, fuel hoses and filters, fuel pressure regulator, and the fuel tank. Fortunately, modern automobiles and trucks now utilize metals and elastomers in the fuel system that are compatible with alcohol blends, since they are already required by law to be compatible with fuel

[3] One example of an aftermarket tuner is the DynoJet Power Commander series of tuning modules. https://www.dynojet.com/power-commander/. Accessed 20 June 2024.

[4] *Survivalist* is a term of American origin that refers to a person committed to having the resources and skills to survive in the event of a total loss of infrastructure and order in society, as might occur in the aftermath of a major disaster, or a civil or world war.

Fig. 7.3 After-market E85 pulse-duration extender for E85 fuel conversion. (Photo from AliExpress https://www.aliexpress.us/item/2251801861809128.html. Accessed June 10, 2023. Noncopyright image)

Fig. 7.4 Nozzle of E10 fuel pump. (Public domain)

containing up to 15% ethanol (or 83% ethanol in flex-fuel vehicles) (Fig. 7.4). The long-established automotive practice of *Terne* (lead-tin or zinc-tin alloy) plating of the interior of a gasoline fuel tank to prevent water corrosion has been mostly discontinued in new cars since ethanol reacts with these coatings. M85 flex-fuel vehicles manufactured in the 1990s are fully compatible with E85. Also, most vehicles for which an E85 flex-fuel option is available, even if purchased without this option, have fuel systems that are E85 compatible, usually requiring only a software modification to accommodate E85. Alcohol compatibility can usually be assumed for any newer US-market car or truck, even if not available with an FFV option, since manufacturing cost considerations favor standardizing components across all vehicle configurations and model lines.

In my experience, the most common components to fail after substitution of alcohol for gasoline are older fuel pumps. Almost all newer cars use submerged "wet" fuel pumps, which circulate fuel through the motor body to cool the motor. As recent as the 2000s, the DC motors in these pumps used brushed commutation, which means that the fuel is in continuous contact with the rotating copper/bronze commutator, subjecting it to corrosion when the pump is not running. This is one of the reasons that racers that use methanol or ethanol fuel always purge the fuel system with gasoline after every race. Newer fuel pumps in recent model cars now use electronic (brushless) DC motors that eliminate this corrosion vulnerability, although all other materials in the pump must also be alcohol compatible [7].

Ignition timing for an ethanol engine has been found to not differ significantly from gasoline, although optimum (MBT) ignition advance is slightly less because of the faster combustion rate of ethanol compared with gasoline. This was discussed in Chap. 6, *Methanol*.

With the higher octane rating of ethanol compared with gasoline, the compression ratio (CR) of the engine can be safely increased without causing combustion knock, discussed in an earlier chapter. The simplest way to do this is to use a thinner head gasket or mill the surface of the cylinder head (up to the point that valves could almost contact the piston surface). This should not done for engines intended to run on either gasoline or alcohol since gasoline has a lower octane rating and is, therefore, less tolerant of high compression ratios. Many (but not all) of the newest automobiles, including almost all hybrids, use *direct cylinder gasoline fuel injection*, as discussed in Chap. 4. Modifying a direct injection engine for any fuel other than gasoline is much more challenging than for a port/manifold injection engine, best left to the vehicle manufacturer.

Figure 7.5 is a photo of a replica of the 1968 Dodge Charger "General Lee" considered by many fans to be the real star of the 1979–1985 TV US television series "Dukes of Hazard" and the 2005 action film based on the TV series. The storyline was that the car ran on homemade ethanol, aka *moonshine*. None of the nearly 300 cars used (and destroyed) in the series actually did. But the popular television exposure helped to increase public awareness of ethanol as a fuel option, inspiring many DIY conversions

Fig. 7.5 Photo of a 1968 Dodge charger customized to look identical to "General Lee" from US TV series Dukes of Hazard. (Image from https://commons.wikimedia.org/wiki/File:1969_Dodge_Charger_SE_(Dukes_of_Hazzard_General_Lee_Tribute) (36671343453).jpg. Creative Commons Attribution 3.0 Unported license)

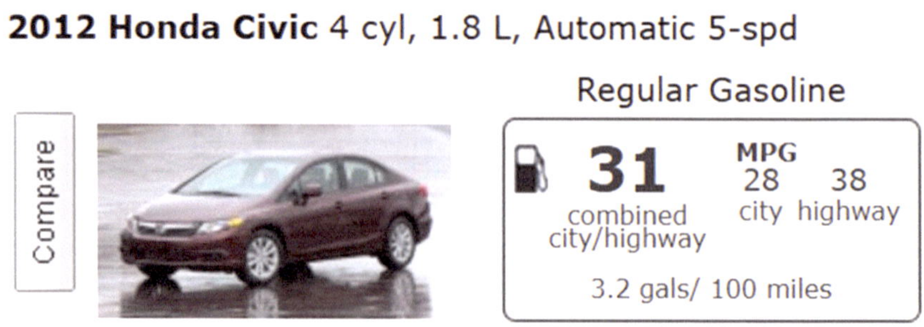

Fig. 7.6 2012 Honda Civic base, manual transmission. (From EPA https://www.fueleconomy.gov/feg/Find.do?action=sbsSelect. Public domain)

CO_2 Emissions of Ethanol Compared with Gasoline

Baseline: Gasoline (Regular Grade, Without Ethanol)

Same 2012 Honda Civic LX that gets 31 MPG = 50 km/gal.

The carbon dioxide produced by combustion of 1 US gallon of gasoline is 8877 g CO_2/gal gasoline (Fig. 7.6).[5]

$$\frac{\left(\frac{8887\text{g CO}_2}{\text{gallon}}\right)}{\left(\frac{50\text{ km}}{\text{gallon}}\right)} = 178\text{g CO}_2\text{ / km}$$

Ethanol (E100)

Density = 0.79 kg/L
Specific energy density = 26.8 MJ/kg

$$\frac{2.36\text{ MJ / km}}{26.8\text{ MJ / kg ethanol}} = 88.1\text{g ethanol / km}$$

Stoichiometric combustion:

$$C_2H_5OH + 3.5O_2 \rightarrow 2CO_2 + 3H_2O$$

$$24 + 6 + 16 = 46\text{ g ethanol} \rightarrow 2(12 + 32) = 88\text{ g CO}_2$$

$$\left(\frac{88.1\text{g C}_2\text{H}_5\text{OH}}{\text{km}}\right)\left(\frac{88\text{ g CO}_2}{46\text{ g C}_2\text{H}_5\text{OH}}\right) = 169\text{g CO}_2\text{ / km}$$ 5.1% reduction compared with gasoline.

[5] Data from 2020 US Code of Federal Regulations (CFR).

Power Output of Ethanol Compared with Gasoline

Baseline for Comparison: Gasoline (Regular Grade, Without Ethanol)

As calculated in Chap. 6,

Power output (manufacturer specifications [8]): 140 HP at 6500 RPM
Thermal efficiency found to be $\eta = 31.6\%$

Ethanol (E100)

Same air volume flow at 6500 RPM (assuming 100% volumetric efficiency): 97.5 L/s

Same air mass flow at 6500 RPM at NTP (20 °C, 1 atm): 117 g/s

Mass air-fuel ratio of stochiometric mixture is 6.45 g air/g methanol (calculated below*)

Fuel mass flow at 6500 RPM:

$$\left(\frac{117\text{g air}}{\text{s}}\right)\left(\frac{1\text{g ethanol}}{9.0\text{g air}}\right) = 13.0\text{g ethanol}/\text{s}$$

Amount of power in this fuel flow:

$$\left(\frac{13.0\text{g ethanol}}{\text{s}}\right)\left(\frac{26.8\text{ kJ}}{\text{g ethanol}}\right)\left(\frac{1\text{ kWs}}{1\text{ kJ}}\right)\left(\frac{1.34\text{ HP}_{\text{US}}}{1\text{ kW}}\right)\eta = 467 \times \eta\,\text{HP}_{\text{US}}{}^{*}\text{at NTP}$$

Assume same thermal efficiency as gasoline: 31.6%
Actual power output:

$$467 \times 0.316\text{ HP}_{\text{US}} = 148\text{ HP}_{\text{US}} \quad 5.7\%\text{ greater than gasoline.}$$

*Calculation of mass AFR for ethanol (also see Appendix 3):

Air contains 21% oxygen by molar volume.
Combustion oxygen/fuel molar ratio:

$$C_2H_5OH + 3O_2 \rightarrow 2CO_2 + 3H_2O \quad \Rightarrow 1\text{ mole of ethanol combines with 1.5 moles of oxygen}$$

Fuel/air ratio:

$$\left(\frac{1\text{mole ethanol}}{3.0\text{moles}O_2}\right)\left(\frac{46\text{g ethanol}}{1\text{mole ethanol}}\right)\left(\frac{0.21\text{mole }O_2}{1\text{mole air}}\right)\left(\frac{1\text{mole air}}{29\text{g air}}\right) = 0.111\text{g ethanol}/\text{g air}$$

$$\Rightarrow \text{AFR} = 9.0$$

Fuel Ethanol Production in the USA

While most countries in the world make fuel ethanol from crops with higher energy yields such as sugar cane, sugar beets, or sorghum, almost all commercial fuel ethanol in the USA today is made by fermentation of corn. The use of corn for the production of ethanol is aligned with its position as the dominant monocrop grown throughout much of the Upper Midwest region of the U.S. As explained by Peter Livingston[6] at the California Polytechnic University in San Luis Obispo, California, the ultimate alcohol yield is roughly proportional to the sugar content of the feedstock. Much of the corn grown for fuel does not quite meet the break-even sugar requirement, so it is beneficial to supplement the corn feedstock with some fraction of a high-sugar crop such as sweet sorghum, which is not itself used as a major fuel feedstock because it is more expensive and weather sensitive. This supplementation is not always considered in energy balance analyses.

On the positive side, traditional fermentation and agricultural methods for fuel ethanol production have seen significant improvements in efficiency since the early 2000s, as reported by the Renewable Fuels Association [9], a US trade association for ethanol and biodiesel producers. This has led to a change in the consensus of neutral evaluators that observed in the 1990s that the production of corn-fermented ethanol was energy negative. (The positive or negative energy status of a fuel is the difference between the additional energy required to produce the fuel and the energy value of the output fuel.) The current consensus is that the typical energy balance is positive due to improved technology. The current debate is now how energy positive it is, with values ranging from 30% to 300% reported by different evaluators with different biases.

Ethanol is currently blended as an additive to gasoline in 64 or more countries, including the Americas, Western Europe, Australia, and parts of Asia. In the USA, 98% of all "gasoline" sold at fuel stations is actually E10 or E15, a volumetric blend of (up to) 10% or 15% ethanol in gasoline. As previously mentioned, the technically sound reason is that the high octane rating of ethanol makes it an effective fuel additive for improving the rather low octane of refined gasoline, replacing previous additives, in particular, tetraethyl lead (TEL), a source of lead particulates, and methyl tertiary butyl ether (MTBE), which is toxic and nondegrading if leaked into groundwater aquifers even in very small quantities.

E10 or E15, being mostly gasoline, does not qualify as an alternative fuel under the Energy Policy Act of 1992 or any of its revisions in 2005, 2007, and 2010. E85 (up to 83% ethanol) is classified as a renewable fuel because of its plant origin and majority ethanol content, which means a *partially closed carbon cycle*. Like M85 that preceded it, this classification is considered by many to be questionable since the 17% or more gasoline content and the energy required to produce ethanol are of fossil origin.

[6] Interview with Peter Livingston, Head of the Bioresource and Agricultural Engineering Department, Cal Poly San Luis Obispo, California, USA. 26 January 2024.

Cellulosic Ethanol

There is a strong incentive to make ethanol from non-food crops. *Cellulosic ethanol* is considered the Holy Grail of ethanol production since it is produced from biomass (e.g., cultivated or waste biomass) with less environmental impact than corn. The obstacle, despite billions of dollars in government and private funds spent on research, is the lack of some mechanism (chemical, biological, radiological, or mechanical) to economically break down cellulosic plant cell walls to facilitate fermentation, albeit at fairly low yields [10, 11]. The most promising and heavily researched solutions have involved specialized enzymes to perform this step, but all are too costly for large-scale ethanol production. If economically sustainable, the use of cellulosic feedstocks for fuel ethanol would make it a much more energy-positive fuel and more justifiably considered a renewable fuel. But despite the enthusiasm of funding agencies and investment, the dream of cellulosic ethanol production remains elusive, economically noncompetitive with the fermentation of corn. As of late 2023, none of the ethanol commercially produced for fuel in the USA is cellulosic.

Cellulosic ethanol can also be produced thermochemically via methanol or syngas as an intermediary, but this approach involves additional energy conversion steps that do not favor efficiency or economics. But this has not stopped enterprising recipients of government-backed loans and subsidies from proposing this or other ethanol fuel production and processing schemes. One case that rose to national attention in 2010 was the startup company *Range Fuels*. Spanning two US presidencies and receiving \$160 million in investor financing, \$162.25 million in government commitments, a \$76 million DOE grant, and a \$6.25 million state grant, the DOE finally suspended payments after the facility produced no ethanol, only a few drums of impure methanol from the first step of their proprietary process [12]. But as is typical in government-funded projects, those involved exonerated themselves by blaming "Washington's failure to impose a tax on carbon via cap and trade" [12, 13].

This story is far from unique and serves as a lesson in politically motivated, optimistic funding of potentially significant but inadequately vetted proposals. Under public and media pressure to "take action" on climate change, billions of dollars of federal and state grants and loans have been dispensed without adequate unbiased technical review to charismatic entrepreneurs seeking to take advantage of easy funding and the investor herd mentality in all fields of alternative energy including biofuels, solar, wind, hydrogen, electric vehicles, energy storage, and even fusion energy.

Efforts continued following the Range Fuels debacle to find ways to utilize cellulosic feedstocks for production of ethanol in a cost-effective and environmentally sound way using syngas gasification methods. Federal, State, and Local political administrations desperate to show progress on energy independence, jobs, and climate change continued to encourage and approve massive taxpayer investments in hopeful projects without adequate oversight. Authorized by the Renewable Fuel Standard (RFS2) in the Energy Independence and Security Act of 2007, hundreds of millions of dollars in direct subsidies, RINs, and government-backed loans were awarded to entities that promised to deliver on hopeful promises of commercial scale cost-effective cellulosic ethanol from waste biomass, despite the lack of any validated advances in the technology. This widespread practice is sometimes called *subsidy harvesting.*

Consider another example of a well-intentioned but doomed-to-fail effort. Between 2007 and 2011, a joint venture between the British-based international chemical producer Ineos Group and the US venture capital firm New Planet Energy received over $125 million in US Federal grants and loan guaranties, and on 9 February 2011 announced groundbreaking for a large cellulosic ethanol facility located adjacent to a refuse disposal facility in Vero Beach Florida. The project would utilize solid waste to produce cellulosic ethanol using a proprietary process, disclosing only that it involved a combination of gasification and fermentation to produce "synthetic ethanol." In a press release on 31 July 2013, Ineos announced successful commercial-scale production with much fanfare [14]. But the announcement was premature as technical issues quietly emerged: the biomass "boiler" was producing syngas that contained cyanide that they attributed to the feedstock wood products containing too much moisture. This contaminated the proprietary bacteria used in the fermentation process that followed it. With additional funding, a large syngas scrubber was installed to remedy the problem. But by 2014, the facility was closed. No ethanol was ever produced. In 2016, the property and equipment were auctioned off. The loss was underwritten mostly by taxpayers. The expected employment promises made to the surrounding community amounted to only a small number of construction and administrative jobs.

The story has been often repeated with other recipients of government biofuel funding. Thirteen years later, commercial scale production of cellulosic ethanol remains an impractical money pit, and US fuel ethanol production continues 100% from corn fermentation.

Fig. 7.7 2010 Ford Escape Flex Fuel (E85) hybrid SUV at 2010 Washington Auto Show. (Photo: Mariordo Ortiz, https://commons.wikimedia.org/wiki/File:Escape_E85_Flex_Fuel_Hybrid_WAS_2010_8941.JPG. Attribution-Share Alike 3.0 Unported license)

E85 Distribution and Utilization in the USA

The slightly reduced energy content of E10 or E15 compared with E0 gasoline is of negligible concern to automobile drivers and manufacturers. It has such a minimal effect on fuel mileage that most consumers are not even aware that it is a component of the gasoline they are buying. It is an issue only in northern cold climates in which the increased ethanol content slightly degrades engine cold startability. Phase separation is a problem with alcohol–petroleum mixtures but rare when alcohol content is 10% or less. For E85, a blend containing up to 83% ethanol, it is a common problem that manifests as the stratification of an ethanol/water fraction to the bottom of a cold fuel tank if the fuel has absorbed small amounts of water. This is discussed below.

At this time (2023), the US government through the DOE and USDA is obligated to purchase more fuel ethanol than the actual demand for blending with gasoline or use as E85. The prediction error for years 2019 and later was based on the trends assumed in the original 2005 mandate, which were problematic because the Energy Policy Act committed the federal government to purchasing specific annual quantities at a fixed price rather than matching annual demand for ethanol. The fixed quantities, although effective for stabilizing agricultural revenues, were overly optimistic because they were based on legacy assumptions about continued exponential growth of demand for gasoline.

But demand for gasoline in the USA has declined well below these pre-EV predictions. This supply–demand imbalance has also been accentuated by significant improvements in gasoline fuel economy mandated by subsequent increases in Corporate Average Fuel Economy (CAFE) standards since 2005. Other factors include the introduction of gasoline–electric hybrids and battery–electric EVs or plug hybrids and the reduction in automobile travel during the 2020–2022 COVID-19 pandemic. Since approximately 2020, the USA has had a glut of subsidized ethanol, and this overproduction is expected to grow [15]. Partially relieving the excess production in the USA are exports of ethanol to Canada, Mexico, and Southeast Asia, which have all increased significantly, e.g., by 27% in just the 10 months prior to May 2023.[7] Consequently, US taxpayers are subsidizing and bearing the environmental and energy cost of crop-based ethanol exported to other countries. This is not a unique situation. The USA is also currently the largest exporter of liquid natural gas (LNG), which is partially subsidized by drilling leases on public land, and public support of the national natural gas infrastructure.

[7] *U.S. Ethanol Export Sales Kick Off 2023 with Large Gains while DDGS* Exports Scale Back.* Renewable Fuels Association, Trade Monitor, Mar 8 2023. Online at https://ethanolrfa.org/media-and-news/category/trade-monitor/article/2023/03/u-s-ethanol-export-sales-kick-off-2023-with-large-gains-while-u-s-ddgs-exports-scale-back. *DDGS are Dried Distillers Grains, a livestock feed supplement that is a byproduct of dry mill ethanol production.

In 2022, following years of pressure from agricultural interests to continue over-subsidizing corn ethanol, the US EPA modified the requirement for E10 allowing and encouraging the use of E15, a 15% volumetric ethanol blend. This increase justified an increase in government contract quantities from ethanol producers to 15 billion gallons per year, which consumes approximately 40% of all corn grown in the USA [16]. As a fuel, E15 is not appreciably different than E10 except in cold climates, where cold starting may be slightly more difficult, and the blend is more susceptible to phase separation.

Manufacturers in the USA introduced flex-fuel technology as an option on existing vehicle models starting in 1986. The first-generation FFVs were designed to run on methanol methanol–gasoline blends containing up to 85% methanol (M85). They would also run just as well on any ethanol/gasoline blend. But in the late 1990s, following intense lobbying and promotion of energy independence in the face of instabilities in Middle Eastern politics, the federal mandate shifted exclusively to ethanol as both the 10% gasoline additive in E10 and the official alternative fuel E85, nominally a volumetric blend of 85% ethanol and 15% gasoline. "Nominal" refers to the actual specifications for E85, defined in US CFR §1090.80 as [17]:

> E85 means a fuel that contains *more than 50 volume percent but no more than 83 volume percent ethanol* and is used, intended for use, or made available for use in flex-fuel vehicles or flex-fuel engines. (italic emphasis added)

Consequently, under the US CFR, E85 could more accurately be named E(51–83), or possibly E67, the mean value in the allowable range. This variability affects the consumer experience in terms of a user's observed mpg and the risk of phase separation, which is much more likely in ethanol/gasoline blends between 20% and 80% ethanol. To assure that phase separation does not occur prior to transferring fuel into the underground tanks at E85 fuel stations, the ethanol is added to the gasoline at the last opportunity: the fuel is "splash blended" at the local distributor or at the fueling station, e.g., added to gasoline already in the underground tank. This also aids in the identification of the exact amount of ethanol delivered to each retailer, assuring the correct subsidy credit to the distributor or retailer.

The justification for this wide blending range in the federal legislation [14] is the allowance of seasonal fuel adjustment, recognizing that cold starting is a known problem with alcohol fuels. In colder months, retailers can reduce the ethanol percentage to avoid engine starting problems [18]. However, as can be seen from the Gibbs phase diagram of Fig. 7.8, a 70/30 blend of ethanol-gasoline containing even a small amount of water has a greater likelihood of phase separation than a 83/17 blend in cold weather [19, 20]. The phase separation thresholds for various gasoline–ethanol–water mixtures at -10, 20, and 60 °C are plotted in Fig. 7.8. A laboratory example of phase separation in an ethanol–gasoline blend is shown in Fig. 7.9. Stratification of the gasoline above the ethanol layer can occur in either the underground fuel storage tank or in the vehicle fuel system, but since the temperature of the underground tank is relatively stable and the tank contents are less exposed to humidity, low-temperature phase separation is more likely to occur in a vehicle that has been sitting unused for a few days than in an underground storage tank. Duration of exposure is also a factor, with E85 stored for several months absorbing more

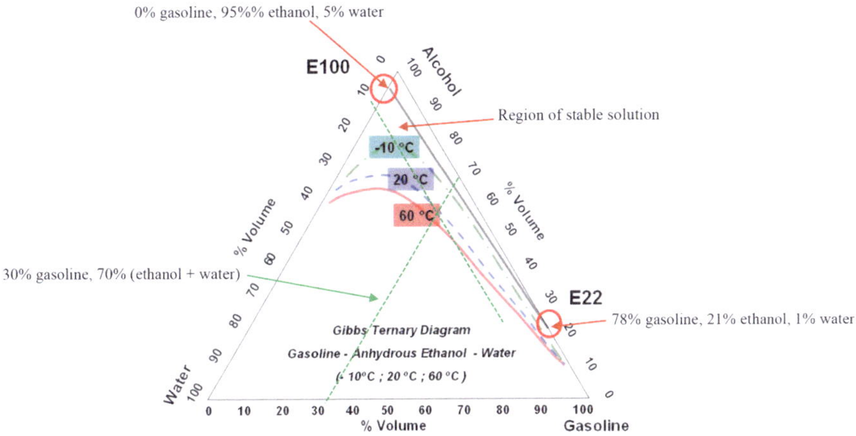

Fig. 7.8 A Gibbs ternary phase stability diagram that identifies phase separation thresholds of ethanol-gasoline-water blends at 60, 20, and −10 °C. The solid line between the small circles at E22 (22% ethanol by mass) and E100 (100% ethanol) represents slightly hydrous ethanol containing 5% water in gasoline/ethanol blends. Green dashed line represents the fuel sample from Fig. 7.9, which shows that the ethanol had absorbed at least 10% of its volume in water from the air. (Gibbs diagram from Hans Keukens. *Hydrous Ethanol in Gasoline: A More Sustainable and Cost Effective Solution*. 2007 (www.e15blends.com) Contained in Orlando Volpato Filhoas, "GASOLINE C MADE WITH HYDROUS ETHANOL", XVI SIMEA 2008. Public domain)

atmospheric humidity. This was previously a common criticism of M85 during the AFUP years. As a practical solution, phase separation can be avoided by not leaving unused E85 in the fuel tank for long periods, and by switching to E10 or E0 (where available) during winter months.

In 2009, a study by Troels Johansen and Jesper Schramm at the Danish Technological Institute [21] observed that as a result of gasoline formulation changes in the approximately 15 preceding years, the phase-separation tolerance for hydrous ethanol–gasoline flex-fuel blends has improved somewhat, and that ethanol, even containing only 5% water by mass, could give significant savings in the production cost and energy use. (A large fraction of the production energy required for fuel ethanol is for the final distillation steps, which are necessary to assure the 1% water limit.) The study concluded that blends containing ethanol with water content above the ethanol/water azeotrope (4.4% water by mass) can be used as blends with gasoline at temperatures above −25 °C (−13 °F), without phase separation occurring. Specifically:

> It was found that an ethanol fuel blend can use ethanol with a purity in the 94–95% purity range and still be acceptable as a flex fuel together with pure gasoline, even as low as −25 °C, and in any conceivable mixture which could occur in the fuel tank of a flex fuel vehicle, without phase separation occurring.

This observation, if borne out in practice, factors favorably for ethanol–gasoline flex-fuel blends but with some risk of unfavorable results due to the high variability of gasoline/ethanol blends worldwide.

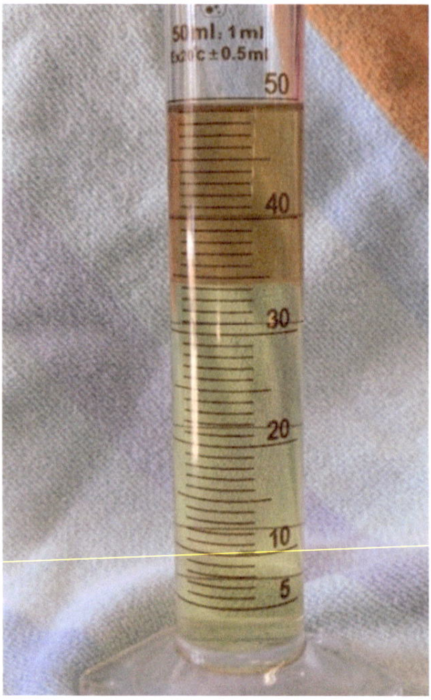

Fig. 7.9 Phase separation of a sample of unused E85 from the fuel tank of a flex-fuel vehicle (FFV) after allowing it to settle for 15 min at 23 °C, 80% RH. As can be seen in the photo, the sample is actually a volumetric blend of 70% (ethanol + water) and 30% gasoline. The lower stratum is ethanol with absorbed atmospheric water. The upper is gasoline. Why the 70:30 ethanol:gasoline ratio instead of 85:15? (1) In the USA, E85 may legally contain any ethanol volume fraction between 50% and 83%. The actual composition need not be disclosed to the buyer. (2) An FFV fuel tank that has previously been filled with E10 will contain some residual gasoline at the time it is refilled with E85, reducing the ethanol fraction. (By author)

Energy Sustainability

As discussed previously, the energy sustainability of corn-based ethanol has been a politically charged topic since the inception of government incentive programs in the late 1980s in the USA. It was a point of debate in the passage of the US Energy Policy Act in 2005. Does corn ethanol provide more fuel energy than the energy required to produce it? Countless studies have been performed by government agencies, commercially funded consulting firms, and advocacy groups on both sides of the question. The conclusions of any given study are consistently correlated with the economic interests or political ambitions of the entity performing or funding the study. The differences in conclusions are easily manipulated by the design and scope of the analysis (what to include and what not to include) and the state of technology at the time.

To be clear, a fuel is considered either *energy positive or negative* by comparing the total amount of energy that is *required to make the fuel*, compared with the usable energy content of the fuel being produced. A negative result does not mean that the fuel provides no useful energy when used in a vehicle; it means that the sum of all forms of energy required to produce the fuel exceeds its fuel value. Therein lies the ambiguity that has led to a wide range of different conclusions: what to include and what not to include as "production energy", and how to factor in the carbon capture or byproduct benefits of the feedstock. This metric came to the public's attention gradually over the past 30 years, relevant only to biofuels, which were portrayed as the epitome of renewable because the energy source is vegetation that captures CO_2 from the atmosphere. But the additional fuel energy (natural gas and diesel fuel) required to grow and convert crops into motor fuel eventually became part of the discussion. For corn ethanol, this ratio has been a contentious topic, predominantly driven by study objectives not related to either energy or environmental sustainability.

It seems obvious that a fuel should be energy positive to justify making it. Otherwise, there is no net energy benefit but all of the environmental and economic consequences. But these are not the only considerations when it comes to government policy and incentives. Economic and other societal benefits that are difficult to quantify are ultimately involved in all US public policy decisions. Initial large-scale production of corn ethanol in the late 1980s was found by almost all independent analyses to be energy negative or at best energy neutral. The resulting ethanol fuel contained no more energy than the additional energy that was required to produce it. But it was still deemed "good for America", justified as supporting farming and ostensibly, energy independence. Following earlier discussion, with process and cultivation improvements over the past 20–30 years, ethanol is now generally agreed to be energy positive by margins varying from 30% to 300%, depending on the study design and included data sources. One prominent metastudy (study of many other studies) in the 1990s concluded that

> … with ethanol about 1.7 gallons of ethanol energy equivalents are required to produce 1 gallon of ethanol. [22]

But this negative energy situation could be explained by the fact that early fuel ethanol production from corn relied on traditional agricultural methods that had been used for food/feed corn, and that production facilities which were designed for beverage ethanol were not optimized for production of fuel ethanol [23]. More recent (post-2010) studies have pointed out that higher per-acre crop yields and more efficient processing methods have reduced ethanol production energy by at least 20% [24] and have included the utilization of biomass for process heat in the newest facilities. As mentioned previously, the current (2022) position of the *Renewable Fuels Association* (a trade association representing ethanol producers) is that ethanol production from corn provides 2.8–3.0 times as much energy as the additional energy required to produce it from corn [9]. The last statement by the USDA that was published online in 2016 concluded a 2.1–2.3 positive energy balance depending on the inclusion of the energy value of the distillers dried grain (DDG) byproduct that is used as a noncaloric supplement in cattle feed [25]. But it is worth noting that almost all of the energy required in the cultivation and production of fuel ethanol, from tractor fuel to heat for drying and distillation, is in the form of fossil fuels, most commonly diesel fuel or natural gas, although dried corn stover biomass now supplements the heat requirements in newer facilities.

Some USDA publications are optimistic that, not including farm cultivation energy requirements,

> There is a significant potential for a 30-fold improvement in energy balance by using biomass (stover)-powered refineries. [26]

But there remains skepticism in the scientific community, especially outside of the USA, about the energy and environmental sustainability of corn ethanol compared with other feedstocks. The 2022 German *Handbook of Fuels: Energy Sources for Transportation* [27] looked at a wide range of crops from which ethanol could be fermented and assessed the full-cycle energy benefits and the GHG emissions relative to gasoline. The results are shown in Fig. 7.10. For climate reasons, corn (maize) ethanol is produced more efficiently in the USA compared with the EU and is energy positive, but corn provides the lowest net energy yields and GHG reduction of all feedstocks except barley and rye. Their analysis concluded a positive energy gain of approximately 34%, but a negative GHG impact, although still 21% better than gasoline. Several documentary videos have been produced that

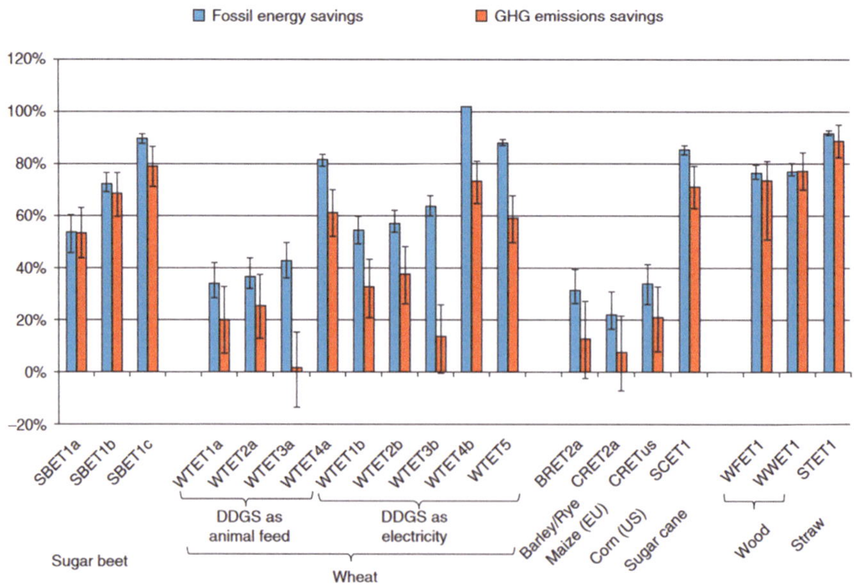

Fig. 7.10 Energy and GHG benefit of various crops for ethanol production. (Chart copied from Handbook of Fuels: Energy Sources for Transportation. Ethanol. 9.12 WTW Analysis. Wiley-VCH, Weinheim, Germany, 2022 (citation 37) Fig. 9.11. Used under STM guidelines)

support concerns about the environmental impact and energy balance of corn ethanol as well as biodiesel for example, the June 2021 Irish documentary *The Problem with Biofuels* [28] available on YouTube.

A concise statement in citation [29] summarized the energy sustainability of corn ethanol in the USA as of 2024:

> The only way ethanol makes sense is as a political issue.

Environmental Sustainability

This topic is even more controversial than the energy balance. Does fuel ethanol, as currently produced, generate more or less environmental harm compared with fossil fuels or other alternative energy options for transportation? Until just the past few years, this question had been largely avoided, since energy security, jobs, and the robustness of agriculture have dominated almost all conversations. But with growing public awareness of climate change (temperature records, drought, floods, extreme weather events), the conversation is now out in the open.

As with the energy balance debate, studies of the overall climate impact of corn ethanol have generated a wide range of conclusions based on the choice of factors included and data sources. The inclusion or exclusion of land conversion impacts of fuel farming and inclusion or exclusion of anticipated cellulosic ethanol production appear to be the largest of these differential factors. A recent study first-authored by Tim Searchinger at Princeton University concluded that the overall carbon balance is 24% negative, while a 20% positive result is emphatically claimed by the Renewable Fuel Association. The positions of both the Princeton research team and the RFA are summarized succinctly in citation [30].

Ethanol contains a higher hydrogen-to-carbon atomic ratio, H:C = 3:1 than the average of the hydrocarbons in gasoline, approximately 2:1. The H:C ratio determines the molar ratio of water to carbon dioxide resulting from combustion, so that compared with gasoline, ethanol exhaust composition is skewed more toward water than carbon dioxide. More water means less CO_2 for a given quantity of fuel energy. Consequently, the combustion of ethanol, like all alcohols (aka oxygenated fuels), contributes less of this primary greenhouse gas to the atmosphere. It is the overall process of creating fuel ethanol from starch crops that leads to conclusions about the lack of environmental sustainability for corn ethanol [31]. One of the major differences in competing analyses of the climate impact of corn ethanol is the *carbon opportunity cost* [32] of the land cleared to grow corn.

A 2022 metastudy conducted by Reuters International concluded [33]:

> Corn-based ethanol, which for years has been mixed in huge quantities into gasoline sold at U.S. pumps, is likely a much bigger contributor to global warming than straight gasoline.
>
> The study, published in the Proceedings of the National Academy of Sciences, contradicts previous research commissioned by the U.S. Department of Agriculture (USDA) showing ethanol and other biofuels to be relatively green.

Also mentioned in citation [29] was a concurrent study from the University of Wisconsin-Madison Center for Sustainability and the Global Environment:

> "Corn ethanol is not a climate-friendly fuel," said Dr. Tyler Lark, lead author of the study.

The research, which was funded in part by the National Wildlife Federation and U.S. Department of Energy, found that ethanol is likely at least 24% more carbon-intensive than gasoline due to emissions resulting from land use changes to grow corn, along with processing and combustion in an engine.

The counterargument from the RFA:

> Geoff Cooper, president and CEO of the Renewable Fuels Association, the ethanol trade organization, called the study "completely fictional and erroneous," arguing the authors used "worst-case assumptions [and] cherry-picked data."

The background behind the differing conclusions was discussed in the Reuters article, excerpts below [32]:

> Under the U.S. Renewable Fuel Standard (RFS), a law enacted in 2005, the nation's oil refiners are required to mix some 15 billion gallons of corn-based ethanol into the nation's gasoline annually. The policy was intended to reduce emissions, support farmers, and cut U.S. dependence on energy imports.
>
> As a result of the mandate, corn cultivation grew 8.7% and expanded into 6.9 million additional acres of land between 2008 and 2016, the study found. That led to widespread changes in land use, including the tilling of cropland that would otherwise have been retired or enrolled in conservation programs and the planting of existing cropland with more corn, the study found.
>
> Tilling fields releases carbon stored in soil, while other farming activities, like applying nitrogen fertilizers, also produce emissions.
>
> A 2019 study from the USDA, which has been broadly cited by the biofuel industry, found that ethanol's carbon intensity was 39% lower than gasoline, in part because of carbon sequestration associated with planting new cropland.
>
> But that research underestimated the emissions impact of land conversion, Lark said.
>
> The U.S. Environmental Protection Agency, which administers the nation's biofuel policy, is considering changes to the program. Under the RFS, Congress set blending requirements through 2022, but not beyond, giving the EPA authority to impose reforms. EPA plans to propose 2023 requirements in May.

As of mid-2024, the most recent development in the ethanol environmental impact debate was the November 6, 2023 public release of a report by the EPA Science Advisory Board [34] which concluded, with abundant political sensitivity:

> …the SAB [Science Advisory Board] finds that resolving the scientific question of whether corn starch ethanol reduces emissions or not, relative to gasoline and diesel, is absolutely central to determining whether the EPA is implementing and enforcing an RFS that has net climate benefits, or one that has neutral climate impacts, or even has net climate damages. Some, though hardly all, recent studies published in peer-reviewed journals, including the *Proceedings of the National Academy of Sciences*, conclude that there are minimal or no climate benefits from substituting corn ethanol for gasoline or diesel.

A politically neutral assessment referenced by the EPA was a metastudy published in the Proceedings of the National Academy of Sciences (PNAS) in 2022 that assessed 20+ years of US experience with fermented ethanol, concluding [35]:

> Agricultural interest groups promote corn ethanol as an environmentally beneficial alternative to gasoline, but many independent scientists have long questioned this view. Nevertheless, the United States has aggressively pursued measures to expand biofuel production. The key policy has been the Renewable Fuel Standard (RFS2) in the Energy Independence and Security Act of 2007, which requires greater use of ethanol up to the current level of 15 billion gallons annually. Ethanol proponents have argued that this reduces greenhouse gas (GHG) emissions, but as Lark et al. [36] show in PNAS, the opposite occurs. The authors find that the life-cycle GHG emissions of the ethanol produced to meet RFS2 are no less than those of gasoline, and are likely even greater. This is because using more corn for biofuel has led to an increase in the intensity and extent of corn farming in the United States. Thus, RFS2 not only fails to mitigate climate change but is actually counterproductive. Furthermore, the authors conclude that RFS2 has exacerbated other environmental problems commonly associated with row crop production, including poor water quality and soil erosion.

The appeal of *energy independence* has always been a powerful theme in the US politics. Proponents and media emphasize local job creation and patriotism as much or more than energy balance or environmental impacts. Consequently, ethanol has traditionally had the support of US voters on both extremes of the political spectrum,[8] No savvy politician will go on record questioning the energy balance or renewability of ethanol since to do so would likely be political suicide.

Economic and Political Sustainability

Perhaps the only actual benefit of subsidized fixed-price fixed-quantity guarantees for fuel corn has been the reduction in traditional USDA programs that pay farmers to NOT grow corn as a means to actively stabilize market prices, the fluctuations of which could be devastating to smaller farming operations. Corn ethanol subsidies and purchase guaranties serve the important function of assuring the critical financial well-being of US agriculture and the food supply. From this point of view, questions about energy or economic sustainability might be considered irrelevant, at least until climate-related issues reach a level of concern that force changes in national and state priorities.

However, an alternative view of the impact of ethanol on food production is the "food vs. fuel" dilemma of using a food/feed crop to make fuel [37]:

> One of the most significant criticisms of ethanol production is its impact on food supply. Using large quantities of corn for fuel can drive up food prices and contribute to food short-

[8] In the USA, the designation *Red* is given to the mostly rural, conservative states in which the US Republican Party dominates electoral results. The *Blue* designation is given to the more urban, liberal Democratic Party dominated states.

ages. Corn that could be used to feed people and livestock is instead diverted to fuel production, exacerbating global food insecurity.

And regarding the completeness of previous studies assessing environmental impacts, e.g., Yi Yang et al. in the Journal of the American Chemical Society [38],

Results show that E85 does not outperform gasoline when a wide spectrum of impacts is considered. If the impacts are aggregated using weights developed by the National Institute of Standards and Technology (NIST), overall, E85 generates approximately 6% to 108% (23% on average) greater impact compared with gasoline, depending on where corn is produced, primarily because corn production induces significant eutrophication impacts and requires intensive irrigation. If GHG emissions from the indirect land use changes are considered, the differences increase to between 16% and 118% (33% on average).

On a gasoline-equivalent energy basis, E85 should cost 20.5% less per gallon than E10. Therefore, subsidies are usually set to allow E85 retail prices to track nearly this percentage less than the retail cost of gasoline per gallon [39]. However, most drivers perceive higher costs per mile if they operate their flex-fuel vehicles on E85. In the USA, the adoption of E85 as a practical motor fuel has been minimal with less than 38% of flex-fuel vehicle owners having ever used E85, and 13% unaware that they own a flex-fuel vehicle [40]. Sales of E85 are so small that many municipalities have no fuel stations that dispense E85, even in some midwestern US states that economically benefit from ethanol use. At the peak of interest in ethanol blends in 2014, there were 3,354 refueling stations in the USA with E85 pumps. This represented approximately 2.4% of refueling stations nationwide according to data from the Oil Price Information Service [41]. According to a 2011 report in Biofuels Digest, USA, E85 sales represented just 0.04% of total US gasoline (E10 or E85) sales, and only 3% of sales at fueling stations that do offer E85 [42]. In 2023, this trend has continued, with now only a trivial amount of E85 sold to consumers. There is growing concern about the inability of corn-derived ethanol fuel to ever reach economic parity with gasoline or other fuel alternatives without continued subsidies [43] (Fig. 7.11).

Traditionally, farming has been the backbone of the US economy. According to [44], family farms accounted numerically for 89% of all US farms. According to the USDA in 2022, just over 60% of farm land is owner operated [45]. The survival of family owned and operated farms is critical to the culture and identity of the nation. This makes it more important than ever that taxpayer support of these small operations actually support them, as opposed to the large agribusinesses that are the dominant recipients of federal and state subsidies. This trend is strongly evident in the corn ethanol industry that has become increasingly consolidated over time. Eight years ago, farmers owned more than 40% of all ethanol production facilities; today their share is just 16–19% [46]. As the largest industry in the USA, agriculture in general has come to rely heavily on government/taxpayer support, and this is particularly the case for corn grown for fuel. The publication and media proliferation of arguments appealing to jobs, energy independence, and even patriotism have helped to cause confusion and entrenched positions by the general public. Hundreds

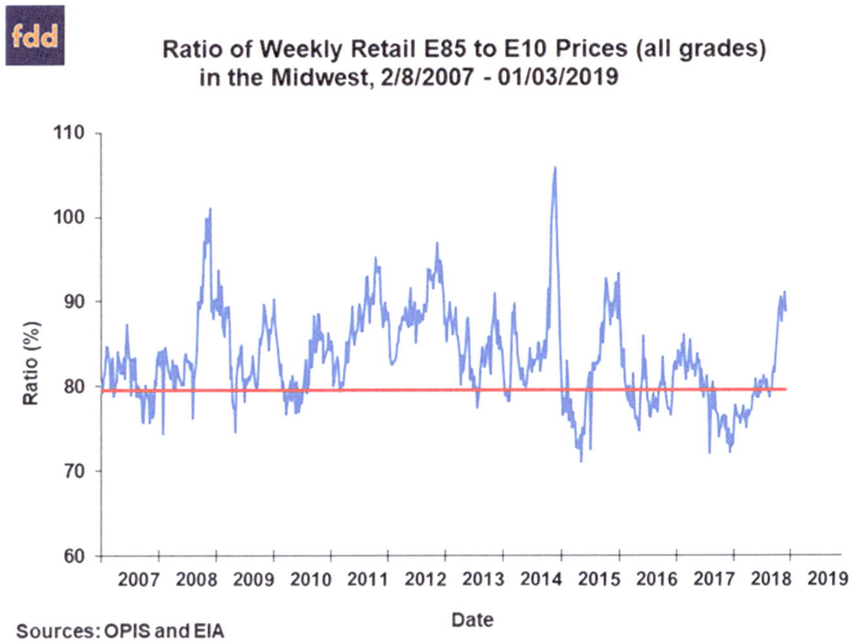

Fig. 7.11 Subsidized retail cost of E85 relative to energy-equivalent cost of E10 gasoline in selected midwestern US states. Target ratio is shown by red line. (Reported by the US EIA, 2019. Public domain)

of expert reports, scientific papers, web sites, and even YouTube videos can be cited. It is impossible to aggregate all arguments into a composite summary.

Most farm equipment is diesel powered, and biodiesel is the direct substitute for D2. But biodiesel blends greater than 20% (B20) are prohibited by manufacturers in all modern diesel engines. It was only in 2023 that a small number of ag equipment manufacturers started building farm equipment that could run on ethanol. In 2023 and 2024, US manufacturer John Deere displayed at Ag Exhibitions a prototype ethanol-powered tractor, and the Dutch manufacturer Mazzotti has demonstrated an ethanol-powered self-propelled sprayer [47].

In the USA, federal government subsidies for corn ethanol have been in place since 1978, starting with the Energy Tax Act of 1978 that allowed a 4 cents-per-gallon reduction in the federal excise tax on gasohol or E10. They have increased continuously over the past 45 years. By 2018, corn ethanol agriculture operations and related business interests had received over $86 billion USD in taxpayer-funded subsidies, with $5.8 to $7.0 billion in 2006 alone [48]. This number does not include state or local subsidies, federal and state incentives to vehicle manufacturers, fuel distributors, and rebates to purchasers of E85 flex-fuel vehicles. In some Midwestern US states such as Minnesota, state subsidies and blending requirements have been even more lucrative than federal expenditures.

The US Renewable Fuel Standard (RFS) program was authorized under the Energy Policy Act of 2005 and expanded under the Energy Independence and

Security Act of 2007. It specified and underwrote the annual purchase of 34 billion liters of fuel ethanol in 2008. In 2024, fuel suppliers were required to blend approximately 57 billion liters of corn ethanol into gasoline to meet mandated targets for expanded use of higher-percentage E15 ethanol blends.

The RFS program directs the Environmental Protection Agency (EPA) to set annual renewable volume obligations (RVOs), which specify the projected demand for ethanol blended into the automotive fuel supply. The accounting process is a bit convoluted: Ethanol producers create renewable identification numbers (RINs) for each gallon of ethanol produced for fuel use. RINs are market commodities that are traded, sold, or used by obligated parties (such as oil refiners) to demonstrate compliance with the RFS. As explained by the EPA in November 2023,

> To implement the RFS program, EPA tracks production and use of qualifying renewable fuel using Renewable Identification Numbers or RINs. These RINs are generated by renewable fuel producers or importers and are bought and sold "attached" to the renewable fuel until the fuel is purchased by an "obligated party" (a refiner or importer of gasoline or diesel fuel) or blended with petroleum-based transportation fuel. At that point the RIN is "separated" from the fuel and may thereafter be independently bought or sold until it is retired to meet an obligated party's renewable volume obligation [49].

From the same EPA source, for D5 and D6 (noncellulosic ethanol, per gallon), the average RIN market price for the 3-year period 2021–2023 was $1.54 USD, varying from $0.84 to $1.88 USD [33]. Like carbon credits, RINs incentivize ethanol production using these credits that buy the right to produce and sell nonrenewable fuels, usually gasoline or diesel fuel. This is a proven method for providing financial incentives without direct cost to the government and taxpayers. But the climate benefit of this incentive is dubious since, in the end, an RIN is just a salable right to pollute.

One point that is not debated by either advocates or detractors is that the fuel ethanol industry in the USA exists only because of government subsidies that, if terminated, would make the enterprise unprofitable regardless of technological or production advances. Subsidies have created an artificial market in which the net energy balance and carbon intensity are only ancillary considerations. If taxpayer subsidies were taken away, the agriculture ethanol industry in the USA would immediately collapse, with significant ripple effects throughout the US economy.

Current Status of the E85 Flex-Fuel Program

The E85 flex-fuel program in the USA remains active today, although only a small number of vehicles are still available in the USA with the flex-fuel option; all are trucks or large SUVs: Chevrolet Silverado, GMC Sierra, Ford Explorer, Dodge RAM 1500, Nissan Titan, and Jeep Grand Cherokee. As of 2022, the USA remained the second largest flex-fuel *vehicle* market in the world (second only to Brazil), yet vehicles capable of using E85 accounted for only 10.8% of all cars on the road, and *most have never been operated on E85*. Only 3% of all gasoline stations in the USA sell E85, and many of these relegate E85 dispensing to a (often nonworking) fuel pump, maintaining it simply for tax benefits despite the lack of demand. In this author's experience in California, the state with the largest flex-fuel vehicle population, it is not possible, even with advanced planning, to drive exclusively on E85 further than the vehicle's single-tank range due to the lack of an adequate distribution of operational E85 stations.

Commentary

Highly polarized opinions dominate public perceptions and government policies associated with corn ethanol in the USA. These cannot be understated. But few would disagree that ethanol production from corn is certainly the most politically blessed of all the alternative fuel options. Like biodiesel fuel, the agricultural biofuel industry, with the support of the USDA, has long benefited from the conflation of agricultural food production and combustion fuel production, with the sympathy of the general population.

As an engine fuel, ethanol has desirable attributes: high octane, energy density less than gasoline but greater than methanol, reduced CO_2 emissions and increased power output compared with gasoline, low toxicity, and improved fire safety. Ethanol in a 10% blend with gasoline (E10) is an excellent replacement for the highly toxic octane additives previously used in gasoline. This application dominates ethanol consumption because it is required by Federal law under the Renewable Fuel Standard. In 2019, the blend limit was allowed to increase to 15%, at the discretion of individual states. As of 2025, eight states, all in the Midwest, have adopted E15 to replace E10. Fuel crop cultivation and processing have contributed to rural employment and the economic stability of agriculture, arguably most important industry in the USA.

But corn ethanol motor fuel in the form of E85 has never been economically sustainable without large taxpayer subsidies. While the same can be said for all cultivated (first generation) biofuels, ethanol has been the largest (by far) recipient of public investment. Initial energy balance concerns have subsided over the past 20–30 years due to advancements in production technology and more efficient agricultural practices, shifting from negative in the 1990s to now significantly positive, although still financially difficult to justify. But there is a nearly uniform consensus from independent evaluations that the net environmental impact remains negative, with impacts for global warming, water use, agricultural chemical pollution, and multiple land use consequences. So as a remediation strategy for climate change, it does more harm than good while continuing to be promoted as an environmental savior. The benefit versus detriment arguments have become too politically charged for rational public discourse. There may still be attainable future improvements, such as the possibility of a breakthrough in cellulosic ethanol production. But at this time the only clear sustainability advantage of corn ethanol appears to be political sustainability.

References

1. T. Thurnell-Read, M. Monaghan, Historical origins and emergence of intoxication, in *Intoxication*, (Palgrave Macmillan, Cham, 2023). https://doi.org/10.1007/978-3-031-19171-8_2
2. S. Osborne, US Department of Commerce, Energy, Assessing the economic effects of commercialization of cellulosic ethanol (2020), https://www.trade.gov/sites/default/files/2020-12/Energy%20in%202020_Assessing%20the%20Economic%20Effects%20of%20Commercialization%20of%20Cellulosic%20Ethanol.pdf
3. California Code of Regulations, Final Regulation Order, Article 5, Approval of systems designed to convert motor vehicles to use fuels other than the original certification fuel or to convert motor vehicles for emission reduction credit or to convert hybrid electric vehicles to off-vehicle charge capable hybrid electric vehicle § 2031, Alcohol or Alcohol/Gasoline Fuels Retrofit Systems, https://ww2.arb.ca.gov/sites/default/files/barcu/regact/2013/altfuel2013/altfuel2013fro.pdf. Accessed 20 June 2024
4. California Air Resources Board, Alternative fuel conversion certification procedures, https://ww3.arb.ca.gov/regact/2013/altfuel2013/altfuel2013.htm. Accessed 20 June 2024
5. California Air Resources Board communication with Rousch Industries Inc (2024), https://roushcleantech.com/wp-content/uploads/sites/all/themes/roushcleantech/pdf/certifications/2014MY_6.8L_CARBretrofit.pdf
6. Industry Week (online edition), Why America has a shortage of skilled workers (2015), https://www.industryweek.com/talent/education-training/article/22007263/why-america-has-a-shortage-of-skilled-workers. Accessed 20 June 2024
7. Snap-on Diagnostics, Fuel pump progress—brushless motor, https://www.snapon.com/EN/US/Diagnostics/News-Center/Fuel-Pump-Progress-Brushless-Motor. Accessed 20 June 2004
8. Edmunds, Used 2012 Honda Civic—specifications and features, https://www.edmunds.com/honda/civic/2012/features-specs/. Accessed 1 June 2024
9. Renewable Fuels Association, Corn ethanol's energy balance continues to improve (2022), https://d35t1syewk4d42.cloudfront.net/file/2214/Ethanol%20Energy%20Balance%20Update%20April%202022.pdf. Accessed 5 June 2024
10. M. Asif et al., Renewable and alternative energy resources, in *Chapter 9: Renewable Energy from Biomass*, (2022), https://www.sciencedirect.com/topics/engineering/cellulosic-ethanol
11. A. Azarpour et al., Chapter 2: Current status and future prospects of renewable and sustainable energy in North America: progress and challenges, in *Energy Conversion and Management*, (2022), https://www.sciencedirect.com/topics/engineering/cellulosic-ethanol
12. Biofuel Digest (2023), https://www.biofuelsdigest.com/bdigest/2011/12/05/the-range-fuels-failure/
13. The range fuels Fiasco, Wall Street Journal (2011), https://www.wsj.com/articles/SB10001424052748704364004576132453701004530. Counterpoint in range fuels is a good energy investment for taxpayers, Wall Street Journal Opinion, (2011), https://www.wsj.com/articles/SB10001424052748703373404576148493142478596
14. Ineos, First facility in the world using new advanced bioenergy technology to convert waste to renewable fuel and electricity, https://www.ineos.com/news/ineos-group/ineos-bio-produces-cellulosic-ethanol-at-commercial-scale/. Accessed 12 Oct 2024
15. H. Zaremba, U.S. doubles down on ethanol despite glut, OilPrice.com, the No. 1 Source for Oil and Energy News (2018), https://oilprice.com/Energy/Energy-General/US-Doubles-Down-On-Ethanol-Despite-Glut.html
16. *EPA Raises Amount of Ethanol that Must be Blended with Gas* (Associated Press, as published in US News and World Report, 2022)
17. US Code of Federal Regulations, Part 1090—Regulation of fuels, fuel additives and regulated blendstocks, https://www.ecfr.gov/current/title-40/chapter-I/subchapter-U/part-1090. Last amended 7 June 2024
18. Velcor Leasing Corporation, *E85 Ethanol Q&A*, https://www.velcor.com/E85EthanolFAQ.pdf. Accessed 10 June 2024

19. R.L. McCormick, National Renewable Energy Laboratory, Letter September 2016. Re: Water uptake of ethanol-gasoline blends in humid environments, https://d35t1syewk4d42.cloudfront.net/file/362/Hydroscopicity-Study-Summary-09-06-16.final_.pdf. Accessed 10 June 2024
20. E.D. Christensen, R.L. McCormick, *Water Uptake and Weathering of Ethanol-Gasoline Blends in Humid Environments* (National Renewable Energy Laboratory), NREL TSA 14-665 for the Renewable Fuels Association, https://d35t1syewk4d42.cloudfront.net/file/589/Evaluation-of-Water-Uptake-by-Ethanol-RFA-09-16.pdf. Accessed 10 June 2024
21. T. Johansen, J. Schramm, Low-temperature miscibility of ethanol-gasoline-water blends in flex fuel applications. Energy Sources A **18**, 1634–1645 (2009). https://doi.org/10.1080/15567030903021897
22. David Pimentel, Ethanol fuels: energy security, economics, and the environment (1996), https://link.springer.com/content/pdf/10.1007/BF02229143.pdf. Accessed 5 June 2024
23. H. Shapouri, J.A. Duffield, M. Wang, *The Energy Balance of Corn Ethanol: An Update*. Agricultural Economic Report No. 813 (U.S. Department of Agriculture, Office of the Chief Economist, Office of Energy Policy and New Uses, 2002)
24. *Ethanol vs. Petroleum-Based Fuel Carbon Emissions* (Office of Energy Efficiency and Renewable Energy, 2022), https://www.energy.gov/eere/bioenergy/articles/ethanol-vs-petroleum-based-fuel-carbon-emissions
25. US Department of Agriculture, Office of the Chief Economist, Office of Energy Policy and New Uses, 2015 Energy balance for the corn-ethanol industry (2016), https://www.usda.gov/sites/default/files/documents/2015EnergyBalanceCornEthanol.pdf. Accessed 15 June 2024
26. PW Gallagher, WC Yee, HS Baumes, Energy balance for the corn-ethanol industry (2015), https://www.usda.gov/sites/default/files/documents/2015EnergyBalanceCornEthanol.pdf
27. B. Elvers, A. Schutze, *Handbook of Fuels: Energy Sources for Transportation. Ethanol. 9.12 WTW Analysis* (Wiley-VCH, Weinheim, Germany, 2022), https://ebookcentral.proquest.com/lib/calpoly/reader.action?docID=6727034&ppg=91
28. B. McManus et al., Real engineering, in *The Problem with Biofuels*, (2021), https://www.youtube.com/watch?v=OpEB6hCpIGMby
29. G. Max Graham, Can ethanol fuel ever live up to its hype? (2024), https://gizmodo.com/can-ethanol-fuel-ever-live-up-to-its-hype-1850414195. Accessed 1 June 2024
30. Inside Climate News, Science, Corn-based ethanol may be worse for the climate than gasoline, a new study finds, https://insideclimatenews.org/news/16022022/corn-ethanol-gasoline-climate-change/. Accessed 1 June 2024
31. M. Cimitile, Corn ethanol will not cut greenhouse gas emissions. Sci. Am. (2009), https://www.scientificamerican.com/article/ethanol-not-cut-emissions/
32. H. Ritchie, Our world in data, What are the carbon opportunity costs of our food? (2021), https://ourworldindata.org/carbon-opportunity-costs-food. Accessed 5 July 2024
33. L. Douglas, U.S. corn-based ethanol worse for the climate than gasoline, study finds, Reuters Environ (2022), https://www.reuters.com/business/environment/us-corn-based-ethanol-worse-climate-than-gasoline-study-finds-2022-02-14/. Accessed 20 July 2022
34. US Environmental Protection Agency, Science Advisory Board, Commentary on the Volume requirements for 2023 and beyond under the renewable fuel standard program (RIN 2060-AV14) (2023)
35. J. Hill, The sobering truth about corn ethanol. Proc. Natl. Acad. Sci. USA **119**(11), e2200997119 (2022). https://doi.org/10.1073/pnas.2200997119
36. T.J. Lark et al., Environmental outcomes of the US Renewable Fuel Standard. Proc. Natl. Acad. Sci. USA (2022). https://doi.org/10.1073/pnas.2101084119
37. N. Kudryavtseva, CCM 7, The petrol mixture with ethanol must be stopped (2024), https://ccm.net/apps-sites/12885-the-petrol-mixture-with-ethanol-must-be-stopped/. Accessed 10 June 2024
38. Y. Yang, J. Bae, J. Kim, S. Suh, Replacing gasoline with corn ethanol results in significant environmental problem-shifting. J. Environ. Sci. Technol **46**(7), 3671–3678 (2012). https://doi.org/10.1021/es203641p

39. S. Irwin, Understanding the price of E85 relative to E10. Farmdoc Daily **9**, 17 (2019) https://farmdocdaily.illinois.edu/2019/01/understanding-the-price-of-e85-relative-to-e10.html. Accessed 5 June 2024
40. S. Pouliot, K. Liao, *Motorists' Willingness to Pay for E85 Versus E10*, Agricultural Policy Review (Center for Agricultural and Rural Development, Iowa State University, 2017), www.card.iastate.edu/ag_policy_review/article/?a=63
41. E85 in the United States, https://en.wikipedia.org/wiki/E85_in_the_United_States
42. Biofuels Digest, https://www.biofuelsdigest.com/bdigest/2011/05/30/e85-ethanol-in-the-us-dead-or-alive/
43. J. Biello, Intoxicated with independence: is domestically produced ethanol worth the cost? Sci. Am. (2011) https://www.scientificamerican.com/article/ethanol-domestic-fuel-supply-or-environmental-boondoggle/
44. The Dairy Alliance, Who owns the American Farming Industry? https://thedairyalliance.com/blog/big-ag-the-truth-behind-the-modern-american-farming-industrys-owners. Accessed 20 June 2024
45. USDA, Farmland ownership and tenure (2022), https://www.ers.usda.gov/topics/farm-economy/land-use-land-value-tenure/farmland-ownership-and-tenure/. Accessed 20 June 2024
46. EWG (Environmental Working Group), Will the real ethanol beneficiaries please stand up? https://www.ewg.org/research/will-real-ethanol-beneficiaries-please-stand. Accessed 20 June 2024
47. Profi, the Farm Machinery Magazine, John Deere ethanol engine at Agritechnica (2023), https://www.profi.co.uk/news/john-deere-ethanol-engine-at-agritechnica/. Accessed 20 June 2024
48. D. Koplow, R. Steenblik, Chapter 4: Subsidies to ethanol in the United States, in *Biofuels, Solar and Wind as Renewable Energy Systems*, (Springer Science + Business Media B.V, 2008)
49. US Environmental Protection Agency, RIN trades and price information, https://www.epa.gov/fuels-registration-reporting-and-compliance-help/rin-trades-and-price-information. Accessed 15 Nov 2023

8 Natural Gas

Natural gas is familiar to almost everyone as the fuel delivered by pipeline to residences and businesses for air and water heating, cooking, and various appliances. Natural gas is methane combined with other minority gases, extracted from the earth's crust by drilling, the same as liquid petroleum. Despite decades of usage, it remains plentiful, considered a largely untapped resource in many areas of the world (including the USA). It is the dominant energy source for electric power generation, accounting in 2023 for 43.1% of all electric generation in the USA [1], and an even larger share in most of the EU. Natural gas is the least expensive portable fuel in the world, with vast deposits in Russia, the Middle East, and the Americas. It is the cleanest burning of all fossil fuels, producing less than half as much carbon dioxide per kWh of electric generation compared with its closest fossil fuel competitor, coal. Worldwide, it is the primary feedstock for hydrogen production for ammonia fertilizer, synthetic fuel production, and hydrogen fuel cell vehicles (HFCVs).

Natural gas is also an excellent IC engine fuel, with a CO_2 intensity (in g CO_2/MJ of combustion energy) 34% lower than gasoline. But its application as a vehicle fuel has been trivial compared with gasoline, diesel, or even electric vehicles. Stored as *compressed natural gas* (CNG) or *liquefied natural gas* (LNG), its main automotive uses have been passenger buses and commercial trucks, all having the benefit of dispatch from central facilities at which fueling can be handled. The lower fuel cost per unit of energy is the main attraction for fleet operators. Its environmental benefit in these applications is secondary.

Over the past several decades, there have been many CNG conversions [2] of gasoline vehicles. In the early 1980s, it seemed like CNG was destined to become a major contender to gasoline, with startup companies investing in the development and sale of vehicle conversion and fueling equipment [3]. But since then, the only major-manufacturer natural gas-powered passenger car was the Honda Civic GX (later badged NGV) offered to the general public 1998–2015 until being discontinued for lack of sales [4]. During its production years, the Honda GX was recognized as one of the most effective carbon-reducing accomplishments in the history of

C. A. MacCarley, *Non-Petroleum Automotive Transportation*,
https://doi.org/10.1007/978-3-031-85028-8_8

Fig. 8.1 2012 Honda Civic GX NGV. (https://upload.wikimedia.org/wikipedia/commons/2/2b/2012_Honda_Civic_GX_CNG_WAS_2012_0823.JPG. CC BY-SA 4.0)

Fig. 8.2 2010 Honda Civic GX Gen 3 NGV photographed at the 2010 Washington DC Auto Show by Mariordo Ortiz. (https://en.wikipedia.org/wiki/Honda_Civic_GX#/media/. CC BY-SA 3.0)

automotive environmental sustainability. It was awarded the Green Car of the Year Award in 2012, and, every year from 2003 to 2011, was on the Greenest Vehicle of the Year list of the American Council for an Energy-Efficient Economy, competing with another groundbreaking innovation by the same manufacturer, the Honda Insight starting in 2001. By any account, it was an exceptional vehicle in terms of its engineering and environmental impact. Now it is just a footnote in the renewable energy vehicle revolution that it helped to usher in (Figs. 8.1 and 8.2).

Natural gas is the simplest hydrocarbon and is classified as a fossil fuel, despite its superior environmental attributes compared with other hydrocarbon fuels. It is unfortunately lumped into the same category as petroleum products because of its extraction using the same hydrofracturing methods as crude oil and the poor record of environmental regulation of well sites, pipelines, and industrial applications. Methane from natural gas leaked into the atmosphere is a potent greenhouse gas, second in climate influence only to carbon dioxide.

Natural Gas Fuel Properties

- A spark ignition (SI) gaseous fuel.
- The USA exports more natural gas than any other country, despite growing concerns over energy security.
- Currently the least expensive portable fuel per unit of energy, excluding gas compression or liquefaction costs.
- Requires minimal engine modifications—technology in place for many years.
- Very low regulated emissions.
- Lowest CO_2 per unit of energy of all combustion fuels except hydrogen.
- Theoretically high octane, but backfire potential at high outputs due to low ignition energy.
- Seven percent reduction in power output compared with gasoline.
- Poor energy storage density—requires high-pressure storage, 25 MPa. Vehicle range limited.
- Readily available anywhere in the USA, but in few public high-pressure fueling stations.
- A fossil fuel unless produced from organic waste.
- Dual-fuel capability is common in conversions.
- Well-developed technology, used in fleet operations (trucks and buses) since the 1960s.
- The most common alternative fuel for transit buses and delivery vehicles.

Fuel Cost

Figure 8.3 provides a snapshot of the US national average retail cost of 3600 psi (25 MPa) CNG in April 2024, 2.90 USD/GGe [5].

The retail price per GGe for CNG is consistently lower and more stable than gasoline as shown in Fig. 8.4 (GGe = gasoline gallon energy equivalent, 1 GGe = 120 MJ).

Vehicle Range Considerations

The combustion energy content (LHV) of geologic natural gas ranges from 40.3 MJ/kg for EU natural gas L to 49.1 MJ/kg for Siberian natural gas H. The US average is 45.0 MJ/kg. For comparison, regular gasoline (EU or USA) contains 41.2–41.9 MJ/kg. CNG exceeds the mass energy storage density of gasoline, but its volume energy density (MJ/L) at 25 MPa is much lower than gasoline. Like hydrogen, methane does not liquefy at room temperature at any pressure. It must be stored either in high-pressure tanks or in a cryogenic dewar (vacuum insulated tank) at −162 °C. Filling a CNG vehicle requires a high-pressure multistage compressor similar to hydrogen, but the standard tank pressure in the USA and Northern Europe is 25 MPa (3600 psi) rather than the 70 MPa (10,000 psi) pressure used in HFCVs. As with all gaseous fuels, the constraining factor for vehicle range is its volume

Compressed Natural Gas (CNG) and Gasoline Average Retail Prices by Region

Region	CNG Prices ($/GGE*)	Gasoline Prices ($/gal)	Price Difference**
New England	$3.78	$3.41	$0.37
Central Atlantic	$2.97	$3.37	-$0.40
Lower Atlantic	$2.44	$3.54	-$1.10
Midwest	$2.64	$3.40	-$0.76
Gulf Coast	$2.76	$3.15	-$0.39
Rocky Mountain	$2.85	$3.10	-$0.25
West Coast	$3.38	$5.27	-$1.89
NATIONAL AVERAGE	**$2.90**	**$3.65**	**-$0.75**

Fig. 8.3 Retail price comparison CNG vs. gasoline by US region, April 2024. GGE is gasoline gallon equivalent energy. (US DOE. Public domain)

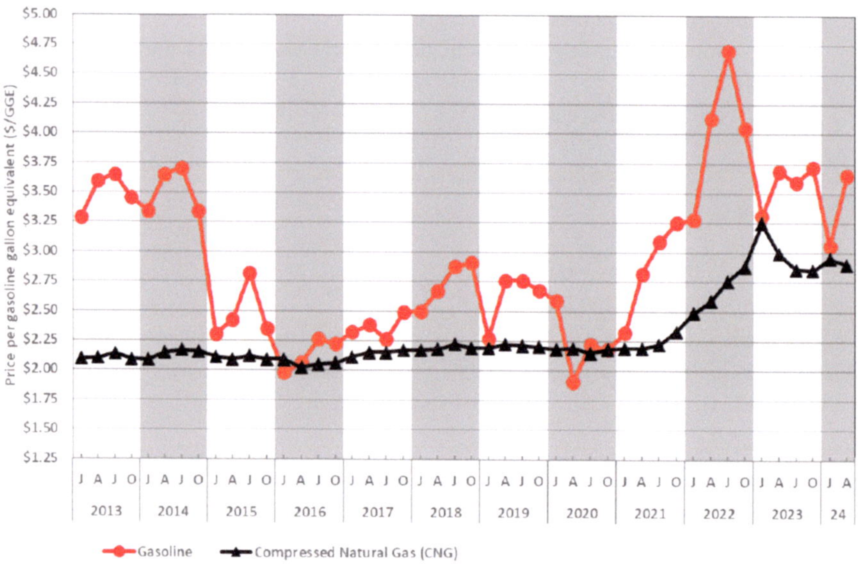

Fig. 8.4 Retail cost of compressed natural gas and regular gasoline in the USA, 2013–2024. (US Dept of Energy, Alternative Fuel Data Center, Clean Cities and Communities Alternative Fuel Prices, April 2024. https://afdc.energy.gov/files/u/publication/alternative_fuel_price_report_april_2024.pdf. Public domain)

energy density rather than mass energy density. CNG at 25 MPa has a volume energy density of 9.2 MJ/L, compared with compressed hydrogen at 70 MPa with a lower density of 6.8 MJ/L. Although CNG at 25 MPa has 0.27 times the volume energy density of gasoline (34.2 MJ/L), this does not include the volume and mass of the required high-pressure tank(s) which is/are significantly larger and heavier than gasoline fuel tanks with similar stored energy.

However, compared with the battery in an electric vehicle, any combustion fuel, including CNG, has superior energy density by mass and volume. The volume energy density of the LNMC battery in a state-of-the-art electric vehicle averages 2.1 MJ/L compared with 9.2 MJ/L for natural gas at 25 MPa. The comparison is even more favorable for CNG when the comparison is on a mass energy density basis, with a typical CNG fuel system (including tanks) storing 54 MJ/kg of fuel energy while the EV battery stores 0.95 MJ/kg of electric energy. The superior efficiency (MJ/km or mile) of the electric vehicle compared with a combustion engine vehicle compensates for the large differences in energy storage density, but not significantly. The mass of an EV is greater than that of a CNG vehicle with the same range, e.g., compare the 1300 kg weight of a 2012 Honda Civic GX having a 382 km range vs. the 1625 kg weight of a 2020 Tesla Model 3 Standard Range having a 402 km range (EPA data).

A comparison of greatest interest to a fleet operator considering replacement of diesel buses with CNG buses is the fuel storage volume for CNG vs. diesel fuel that has the highest volume energy density of any engine fuel (38.6 MJ/L): Roughly speaking, a diesel vehicle would have a range over four times that of a CNG vehicle with the same fuel storage volume. The large storage tanks enclosed in the elevated roof of a CNG transit bus provide a visual indication of this difference.

CNG ICVs vs. HFCVs

In terms of practicality, the closest alternative fuel comparison with a CNG vehicle is an HFCV, both having limited energy storage capacity onboard and both requiring refueling with a high-pressure gas at specialized fueling stations. The combined fuel cell, battery round trip efficiency (RTE) and electric motor efficiency (not including hydrogen production) of a HFCV is typically 40–50%, while that of an ICV fueled by natural gas is typically 35%. A CNG vehicle will have a range slightly greater than a HFCV with the same volume of fuel storage system, even at the higher pressure 70 MPa for hydrogen vs. 25 MPa for natural gas. At the same storage pressure and volume, the range of a natural gas vehicle exceeds that of an HFCV.

Both CNG and hydrogen require significant energy for gas compression, although 70 MPa hydrogen requires much greater energy than 25 MPa CNG. In either case, the compression energy must be subtracted from the stored energy, decreasing the effective fuel efficiency. Other factors differentiating pressurized fuel storage from liquid fuels include the following:

- Fueling time
- Energy required for gas compression
- Energy loss due to adiabatic compression heating
- Safety issues due to high pressure gas storage on vehicle

A more subtle but important factor is the form of the energy used for gas compression, electricity, which incurs its own system losses on the way from grid generation (typically 43% if produced natural gas) to use by the multistage compressor at the refueling facility. The performance specs vary quite a bit for different compressors. For hydrogen, a typically quoted value for the energy required to precool and compress from 2 to 70 MPa is 6.0 kWh/kg, which amounts to about *18% of the fuel energy*. The energy to compress CNG from 2 to 25 MPa is lower—about 1 kWh/kg of natural gas at a refueling facility, which amounts to *8% of the fuel energy*. These losses are significant considering that liquid fuels such as gasoline, LPG, biodiesel, E85, or M85 require only trivial electric energy to dispense because no local compression is required.

Another consideration is worth mentioning regarding the refueling of a natural gas vehicle. The effect of adiabatic heating during isentropic (no heat loss) compression approximately follows the ideal gas pressure vs. temperature relationship—if a gas is compressed, its temperature increases. The adiabatic temperature rise during the filling of the 25 MPa tank reduces the mass of methane that can be stored at that pressure. During the roughly two hour period after a CNG fill, the gas cools, and therefore, the pressure decreases. Figure 8.5 from the US DOE Alternative Fuels Data Center illustrates the phenomenon: an effective loss of 10–15% of the tank capacity due to the temperature rise. Fortunately, under DOT/ICC standards, the vehicle CNG tanks must be capable of handling 125% of their rated pressure, so they can safely be overfilled to as high as 28 MPa (4100 psi), with the pressure after cooling decreasing to 25 MPa, allowing the storage of the intended mass of fuel.

The equipment required at a CNG refueling station is depicted in Fig. 8.6.

An EU-standard 25 MPa fill port on a CNG car is shown in Fig. 8.7.

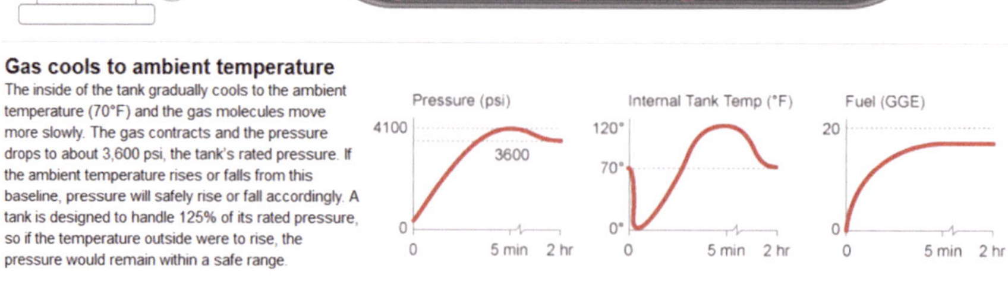

Fig. 8.5 Pictorial on adiabatic heating of natural gas during filling of a 25 MPa (3600 psi) tank. (From DOE AFDC, https://afdc.energy.gov/vehicles/natural-gas-filling-tanks. Public domain)

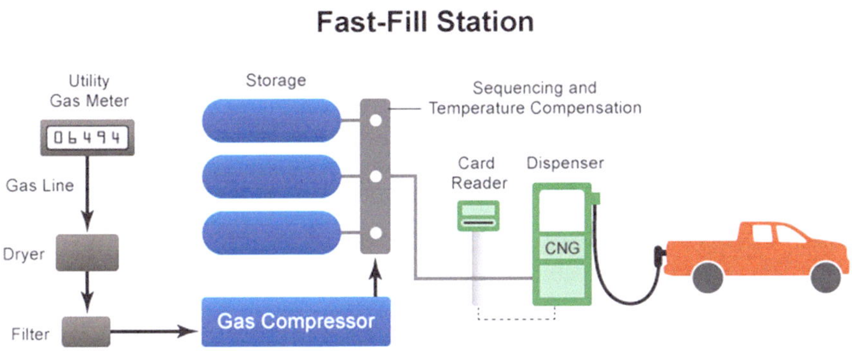

Fig. 8.6 Fast-fill dispensing of CNG at a refueling station. (From DOE AFDC. https://afdc.energy.gov/fuels/natural-gas-cng-stations. Public domain)

Fig. 8.7 Filling a CNG Fiat Multipla, 2012. (Photo by Andreas Geick. https://en.wikipedia.org/wiki/Natural_gas_vehicle#/media/File:FillingUpCNG.jpg. CC-BY-SA 2.0)

The Technology of Natural Gas-Powered Vehicles

The main fuel system components of a purpose-built CNG vehicle are shown in Fig. 8.8.

The modification of a gasoline engine to run on natural gas is a fairly simple matter using a gaseous fuel carburetor (gas mixer) available from several companies worldwide [6], example shown in Fig. 8.9.

A gas mixer (aka gaseous fuel carburetor) is designed to maintain a constant mass flow ratio between the intake air and the fuel. But even more so than gasoline carburetors, gas mixers are imperfect at this task, resulting in regimes of lean and rich engine operation. Lean (excess air) operation can lead to misfire and low power which is easily noticed by the driver, so in practice, gas mixers are often set up by mechanics to avoid lean spots by allowing rich operation at some flow rates. This is not a trivial problem. Since CNG vehicles are usually exempt from periodic emission inspections in most US states, unburned methane in the exhaust is not penalized even though it is a much more harmful greenhouse gas than carbon dioxide.

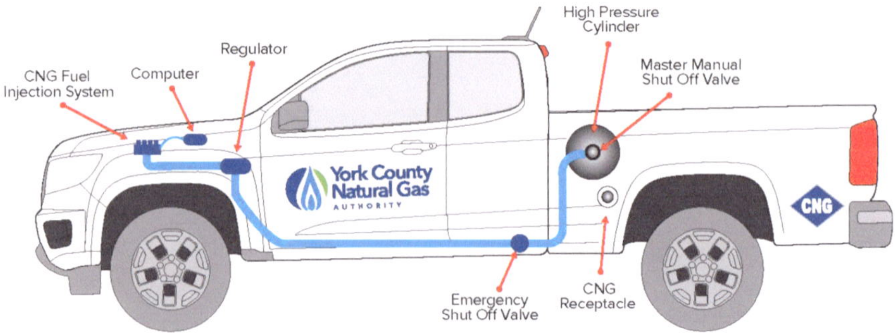

Fig. 8.8 Phantom diagram showing the fuel system components of a purpose-built CNG-powered pickup truck used as service vehicle. (Image courtesy of Hannah Smith, York County Natural Gas Authority, SC)

Fig. 8.9 Impco natural gas carburetor for retrofitting a medium-displacement gasoline engine. (Image from https://store.nashfuel.com. Used with permission)

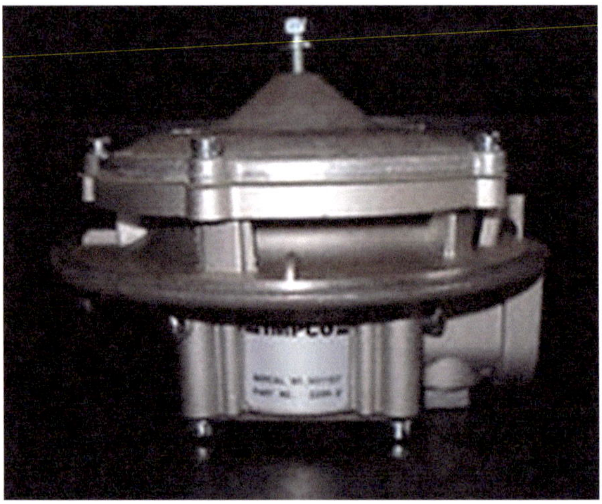

A more precise fuel management solution is electronic port fuel injection similar to that used on most gasoline vehicles (see Chap. 4 *Engines and Fuels*). However, the very low volumetric energy density of gaseous fuels (especially natural gas or hydrogen) requires much higher injector flow rates compared with liquid fuels [7]. Figure 8.10 shows the HANA H2001 natural gas fuel injector used by Honda on the Japanese-market Civic NGV. Natural gas electronic fuel injectors are the same type of solenoid valves as liquid fuel injectors but with higher flow coefficients and no need for a spray nozzle since gaseous fuel disperses without the need for atomization. Mass fuel delivery is approximately linear with the injection pulse duration, which typically varies from 1.0 to 12.0 ms or less for high-speed automotive engines.

Fig. 8.10 HANA H2001 natural gas fuel injector used by Honda on Civic GX NGV. (Image from http://hanaems.com/h2001/, not copyright)

Fig. 8.11 1982 natural gas fuel-injected, turbocharged automobile constructed by author for evaluation of CNG fuel by Colorado State Agencies, 1982

Figure 8.11 is a photo of a CNG-converted 1982 Ford Mustang GT that incorporated a prototype fuel injection system and turbocharging, built in 1982 for evaluation by the Colorado Highway Patrol, that was considering natural gas fuel due to its lower cost and reliable supply compared with gasoline at that time. Previous CNG conversions using gas mixer kits provided inadequate power and excessive methane emissions. The precise fuel control maintained a consistent AFR across the engine operational range compared with a gas mixer, allowing the operation of the engine closer to a stoichiometric ratio (lambda = 1) without the risk of exceeding this limit.

Fig. 8.12 Washington DC metro bus converted to run on natural gas. (https://en.wikipedia.org/wiki/Natural_gas. Photo by Adam E. Moreira, https://commons.wikimedia.org/w/index.php?curid=1347035. CC BY-SA 3.0)

This improved the engine power output without the emission of unburned methane. It also made possible forced induction (turbocharging) of the engine, which required higher fuel pressures to compensate for the above-atmospheric intake air pressure. The performance of the vehicle exceeded that of the unconverted Mustang GT, making it acceptable as a law enforcement or emergency services vehicle.

Like hydrogen engines, the low ignition energy and lack of fuel vaporization cooling make natural gas engines prone to intake backfiring, the same phenomenon but not as severe as hydrogen engines. The delayed fuel injection method developed for hydrogen fuel injection (see Chap. 9, *Hydrogen*) is equally effective at eliminating backfiring in natural gas engines, allowing turbocharge boost to compensate for the 7% volumetric power reduction intrinsic to natural gas.

Probably the most successful deployments of natural gas as a motor fuel have been passenger transit buses, since they are fueled and maintained at a central facility where CNG and qualified staff are readily available. The emissions of the buses are much cleaner than the status quo, diesel, even considering the complex exhaust aftertreatment of modern diesel engines. Natural gas also reduces fleet fuel costs by as much as half compared with diesel operation. Figure 8.12 is a 2014 photo of a CNG bus in regular service in Washington DC.

Methane Is a Potent Greenhouse Gas

Natural gas is mostly methane (CH_4). Depending on the source and/or location, methane constitutes 65% [8] to 98% [9] (by mass) of extracted natural gas. While not generally appreciated as a greenhouse gas prior to about 2000, methane is now known to be a much more potent greenhouse gas than carbon dioxide in the atmosphere. According to a recent MIT study, the severity of its impact may actually be understated by the use of a 100-year impact when calculating its global warming potential (GWP), instead of a more immediate 10- or 20-year impact horizon. The choice of analysis period is primarily responsible for the wide range of GWPs commonly quoted for methane. For a 100-year horizon, the US EPA [10] reports a GWP for methane of 27–30, while the MIT study and other credible sources that use shorter analysis horizons report GWP as high as 80. According to the study [11]:

> Over 20 years, the methane would trap about 80 times as much heat as the CO_2. Over 100 years, that original ton of methane would trap about 28 times as much heat as the ton of CO_2.

Looking at the global warming harm of unburned fuel leakage or exhaust emission is a departure from common practice in the energy industry since leakage is considered a maintenance problem rather than an intrinsic property of the fuel. The primary sources of natural gas in the atmosphere are organic: decaying organic material and agriculture. But natural gas that is deliberately extracted from the earth's crust is arguably an even larger contributor to global warming [12]. As discussed in Chap. 9, it is conservatively estimated that 2% of the entire anthropogenic greenhouse effect is due exclusively to carbon dioxide emissions from the production of hydrogen from natural gas for manufacturing ammonia fertilizer, not even including methane leakage from this worldwide industry. Production and combustion of natural gas as a transportation fuel are only a tiny fraction of these methane emitters.

In large areas of the world including Russia, the Middle East, and the Americas, natural gas is plentiful and by far the least expensive form of portable energy in cost per energy unit. Even if natural gas is not the objective of a drilling operation, it is often encountered as a byproduct of petroleum extraction. In the quest for liquid petroleum, large amounts of natural gas are flared (or illegally released unburned), not only wasting its energy value but contributing to global warming in a much more potent way than carbon dioxide. Byproduct natural gas is a particular problem at remote drilling locations, where the capture and transport of natural gas is uneconomical, exacerbated by the low market value of natural gas and lack of pipeline networks. In assessments of the GHG harm of natural gas for transportation, the flaring, venting, leakage, compression, liquefaction, and distribution of natural gas are often ignored, assuming that these are negligible because of the minimal refining required for natural gas prior to combustion use. But this is not true. According to Desirée Plata, Associate Professor and Director of the MIT Methane Network [13]:

> We're definitely under-accounting natural gas's contribution to global warming relative to coal," says Official government. "Calculations of natural gas's contribution to warming don't account for leaks, says Plata. And in fact, she says, if methane emissions from natural gas are on the high end of current estimates, natural gas may be no improvement over coal at all.

The connection between petroleum extraction and natural gas is poorly quantified and not likely to change in the near future. Why?

- Most oil in the USA and some other countries is now extracted using hydrofracturing methods. Natural gas is very often released from new oil wells, and hydro-

fracturing is particularly adept at releasing natural gas from wells intended for oil extraction. Exact numbers are not available, but the environmental harm done by venting or flaring of unwanted gas should be included among the greenhouse gas consequence of petroleum-fueled vehicles. Currently, it is not.
- 43+% of electric power generation in the USA, and an even larger percentage in some EU countries is fueled by natural gas. And despite the rapid growth of intermittent renewable electricity sources (solar, wind), it is unlikely that this percentage will drop in the near future due to the continuing need for 24/7 dispatchable electric power that cannot be provided by solar or wind resources. The proliferation of electric vehicles increases the need for consistency of electric power generation, while local or grid electric energy storage lags far behind the time-dependent demand for electric power due to the misalignment between periods of renewable availability and vehicle charging times.

The USA is currently a net energy exporter, with the large majority of exported energy in the form of LNG transported by specialized tanker ships [14]. In early 2024, natural gas exports were finally acknowledged as an important climate issue, and the construction of two LNG "mega terminals" in the Gulf of Mexico and California was temporarily suspended by an executive order to block LNG exports to non-free-trade-agreement (FTA) countries and consider climate impacts [15]. As of November 2024, the suspension remains in litigation, with exports expected to resume in 2025 under a different federal administration.

Natural Gas Vehicle Safety

A CNG vehicle has extra risks related to its high-pressure composite fuel tank(s), fuel lines, pressure regulators, and fuel delivery components. The risk of fuel leakage is higher than liquid fuels or LPG because of the high-pressure fuel system. Of greater concern than fire due to fuel leakage is the kinetic energy release from the catastrophic rupture of the tank in the event of a collision. Arguably, the lack of a gasoline fuel tank makes CNG potentially safer for the vehicle occupants because natural gas, like hydrogen, is lighter than air. If released, it rises and dissipates quickly rather than pooling around the vehicle, blocking passenger evacuation and access by first responders.[1] The exception, of course, occurs when the vehicle is operating indoors or in a tunnel.

The design and safety requirements for natural gas fuel storage and delivery systems are regulated by multiple overlapping codes and jurisdictions. A compendium of resources and regulations can be found in citation [16]. Almost all vehicles smaller than a Class 5 commercial truck or bus are fueled by CNG, not by LNG.

[1] For example, a video clip of a CNG tank rupture in a transit bus in Italy: Motor1.com. 25 April 2022. *Bus Turns Into Flamethrower When CNG Tanks On Roof Explode.* https://www.motor1.com/news/581962/bus-flamethrower-cng-tanks-explode/. Accessed 15 June 2024.

US National Regulations and Standards Related to CNG Vehicle Safety

National Fire Protection Association (NFPA):

- NFPA 52, the Vehicular Natural Gas Fuel Systems Code, defines specific safety requirements for NGVs and their fueling facilities.
- NFPA 30A applies to facilities that perform maintenance and repair of NGVs.
- NFPA 88A applies to parking garages.

Federal Motor Vehicle Safety Standards:

- (FMVSS) No. 304, CNG Fuel Container Integrity

Society of Automotive Engineers:

- J1616_201605 Standard for CNG Vehicle Fuel
- J1616_201703 Recommended Practice for CNG Vehicle Fuel

American Society for Testing and Materials (ASTM):

- ASTM D8080-21 Standard Specification for CNG and LNG Used as a Motor Vehicle Fuel

US Dept of Transportation:

- Natural Gas CNG Fuel System and Tank Maintenance, https://afdc.energy.gov/vehicles/natural-gas-cylinder

US National Highway Transportation Administration:

- CNG and LNG Vehicle Fuel Systems: First Responder Guide ENP-084 Rev. D, https://static.nhtsa.gov/odi/tsbs/2020/MC-10177251-9999.pdf

GHG Emissions of CNG Compared with Gasoline

Appendix 2 provides detailed calculations of the CO_2 emission resulting from the combustion of natural gas. The method is shown below using two models of the same vehicle, one deigned to run on gasoline and the other exclusively on natural gas. Vehicle specifications for both are from US DOE fueleconomy.gov website [17] (Fig. 8.13).

Fig. 8.13 EPA fuel economy specifications for 2012 Honda Civic 1.8L car. (Public domain)

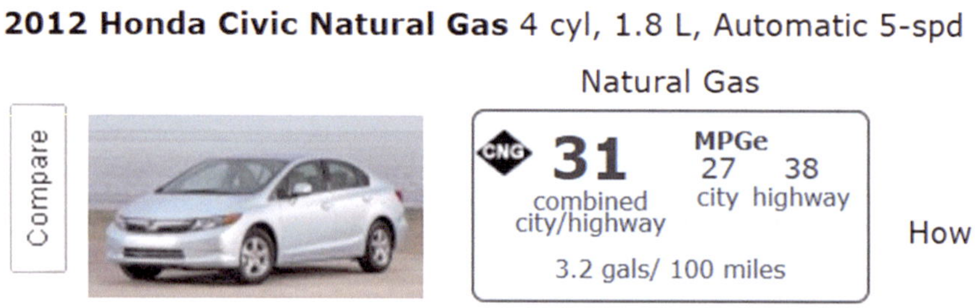

Fig. 8.14 EPA fuel economy (MPGe) specifications for 2012 Honda Civic 1.8L natural gas car. (Public domain)

Gasoline (Without Ethanol)

Baseline: *2012 Honda Civic LX* that gets 31 MPG = 50 km/gal

The carbon dioxide produced by combustion of 1 US gallon of gasoline is 8877 g CO_2/gal gasoline[2] (Fig. 8.14):

$$\frac{\left(\frac{8887\,\text{g CO}_2}{\text{gallon}}\right)}{\left(\frac{50\text{ km}}{\text{gallon}}\right)} = 178\,\text{g CO}_2/\text{km} \quad \left(285\,\text{g CO}_2/\text{mile}\right)$$

Natural Gas

The CO_2 emissions per km of a *2012 Honda Civic GX/NGV* that gets 31 MPGe = 50 km/GGe

[2] Data from 2020 US Code of Federal Regulations (CFR).

For simplicity, assume that natural gas is entirely methane. The combustion energy of CH_4 = 50.0 kJ/g and 1 GGe = 121 MJ.

The amount of energy used to travel 1 km:

$$\left(\frac{1\,\text{GGe}}{50\,\text{km}}\right)\left(\frac{121\,\text{MJ}}{1\,\text{GGe}}\right) = 2.42\,\text{MJ}/\text{km}$$

The amount of methane that contains 2.42 MJ is

$$(2.42\,\text{MJ})\left(\frac{1\,\text{kg}}{50\,\text{MJ}}\right) = 0.0484\,\text{kg} = 48.4\text{g methanol}$$

The amount of carbon dioxide produced by the combustion of 1 g of methanol is found from the combustion equation:

$$CH_4 + 2O_2 \rightarrow CO_2 + 2H_2O$$

$$16\text{g}\,CH_4 + 64\text{g}\,O_2 \rightarrow 44\text{g}\,CO_2 + 36\text{g}\,H_2O$$

$$\frac{44\text{g}\,CO_2}{16\text{g}\,CH_4} = 2.75\text{g}\,CO_2/\text{g}\,CH_4$$

The amount of CO_2 emitted per distance for the Honda GX is

$$(48.4\text{g}\,CH_4)(2.75\text{g}\,CO_2/\text{g}\,CH_4) = 133\text{g}\,CO_2/\text{km}\,(213\text{g}\,CO_2/\text{mile})$$

Comparing natural gas with gasoline…

$$\frac{178\text{g}/\text{km}}{133\text{g}/\text{km}} = 1.34 \Rightarrow \text{Gasoline produces 34\% more } CO_2 \text{ per distance traveled.}$$

Power Output of Natural Gas Engines

Conversion or replacement of fleets of service vehicles to natural gas (CNG) operation is attractive to fleet operators due to the lower cost of natural gas compared with gasoline on a per-mile basis. But methane/CNG intrinsically provides a lower power output than gasoline or diesel fuel. This can be a concern for law enforcement or first responder vehicles [3]. The previously discussed vehicle shown in Fig. 8.11 was commissioned specifically as a counterpoint to the low-power reputation of CNG. It incorporated natural gas fuel injection and turbocharging and exceeded the performance requirements for CHP pursuit vehicles.

Appendix 4 includes detailed calculations to determine the power output of a natural gas engine compared with the same engine fueled by gasoline. Below is an example of a comparison between two otherwise-identical Honda Civic models, one fueled by gasoline, the other, natural gas.

Gasoline (Without Ethanol) Baseline

2012 Honda Civic, 31 MPG = 50 km/gal

Manufacturer specifications [18]: 140 HP at 6500 RPM

Compare this with the calculated HP to determine the thermal efficiency of the engine:

Air volume flow at 6500 RPM (assuming 100% volumetric efficiency):

$$\left(\frac{1.8\text{ L}}{2\text{ revs}}\right)\left(\frac{6500\text{ revs}}{\text{min}}\right)\left(\frac{1\text{ min}}{60\text{s}}\right)=97.5\text{ Lair / s}$$

Air mass flow at 6500 RPM at NTP (20 °C, 1 atm):

$$\left(97.5\frac{\text{L air}}{\text{s}}\right)\left(\frac{1.204\text{g air}^*}{\text{L air}}\right)=117\text{g air / s}^*\text{from Ideal Gas Law, Appendix}$$

Fuel mass flow at 6500 RPM, for gasoline AFR = 14.6 g air/g fuel [19]:

$$\left(\frac{117\text{g air}}{\text{s}}\right)\left(\frac{\text{g gasoline}}{14.6\text{g air}}\right)=8.01\text{g gasoline / s}$$

Amount of power in this fuel flow:

$$\left(\frac{8.01\text{g gas}}{\text{s}}\right)\left(\frac{41.5\text{ kJ}}{\text{g gas}}\right)\left(\frac{1\text{ kWs}}{1\text{ kJ}}\right)\left(\frac{1.34\text{ HP}_{\text{US}}}{1\text{ kW}}\right)\eta=446\times\eta\text{HP}_{\text{US}}{}^*\text{at NTP}$$

where η is the thermal efficiency of the engine. It is typically about 30% for ICEs of this type. Solve for η using manufacturer-specified horsepower, 140 HP_{US}:

$$\eta=\frac{140\text{ HP}_{\text{US}}}{446\text{ HP}_{\text{US}}}=31.4\%.$$

Natural Gas

2012 Honda Civic GX/NGV, 31 MPGe = 50 km/GGe

Like hydrogen, the effective engine displacement is reduced due to the volume of the fuel in the intake air-fuel charge.

What volume and mass of fuel must be inducted for a given mass of air to form a stoichiometric mixture?

16 g of methane reacts with 64 g of oxygen:

$$CH_4+2O_2\rightarrow CO_2+2H_2O$$

Air is 21% oxygen and 79% nitrogen by mole or volume:

$$AFR_{\text{vol}}=\frac{2\left(1\text{mole}O_2+\dfrac{0.79}{0.21}\text{moles}N_2\right)}{1\text{mole }CH_4}=9.52\text{ Lair / L }CH_4$$

Percentage of intake volume displaced by the gaseous fuel:

$$1-\frac{9.52\text{L air}}{9.52\text{L air}+1\text{L CH}_4}=9.51\%$$

Effective displacement of 1.8 L engine at full power:

$$(1.8\text{L})(1-0.095)=1.63\text{L}$$

Air flow at 6500 RPM:

$$\left(\frac{1.63\text{L}}{2\text{ revs}}\right)\left(\frac{6500\text{ revs}}{\text{min}}\right)\left(\frac{1\text{ min}}{60\text{s}}\right)=88.3\text{L air / s}$$

Fuel flow at 6500 RPM:

$$\left(88.3\frac{\text{L air}}{\text{s}}\right)\left(\frac{\text{L CH}_4}{9.52\text{L air}}\right)=9.28\text{L CH}_4\text{ / s}$$

Amount of power in this fuel flow:

$$(9.28\text{L CH}_4\text{ / s})\left(\frac{0.667\text{g CH}_4^{*}}{\text{L CH}_4}\right)\left(\frac{50.0\text{ kJ}}{\text{g CH}_4}\right)\left(\frac{1\text{ kWs}}{1\text{ kJ}}\right)\left(\frac{1.34\text{ HP}_{\text{US}}}{1\text{ kW}}\right)\eta$$
$$=415\times\eta\text{HP}_{\text{US}}{}^{*}\text{at NTP}$$

Since all of the specifications and MPGe are identical for the gasoline and the natural gas versions of the Civic, we can use the value found for gasoline above of $\eta=0.315$:

$$414\times0.315\text{ HP}_{\text{US}}=130\text{ HP}_{\text{US}}$$

The percentage power reduction from gasoline to natural gas is $1-\frac{130\text{ HP}_{\text{US}}}{140\text{ HP}_{\text{US}}}=7.1\%$.

Renewability

Natural gas extracted from the earth's crust is a fossil fuel. Among fossil fuels, however, it is the least environmentally harmful provided that when leakage occurs, it is fully combusted rather than released to the atmosphere, since methane is as much as 80 times more potent a GHG as carbon dioxide. As calculated above, the direct operation of an ICV on natural gas can generate lower CO_2 emissions than any fuel except hydrogen. But emission of methane to the atmosphere due to over-rich fuel aspiration of leakage from fuel extraction and processing equipment could easily negate this potential benefit. These problems are technically solvable. The problem mostly exists now because only a few years ago, CNG vehicles were unregulated, so there was little incentive to develop more accurate fuel and emission controls.

For perspective, much larger methane emissions can be attributed to agricultural livestock operations, landfill emissions, and natural seepage, in addition to fuel extraction, refining, and distribution.

Biomethane

There has been much speculation in recent years that biomethane from landfills or livestock operations has the potential to offset some of the extracted natural gas. These sources are envisioned as carbon-neutral because the methane would naturally be released into the atmosphere anyway, so any collection and use as an energy source is a plus for environmental and energy sustainability. This is a promising possibility, considering the sheer magnitude of these methane emissions. According to a 2021 United Nations report, livestock emissions—from manure and gastroenteric releases—account for roughly 32% of human-caused methane emissions [20]. Worldwide, the cost of capturing biomethane is more expensive than the production of fossil natural gas, estimated in the cited UN report to be \$19/MBTU, while the average price of fossil natural gas is \$2.58/MBTU (June 2024).[3] The only cost-effective scenario for biomethane seems to be the internal use of the collected unpurified gas as part of the agricultural operations, usually to power ICE generators that create greater CO_2/kWh than utility electricity.

But in the creative accounting world of carbon credits, in California, biomethane captured from Ag operations, landfills, or natural swamps is allowed carbon offset credits. These "renewable natural gas" (RNG) tax credits are free market sold to existing hydrogen producers that use carbon-intensive SMR methods, allowing them to claim that their production is carbon-neutral and, therefore, legally receive "green hydrogen" carbon offset credits for their gray, brown, or black hydrogen.

From Canary Media, September 2023 [21]:

> The RNG [Renewable Natural Gas] industry is pressing the [US] Treasury Department to adopt a controversial emissions accounting practice that has its roots in California's Low-Carbon Fuel Standard (LCFS). That scheme, which watchdogs say is already starting to be abused in California, allows producers of hydrogen and other fuels to cancel out their emissions by purchasing 'carbon-negative' credits from commercial dairies and livestock operations that capture the planet-warming methane bubbling out of their manure lagoons.
>
> California's rules, critics say, are much more effective at turning factory-farm manure lagoons into subsidy gold mines than they are at actually reducing the carbon intensity of fuels sold in the state. In fact, many analyses indicate that they increase, rather than decrease, greenhouse gas emissions and air pollution.
>
> That runs the risk of making 45 V 'the single greatest waste of climate money in U.S. policy history,' with the possible exception of the massive, decades-long subsidization of corn ethanol that has worsened climate and food crises.

[3] Henry Hub Spot price, which is the pricing benchmark for natural gas.

As of early 2024, this and other loopholes in the IRA remain under debate in Congress, subject to intense lobbying efforts.

Too Much Natural Gas, at Least for the Next Decade

Since 2013, oil and gas companies in the USA have overextracted natural gas, almost entirely because of highly successful hydrofracking operations in North Dakota, Pennsylvania, Texas, and Alaska. This is the consequence of the largely unregulated energy market in the USA. The extraction of natural gas has progressed at a faster rate than expected before enhanced oil and gas recovery methods were first deployed in the early 2000s. Domestic fossil fuel consumption has declined, leading to the USA becoming a net energy exporter, with the large majority of the exported petroleum-based energy in the form of LNG transported by specialized tanker ships. Gas exports don't need federal approvals if they go to the 17 countries, from Australia to Singapore, which have free trade agreements (FTAs) with the USA [22]. US natural gas resources that were previously expected to peak in 2022 [23] are now expected to peak between 2024 and 2030 [24], and last between 10 and 50 years, depending upon the predictor and criteria. But thankfully from a GHG point of view, the current glut of natural gas is expected to be short-lived. These predictions suggest that natural gas may not continue to be the least expensive energy option in a decade or two [25], although this statement is based only on known reserves and the current regulatory framework.

The creation of a national-scale fuel storage reserve (in ready-to-use form) has been recurrently discussed since the National Energy Policy of the Carter Administration prior to 1980 [26]. The cost of storing more than a few days of energy for the US transportation sector alone is staggering—20 million barrels of oil equivalent per day [27], with each barrel providing 5.4 GJ [28] (LHV), for a daily energy use of 108×10^{15} or 108 million GJ. This would be energy equivalent to 5.4 billion liters of LNG. As mentioned above, recent efforts to site off-shore or on-shore LNG storage facilities have usually met with public concerns over safety and environmental harm, slowing or preventing projects from moving forward [29, 30].

References

1. US Energy Information Administration, Net generation by energy source: total (all sectors), 2013—November 2023, Table 1.1, https://www.eia.gov/electricity/monthly/epm_table_grapher.php?t=table_1_01. Accessed 1 June 2024
2. The Transport Project, Vehicle availability, https://transportproject.org/vehicles/availability/. Accessed 5 June 2024
3. C.A. MacCarley, *Acceleration of Automobiles Fueled by Natural Gas and Gasoline: A Comparison*, Report to the National Science Foundation (Denver Urban Observatory and City of Denver, University of Denver Research Institute, 1982)
4. American Honda Corp., Natural gas-powered 2011 Civic GX remains smart, green choice, HondaNews (2010), https://hondanews.com/en-US/releases/natural-gas-powered-2011-civic-gx-remains-smart-green-choice. Accessed 1 June 2024

5. US Dept of Energy, Alternative fuels data center, in *Fuel Prices*, https://afdc.energy.gov/fuels/prices.html
6. SkyCNG Conversion Kits, https://www.skycng.com/technology. Accessed 1 June 2024
7. C.A. MacCarley, Development of a high speed injection valve for electronic hydrogen fuel injection. Proc. Third World Hydrogen Energy Conference, Tokyo, Japan (International Association of Hydrogen Energy, Coral Gables, Florida, 1980)
8. Penn State University, College of Earth and Mineral Science, Natural gas composition and specifications, https://www.e-education.psu.edu/fsc432/content/natural-gas-composition-and-specifications. Accessed 15 May 2024
9. Robert Bosch GmbH, *Automotive Handbook*, 10th edn. (SAE International, 2018)
10. US Environmental Protection Agency, Understanding global warming potentials (2024), https://www.epa.gov/ghgemissions/understanding-global-warming-potentials. Accessed 1 May 2024
11. Ask MIT Climate, Why do we compare methane to carbon dioxide over a 100-year timeframe? Are we underrating the importance of methane emissions? https://climate.mit.edu/ask-mit/why-do-we-compare-methane-carbon-dioxide-over-100-year-timeframe-are-we-underrating. Accessed 1 May 2024
12. Our World in Data, Global warming contributions by gas and source, World, 1851 to 2022, https://ourworldindata.org/grapher/global-warming-by-gas-and-source. Accessed 20 May 2024
13. Ask MIT Climate, How much does natural gas contribute to climate change through CO_2 emissions when the fuel is burned, and how much through methane leaks? (2023), https://climate.mit.edu/ask-mit/how-much-does-natural-gas-contribute-climate-change-through-co2-emissions-when-fuel-burned. Accessed 1 May 2024
14. NRDC (Natural Resources Defense Council), DOE shrugs as U.S. natural gas exports surge (2023), https://www.nrdc.org/bio/gillian-giannetti/doe-shrugs-us-natural-gas-exports-surge-0. Accessed 15 May 2024
15. NPR (online), U.S. pauses build-out of natural gas export terminals to weigh climate impacts (2024), https://www.npr.org/2024/01/27/1227292600/u-s-pauses-build-out-of-natural-gas-export-terminals-to-weigh-climate-impacts. Accessed 15 May 2024
16. US DOE Alternative Fuels Data Center, Natural gas fuel safety, https://afdc.energy.gov/vehicles/natural-gas-safety. Accessed 1 May 2024
17. US Dept of Energy/US Environmental Protection Agency, Fuel economy of 2012 Honda vehicles, https://www.fueleconomy.gov/feg/bymake/Honda2012.shtml. Accessed 1 June 2024
18. Edmunds, Used 2012 Honda Civic—specifications and features, https://www.edmunds.com/honda/civic/2012/features-specs/. Accessed 1 June 2024
19. Robert Bosch, *Automotive Handbook*, 10th edn. (SAE International, 2018), p. 345
20. United Nations Environment Programme, Methane emissions are driving climate change. Here's how to reduce them (2021), https://www.unep.org/news-and-stories/story/methane-emissions-are-driving-climate-change-heres-how-reduce-them. Accessed 6 June 2024
21. J. St. John, The biomethane boondoggle that could derail clean hydrogen, Canary Media (2023), https://www.canarymedia.com/articles/hydrogen/the-biomethane-boondoggle-that-could-derail-clean-hydrogen?_hsmi=273724509. Accessed 15 May 2024
22. D. Gilbert, T. Fowler, Natural gas glut pushes exports, Wall Street J. (2012), http://online.wsj.com/article/SB10000872396390444223104578036403362012318.html, Accessed 7 Mar 2013
23. US EIA, U.S. marketed natural gas production forecast to rise in 2022 and 2023 (2022), https://www.eia.gov/todayinenergy/detail.php?id=51319. Accessed 1 June 2024
24. University of Texas News, New, rigorous assessment of shale gas reserves forecasts reliable supply from Barnett shale through 2030 (2013), http://www.utexas.edu/news/2013/02/28/new-rigorous-assessment-of-shale-gas-reserves-forecasts-reliable-supply-from-barnett-shale-through-2030/. Accessed 7 Mar 2013
25. EIA, Natural gas explained. How much natural gas is left, https://www.eia.gov/energyexplained/natural-gas/how-much-gas-is-left.php. Accessed 15 May 2024

26. UC Santa Barbara, The American Presidency Project: Jimmy Carter, 1977–1981, Fact Sheet on the President's National Energy Program (1977), http://www.presidency.ucsb.edu/ws/index.php?pid=7373. Accessed 7 Mar 2013
27. US Dept of Energy, *Transportation Energy Databook*, 31st edn. (2012), http://cta.ornl.gov/data/index.shtml. Accessed 7 Mar 2013
28. Wikipedia, Barrel of oil equivalent, http://en.wikipedia.org/wiki/Barrel_of_oil_equivalent. Accessed 7 Mar 2013
29. J. de Baan, M.H. Krekel, R. Leeuwenburgh, M.M. McCall, Offshore transfer, re-gasification and salt dome storage of LNG, in *Paper OTC 15301, 2003 Offshore Technology Conference, Houston, Texas, USA*, (2003)
30. US DOE, North American LNG export terminals approved, not yet built (2022), https://www.energy.gov/sites/default/files/2022-03/FERC%2C%20N.%20American%20LNG%20export%20terminals_0.pdf. Accessed 1 Oct 2024

9 Hydrogen

Inflammable Air

Before it was given the name *hydrogen* by Antoine Lavoisier in 1783 [1], hydrogen was referred to as a type of *Inflammable Air*. Hydrogen is both a combustion fuel and a fuel cell reactant. It contains no carbon, which means no carbon dioxide emissions when combined with oxygen from the air to produce mechanical power or electricity. It is the most common substance in the universe but is rarely found on Earth as a free gas. It is the basic building block of the Earth's oceans, soil, and all life forms. It is in the atomic makeup of almost every combustion fuel.[1] It can be isolated from thousands of substances using electricity, heat, or chemical reactions. As a fuel for transportation, it shares the dual distinctions of having the highest mass energy density and the lowest volume energy density of any fuel at a given temperature or pressure.

Hydrogen has been used since the early nineteenth century as a source of heat or illumination, most often blended with carbon monoxide in coal gas, water gas, or town gas. It was used as an automotive fuel at a time of scarce petroleum in Europe 1940–45. Worldwide, over 95% of hydrogen is produced solely to make ammonia for agricultural fertilizer, generating large amounts of carbon dioxide [2].

More is known about hydrogen's elemental properties than any other element. Still, it remains somewhat mysterious to the general public, perennially the subject of as much science fiction as science fact. And now, facing the environmental consequences of our exploitation of fossil fuels for more than a century, we turn to hydrogen as a possible alternative path to allow us to maintain the energy-hungry standard of living that has come to define human civilization. This chapter discusses hydrogen's role in transportation as a possible means to mitigate the impact of this sector on climate change (Fig. 9.1).

[1] Exceptions to this statement include coal (carbon), sulfur and reactive metals such as lithium and magnesium, although none of these would be considered motor fuels in the modern sense.

C. A. MacCarley, *Non-Petroleum Automotive Transportation*,
https://doi.org/10.1007/978-3-031-85028-8_9

Fig. 9.1 Hydrogen Fuel Cell Vehicle refueling at Shell Hydrogen station in Southern California, 2016. Photo: US Dept. of Energy. Public domain

The Story of Hydrogen

Hydrogen is the first element on the Periodic Table of Elements, a single proton and electron. As a free gas, it is always found as a diatomic molecule: two hydrogen atoms sharing a covalent bond, thus the formula H_2. Its primary applications are not related to energy but rather the production of ammonia for agriculture and the synthesis of other hydrocarbons. The overwhelming majority of hydrogen (98% worldwide) is produced from fossil fuels, specifically natural gas and coal. Carbon-free (green) hydrogen is usually made by water electrolysis using electricity generated entirely from carbon-free sources, i.e., solar, wind, hydroelectric, nuclear, or geothermal. But there are two, possibly three, exceptions for carbon-free hydrogen, all in the research phase. One is Solar-to-Hydrogen (STH), in which water is decomposed directly by solar insolation via *photolysis*. Another is direct water splitting at very high temperatures found only in a nuclear reactor or at the focal point of a large concentrating solar collector. Lastly, there is the hopeful possibility of naturally occurring hydrogen in deep geologic formations. A variation of this method is the injection of water into these rock formations to stimulate natural hydrogen production. Both are referred to as *geologic hydrogen*. This last development is still uncertain, but if successful, it has the potential to change the role of hydrogen from exclusively an energy carrier to an energy source, with a profound impact on the utility of hydrogen to aid in decarbonizing transportation energy.

Recently, the potency of hydrogen as an indirect greenhouse gas with a 20-year GWP of 37.3 ± 15.1 (37.3 times worse than carbon dioxide) [3] has been recognized, leading to concerns that leakage of either geologic or synthetic hydrogen into the atmosphere could potentially contribute to global warming. Could this be a problem? Experience at this time is lacking, being mostly extrapolated from natural gas. Hydrogen is a tiny molecule, just two protons in diameter, with a greater propensity to leak or permeate through even dense ground formations, enclosures, processing equipment, pipes, valves, and storage tanks than any other gas. Fugitive (leaked) hydrogen, like methane, could nullify some (or even all) of its benefits as a zero-carbon fuel, especially in transportation applications. Leakage is typically

Fig. 9.2 Hydrogen fuel cell bus, London, England, 7 April 2004. Photo by Spsmiler. https://commons.wikimedia.org/wiki/File:London-fuel-cell-bus-route25.jpg. Public domain

expected during hydrogen extraction, transport, and compression. However, this limitation may be optimistic when it comes to hydrogen FCVs, which contain complex assemblies of ultra-high-pressure hydrogen plumbing and components. The current recall of all Hyundai Nexo FCVs in Korea and the USA for potential hydrogen leakage, possible at a very high rate, is discussed in the later section Hydrogen Fuel Safety (Fig. 9.2).

Hydrogen used as a fuel cell reactant produces only water as a byproduct. Hydrogen burned in an excess of air produces only water, but if burned in a nearly stoichiometric mixture in a combustion engine, it produces levels of NOx similar to or greater than those from gasoline.

But by far the most prominent questions about the benefits or detriments of hydrogen are the GHG emissions and energy efficiency from its production. Given popular perceptions and massive public and private investment in large-scale hydrogen deployment, it is essential to examine these details closely [4].

Is hydrogen the ultimate solution to climate change or a distraction from better options? Hydrogen has become an almost magical topic that has been particularly susceptible to misinformation, some of which has created unrealistic expectations, one example in citation [5]. While this book focuses on transportation energy, in the case of hydrogen, the breadth of these expectations across energy and the environment warrants more extensive discussion.

Making and Burning Hydrogen

The simplicity of hydrogen production, combustion in an engine, or reaction in a fuel cell is unique among all fuels:

$$H_2O + \text{electricity} \rightarrow 2H_2 + O_2 \, (\text{electrolysis})$$
$$2H_2 + O_2 \rightarrow H_2O + \text{heat} \, (\text{combustion})$$

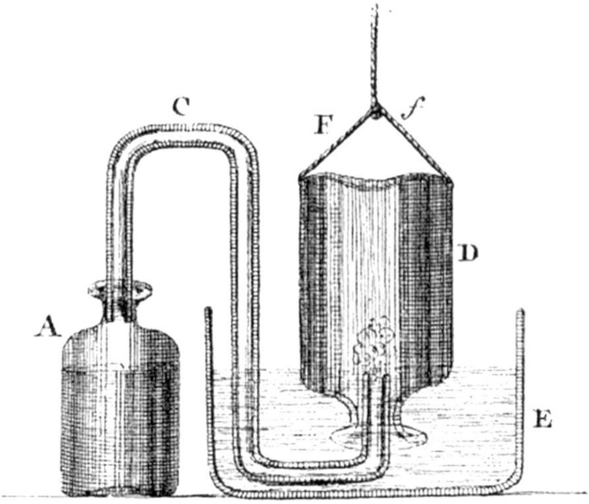

Fig. 9.3 Illustration of the 1766 Cavendish apparatus for collection of "Inflammable Air" from reaction of iron and sulfuric acid. From: H. Cavendish. *Cavendish's pneumatic trough.* Philosophical Transactions of the Royal Society, 1766, 56, p. 141, Fig. 9.1. Public domain

In 1520, the Swiss physician Paracelsus made the first recorded observation of hydrogen liberated from the reaction of metals (iron, zinc, and tin) with sulfuric acid. Hydrogen collection from these reactions, and generation by electrolysis of water would not be published until centuries later by English scientist Henry Cavendish in 1766 and William Nicholson and Anthony Carlisle in 1800 [6] (Fig. 9.3).

Arguably, the first practical use of hydrogen was as a component of various fuel gases discovered or invented in the late eighteenth and early nineteenth centuries. The common components were primarily carbon monoxide and hydrogen, with minority components including methane, ethylene, and carbon dioxide. Coal gas and town gas were common names for early fuel gases made from coal of various qualities, combusted in the absence of air, a process now referred to as *coal gasification*. With the later addition of steam, the resulting *water gas* contained a greater fraction of hydrogen making it cleaner burning, but was less desirable for illumination due to its less luminous flame. Carbonaceous forms of biomass (e.g., wood, crop residues, peat) sometimes replaced coal, and the resulting product was more broadly referred to as *producer gas*. Its energy content is primarily from carbon monoxide, plus smaller fractions of hydrogen and unreacted nitrogen, depending on the hydrocarbon content of the coal or biomass:

$$2C + O_2 \rightarrow 2CO$$

With the addition of superheated steam, some of the carbon monoxide from the coal oxidation combines with the water via the *water gas shift reaction* [7]:

$$H_2O + CO \rightarrow H_2 + CO_2$$

A concise history of early fuel gases can be found in citation [8].

Fuel gases containing carbon monoxide and hydrogen were initially used for illumination, and later for cooking and heating. By the late nineteenth century, almost all major European cities operated "gas works" that used coal to manufacture fuel gas. In some locations in England and continental Europe, town gas remained in use until the early 1960s. Its toxicity was well known. The 1963 suicide of the famous English/American poet Sylvia Plath in London was by carbon monoxide poisoning from a kitchen oven [9]. By the 1960s, the use of manufactured gas for domestic service had almost all (except, unfortunately, for Ms. Plath) been replaced by natural gas obtained from wells, and electric lighting had long before replaced gas lamps. Urban "gaslight districts" in affluent cities persist to this day for their decorative charm, but they use natural gas.

Manufactured H_2/CO mixtures are referred to as *syngas (synthetic gas),* which is the starting point for producing a variety of other fuels and commodities that require hydrogen, for example, ammonia as an agricultural soil amendment. Worldwide, hydrogen production from coal remains the second most common method for industrial and agrarian hydrogen production.

With the availability in the early twentieth century of plentiful natural gas in some countries, much of the hydrogen production shifted from coal to less carbon-intensive production by reacting methane in natural gas with superheated steam above 1000 °C:

$$CH_4 + H_2O \rightleftharpoons CO + 3H_2$$

Additional steam beyond the stoichiometric requirement then facilitates a reaction (water gas shift) that generates additional hydrogen by reacting steam at high temperature with the carbon monoxide (CO):

$$CO + H_2O \rightleftharpoons CO_2 + H_2$$

This overall process is referred to as *Steam Methane Reformation* (SMR), which (ideally) yields a mixture of only hydrogen and carbon dioxide from the natural gas feedstock:

$$CH_4 + 2H_2O \rightleftharpoons CO_2 + 4H_2$$

On a mass basis, this means that 16 kg of methane from natural gas produces 8 kg of hydrogen and 44 kg of carbon dioxide. The generation of 5.5 times greater mass of carbon dioxide than hydrogen is the main climate impact concern for hydrogen produced by SMR, which dominates hydrogen production in the United States and Europe. The use of electrolysis to make so-called *green hydrogen* is much more energy intensive and expensive, and its relative *greenness* depends entirely on the source of the electricity.

At this time, approximately 98% of hydrogen worldwide is produced from either natural gas or coal via these processes, with the majority from natural gas using SMR. The largest market for hydrogen is the production of ammonia, which is critical to the high productivity of modern agriculture, replacing depleted nitrogen in the soil.

Hydrogen production from fossil fuels is just a surrogate for these fuels, specifically coal and natural gas (Fig. 9.4). In a 2023 report by the International Energy Association (IEA), the average GHG emissions from hydrogen production is 12–13 kg CO_2e/kg H_2, with the range reflecting different accounting methods for byproduct hydrogen production in oil refineries [10]. As discussed in citation [11], hydrogen production from natural gas to make ammonia alone is responsible for 0.93% of all anthropogenic greenhouse gas emissions [12], releasing an average of

Fig. 9.4 Entrance to Shell Oil Corp. hydrogen SMR Production Facility. Photo: US Dept. of Energy, http://www.publicdomainfiles.com/show_file.php?id=14024560215339. Public domain

two to three tons of CO_2 into the atmosphere to manufacture one ton of ammonia produced for fertilizer. As of 2012, 72% of ammonia production worldwide was from natural gas [13], with the remainder produced almost entirely from coal.

Hydrogen as a Transportation Fuel

Hydrogen's potential as a transportation fuel dates back several centuries. It has been "rediscovered" every generation, starting in the 1920s with the conversion of several trucks to hydrogen by Rudolph Erren [14] in Germany. Aside from the natural aversion of over-empowered entrepreneurs to literature searching, these recurrent epiphanies are enabled by the ease by which any spark ignition (gasoline) internal combustion engine can run on hydrogen. Simply dumping almost any continuous flow of hydrogen into the intake manifold of a spark ignition engine will allow the engine to at least run due to the unusual flammability limits of a hydrogen-air mixture. In terms of the normalized air-fuel mass ratio *lambda*, hydrogen will burn in a range from more than $\lambda = 5$ (extremely lean) to 0.2 (extremely rich); i.e., *almost any mixture* of hydrogen and air will burn. An unfortunate concomitant of the numerous unscientific demonstrations of hydrogen engines is the ease by which the general public is duped into believing conspiracy-laden claims connected to hydrogen, such as engines that are claimed to run on water [15, 16].

Hydrogen's combustion characteristics are different than almost any other fuel, gaseous or liquid. Engine control by fuel quantity rather than air throttling is uniquely possible due to its extremely wide flammability limits. This allows the engine to be controlled without an air throttle, eliminating part-throttle losses caused by the engine acting as a vacuum pump. This improves part load efficiency, which, in conjunction with its fast stoichiometric combustion rate, gives hydrogen the highest potential thermal efficiency of all combustion fuels. However, its extremely low ignition energy requirement leads to the unique problem of intake manifold backfiring, usually discovered by new advocates immediately after attempting to achieve an acceptable power output from a hydrogen-fueled engine. Because of the extremely low density of hydrogen, it can occupy as much as 30% of the intake fuel-air charge, significantly reducing the power output of the engine. Emissions of oxides of nitrogen (NO_x) can be greater than gasoline or diesel fuel.

After extensive research and development of hydrogen combustion engine technologies in the 1970s and 80s, interest in hydrogen fuel shifted from IC engines to its use as a reactant in electrochemical fuel cells for electric vehicles. The original excitement about hydrogen Fuel Cell Vehicles was due to the higher energy storage capacity (e.g., range) of a hydrogen FCV compared with that of battery electric vehicles (BEVs) that relied on the state-of-the-art batteries of the 1990's—all objectionably heavy for the electric energy they could store. At least six major and many more minor manufacturers have demonstrated prototype or production FCVs within the past approximately 25 years: Toyota, Honda, Hyundai, BMW, Mercedes-Benz, and Renault. At this time, three still have FCVs in production: Toyota, Hyundai, and Honda.

This chapter covers both hydrogen combustion and fuel cell vehicles since both rely on the same fuel and fueling infrastructure and face most of the same technical obstacles. Respectful of the historical sequence, I will first discuss hydrogen as a combustion fuel for automotive ICVs. Then, I will discuss hydrogen fuel cell vehicles. Interspersed through both applications, I will discuss hydrogen production and storage and hydrogen applications for aircraft, rail, marine transport, and heavy trucks.

Advantages of Hydrogen as a Vehicle Fuel

Hydrogen Can Be Made from Almost Any Energy Source

If any fuel could be called a universal fuel, it is hydrogen. It can be used in virtually any combustion engine (internal or continuous) and is the only practical fuel for fuel cells operating at ambient temperature.

The possible energy or chemical sources for making hydrogen seem nearly limitless. It can be trivially made from water by electrolysis using electricity. However, the multiple energy conversion steps from energy source to hydrogen fuel greatly reduce its net energy efficiency compared with other fuels and energy carriers Except for electrolysis using exclusively zero-carbon energy, the byproducts of its production all include greenhouse gases.

High Combustion Energy Efficiency

Hydrogen is well known as a clean-burning fuel for ICEs and a clean reactant for fuel cells. In both combustion and electrochemical reactions, the essential converstion step is

$$H_2 + \frac{1}{2}O_2 \rightarrow H_2O + \text{energy}$$

where energy could be heat (from combustion) or electricity (from fuel cell output).

In reciprocating IC engines, thermal efficiencies of 40% are possible for lean H_2-air mixtures [17]. These efficiencies are the highest of all engine fuels. For comparison, gasoline ICEs achieve up to 32%, with 28% a typical thermal efficiency.

Hydrogen Is the Only Viable Fuel for Room-Temperature Fuel Cells

Other than hydrogen, the only potential competitors as fuel cell reactants are methane from natural gas, used in high-temperature fuel cells, or methanol, which has a much lower conversion efficiency and cannot be fueled continuously. Consequently, all FCVs use hydrogen fuel.

Hydrogen polymer electrolyte membrane or proton exchange membrane (PEM) fuel cell efficiencies as high as 75–80% have been achieved, although, in practice, production-size fuel cells typically operate at 60–65% efficiency at ambient temperature. The high-temperature operation can improve efficiency but has not been used in FCVs [18] for safety and complexity reasons.

Highest Energy Mass Density of All Combustion Fuels

When our objective is mobile storage onboard an FCV, hydrogen presents a strange conundrum. Its mass (gravimetric) *energy density* is the highest of all fuels. This would be great news, except that the volumetric energy density (energy per unit volume) at normal temperature and pressure (NTP) is the worst of any combustion fuel. Hydrogen must be compressed to extremely high pressure (70 MPa) to store enough energy to provide a reasonable vehicle range in an FCV or combustion hydrogen vehicle. To make matters worse, hydrogen, unlike butane, propane or DME, does not spontaneously liquefy at normal temperature under any pressure. It can only be forced into liquid form with extreme refrigeration to –253 °C (20 K). This is the second-lowest is more accurate boiling point of any substance. Only liquid helium at –271 °C (2 K) has a lower liquefaction temperature.

Other potential methods for vehicular hydrogen storage include storage in interstitial metal hydrides or chemical hydrides, or onboard generation from methanol or ammonia. Hydrogen storage methods will be discussed later in this chapter.

Despite its low volume energy density, the exceptionally high mass energy density of hydrogen makes it a preferred rocket fuel, used in the Saturn C5 booster stages of the Apollo moon missions and for the main engines of the NASA Space Shuttle. This also makes it a potential fuel for commercial jet transport. These are among the few applications for which *mass* energy density matters more than *volume* energy density.

Greater Range Compared with BEVs

Before the development of practical lithium batteries, hydrogen fuel cells were a mature technology that could provide more electric energy on board at a much lighter weight than the best available batteries. In either compressed or liquid form, hydrogen's mass energy density was superior to that of an electrochemical battery. Hydrogen fuel cells were a technically superior solution for vehicles at that time.

For example, the 2024 Toyota Mirai XLE FCV [19] has an EPA range of 402 miles, while a 2024 Toyota bZ4X Electric offers a range of "up to 252 miles" [20].

But since the late 1990s, batteries have radically improved. The range of a 2024 electric passenger car varies from 149 miles for a Nissan Leaf S, to 520 miles for the Lucid Air Touring [21, 22]. The ranges of the Toyota Mirai XLE and the Hyundai Nexo HFCV are 402 and 380 miles respectively [23, 24]. The time required for rapidly charging lithium batteries also reduced significantly, although it is still longer than the refill time of FCVs. A high-rate 60% battery recharge of a 100 kW battery can now be accomplished in as little as 30 min. Post-2020, the original advantages of FCVs over BEVs have become mostly moot.

An FCV can be thought of as an electric vehicle with an onboard fuel cell that continuously charges the battery that powers the electric motor. This requires a fairly large "ballast" battery to match the power production from the fuel cell with the rapidly changing peak power demands of the vehicle. The battery in an FCV is subject to RTE (Round Trip Efficiency) losses the same as in a BEV. For an LNMC battery in this application, the high-rate cycling is similar to or greater than that of a typical gasoline-electric hybrid. By comparison, as long as there is no leakage, compressed hydrogen incurs no losses related to the vehicle's use or non-use. For liquid hydrogen, there is continuous fuel loss when not in use, as the cryogenic liquid continuously boils due to the small amount of heat that penetrates the tank insulation, and this gas must be vented, usually through a catalytic converter to transform it into water vapor. Nevertheless, the storage of a liquid hydrogen vehicle indoors always involves some risk of explosion of the slowly vented hydrogen gas.

Large Vehicles Favor FCVs over BEVs

For BEVs, the range limit is proportional to the battery mass. For an FCV, the limit is typically the *space* available onboard for the compressed or liquid hydrogen tanks. As the range of a BEV increases, the vehicle weight and cost increases significantly since the primary source of the vehicle weight and cost is the battery, for which kg/kWh and $/kWh are nearly constant. The mass of the vehicle structure required to safely carry additional batteries also increases. This range/mass relationship does not apply to a hydrogen FCV for which the range is determined by the capacity of the compressed or liquid hydrogen tank(s) alone. Double the range requires double the tank volume, but since the tanks are a relatively small fraction of the overall vehicle weight, and the fuel mass is almost trivial, twice the range requires only a small incremental vehicle weight.

For example, compare a 2023 Toyota Mirai with a 2024 Nissan Leaf S:

The 2023 Toyota Mirai LXE has a *curb weight*[2] of 1930 kg [25]. The total weight of its two high-pressure fuel tanks is 87.5 kg. To increase its range by 50% would require an increase in the tank capacity and weight by about 50% (44 kg). The

[2] *Curb weight* is the mass in kg or lbs of a car ready to drive with a full fuel tank, but not including passengers or cargo. It is the most accurate indicator of practical vehicle mass, somewhat greater than the *dry weight* that is usually quoted in manufacturer's specifications.

increase in fuel mass would be exactly 50% (2.5 kg). Therefore, a 50% range increase would require only a **2.3%** increase in the vehicle weight.

For comparison, the curb weight of a 2024 Nissan Leaf S with its 40 kWh battery and 149-mile range is 3509 lbs. A Nissan Leaf S-*Plus* has a 50% higher capacity 60 kWh battery and a 226-mile range. The *Plus* weighs 3901 lbs [26], an **11%** increase in vehicle weight for a 52% increase in range.

Possibly the only ground vehicle application that favors H_2 FCVs over BEVs is large vehicles with long-range requirements, e.g., long-haul trucks, railroad engines, ships, and aircraft. For these applications, cryogenic liquid hydrogen may be a better option than batteries due to its reduced mass compared with batteries.

Here is a back-of-envelope comparison of the fuel system mass required to provide the same range for a long-haul semitruck powered by each fuel option: diesel fuel, liquid hydrogen, or an LNMC battery:

Diesel: Dual 120-gallon side-saddle diesel fuel tanks on a long-distance freight hauler would weigh when full about **2000 lbs** and contain about 35,000 MJ of fuel energy. At an average of 6 MPG (23 MJ/mile), this fuel quantity provides a range of about 1500 miles.

LH_2 fuel cell electric: Adjusting for the slightly higher efficiency of a fuel cell electric drivetrain, the equivalent energy would require about 1200 gal of liquid hydrogen weighing, including the cryogenic tanks, about **1,000 lbs**.

LNMC battery electric: For state-of-the-art batteries having an energy density of 1.0 MJ/kg, adjusting for the higher efficiency of an electric drivetrain, an LNMC battery weighing about **80,000 lbs**. would be required to provide the equivalent range. Considering that the maximum legal weight of a tractor-trailer rig in most states in the USA is 80,000 lbs, the truck would be illegal to operate even if it carried no freight.

Reduced Refueling Time Compared with BEVs

The shorter refueling time of FCVs compared with EVs remains an incentive favoring FCVs. Toyota and Hyundai both advertise that their FCVs can be 80% refueled in 5–7 min, compared with a Tesla S with a 100 kWh battery that can be Level-2 home charged to 80% in 6–12 hours, or fast charged to 60% capacity at a 200 kW rate in 15–20 min. Charge times depend on available charger power. The difference is not very significant for a fast-charged BEV, but most EV charging is done with Level-2 chargers at home or near business.

Disadvantages of Hydrogen as a Vehicle Fuel

Hydrogen Fuel Storage

Onboard hydrogen storage has been a research challenge since the first experimental hydrogen vehicles in the mid-1930s. At standard temperature and pressure, hydrogen has the lowest volume energy density of any fuel. Large volumes are

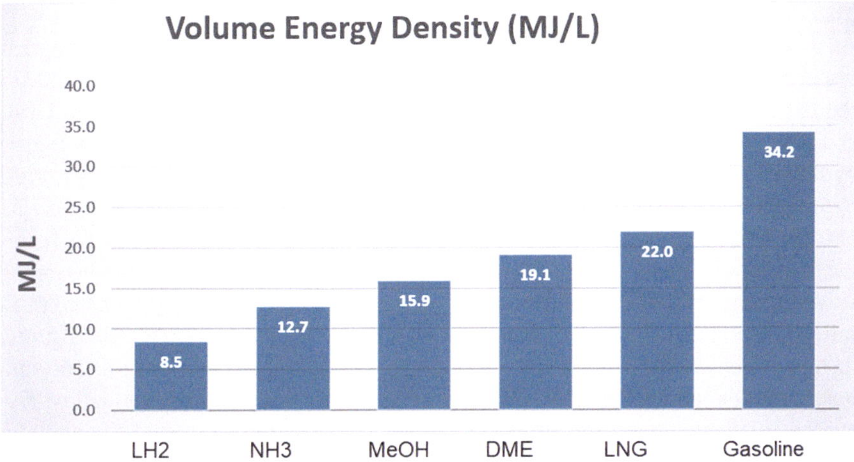

Fig. 9.5 Comparison of volume energy density of liquid hydrogen (LH_2) with other alternative fuel options. Data from standard chemistry references. Chart by author

required to store even small amounts of energy in the form of hydrogen, despite the fact that the fuel mass is quite small. Volume energy density at 70 MPa (10,000 psi) is about the same as a lithium/cobalt/manganese battery.

As discussed earlier, four storage methods have been demonstrated for hydrogen vehicles: High pressure tanks, cryogenic (liquid hydrogen), metal hydrides, or chemical hydrides. High-pressure hydrogen storage has emerged as the winner for passenger HFCVs, but no method for onboard hydrogen storage is as compact or convenient to refuel as gasoline or even CNG.

Figure 9.5 compares the volume energy density of liquid hydrogen with other alternative fuels. Its volume energy density is about one-quarter that of gasoline and half that of DME (*Dimethyl Ether*, to be discussed in Chap. 13).

Reduced Power Output in Hydrogen ICEs

For combustion engine hydrogen vehicles only: Compared with a gasoline engine, the power output of a naturally aspirated hydrogen engine is limited by the large 30% volume fraction of the intake fuel-air charge that the hydrogen must occupy to form a stoichiometric mixture. This effectively reduces the displacement of the engine to only 70% compared with liquid fuel. Maximum power output is reduced proportionally. Charge air pressurization (supercharging or turbocharging) is almost necessary for hydrogen engines to restore the power reduction caused by this 30% volumetric efficiency reduction. The power reduction is even greater in practice due to concerns about intake backfiring and NO_x emissions, which usually require that the hydrogen engine run with a very lean mixture [27].

Intake Manifold Backfiring in Hydrogen ICEs

As discussed earlier in the context of hydrogen engine emissions, one of the operational problems recognized as early as Erren's converted lorries (trucks) was due to the extremely low ignition energy requirements of hydrogen, as little as 0.1 mJ, compared with the typical 100–300 mJ energy output of a spark plug. This makes hydrogen-air mixtures extremely easy to ignite. Consequently, when a pre-mixed hydrogen-air mixture is inducted during the intake stroke of the engine, it may be unintentionally ignited by even the slightest hot surface or hot particle in the residual exhaust gas after the end of the exhaust stroke. Egregious hot spots occur on the valves, spark plug, or piston. All early hydrogen conversions of ICEs that used mixture formation upstream of the intake valve (i.e., used a gas carburetor) were prone to this unintended pre-ignition of the intake charge, which results in a loud and potentially dangerous *backfire*. (Means for managing this phenomenon will be discussed in the later section, *Backfire suppression*.) Consequently, the maximum equivalence ratio for hydrogen engines that use a manifold-mixed air-fuel charge is typically $\Phi = 0.6$, which is also the threshold of NO_x production. This limits the engine power output even further. For example…

The infamous Hindenburg airship that was destroyed while landing in 1937 used six large 16-cylinder Daimler diesel engines, each producing a maximum of 1320 HP on diesel fuel. The fuel alone for a transatlantic crossing weighed 58,880 kg. The original design had called for the engines to be powered by hydrogen from the airship's huge envelope. Plans had been made to add two more engine nacelles to compensate for the anticipated reduction in power. But when one of the engines was converted to operate on hydrogen and tested on a dynamometer, the maximum power that could be obtained (without backfiring) was 300 HP. The hydrogen engine plan was abandoned since even with the additional two engines, the total power would be only one-third of the requirement for the airship to achieve its specified 135 kph (76 mph) cruising speed [28].

Details of the D-LZ129 Hindenburg are covered in the *Hydrogen safety* section of this chapter.

Risk of Rupture of a High-Pressure Hydrogen Tank in Collision

With hydrogen stored at ultra-high pressure (70 MPa) on board an FCV, the risk of the rupture of the specialized tank in a collision is a major concern. But the potential harm isn't due so much to the combustibility of the fuel; it is due to the kinetic energy released from the rupture, which can be devastating to the FCV as well as proximate vehicles and pedestrians, or worse, an enclosed garage or tunnel.

The safety record of FCVs to date is excellent. But this potential risk is obviously greater than that of a conventional gasoline or diesel vehicle. The risk has factored into the success or failure stories of many startup companies that have claimed to "soon" offer hydrogen vehicles (e.g., Apricale, Nikola, Hyperion, Hyson) or have actually demonstrated vehicles (BMW, Volvo, Daimler). Almost all accidents related to H_2 FCVs have occurred at refueling facilities, for example, an explosion at a hydrogen fueling station in Norway in 2019 [29], discussed later in this chapter.

Complexity and Reliability of FCVs compared with BEVs

The level of technology onboard an H_2 FCV is incredibly advanced, more worthy of a spacecraft than an automobile. And this is in addition to the conventional components found in a BEV (including a battery). The hydrogen-specific components are the ultra-high-pressure hydrogen tanks, intake air scrubber (for removal of CO_2 from the intake air), the hydrogen-oxygen fuel cells, and specialized control and safety equipment. For the 2023 MY, the $60,135–$63,585 MSRP of a Hyundai Nexo or the $50,595–$67,095 MSRP of the Toyota Mirai, while seemingly high, were probably less than their actual production and amortized development costs, a common situation for radically new automotive technologies. This reality suggests that FCV manufacturers have been willing to accept a financial loss on hydrogen FCVs, looking hopefully for the niche-market segment to grow in the future, and/or to support a "green" corporate image which aids in the sale of mainstay ICV and BEV products. All three remaining FCV manufacturers have been willing to cover extended warranty costs of the vehicles, which for the hydrogen fuel-cell and electric power systems are beyond the experience of dealer service personnel and require a large investment in training, specialized tools, and documentation for a very-low-production vehicle.

Poor Well-to-Wheel Efficiency and Climate Impact

From a climate solution point of view, the main problems associated with hydrogen are its poor *well-to-wheel* (WTW) energy efficiency and the carbon footprint of its production. The unusually large number of energy conversion steps, need for gas compression or liquefaction, and fuel transport requirements, turn a fuel with exceptionally high efficiency onboard a vehicle into one of the worst for WTW efficiency. Its net carbon footprint for existing production methods is also poor, although this depends on the "color" assigned to it in various production scenarios, to be discussed later.

The promotion of hydrogen as a transportation fuel can be considered an example of the consequences of a narrow focus on its sound-bite energy and environmental benefits, while diminishing or ignoring the immutable limits of its physical properties. From Politico, Feb 1, 2024 [30]:

> There are several reasons hydrogen hasn't taken off as a passenger fuel. One is the high cost—about $25 per kilogram in California—which car companies have had to heavily subsidize to make driving their vehicles feasible for customers. At that price, it costs about $135 to fill up a standard car, whereas charging an EV at a commercial station costs $10–30, which makes that option far more economical, even if it requires more frequent charging.
>
> The Environmental Defense Fund calculates that it requires three to nine times the energy to power a hydrogen fuel cell vehicle as it does a battery electric one. The stations are also expensive: The Energy Commission notes that building a hydrogen fueling station costs $6.5 million, while an EV charging station costs about $110,000.

As of March 2024, hydrogen at 70 MPa is being dispensed at most of the remaining hydrogen refueling stations in California for $36/kg USD, greatly exceeding assurances given to FCV owners of $7/kg and the DOE target of $1/kg by 2030. The positive climate impact of the hydrogen fuel cell vehicles in the US is also a ruse, since all of the hydrogen currently dispensed at hydrogen highway stations is produced from natural gas using SMR despite CO_2 emissions that are no better than the direct combustion of the natural gas. While government funding agencies and investors continue to remain hopeful, hydrogen has not yet fulfilled its promise to reduce the carbon footprint of transportation by transitioning to green (carbon free) hydrogen, or even *blue* hydrogen, to be discussed later.

Hydrogen Internal Combustion Vehicles (ICVs)

As mentioned previously, hydrogen is both a combustion engine fuel and a reactant for fuel cells. In this section, its history, technology, and the current state of the art as an ICE fuel are discussed. Its application to fuel cell vehicles is discussed in the following section.

Here is a bullet summary of its attributes as a combustion fuel:

- SI engine fuel, like gasoline.
- Energy carrier, NOT an energy resource unless obtained directly from geologic sources.
- A carbon-free fuel for combustion, but overall carbon intensity depends upon the energy source.
- Hydrogen is more prone to leakage than any other gas, and special care must be taken to assure that it is not allowed to escape from pressurized fuel systems.
- One of only two non-hydrocarbon combustion fuels for IC engines.
- Emission from lean combustion is only water vapor.
- Emissions from fuel-air mixtures above 50% of stoichiometric include oxides of nitrogen at levels potentially greater than gasoline.
- Highest combustion energy per unit mass of any ICE fuel.
- Lowest combustion energy per unit volume at atmospheric pressure and temperature.
- Very wide limits of flammability make it easy to ignite in almost any mixture with air.
- Most gasoline engines can be converted to run on hydrogen with fuel system modifications.
- Engine power output of a converted engine is typically less than half that of the gasoline performance of the engine.
- Theoretically very high octane rating, but not necessarily amenable to very high compression ratios due to its very low ignition energy.
- Intake manifold backfiring is a problem if a conventional gas mixer (e.g., natural gas carburetor) is used. Problem reduced or eliminated by timed fuel injection.
- Non-toxic. Not detectable by sight or smell due to lack of color or odor.

- Combustion produces no luminescence.
- Low density assures leaks dissipate quickly, but note the GHG consequences mentioned above.
- Due to low density, less severe consequences from fuel tank rupture compared with other fuels.
- Several methods for onboard fuel storage or generation, but all have lower energy density than any other fuel and involve exotic technologies:
  - Ultra-high-pressure tanks (70 MPa) dominate passenger car applications.
  - Cryogenic (20 K) liquid hydrogen storage is expensive due to liquefaction and storage.
  - Solid-state storage in metal hydrides is safest, but the hydrogen mass density in existing hydrides (1–5%) is currently lower than compressed or liquid hydrogen.
- Production energy efficiency varies greatly with energy source.
  - Best: synthesis from coal or natural gas (black or gray hydrogen).
  - Worst: electrolytic (green) hydrogen.
- Climate impacts of production varies with energy source:
  - Best: electrolytic (green) hydrogen.
  - Worst: Synthesis from coal or natural gas (black or gray hydrogen).
- In general, cleaner hydrogen production always comes with a higher cost.
- Carbon capture and storage (CCS) could mitigate some of the CO_2 emissions for hydrogen made from natural gas or coal, but not nearly enough to consider the resulting hydrogen to be carbon free or even low carbon (Fig. 9.6).

Early Hydrogen ICE Vehicles

In the late 1920s and early 1930s German engineer Rudolf Erren converted several trucks, buses, and even a submarine to use hydrogen or hydrogen-rich producer gas mixtures [31]. While probably not the first to fuel an ICE with hydrogen, his work became highly relevant during the petroleum-starved years of World War II.

Fig. 9.6 Russian hydrogen powered GAZ-AAA truck, 1941. From Hydrogen Cars Now, "Tense Times in Hydrogen and Fuel Cell History" 28 June 2017 https://www.hydrogencarsnow.com/index.php/history/tense-times-in-hydrogen-and-fuel-cell-history/ Article copyright George Wand. Historical image public domain

The UCLA Hydrogen Cars, 1972–1978

Among the first modern ICVs to receive national exposure were two vehicles constructed by engineering students at UCLA 1972–78 [32]. The first was a Barris-customized AMC Gremlin that stored hydrogen in two steel tanks, together weighing 182 kg (400 lbs) and storing 2.7 kg (6.0 lbs) of compressed hydrogen at 41 MPa (6000 psi). Since 6000 psi hydrogen was not easily obtained at the time, most of its competitions and appearances relied on readily available 2000 PSI hydrogen, giving it a range of only about 26 km (16 miles) (Figs. 9.7 and 9.8).

In 1975, a UCLA team constructed an experimental liquid hydrogen-powered mail delivery vehicle under contract to the United States Postal Service (USPS). It used a 190 L (50 gallon) spherical cryogenic dewar custom manufactured by Minnesota Valley Engineering to store approximately 13.6 kg (30 lbs) of LH_2 at 20 Kelvins, giving it a useful range of 400 miles. Liquid hydrogen donated by Union Carbide Corp. could only be obtained at its hydrogen production plant, 65 miles distant from UCLA. Fortunately, with a 400-mile range, this distance was not an issue (Fig. 9.9).

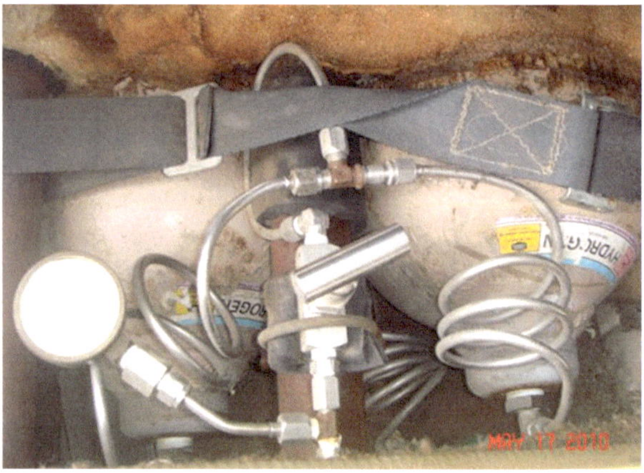

Fig. 9.7 High pressure fuel storage tanks in UCLA-modified AMC Gremlin. Two steel pressure tanks weighing 91 kg each stored a maximum of 2.7 kg (6.0 lbs) of hydrogen at 41MPa (6000 psi)

Fig. 9.8 The UCLA converted hydrogen powered AMC Gremlin. Photo: UCLA Hydrogen Car Team, 1973

Fig. 9.9 The 1974/5 UCLA Liquid Hydrogen mail service vehicle. Built under contract to the US Postal Service as an evaluation platform. Photo: author, 1975

Fig. 9.10 Liquid hydrogen converted Datsun B210 constructed by students at Musashi Institute, Japan. Photo from the 1975 SEED (Student Engineered Economy Design) Rally. Public domain

In 1976, this vehicle was involved in a major freeway accident while being towed to a competition but survived without damage to the LH_2 tank or fuel system. This is believed to be the first and possibly only case of a liquid hydrogen passenger vehicle on-road accident.

Musashi Institute, Japan, 1975

In the 1970s, hydrogen in compressed or liquid form was used in several experimental vehicles constructed by university student teams. A 1974 Datsun Sunny (B210 in the USA) was converted in 1975 to liquid hydrogen by students at the Musashi Institute of Technology in Japan. The photo of Fig. 9.10 was taken during the 1975 Student Engineered Economy Design (SEED) rally, a 1250-mile alternative fuel vehicle rally between Western Washington State College in Bellingham, WA, and the University of California at Los Angeles (Fig. 9.10).

Hydrogen Passenger Buses, 1982–84

In 1976, the dual-fuel hydrogen-gasoline 11-passenger van shown in Fig. 9.11 was constructed and used in limited service in Moscow [33]. It was probably the first operational hydrogen bus since Rudolf Erren's conversions, 1920–1945.

In 1978, Roger Billings of Provo, Utah conducted several public demonstrations of a Winnebago shuttle bus modified to run on compressed hydrogen [34]. It was reported in local newspapers that it successfully made several 13-mile trips between Orem and Provo Utah (Fig. 9.12).

Mr. Billings was a charismatic entrepreneur who founded Billings Energy Research and many other short-lived enterprises, as well as a polygamous church [35], a religious cult [36], and three unaccredited educational institutions, one for the sole purpose of awarding himself a doctorate degree [37]. His many claims of "firsts" in the hydrogen area have all been without merit, but his early promotion of hydrogen as a vehicle fuel deserves recognition.

Fig. 9.11 Hydrogen/gasoline dual-fuel van demonstrated in the USSR in 1976. Photo public domain, https://www.greencarreports.com/news/1097672_more-olympics-green-cars-soviet-hydrogen-gasoline-minibuses-of-1980

Fig. 9.12 Winnebago shuttle bus modified to run on compressed hydrogen by Billings Energy Research, possibly as early as 1976. Photo and information from https://www.rogerebillings.com/1976-the-hydrogen-bus/. Public domain

In 1979, another hydrogen shuttle bus was constructed in Latvia, USSR, powered by fuel cells using compressed hydrogen storage. The *Kvant-RAF hydrogen fuel cell van* was demonstrated at the 1980 Summer Olympics in Moscow [38].

Another noteworthy early hydrogen transit bus was a 24-passenger transit bus constructed in 1981 at the University of Denver Research Institute, co-funded by the US Dept of Energy and the Denver Regional Transit District [39]. It incorporated a prototype hydrogen fuel injection system and was turbocharged to restore the power loss characteristic of hydrogen engines. It also incorporated a system that purged the engine and exhaust system of residual hydrogen following shutdown to prevent startup exhaust explosions that often occurred in previous hydrogen vehicles.

One of the few other early hydrogen buses for which published information is available was a Mercedes-Benz TN 310 10-passenger shuttle bus converted by the manufacturer to hydrogen in 1984. The van used an internal combustion engine delivering 75 kw at 5600 rpm and a metal hydride hydrogen storage system. A small number of these prototypes were deployed in medical and passenger service in Berlin, Germany [40] (Fig. 9.13).

Specialized Vehicle Applications

Many other examples of hydrogen vehicles appeared from 1980 through approximately 2005. None have been more than one-off experiments, not put into production or useful service.

However, one specialized application that has seen actual use has been indoor-operated forklift trucks which must be free of harmful emissions. Hydrogen provided a clean alternative to propane or battery-powered forklifts, with greater operational time compared with lead-acid batteries—refueling once a day rather than recharging or exchanging batteries two or three times each day. The excessive weight of a room-temperature metal hydride tank such as iron-titanium alloy is actually a plus in this application since it replaces the heavy counterweight on the back of the lift truck.

In 1979 an ACC50 Allis Chalmers forklift truck was converted to hydrogen at the Denver Research Institute as a demonstration for sponsor International Nickel Company (Inco), a producer of iron-titanium metal hydride alloys for hydrogen

Fig. 9.13 Modified 1984 Mercedes TN 310 passenger van operated in Berlin. From citation [40]. Public domain

storage systems. Allis Chalmers was possibly the earliest company to embrace hydrogen as a vehicle fuel, but never put this prototype into production and became insolvent in 1999. As of 2023, a hydrogen fuel cell electric forklift truck was still available by special order from Hyster-Yale [41]. Indoor-operated industrial vehicles have been the only successful industrial equipment market for hydrogen vehicles [42].

Agricultural equipment has seen the least penetration of alternative fuels in general (ironically, even ethanol), but there have been some exceptions. Figure 9.14 is a 2009 photo of a prototype hydrogen-fueled New Holland farm tractor that uses anhydrous ammonia which is reformed onboard to generate hydrogen that (apparently) partially powers it. The ammonia-hydrogen connection is logical in agriculture: ammonia is used as a nitrogen fertilizer in almost all large farm operations and is readily available for this application.

In the years immediately before automotive hydrogen fuel cells, a few hydrogen combustion vehicles were constructed by major manufacturers, mostly as public relations exercises. One exceptional example was a 2005 BMW 750i modified by BMW to operate on liquid hydrogen. It now resides in the Deutsches Museum von Meisterwerken der Naturwissenschaft und Technik (German Museum of Masterpieces of Science and Technology) in Munich, Germany, as shown in Fig. 9.15. Its liquid hydrogen tank is shown separately in Fig. 9.16.

The Peculiar Combustion Characteristics of Hydrogen

NO_x Formation

Hydrogen burns in air, theoretically producing only water (thus the ZEV designation, which refers only to tailpipe emissions). However, for air-fuel mixtures even as lean as 60% of stoichiometric (Φ=0.6, λ=1.7), the peak combustion temperature of a stoichiometric hydrogen-air mixture is higher than that of other spark ignition fuels such as gasoline or natural gas, and oxides of nitrogen are formed by reaction of small amounts of the nitrogen and oxygen in air. For any combustion process, the

Fig. 9.14 New Holland prototype farm tractor powered by hydrogen generated onboard by reformation of ammonia. https://commons.wikimedia.org/wiki/File:New_Holland_NH2_hydrogen_tractor_at_Agritechnica_2009.jpg. CC Attribution-Share Alike 2.0 Generic license

Fig. 9.15 The 2005 BWM 750i Liquid Hydrogen "Limited Production" Car, Deutsches Museum, Munich, Germany. Photo: author, 2016

Fig. 9.16 Linde Liquid hydrogen tank used by BMW in 750i hydrogen conversion of Fig. 9.17. Deutsches Museum, Munich, Germany. Photo: author, 2016

formation of NO and NO_2 (aka NO_x, oxides of nitrogen) is unrelated to the carbon content of a fuel. It is just as easily formed from hydrogen as from any other combustion fuel. It is generated any time air is heated to a high temperature, even in home cooking or heating with natural gas. NO_x is highly toxic and a precursor to photochemical smog even in low-ppm (parts per million) quantities. It is a regulated automotive emission in all Paris Accord countries.

The oft-quoted clean combustion reaction of a hydrogen-air mixture is

$$H_2 + \frac{1}{2} O_2 \rightarrow H_2O + \text{heat}$$

Oxygen is only 21% of air, the balance being mostly nitrogen. At high temperatures, some of nitrogen reacts with the oxygen in the combustion chamber, producing NO and NO_2. Accounting for atmospheric nitrogen, the complete combustion reaction in the engine is

$$H_2 + \frac{1}{2}\left(O_2 + \frac{78}{21} N_2\right) \rightarrow H_2O + \text{heat} + 1.86 N_2 + \text{trace}\, NO_x$$

Trace NO_x means a numerically small but non-zero fraction that is dependent upon the combustion temperature and residence time in the cylinder at that temperature. Combustion peak temperature in an engine is most closely related to the fuel/air ratio. NO_x formation is also a function of the duration of the high-temperature exposure, which is referred to as the *residence time* at that temperature. Residence time

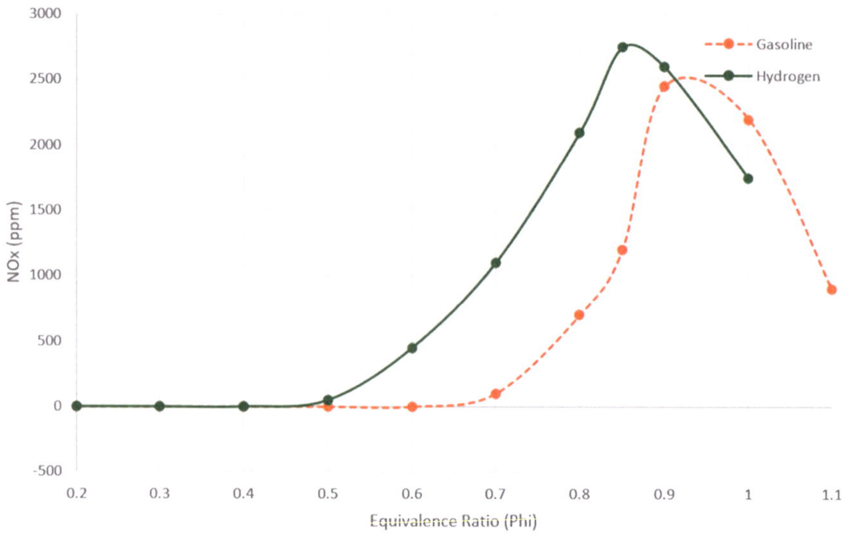

Fig. 9.17 Example of NO_x emissions related to equivalent ratio (inverse of normalized AFR). Different but very similar engines were used for the hydrogen and gasoline plots. Both tests were performed at 2000 RPM. Dashed line: Gasoline engine. Solid line: Hydrogen engine using direct cylinder injection. Gasoline data from [43]; Hydrogen data from [44]

is inversely proportional to the engine speed (RPM) and flame speed, such that at the same engine torque, NO_x has been observed to remain constant with RPM, or even increase at low engine speeds.

The dependency between NO_x formation and fuel-air equivalence ratio is exponential up through approximately $\Phi = 0.9$ equivalence ratio, as illustrated in Fig. 9.17. The decline of NO_x at $\Phi = 1.0$ might be explained by a less-than-perfectly-mixed fuel-air mixture, a known problem with hydrogen engines with richer AFRs. However, this explanation is just intuitive and has not yet been confirmed experimentally.

While the threshold of NO_x emissions varies between engines, NO_x is nearly zero for very lean fuel-air mixtures below the equivalence ratio Φ=0.6. This means that to prevent NO_x formation, the engine can be operated at a reduced output compared with other fuels. Further reducing the power output is hydrogen's low density, which occupies 30% (by volume) of a stoichiometric intake fuel-air charge. Consequently, a naturally aspirated hydrogen engine will always be power-limited to less than half the output of the same engine running on gasoline.

As can be seen in Fig. 9.17, the zero-emissions designation for hydrogen is only true for hydrogen ICEs operated at very lean mixtures ($\Phi < 0.5$, $\lambda > 1.7$) which keep NO_x emissions to nearly zero, but at the cost of the power output of the engine. Exhaust gas recirculation (EGR) has been used but is inadequate to reduce NO_x emissions for equivalence ratios that are not lean. Since NO_x is generated for fuel-air equivalence ratios Φ between 0.5 and 0.9, in the middle of the power band, exhaust after treatment and/or various means to reduce combustion peak temperature are usually employed, e.g., water induction or injection, but with only limited success.

Conventional catalytic converters designed for NO_x, HC, and CO reduction in gasoline engines are effective only in a narrow range of Φ centered at stoichiometric. They have been found to be less effective at reducing NO_x in quality governed hydrogen engines (see below), presumably due to the excess oxygen content in the exhaust compared with throttle-controlled engines. Experimental catalytic converters have been developed and tested with limited improvement in NO_x reduction [45]. But little research has been published in this area since the early 1980s due to the decline of interest in alternative fuels in general, and the introduction of hydrogen fuel cells that power modern FCVs.

Quality Governing

Because of its unusually wide flammability limits, hydrogen SI engines can be controlled similarly to diesel engines by varying the fuel quantity directly rather than controlling the manifold air pressure with a throttle and matching the fuel delivery to the air density. This control method is referred to as *quality governing*. It eliminates engine vacuum "pumping" losses, allowing greater mid-range efficiency than a throttle-controlled engine. In terms of engine thermal efficiency, a quality governed hydrogen engine is capable of 40+% efficiency [46], compared with the typical value of 30–32% for a high-efficiency gasoline engine.

However, a quality governed hydrogen engine is prone to poor mixing properties of the fuel and air, at least partially because of hydrogen's low gas inertia. This is particularly true if fuel is injected directly into the cylinder after intake valve closure. And since the combustion flame speed varies radically with the local fuel-air ratio Φ in the combustion chamber, (MBT) ignition timing may require an unusually wide range of advance. With an extremely high flame speed at stoichiometric, only a few degrees of ignition advance is required even at high RPM, while with an extremely low flame speed near the lean limit, combustion in local lean areas may not be complete even with as much as 60 degrees BTDC advance.

Another alternative method for partial-load control of hydrogen engines was made possible by the introduction of variable intake valve timing and electronic throttle control in the mid-1990s. The *Atkinson Cycle* [47] was initially used in high-efficiency hybrid EV engines, e.g., the Gen 2 and later Toyota Prius starting in 2003. This control method is now used in many high-efficiency and/or high-performance spark ignition engines, especially those with direct injection spark ignition (DISI) to implement dual-mode (homogeneous charge and stratified charge) control [48]. By controlling the intake valve closure angle, the usual throttle control of engine torque is supplemented by the variable closure angle of the intake valve. The intake valve closure is delayed to various degrees at the start of the compression stroke. During this period, intake air is allowed to backflow through the intake port as the piston rises. This effectively results in a variable cylinder displacement and a variable compression ratio, since the effective cylinder volume at the crank angle at which compression actually begins is reduced, while the volume at TDC remains the same. Atkinson control combined with timed hydrogen fuel injection can afford the efficiency of quality governing without having to span such a wide fuel-air ratio and ignition timing range. The 2004 BMW H2R prototype hydrogen land speed record attempt was reported to utilize timed hydrogen fuel injection along with *Valvetronic* variable valve timing, the combination of the two suggesting the use of this engine control method (though this cannot be confirmed).

Supercharging to Recover Power Loss

Regarding the power output of a hydrogen ICE: Forced induction with aftercooling, i.e., supercharging or turbocharging, is effective in compensating for the reduced volumetric efficiency of hydrogen ICEs, by increasing the cylinder energy content, although the increased energy density in the cylinders increases the peak combustion temperature thereby increasing NO_x and the tendency for intake backfire. For a naturally aspirated hydrogen engine to match the power output of a gasoline engine requires approximately twice the engine displacement, or supercharging at greater than 1 Bar (14.5 psi, 100 kPa).

Backfire Suppression

As mentioned previously, fuel-air mixtures approaching stoichiometric $\Phi > 0.5$ lead to higher cylinder surface temperatures that, due to the extremely low ignition energy requirement of hydrogen, can cause the unintended ignition of a pre-mixed fuel-air mixture as it enters the cylinder during the intake stroke, a phenomenon referred to as *intake backfiring* (as distinguished from *exhaust backfiring* aka *afterfiring* due to presence of a flammable mixture in the exhaust system). This pre-ignition problem can also be aggravated by residual hot gases or particulates at the end of the exhaust stroke that may function as combustion nuclei for the incoming flammable charge [49]. Backfiring is particularly a problem for supercharged pre-mixed intake mixtures at high loads.

The problem of intake backfiring was rigorously investigated and substantially reduced in 1978 at UCLA [50] by the delay of fuel delivery to the intake port until well after the opening of the intake valve. This allows the induction of cold air during the initial intake stroke, which can help to quench low-thermal-mass surfaces and combustion nuclei remaining from the preceding exhaust stroke. The delayed fuel introduction was accomplished by timed electronic fuel injection, with the start of injection delayed typically to 45 degrees ATDC. This approach also eliminated the presence of a flammable pre-mixed fuel-air charge in the intake manifold, such that if unintended ignition were to occur while the intake valve was open, the physical consequences would be small compared with the ignition of a large manifold volume of pre-mixed hydrogen and air.

Hydrogen, having extremely low volumetric energy density at NTP (normal temperature and pressure, 20° C and 1 atm), requires a higher volume fuel flow rate than any other fuel, liquid or gaseous. Implementing electronic fuel injection for hydrogen requires a high-speed fuel injection valve with a much higher flow coefficient than that required for any other fuels, liquid or gaseous. Figure 9.18 is a

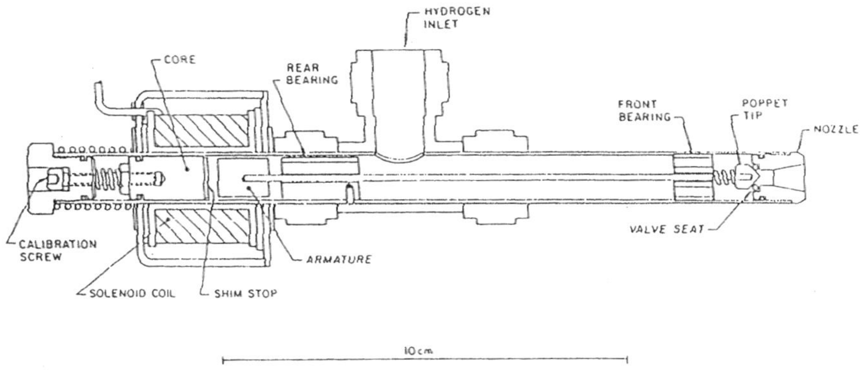

Fig. 9.18 Electrically-actuated hydrogen fuel injection valve developed in 1979, used to implement time-delayed port fuel injection as a means to reduce/eliminate intake backfiring in high power hydrogen engines. Diagram by author, 1979

cross-sectional diagram of an electrically actuated gaseous fuel injection valve developed at the University of Denver Research Institute in 1979 to implement time-delayed hydrogen fuel injection [51, 52].

EGR or water induction with a pre-mixed air-fuel intake charge has been found to be only trivially effective in suppressing backfiring. But it was found that if water injection was timed to occur during the fuel injection delay interval in the early intake stroke, it further enhanced the pre-cooling effect prior to entry of fuel into the cylinder by increasing the quenching of small-thermal mass pre-ignition sources and cooling of cylinder hot spots, especially the valves and the spark plug electrodes.

A completely different phenomenon can also be related to hydrogen intake backfiring: electromagnetic inductive coupling between proximate ignition cables which are usually routed together on and around the engine. This is not a problem with any liquid or heavier gaseous fuels due to their much higher ignition energy requirements, but it has been observed in high-power natural gas as well as hydrogen engines. The small current induced electromagnetically in a proximate ignition lead for a non-firing cylinder can be sufficient to allow its spark plug to ignite the cylinder's hydrogen air mixture. The most expedient solution is to prevent inductive coupling by keeping ignition leads well separated and allowing them to cross only at right angles, or use a coil-on-plug ignition system. While not rigorously tested, it was observed informally that cross-induction ignition was actually a *less* likely cause of backfiring at atmospheric or higher (supercharged) intake pressure, because the dielectric constant of air increases with pressure, providing a higher voltage threshold that inhibits errant spark plug firing.

In dynamometer tests, the concurrent use of all of the backfire prevention methods just discussed was found to eliminate backfiring at BMEPs (brake mean effective pressure or normalized torque) as high as ten bars, with a turbocharger boost pressure of 1.0 bar. Recently (2023) hydrogen fuel injection has been utilized in prototype hydrogen combustion engines developed by Honda, Toyota, and AVL, each reporting power output levels equivalent to or greater than high-performance naturally aspirated gasoline engines [53, 54].

At this time (2024), there seems to be a resurgence of interest in hydrogen combustion engines, although no production examples have yet appeared or are planned.

Hydrogen Fuel Cell Vehicles (FCVs)

While the principles of fuel cells have been known since first documented by Welsh inventor Sir William Robert Grove in 1839, it has only been since the 1960s with the development of practical "hydrox" (hydrogen-oxygen) fuel cells by General Electric Co. (USA) that their potential for useful levels of electric power generation has been demonstrated.

The first documented fuel cell powered land vehicle was a converted electric golf cart demonstrated in 1959 by Allis Chalmers (USA) [55], shown in Fig. 9.19 and discussed later in this chapter. In the late 1969s and early 1970s, fuel cells developed by United Technologies Corp. (USA) provided electric power for all of the Apollo lunar missions. Fuel cells opened the door to the production of electricity on manned spacecraft without the need for heavy batteries or a mechanical generator. They also found applications in which even the most advanced batteries were too heavy, such as the AeroVironment Global Observer High-Altitude Long Endurance remotely piloted aircraft in 2005 [56]. Circa 2020, bidirectional hydrogen fuel cell/electrolyzers have been proposed for utility-scale electric power storage, with excess electric energy stored as compressed hydrogen in impervious underground formations such as salt caverns or

Fig. 9.19 Allis Chalmers PR photographs of an experimental hydrogen fuel cell tractor and golf cart from 1959. Original prints provided by Allis Chalmers Corp. to author for a school science project in 1966

depleted oil formations. But the best-known application of fuel cells has been as the onboard power source for electric vehicles, replacing some of the batteries with high-pressure hydrogen tanks. This variation of an electric vehicle is referred to as a FCV.

Although methane and methanol have been used as fuel cell reactants in stationary applications, hydrogen remains the only high-efficiency fuel source for vehicular fuel cells. Methane has been used in high-temperature fuel cells to be discussed, but with significant carbon dioxide emissions that negate the objective of a ZEV. While hydrogen ICVs first appeared in the 1930s, fuel cells were not used to power electric vehicles until 1959 (Allis Chalmers, above). Since then, auto manufacturers including General Motors, BMW GmbH, Daimler Benz, and Toyota Motor Co have experimented with fuel cell electric vehicles, and current FVCs are sold by Toyota, Hyundai, and Honda.

Hydrogen ICEs are less efficient than fuel cells, but the simplicity and lower cost of an ICE compared with the fuel cell stack, batteries, motor, and controls of an FCV have allowed the idea of hydrogen combustion vehicles to persist as a second-choice option even today. Only battery electric and hydrogen fuel cell vehicles are granted the "zero-emission" classification in the USA and Europe. The advantages are:

- Hydrogen provides the highest efficiency of any reactant.
- Exceptionally highest mass energy density.
- Non-toxic. No color or odor.
- Multiple production paths, both renewable and non-renewable.
- Faster refueling time and potentially longer range than BEVs.

However, even in fuel cell vehicles, hydrogen's properties are responsible for technical challenges:

- Exceptionally low volume energy density.
- Onboard storage is a challenge. Compressed storage at pressures as high as 70 MPa are required to provide sufficient vehicle range. Cryogenic liquid hydrogen storage is possible but considered unfeasible except possibly for aircraft, interstate trucks, rail, or marine applications.
- High-purity hydrogen is required for fuel cells, limiting the use of hydrogen generation onboard by reforming alcohol or hydrocarbon fuels.
- Almost all hydrogen is currently made from natural gas or coal, which generate CO_2 emissions.
- Currently, the only source of carbon-free *green hydrogen* is the electrolysis of water using carbon-free electricity, a path that involves many energy conversion steps and results in questionable efficiency gains compared with the direct use of electricity in an electric vehicle.

All FCVs use hydrogen as a reactant with oxygen from ambient air in a series connection (stack) of hydrogen-oxygen (hydrox) fuel cells. Hydrogen at the anode and air at the cathode interact through an ion-exchange membrane. Hydrogen

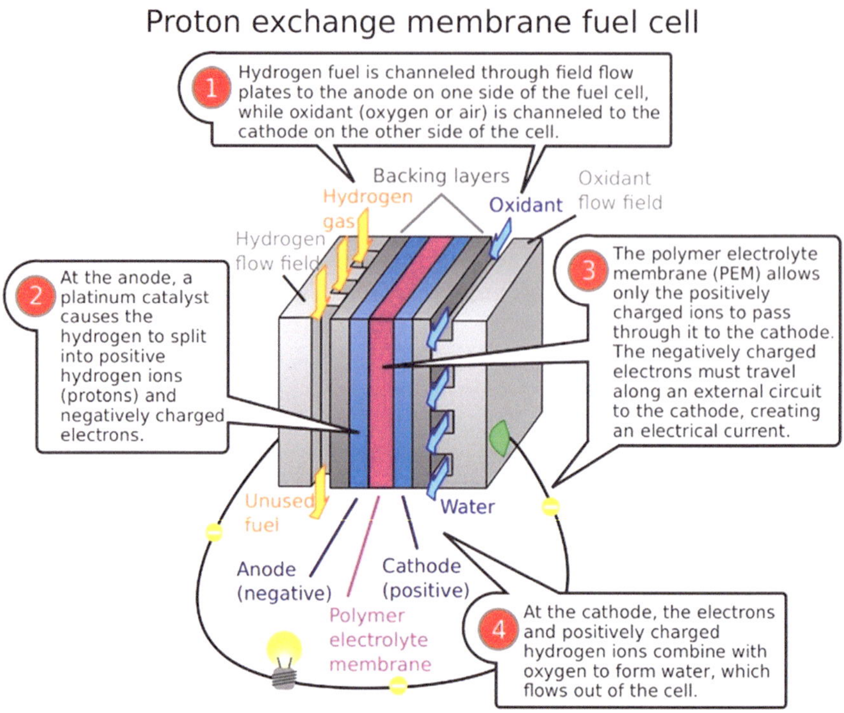

Fig. 9.20 Diagram of a typical polymer electrolyte membrane (PEM) fuel cell, from https://www.energy.gov/eere/fuelcells/types-fuel-cells. Public domain

combines with oxygen to create water, as illustrated in Fig. 9.20. Waste heat is generated but it is small compared with exhaust heat from an equivalent ICE.

Main Components of a Hydrogen FCV

Referring to Fig. 9.21, power flow in hydrogen FCV is as follows: High pressure H_2 tank → reduction regulator → fuel cell → electricity to charge controller → main traction battery → electric motor controller → motor → driveline/wheels

An intermediate battery (usually 3–5 kWh) is necessary for all FCVs because a fuel cell cannot be throttled in real-time, and the fuel cell cannot generate the instantaneous power demands of the traction motor. The battery also serves as an energy reservoir to absorb power from regenerative braking, which cannot be reabsorbed by a fuel cell [57].

Milestones in the Development of Hydrogen Fuel Cells and FCVs

The earliest examples of vehicles powered by hydrogen fuel cells are shown Figs. 9.19 and 9.22.

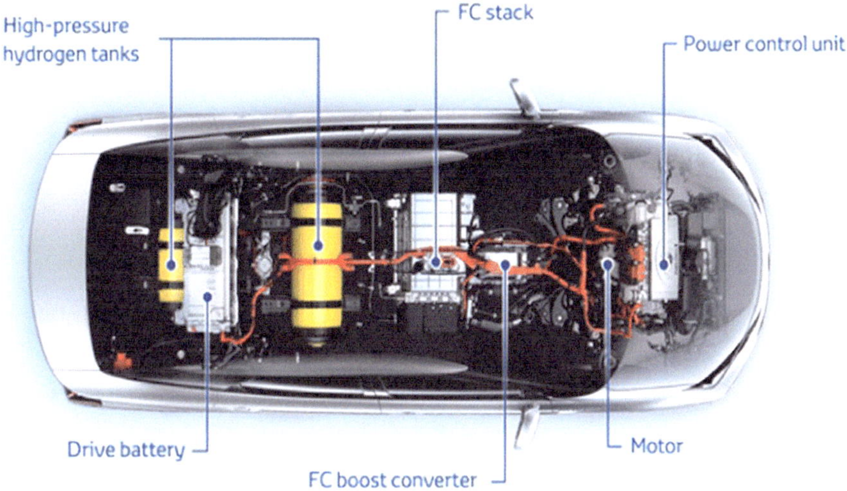

Fig. 9.21 Location of hydrogen fuel cell, drivetrain and storage components in Toyota Mirai. From https://www.hydrogen.energy.gov/docs/hydrogenprogramlibraries/pdfs/htac_feb18_07_scott.pdf?Status=Master 2018. Public domain

Fig. 9.22 (Right) GM fuel cell "Electrovan" concept vehicle, 1966. The first road-legal vehicle to use fuel cells or liquid hydrogen storage. The fuel system occupied almost the entire cargo area of the van. On permanent display at the GM Heritage Center, Grand Blanc Township, Michigan, USA Photo: GM HeritageCenter. Public domain

Allis Chalmers, 1960

In the early mid-1960s, fuel cells were a hot R&D topic with the promise of conversion of fuels to electric power without combustion, and at high efficiency. Compared with the best batteries of the time, they provided considerably higher energy mass density. Figure 9.22 shows publicity photos of two experimental Allis Chalmers (US tractor company) fuel cell vehicles constructed in 1959–60: a fuel cell golf cart and a modified farm tractor. Unfortunately, Allis Chalmers never monetized these innovations.

Fig. 9.23 Hydrogen-oxygen fuel cells in Apollo command module. NASA photo. Public domain

United Technologies, 1967

The potential of fuel cells to replace heavy batteries in spacecraft applications was immediately apparent from their higher specific (per unit mass) energy density.

Figure 9.23 is a photograph of the United Technologies fuel cells on one of the NASA Apollo spacecraft, 1965–71.

Fueled by liquid hydrogen and liquid oxygen, they provided a lightweight solution to the need for electric energy for long missions. Since then, all US manned missions and all space stations from 1973 Skylab through the 1998 International Space Station have relied on hydrogen-oxygen fuel cells for electric power and even the production of potable water for the crew. Both the hydrogen and oxygen are stored in cryogenic liquid form, which is less of a challenge in space than on the Earth because of the total vacuum which is a perfect thermal insulator for the tanks if shielded from solar infrared radiation.

The near tragedy of the failed Apollo 13 mission was caused by the explosion of one of the two liquid oxygen tanks due to a broken or excessively hot electrical connection inside the tank. In a saturated oxygen environment, any flammable material becomes an explosive. The in-tank wires that connected the tank heating element were insulated with a flammable plastic that ignited explosively in the presence of pure oxygen. The tank was subsequently redesigned for Apollo 14 with no interior flammable materials, as shown in the NASA publication photo in Fig. 9.24. The right diagram labeled Apollo CSM (Command Service Module) Oxygen Tank identifies which materials were replaced to eliminate any possibility of combustion.

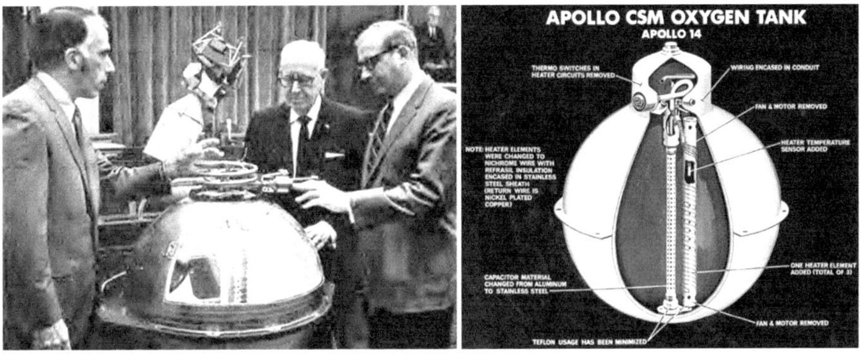

Fig. 9.24 Redesigned liquid oxygen tank used on NASA moon landing missions following the explosion of a liquid oxygen tank on Apollo 13. NASA photos, public domain

Unlike spacecraft applications, vehicular fuel cells use air, not oxygen as the oxidizer, but at lower efficiency compared with pure oxygen. The nitrogen in air requires periodic purging to prevent accumulation, but otherwise does not damage the cell. However, even very small amounts of carbon dioxide (CO_2), oxides of nitrogen (NO and NO_2), sulfur dioxide (SO_2), ammonia (NH_3) or volatile organic compounds (VOCs) can irreversibly poison platinum catalysts and damage the ion exchange membrane, requiring that the intake air be pretreated and scrubbed of these contaminants.

Ballard Power Systems, 1979

Worthy of special mention is the Canadian company Ballard Power Systems, founded in 1979, that was the first to successfully commercialize hydrox fuel cells at a low enough cost to be considered for ground transportation applications [58]. Ballard invented and controls a large portfolio of patents, and produces fuel cells for a wide range of commercial buses and trucks, rail, automobiles, aircraft, and marine applications (Figs. 9.25 and 9.26).

Fuel Cells Require Pure Hydrogen

Fuel cells require high-purity hydrogen to prevent poisoning of the electrolyte, anode or cathode materials. This restricts their use to "green hydrogen" that is produced by electrolysis of water, or by purified hydrogen produced from natural gas with removal of all carbon dioxide and other contaminants.

The need for high-purity hydrogen for fuel cells is an important distinction between hydrogen FCVs and hydrogen ICVs. By comparison, combustion in a hydrogen ICE is tolerant of almost any contaminant or co-fuel, including methane, although these undesired components are invariably greenhouse gases. The purity requirement also effectively prevents consideration of onboard reformers that generate hydrogen from denser, easier-to-store alcohols or ammonia, since these methods produce large quantities of carbon monoxide from reformed methanol, or residual ammonia from ammonia reformation. Despite their cost and limitations, hydrogen fuel cells are still favored over hydrogen IC engines due to their higher efficiency and zero NO_x emissions.

Fig. 9.25 1 kW Ballard fuel cell used in Cal Poly student project, 2010. Photo from product manual

Fig. 9.26 2022 Ballard self-contained fuel cell system. Image from Ballard Power Systems. Used with permission

Other Mobility Applications

Modern Hydrogen Transit Buses

Municipal transit buses seem to have become one of the last automotive applications for which hydrogen can be justified, in direct competition with electric buses. In the past 10 years, a number of startup companies have entered the hydrogen fuel cell market, almost all relying on the expected continuation of large government grants and carbon offset credits for municipal bus fleets to deploy hydrogen vehicles in an effort to decarbonize. One example, shown in Fig. 9.27, is from American Fuel Cell, now a division of Plug Power, Inc., a US manufacturer of hydrogen electrolyzers and hydrogen fueling infrastructure. There is, however, growing skepticism of hydrogen even for this application for which capital and operating costs are much higher than those of battery electric buses, at least in the deployments to date [59].

Fig. 9.27 Transit bus equipped with hydrogen fuel cell drivetrain produced by American Fuel Cell Inc. operating in Palm Spring, California, 2019. Photo by Leslie Eudy, NREL. https://www.nrel.gov/state-local-tribal/blog/posts/fuel-cell-electric-buses-in-the-usa.html. Public domain

Railway Applications

The use of hydrogen either in compressed or liquid form has met with limited success in railroad applications in Europe and Asia. The transition from diesel electric to hydrogen electric is straightforward; a fuel cell replaces the large diesel generator that charges the locomotive ballast batteries and powers the motors. Since the size and mass of the locomotive are not major concerns, the fuel storage volume required for liquid hydrogen is less of a problem compared with a huge battery that would be required to provide an equivalent range for a (hypothetical) battery-electric locomotive. In stretches of rail lines that are not currently electrified, the quiet operation and zero local emissions of hydrogen powered light rail are clear advantages.

Starting limited service in 2018, the first demonstration of a hydrogen commuter train is shown in Fig. 9.28. Designed and built by Alstom, five of these trains are currently in service, with another 27 under construction for the Frankfurt metro area, and an unstated number to be delivered in Italy and France [60]. A similar hydrogen rail deployment is currently in progress in Switzerland, with passenger trains manufactured by Swiss rail vehicle maker Stadler Rail. Originally built for California's San Bernardino County Transportation Authority but ultimately rejected, its top speed is reported to be 127 km/h (79 mph), and its range is 460 km (286 miles) between refueling stops. However, in a special configuration with a liquid hydrogen tank occupying an entire rail car in the middle of the train, it was reported that a new Guinness World Record for the range was established: traveling for almost 2 days around the clock for a distance of 1,741.7 miles at the ENSCO test circuit in Pueblo, Colorado, USA. Only one unit was placed in passenger service in San Bernardino, California [61]. Stadler has announced that two Italian rail operators have signed contracts for the supply of 15 hydrogen-powered trains [62]. Similar developments have been recently announced in Spain and China.

Fig. 9.28 The Coradia iLint, the first fuel cell hydrogen commuter train, shown in service in the Lower Saxony region of Germany, 2023. From https://cdn.railuk.co/wp-content/uploads/2016/09/24120435/IMG_5056_iLint_D-1068x712.jpg UK government agency, public domain

The cost of operation as well as the acquisition cost have been criticized as being noncompetitive compared with less costly diesel-electric, or even shorter range battery electric propulsion. Safety was also an issue. As reported in citation [62]:

> Storing large quantities of hydrogen on a fast-moving train is potentially hazardous. The UK's Rail Safety and Standards Board has published a report that addresses questions of safety, technical standards and regulation required to facilitate the safe introduction of hydrogen trains into regular service. It identifies fire as the main concern and explores several ways to mitigate the risks, including the location of hydrogen tanks on rolling stock and the maintenance requirements to ensure safe operation.

As of early 2024, the realities of hydrogen in transportation were starting to overtake the facade of "green energy," even in media that have always been dedicated promoters of hydrogen. Regarding the hydrogen rail system mentioned above, from *Accelerate Hydrogen,* October 3, 2024 [63]:

> Fed-up authorities in the Frankfurt region have threatened to cancel contracts at the world's first hydrogen-only railway and return to diesel engines after 2 years of almost constant technical problems with the H_2 trains.
>
> The RB15 line on the mountainous Taunus network was supposed to make a complete switch from diesel to hydrogen on 11 December 2022, when the 27 Coradia iLint fuel-cell trains ordered from Alstom for €500 m ($552 m) were scheduled to start operating, but only six of them were delivered in time and proved to be defective, requiring retrofitting of new hardware and software components.
>
> Six months later, only 12 of the trains were in operation....

In October 2024, all except one hydrogen train were taken out of service due to excessive capital and operating costs.

High-Temperature Solid Oxide Fuel Cells

While not currently used for vehicles, a special class of fuel cells was originally productized in the 1960s but used only in military and emergency operations due to cost. High-temperature fuel cells can be fueled directly by purified methane from natural gas using a ceramic (solid oxide) electrolytic membrane in the cells. Operating at 1000 °C (1800 deg F), they were never considered suitable for FCV applications, but they could serve as compact, quiet energy sources which made them attractive in some environments such as battlefields. The ability to operate directly on natural gas was a huge benefit compared with room-temperature hydrogen fuel cells, bypassing the need to first make hydrogen from natural gas, pressurize, and store it [64].

A high-temperature solid oxide fuel cell (SOFC) uses a ceramic electrolyte instead of liquid or polymer, typically zirconium oxide and calcium oxide pressed into a crystalline lattice. The ceramic electrolyte is coated on both sides with porous non-corroding conductive materials which serve as the anode and cathode. The highest performance material would be platinum, but in commercial applications for which cost is more important, less-expensive high-nickel stainless steel is used—see Bloom Energy below. Oxygen ions (with a negative charge) migrate from the cathode toward the anode through the crystal lattice. These combine with positively charged hydrogen atoms at the anode, a continuous process that causes electrical current to flow through an external circuit. Some of the fuel energy is lost in maintaining the fuel cell temperature. Energy conversion efficiencies as high as 60% have been demonstrated in laboratories, but practical efficiencies of 45–50% are typical. As a climate change solution, however, their impact is about neutral compared with grid electricity since all of the carbon from the methane is released into the atmosphere as carbon dioxide.

SOFCs have been a laboratory curiosity since the 1930s. The method was originally patented and developed into a modular package by Westinghouse in the late 1950s, followed by General Electric and Siemens in the early 1960s. The only applications at that time were military field power units for which cost was not an obstacle.

In 2000, the U.S. Department of Energy funded the development of an experimental SOFC power generation facility built by Siemens Westinghouse. It reported that the 220 kw SOFC operated for nearly 3400 h, and achieved an electrical efficiency of about 53% [65].

There seems to be only one major commercial entity in the USA that has successfully commercialized SOFCs, *Bloom Energy* of Sunnyvale, California. The company refers to these as "energy servers" (Fig. 9.29), but they are commonly known as "Bloom Boxes" that produce electric power from utility-sourced natural gas on-site. They are most useful for backup power rather than primary electric

Fig. 9.29 Installation of "Bloom Box" solid oxide high temperature natural gas fuel cells produced by Bloom Energy (USA) circa 2015. Photo from Wikipedia Commons. 9 March 2020 https://commons.wikimedia.org/wiki/Category:Bloom_Energy_Server. Attribution CC BY 2.0

power, since the cost of the electricity that they produce considerably exceeds that of utility power. With an installed cost of $1.2M per module (2020), Bloom Energy fuel cells have been deployed almost exclusively by high-profile corporate campuses and data centers such as those operated by Google, Amazon, and eBay, and at a few hospitals for backup electric power.

Bloom Energy originally marketed their energy servers as a "green" energy solution, which was embraced by corporate customers sensitive to the optics of diesel or natural gas backup generators. The quiet operation and attractive enclosures supported this image. But what may not have been clear to adopters and investors is that since the cells are fueled by methane from natural gas rather than hydrogen, *their CO_2 emissions are nearly the same as those from combustion of the fuel in the engine of a natural gas-fueled generator*, although there is some efficiency advantage compared with a natural gas-fueled ICE generator. Compared with utility electricity, this difference could theoretically amount to a reduction in CO_2 emissions per kWh if the energy conversion efficiency of the Bloom Box was significantly higher than a modern natural gas power plant. But that is not the case. An independent study of deployed Bloom Boxes reported an average efficiency of 45% [66], which compares less-than-favorably with a combined cycle natural gas power plant that typically operates with an efficiency above 50%. In both the natural gas SOFC fuel cell and direct combustion or natural gas, the mass of CO_2 emitted per kWh is directly proportional to the fuel consumption. The same study also found that the useful lifetime of the ceramic electrolyte material is 2 years, requiring major service at the end of each period.

Like many popular startups in the renewable energy field, Bloom raised more than $1 billion in venture capital funding even before going public in 2018. Its fuel cells have and continue to be subsidized by government green energy programs, despite their lack of actual benefit to the climate. As of 2020, Bloom reports that they had installed about 600 megawatts worth of fuel cells, although the company has never made a profit [67]. Its business practices have been called into question in several investor lawsuits, and it was fined $1 million USD by the US EPA in 2015 for hazardous waste disposal violations [68, 69]. As of 2023, Bloom Energy diversified into hydrogen fuel cells and electrolyzers and continues to earn renewable energy subsidies from the state of Delaware, as well as Federal carbon replacement credits [70].

There are certainly specialized applications for SOFCs in which their quiet operation and compactness prevail over their relatively high carbon dioxide emissions, low efficiency, and high cost. But other than such specialized applications, the marginal benefits are difficult to justify. Their proposed application for emergency backup power is questionable since the warm-up time is typically about one hour for Bloom Energy Servers. SOFCs have been experimentally used as motive power sources, but no major FCV manufacturer has used this type of fuel cell, despite their obvious advantage of fueling with natural gas rather than hydrogen. The CO_2 emissions of this type of cell are a stumbling point if the objective is a zero-carbon vehicle. But there is still room for innovation, including integration with carbon capture technologies that could change the landscape for FCVs by allowing them to be fueled by natural gas or some other hydrocarbon having a higher energy density than hydrogen.

Fuel Cells Adapted to Vehicles

FCVs are battery electric EVs with an onboard hydrogen fuel cell that continuously charges the battery. The fuel cell cannot be simply directly coupled to the motor drive system because it cannot be throttled as required for vehicle power control, and its continuous output is too small to meet the instantaneous needs of the electric motor. Fuel cell vehicles might, therefore, be more accurately described as fuel cell hybrid EVs, having a smaller battery than a full BEV and a hydrogen fuel cell range extender.

Figure 9.30 is a diagram depicting the production and automotive use of "Green Hydrogen" from a zero-carbon electricity source: solar PV, wind, hydroelectric, or geothermal. Unfortunately, the multiple energy conversion steps and gas compression energy requirements are equivalent to as much as 70% of the energy value of the hydrogen produced. These losses are considerably greater than the RTE losses of a lithium battery in a BEV that could use the original electricity directly.

Overlooking the inefficiency of green hydrogen production, FCVs still offer a range of advantages compared to BEVs since the FCV range is determined by the size of the compressed hydrogen tank(s), which weigh considerably less than the batteries needed for that range in a BEV. Also, the refueling time is faster than all except the highest power DC rapid EV chargers. One popular media source concluded that the biggest deal-breaker for potential FCV buyers is just the lack of refueling facilities [71].

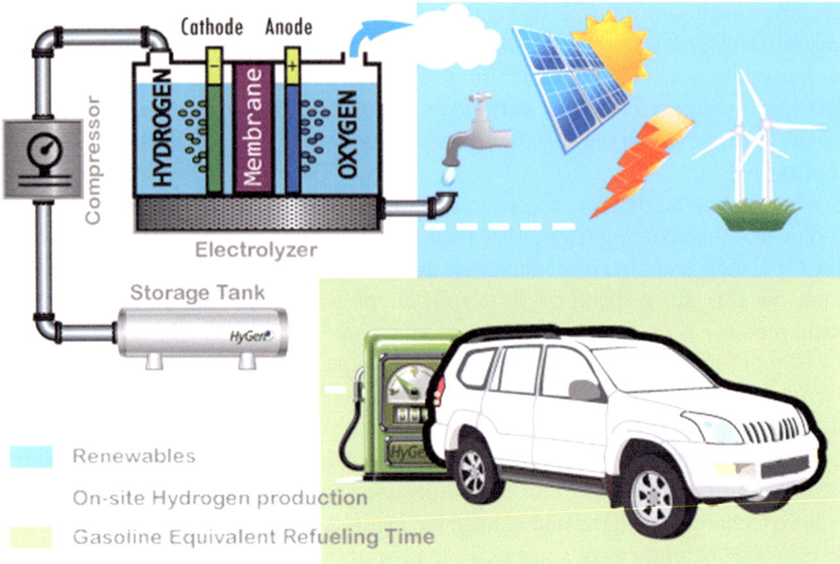

Fig. 9.30 Diagram of hydrogen fuel cell vehicle with fuel produced by electrolysis. From Hydrogen Fuel Cell Partnership, http://cafcp.org/stations/buildingastation. Public domain

Onboard Hydrogen Storage

Applicable to both hydrogen ICVs and FCVs. For a hydrogen ICE, we can calculate the energy equivalent gasoline fuel consumption, starting with the lower heating value (from 2020 Bosch Handbook data), one gallon of gasoline has a specific calorific value (lower heating value) of 117 MJ/gal (US) [72]. Hydrogen has a density of 120.0 MJ/kg. These numbers are close enough to conclude that approximately

$$1\,\text{kg}\,H_2 = 1\,\text{GGE}\,(\text{gallon gasoline equivalent})$$

This equivalence does not take into account the higher thermal efficiency of a lean-burn hydrogen engine, typically 40% compared with a gasoline engine at 30%. With these factors considered in a practical hydrogen ICE, we get an adjusted GGE equivalence that provides a more realistic estimate of the range of an H_2 ICE vehicle compared with the same vehicle fueled by gasoline.

$$1\,\text{kg}\,H_2 \text{ is range equivalent to } 1.33\,\text{GGe}$$

An early example from my experience: The original 3.8L six-cylinder engine of the 1972 UCLA hydrogen AMC Gremlin was replaced by a larger Ford 5.7L V8 engine in an attempt to restore some of the reduced power output. The EPA rating of the Ford engine was 12 mpg. Assuming that the hydrogen Gremlin could achieve the same fuel economy, the 0.9 kg of hydrogen that its two tanks could carry at 2000 psi gave it a range of only 16.0 miles. If filled to the maximum pressure of 6000 psi, the tanks could store a maximum of 2.7 kg of hydrogen, giving it a maximum range of

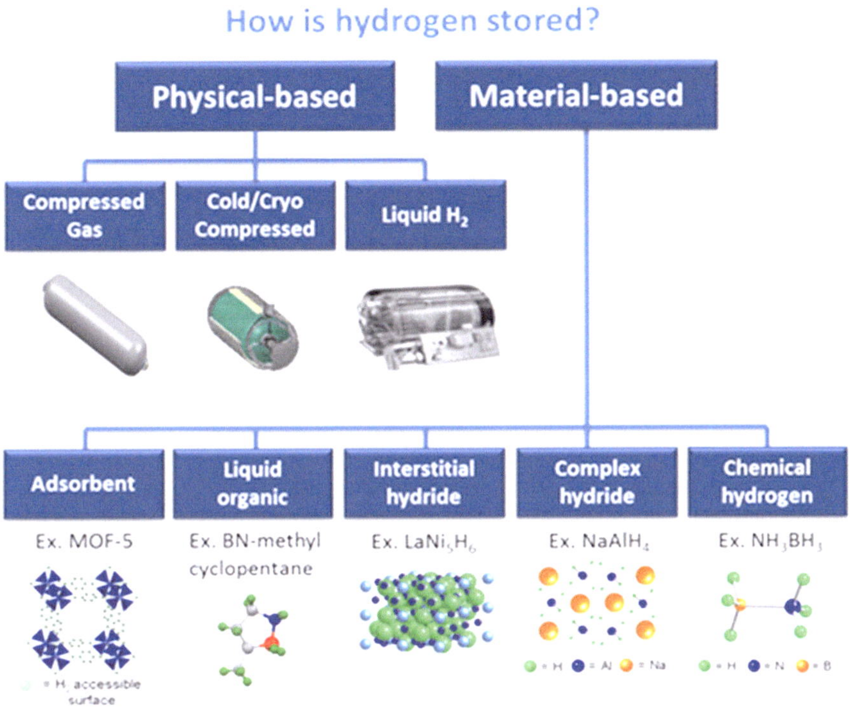

Fig. 9.31 Methods for vehicular hydrogen storage. US DOE. Public domain

approximately 48 miles. This was adequate for the alternative fuel competitions it competed in, but would certainly not be practical for daily driving.

Onboard hydrogen storage is probably the most problematic issue for hydrogen-fueled vehicles (both ICEs and FCVs). The very low volumetric energy density of hydrogen, even at high pressure or in cryogenic liquid form, limits the vehicle range. Were it not for the even worse electric energy density of electrochemical storage in batteries, there would probably be little interest in the use of hydrogen for mobile applications.

Several methods for the storage of hydrogen have been demonstrated, and additional methods have been evaluated in laboratories but never reduced to practice [73] as illustrated in the US Department of Energy graphic of Fig. 9.31. But of these, only high-pressure compressed hydrogen storage has been used in hydrogen fuel cell vehicles sold to the public. Improvement of hydrogen storage remains an area of research interest [74].

As previously mentioned, despite having the highest mass energy density (J/kg), hydrogen has the lowest volume energy density of any fuel at standard temperature and pressure. Even at high pressure or liquified at 20K, its volume energy density isn't even close to that of gasoline or diesel fuel.

Liquid hydrogen at −253 °C is the densest form of pure hydrogen. The density of liquid hydrogen is 70.85 g/L, compared with water 1000 g/L or gasoline 875 g/L. In fact, since the hydrogen mass fraction of gasoline is about 14%, there is a 71% greater mass of hydrogen in a liter of gasoline than in a liter of liquid hydrogen.

A quick comparison of fuel energy content (LHV) and vehicle range:

> 85 L of liquid hydrogen (not including the cryogenic tank volume) contains 6 kg of LH_2 having 720 MJ of fuel energy. This is approximately the same mass of hydrogen stored onboard a Toyota Mirai, which has a stated range of 402 miles (647 km).
>
> 85 L of regular gasoline without ethanol weighs 64 kg and contains 2656 MJ of fuel energy. For a 25 mpg vehicle, this would provide a range of 561 miles (903 km).

High-Pressure Storage and Compression

In the USA, the most common method for commercial distribution of small quantities of hydrogen (for welding or laboratories) is a standard "K" cylinder at 14 MPa (2000 psi), each cylinder weighing 45 kg and storing 0.41 kg of hydrogen as shown in Fig. 9.32. The hydrogen mass content constitutes <1% of the tank+fuel total mass. As a rough figure of merit, one cylinder could provide a

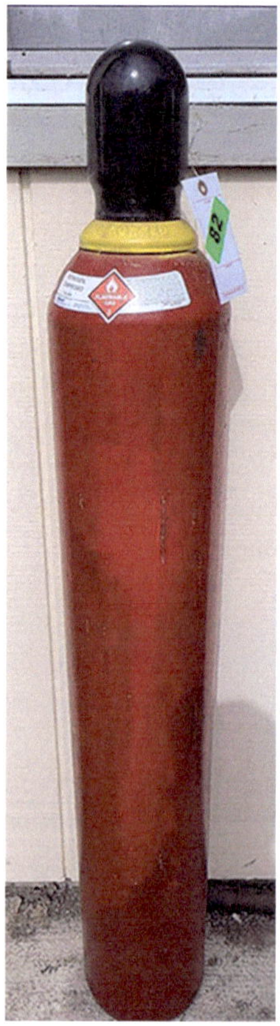

Fig. 9.32 US Standard size K steel cylinder weighs 50 kg and stores 0.4 kg of hydrogen at 14 MPa, 20 ?C. Photo from auction, public domain

range of approximately 21 miles in a Toyota Marai FCV that can travel an average of 50 miles per kg of H_2. This limited range makes clear why much higher storage pressures were necessary for modern FCVs, despite the higher efficiency of fuel cells compared with ICEs. Hydrogen used in fuel cell electric vehicles is compressed and dispensed at either: 35 MPa (5,000 psi, designated by Shell Oil Co. as H35) or 70 MPa (10,000 psi, designated H70).[3]

Energy Required for Gas Compression

As with natural gas discussed in Chap. 8, *Natural Gas*, when considering the net energy content of compressed hydrogen, we must take into account the energy required for gas compression in any *well-to-cell* efficiency analysis. As reported by the US Dept of Energy Hydrogen and Fuel Cells Program regarding the energy requirements for hydrogen gas compression and liquefaction [75],

> The theoretical energy to compress hydrogen isothermally from 20–350 bar (5000 psi or ~35 MPa) is 1.05 kWh/kg H_2 and only 1.36 kWh/kg H_2 for 700 bar (10,000 psi or ~70 MPa). Greater compression energies are required to fill vehicles in practice due to compressor inefficiencies and heating during fast fills. DOE Technology Validation Project data for compression from on-site H_2 production is 1.7–6.4 kWh/kg H_2. Additional energy required for pre-cooling (as cold as –40 °C) to ensure on board fast fill temperatures are 85 C or lower can be modest (0.15 kWh/kg H_2).
>
> Precooling the hydrogen to between –40 °C and –100 °C prior to dispensing is necessary to counteract the adiabatic heating of the gas compression, that would leave the gas at an excessively high temperature in the vehicle tank. Even if this were not dangerous for the integrity of the composite tank, after the gas cools to ambient temperature there would be a significant decrease in pressure.

As mentioned above, the energy required for 70 MPa gas compression and pre-cooling has been reported to be in the range of 1.7–6.4 kWh/kg H_2 [197]. When the pre-cooling energy is added, this amounts to $\frac{1.7+6.4}{2}+0.15\text{kW}=4.2\ \text{kWh}/\text{kg}\ H_2$. This energy investment represents[4]

$$\frac{4.2\,\text{kWh}/\text{kg}}{120\text{MJ}/\text{kg}}\times\frac{3.6\text{MJ}}{\text{kWh}}\approx 13\%\ \text{loss of fuel energy}\left(\text{process efficiency}=87\%\right)$$

The energy invested in gas compression and pre-cooling is completely lost as the hydrogen leaves the vehicle tank(s) during use, passes through multiple pressure

[3] H35 and H70 are ad hoc designations used in hydrogen refueling stations to separately dispense fuel at either 35 MPa of 70 MPa respectively. The compression and precooling energy required for H70 is disproportionately higher than H35, justifying a higher cost per kg at the "pump".

[4] Most medium size compressors have efficiencies between 50–70%, and large size compressors in general have efficiencies of 60–80% at full load, data from: https://lambdageeks.com/compressor-isentropic-efficiency/.

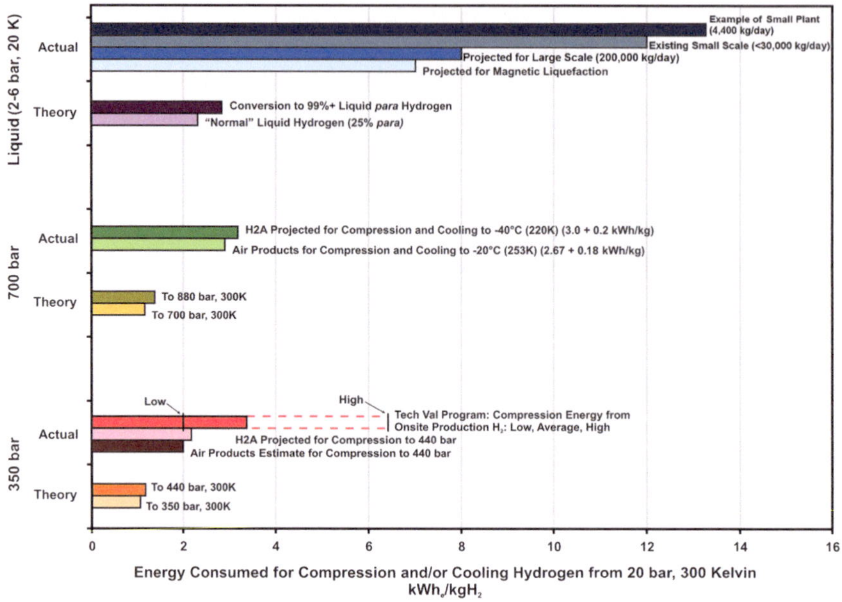

Fig. 9.33 Energy required to compress hydrogen at either 35 MPa or 70 MPa, compared with liquefaction of hydrogen. Chart copied from "Energy requirements for hydrogen gas compression and liquefaction as related to vehicle storage needs" DOE Hydrogen and Fuel Cells Program Record, Record: 9013, July 7th, 2009. Public domain

reduction stages and ultimately is supplied to the fuel cell at low pressure, usually under 1 MPa (145 psi). But more importantly, the efficiency measure does not consider *energy quality* (See chapter "Introduction"). The lost energy is in the form of electricity that is used to precool and compress hydrogen at the production and refueling facilities. Electricity is the highest quality form of energy, so the loss of energy usefulness is even more consequential than 13% might suggest. The US DOE chart of Fig. 9.33 compares the electric energy required for one kg of 35 MPa and 70 MPa compressed hydrogen, as well as one kg of liquid hydrogen.

Hydrogen Leakage and Metal Embrittlement

Hydrogen is more prone to leakage than any other gas, especially at high pressures, due to its very small molecule size (two protons) that allows leakage or diffusion through structural and plumbing metals, pressure vessels, and elastomeric seals that would be hermetic for any other gas. Leakage becomes worse with temperature. Aside from problems with external leakage to the atmosphere, hydrogen is extremely difficult to internally retain with sliding or rotating seals in compressors, making it one of the most difficult and expensive conventional gases to compress, even compared with many highly toxic gases for which leakage could be disastrous.

But possibly even more important is that it is readily absorbed into the crystalline lattice of many metals and alloys, causing them to become more brittle or fracture over time. This is another unique characteristic of hydrogen that requires different handling and processing procedures. The problem of "hydrogen embrittlement or hydrogen corrosion" can lead to catastrophic failure at high pressures or under mechanical stress, a risk that grows with time. Iron, many high-strength steels, nickel, titanium, and cobalt are the most susceptible metals, but even aluminum and magnesium alloys can be affected at high temperatures. The number of materials that can be safely used in pressurized or cryogenic hydrogen systems is limited. Even highly abrasion-resistant protective coatings such as titanium nitride are affected, and the failure of these coatings has led to product failures [76]. Interesting to note that these are the same metals that are candidates for use as interstitial hydrides for hydrogen storage because of their absorption properties, to be discussed later in this section. Hydrogen embrittlement can occur over time to some degree in most metals, especially at very high temperatures and pressures. Even non-ferrous Inconel® stainless steel used for the fuel rods of nuclear reactors is of some concern after a few years of operation. Hydrogen embrittlement is a long-term inspection and maintenance item in nuclear power plants for components subject to hydrogen that is formed by unintended water splitting due to exposure to high radiation levels and high temperatures.

The preferred metals for high-pressure hydrogen plumbing are 304 or 316 stainless steel, both non-magnetic and classified as austenitic that is less susceptible to hydrogen permeation. Multi-stage reciprocating (piston-type) high-pressure hydrogen compressors must use embrittlement-resistant metals for hydrogen-exposed surfaces, or non-metallic composites for multi-stage diaphragm-type compressors. In general, the maintenance intervals for hydrogen compressors are shorter and cost more than those of air or natural gas compressors. Unconventional compression methods such as the use of interstitial metal hydrides have been investigated for hydrogen compression and may hold future promise, but are currently considered impractical for pressures above about 3 MPa (435 psi).

Hydrogen embrittlement and leakage problems are also a concern if existing natural gas distribution pipeline networks and equipment are ever to be used for hydrogen distribution. Current industry standards limit natural gas/hydrogen blends to a maximum of 5% hydrogen. This is a serious obstacle to plans to transition existing gas infrastructure to hydrogen since the cost of replacing or retrofitting the vast legacy gas distribution network with hydrogen-compatible materials would be prohibitive.

While the technical and safety issues associated with compressed hydrogen storage onboard vehicles are challenging, all alternative low-pressure hydrogen storage methods are considered even more problematic, for example, cryogenic liquefaction, reversible metal hydrides, or onboard reformation of methanol or ammonia [77]. Consequently, compressed storage has been adopted in all current and past FCVs.

Specialized ultra-high-pressure tanks are required to contain hydrogen at 70 MPa (10,000 psi). Several types of filament-wound composite tanks have been used in FCVs, each capable of different maximum pressures: Aluminum/fiberglass: 305 bars (30.5 MPa; 4420 psi), aluminum/aramid fiber: 438 bars (43.8 MPa; 6,350 psi), aluminum/carbon-fiber: 700 bars (70 MPa; 10,000 psi). All tanks incorporate a non-structural liner, usually made of aluminum or a high-density polymer or fluorocarbon that is impervious to hydrogen diffusion.

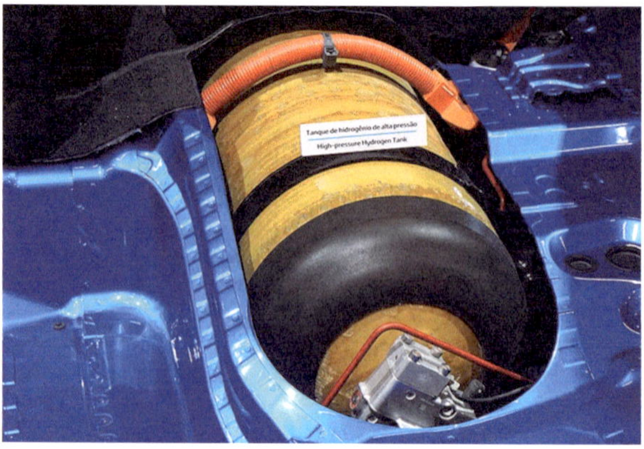

Fig. 9.34 Composite high-pressure hydrogen tank in Toyota Mirai. Photographer: Mario Roberto Duran Ortiz, 2016. https://commons.wikimedia.org/wiki/File:Toyota_Mirai_high-pressure_hydrogen_tank_SAO_2016_9036.jpg. Wikimedia Commons CC-BY-2.0

Fig. 9.35 Carbon fiber over aluminum high pressure cylinder after overpressure failure. From report by William P. Schonberg. *Will it Leak? Will it Burst? COPV Perforation and Rupture after a MMOD Impact* NASA publication, 2023. https://ntrs.nasa.gov/api/citations/20230012105/downloads/8-16-23%2020230012105.pdf. Public domain

As shown in Fig. 9.34, the main fuel tank of a Toyota Mirai can store 6.0 kg of hydrogen at 70 MPa. Converted to electricity by Mirai's fuel cell stack, it provides an EPA-rated vehicle range of 483 km (300 miles). To date, the Mirai has the largest capacity tank and the longest range of available FCVs.

High-pressure storage of any gas always entails some risk of a tank rupture in a collision or due to a manufacturing or installation defect. Hydrogen is no different, and the extremely high pressure increases the risk. Filament-wound tanks of this type are extremely strong under evenly applied force, but a relatively minor point impact can initiate a rapid disintegration of the shell, leading to catastrophic failure. I could find no documented cases of hydrogen tank ruptures in FCVs, but there have been many failures of 25 MPa (3600 psi) composite natural gas tanks. The result of the rupture of a composite CNG tank is shown in Fig. 9.35.

It is logical to assume that the casualties from a hydrogen tank rupture would be due to the combustion of the fuel. But the most dangerous consequence is the huge kinetic energy release due to the instantaneous gas expansion, which can scatter debris at lethal velocities hundreds of meters from the site of the rupture. This does not deprecate the damage that can be done by the combustion of the hydrogen fuel, but it is an extremely light gas that rises upward and dissipates quickly, as opposed to vehicular fires with liquid fuels that spread across the ground and engulf the vehicle. This safety advantage, however, is only valid outdoors, not inside a structure or tunnel.

High-Pressure Refueling Requires Pre-cooling

70 MPa (10,000 psi) is the highest pressure used for any commercial gas distribution, five times the pressure of a SCUBA tank or a welding gas cylinder. Hydrogen is one of only a few common gases that cannot be considered ideal and therefore does not behave according to the Ideal Gas Law T=PV/nR. During refueling a 70 MPa hydrogen tank, it exhibits a negative Joule-Thompson coefficient, causing the temperature of the vehicle tank to rise during rapid filling [78, 79], which is necessary to provide refueling times acceptable to consumers. Pre-cooling hydrogen to approximately –40 °C is standard practice at hydrogen refueling stations to avoid temperatures high enough to damage the vehicle's composite pressure tank. Safety standards for hydrogen refueling have been established by the Society of Automotive Engineers (SAE) [80], the International Standards Organization (ISO) [81], and other safety standard organizations. A tutorial about the related physics can be found online at citation [82]. This pre-cooling step adds to the energy investment in the fuel beyond the energy required for gas compression.

Slow refueling is possible without pre-cooling, but it would require a large increase in the refueling time, of the order of an hour. From a sampling of experiences at California hydrogen fueling stations, a typical *half fill* (H35) to 35 MPa (5000 psi) typically requires 3–5 min. A *full fill* (H70) to 70 MPa (10,000 psi) requires typically 5–10 min, although faster fill times are possible (Fig. 9.36).

Cryogenic (Liquid) Hydrogen

Hydrogen is cooled through multiple refrigeration stages to 20 Kelvins (–253 °C). It is stored in a cryogenic dewar (example shown in Fig. 9.37), essentially a super-insulated thermos bottle, which keeps the liquid hydrogen at this temperature for long periods. Some always continue to boil due to heat conduction through the super-insulated tank walls and fittings. This requires some provision for venting this leakage, either directly to the atmosphere or through a catalytic converter that converts it to water vapor.

Venting of H_2 from an LH_2 tank will occur no matter how good the insulation, as ambient heat eventually creeps into the dewar causing the liquid hydrogen to boil. The tank pressure will increase, opening a pressure relief valve. To prevent the release of hydrogen into a closed garage or storage area, the vented hydrogen is passed through a selective catalytic converter to "burn" it into water vapor. But it is still a bad idea to park any vehicle with a cryogenic storage tank (H_2 or LNG) in a closed garage—it takes very little hydrogen to form a flammable mixture in air, and the required ignition energy of the mixture is very low. Possibly the earliest design descriptions for an automotive-size liquid H_2 can be found in citation [83].

Fig. 9.36 Hydrogen "gas pump" at fuel H_2 station in Torrance California, 2011. Photo by haymarketrebel. https://www.flickr.com/photos/26557483@N08/6269926869 Attribution 2.0 Generic, CC BY 2.0

For safety reasons, any vehicle that uses cryogenic storage (LH_2 or LNG) should never be parked in an enclosed space. This is usually not a problem for large trucks or buses that are parked outdoors, but it is a safety limitation for passenger automobiles.

As discussed above, another consideration for hydrogen suppliers and fueling stations is the large amount of energy required for the liquefaction (Kelvin) process, typically 41% of the hydrogen fuel value [84]. And prior to liquefaction, all contaminant gases must be removed. And to reduce the energy required as the temperature approaches its liquefaction point during cooling, the hydrogen must be catalytically converted to entirely "para" form, from the naturally-occurring mix of ortho and para forms defined by the electron spin direction [85]. The total energy investment required for liquifying hydrogen is much greater than the energy required for compression to 70 MPa.

All requirements and limitations considered, it is unlikely that liquid hydrogen storage will ever be used in production passenger vehicles, but it has likely applications in commercial aviation, large truck transport, or rail vehicles for which cost, GHG impact, and overall efficiency of the fuel are not as important as mass energy density.

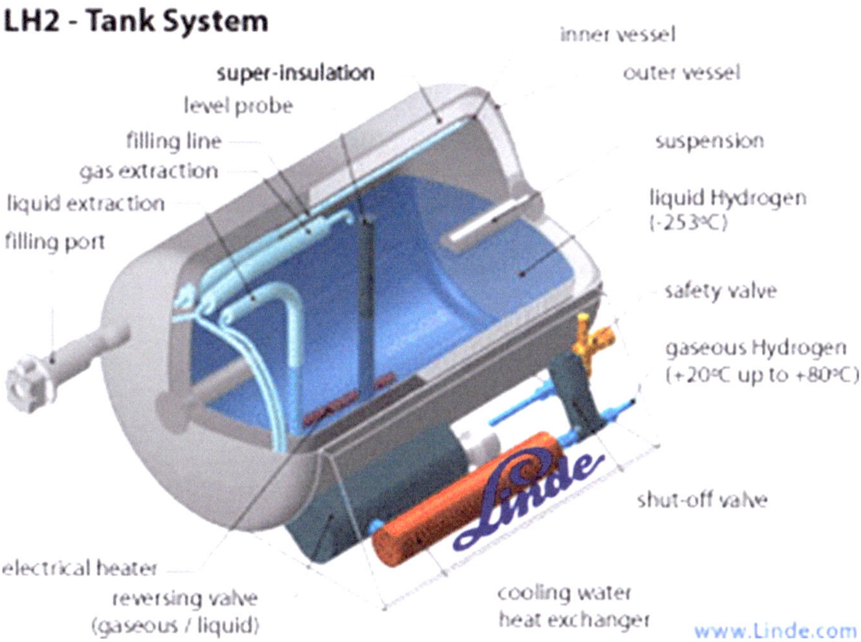

Fig. 9.37 Cross-section diagram of automotive liquid hydrogen tank, as used in the BMW hydrogen 750i concept car of Fig. 9.16. Illustration from *HFCIT Hydrogen Storage: Gaseous and Liquid Hydrogen Storage*. US DOE 2007. http://www1.eere.energy.gov/hydrogenandfuelcells/storage/hydrogen_storage.html. Public domain

Cryo-Compressed Hydrogen

Cryo-compression is a compromise between liquid hydrogen and high-pressure hydrogen storage that combines both the benefits and limitations of each method, allowing storage above the 20K temperature of liquid hydrogen and pressures below 70 MPa of compressed hydrogen. In one prototype system, gaseous hydrogen is cooled to 77K (−196 °C) by liquid nitrogen and stored at 15 MPa (2,200 psi). At this temperature and pressure, a 100L tank will hold 4.1 kg of hydrogen [86]. By comparison, 4.1 kg of liquid hydrogen would occupy 58L, and 4.1 kg of 70 MPa compressed hydrogen would also occupy about 58L not including the tank. Obviously, cryo-compressed hydrogen requires a specialized tank that is both a super-insulated dewar and a medium pressure (15 MPa, 2,200 psi) compressed gas tank. The concept is sound but complex and expensive since both gas supercooling and compression are required. It may have promise for specific applications, but it is not in current use for vehicular hydrogen fuel storage [87].

Fig. 9.38 Iron titanium hydride granules. From https://wikivisually.com/ CC-BY-3.0

Metal (Interstitial) Hydrides

Diatomic (common) hydrogen is the smallest stable molecule in existence. Its size and physical characteristics give it properties not found in any other gases. It has the ability to diffuse and be absorbed into the crystalline lattice of some metals and metal alloys. This is the principle behind hydrogen embrittlement of ferrous metals such as iron and carbon steel, a serious problem if transporting hydrogen via pipeline or pressure tank. But this unusual property also makes it possible to reversibly store a significant mass fraction of hydrogen in certain metal alloys referred to as interstitial metal hydrides, a subset of the broader class of absorption metal hydrides. An example of a low-temperature interstitial metal hydride, iron-titanium alloy, is shown in Fig. 9.38.

A large number of metal alloys share this property, as shown in Fig. 9.39 that identifies hydrides that have been examined in three US DOE centers of excellence [88]. The brackets in the figure enclose the small number of materials that meet acceptable mass hydrogen density requirements and operate within a reasonable range of temperatures and pressures[5] for practical hydrogen storage. The topic is well-researched, spanning several decades since the 1970s. Further information is available in citations [89, 90].

Interstitial metal hydrides are stored in granular form in a sealed metal tank, usually made of aluminum or nonferrous stainless steel, capable of withstanding the pressure and temperature that would be encountered during recharging, and non-reactive with hydrogen. The absorption reaction is exothermic, occurring at elevated pressures. The hydrogen desorption reaction is endothermic, requiring external heat to release the hydrogen, increasing its pressure to a level sufficient to meet the fuel supply needs of either a combustion engine or a hydrogen fuel cell. The nominal pressure target chosen by the US Department of Energy's Metal Hydride Center of Excellence is five bar (73 psia) [88], although this is somewhat arbitrary. As long as the desorption pressure is greater than atmospheric pressure at the temperature attainable with the exhaust or coolant heat, hydrogen will flow out of the tank. At a given temperature, the hydrogen mass fraction vs pressure follows a hysteresis curve, shown in Fig. 9.40 for a generic metal hydride.

[5] In this chart, MOF stands for Metal Organic Frameworks, which are porous metal clusters containing organic ligands, which increase the surface area and absorption properties.

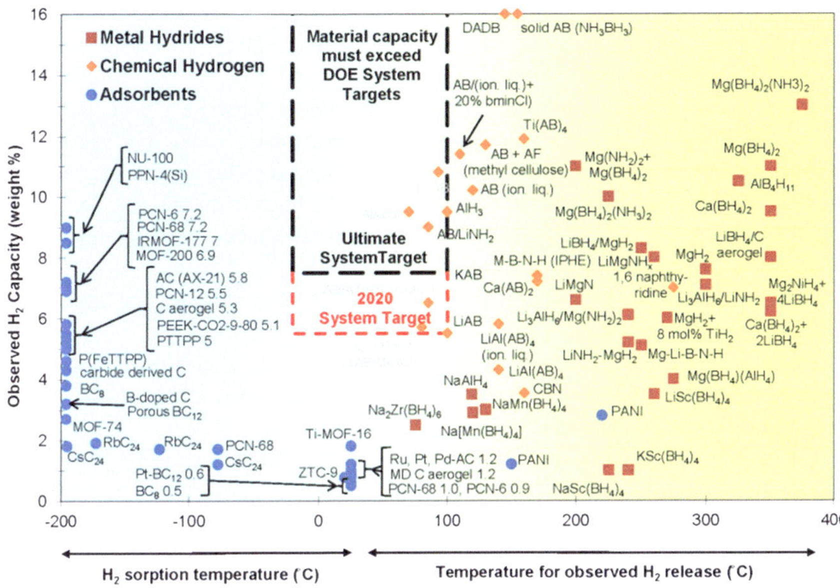

Fig. 9.39 Metal hydrides identified in Final Report for the DOE Metal Hydride Center of Excellence, 2012. From https://www.energy.gov/eere/fuelcells/articles/final-report-doe-metal-hydride-center-excellence. Public domain

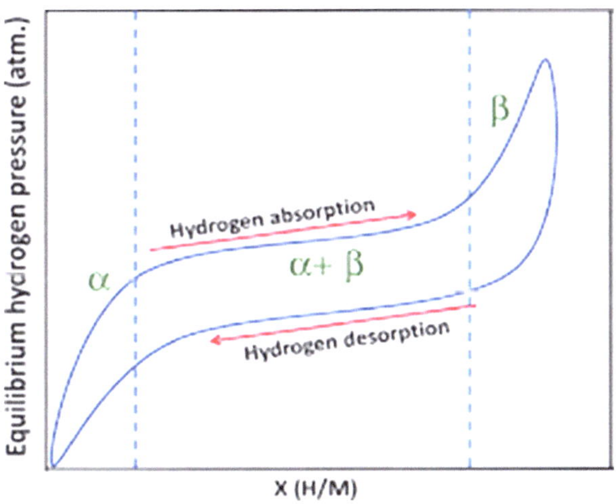

Fig. 9.40 Hydrogen absorption/desorption hysteresis curve for a generic metal hydride. The abscissa scale H/M is the mass ratio of hydrogen to metal. It is sometimes normalized such that the maximum ratio or saturated state is assigned a value of 1. Diagram from Banerjee, S., Ruz, P. (2021). Synthesis and Characterization of Metal Hydrides and Their Application. Handbook on Synthesis Strategies for Advanced Materials. Indian Institute of Metals Series. Springer Nature, 2021. Used under STM guidelines

For a reversible metal hydride to be useful for vehicular hydrogen storage, its properties must usually fall within the following constraints:

- The desorption temperature and heat requirements must be within the range available from the waste heat of either the combustion engine or the fuel cell that uses hydrogen as fuel. This may not be an obstacle for combustion engines that generate large amounts of exhaust or coolant heat, but it is an obstacle for the use of metal hydride hydrogen storage for fuel cells, which do not generate as much waste heat because of their higher efficiency.
- If the storage system is to be recharged onboard the vehicle, the exothermic absorption temperature must not be so high as to create a safety problem.
- The hydrogen storage mass energy density must be at least competitive (not necessarily higher) with that of compressed hydrogen, which is the status quo for vehicular hydrogen storage at this time.
- The hydrogen storage volume density is usually not a primary concern, because the density of the metal in the tank is very high. For example, a metal hydride with a 2% hydrogen storage density would weigh 50 times the mass of the hydrogen, but would occupy a relatively small volume compared with other hydrogen storage options.
- The material cost must be reasonable and economically sustainable. Considering the advantages of metal hydrides, the cost need not necessarily be comparable with the cost of high pressure of cryogenic hydrogen storage tanks.
- Hydride metals must not be dependent on scarce, politically sensitive, or environmentally damaging supply chains. Unfortunately, the exotic materials required do have these associations.
- The recharge (absorption) time for hydrides must not be excessive. Reasonable expectations are longer than the refill times for compressed or liquid hydrogen, but less than high-rate EV battery charging.
- Metal hydride granules repeatedly fracture during charge/discharge cycling, which can ultimately reduce them to fine dust that can become densely packed and resistant to hydrogen absorption. The gas passages in the hydride tank must be engineered to allow for this degradation without excessive loss of capacity.

The heat generated during refueling a metal hydride storage tank on a vehicle is usually wasted, but creative uses for this heat such as building air heating near the recharging facility are possibilities. The use of room-temperature metal hydrides for absorption heat pumps have been proposed for HVAC, utilizing hydrogen transferred between interior and exterior hydride tanks.

These practical constraints and the generally poor mass storage density have limited the use of metal hydrides in road-going vehicles. Two of the few hydrides that meet the temperature and pressure range requirements are iron-titanium (FeTi) and Lanthanum-Nickel ($LaNi_5$) alloys. The maximum hydrogen density for FeTi is 1.2%. It has been the most commonly used metal hydride for hydrogen storage, and was used in experimental hydrogen ICE vehicles as early as the 1970s. Lanthanum-Nickel can theoretically store 1.4% of its mass in hydrogen at room temperature and has slightly better hydrogen volume density compared with FeTi. It has most commonly been used in nickel metal hydride batteries rather than for vehicular hydrogen storage [91].

The relative safety of metal hydride hydrogen storage is superior to any other vehicular storage method. A ruptured metal hydride tank is much less consequential than a ruptured high-pressure hydrogen tank or an LH_2 dewar that will fail catastrophically rapidly releasing hydrogen that will likely be ignited. A metal hydride tank, being a low-pressure tank filled with granular metal, releases hydrogen slowly, and the endothermic adsorption reaction cools the tank during the release, reducing the risk of ignition [92].

Despite the poor hydrogen mass storage density, materials cost, and recharge time limitations of known metal hydrides, their intrinsic safety and low-pressure operation make this technology advantageous for some specialized applications in which these limitations are not objectionable. Examples are indoor-operated forklifts, underground mining vehicles, military vehicles, or stationary electric generators.

Chemical Hydrides

A large number of chemical compounds have large hydrogen mass fractions, and the hydrogen can be disassociated via hydrolysis reactions with water or acids, allowing them to be used for high-density hydrogen storage. This makes them potential candidates for vehicular hydrogen storage. With only one exception (sodium borohydride) there have not yet been any such deployments other than experimental. Most are not reversible in the same sense as an absorption metal hydride. The classic aluminum + sodium hydroxide reaction of countless school projects comes to mind. The practicality of hydrogen storage depends mostly on the controllability of hydrogen generation, but also on the compactness of the reactants, safety, fuel handling considerations, recyclability of the reactants, and ultimately the cost per unit of hydrogen produced. While reversibility of the reaction is desirable, it is not possible onboard a vehicle since *rehydrogenation* requires more than just a change of pressure and temperature—usually other chemicals are required. The reactants must be treated as fuels themselves rather than as reversible hydrogen storage materials.

Sodium Borohydride

One of the few examples of chemical hydrides that have actually been used for hydrogen generation onboard a vehicle is sodium borohydride ($NaBH_4$). Sodium borohydride ($NaBH_4$) undergoes hydrolysis (hydrogen generation) at room temperature in the presence of water and a suitable catalyst, or almost any acid, liberating hydrogen gas from both the hydride *and* the water, leaving sodium metaborate $NaBO_2$ as a recyclable waste product. The reaction requires a heterogeneous catalyst (such as Pt or Co) that is not consumed in the reaction, or an acidic solution, usually phosphoric acid due to its relatively low volatility. Schlesinger et al in 1953 first observed that an aqueous solution of $NaBH_4$ in a basic solution is very stable and could be treated like a liquid fuel. Hydrogen generation can then be initiated and controlled on demand by bringing the stabilized solution into contact with the catalyst or by injecting an acid into the solution. The reaction is mildly exothermic, requiring no additional heat or other form of energy. It produces a mass of hydrogen that is twice the hydrogen content of the hydride [93]. Essentially, the reaction releases the entire hydrogen content of the hydride, as well as hydrogen from the water that it reacts with:

$$\mathrm{NaBH_4 + 2H_2O \rightarrow NaBO_2 + 4H_2}\ \Delta H = -75\,\mathrm{kJ/mole\,H_2}$$

1 g of $NaBH_4$ will produce 2.37 L of hydrogen at standard temperature and pressure (STP). Not including the required water, the gravimetric hydrogen storage capacity is 10.8 wt%, which is superior to both compressed and liquified hydrogen when the weight of the respective enclosures is factored in.

In solid form, $NaBH_4$ is a white powder with the consistency of gypsum which can be pressed into standardized shapes, allowing the possibility of fueling hydrogen vehicles with brick-like solid cartridges at room temperature and moderate pressure. Since contact with any environmental acid can initiate the liberation of hydrogen, a small amount of a common base such as sodium hydroxide may be mixed with the solid $NaBH_4$ to assure long-term stability. The onboard hydrogen generator can be as simple as a *Kipp Apparatus* [94], which controllably exposes the stabilized $NaBH_4$ to a low-volatility acid or an appropriate catalyst. It's worth noting that unlike hydrogen generation by onboard reforming of LOHCs (see next section), hydrogen generated from sodium borohydride is exceptionally pure, with the only contaminant being water vapor. In fact, sodium borohydride has been used for generation of small quantities of pure hydrogen in laboratories since the 1960s [93].

The hydrogen mass and volume density for $NaBH_4$ storage of hydrogen is much greater than compressed or liquid hydrogen or any known interstitial hydrides. Like absorption metal hydrides, the hydrogen produced is pure, eliminating the need for gas-purifying steps. The hydrolysis reaction does not require external heat. This would theoretically make it an excellent candidate for hydrogen storage for fuel cell vehicles.

Unfortunately, the reaction is not reversible. It is not possible to regenerate $NaBH_4$ directly from the recycled $NaBO_2$ end-product. The synthesis of $NaBH_4$ requires multiple steps and significant energy, making it unacceptably costly except in situations such as emergency APUs, such as those used for remote military field operations. Nevertheless, in the early 2000s, Siemens GmbH and Daimler Chrysler promoted $NaBH_4$ as a (questionably) proprietary hydrogen storage medium for hydrogen FCVs under the trademark name *Natrium* [95].[6] The Chrysler Town &

[6] The trademarked name "Natrium" is Latin for Sodium. Similar genus as "Vulcanol", used by Carbon Recycling International to refer to methanol produced from CO_2 sequestered by its direct air capture facility in Iceland, ref: CRI, RENEWABLE METHANOL - VULCANOL (TM). https://carbonrecycling.com/industries/co2-methanol/products.

Country *Natrium* concept minivan was claimed to have a range of 300 miles (482 km) [96]. It never came close to production for undisclosed reasons, almost surely related to the hydride cost and the need for an entirely new fueling infrastructure [97].

One other borohydride has received some interest recently: lithium borohydride $LiBH_4$. It is similar in molar hydrogen storage density but lighter weight, with an effective hydrogen production density of 9% [98]. This makes it arguably the highest-density chemical hydride usable for practical hydrogen storage. Unfortunately, it is more costly than even $NaBH_4$, and like $NaBH_4$, is not easily reversible.

The search for a stable and easily reversible chemical hydride that can be produced at a reasonable cost remains a topic of continued, albeit low-priority, research interest.

Silanes

A relatively new area of investigation, compounds of silicon and hydrogen, i.e., silanes, polysilanes, or organosilanes, hold some promise for portable hydrogen storage. It is well known that exposure of these compounds to water, alcohol, or heat can rapidly generate large quantities of hydrogen, requiring special precautions for storage of these chemicals. However, it has only been since approximately 2005 that the ease with which they can generate hydrogen at NTP on demand has been considered for the task of mobile hydrogen storage [99]. For example, the reaction of polysilyl organics such as $HC(SiH^3)^3$ can produce > 6% hydrogen by mass [100], a superior hydrogen mass and volume density compared with conventional hydrogen storage methods.

In citation [101] 12 silicon-hydrogen compounds were identified as candidates for service as hydrogen storage media for hydrogen vehicles. The study authors observed that the general catalyzed reaction of organosilanes with water could produce hydrogen on demand in a fully controllable, environmentally friendly, rapid, and safe process. With no carbon involved, CO_2 emissions from the vehicle (but not necessarily the production process) would be zero. As liquids or dual-phase gas/liquids, they would be familiar to users of liquid fuels. And unlike interstitial metal hydrides, no external heat is required for the release of hydrogen, as the hydrolytic oxidation reaction is slightly exothermic.

However, the highest H_2 density silanes have handling issues that would preclude their use as a vehicle fuel, i.e., the root compound Silane SiH_4 is a highly flammable and toxic gas that is prone to autoignition during handling, recalling unstable boranes such as pentaborane B_5H_9, a borohydride that in the 1950s was investigated as a possible rocket fuel because of its very high specific energy density [102].

The design of an apparatus required to controllably generate hydrogen on demand has not been discussed in the literature, but it would logically be similar to the apparatus used for hydrogen generation from sodium borohydride, discussed above.

Like other chemical hydrides, silanes cannot be "recharged" on board the vehicle. Residuals from the depleted silane/hydrogen reaction would have to be drained upon refueling, to be recycled and processed. But in general, the catalyst materials can be recovered and reused indefinitely.

A 2011 study reported success in regeneration of methoxysilanes, $(CH^2Si(OMe)2)^3$ and $(CH_2Si(OMe)_2CHSi(OMe)^3)^2$ which were experimentally regenerated in high yields by $LiAlH_4$ reduction [99]. This remains an area for further research.

A few organosilanes are dry-air-stable and non-toxic. One citation [103] notes that two candidate organosilanes, polymethylhydrosiloxane ($Me_3Si(OSiMeH)_n OSiMe_3$, PMHS) and tetramethyldisiloxane ($Me_2SiHOSiHMe_2$, TMDS), are currently unused byproducts of the semiconductor industry and are therefore cost-effective, and that the residual silanols after dehydrogenation are useful in the production of silicone rubber. Regarding control of the reaction, the use of recoverable heterogeneous catalysts is reported to allow real-time metering of H_2 production. Several heterogeneous catalysts for the hydrolytic oxidation of organosilanes have been reported, e.g., citation [104].

Production costs and life-cycle energy efficiencies for the widespread use of silanes as practical hydrogen carriers have not been published, and no use of silanes for automotive hydrogen storage has been reported in publications as of 2024. However, production-oriented research in this area is still in its infancy. Silanes have potential for hydrogen storage for specialized fuel cell vehicles, but it seems unlikely that they will ever be competitive for general automotive hydrogen storage.

Onboard Reforming of Liquid Organic Hydrogen Carriers (LOHCs)

Methanol (CH_3OH)

Methanol has a volume and mass energy density about half that of gasoline. The number of potential feedstocks for making methanol is probably second only to hydrogen. It is most commonly made from natural gas via a process in which hydrogen is an intermediate product. But it can be made from almost any carbonaceous material, including most forms of dry biomass and wood residues.

Methanol is usually synthesized by an exothermic reaction between hydrogen and carbon monoxide from syngas in the presence of an exogenous catalyst—typically metallic copper and/or zinc. But the synthesis process can be easily reversed in an endothermic reaction to produce hydrogen, carbon monoxide, and/or carbon dioxide. This makes it probably the best candidate as a LOHC, a potential replacement for hydrogen storage onboard a hydrogen-fueled vehicle. It is the only LOHC that has seen use as a hydrogen carrier in anything other than an experimental or demonstration environment [105].

As a bulk hydrogen carrier, methanol has the potential to replace the hazardous and energy-intensive storage of hydrogen as a compressed gas or cryogenic liquid. This is a particularly attractive option for large-scale marine transport [106].

Possibly the only automotive application in which onboard methanol reformation is currently used is the combination of an onboard methanol-to-hydrogen reformer and a hydrogen fuel cell, which has been referred to as an *Indirect Methanol Fuel Cell (IMFC),* even though it is not actually a methanol fuel cell. This technology is discussed in the *Methanol Fuel Cell* section of Chap. 6, *Methanol.* In that application, a self-contained apparatus onboard a vehicle reforms methanol into syngas using external heat, then scrubs the CO_2 and CO from the gas and uses the purified hydrogen in a PEM hydrogen fuel cell. The electric output continuously charges a battery, which powers an electric drivetrain.

One of the drawbacks to any "green" LOHC is the total energy required to form it from hydrogen, and then reform it back into hydrogen onboard a vehicle, both processes penalizing the overall efficiency of the vehicle. Methanol is most commonly made from natural gas with 61–66% energy efficiency, although at least one source reports optimized efficiencies as high as 75% [107]. But truly "green" methanol must be made from "green" hydrogen which is made from renewable electricity, a process that is energy-intensive. The combined efficiency of an onboard reformer and a hydrogen fuel cell is only 35–40% as reported by Blue World Technologies [108], the manufacturer of the integrated reformer/fuel cell module referred to in the example above. The advantage, of course, is the convenience and high volume energy density of methanol compared with an equivalent amount of hydrogen stored as a compressed gas or a cryogenic liquid.

Onboard a vehicle, methanol can be directly reformed in a single step using external and/or recycled heat to react methanol over a catalyst to produce hydrogen and carbon monoxide:

$$CH_3OH + \text{heat} \rightarrow CO + 2H_2$$

Or, with the addition of steam, the CO and H_2O can be converted into CO_2 and hydrogen, leading to an overall reaction:

$$CH_3OH + H_2O \rightarrow CO_2 + 3H_2$$

In this latter process, a mixture of methanol and water is vaporized at 250–360 °C and 20 bar (300 psig), then passed through a catalyst bed. Exogenous catalysts as simple as copper and zinc can be used, but higher conversion rates can be achieved with the addition of aluminum oxide or the much more expensive noble metal catalysts palladium or platinum. The reaction is endothermic, meaning that heat input is required to drive the reaction. If the reaction goes to completion, the output is just hydrogen and carbon dioxide, but usually small fractions of carbon monoxide, water, and methanol vapor remain from the steam reforming process. Water and methanol vapor is easily condensed out, but if pure hydrogen is needed for a fuel cell, the carbon monoxide and dioxide contaminants must be separated by either pressure swing adsorption (PSA) or filtering via a hydrogen-permeable polymer or ceramic membrane.

The heat generated by the fuel cell can be recycled to supplement the heat required by the methanol reformer, but it is inadequate to provide more than a minor fraction of the heat requirement. However, in a combustion hydrogen vehicle, most or all of the required heat can be provided by the (much more wasteful) engine exhaust heat, making this the more logical application of this technology. The heat balances of a completely reformed methanol fuel system for a hydrogen combustion vehicle were explored in the early 1980s at the DOE's Solar Energy Research Institute (SERI), in which a Chevrolet Citation was converted to run on hydrogen and CO generated by an onboard methanol reformer. The vehicle was successfully demonstrated, with all of the reformer heat provided by the engine exhaust after the engine warmed up [109].

Using a methanol + water feedstock is referred to as *methanol steam reforming* (MSR, as opposed to SMR), as described above. The CO_2 co-product is released into the atmosphere, reducing or possibly eliminating any advantage that can be claimed by the hydrogen-powered vehicle as a ZEV. CCS isn't practical at the scale of a small automotive methanol-to-hydrogen reformer. As a result, methanol as an LOHC isn't really any better from a GHG point of view than the direct combustion of methanol in the engine. Consequently, despite its advantages compared to other LOHCs under consideration, onboard methanol reforming is not currently used by any major manufacturer, with the possible exception of Fraunhofer GmbH, which has developed a small exhaust-heated methanol-hydrogen reformer to produce a hydrogen-rich supplemental fuel for a diesel engine. This novel application has been found to improved combustion characteristics and reduce particulate emissions. It is also advantageous for cold-starting the diesel engine [110].

Also demonstrated in the above-referenced SERI project were the challenges of matching the hydrogen generation rate with the instantaneous demand of an ICE or fuel cell. Storage of even a small amount of hydrogen to serve as a buffer between instantaneous supply and demand requires yet another hydrogen storage mechanism, adding further complexity to the hydrogen production plant onboard the vehicle.

Ammonia (NH_3) [111]

As discussed in Chap. 11, *Ammonia*, NH_3 is itself a usable fuel for internal combustion engines. But its combustion quality is poor, NO_x emissions high, and its hazardous physical properties make it challenging to handle safely [111]. As a potential hydrogen carrier, ammonia's hydrogen density is excellent, with a hydrogen content of 121 g H_2/L of liquid ammonia (under pressure), even higher than methanol.

As discussed previously, most commercial ammonia is produced for fertilizer, starting with hydrogen made from natural gas or coal, followed by the Haber-Bosch synthesis that reacts hydrogen with nitrogen in the presence of an iron catalyst at high temperature and pressure—typically 200 to 400 bar and 400° to 650° C. The process is very energy intensive, and the invested heat energy required to make ammonia from hydrogen is not recoverable.

Ammonia is stored and transported in liquid form at ambient temperature (20 °C) under moderate pressure, typically 7.5–10 bar (109–145 psig) [112]. Its physical properties are similar to butane or propane, but ammonia is highly toxic and caustic, readily attacking exposed flesh. Full-body personal protection equipment (PPE) is required if there is any risk of contact or inhalation of anhydrous ammonia [113].

Essentially the same Haber-Bosch process used to make ammonia from hydrogen can be reversed to reform ammonia to hydrogen onboard a vehicle. Ammonia has one major advantage over methanol as a hydrogen carrier: its disassociation does not produce carbon dioxide, so the direct carbon emissions of the hydrogen vehicle remain zero. Dehydrogenation of ammonia to produce hydrogen requires moderately high temperatures. The primary byproduct is gaseous nitrogen.

Although ammonia is recurrently suggested as a potential portable hydrogen carrier, its hazardous handling issues have prevented its use in any except a few experimental vehicle deployments to date.

MCH (Methylcyclohexane, C_7H_{14})

One other LOHC that has gained some traction recently is Methylcyclohexane (MCH). It could equally be classified as a chemical hydride, although its ease of recyclability sets it apart from most chemical hydrides that require more extensive reprocessing to regenerate. As a portable hydrogen storage medium, it has found its way into at least one full-scale demonstration project—the conversion of a small collection of diesel-electric rail locomotives to hydrogen fuel cell electric locomotives [114]. Its use as an automotive fuel seems unlikely, however.

There are multiple paths for the production of MCH, all involving catalytic reactions of toluene (C_7H_8) and hydrogen at elevated temperatures. Toluene is a common industrial reagent, paint solvent, and the basis of the high explosive trinitrotoluene (TNT). It is inexpensive since it is a byproduct of gasoline and coke production. The innovation that has made it popular recently as a hydrogen carrier is referred to as Direct-MCH® by the Japanese company Eneos [115]. This is a production process very similar to the electrolysis of water, except that the hydrogen normally liberated as a gas at the cathode reacts directly with a toluene solution to form MCH. This process bypasses the separate generation of hydrogen by electrolysis and then the conversion of toluene to MCH, accomplishing both in a single step. But as with most other carbon-free energy production schemes, the electrical energy input to the process must be "renewable electricity"; otherwise, its generation could be carbon intensive to a degree greater than the carbon intensity of the electric energy source.

MCH is a liquid at room temperature and pressure, making handling similar to gasoline—with one major difference: after it is reformed onboard to hydrogen and toluene, the toluene is recycled by removal at the time of refueling and returned to the MCH production facility for processing. Onboard a hydrogen-fueled vehicle, hydrogen would be extracted from MCH by dehydrogenation involving a catalyst and heat of around 250 °C, which could be partially provided by the exhaust of a hydrogen ICE. The remaining toluene is later removed during refueling and recycled to become MCH at a reprocessing facility. Hydrogen reformed from MCH is not pure enough to be used directly in hydrogen fuel cells, but it is acceptable as a combustion fuel in hydrogen ICVs. For fuel cell use, an additional purification stage is required, which requires additional energy. The hydrogen mass fraction of MCH is 7.14%. But not all of the hydrogen in MCH is released since the byproduct is toluene which has a hydrogen mass fraction of 8.7%. MCH reforms into 47.3 g H_2/L of MCH. For comparison, the hydrogen mass fraction of methanol is 12.5% and it reforms into 98.8 g H_2/L methanol. For ammonia, the hydrogen mass fraction is 17.7% and it reforms into 121 g H_2/L of liquid ammonia. Both methanol and ammonia have a higher hydrogen volume density than liquid hydrogen (71 g H_2/L) while MCH is lower than liquid hydrogen. For perspective, if hydrogen were to be generated onboard from methane via MSR, the H_2 mass/methane volume at 25 MPa would be 36.0 g H_2/L of compressed methane, lower than any of the above (Fig. 9.41).

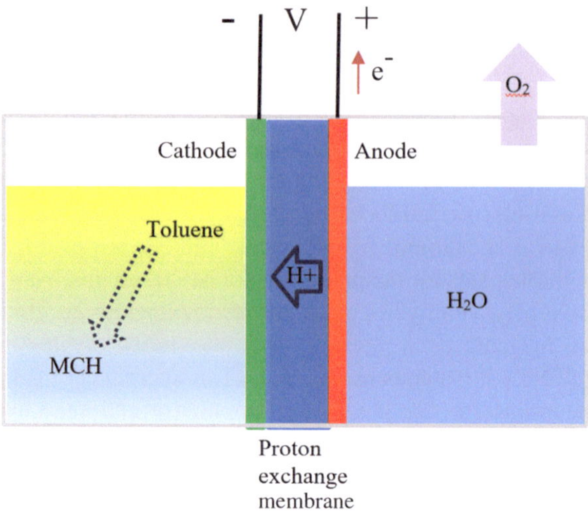

Fig. 9.41 Electrolytically recycling toluene to MCH. Diagram by author, based on similar diagram in citation [115]

Table 9.1 H_2 storage or carrier types in order of increasing hydrogen volume storage density, excluding the storage tank or reformer

Hydrogen storage medium	H_2 Density (g H_2/L storage)	Notes
Methane (onboard SMR)	36.0 (at 25 MPa, 3600 psi)	Requires large amount of heat
Methylcyclohexane (MCH)	47.3	Toluene byproduct recycled
Compressed hydrogen (70 MPa)	52.3	Bulky tank mass
FeTi hydride	53.1 (1.2% mass)	Room temp hydride
$LaNi_5$ hydride	115 (1.4% mass)	Room temp hydride
Mg-Al hydride	54.1 (6.0% mass)	High temp hydride
Liquid hydrogen (20?K)	72.1	Exotic fuel handling
Methanol (CH_3OH)	98.8	CO_2 byproduct
Ammonia (NH_3)	121.0 (liquid ammonia)	Hazardous gas at NTP
Sodium Borohydride ($NaBH_4$)	126.2 (10.8% mass)	Sodium borate byproduct recycled
Silanes, e.g., silicon tetrahydride	12.5% mass % (excl. water)	Hazardous gas

Silanes listed as a typical percentage mass density

While sharing the advantage of methanol in its compatibility with existing fuel transport and dispensing infrastructure, and being fully recyclable, MCH is limited by its relatively low hydrogen density, high energy losses during production and reforming, and expected high production cost compared with other portable hydrogen storage options.

A summary of the various methods discussed above for vehicular hydrogen storage is shown in Table 9.1.

Hydrogen as an Aviation Fuel

While the focus of this book is ground transportation, it is worthwhile to look at commercial aviation, a related transportation application which could potentially benefit from the use of hydrogen fuel. If the hydrogen is produced entirely from carbon-free electricity, hydrogen qualifies as a "Sustainable Aviation Fuel" (SAF) and an *eFuel*, to be discussed in a later chapter. In fact, it is the root of *all* eFuels, which are made by combining carbon compounds with hydrogen to synthesize bespoke hydrocarbons.

Commercial aviation is responsible for about 13% of transportation GHG emissions, and 4% of all anthropogenic atmospheric warming [116]. In 2021, US commercial air carriers have committed to carbon-neutral international aviation beginning in 2021, and a 50% reduction in carbon dioxide (CO_2) emissions in 2050 compared to 2005 levels [117]. Jet or turboprop engines require a combustion fuel. Green hydrogen is currently the only carbon-free solution for powering these types of aircraft engines.

The use of hydrogen in aerospace is well established, considering its use as the fuel of choice for large rocket boosters, such as in the US satellite and manned space programs. The idea of liquid hydrogen as a commercial or military aviation fuel has been around since the 1950s, but it has only been used with marginal success in a few military applications [118]. Three unpiloted exceptions were the Centurian, Pathfinder, and Helios remotely piloted ultralight solar/electric aircraft, built for NASA by AeroVironment of Monrovia, California. In the late 1990s, these experimental UAVs were powered solely by solar energy, hydrogen fuel cells, and electric motors, and had the ability to fly multi-day missions at altitudes as high as 100,000 ft (30,000 m). They continued operation day and night by storing excess energy during the day as hydrogen, using reversible hydrogen fuel cells. A NASA photo of the Helios in flight is shown in Fig. 9.42.

Despite its much higher mass energy density compared with jet fuels (120 MJ/kg vs 43.5 MJ/kg for Jet A), its *much lower* volume energy density (8 MJ/L vs 35 MJ/L for Jet A), and its exotic fuel handling and containment are barriers to commercial deployment. The theoretical advantage of liquid hydrogen over other aviation fuel options is illustrated in Fig. 9.43. For a given total aircraft mass, the lower mass density of hydrogen allows more energy to be carried on board within the maximum allowable fuel fraction of the aircraft weight. This could provide a longer range than conventional jet fuel or other alternative aviation fuels, *but only if the huge increase in the aircraft fuel tank volume can be accommodated within the airframe*. Industry estimates are highly variable based on the configuration of the aircraft, but a visual estimate of the total volume displaced by liquid hydrogen tanks suggests that it could reduce passenger and cargo space by as much as 50%. As shown by the

Fig. 9.42 2001 NASA photo of Helios remotely piloted ultralight aircraft, designed for nearly perpetual flight by use of solar/electric propulsion and reversible hydrogen fuel cells for storage of energy for night flight. https://ia601309.us.archive.org/27/items/NIX-ED01-0230-4/ED01-0230-4.jpg Public domain

dashed red line in Fig. 9.43, the rule-of-thumb maximum allowable fuel fraction of the total weight of an aircraft is 44%.

Most manufacturers of commercial aircraft as well as a number of recent startup firms have been dabbling with LH_2 concept aircraft; some have even announced major investments in their development [119]. Large aircraft would most likely use LH_2 as a fuel for conventional turbojet engines. Commuter-scale and smaller propeller-driven aircraft would more likely use hydrogen fuel cells to power electric motors. An excellent review of the current state of the art in hydrogen-powered aviation can be found in citation [120]

Figure 9.44 is a photo of the Airbus SAS ZEROe Turbofan concept liquid hydrogen aircraft. Two hydrogen-fueled but otherwise conventional high-bypass turbofan engines provide thrust. Liquid hydrogen sufficient for transcontinental range is stored in the aft (windowless) section of the fuselage. The very low mass of liquid hydrogen does not upset the weight and balance of the aircraft.

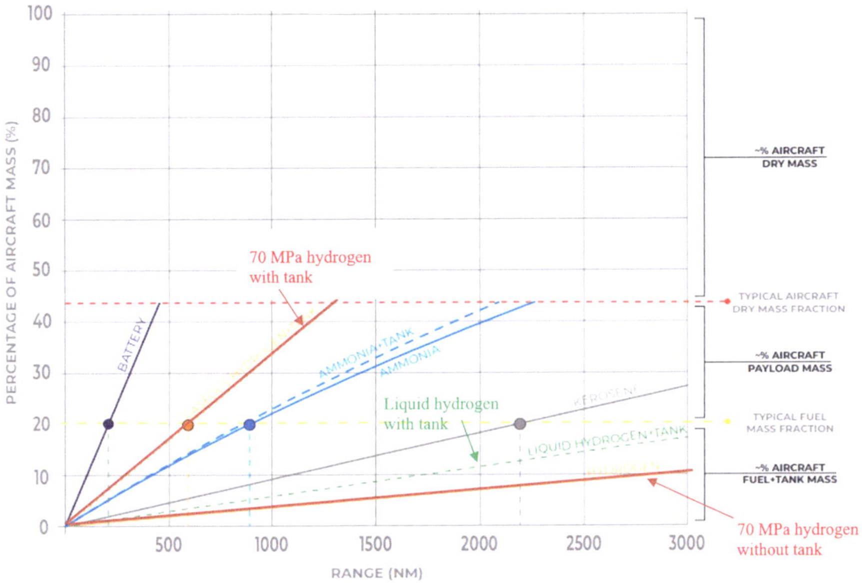

Fig. 9.43 Comparison of fuel mass percentage of aircraft with practical range: Batteries, compressed hydrogen, ammonia, kerosine, and liquid hydrogen, including tank. Source: Saurav Tiwari, Michael J. Pekris, John J. Doherty. A review of liquid hydrogen aircraft and propulsion technologies International Journal of Hydrogen Energy, Volume 57, 29 February 2024, Pages 1174-1196. Used under STM guidelines

Fig. 9.44 Rendering of the Airbus SAS ZEROe Turbofan concept liquid hydrogen aircraft. ? Airbus SAS 2020. All rights reserved. Used with permission

Concerns About High-Altitude Water Vapor Emissions

As with all efforts to quickly implement solutions to climate change, the overall impact of hydrogen in commercial aviation is still uncertain as we learn more about the atmospheric effects of aircraft emissions at high altitudes. Previously focused on just carbon dioxide as the GHG of concern, it has been reported for some time that the aviation industry is responsible for 2.5% of global carbon emissions, and an even greater percentage effect on global warming. This has logically motivated interest in developing low-carbon synfuels, and carbon-free hydrogen fuel for commercial aviation.

But would the transition to hydrogen actually be beneficial? Only since about 2019 have climate impact studies addressed *all* aviation emissions, not just carbon dioxide, and identified water vapor emitted at high altitudes in the form of contrails (condensation trails, as shown in Fig. 9.45) as a serious concern [121, 122]. From citation [123]:

> Taking all of these effects into account, the authors estimate that aviation has accounted for approximately 3.5% of effective radiative forcing to date. Another study estimates that it has been responsible for 4% of global temperature rise since pre-industrial times [124].
>
> Although CO_2 gets most of the attention, it accounts for less than half of this warming. Two-thirds come from non-CO_2 forcings. Contrails—water vapor from aircraft exhausts—account for the largest share. This explains why aviation contributes 2.5% of annual CO_2 emissions but more when it comes to its total impact on warming.

With the exception of NO_x, the sole combustion product of hydrogen is water vapor, *more than twice the mass emitted by hydrocarbon aviation fuels*, since all of its energy content obviously comes from its hydrogen content, rather than the carbon and hydrogen content of hydrocarbon fuels. Water vapor from combustion is of little consequence to global warming at or near the Earth's surface. But if released into the stratosphere at 10–13 km (33,000–42,000 ft), its effect is that of a greenhouse gas more potent than CO_2, nullifying any advantage of the use of reduced-carbon fuels compared with conventional fuels.

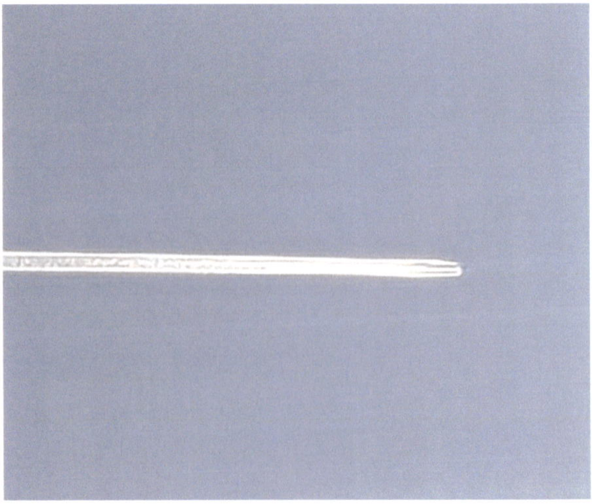

Fig. 9.45 Example of high-altitude contrails from a commercial jetliner. Photo: author

Few If Any Clean Fuel Options for Aviation

If indeed hydrogen is not the panacea for commercial aviation, what are the alternatives? Hydrogen for fuel cells may still be viable for low-altitude prop aircraft driven by electric motors. But there are few replacements for jet fuels used for transcontinental flights. Continued use of kerosine-like jet fuels is obviously an unacceptable path to climate degradation. At this time, the most common replacements for conventional fossil-based jet fuel are the various SAFs (sustainable [or synthetic] aviation fuels), some of which are eFuels that can be made from hydrogen and simpler hydrocarbons. The hydrocarbon feedstocks for SAFs are as varied as those for biodiesel, but most are currently made from plant oils or biomass, the same as biodiesel feedstocks. SAFs are synthetic hydrocarbon fuels, and as such, their full-cycle CO_2 emissions are typically only slightly better than fossil fuels, and the designation "renewable" assigned to them derives only from the arguable renewability of the agricultural or waste biomass feedstocks and hydrogen used to synthesize them. This subtlety may be glossed over in public promotion of SAFs. According to the US Dept of Energy Bioenergy Technologies Office (BETO) [125] ,

> SAF made from renewable biomass and waste resources have the potential to deliver the performance of petroleum-based jet fuel but with a fraction of its carbon footprint, giving airlines solid footing for decoupling greenhouse gas (GHG) emissions from flight.

SAF producers claim that it has the potential to reduce CO_2 from commercial aviation emissions by up to 85% compared to conventional Jet A1 fuel over its life-cycle But at this time, almost all of the hydrogen used to synthesize SAFs does not come from carbon-free sources, and the "waste resources" have questionable renewability. All SAFs are more costly and consume more source energy than conventional aviation fuel. But the science continues to evolve, and lacking any other alternative that meets aviation requirements, eSAF remains the only hope for reducing the carbon emissions of aircraft. The US DOE is currently sponsoring the *Aviation Fuel Grand Challenge* [126] competition in the hope of a game-changing solution.

SAFs made from electrolytic hydrogen are in the class of eFuels, and are designated eSAFs. As with all eFuels, the production path begins with renewable electricity (from sources such as solar, hydro, or wind), which is used to make hydrogen by electrolysis. Carbon dioxide can come from many different sources, but ideally would come from CCS at SMR hydrogen or electric power generation facilities, or in the rare situations in which it is cost-tolerable, directly from the air (direct air carbon capture and storage). Renewable hydrogen and captured CO_2 are converted to water and carbon monoxide (CO), via the equivalent of the water gas shift process in reverse. Then the CO and additional hydrogen undergo Fischer-Tropsch (FT) synthesis (also in reverse), to produce synthetic crude oil that can be further processed with selective catalysts to produce boutique hydrocarbon fuels, including eSAF. The US FAA and jet engine manufacturers currently limit the maximum blend of eSAF with fossil JP-1A to 50%, although, like biodiesel, these

Table 9.2 Comparison of existing and potential future aviation fuel properties

	JP-1A (ASTM D1655) [73]	SAF (ASTM D7566) [129]	Liquid Hydrogen [73] (LH_2)	Dimethyl Ether [73] (DME)
Storage form [130]	Liquid. Regular fuel tank	Liquid. Regular fuel tank	Cryogenic liquid at ?253 °C and 6 bar	Liquid at 20 °C and 6 bar. Low-pressure tank
Density (kg/L)	0.720	0.786	0.07085	0.667
Mass energy[a] density (MJ/kg)	43.5	46.7	120.0	28.8
Volume energy[a] density (MJ/L)	31.3	36.7	8.5	19.2
Mass of fuel[b] for 10,000 km range (kg)	86,400	80,500	31,300	130,600
CO_2 combustion[c] emissions (kg)	269,000	251,000	0	250,000
H_2O combustion[c] emissions (kg/km)	117,000	109,000	282,000	153,000
Volume of fuel[d] for 10,000 km range (L)	120,000	102,000	442,000	196,000
Range of aircraft if fuel mass restricted to 86,400 kg and fuel volume restricted to 120,000 L (km)	10,000	11,800	2715	6122

[a]Energy densities are lower heating values, appropriate for aircraft turbojet engines
[b]The fuel energy burn rate of the Boeing 747 on JP-1A (aka Jet A) is 376 MJ/km. This energy use rate is assumed to be the same for all fuels. The actual burn rate would be slightly higher for SAF, and slightly lower for LH_2 and DME, compared with JP-1A
[c]Direct emissions for fuel combustion, not including fuel production/distribution
[d]Tank structural volume or mass not included for any fuel. Compared with existing JP-1A tanks, the tank structural volume would be slightly larger for DME at six Bar, but much larger for LH_2 to accommodate the vacuum and superinsulation space of the cryogenic tank

requirements are probably overly cautious. The first commercial airliner to use 100% SAF was a Boeing 777 operated in freight service by FedEx in 2018 [127]. More recently in 2022, a Boeing 737 Max and an Airbus A380 were flown in actual passenger service with one engine running on 100% SAF [128].

Table 9.2 compares JP-1A, SAF, LH_2, and DME (dimethyl ether) for a hypothetical transcontinental flight in a Boeing 747. For perspective, I have included in the fourth column the largely ignored synthetic fuel, DME, which like hydrogen is currently produced from natural gas, but can be produced from renewable electricity or directly from waste biomass. It is much easier to store than hydrogen, requiring only modest pressurization to liquify, very much like butane. At atmospheric pressure, it liquifies at –24 °C (–11 °F). Or at 20 °C (room temperature), DME remains liquid under only 4.5 bar (66 psig) pressure, so that low-pressure uninsulated tanks no stronger than a cigarette lighter refill cartridge can be used to store it in the liquid phase.

Some Observations from Table 9.2

- The range on SAF would be approximately[7] 118% of the JP-1A range. (The mass energy density of SAF is from a different citation than JP-1A; the comparison may be overly optimistic.)
- The CO_2 emissions of SAF for a transatlantic flight are approximately 7% lower than JP-1A.
- The range on DME would be 61% of the JP-1A range. It would take 64% more fuel volume to match the range provided by JP-1A.
- The CO_2 emissions of DME for a transatlantic flight are approximately 7% lower than JP-1A.
- The range on LH_2 would be 27.5% of the JP-1A range. It would take almost four times the fuel volume to match the range provided by JP-1A.
- The CO_2 emissions of an LH_2 transatlantic flight are zero. But its water vapor emissions are 2.4 times those of JP-1A and 1.8 times those of DME.
- Water vapor at 12,000 m is estimated to have a GWP > 2.0. A crude comparison of CO_2 and H_2O effects at high altitude suggests that LH_2 would be responsible for at least 2.4 times the global warming impact of JP-1A. Of course, this depends upon the GWP of water vapor at 25,000 m, which is not yet agreed upon. But even if the GWP was 1.0 the same as CO_2, LH_2 would still be more harmful than JP-1A.
- If the rated 10,000 km range of the aircraft was a requirement, a liquid hydrogen airliner would have to compromise a large fraction, possibly more than 50%, of its passenger or cargo space to carry the additional fuel volume.
- The replacement of JP-1A with liquid hydrogen would require radical modification of the aircraft fuselage, and much more hazardous and expensive airport refueling infrastructure.
- A DME airliner might be able to retain the existing cabin space but lose most of the cargo space to accommodate the range requirement. But the modifications would be less intrusive and less expensive that those required for liquid hydrogen.

The Hydrogen Rainbow

How is hydrogen produced? There are probably more ways to produce hydrogen from other energy sources than any other fuel. And almost all synthetic fuels are made starting with hydrogen produced from a variety of sources. In an effort to create soundbite names for the various methods of hydrogen production, colors have been assigned to each, apparently by popular consensus. Note that since there is no actual standard for color assignment, the colors referred to in various literature may differ. The assignments below represent the author's informal survey of most available publications. There are now so many hydrogen color designations that we may

[7]The H-C ratio of typical eSAF is about the same as that of JP-1A. The 18% difference in energy density is probably due to the use of two different references for this data.

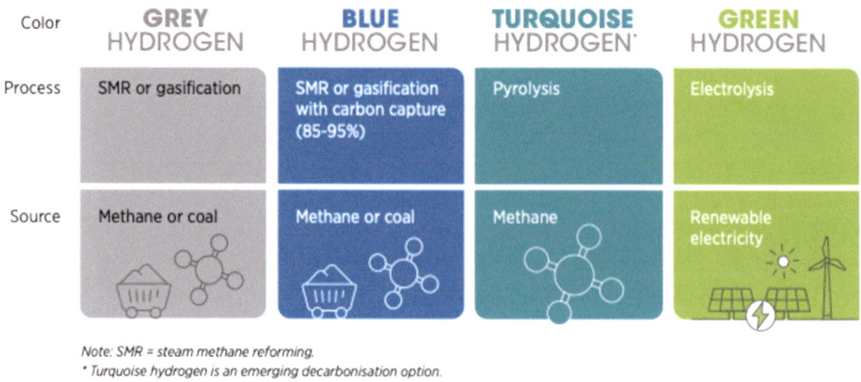

Fig. 9.46 Characteristics of some of the most popular colors of hydrogen. Graphic from the International Renewable Energy Agency, "Green Hydrogen: A Guide to Policy Making". Available online at https://www.irena.org/-/media/Files/IRENA/Agency/Publication/2020/Nov/IRENA_Green_hydrogen_policy_2020.pdf. Public domain

soon need a bigger box of crayons. Figure 9.46 mentions just the most commonly referenced colors of hydrogen.

Figure 9.47 from S&P Global Oct 2021 shows industry-provided data on the relative carbon intensities of these methods. Carbon capture (CCS) rates above 68% are possible but require exponentially larger energy expenditures and have never been achieved in large scale.

What follows below are descriptions of each hydrogen color, starting with the most carbon intensive. To this list of generally agreed-upon designations, I have added "drab green hydrogen" to designate hydrogen made by electrolysis of water using grid electricity, the most common source of electricity for electrolysis at this time.

Black, Brown, and Gray Hydrogen

Overwhelmingly, the majority of hydrogen produced worldwide is gray, black, or brown.

Gray hydrogen is generated from the methane content of natural gas or biomethane by Steam Methane Reforming (SMR), discussed previously, without processing of the exhaust products to remove or reduce carbon dioxide. Gray hydrogen from methane is a huge emitter of carbon dioxide, although less so compared with black or brown hydrogen, which uses black (anthracite or bituminous) or brown (lignite) coal for hydrogen production.

An informative animation created by Mahler AGS GmbH of the SMR process for the generation of hydrogen from natural gas is available as a YouTube video at citation [131].

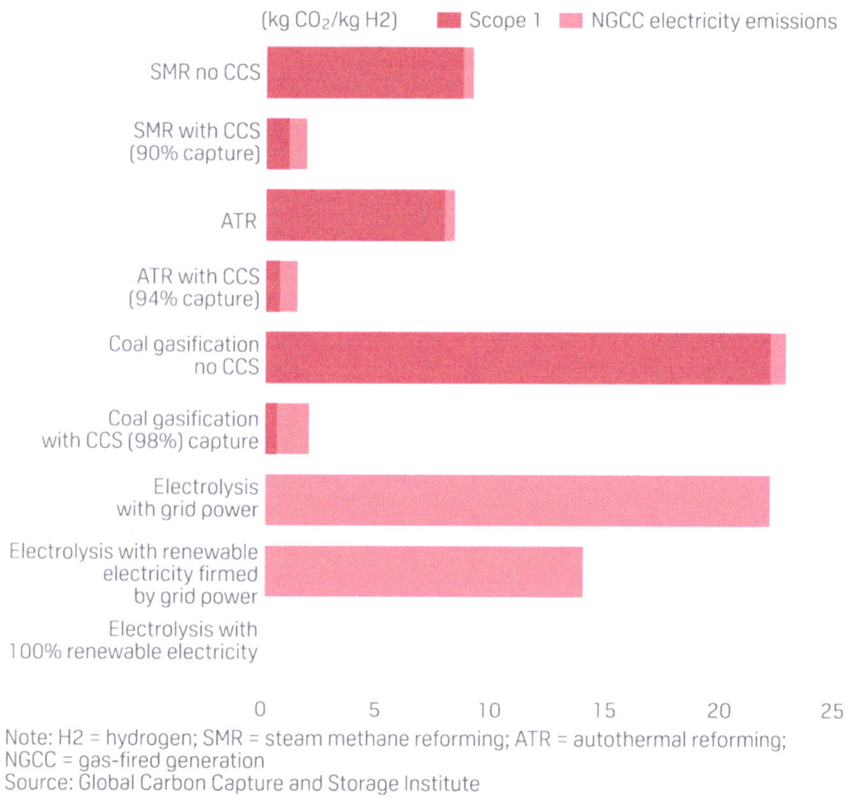

Fig. 9.47 Carbon intensity of hydrogen production methods, current and planned. From https://www.spglobal.com/commodityinsights/en/market-insights/latest-news/electric-power/100821-feature-blue-hydrogen-lifespan-raises-questions-over-carbon-offsets. Accessed 10 April 2024. Used under STM guidelines

Black or brown hydrogen is the most environmentally damaging classes of hydrogen, since not only are larger masses of carbon dioxide and carbon monoxide released into the atmosphere compared with gray hydrogen, but also large masses of particulates that are directly hazardous to health. Worldwide, coal (black and brown) hydrogen is the second most prevalent method for merchant hydrogen production, exceeded only by gray hydrogen produced from natural gas. Collectively, these methods account for over 95% of all hydrogen produced worldwide, ranking hydrogen production among the worst drivers of climate change, second only to agricultural GHG emissions.

As of 2022, SMR and coal gasification (CG) are the most cost-effective but dirtiest processes for hydrogen production, with a levelized cost of hydrogen (LCOH) of USD \$1.03–2.16 per kg for SMR or \$0.96–2.16 per kg for CG [132].

Blue Hydrogen

Hydrogen is labeled as blue whenever it is produced from natural gas by SMR, CG, or pyrolysis of biomass, followed by some method for carbon dioxide capture from the effluent gas. There are a number of processes in small-scale use, and several other methods proposed or in development. These processes are referred to as CCS, *carbon capture and storage, or carbon capture and sequestration.*

CCS involves the absorption and separation of CO_2 from industrial and energy-related stack emissions and utilization for some beneficial purpose or storage below ground in storage voids of various kinds. The goal is to prevent the release of CO_2 into the atmosphere. The mass rate of carbon captured can ideally be as high as 85–95% for the capture from concentrated streams such as power generation stack emissions, but is more typically in the range of 60–70%. For the capture of CO_2 from the atmosphere, known as direct-air carbon capture and sequestration (DCCS), the maximum recovery percentage is lower and energy requirements higher. A comprehensive report on DCCS can be found in citation [133].

A number of innovative CCS processes are currently in the research or pilot plant phases, three of which (e.g., [134]) are currently receiving major funding from the US DOE Hydrogen EarthShot program in pursuit of the DOE's goal of 95% effective carbon capture [135]. In the case of blue hydrogen, the purpose of CCS is to remove CO_2 produced in the SMR process from the output product hydrogen. All current methods involve the separation of CO_2 from the CO_2 + hydrogen output stream (or during the SMR reaction) using selective absorbing materials and/or filtration which are regenerated using liquid solvents from which the CO_2 is later separated. These processes all require significant heat and electrical energy which are usually provided by additional combustion of natural gas, producing additional CO_2 [136] which also must be removed by the same or a separate CCS process.

The designation "blue" implies that the separated CO_2 from the SMR process is either pipelined off to be stored underground or *used for some beneficial purpose*, in which case it is sometimes referred to as *carbon capture utilization or storage* (CCUS). In some international reports and US legislative documents the "S" represents *sequestration* [137] which is technically a stronger constraint than *storage*, implying that the isolated carbon dioxide is converted to a form that cannot re-enter the atmosphere, such as biochar buried in the ground or reacted with underground minerals. An example of *utilization* of captured carbon dioxide is addition to the air in high-tech sealed agricultural greenhouses, which enhances plant photosynthesis resulting in faster and larger plant growth [138]. Another example, demonstrated to much fanfare in Iceland, is the production of methanol from captured CO_2 and excess electrical energy which is almost free on this volcanic island due to the abundance of geothermal and hydroelectric power. The electric energy required for the CCS process applied to SMR hydrogen production is typically 30% of the ultimate hydrogen fuel value. If the energy required for the CCS process is not carbon-free, the system efficacy in reducing net CO_2 emissions does not even break even compared with the obvious solution: don't use fossil fuels to make hydrogen.

Lacking immediate and proximate opportunities for utilization of the CO_2 captured from CCS-equipped SMR hydrogen production, there seems to be a consensus that the best destination for the CO_2 is deep underground geologic rock formations that lie below some impermeable strata layer (e.g., dense rock) to reduce leakage to the surface. Under the ideal geologic conditions, carbon dioxide will be absorbed into deep groundwater to form carbonic acid (H_2CO_3), which will eventually (possibly over hundreds of years) react with minerals to be effectively sequestered in solid form. But this is an optimistic assumption. Even for well-sealed wells, CO_2 leakage can be substantial and can become a serious threat to groundwater quality within a radius of at least 0.4 km (¼ mile) [139, 140]. In a proprietary variation of this method being tested by Carbfix [141] in Iceland, the CO_2 is pre-absorbed into fresh water and the water is injected into wells for reaction with basaltic (Volcanic) rock, claimed to form carbonate minerals in <2 years. This process would seem to be best deployed in locations with the appropriate subsurface geology and adequate fresh water, as well as the availability of carbon-free electricity, conditions similar to those found in Iceland.

As mentioned above, the scientific consensus (not DOE target) at this time is that CCS could potentially remove 60–70% of the carbon dioxide from the effluent stream before releasing it to the atmosphere.[8] According to the Institute for Energy Economics and Financial Analysis (2022), only two commercial plants in the world (Quest in Alberta Canada and Air Products in Texas USA) are producing blue hydrogen from natural gas with CCS [142] , although the support in the USA of the 2022 IRA will soon increase this number. The CCS performance of these facilities is listed in Fig. 9.48, which reveals that the best CCS technology in use at this time captures 68% of the emitted CO_2 including the CO_2 emissions generated by the energy used for the CCS is included. (These accessory emissions are often deprecated in study results by the optimistic assumption of carbon-free electricity). As of 2023, CCS is not used by any of the US hydrogen suppliers for FCV refueling. All stations actually use traditional gray hydrogen, although one supplier, *FirstElement*, advertises that 1/3 of their gray hydrogen is actually green because it is produced from the incineration of renewable biomass [143] in the form of municipal refuse. This is not an uncommon but a specious justification that equilibrates municipal trash with natural biomass having a closed carbon cycle, simply because it could have been burned instead of buried in a landfill.

In much of the commercial and even in some government publications, the term *blue hydrogen* is proffered as nearly equivalent to zero-carbon or carbon-neutral hydrogen. The US DOE *Earthshot* objective is CCS/CCUS which aims to capture 95% of the effluent carbon dioxide, a target that most researchers consider to be unjustified due to its energy cost per unit of carbon recovered. The energy required for high-percentage capture of CO_2 is much larger than for lower capture percentages, an epiphany that has led just recently to a possible rethinking of the official 95% capture goal, with a new target of 90% capture replacing the 95% stated in some DOE news releases. The solution of using the hydrogen produced by SMR

[8] Hui Song, et al. ibid.

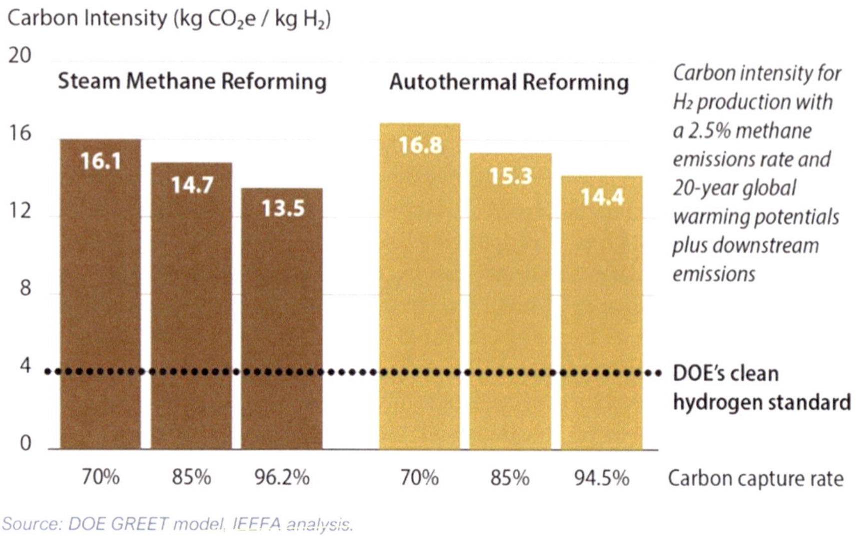

Fig. 9.48 Carbon intensity (kg CO2e / kg H_2) for hydrogen production using various capture percentages and two methods for hydrogen production. Data from DOE GREET model, graphic by IEEFA, published in David Schlissel, Blue Hydrogen: Not Clean, Not Low Carbon, Not a Solution. https://www.energy.gov/sites/default/files/2024-02/090.%20David%20Schlissel%20and%20Anika%20Juhn%2C%20IEEFA%2C%20Blue%20Hydrogen%20Not%20Clean%20Not%20Low%20Carbon.pdf. DOE publication, public domain

with CCS as the energy source for the CCS process is almost pointless and energy-wasteful.

There is also the reality that even if carbon capture can reduce the CO_2 emissions from SMR hydrogen at the current technology limit of 68%, the 32% remaining CO_2 is released into the atmosphere, reducing the carbon reduction efficacy of hydrogen by at least that same percentage (actually worse due to second law considerations). Worth noting is that hydrogen intended for use in fuel cell vehicles must be of high purity. Even a small fraction of CO_2 in the hydrogen fuel can poison and deactivate the ion-exchange membrane or solid electrolyte that is the core of the fuel cell's ability to generate electricity. Therefore hydrogen produced by SMR with or without CCS already requires carbon dioxide removal to isolate the hydrogen output product, partially justifying the CCS process.

Most non-advocate independent analyses conclude that "blue" hydrogen, even with the best available CCS technology, would be questionable as to any net positive impact on climate change. As summarized by Hydrogen Insight, Dec 1, 2022 [144]:

> Ramping up the use of blue hydrogen to just 30% of the H_2 mix by 2050 could wipe out the climate benefits from the hydrogen economy over 20 years on account of fugitive methane emissions and the fact that not all the CO_2 in the production process can be captured, climate scientists write in a new study.

The arguments from most environmental organizations are that at most, "reduced carbon" would be a more accurate description for blue hydrogen, while some calculate that the net carbon dioxide emissions may exceed those that would have been generated if the natural gas had been burned directly rather used to make blue hydrogen [147]. As mentioned above, all of the operational hydrogen highway stations in California currently dispense *gray* hydrogen, without CCS, while originally proposing to upgrade to green or at least blue hydrogen if/when it becomes available. But this is not the perception of owners of FCVs who are motivated by the belief that they are leading by example to help mitigate climate change [145].

There are apparently only a few self-contained electrolytic (labeled green) hydrogen stations in the world, some in the interior areas of Australia that use grid electricity to produce *drab green* hydrogen (see below) [146]. The use of grid electricity for electrolytic production of hydrogen is counterproductive due to the high carbon intensity of the grid in most countries. It has been a major cause of concern for watchdog agencies. For example, the environmental organization Sierra Club [147]:

> Upwards of 95% of hydrogen is produced from methane (known as gray hydrogen) or from coal (brown hydrogen). Fossil fuels provide both the feedstock and the power for the energy-intensive process. Dirty hydrogen production accounts for 6% of global methane gas use and 2% of coal use, with massive carbon dioxide emissions as a byproduct.
>
> A 2021 peer-reviewed article in Energy Science & Engineering found that the life-cycle greenhouse gas emissions of blue hydrogen were more than 20% greater than those from burning straight methane gas. "The use of blue hydrogen," the authors concluded, "appears difficult to justify on climate grounds."

Specifically regarding CCS, David Schlissel et al. of the Institute for Energy Economics and Financial Analysis (Canada) stated in citation [148]:

> CCS technology has been around for decades, yet its actual, real-world implementation in either the large commercial hydrogen production sector or the utility-scale power production sector has been unreliable and far below the 90–95% capture rate that is considered the industry's prime objective for CCS. Not only that, but among the projects that have been built, substantial failures have occurred. This might have been understandable in the 1970s, 1980s, and possibly even the 1990s. But the fact that the problem persists into the 2020s makes CCS a highly risky investment.

In the USA, the distinction between blue hydrogen and truly carbon-free green hydrogen has become a point of contention for qualifying for government incentives under the 2022 Federal IRA. The fossil fuel interests, with the benefit of massive government subsidies that previously promoted hydrogen as a means to sustain their industry in the face of electrified transportation, were essentially shortchanged by provisions in the IRA that required that hydrogen qualifying for the \$3/kg USD subsidy *actually had to be truly green (zero carbon), not SMR with or without CCS*. Aggressively lobbying efforts are ongoing to have blue hydrogen considered to be fully or partially synonymous with green hydrogen for purposes of subsidies. The arguments take the form of "how clean does the hydrogen have to be to qualify as a climate solution?" According to the non-profit political watchdog group *Open Secrets* [149]:

> The fossil fuel industry has aggressively lobbied the White House, Congress, and Energy and Treasury departments to ensure gas-based hydrogen qualifies for federal subsidies.
>
> Thirty-two oil and gas producers reported lobbying on hydrogen, among other issues, and spent a combined $41.3 million on federal lobbying efforts this year, as of Sept. 30 [2023].

Julie McNamara, the deputy policy director for climate and energy at the Union of Concerned Scientists, is quoted in citation [150]:

> The government risks "aiding and abetting fossil fuel" interests. There are so many ways that hydrogen can go that it just perpetuates the status quo. That is an extremely lucrative place for the fossil fuel industry to be.
>
> If we have weak standards, it can mean more use of natural gas for longer with more profit along the way.
>
> Getting hydrogen wrong would be a catastrophic waste of time.

To date, CO_2 emissions of any operational blue hydrogen facility have never been justified compared with the emissions from direct combustion of the natural gas feedstock [151]. This implication is that if fossil hydrogen with or without CCS is allowed the same subsidy status as more expensive electrolytic hydrogen, the environmental benefit would be insignificant, and the motivation to produce hydrogen that is actually carbon-free would be diminished.

Turquoise Hydrogen

Similar to one of the two popular meanings of pink hydrogen (below), turquoise refers to the production of hydrogen by the process of methane pyrolysis, for which the residual carbon effluent is mostly char (solid carbon) rather than carbon dioxide. This solid form of carbon is periodically or automatically removed from the reactor and either utilized in another application or buried, where it is very stable and potentially a beneficial soil supplement. In other words, much of the carbon is already sequestered to a solid form that produces no environmental harm. The process requires higher temperatures than SMR, and is slightly less efficient. Although not yet beyond the demonstration stage, if eventually scaled to production levels, this distinction would make turquoise hydrogen more environmentally sound than gray, brown, or black hydrogen. In combination with CCS to capture residual CO_2 effluent, it could easily yield lower net GHG emissions than current blue hydrogen, especially if the feedstock is particularly carbon-rich such as biomass.

Pink Hydrogen

An ambiguous designation, with two very different meanings depending on the source publication.

Pink was originally assigned to hydrogen made by pyrolysis of biomass followed by separation of the hydrogen from the carbon monoxide and other compounds in the dirty syngas.[9] This is similar to turquoise hydrogen, except that the feedstocks are a wide range of cellulosic biomass. While a compelling argument can be made for the energy sustainability of biomass-derived fuels such as pink hydrogen, the high carbon fraction and variability of different forms or batches of biomass usually make pink hydrogen nearly as carbon intensive as black or brown hydrogen.

Second and arguably more common, an entirely different definition of pink hydrogen made by electrolysis using electricity *from nuclear energy rather than renewables* [152]. As such, it would be carbon free, so that it could be included in the green classification. Within this latter class is high-temperature high-pressure electrolysis using the waste heat of a nuclear reactor to improve the electrolysis efficiency. And looking further into the future, the possibility of direct thermal decomposition of water at the extreme temperatures present in the core of a high-temperature gas-cooled (HTGR) nuclear reactor. As mentioned in the chapter on electric vehicles, the use of nuclear fission for energy has been and probably always will be a point of strong differences, sometimes pitting environmental advocates against each other. So it is also if nuclear is the energy source for hydrogen. But it is important to note that hydrogen produced electrolytically from nuclear electricity is indeed carbon free, notwithstanding the small but highly consequential risks associated with nuclear power plants.

Yellow Hydrogen

Yellow has been attached to a wide range of hydrogen generation methods, and a consensus has not yet been reached. Circa 2020 it was used to designate the hopeful new field of Solar Thermal Hydrogen (STH), but it was also used to designate hydrogen generated by electrolysis using nuclear-generated electricity (aka pink hydrogen), and most recently its association has been expanded to include all hydrogen made using grid electricity.

While certainly arguable, it seems most intuitive to retain the original association of yellow with STH, simply because 'solar' and 'yellow' have an intuitive connection, while there is no logical argument to associate the color yellow with nuclear or the entire mix of coal, natural gas, hydroelectric, PV, wind and biomass combustion that power the grid. (If all were figuratively 'mixed' together on a canvas, the result would be "drab green", the color I assign in this book.)

[9]This is the definition of pink hydrogen provided by ChatGPT in response to the prompt "Pink Hydrogen". This unreferenced response is relevant because we are discussing *popular* color designations for hydrogen, all motivated by the original assignment of "green" to zero carbon hydrogen. None have any precedent or basis in physical laws.

Narrowing the definition of yellow hydrogen in this discussion to STH, this is a promising area of intense research with the potential to effectively integrate solar PV electricity and electrolysis into a single energy conversion step via STH modules. The term "STH" actually covers several solar H_2 generation technologies: photocatalytic (PC) water splitting, photoelectrochemical (PEC) water splitting, photovoltaic-electrochemical (PV-EC) water splitting, solar thermochemical (STC) water splitting, photothermal catalytic (PTC) H_2 production from methane, and photobiological (PB) H_2 production. Efficiencies in the range of 13.1–19.3% have been reported (see Footnote 7). These numbers are comparable to the 8–13.2% combined efficiency of state-of-the-art solar modules (20–22%) followed by PEM electrolyzers (40–60%). Development is still immature compared with existing technologies, with commercial products expected by 2030, if progress continues at the current rate.

Green Hydrogen

The best-known designation, *green hydrogen* refers to hydrogen made or recovered by any process that overall does not produce airborne GHGs (most notably carbon dioxide). Up until recently, green hydrogen was assumed to be produced exclusively by electrolysis that directly uses carbon-free electricity: solar, wind, geothermal, tidal, or (arguably) hydroelectric or nuclear power. However, regulatory provisions in the US inadvertently allowed the purchase of US and International carbon offset credits to claim that almost any form of hydrogen generation could be referred to as "green." In California, following the approval of California Energy Commission Rulemaking 20-05-012, the green designation was and remains restricted to hydrogen produced electrolytically using any zero-carbon electricity source except nuclear generation (the non-nuclear provision was a nod to public polarization over nuclear energy). The use of "carbon-producing electricity offsets" is now disallowed.

Another ploy was hydrogen produced using electricity generated at a time of excess grid electricity, and *stored in the grid rather than local storage*, which exacerbates major problems with grid stability. It is now also disallowed. And finally, hydrogen derived after multiple processing stages entirely from seasonally renewing crops is disallowed, although this provision remains a topic of debate based on the assertion by agriculture advocates that agricultural sources are intrinsically carbon neutral and should fall under an exception in the regulations allowing the generation of electricity from "renewable biomass." Biomass is broadly defined to potentially include almost anything that began as plant life and ended up as a combustion material, even if processed through the digestive tract of an animal [153]. This interpretation of *green* justifies the production of hydrogen made by high-CO_2 pyrolysis or SMR using dirty feedstocks that can be considered "biological" because they originally obtained some or all of their carbon content from the atmosphere. This includes most forms of solid waste.

Unexpectedly, in the brief time since the 2021 regulation, two other means for obtaining or producing *green* hydrogen have gained considerable media attention and traction with funding agencies and investors. While these meet the definition of green hydrogen, the completely different origin merited two new colors in the hydrogen spectrum. The first is white *(aka gold) hydrogen*, extracted from naturally occurring underground pockets of hydrogen gas, usually trapped beneath impervious salt domes or igneous rock "lids." Also it is hypothesized that the injection of acidic water into iron/magnesium olivanic rock formations can stimulate the formation of geologic hydrogen that can be extracted from deep wells. This has been assigned the color *orange hydrogen*. Terminology in the relatively new field or geologic hydrogen is still in flux [154].

The other innovation, previously discussed, is *yellow* hydrogen produced directly from solar insolation via STH modules as. If sunlight is used *directly* to split water into hydrogen and oxygen, then it can be considered an energy source rather than an energy carrier created by electrolysis using solar PV electricity.

If these technologies are scalable to a level that could significantly contribute to global hydrogen demand, they could eventually become game-changers for the future of hydrogen in transportation. But at this time they have only been demonstrated in laboratories or small pilot plants.

Truly green hydrogen is the only form of hydrogen that is actually climate-neutral, meaning that it could play a significant role in at least stabilizing atmospheric carbon content. But despite legal description restrictions, in the current environmental vernacular the term "green hydrogen" has become an inexplicit descriptor for almost whatever type of hydrogen an entity or agency is promoting. The term is used loosely in business and media to promote hydrogen as a sweeping climate solution. For instance, the misleading statement "hydrogen is carbon-free." Hydrogen generated by SMR with CCS, or electrolysis using grid electricity is not carbon-free and has a very wasteful energy balance. Claims have been made that SMR hydrogen should be classified as green if the methane used to make it was surplus that would otherwise have been flared or if it originated from biological sources, overlooking the CO_2 emissions required to create those biological sources. And as will be explained further in the next section, hydrogen generated by electrolysis using electric grid power is inevitably less green than the grid electricity used to make it, due to the inefficiency of converting the electricity into hydrogen.

As of January 2024, lacking solar fields or wind farms built exclusively for hydrogen generation, the few electrolyzers in the world that generate hydrogen for transportation are all powered from the electric grid, not from dedicated renewable electricity sources. In the USA, this situation was expected to change starting in 2025 since the subsidies provided under the amended provisions of the 2022 US Federal IRA [155] require that no existing grid-connected solar or wind facilities could be "repurposed" for this purpose, which would simply transfer green electricity from the grid to hydrogen production. Any qualifying facilities must be newly built specifically to power electrolytic hydrogen generation, and that either the electricity or hydrogen be stored on-site, eliminating the grid-destabilizing practice of banking solar or wind electricity in the electric power grid, then powering the electrolyzers from the grid. As expected, this requirement has seen strong objections from entities hoping to qualify for green hydrogen subsidies with minimum capital expenditures. These concerns, however, may become moot with the change in the US federal administration starting in 2025, which could result in the cancellation of many/most of the provisions of the IRA.

Considering the energy losses of the electrolyzer and the gas compression required for storage, the energy content of any *green* hydrogen hypothetically dispensed at a fueling station is only a minor fraction, estimated between 9% and 25% in various reports, of the energy content of the original sources of energy used to generate the electricity, e.g., natural gas, solar or wind.

In 2021, the US Department of Energy's "Hydrogen Energy Earthshot" program set an objective of having green hydrogen available throughout the USA for USD $1/kg by 2031 [156]. This ambitious goal was modeled after subsidized investment in advanced lithium batteries starting in the early 2000s, which has paid off today as modern EVs. But the science is different in this case. Proponents extrapolate from the lowest production cost achievable to date, $4/kg at laboratory scale, which justifies the $3/kg subsidy provided by the IRA to spur green hydrogen research and development. But at this time, the net energy efficiency of electrolytic hydrogen production is among the lowest of the available production paths, motivating the question "Is the expectation of a 4:1 reduction in the cost of "green" hydrogen a worthwhile strategy to address climate objectives?

Drab Green Hydrogen

As discussed above in the *Yellow Hydrogen* section, *drab green* is my own designation for hydrogen produced by electrolysis *using utility grid electricity*. It is necessary to distinguish electrolytic hydrogen made using grid electricity from carbon-free or renewable electricity. Worldwide, hydrogen produced by electrolysis almost entirely uses grid electricity, even in cases of colocation with intermittent renewable energy facilities for which a grid power connection is necessary to operate the facility continuously. In early 2024, the average carbon intensity of the US electrical grid was reported to be approximately 400 g CO_2/kWh including all energy sources[10]. Worldwide, the IEA estimates 475 g CO_2/kWh, at least partly attributed to greater use of coal compared with the USA.

I am aware of only three (questionable) exceptions other than purely experimental demonstrations in the USA. Plug Power's hydrogen plant in Latham, New York claims this status because it uses grid electricity from the Quebec grid that is almost all from hydropower, and Air Liquide's hydrogen plant in Las Vegas, Nevada USA, that uses solar energy during daylight hours but is connected to the utility power grid to allow continuous operation. The Las Vegas power grid derives much of its power from Hoover Dam on the Colorado River. The soon-to-be operational SoHyCal plant in Fresno, California [157] is the first of the expected new wave of IRA-subsidized green hydrogen

[10] Estimates of the carbon intensity of the average USA grid vary from 350 to 450 kg CO_2/kWh depending on the source and region. US EIA: 388 kg CO_2/kWh in 2020, 417 kg CO_2/kWh in 2021, 440 kg CO_2/kWh in 2022 https://www.eia.gov/todayinenergy/detail.php?id=53819. But it is unclear if these numbers include all distributed generators, in particular, home solar PV. As of the end of 2023, *Our World in Data, Carbon intensity of electricity generation, 2023* reports **368 g CO_2/kWh.** https://ourworldindata.org/grapher/carbon-intensity-electricity. Accessed 10 January 2024. But 400 g CO_2/kWh seems to the consensus as of early 2024.

production plants that use dedicated solar PV electricity when available and grid electricity when not. None of these facilities can exclusively use renewable electricity since, at the very least, they operate 24/7 in the interest of cost-effectiveness.

The method for hydrogen production is particularly relevant as the US (primarily California) and some other countries are still expanding their *hydrogen highways*, which are chains of periodically spaced hydrogen refueling stations on major travel routes. Hydrogen dispensed at these stations must be delivered by truck in high pressure or liquified form, or produced on-site with a grid-connected electrolyzer. At this time, no California hydrogen refueling stations use electrolytically produced hydrogen. Electrolytic hydrogen is dispensed at a small number of refueling stations in Europe and Australia, although in no case is the electricity source entirely renewable (Fig. 9.49). Many hydrogen stations have token solar panels on the canopy above the fuel pumps, but this tiny surface area would only be capable of providing a trivial percentage of the power needed by an electrolyzer.

As will be discussed in greater detail below in the *hydrogen highways* section, there is no way to avoid concluding that hydrogen refueling stations are misleading in their stated or implied environmental benefits, since they exclusively utilize grey hydrogen. For the few electrolytic hydrogen producers, there is a wide (0–50 kg CO_2/kg H_2) range of carbon intensities. The zero emission end of the range would ideally be electrolysis plants collocated with solar or wind generation facilities. Worldwide, truly zero emissions facilities are probably nonexistent, since even in Iceland with its overabundance of carbon-free electricity from hydroelectric and geothermal sources, a small (approximately 1%) of electricity still comes from fossil fuels, and peaking plants are still required for load leveling. The dirty extreme 50 kg CO_2/kg H_2 for grid electricity would be exemplified by the US states of West Virginia where 88.6% of the electric power is generated from coal, or Wyoming where 71% comes from coal. In these cases, drab green hydrogen made from grid electricity is considerably more carbon intensive than even gray hydrogen produced directly from coal gasification (CG) [158].

Fig. 9.49 Photo of Toyota Mirai at hydrogen fueling station in UK, 2016 (pre-Brexit). Hydrogen at this station is reported (unconfirmed) to be produced by local electrolysis using grid power. Photo by Alister Thorpe. https://commons.wikimedia.org/wiki/File:ITM_Power_Hydrogen_Station_and_Toyota_Mirai.jpg. CC-BY-SA-4.0

Since the US overall carbon intensity of grid electricity (2023) is approximately 400 kg CO_2/kWh, drab green hydrogen is about as carbon intensive as state-of-the-art blue (68% CCS) hydrogen.

Biohydrogen and BECCS

Two other classes of hydrogen production are also allowed the "green" designation:

The first is biohydrogen, which can be produced by many possible biological processes, similar to biomethane, fermented ethanol, and crop-derived biofuels. Among the feedstocks are renewable biomass that can be used in "dark fermentation" processes, selective algae capable of biophotolysis, and enzyme-assisted fermentation of starch crops. Many processes have been proven in laboratories, but at this time, few, if any, have demonstrated scalability to a level that could provide economically sustainable hydrogen production.

The other class is *Bioenergy with Carbon Capture and Storage* (BECCS). BECCS involves the same carbon-intensive gasification processes used for the production of hydrogen from fossil fuels (coal, in particular, described previously), but it is treated differently because the source hydrocarbon is biomass or ag residues, and it includes CCS to reduce CO_2. The "green" justification is earned because the biomass feedstock is assumed to have been recently formed by plant photosynthesis that removed carbon dioxide from the air. So except for the energy losses, the return of the CO_2 to the atmosphere as a result of SMR or pyrolysis should theoretically balance the rate of biological uptake. Then even if only a fraction of the carbon dioxide emissions are captured and stored via CCS, the *overall* cycle can be considered carbon negative, actually restoring more carbon to the Earth since the original plant source of the biomass sequestered the carbon during its growth cycle. Of course, this arithmetic is dependent upon the type of biomass, ignores the energy required to dry and prepare wet biomass, the energy required for the processing, and the CCS recovery energy required.

The basic idea is sound. In fact, it falls within the original definition of a "renewable fuel". But biofeedstocks that are true to the original concept are almost nonexistent. The ambiguity of what feedstocks should qualify for subsidies and/or regulatory carbon credits. Should "biomass" include municipal refuse, residuals from forest harvesting, landfill methane, or manufacturing waste? For example, in California (2023) nearly three-fifths of the state's utility-scale biomass fueled electric generation capacity came from 28 power plants *fueled by wood and wood waste*, and 77 fueled by *landfill gas, or municipal solid waste* that includes waste plastic, paper, metals and even toxic chemicals. And if the biomass is intentionally cultivated for fuel production, BECCS can be highly land and water intensive [159].

Extrapolating from limited experience, it appears that BECCS using anything other than truly renewable biomass would probably not be carbon negative or even neutral. But future developments can be expected if hydrogen continues to proliferate.

White or Gold Hydrogen

White (aka gold or *geologic*) hydrogen is the most recent color designation. While natural seepages of geologic hydrogen have been known for years, their occurrence was never considered a potential energy source. However, recent announcements of accidental discoveries of large reservoirs of geologic hydrogen have led to speculation that naturally occurring hydrogen could supply a significant part of the current and future demand for hydrogen [160]. Since 2022, interest in this potential resource has quickly grown from zero to a major investment focus, the target of numerous startup exploration and drilling companies and more than one sensational claim.

White hydrogen can be extracted from deep wells in areas that have iron-rich (olivine) underground strata. While isolated pockets of high-purity natural hydrogen have been found, the limited experience to date suggests that for the quantities necessary to become a useful energy source, white hydrogen is still speculative [161]. A small percentage of exploratory wells have produced high-purity hydrogen, but most have found hydrogen mixed with other geologic gases, especially nitrogen, methane, and carbon dioxide. One source estimated that a reasonable mean expectation for the hydrogen content would be 75–85% hydrogen, with a balance of nitrogen, methane, and inert gases such as argon and helium. This is partly because hydrogen is a reactive gas that combines with many minerals, and is readily metabolized by a wide range of subsurface bacteria that consume hydrogen and release methane [161]. The level of purity is similar to that of natural gas, which is predominantly methane with minor percentages of other light alkanes and carbon dioxide. But unlike natural gas, which is tolerant of almost all compositions, high-purity hydrogen is required for use as a fuel for FCVs. Separation of geologic hydrogen from its contaminants requires external energy and its associated CO_2 emission. And only if minimal GHG contaminants are released into the atmosphere during its recovery and purification can white hydrogen be truly considered green. If all works out, white hydrogen joins the only other green hydrogen alternatives: electrolysis using completely carbon-free electricity, solar STH, and nuclear thermochemical.

The newness of this energy opportunity leaves many questions unanswered. An online presentation by the USGS [162] observed that global demand for hydrogen is expected to be >500 MT/yr by 2050, while annual global production of geologic hydrogen is estimated to be only 23 MT/yr, less than 5% of demand. And there have emerged questions about the relative climate benefit of geologic hydrogen. One reference cited other studies estimating that natural hydrogen could have an emission intensity of 0.37kg of CO_2e per kg of hydrogen, based on modeling [163]. While another analysis concluded that "even for 70% pure H_2 with 30% other carbon-bearing gases, the net carbon impact would be negative" [164].

There are significant financial and environmental costs for drilling and operating the well, purifying the effluent gas, and disposing of the undesired gases. Hydrogen leakage from the well or processing and storage equipment is expected to be greater than for natural gas but is not yet quantified. Disposal of the unwanted impurity gases could end up being a major environmental concern. And the potential for seismic consequences has been a growing concern, extrapolating from experience with natural gas and oil hydrofracturing [165]. Only two realistic methods for impurity gas disposal have been proposed: flaring flammable unwanted gases as is the practice at natural gas wellheads and refineries, or injecting nonflammable gases back into the ground through other wells. A disposal well must be located far from the extraction well to avoid the reintegration of the waste gases with the extracted hydrogen, and impacts on ground water are known problems.

The underground formations that can both produce and trap geologic hydrogen may be more common than currently assumed. The reaction of ground water with iron-bearing minerals is the prime candidate for hydrogen formation. Wells may not need to be as deep as originally expected. But trapping requires some type of impervious "lid" strata such as a salt dome or layer of impervious rock above the formation strata. At this time, little is currently known about where to find this combination of underground requirements, but that situation is changing due to the level of investment interest. As of the end of 2023, naturally occurring hydrogen had not yet been captured in any commercially viable scale in the USA or EU, although new "finds" are being reported with increasing frequency, some accompanied by major media attention. The first-cited example of high-purity geologic hydrogen that seems to be driving much of the speculation was in Mali (West Africa) in 2011 [166]. Exploration wells were drilled in Nebraska in 2021, and in many states since then in the USA. More recently, in 2023, a large natural hydrogen reservoir was reported near the German-French border by geoscientists at the University of Lorraine [167]. An exploration well in Australia found "significant concentrations of up to 73.3% H_2" in February 2023 [168]. And "one of the largest volumes of natural hydrogen that has ever been measured" was reported in an Albanian chromium mine on February 8, 2024, although this has not been independently confirmed [169].

A December 2024 article by Geoffrey S. Ellis and Sarah E. Gelman of the USGS, published in *Science Advances* speculated that [170]:

> Geologic hydrogen could be a low-carbon primary energy resource; however, the magnitude of Earth's subsurface endowment has not yet been assessed. Knowledge of the occurrence and behavior of natural hydrogen on Earth has been combined with information from geologic analogs to construct a mass balance model to predict the resource potential. Given the associated uncertainty, stochastic model results predict a wide range of values for the potential in-place hydrogen resource [10^3–10^{10} million metric tons (Mt)] with the most probable value of ~5.6 × 10^6 Mt. Although most of this hydrogen is likely to be impractical to recover, a small fraction (e.g., 1 × 10^5 Mt) would supply the projected hydrogen needed to reach net-zero carbon emissions for ~200 years. This amount of hydrogen contains more energy (~1.4 × 10^{16} MJ) than all proven natural gas reserves on Earth (~8.4 × 1015 MJ).

Recent reports with similar exciting but speculative conclusions are appearing in technical and popular periodicals. Considering the potential value of the product and the maturity of gas drilling technology, it is surprising that it hasn't been investigated and exploited earlier.

With current research interest and the exuberant investment that follows announcements of this kind, there is a real possibility that our entire perspective of hydrogen could change from solely an energy carrier to that of an actual energy source. Drilling for naturally occurring hydrogen is already trending toward the wildcat levels of major oil and gas discoveries of the past, i.e., a new *gold rush*. But with the limited state of actual experience, it would probably be best to temper any expectations about the future of white hydrogen and its impact on transportation until more solid data is available.

Orange Hydrogen

There is not yet uniform acceptance of this new color designation, or whether this designation is even needed. This classification is a variation of white/gold hydrogen in which surface water is injected at high pressure into deep iron-rich olivine rock formations at temperatures above 200 C. The hoped-for reaction is the anaerobic oxidation of iron (aka rusting) which liberates hydrogen from the water as its oxygen combines with the iron. Considering that much of the Earth's mantle and core is hot or molten iron, this resource could be potentially vast, but not without consequences. One could easily imagine hot water injection combined with hydrofracturing in olivine formations that lack the required impervious "lid." Hydrogen seepage through porous layers would simply have to be considered a tolerable production loss factor. Also, the reactivity of fugitive hydrogen or the injected acidic water with minerals above the hydrogen formation depth could potentially generate methane or carbon dioxide, which could corrupt the recovery process or contribute to atmospheric warming [171]. However, this is just speculation at this point. If the orange hydrogen process is found to be viable and scalable, the hydrogen extracted from these water-injected formations could be very significant. The prospects of this new avenue for hydrogen supply have already (as of February 2024) garnered massive investment and grant funding, but there have not yet been any actual breakthrough announcements of practical hydrogen production.

Some Comments on Hydrogen Colors

Hydrogen colors invite vague and possibly misleading interpretations. When referring to hydrogen sources, it is important to clarify the exact details of the hydrogen production method, *especially if the green designation is claimed.*

In practice, no form of hydrogen production is completely free of carbon emissions. Citation [172] compares the carbon impacts of the five most prevalent methods for hydrogen production worldwide as of early 2024. The carbon release to the atmosphere in kg CO_2/kg H_2 for SMR with CCS can vary from 0.5 kg CO_2/kg H_2 for 95% carbon capture (which has never been implemented outside laboratories) to 8.8 kg CO_2/kg H_2 for the more realistic 50% carbon capture percentage (extrapolated from CCS at electric power plants). Electrolytic hydrogen can have a highly variable climate impact depending on the source of the electricity. The minimum carbon footprint (zero CO_2) assumes entirely carbon-free electricity from solar, wind, hydro, or nuclear. At the other extreme is electric power generated using anthracite coal, yielding drab green hydrogen with 50 kg CO_2/kg H_2 greater than that of black hydrogen from anthracite coal.

Hydrogen's Indirect Greenhouse Effect

As mentioned in the previous discussion of hydrogen as an aviation fuel, there is growing evidence that hydrogen, when leaked into the atmosphere acts as an "Indirect Greenhouse Gas" multiplying the GWP of methane and other recognized GHGs. Conservative estimates of the effective 100-year GWP of hydrogen leaked into the lower atmosphere have been reported in the range of 8.8 to 14.4 (relative to CO_2). As reported in [173]:

> Hydrogen is considered an indirect greenhouse gas because although it does not cause a warming effect on its own, it interacts with airborne molecules called hydroxyl radicals to prolong the lifetime of atmospheric methane — a highly potent greenhouse gas—and increase the production of ozone, another greenhouse gas.

Natural hydrogen in the atmosphere is rare because it is a very reactive gas. A recent multimodal study [174] considered natural absorption mechanisms. Soil uptake is estimated to account for 65–85% of the total hydrogen sink. The atmospheric lifetime of hydrogen, defined as the total atmospheric burden divided by the total sinks, was reported as about 2 years One other report [175] has emphasized the critical importance of containing hydrogen leaks to prevent the atmospheric release of hydrogen, especially if the scale of its use as a fuel increases as hypothesized in various scenarios for hydrogen economies. The analysis in the paper used the most common 10-year GWP impact of hydrogen, concluding that by this metric, its equivalent GWP is 100 times that of CO_2.

Considering that hydrogen is more prone to leakage than any other substance on Earth, the conclusions of these analyses certainly warrant more serious consideration since they anticipate that the escape of even small amounts of hydrogen from production facilities, pipelines, storage systems, and hydrogen vehicles could potentially undermine the entire climate benefit of hydrogen.

FCV Efficiency

Like any energy conversion device, the thermodynamic efficiency of a hydrogen fuel cell is simply the ratio of the energy output to the energy input. Hydrogen fuel cell efficiency is the ratio of electric energy produced to the fuel energy of the hydrogen consumed.

The low-pressure fuel cells used in FCVs have improved considerably in recent years following advances in polymer ion-exchange membranes and anode and cathode innovations. The National Renewable Energy Laboratory (USA) reported that current-technology automotive fuel cell stacks have a mean peak system efficiency of 58%, while a range of 40–60% is reported by manufacturers [176]. A complete efficiency analysis of FCVs must include not only the vehicle fuel cell and drivetrain, but the efficiency of the electrolyzer (72%), gas compression and pre-cooling prior to dispensing (87%), the onboard battery RTE (90%), and the electric power distribution and conversion losses (assuming entirely renewable energy). The elements of a complete efficiency analysis of an FCV and its fueling infrastructure are illustrated in Fig. 9.50.

Aside from the hydrogen-related efficiency issues, FCVs share all of the same electrical and mechanical losses as a BEV. The entire energy investment in gas compression or liquefaction for storage is lost. For all fuels burned in internal combustion engines (including hydrogen for combustion), the lower heating value (LHV) is used because the water created leaves the engine as a vapor, and no useful energy can be extracted from its condensation. For fuel cells, the distinction is less clear. The use of the LHV seems to be more common in North American publications, while the HHV (higher heating value) is more common in Europe.

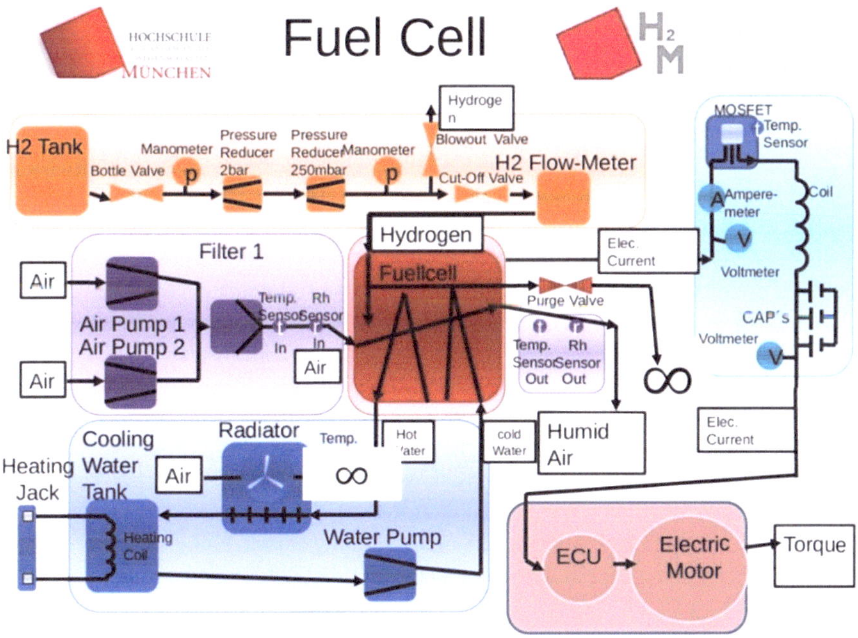

Fig. 9.50 Diagram from HydroMotion project at Munich University of Applied Science. https://ptrace.getkotori.org/2016-03-07_h2m-fuel-cell-schematics-150.jpg With permission

This distinction is particularly important when comparing the efficiency of hydrogen fuel cells with hydrogen ICEs, because the HHV of hydrogen is 142 MJ/kg and the LHV of hydrogen is 120 MJ/kg. These values have the highest spread among all combustion fuels because the entire waste product of the combustion is water, while hydrocarbon fuels produce water vapor and non-condensing carbon dioxide.

A Comparative Analysis of FCV, BEV, and ICV Emissions for Various Hydrogen Production Scenarios

In the following pages, I attempt to calculate the CO_2 emissions of a hydrogen FCV with three methods of hydrogen production. These results are compared with an equivalent BEV, CNG vehicle, and gasoline vehicle. The analysis in each case works backward from the energy required by a vehicle to travel 60 km, tracing this requirement back to either a common or equivalent common energy source. Overall results are reported as grams CO_2 produced to travel 60 km, and also a (US) g CO_2/mile rating.

My objective is a fair comparison between motive power options using the same experimental assumptions since the conclusions depend significantly on the assumptions. The choice of energy conversion steps or process losses, and the efficiencies and carbon intensities for each come from sources cited below.

Notes on Data Sources

- Grid electricity is the USA average mix of renewable and non-renewable energy sources in 2023. This value has increased each year, despite increased renewable energy sources.
- Hydrogen is delivered to hydrogen refueling stations (HRSs) in either high pressure or cryogenic form by specialized truck-transport trailers. Compressed hydrogen delivery requires three times the number of round trips for delivery of a given mass of hydrogen, but requires about one-third of the energy compared with liquefaction. 53 MPa-rated (7,500 psi) "tube trailers" are the most common transport vehicle. The largest of the hydrogen suppliers, Air Products Corp, operates 12 purpose-built "drop-and-swap" hydrogen trailers that are self-contained with gas purification, drying, and compression equipment that can be left on site. These are exchanged at the fueling stations rather than transferring hydrogen from the trailer to a ballast tank at the station [177].
- Comparison vehicles and their energy storage media:
  - FCV: 2023 Toyota Mirai (Compressed H_2 at 70 MPa (10,000 psi))
  - EV: 2023 Tesla Model 3, base (82 kWh battery)
  - CNG: 2015 Honda Civic GX NGV (CNG at 25 MPa (3,600 psi))
  - Gasoline: 2023 Honda Civic EX (12.4 gallon gasoline tank)
- All vehicle efficiencies and emissions from US EPA https://www.fueleconomy.gov/feg/.
- Hydrogen and methane leakage NOT included in Cases 1, 2, 3, and 5.
- Data sources are documented in Table 9.3, below. Some numbers are averages from several sources. In some cases, the numbers used to find the average differed significantly.

Table 9.3 Data and physical properties used for the CO_2 emission calculations in the six cases to follow

Process or energy conversion	Efficiency	CO_{2e} emitted	Energy required/kg H_2	Reference
Electric power generation				
Electric energy, USA grid avg.	40%	0.368 kg CO_2/kWh		[178, 179]
Fuel energy				
Natural gas, USA avg. (45 MJ/kg)		2.78 g CO_2/kg NG	12.5 kWh/kg	[180, 181]
Gasoline, USA avg. (41.5 MJ/kg)		2.3 kg CO_2/L gasoline	11.5 kWh/kg	
Fuel production				
H_2 Electrolysis (kWh_{H2}/kWh_e %)	61%		55 kWh_e/kg H_2	[182, 183]
H_2 SMR w/o CCS	3.15 kg/kg H_2	9.5 kg CO_2/kg H_2	0.80 kWh/kg H_2	[184, 185]
H_2 SMR w/ 68% CCS	3.35 kg/kg H_2	2.85 kg CO_2/kg H_2	5.50 kWh/kg H_2	[186–188]
NG extraction, refining	90%	0.20 kg CO_2/kg NG	0.67 kWhe/kg NG	[189]
Petrol (Gasoline) extraction, refining	88%	0.33 kg CO_2/kg gasoline	0.73 kWhe/kg petrol	[190, 191]
Electric distribution				
Transmission & Distribution (USA avg)	94.1%			[192–194]
Fuel transport (zero leakage)				
H_2 compression for transport, 53 MPa		0	1.5 kWh_e/kg H_2	[75, 84, 195]
H_2 transport, 2000 kg, 100 km		0.069 kg CO_2/kg H_2	Negligible	
Gasoline transport, 12,000 gal, 100 km		0.023 kg CO_2/kg gas	Negligible	
Natural gas transport (pipeline)		0	Negligible	
Fueling or charging station				
EV, mean of L2 (AC) and L3 (DC) charge rates	88%			[196]
H_2 precool/compr, –40 °C, 70 MPa			4.2 kWhe/kg H_2	[197]
CNG compression, 25 MPa			0.39 kWh_e/kg NG	[198, 199]
Vehicles				
FCV Toyota Mirai, hydrogen	102 km/kg H_2	0		[200]
EV Tesla M3, base, XLE, electric	6.24 km/kWh	0		[201]

Table 9.3 (continued)

Process or energy conversion	Efficiency	CO_{2e} emitted	Energy required/ kg H_2	Reference
ICV, 2015 Honda Civic NGV, CNG	19.4 km/kg	0.143 kg CO_2/km		[202, 203]
ICV, 2023 Honda Civic 4-DR, gasoline	14.1 km/L	0.164 kg CO_2/km		[204]

Notes:
SMR H_2 without CCS:
Natural Gas Required: 3.15 kg/kg H_2
Electricity Required: 0.80 kWh/kg H_2
CO_2 Emissions: 9.5 kg CO_2/kg H_2
SMR H_2 with 68% CCS:
Natural Gas Required: 3.35 kg/kg H_2
Electricity Required: 5.50 kWh/kg H_2
CO_2 Emissions: 2.85 kg CO_2/kg H_2

Test Cases

Case 1: FCV, Gold (white) geologic hydrogen, or Orange stimulated geologic hydrogen. Based on available published information. Note: Original name "gold" given to geologic hydrogen was changed to "white" in later publications.

Case 2: FCV, Green, or Drab Green hydrogen. On-site electrolysis, renewable or grid electricity respectively.

Case 3: FCV, Gray or Blue hydrogen. SMR without or with 68% CCS, respectively. Transport to refueling stations as LH_2.

Case 4: BEV, Average of Level 2 and Level 3 charging using grid electricity. This is the the norm for almost all BEVs.

Case 5: ICV, Compressed natural gas, 25 MPa 2015 Honda NGV.

Case 6: ICV, Gasoline. 2023 Honda Civic EX sedan.

Case 1: FCV using geologic hydrogen: Gold (white or natural) or Orange(hydro-induced).

Geologic hydrogen is assumed to be accessible directly from wells and requires an estimated electric energy input for extraction of an estimated 2 kWh_e/kWh H_2 for gold hydrogen (natural) or 3 kWh_e/kWh H_2 for orange hydrogen (stimulated by water injection). Lacking actual data from commercial operations, these numbers are estimated based on typical practice with natural gas wells. The other electricity inputs are for pre-cooling and compression at the refueling station 4.2 kWh_e/kg H_2. The only CO_2 emitters in this chain are the electric power and the transport of compressed H_2 to dispensing facilities. The Toyota Mirai energy use rate, according to EPA testing, is 102 km/kg H_2 so a travel distance of 60 km consumes 0.59 kg of H_2 with an energy value of 19.6 kWh.

Including the electrical energy inputs, the production of the 0.59 kg H_2 required for the car to travel 60 km will result in the net CO_2 emission of 1.47 kg CO_2 (39.4 g CO_2/mile) for gold hydrogen (cleanest except for completely green H_2) or 1.70 kg CO_2 (45.6 g CO_2/km) for orange hydrogen (Fig. 9.51).

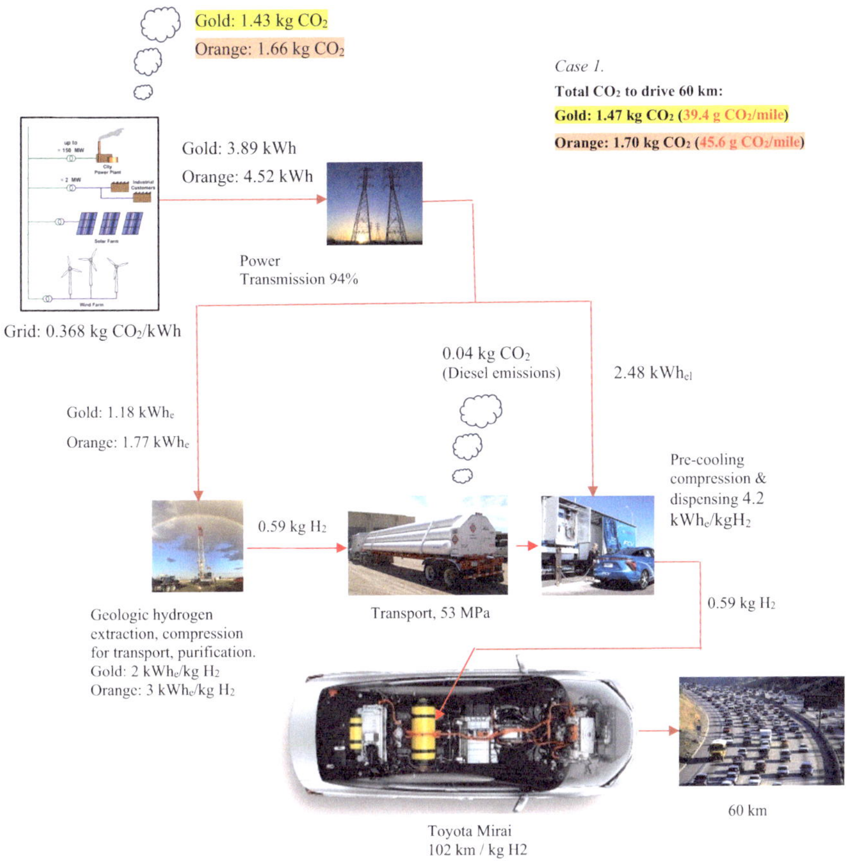

Fig. 9.51 **Case 1, Gold or Orange Hydrogen**. *Same idea of the other five Cases in Figs. 9.52, 53, 54, 55, 56* where the Case name appears following the Case number

Case 2: FCV using Green or Drab Green hydrogen. On-site electrolysis.

Hydrogen is made by electrolysis of water at the refueling facility using either completely renewable *green* electricity or *drab green* electricity from the utility grid. For the green H_2 case, carbon-free electricity is used exclusively for every electrical load from electrolysis to fuel dispensing.

For the grid electricity case, the average carbon intensity of the US electric grid was used, which includes both non-renewable and renewable sources. This means that the conclusion could be location-specific. If the hydrogen station was located in an area where coal makes up a larger fraction of the fuel for electric power generation, the local grid CO_2 emissions would be (much) larger.

As in Case 1, the criterion is a 60 km travel distance in a Toyota Mirai that achieves 102 km/kg H_2.

For green hydrogen, no CO_2 is emitted at any point in the production or use processes so the net CO_2 emissions are zero.

For drab green hydrogen using grid electricity, 13.7 kg of CO_2 is generated (228 g CO_2/km) during the 60 km trip. *This is the highest CO_2 emission rate of all options examined here, even slightly worse than gasoline.* Why so high? Electrolysis is not a particularly efficient energy conversion, requiring a large amount of electric energy compared with the fuel value of the hydrogen produced. This electricity incurs CO_2 emissions of 368 g CO_2/kWh. Unless fully renewable electricity is available to power the electrolyzer as well as the hydrogen compression/dispensing equipment, the hydrogen is far from green. *If even a small percentage of the electrical power is drawn from the grid, the hydrogen could be more carbon intensive than most of the cases examined here* (Fig. 9.52).

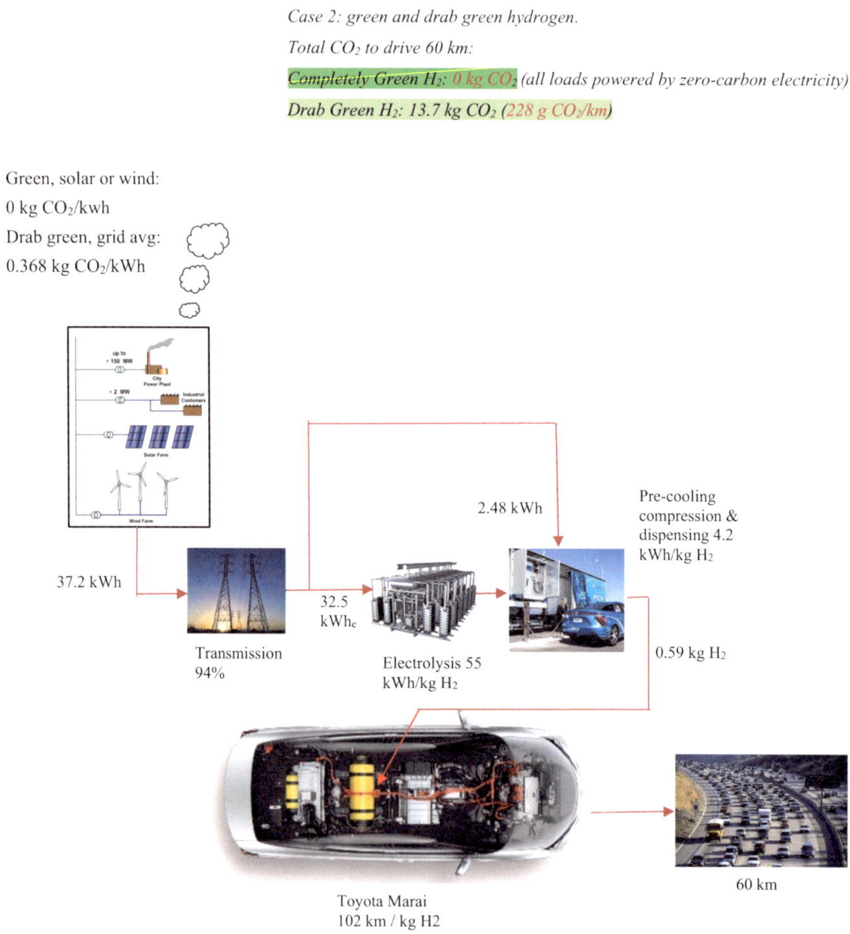

Fig. 9.52 Case 2: green and drab green hydrogen
Total CO_2 to drive 60 km: Completely Green H_2: 0 kg CO_2 (all electricity from zero-carbon sources), the lowest emissions of all cases; Drab Green H_2: 13.7 kg CO_2 (228 g CO_2/km), the highest emissions of all cases

Case 3: FCV, Gray (SMR without CCS), or Blue (SMR with CCS) hydrogen

Hydrogen is made from natural gas via SMR either without (gray) or with (blue) CCS. The extraction, purification, and pipeline transport of the natural gas from the well head to the SMR facility is highly variable depending on the well and facility characteristics; in this case, it is conservatively estimated to be 0.5 kWhe/kg NG. The final result is only minimally affected by the variability of this estimate. No leakage of either NG or H_2 is assumed, which is optimistic.

As in all cases, the basis is a 60 km travel distance in our Toyota Mirai that achieves 102 km/kg H_2.

For gray hydrogen, the net CO_2 emissions are 7.13 kg CO_2 (119 g CO_2/km). For blue hydrogen that employs CCS with a capture percentage of 68%, including the additional natural gas and electricity required by the CCS process, the CO_2 emissions would be 4.33 kg CO_2 (72 g CO_2/km) (Fig. 9.53).

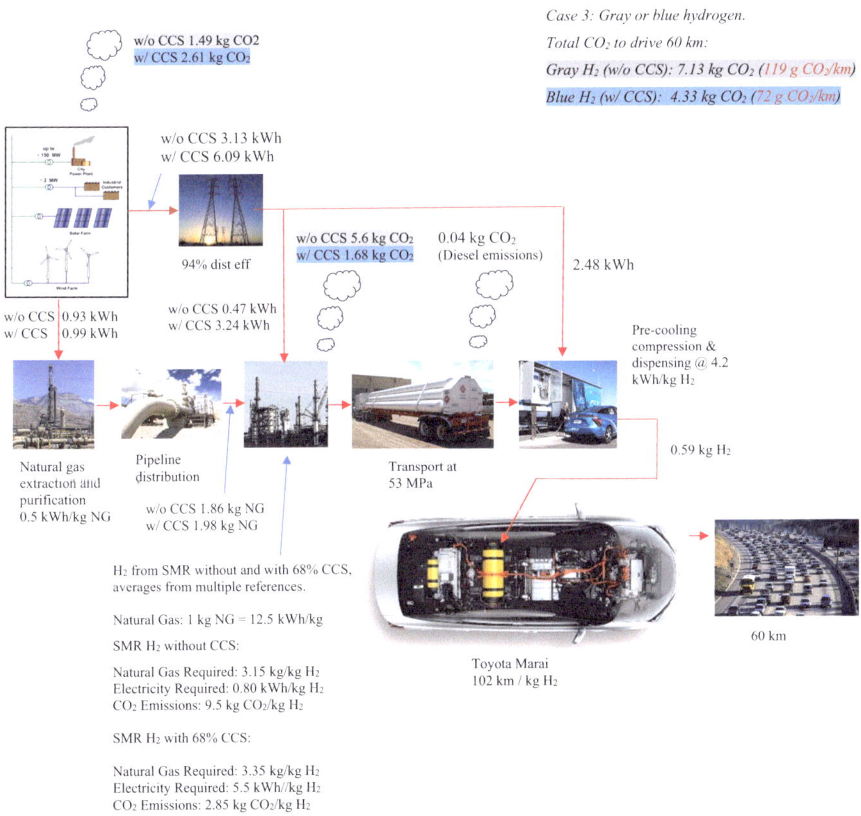

Fig. 9.53 Case 3: Gray or blue hydrogen
Total CO_2 to drive 60 km: Gray H_2 (SMR w/o CCS): 7.13 kg CO_2 (119 g CO_2/km); Blue H_2 (SMR w/68%CCS): 4.33 kg CO_2 (72 g CO_2/km)

Case 4: ICV, Compressed Natural Gas, 25 MPa.

Q: Why compare hydrogen FCVs with natural gas combustion vehicles, when natural gas vehicles are not actually being considered as alternatives to hydrogen FCVs?

A: Because both vehicles actually run on natural gas. The difference is that for the hydrogen vehicle, the natural gas is converted into hydrogen and then "burned" in a fuel cell. The NGV just burns the natural gas directly in an IC engine

We have a large experience database for natural gas-powered vehicles, although only one (now discontinued) production natural gas vehicle, the Honda Civic NGV, which is in the same size/use class as the Toyota Mirai and Tesla Model 3. It has an EPA fuel economy rating of: 19.4 km/kg natural gas. Note that natural gas is quite variable in its composition as energy content, ranging from 40 MJ/kg (North Sea NG-L) to 53 MJ/kg (Russian NG-H, the "good" stuff). I used the median value 45 MJ/kg in this analysis, which approximately represents the average for natural gas in the USA.

The analysis includes the electrical energy from the grid required for natural gas extraction and processing and distribution by pipeline. Also, the electric energy needed to compress and dispense natural gas at 25 MPa at NG refueling facilities.

Like cases 1–3, leakage of hydrogen or natural gas is not considered, aware that it can potentially be a significant source of GHGs from natural gas distribution and hydrogen production.

Subject to these assumptions, the net CO_2 produced by the NGV to drive 60 km is 10.9 kg CO_2 (182 g CO_2/mile), more than twice the CO_2 emissions of the FCV using SMR hydrogen with 68% CCS. To travel 60 km, the Honda NGV ultimately uses 3.1 kg of NG and 2.84 kWh of electricity, whereas the Toyota Marai with SMR/CCS hydrogen uses 1.98 kg NG and a total of 7.1 kWh of electricity. In terms of total energy consumption per km, the NGV is actually more efficient (Fig. 9.54).

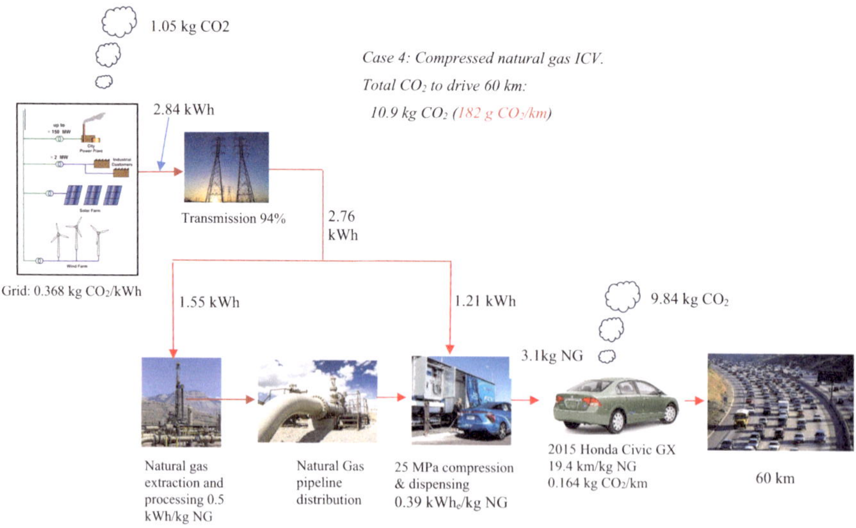

Fig. 9.54 Case 4: Compressed natural gas ICV
Total CO_2 to drive 60 km: 10.9 kg CO_2 (182 g CO_2/km)

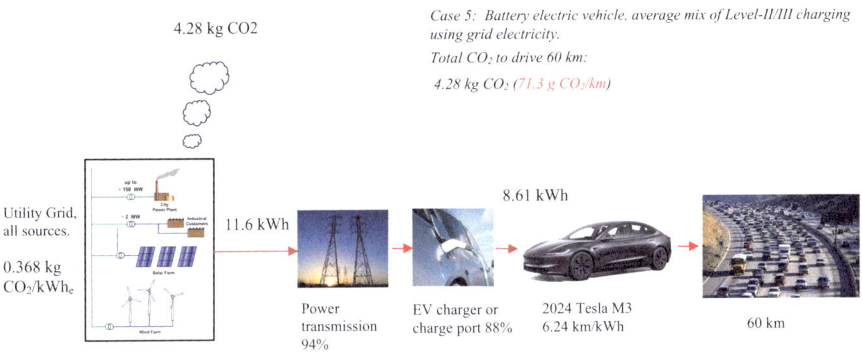

Fig. 9.55 Case 5: Battery electric vehicle, average mix of Level-II/III charging using grid electricity
Total CO_2 to drive 60 km: 4.28 kg CO_2 (71.3 g CO_2/km)

Case 5: BEVs, for a mix of Level II/III charging using grid electricity.

For perspective, this case estimates the *net* CO_2 emissions of a Tesla Model 3 electric vehicle, including the CO_2 emissions of utility grid electricity. As in all cases, the basis is a 60 km travel distance in the 2024 Tesla that achieves an EPA-rated mileage of 3.9 miles/kWh.

For emission purposes, electricity is assumed to be entirely from the US utility grid, which is the norm for all EVs in the world at this time, so I have not considered the case of green electricity solely generated by solar or wind. Obviously, if the electricity did entirely come from a carbon-free source, the net CO_2 emissions would be zero.

The net emissions for this travel distance in our Tesla are 4.28 kg CO_2. The indirect emissions related to vehicle or battery manufacturing are not considered, nor were they in any other test cases. This CO_2 level is about the same as blue hydrogen with 68% CCS (as long as all of the captured CO_2 from the CCS is fully sequestered).

I should note that the 368 kg CO_2/kWh carbon intensity of the US electric grid used in this analysis may be conservative. The data spread for 2022–3 from authoritative references is as high as 440 kg CO_2/kWh [205], 20% higher, which would proportionally increase the carbon intensity of electric vehicles (Fig. 9.55).

Case 6: ICV, gasoline.

And last, the expected comparison with an equivalent gasoline vehicle. For this case I chose the gasoline version of the base model Honda Civic, almost exactly equivalent to the Civic NGV of Case 4 and very similar to the Toyota Mirai.

Like the other vehicles, we use EPA certification values for the fuel use and the CO_2 emission of the vehicle on a per km basis, and include the emissions of the gasoline supply chain. The energy cost and CO_2 emission consequences of the oil extraction, refining, transport, and dispensing are included, although some assumptions were required for the highly variable data related to oil extraction and refining. The US (conservative) average electric grid CO_2 intensity was used for the power requirements of extraction, refining, and dispensing of gasoline at a typical gasoline station.

Subject to these assumptions, for the 60 km travel distance this gasoline car would generate 11.7 kg CO_2 (195 g CO_2/mile), higher than any of the other cases except for drab green hydrogen (electrolysis using grid electricity) (Fig. 9.56).

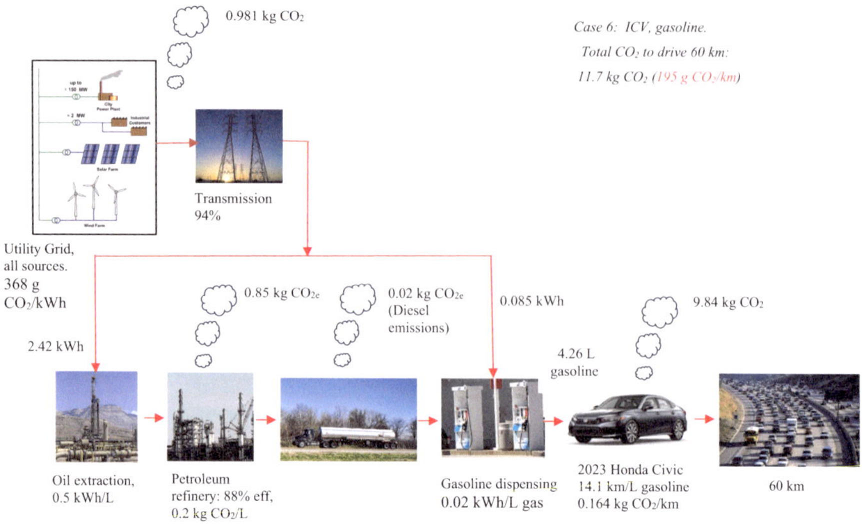

Fig. 9.56 Case 6: ICV, gasoline
Total CO_2 to drive 60 km: 11.7 kg CO_2 (195 g CO_2/km)

Observations from These Six Scenarios

The simplified analyses here are based on mean values from published data in a broad literature search. The large variability in the source data suggests that the results are equally variable. Aside from this concern, some results were a bit surprising. Except for green hydrogen made from a co-located solar or wind farm, gold or orange hydrogen are obviously worth holding out for. Blue hydrogen with 68% CCS may not be as bad as some recent analyses report, only slightly worse than grid-charged BEVs, although it would be much worse with lower CCS recoveries. But the carbon intensity of drab green electrolytic hydrogen using grid electricity *was worse than gasoline*. This can be traced to the carbon intensity of the electric grid, and at this time, almost all electrolytic hydrogen uses grid electricity. These observations have profound implications for current climate mitigation strategies, including the "Hydrogen Highways" in the USA and Europe, with hydrogen ranking both the best (gold) and the worst (drab green) in climate impact. The severity of the drab green hydrogen case is especially worth noting since at this time, almost 100% of all electric vehicles on the road are charged with grid electricity.

Hydrogen Fuel Safety

The safety record of the approximately 72,000+ hydrogen FCVs on the road worldwide [206] has generally been excellent. But since the question of safety is always raised when hydrogen is mentioned, it is worthwhile to look at a sample of the few situations in which things went wrong and the consequences.

Compressed Hydrogen Vehicle Accidents

In the USA, there have been very few accidents among hydrogen fuel cell vehicles on the road. A synopsis of safety issues unique to hydrogen FCVs can be found in citation [207]. Fortunately, there are enough FCVs worldwide to have warranted full-scale safety testing by the European New Car Assessment Program (NCAP). A group of 2021 Toyota Mirai's were submitted for testing in September 2012. The Mirai received the maximum five-star rating, with only a cautionary request that the hydrogen tanks should be shielded from impact to the same degree as the human occupants [208]. In 2018 the Hyundai Nexo FCV was tested and also awarded a five-star safety rating, although the NCAP report noted that the tests were not comparable due to different standards and procedures. It is worth noting that in none of the tests was the high-pressure hydrogen tank ruptured, as it was well protected within the vehicle's unibody structures. Had the tank ruptured, the results would have certainly been different; but it is also true that fuel tank ruptures are also rare when gasoline or diesel-fueled vehicles are tested with the same protocols.

Hydrogen leakage from an FCV parked in a closed garage or a parking structure could have potentially catastrophic consequences, warranting aggressive preventative measures. In mid-2024 the Korea Transportation Safety Authority reported that 1,463 out of 9,482 MY 2019–2024 Hyundai Nexo FCVs tested in South Korea failed tests for pressure-resistant containers since 2021, with 91% of the defective vehicles leaking from a particular overpressure relief valve. No incidents have occurred in the USA, but in October 2024 the NHTSA proactively ordered the recall of all 1,545 Nexo US models, with the 2020 MY recall classified as "Urgent, fire risk when parked" [209].

Nevertheless, it is probably safe to conclude that hydrogen FCVs currently in North America are well-engineered, safe cars, exceeding the safety ratings of most gasoline ICVs.

Liquid Hydrogen Vehicle Accidents

Probably the first reported incident involving a liquid hydrogen vehicle fire occurred with a hydrogen conversion by Roger Billings of Provo Utah, circa 1976. A 40L cryogenic LNG tank manufactured by Beech Aircraft had been used to store liquid hydrogen, located in the trunk of a full-size car. A fire broke out while parked, destroying the vehicle. The vehicle was not occupied at the time, and no one was injured. No causality was established, but it is reasonable to assume that a leak in the low-pressure fuel system allowed hydrogen to escape into the closed truck and come into contact with an unidentified ignition source. A copy of a since-redacted news photo of what is presumed to be the vehicle and tank prior to the incident is shown in Fig. 9.57.

While several accidents have occurred during the transport of liquid hydrogen by truck, the first and possibly only recorded traffic accident involving a liquid hydrogen powered automobile or light truck occurred in June 1975 when the UCLA Hydrogen Postal Service vehicle shown previously in Fig. 9.42 was involved in a freeway accident in which it rolled over while being towed to the start of the 1975 SEED Rally, a clean air vehicle competition. This was the first known opportunity to assess the safety of a liquid hydrogen-powered automobile in a roadway traffic accident. Fortunately, the 190L (50 gal US) cryogenic hydrogen tank located in the back area of the vehicle interior was not damaged and there was no fire, even though the vehicle came to rest inverted on an embankment. Among the lessons learned from the incident was the need to modify protocols for first responders at liquid hydrogen (and LNG) incidents: fire department personnel dispatched to the scene instinctively doused the vehicle and its liquid hydrogen fuel tank with water, which immediately froze the cryogenic fuel valves preventing the fuel supply from being shut off. Since the vehicle was inverted, it was expelling liquid hydrogen downward from the roof vent. Fortunately, since no ignition source was present near the vent, the only harm was the loss of about a kilogram of liquid hydrogen before the vehicle was manually rolled back over by the student team (responders stayed clear of the vehicle). Despite major body damage, it was safely driven powered by hydrogen 200 miles back to the UCLA campus, leaving behind the heavily damaged tow vehicle and trailer.

While serious accidents involving FCVs have been rare, a number of major accidents have occurred involving hydrogen transport trucks, both pressurized and liquid. Unlike the small well-protected compressed H_2 tanks in all modern FCVs, LH_2 transport trucks typically carry 10,000 gallons of LH_2, and typical high-pressure

Fig. 9.57 Circa 1976 LH_2 conversion using a Beechcraft liquid natural gas tank for fuel storage in the trunk. News coverage was scant with no details reported, but noted that the car was destroyed when a fire broke out while parked. Photo reproduced from https://simanaitissays.com/2012/10/07/hydrogen-i-c-part-1-billings/. Public domain

"Tube Trailers" about a third of this hydrogen mass at 25–30 MPa. Both of the accidents cited in the hydrogen fuel transport discussion below-involved fire, property damage, and minor injuries, but none resulted in fatalities [210]. Considering the size of the vehicles involved, this was a remarkable testament to the relative safety of hydrogen transport.

Fire Safety of Hydrogen Vehicles

Hydrogen vehicle safety in the event of a crash has always been a topic of concern. But as of 2021, I could find no public records of injuries from incidents involving FCVs [211]. Figure 9.58 shows a staged test of the comparative consequences of the release and combustion of fuel from a hydrogen-fueled vehicle vs a gasoline vehicle fire, published by the US Dept of Energy in the early 2000s. The overpressure safety valve of the H_2 vehicle released pressurized hydrogen which vented upwards with minimal damage. The gasoline vehicle was engulfed in fire as the fuel spread out on the ground beneath the vehicle. If the hydrogen tank had catastrophically ruptured in a collision, the consequences would have been more severe, but probably still not as life-threatening as the spillage and burning of the contents of a gasoline tank. As will be illustrated using the 1937 crash of the Hindenburg later in this chapter, the physical characteristics of hydrogen after a release reduce potential harm in three ways: low energy per volume, minimal infrared radiation from the combustion, and the low density of hydrogen that causes it to immediately rise above the vehicle and dissipate quickly rather than pooling around the vehicle.

Fig. 9.58 Demonstration of a hydrogen vs gasoline vehicle fire. DOE-funded 2001 demonstration conducted by Michael Swain at the University of Miami in Florida, USA. Photo from https://ecv.hyundai.com/upload/newsroom/BL00200514/Green-Tech-010-Is-Hydrogen-Safe-Sep2023.pdf. US DOE copyright in public domain

Hydrogen Fuel Transport Safety

With the increase in the delivery of hydrogen to public hydrogen fueling stations, there are reasons to be concerned about the safety of transport trucks carrying large quantities of hydrogen in liquid or pressurized form. But some perspective is in order. Figure 9.59 shows an example of the crash of a *gasoline* tank truck on a freeway in Connecticut. This can be compared with Fig. 9.60, a crash of similar severity between a high-pressure *hydrogen* tube trailer and a pickup truck at an intersection in Ohio. The gasoline fuel energy content was of course, greater. But the photos illustrate the ramifications of similar traffic incidents. The gasoline crash resulted in a major conflagration as burning fuel leaked onto the freeway. The road surface and a nearby overcrossing were destroyed. By comparison, the hydrogen truck crash resulted in a succession of explosions as individual tubes of the trailer ruptured from the heat of proximate tubes. Combusting hydrogen from the ruptured pressure tanks in Fig. 9.60 rose entirely into the air without spreading laterally to the surrounding area. Other than the destroyed vehicles, the highest value property damaged was the traffic signal light immediately above the collision site. A proximate structure was untouched.

This is only a single comparison, but it is representative of the relative risks of hydrogen vs petroleum transport. It is probably fair to say that hydrogen transport is no more dangerous than gasoline, and in most cases to date, less dangerous.

Hydrogen Fueling Station Incidents

In perspective over the years, it appears that the greatest risks of FCVs are not with vehicles, but with the distribution and dispensing of hydrogen for FCVs. One of the better-known examples: In 2019, a Uno-X hydrogen refueling station near Oslo, Norway caught fire and exploded, aerial photos in Fig. 9.61. Two minor injuries

Fig. 9.59 Gasoline fuel truck collision on freeway in Connecticut USA. Image is a frame from video newscast by NBCCT titled "Car merging onto I-95 South hit fuel truck before fiery crash in Norwalk: state police", 4 May 2024. https://www.nbcconnecticut.com/news/local/car-merging-onto-i-95-south-hit-fuel-truck-before-fiery-crash-on-i-95-in-norwalk-csp/3282690/

Fig. 9.60 Collision of hydrogen tube trailer transport in Ohio, USA. 7 February 2023. Still frame link to video newscast, Alex Barry, WSYX. https://www.hydrogeninsight.com/transport/explosion-after-explosion-three-injured-after-truck-towing-420kg-of-hydrogen-crashes-and-explodes-in-ohio/2-1-1400104. Public domain

Fig. 9.61 Shell hydrogen fueling station explosion in Sandvika, Norway, 2019. Images from https://uk.motor1.com/news/354304/hydrogen-fuelling-station-explodes-norway/. Public domain

were reported. This incident, and a similar one in Sweden, are believed to have motivated Toyota to discontinue the sale of the Mirai FCV in 2019 [212], although it was reintroduced in 2021.

In the USA in 2023, the roof-mounted hydrogen pressure tanks of a fuel cell transit bus caught fire while refueling at a central facility in the city of Empire, near Bakersfield, California, as shown in Fig. 9.62. Despite the catastrophic appearance, no injuries were reported, and the fueling facility sustained only modest fire damage.

What About the Hindenburg?

Any mention of hydrogen safety in non-technical conversation usually brings up the German Hindenburg airship that caught fire and burned while landing at Lakehurst, New Jersey, USA, in 1937 at the conclusion of the first leg of its second successful round-trip Atlantic Ocean crossing. *This incident did not involve*

Fig. 9.62 Hydrogen Fuel Cell bus in Bakersfield California caught fire while refueling at fleet facility 19 July 2023. Frame from video news report, Eyewitness News, CBS affiliate KBAK. Video by Brandon Barazza. https://www.youtube.com/watch?v=-toRJseb_o0. Public domain

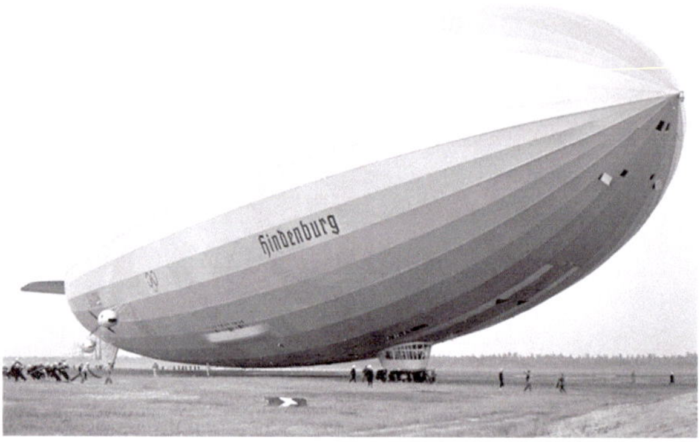

Fig. 9.63 Photo of LZ-129 Hindenberg Airship, 1935 at flight test. Photo: US Library of Congress, historical archives. Public domain

hydrogen as a fuel and therefore is not of direct relevance to hydrogen-powered vehicles, but it revealed much about the relative risk and safety of hydrogen roadway vehicles. To this day, the incident serves as an example of the relative safety of hydrogen compared with most liquid vehicle fuels due to its extremely low volumetric energy density at atmospheric pressure and temperature, despite its ease of ignition.

The LZ-129 Hindenburg rigid airship used hydrogen as a lifting gas. Extensive precautions were in place to prevent its accidental ignition. Nevertheless, the risk of accidental hydrogen ignition was so high and the volume was so large that such an incident was inevitable.

Shown in Fig. 9.63 during preliminary testing in 1935, the LZ-129 was the largest lighter-than-air vehicle ever since built. Its sheer volume was unprecedented: 200,000 cubic meters of hydrogen at atmospheric pressure

Fig. 9.64 Launch of LZ-129Hindenburg at Friedrichshafen, Germany, March 4, 1936. US Library of Congress, historical archives. Public domain

(approximately 1 bar). It was specifically designed for transatlantic ocean crossings and was promoted to affluent passengers as the pinnacle of prestige travel, on par with the RMS Titanic in 1912, which coincidentally had nearly identical exterior dimensions. It was the ultimate conveyance for an Atlantic crossing. With a maximum speed of 135 kph (84 mph), it could complete the trip in less than a quarter of the time required by the fastest ocean liner, and with a better view and more elite passenger capacity (Fig. 9.64).

The Hindenburg was powered by four Daimler Benz LOF-6 (DB-602) 16-cylinder diesel engines, each with a maximum output of 1320 hp @ 1650 RPM, and 900 hp @ 1480 RPM continuous power. Original plans had called for the engines to be fueled by hydrogen, but preliminary engine tests showed that the engines could only produce a maximum of 300 hp [213] making them inadequate for the airship's requirements.

Specifications for the LZ-129 are listed below, reproduced from a detailed history of the Hindenburg in citation [214].

LZ-129 Hindenburg statistics:

- Length: 245 m / 803.8 feet *(Nearly three American football fields).*
- Diameter: 41.2 m / 135.1 feet
- Gas capacity: 200,000 cubic meters / 7,062,000 cubic feet
- Lift: 511,500 lbs
- Cruising Speed: 125 km/h (76 mph)
- Maximum Speed: 135 km/h (84 mph)
- Main Powerplant: 4 Daimler-Benz 16-cylinder LOF 6 (DB 602) Diesels
- Crew: 40 flight officers and men; 10-12 stewards and cooks
- Passengers: 50 sleeping berths (1936); 72 sleeping berths (1937)
- First flight: March 4, 1936
- Final flight: Crashed, May 6, 1937

LZ-129 Hindenburg	kg	lbs
Dead weight	118,000	260,145
Crew	5,400	11,905
Provisions	3,000	6,614
Fuel	58,880	129,808
Oil	4,000	8,818
Ballast	7,950	17,527
Misc.	9,120	20,106
	206,350	454,924
Gross lift/hydrogen (68lbs/1,000 cu. ft.)	215,910	476,000
Payload for passengers, mail, freight w/ hydrogen	9,560	21,076

A Commercial Airliner Crash Would Have Been Much Worse

The energy release calculations below are not intended to diminish the tragedy of the deaths of the 36 people who died in the 1937 accident (35 onboard, 1 ground crew member). But of the 97 passengers and crew onboard, 62 survived. Considering the size and horrific imagery of the incident, it surprises people even today that everyone on board was not incinerated, as expected from new reports such as the Los Angeles Times front page shown in Fig. 9.65. In fact, burns from

Los Angeles Times

ZEPPELIN BLAST KILLS THIRTY-FIVE

Giant Dirigible Blazing Wreck

BLASTS SEND GIANT BALLS OF FLAME SPURTING FROM AIRSHIP

C.I.O. Joins Film Strike

JURY QUIZZES MRS. SHELBY IN TAYLOR DEATH MYSTERY

Survivor Tells Leap to Safety

TENTATIVE LIST OF MISSING AND KNOWN CRASH SURVIVORS

AGE PENSION MEASURE VOTED

LIST OF KNOWN SURVIVORS

Figure 9.65 Los Angeles Times front page, the day after crash at Lakehurst New Jersey, May 6, 1937. US Library of Congress newspaper archive. Public domain

Fig. 9.65 Los Angeles Times front page, the day after crash at Lakehurst New Jersey, May 6, 1937. US Library of Congress newspaper archive. Public domain

hydrogen combustion were few since the burning gas rose above the airship rather than engulfing the gondola. Most fatalities occurred from jumping or the falling airframe [215]. Morbidly, thc 36% fatality rate compares favorably with the usual 100% fatality rate of a commercial airliner crash. The reason has to do with the physical properties of hydrogen that made the accident far less consequential than appeared to observers and media. These properties are also relevant to automotive uses of hydrogen.

The airship envelope contained 200,000 cubic meters of hydrogen, which might be assumed to constitute a huge amount of combustion energy. But due to the extremely low volumetric energy density of hydrogen, that was not the case. This can be easily calculated using the Ideal Gas Law, which approximately applies to this NTP condition despite the fact that hydrogen is not actually an ideal gas.

The Ideal Gas Law states that PV = nRT, where n = number of moles, P = absolute pressure, V = volume, T = absolute temperature, and R is the ideal gas constant, all in consistent units.

1 mole of hydrogen at 1 atm and 22 °C occupies 1/0.0413 = 24.2 L.
One mole of diatomic hydrogen (H_2) has a mass of 2.0 g.
The specific combustion energy (LHV) of hydrogen = 120 MJ/kg.
The combustion energy contained in the lifting gas can be calculated:

$$\frac{2\,\mathrm{g}}{\mathrm{mole}}\frac{1\,\mathrm{mole}}{24.2\,\mathrm{L}}\frac{1000\,\mathrm{L}}{1\,\mathrm{m}^3} = 82.6\,\mathrm{g/m}^3$$

$$82.6\,\mathrm{g/m}^3 \times 200{,}000\,\mathrm{m}^3 = 16.6\times10^6\,\mathrm{g} = 16.6\times10^3\,\mathrm{kg\,hydrogen}$$

$$\frac{120\,\mathrm{MJ}}{\mathrm{kg}}16.6\times10^3\,\mathrm{kg} = 2.0\times10^6\,\mathrm{MJ}\ \text{total hydrogen energy (LHV)}$$

For transatlantic crossings, the airship carried 58,880 kg of diesel fuel, with an energy value (LHV) of 45 MJ/kg with a total energy content of:

$$58{,}880\,\mathrm{kg\,diesel\,fuel} \times 42.6\,\mathrm{MJ/kg} = 2.51\times10^6\,\mathrm{MJ}$$

The diesel fuel onboard the LZ-129 had *greater combustion energy than the energy of the hydrogen lifting gas.* For comparison, the fuel capacity of a short-body Airbus A320 having a 3000 NM (nautical mile) range is 34,000 kg of jet fuel [216], having a combustion energy value of 1.50×10^6 MJ.

There are no published estimates of the remaining diesel fuel onboard at the completion of the transatlantic crossing, any fire on the tarmac would have been caused by the diesel fuel and burning envelope rather than the hydrogen.

It would be misleading to compare a fire involving 16,600 kg of hydrogen in an airship with vs 6 kg of hydrogen in an FCV, but the 2,770:1 scaling alone illustrates the relative safety of hydrogen if released in a vehicle collision.

A Roadmap for Green Hydrogen in Transportation

Figure 9.66 is a copy of the Hydrogen Roadmap created in 2022 by The International Renewable Energy Agency (IRENA). The "green" electric energy sources for the hydrogen assumed in the roadmap are solar PV and wind. The intermittency of these resources would certainly be a factor in the actual implementation of the proposed Hydrogen Roadmap.

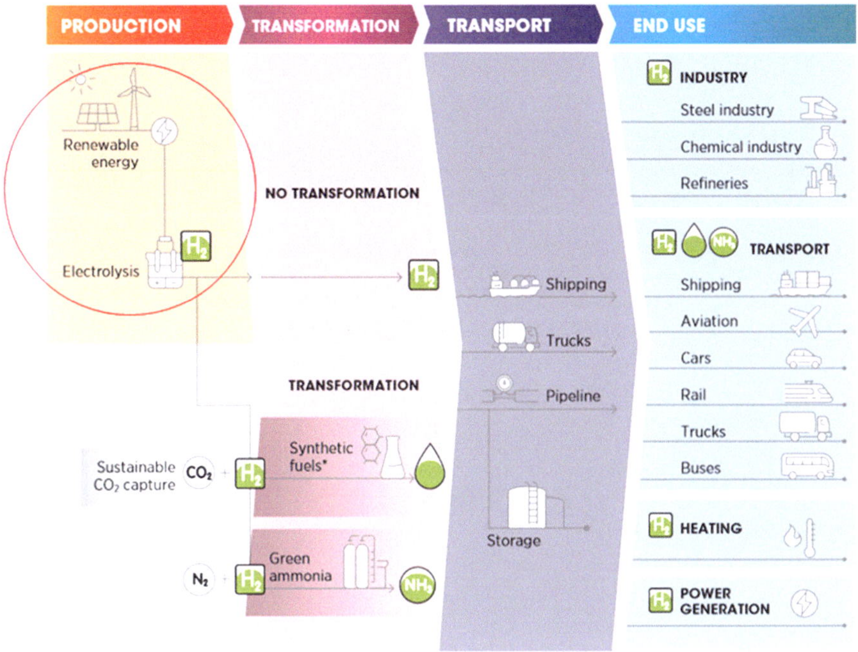

Fig. 9.66 IRENA Roadmap for hydrogen economy, 2022. Public domain

The Hydrogen Highways

The deployment of a refueling infrastructure to support hydrogen vehicles has been a much greater challenge than, for example, public EV charging networks. A map of hydrogen refueling locations in Europe is shown in Fig. 9.67. Possibly the first such network in the world was implemented in California, USA to support the small population of FCVs, all located in the urban areas of Coastal California, as shown in Fig. 9.68. Several other extensive hydrogen refueling networks have also been deployed with government assistance in Europe, with locations marked in Fig. 9.67. As of March 2024, the US Department of Energy Alternative Fuels Data Center reported that there were 58 operating public hydrogen refueling stations (HRS) in California, and one station in Hawaii. These involved a capital expenditure of $180 million, funded by the California Energy Commission that provided 69% of that investment, while private operators provided 31%. Collectively, these are the only publicly accessible hydrogen stations in the USA [217]. In 2023, this number was reported by the California Energy Commission as "66 public-accessible *Hydrogen Highway* fueling stations between San Diego and Sacramento" [218], suggesting that eight of the facilities have not yet been constructed. The California Energy Commission further announced plans to fund the construction of an additional 124 more stations by 2027, including 13 that can service trucks and buses. More complete information about the technologies and operation of HRSs in the USA and Europe can be found in citations [219] and [220], respectively. A concise assessment of the current situation in California can be found in citation [221].

Fig. 9.67 Map of EU and UK Hydrogen Stations, 2022. https://hydrogencarsnow.com. Public domain

Fig. 9.68 Map of California Hydrogen Highway Stations. From US Dept of Energy Alternative Fuels Data Center. https://afdc.energy.gov/fuels/hydrogen_locations.html#/find/nearest?fuel=HY. Public domain

What happens at a hydrogen refueling station? In all of the California Hydrogen Highway stations, hydrogen is produced from natural gas at existing commercial SMR hydrogen production facilities. It is processed to remove contaminants (mostly CO_2) and then either pressurized or liquified for truck transport to nearby (within a 100-mile radius) HRSs. At the station, the hydrogen is either transferred from the trailer to a large local storage tank, or more commonly, the transport trailer is left on-site, to be exchanged when empty with another full trailer. At the station, the hydrogen is compressed to a final pressure above 70 MPa (10,000 psi) and cooled to –40 °C, in preparation for dispensing to vehicles. Figure 9.69, illustrate the sequence of events.

a. Liquid hydrogen delivery to station in California, 2018.

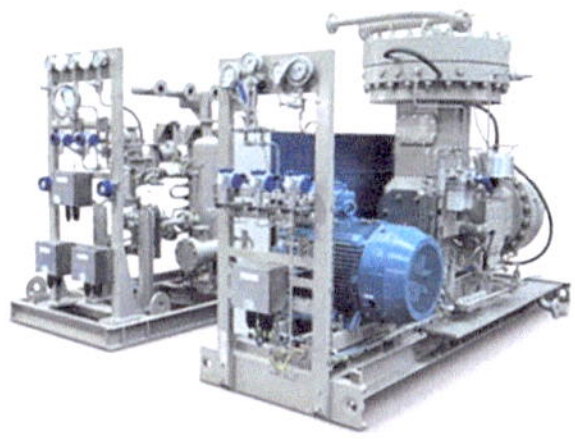

b. Multistage hydrogen compressor followed by gas cooling equipment.

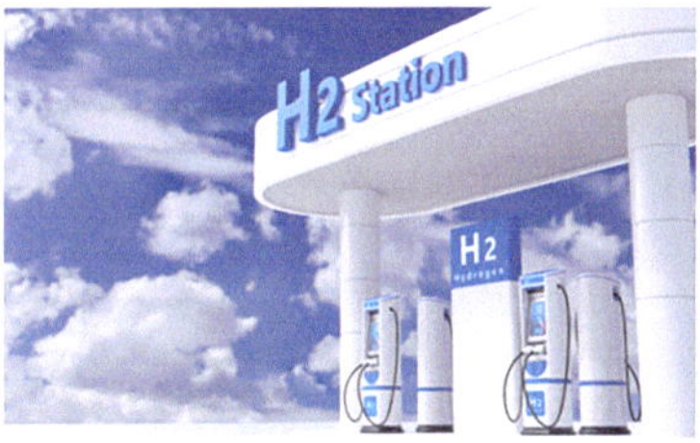

c. Fuel pumps at Hydrogen Refueling Station dispensing hydrogen at 70 MPa and -40°C.

d. Self-fueling hydrogen FCV. In this photo, California Governor Arnold Schwarzenegger and a hydrogen-converted GM Hummer H2

Fig. 9.69 Sequence of events that occur at a Hydrogen Refueling Station. (**a**) https://upload.wikimedia.org/wikipedia/commons/8/88/Linde_hydrogen_station_Mountain_View.jpg CC BY-SA. (**b**) https://www.linkedin.com/pulse/hydrogen-compressor-market-unveils-comprehensive-report-shubham-yadav-lhndf/?trk=public_post_main-feed-card_feed-article-content Product sales image. (**c**) https://depositphotos.com/photos/hydrogen-fuel.html. Public access computer-generated image. (**d**) Photo from California Air Resources Board, 2007, Public domain. Accessible at https://afdc.energy.gov/fuels/hydrogen_locations.html#/find/nearest?fuel=HY

Hydrogen Color Blindness

A 2021 feature article by the Los Angeles Times [222] made public the fact that all refueling stations that form the Hydrogen Highway Network in California use SMR (gray) hydrogen, *not carbon-free "green" hydrogen or even blue hydrogen.* The exclusive use of gray hydrogen appears to also be true for hydrogen stations in Europe, Southeast Asia, and the Middle East [223], although inadequate data is available to confirm numbers. As previously mentioned, few stations worldwide use electrolytic hydrogen, with notable exceptions at locations in central Australia that uses grid electricity [224] to generate drab green hydrogen. In California, gray hydrogen dispensed at these fueling stations is responsible for GHG emissions considerably worse than those produced by equivalent gasoline-fueled cars. And in the few cases worldwide where hydrogen is produced electrolytically using utility grid electricity, the equivalent per-vehicle-km GHG emissions are possibly even greater than those of gray hydrogen, as calculated in the preceding six-case comparative analysis. Figure 9.47 listed the relative CO_2 contributions of these hydrogen production methods.

Status of the California Hydrogen Highway, 2024

Despite California's investment of over $250 million USD in a hydrogen vehicle refueling network, with larger investment planned, and with Federal and State subsidies and carbon offset credits that continue to underwrite the majority of the fuel supply and operating costs, the hydrogen highway is in decline. As reported in CleanTechnica [225]:

> A Shell spokesman told Hydrogen Insight on February 9, 2024, "Shell discontinued the build out of its light duty hydrogen station network in California in 2023, and after temporary closure of five of its seven light duty stations, made the decision to permanently close its light duty station network in California in early 2024.
>
> California's largest H_2 fuel retailer, True Zero, operates 37 of the 53 hydrogen filling stations in the California. It recently hiked the price of hydrogen at all its pumps to *$36 per kg*, up from around $30/kg. As recently as April 2021, it was charging just $13.14 per kg.

Note that the $36 USD/kg price to consumers already includes the cost reduction from large state and federal subsidies to hydrogen highway fueling station operators as well as the hydrogen producers. At the launch in June 2021 of the *Hydrogen Energy Earthshot* program, a DOE spokesperson maintained that [226]

> The overall challenge to hydrogen production is cost. DOE's Hydrogen and Fuel Cell Technologies Office is focused on developing technologies that can produce hydrogen at $2/kg by 2026 and $1/kg by 2031 via net-zero-carbon pathways, in support of the *Hydrogen Energy Earthshot* goal of reducing the cost of clean hydrogen by 80% to $1 per 1 kg in 1 decade ("1 1 1").

An 80% cost reduction? That would imply that the current consumer price of hydrogen is *\$5/kg*. But the current retail cost of hydrogen at California Hydrogen Highway stations is \$36/kg, such that a target retail price of \$1 per 1 kilogram would require a cost reduction of 97% which is highly unlikely.

Possibly not obvious is that the *clean hydrogen* referred to in the DOE Energy Earthshot program isn't low-cost SMR gray hydrogen from existing commercial facilities. It is much more expensive electrolytic hydrogen made from zero-carbon renewable electricity, assumed to be solar or wind with storage. Even if a truly green hydrogen production facility comes online, the hydrogen it will generate is expected to cost *four times* as much as the gray hydrogen that is currently being dispensed at California hydrogen stations [227]. And even if renewable electricity becomes ubiquitous at low or zero cost, the question remains, is conversion of renewable electricity to hydrogen really its best use compared with the many other applications that depend entirely on it to earn the renewable designation and concomitant carbon credits?

Fueling issues for FCVs are far more daunting than even those of BEVs. Bloomberg Green [228] on 6 April 2024 to report:

> On April 2, for instance, California had just 54 hydrogen fueling stations and 16 of them were closed, according to real-time data from the Hydrogen Fuel Cell Partnership.
>
> Complaints about fuel prices are also common: Filling up a Mirai, an FCX Clarity or a Hyundai Nexo runs a California driver about \$200 — the equivalent of paying \$14.60 for a gallon of gas.... On Reddit, a hydrogen fuel cell car owner in Southern California vented that, "The cost is the equivalent of driving a monster truck that gets 9 mpg".

The FCV owner is correct. As of October 2024, the average retail cost for regular E10 gasoline in Los Angeles was \$4.38/gal, while the average retail cost at the same time and place for H70 hydrogen was \$34.55/kg. Comparing a Toyota Mirai FCV (68 miles/kg H_2) with a Toyota Camry LE non-plug hybrid (50 miles/gal gasoline), the comparative cost per mile is 5.8 times more for the Mirai compared with the hybrid. Indeed, *a monster truck would only need to get 9.6 mpg* to match the fuel cost of the FCV. And if the hydrogen was not so heavily subsidized, even the "Grave Digger"[11] at 3 mpg could probably match the FCV in cost per mile.

Combined with the lack of working refueling facilities, there is little incentive for buyers of hydrogen FCVs.

Late-breaking news: Toyota and Hyundai are currently selling MY 2024 hydrogen FCVs at discounts of \$33,000 for the Mirai and \$35,000 for the Nexo, both discounts of over 50%. And both are throwing in \$15,000 of free fuel. [229]

A Novel Approach to Hydrogen FCV Introduction

As with any new technology, especially those involving energy, early adopters face cost and inconvenience barriers. In the case of automotive technologies still in their

[11] *Grave Digger* is the name of a popular monster truck in the USA that has consistently won competitions since its debut in 1981.

infancy, it is prudent for manufacturers to maintain control of vehicles placed with consumers rather than release them for sale in the open market. This has been the genesis of most modern electric cars, originally offered under closed-end fixed-term leases, requiring that the vehicle be returned to the manufacturer at the end of the lease. Probably the most media-exposed examples were the 1997 GM EV-1 and the 2012 Electric Honda Fit. In addition, the operation and refueling of complex, ultra-high-pressure hydrogen FCVs pose a much greater risk to manufacturers than did the first generations of BEVs.

Recognizing an opportunity to gain much-needed experience and user feedback while aware of the potential risks of the outright sale of post-prototype FCVs to car buyers, a short-term rental (car sharing) program was established by Linde Gas Products GmbH based at their (gray) hydrogen production plant in Munich Germany. From 2016 to 2022, they offered public short-term rentals of hydrogen FCVs at very reasonable rates. The "BeeZero" single-day FCV rental program avoided the involvement of customers in refueling, assuring safety, and overcoming the lack of public hydrogen refueling facilities in Germany at the time. A fleet of 20 Hyundai Tucson [230] SUVs were converted in 2016 by Hyundai to FCVs for this program. The visibility of BeeZero cars on the streets of Munich and the surrounding area provided great exposure, with none of the complaints associated with refueling FCVs. Any problems that the vehicles might encounter were immediately attended to by a field service team.

Potential renters used a phone app to reserve a fully-fueled FCV that could be picked up at several designated pickup locations in Munich and returned either 6, 12, or 24 hours later. The cars could be driven up to 200 km for a 24-hour "adventure," at not much more than the cost of a rental car in Munich. And no cost for gasoline (Fig. 9.70).

As with the "Hydrogen Highway" experiments, any actual climate benefits were dubious because of the fossil fuel origin of the hydrogen. But the program was a success in terms of the exposure it provided for hydrogen FCVs while minimizing the infrastructure requirements and fuel handling risks. And the development work

Fig. 9.70 Linde Gas Products hydrogen FCV demonstration program, Munich Germany, 2016. (Program discontinued in 2022.) Image from https://beezero.com/en/beezero. Public domain. Accessed July 1 2018

invested by Hyundai in the Tucson FCVs constructed for Linde clearly presaged the Hyundai Nexo FCV that was offered for public sale starting in 2019.

FCVs Available 2022–24 for Public Purchase

2021 Toyota Mirai ($67,095 USD)

The Toyota Mirai has been the best selling of the three major-manufacturer FCVs. It was discontinued in 2021 almost immediately after the explosion of a hydrogen fueling station in Norway although it is unclear if the events were correlated. The Mirai was reintroduced in 2023, coincidentally with media criticism of a withdrawal of commitment to renewable fuels (Fig. 9.71).

A trend toward sunsetting hydrogen FCVs is in progress: As announced in Cars Direct (online) February 15, 2024: *Toyota Offering $40,000 Off Mirai Hydrogen Fuel Cell EVs:*

> Starting today, 2023 Toyota Mirai Limited models offer $40,000 in APR Subvention Cash, a rebate when you qualify for promotional financing. The car has a sticker price of $67,095, making this a 60% discount off MSRP and potentially bringing the price down to just $27,095. That's less than a well-equipped Toyota Corolla. Toyota also offers a $15,000 hydrogen fuel card with the Mirai. ... Is Toyota basically giving away the Mirai?" ... Inventory data only shows around two dozen Mirai Limited models left in stock.

Fig. 9.71 2023 Toyota Mirai hydrogen FCV. Photo by Toyota Motor Co., copied from nanalyze.com. Public domain

Fig. 9.72 2023 Hyundai Nexo hydrogen FCV. Photo by Arthur Braun, https://upload.wikimedia.org/wikipedia/commons/7/7b/Hyundai_NEXO_Hydrogen_Fuel_Cell_Car_beim_Autosalon_Genf_2018_y.jpg. Creative Commons, CC-BY-SA-4.0

Fig. 9.73 2021 (last year of issue) Honda Clarity Hydrogen FCV. Image from https://upload.wikimedia.org/wikipedia/commons/c/c0/Honda_Clarity_FCV_1_-_Vienna_Autoshow.jpg. Creative Commons, CC-Zero license

2023 Hyundai Nexo ($60,135 USD)

As announced by Hyundai in 2023:

> When you drive home a new 2023 NEXO, we'll give you NEXO Cards worth up to $15,000 that you can use like a debit card at hydrogen fueling stations for up to six years with a purchase or three years with a lease (Fig. 9.72).

2016–2021 Honda Clarity FCV

The first mass production hydrogen fuel cell vehicle sold to the general public. It was discontinued after 2021, with Honda citing continued losses on the vehicle as reported by *Inside EVs* [231] (Fig. 9.73)

"Honda has officially confirmed to *InsideEVs'* sister site *Motor1* that it will conclude the production of the Clarity Plug-in Hybrid and Clarity Fuel Cell in August 2021. … we believe that the Clarity Fuel Cell sales were very low."

Fig. 9.74 2025 Honda CR-V Fuel Cell Touring FWD CR-V, a plug hybrid hydrogen FCV. Non-copyright photo from Culver City Honda, California, online at https://www.cargurus.com

2025 Honda CR-V FCEV

After an absence of 4 years from the Honda car line, a descendant of the Honda Clarity was introduced for MY 2025, the first battery/fuel cell plug hybrid FCV. A larger battery than previous FCVs provides enough range to allow electric-only travel when hydrogen in unavailable. According to American Honda Motor Company, 2024 [232] (Fig. 9.74):

> As the first plug-in Fuel Cell EV in North America, the CR-V e:FCEV combines plug-in charging capability with fast hydrogen refueling for more flexible power for every drive.

Toyota Corolla Cross H_2 Concept (Japan)

An unexpected return to combustion hydrogen vehicles. From a Toyota Motor Co. press release, summarized in citation [233]:

> While the Mirai fuel cell electric vehicle has experienced a slow roll-out to limited adoption – for a spectrum of reasons – Toyota's prototype for an H_2 internal combustion engine (ICE) in the form of the *Corolla Cross H_2 Concept* opens up a new category for these cars.

A hybrid electric vehicle that uses a *hydrogen-fueled ICE*. The departure from fuel cells in favor of a hydrogen combustion engine makes it the first-ever mass-market hydrogen ICV sold to the public. It uses a 1.6L 3-cyl engine developed for its rally racing team that competed in the Super Taikyu endurance series in 2021 and the World Rally Championship (WRC) in 2021, running on liquid hydrogen - another first. The production version will use compressed hydrogen storage like all current FCVs and is claimed to be available in the USA in late 2024, although announcements have been vague (Fig. 9.75).

Fig. 9.75 Prototype of Toyota Hydrogen ICE Hybrid Corolla, announced in 2022. Photo by KKPCW (Kyu3) in Nagoya, Japan, 2022-10-23. CC-BY-SA-4.0 https://commons.wikimedia.org/wiki/File:TOYOTA_COROLLA_CROSS_H2_CONCEPT_-_1.jpg

Hydrogen for Grid Electric Energy Storage

While not directly related to its use as a vehicle fuel, hydrogen is often proposed as a medium for grid energy storage. The process is depicted in Fig. 9.76. Electric energy storage is critical if intermittent energy sources such as solar and wind will be able offset the increased grid demand of EVs.

While countless schemes have been suggested for storing excess grid electricity, only three methods have been seriously considered for large-scale deployments: Utility-scale batteries, underground storage as electrolytic hydrogen, and the well-established status quo: pumped hydroelectric storage (aka pumped storage hydroelectric, PSH). PSH will be discussed further in Chapter 14 *Electric Vehicle*s. The idea of hydrogen-based electricity storage is to make hydrogen by electrolysis during peak solar or wind generation periods when electric demands are low, inject the hydrogen under moderate pressure into underground voids such as salt caverns, and when peak electric power is demanded, recover the hydrogen and feed it to a fuel cell to generate DC electricity, then convert the DC power to line-synchronized AC and feed this back into the grid. The use of above-ground high-pressure storage in huge tanks has also been suggested, but the additional energy required for gas compression, as well as the storage tank costs, are prohibitive.

Despite obvious red flags due to the energy losses from multiple energy conversion steps, gas storage and transport, the proposed method continues to attract strong support from funding agencies and investors. Based on a review of published literature, hydrogen-based electric energy storage has not yet been deployed anywhere in the world except in small-scale demonstrations funded entirely by government grants [234]. Each of the technologies and equipment requirements is well established. These include the AC/DC power conversion, electrolyzer, gas compression, fuel cells, and DC/AC power conversions, each incurring significant energy losses. But the low round trip efficiency (RTE) and lack of suitable underground storage locations where electric load balancing is needed make this proposed electric energy storage method infeasible for any except specialized situations. A study by the US Dept of Energy yielded the cost/capacity comparisons of Fig. 9.77,

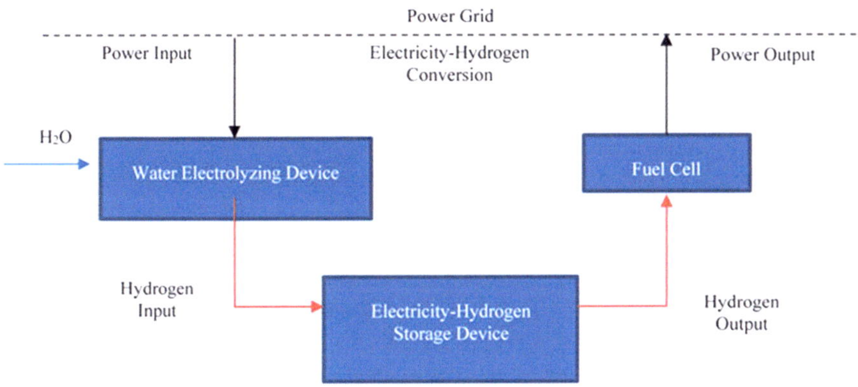

Fig. 9.76 Grid electric energy storage using hydrogen as storage medium. From Qing Xue et al. *Environmental Benefit and Investment Value of Hydrogen-Based Wind-Energy Storage System.* Frontiers of Energy Research, Volume 9, 09 March 2021 https://doi.org/10.3389/fenrg.2021.629136. Creative Commons Attribution license (CC-BY)

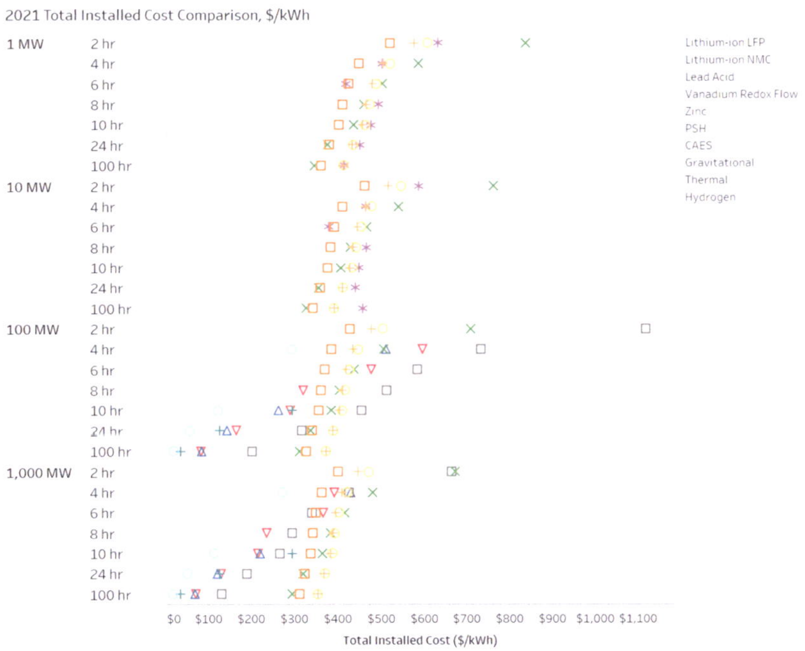

Fig. 9.77 Cost/capacity scatterplot for various grid electric storage methods. Graphic from Vilayanur Viswanathan, et. al, Energy Storage Grand Challenge Cost and Performance Assessment August 2022 https://www.energy.gov/sites/default/files/2022-09/2022%20Grid%20Energy%20Storage%20Technology%20Cost%20and%20Performance%20Assessment.pdf. Public domain

concluded that hydrogen energy storage may be competitive, but only for facilities of 100 MWh or larger, the lower left quadrant of the scatter plot.

The capital costs of a hydrogen fuel cell/electrolysis electric energy storage facility are estimated to be in the same ballpark as a new PSH facility if an upstream reservoir must be constructed, but hydrogen is considerably more costly if compared to a geographically favorable PSH location or an existing PSH facility. The many energy conversion steps that take a toll on the RTE of this method are further compounded by hydrogen storage losses. Hydrogen has a propensity to leak through almost any common containment materials and underground geologic formations. Even the best-formed salt domes or depleted aquafers under solid strata have significant leakage rates that vary for different subsurface geologies and can change over time.

Possibly the only appealing scenario may be if the hydrogen generated from the excess electricity is used directly or concurrently as a fuel source for vehicles. This is a viable proposition if the hydrogen fuel station is co-located with the grid energy storage facility, which would eliminate the need for hydrogen transport. However, considering that the locations of the storage facilities would have to be at least proximate to acceptable underground storage geology and that these locations would not likely be convenient for retail hydrogen vehicle fueling facilities, colocation would be unlikely. Hydrogen pipelines or transport arrangements would be necessary.

When assessing the actual efficiency of electric energy from intermittent generators (solar and wind), the RTE losses of the grid energy storage method must be included since the usefulness of these renewable electricity source requires such storage. More is said about grid energy storage in Chap. 14 *Electric Vehicles*, but for purposes of assessing the efficiency of hydrogen used for grid energy storage, it is worthwhile to consider some typical efficiency for large electrolyzers and fuel cells:

- Manufacturer-published ambient temperature electrolyzer efficiency values vary by type and size, from a minimum of 54.9% [235] to a maximum of 80% [236], or a mean value of 67.5%.
- In 2023, for a large-scale PEM hydrogen fuel cell, a 60% average efficiency was reported by the US DOE [237].

At this time, large-scale hydrogen storage for grid energy storage or other purposes remains more speculative than proven, although aggressive research is in progress that may yield promising conclusions [238]. Underground storage in rock salt caverns, depleted oil or gas formations, partially drained aquafers, or yet-to-be-discovered impervious geological formations are candidates for underground storage (Fig. 9.78). There is already extensive experience with underground natural gas storage, especially in China [239], which imports and stores the largest quantities of natural gas worldwide. However, even the best deep formations have a finite leakage rate that must be considered a form of energy loss with time. Lacking actual data, no values for the expected losses for underground hydrogen are known yet. Industry estimates range from a best case scenario of 1% to a worst case of 10% for established storage formations, hydrogen pipelines and processing systems [240].

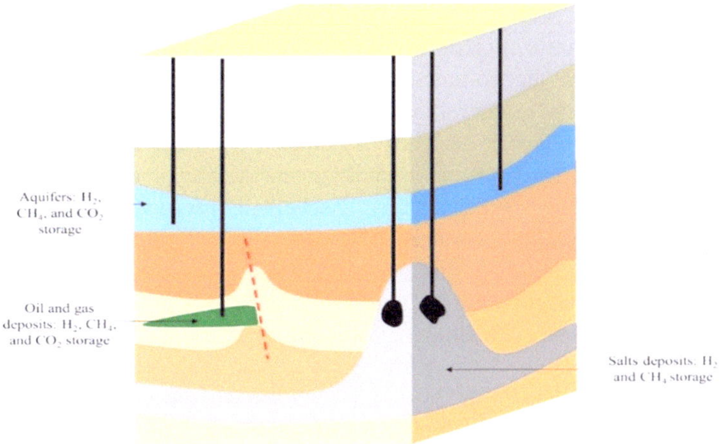

Fig. 9.78 Geological formations suitable for storage of hydrogen, methane or carbon dioxide. Graphic from Tarkowski, R., Uliasz-Misiak, B., Tarkowski, P., *Storage of hydrogen, natural gas, and carbon dioxide: geological and legal conditions*. Int. J. Hydrog. Energy 46 (38), 13 March 2021. https://doi.org/10.1016/j.ijhydene. Used under STM guidelines

Putting just these efficiency numbers together and assuming no leakage, underground energy storage using hydrogen at low/medium pressure:

87% (compression and pumping) × 65% (electrolyzer) × 60% (fuel cells) = 35%

RTE from electricity into electricity out. By comparison, extensive data from existing pumped hydroelectric storage (PSH) facilities show a range of RTEs of 70–87% [241].

Commentary: Why or Why Not Hydrogen?

Arguments abound, most favorable, but some highly critical of hydrogen as a potential alternative to fossil fuels for transportation. Hydrogen continues to be the darling of both diehard environmentalists and financially motivated opportunists, all ostensibly committed to it as a means to reduce the carbon footprint of automotive transportation, especially in California, USA [242].

A compendium of relevant background publications can be found in citation [243] covering a range of environmentally beneficial opportunities in which hydrogen plays a critical role. Most of the predicted favorable assessments are predicated upon the use of *green electricity* from utility-scale solar photovoltaic or wind turbine facilities to power electrolyzers to make hydrogen. Carbon-free hydrogen requires carbon-free electricity. Meanwhile, demand for carbon-free electricity continues to increase exponentially to support the all-electric future envisioned by many: a future without any fossil fuel combustion, including gasoline, diesel, coal, or natural gas. Yet, the use of electricity to produce *any other form* of energy requires multiple lossy energy conversion steps, the efficiency of each multiplying, resulting in overall inefficient utilization of this highest quality form of energy.

It is understandable that commercial and government agencies and consortia of fossil fuel industries are empowered to promote hydrogen energy, e.g., see citation [244]. But a mythology has evolved, perpetuated by these entities and popular media, that hydrogen by definition implies "green." Reality check: If the source of the claimed "green" hydrogen is not clearly specified, *it is almost surely not green*. In fact, there are currently no commercially competitive *green hydrogen* production facilities in the USA, although this situation will soon change as *subsidy harvesters*[12] take advantage of the US IRA and similar initiatives in other countries. This critical view is supported by almost all non-political science-based organizations including the Sierra Club [245], Earth Justice [246], the Environmental Defense Fund [247], Friends of the Earth [248], and the Union of Concerned Scientists [249].

Admittedly, energy efficiency and CO_2 emissions from energy production are not the only criteria for public investment decisions. Nor are fuel or technology costs, which could decline over time, as anticipated by the US DOE Hydrogen Energy Earthshot Program [250]. But at this time, the unsubsidized cost of green hydrogen is much greater than the cost of the energy used to make it [251]. The net benefit or harm to the environment is almost entirely dependent on the production method. Hydrogen from fossil fuel (gray, brown, black, or blue) sources remain the overwhelmingly dominant methods, and in all these cases remain environmentally a negative proposition. Drab green hydrogen (from electrolysis using grid electricity) depends on the average CO_2 emissions of the electric grid, which are currently far from net zero. Truly green electrolytic hydrogen using entirely renewable electricity is environmentally sustainable, but a disaster for energy sustainability compared with the use of electric energy for almost anything else. Only the speculative future options of geologic (gold/white/orange) hydrogen, nuclear-electric (pink), or direct STH present solutions that are both environmental and energy sustainable at this time.

Regarding green electrolytic hydrogen: A compelling argument can be made that since electricity from solar or wind is free, it doesn't matter how efficiently it is used or what it is used for. This would be a legitimate argument if there were no other needs for the electricity at the time and place it is generated, and there was adequate provisions to store the hydrogen without leakage during the periods in which the electricity is available. The ability to store unneeded electricity in the form of electrolytic hydrogen makes sense if the objective is to make the hydrogen itself, which can be pipelined or transported in liquid form to locations in which it is needed. But the RTE of this strategy for electric load leveling is very low, much lower than the status quo pumped hydroelectric storage (PSH).

Gray or blue hydrogen, made from natural gas, is much less costly than electrolysis, but neither has ever demonstrated an environmental benefit. Gray, brown,

[12] Rachel Parkes, Hydrogen Insight. 28 March 2024. *"Subsidy harvester" | Hydrogen-derived e-methane import terminal moves a step closer to FID after exemption.* The term is borrowed from agriculture, referring to USDA payments to farmers to *not grow* certain crops in an effort to stabilize market prices.

or black hydrogen accounts for an estimated 98% of hydrogen produced worldwide solely because of its low cost. Efforts to perpetuate fossil hydrogen with the addition of CCS have proven inadequate for reducing CO_2 from the overall process, but remain strongly incentivized as a way to perpetuate dependency on fossil fuels.

Why do we need hydrogen? The most legitimate justification for hydrogen is that it meets a need for a standardized energy carrier for applications that cannot be met directly by electricity. This is certainly true for industries such as steel, cement, or ammonia that are traditionally huge consumers of natural gas, and large contributors to global warming [252]. Or specialized applications that require a combustion fuel, regardless of efficiency or GHG emissions. Commercial aviation is the low-hanging fruit here, since the mass energy density of the fuel is more important than volume energy density, and hydrogen has a big advantage with respect to that metric. But recent research has nearly uniformly suggested that its global warming effect could be worse than existing fossil aviation fuel when its water vapor emissions at high altitudes are considered. This leaves marine, rail, or heavy truck-transport applications which simply need greater range than can be provided by battery electric propulsion [253]. These applications may indeed benefit from *hydrogenification*[13] since their requirements cannot be met by electricity stored in batteries. But there are other fuel options (discussed in other chapters of this book) more practical than hydrogen.

Specifically regarding hydrogen as an automotive fuel: This is an application where energy, environmental, and economic sustainability all matter, if for no other reason than its dominant role in both GHG emissions and energy use. Add to this the infrastructure investment requirements needed to support hydrogen fueling on a scale larger than a small number of heavily subsidized fueling stations.

So why is this juggernaut still in motion? Some insight is possible by examining who benefits, and what political and economic forces are driving massive government and private investments in hydrogen technology and infrastructure.

A convenient observation… the list of sponsors of the 2024 World Hydrogen Technology Expo, from citation [254]: At the top of the hydrogen leaders list are companies that already produce and distribute or provide equipment for fossil fuels or hydrogen from fossil fuels. Major fossil fuel interests seek the preservation of their existing business models that are based on petroleum or natural gas [255]. Legacy firms producing gray hydrogen for industrial uses and ammonia synthesis see an opportunity for a larger and more profitable market for their product [256]. After years of promoting hydrogen in an effort to stave off the electrification of transportation, these entities were counting on being able to simply add token CCS to their fossil fuel hydrogen production facilities, and the resulting "blue" hydrogen would be considered "green enough" to receive the same government incentives as carbon-free electrolytic hydrogen. But, to the credit of the authors of the 2022 US IRA, they excluded gray or blue hydrogen, aware that at any reasonable cost, CCS can capture only between 50% and 68% [257] of the generated CO_2. Considering that the mass of CO_2 produced by state-of-the-art SMR is 5.5 times the mass of the

[13] A contrived word. But its meaning should be clear in context.

hydrogen produced, the net GHG emissions with even the most advanced CCS technology are still more climate-harmful than other options.

As is always the case when large government subsidies are available with the best of intentions, those that benefit the most are riding the incentive and/or speculation train until the last stop when it ends due to lack of promised results (or by Federal Trade Commission investigations). But it can be a long ride. During the 2–3 years that it usually takes for a meaningful assessment, the continued receipt of government funding or carbon credits are portrayed to private and institutional investors as endorsements, which encourages the herd mentality of investors seeking short-term gains from initial stock offerings. Selective reporting of breakthroughs is plentiful, while insurmountable obstacles are rarely revealed. Non-technical media seeking uplifting stories are regularly duped by sensational announcements. Dollars/Euros flow, but the exuberance is usually short-lived. This phenomenon applies to the entire FiTech field, but it is particularly common with ever-mysterious hydrogen. For example, it was reported in a popular financial publication [258] in September 2023 that there was *a huge unserved market for hydrogen electrolyzers*. But in March 2024, even a leading hydrogen advocacy website reported serious overcapacity in electrolyzer manufacturing [259].

The illusion of large and/or reliable sources of *unneeded* "green electricity" is not unique to hydrogen, but when assumed to be a reasonable source of energy for green hydrogen production, it is particularly misleading. Whenever a media sound bite, business prospectus, proposal for agency funding, carbon credit application, or promotional article begins with the words "using green electricity" it's usually a red flag. Yet this phrase appears even in highly credible "green hydrogen" proposals and policy statements [260, 261]. The reality is that *making almost anything exclusively with green electricity (solar, wind, hydro) can allow it to qualify as green and/or energy positive,* since the energy source is essentially zero-cost and carbon-free. And lost in the sales pitch is thermodynamics: Electricity is at the top of the energy quality[14] "food chain." The use of electricity to make combustion fuels, especially hydrogen, is a wasteful use of this highest-quality energy carrier. When combined with the efficiency of the combustion process that the fuels will be used for, hydrogen incurs large energy losses enroute to the mechanical work it was ultimately intended to do. It can be justified only if there is no other need for this electricity, a situation that exists in only a few places in the world, notably Iceland with geothermal power to spare, or for nuclear-generated electricity in which the incremental fuel energy cost is nearly inconsequential and its use is carbon free. Current studies are reconsidering the role of nuclear as a possible solution to climate change, e.g., [262].

The intermittency of solar or wind is often ignored when they are prescribed as the electricity sources for "green" hydrogen. Operation of electrolyzers only during mid-day solar peak insolation and storing hydrogen on-site for delivery when needed is almost always a financial non-starter. This explains why (from every published source that discloses this information) projects that initially proposed to

[14] Please refer to Chap. 2 *Transportation Energy Realities* for the definition of *Energy Quality*.

use renewable electricity or renewable hydrogen have ended up fully or partially using grid (non-renewable) electricity or gray or blue hydrogen, while still promising to switch to "green hydrogen when it becomes available". This may also explain why all hydrogen fueling stations in the USA use gray (SMR) hydrogen from existing suppliers, even though this path produces significant net GHG emissions.

Geological hydrogen is still a large unknown, but its incredible prospects provide motivation to be hopeful. Indeed, if hydrogen were to come from the Earth's crust like natural gas, most criticisms about its production become moot. But this remains only speculation at this time, and past experience strongly suggests a healthy dose of pessimism.

Overall, *green hydrogen* seems to be justifiable only for specialized applications in which efficiency and cost are not consummate considerations. This may include steel and cement production (many sources can provide more info about these), or transportation applications in which battery energy storage is simply not feasible, e.g., heavy truck or rail transport. There will always be the 95th percentile need for hydrogen, production of ammonia for agriculture, which is so environmentally harmful than any alternative that provides even a small improvement over coal-to-gas or natural gas SMR and Haber-Bosch would be a positive contribution.

But for automobiles and light trucks, at least at this time, there are options with superior energy, environmental and economic sustainability prospects discussed in other chapters of this book.

Selected Historical News Releases About Hydrogen for Transportation

US Environmental Protection Agency, Hydrogen in Transportation, 2023

> Most fuel cell electric vehicles (FCEVs) are powered by hydrogen (H_2) and considered zero-emission vehicles (ZEVs). … Hydrogen can be used as fuel in a variety of fuel cell electric applications to generate power, emitting only water and heat as byproducts. … Hydrogen has the potential to meaningfully reduce GHG emissions in the transportation sector.
>
> https://www.epa.gov/greenvehicles/hydrogen-transportation

Sierra Club, Hydrogen: Future of Clean Energy or a False Solution? January 4, 2022

> At the core of much of the debate is the question does the term "green" or "clean" hydrogen mean only hydrogen generated electrolytically (green hydrogen), or should it be allowed to include blue hydrogen that is made from MSR followed by CCS? This latter argument is aggressively lobbied by all major oil companies in the USA and EU. If blue hydrogen was to be reclassified as green hydrogen for purposes of government incentives and carbon credits, it would allow the existing 99+% hydrogen production from fossil fuels to con-

tinue. These facilities generate huge amounts of CO_2. Consequently, hydrogen FCVs would continue to generate greater Fuel to Wheel CO_2 even when compared with the combustion of the original fuel used to generate the electricity.

https://www.sierraclub.org/articles/2022/01/hydrogen-future-clean-energy-or-false-solution

Elon Musk on Hydrogen Fuel Cell Vehicles, 2021

"Extremely silly."

https://www.cnbc.com/2021/12/06/elon-musk-has-strong-views-on-hydrogen-and-not-everyone-agrees.html

Toyota-Sponsored YouTube Video About H_2 Cars and Safety, 2018

A balanced summary FCV pros and cons, although the speaker understates the safety issues of a 70 MPa H2 tank failure, assuming that the only hazard is due to combustion of the fuel.

https://www.youtube.com/watch?v=YJBzEYduKK8

A Celebrity Hydrogen Land Speed Record Attempt, 2009

A media stunt by Jesse James, host of the popular (at the time) television program "West Coast Choppers" [263] (married at the time to American actress Sandra Bullock, divorced in 2010). Mr. James claimed an unofficial hydrogen land speed record of 199 MPH. Unaware of hydrogen fuel injection, Mr. James complained in an interview that the maximum speed was limited by engine intake backfiring.

https://www.paramount.com/press/jesse-james-sets-land-speed-world-record-for-hydrogen-powered-vehicle-in-season-finale-on-spike-tv

References

1. *The Chemical Revolution of Antoine-Laurent Lavoisier*. Commemorative booklet published by the American Chemical Society, 1999. https://www.acs.org/education/whatischemistry/landmarks/lavoisier.html. Accessed 20 June 2024
2. Trevor Brown. *Renewable Hydrogen for Sustainable Ammonia Production,* AIChE Special Section, August 2019. https://www.aiche.org/resources/publications/cep/2019/august/renewable-hydrogen-sustainable-ammonia-production. Accessed 15 April 2024
3. M. Sand, R.B. Skeie, M. Sandstad, et al., A multi-model assessment of the global warming potential of hydrogen. Nat Commun Earth Environ **4**, 203 (2023) https://www.nature.com/articles/s43247-023-00857-8. Accessed 20 April 2024
4. Hydrogen Insight. *California places near-$130m order for six more hydrogen trains for intercity lines. February 2024.* https://www.hydrogeninsight.com/transport/california-places-near-130m-order-for-six-more-hydrogen-trains-for-intercity-lines/2-1-1599032. Accessed 1 March 2024

5. John Hyatt, Forbes, 23 April 2024. *Anatomy of a $46 million renewable energy swindle.* https://www.forbes.com/sites/johnhyatt/2024/04/23/anatomy-of-a-46-million-renewable-energy-swindle/. Accessed 24 April 2024
6. Wikipedia. *Timeline of hydrogen technologies.* https://en.wikipedia.org/wiki/Timeline_of_hydrogen_technologies
7. Wikipedia. *Water-gas shift reaction.* https://en.m.wikipedia.org/wiki/Water%E2%80%93gas_shift_reaction. Accessed 1 June 2023
8. Wikipedia. *History of manufactured fuel gases.* https://en.wikipedia.org/wiki/History_of_manufactured_fuel_gases. Accessed 15 October 2024
9. Wikipedia. *Sylvia plath.* https://en.wikipedia.org/wiki/Sylvia_Plath. Accessed 20 April 2024
10. International Energy Agency, *Towards Hydrogen Definitions Based on Their Emissions Intensity. Report from the 2023 G7Hiroshima Summit* (IEA Publications International Energy Agency) www.iea.org
11. Y. Bicer, I. Dincer, G. Vezina, F. Raso, Impact assessment and environmental evaluation of various ammonia production processes. Environ Manage **59**, 842–855 (2017). https://doi.org/10.1007/s00267-017-0831-6
12. P. Gilbert, P. Thornley. *Energy and carbon balance of ammonia production from biomass gasification*. Poster at Bio-Ten Conference, Birmingham, England, 2010.
13. International Energy Agency. *Energy Technology Perspectives 2012, Pathways to a Clean Energy System,* https://www.iea.org/publications/freepublications/publication/ETP2012_free.pdf. 2012
14. Rudolph Erren, Wikipedia, https://en.wikipedia.org/wiki/Rudolf_Erren. Accessed 1 May 2023
15. CUS. *Gotta do it yourself. Engine That Runs On 80% Water – Fact or Fiction?* 9 February 2015. https://www.gottagodoityourself.com/engine-that-runs-on-80-water-fact-or-fiction/. Accessed 15 April 2024
16. Anthonycut. YouTube video. *Water engine.* https://www.youtube.com/watch?v=x1UrlDNkYSo posted 26 June 2007, 557,346 views. Accessed 1 May 2024
17. C.A. MacCarley, W.D. Van Vorst, Electronic fuel injection techniques for hydrogen powered I.C. engines. Int J Hydrogen Energy **5**(2) Pergamon Press, Oxford, England (1980)
18. The only exception I could find was *Volkswagen reveals high temperature fuel cell technology,* Nov 1, 2006 https://www.vwpress.co.uk/en-gb/releases/842. Accessed 1 March 2024. This technology never entered production
19. Toyota Motor Co., *2024 Mirai full specs* https://www.toyota.com/mirai/features/mpg_other_price/3002/3003. Accessed 20 March 2024
20. Toyota Motor Sales USA. 2024 *bZ4X Specs* https://www.toyota.com/bz4x/. Accessed 1 March 2024
21. Ride and Drive Clean, *Vehicles.* https://rideanddriveclean.org/electric-vehicle-makes-and-models/?gad_source=1. Accessed 1 May 2024
22. *Cars.com online*, 29 March 2024. https://www.cars.com/articles/electric-vehicles-with-the-longest-range-422227/. Accessed 1 May 2024
23. John Bogna, PC. Magazine, June 22, 2022. *Hydrogen-powered cars: fuel cell electric vehicles explained.* https://www.pcmag.com/how-to/hydrogen-powered-cars-fuel-cell-electric-vehicles-explained. Accessed 2 May 2024
24. Hyundai USA. *2023 Nexo fuel cell, our hydrogen powered SUV.* https://www.hyundaiusa.com/us/en/vehicles/nexo. Accessed 2 May 2024
25. Toyota Motor Sales USA. *2024 Mirai full specs* https://www.toyota.com/mirai/features/mpg_other_price/3002/3003. Accessed 1 March 2024
26. Nissan USA. *2024 Nissan leaf specifications* https://www.nissanusa.com/vehicles/electric-cars/leaf/specs-trims.html#modelName=S|40%20kWh%20lithium-ion%20battery. Accessed 1 May 2024

27. C.A. MacCarley, Development of a high speed injection valve for electronic hydrogen fuel injection, in *Proceedings of the Third World Hydrogen Energy Conference, Tokyo, Japan, September, 1980*, (International Association of Hydrogen Energy, Coral Gables, Florida, 1980)
28. Wikipedia. *Hindenburg disaster.* https://en.wikipedia.org/wiki/Hindenburg_disaster. Accessed 15 May 2024
29. Chris Randall. *Norway: Explosion at hydrogen filling station.* Electrive, 2019, updated 2024. https://www.electrive.com/2019/06/11/norway-explosion-at-fuel-cell-filling-station/. Accessed 15 March 2024
30. Eriko Shrestha, Tianyi Sun. Is this the end of the hydrogen highway? Politico, 2024. https://www.politico.com/news/2023/08/15/california-hydrogen-cars-00111323. Accessed 10 Feb 2024
31. Author's *interview with Rudolph Erren*, Sept 1978, at World Hydrogen Energy Conference, Zurich Switzerland. Also, chronicled in Alternative Energy Magazine, April 2009. https://www.altenergymag.com/article/2009/04/the-history-of-hydrogen/555/
32. G. Joseph, F.E. Lynch, N.R. Baker, R. Takahashi, A.F. Bush, The UCLA hydrogen car: design, construction, and performance. SAE Trans **82**(2), 1626–1637. SAE International (1972)
33. Stephen Edelstein, Green car reports, 10 April 2015. *More Olympics Green cars: Soviet hydrogen-gasoline mini buses of 1980.* https://www.greencarreports.com/news/1097672_more-olympics-green-cars-soviet-hydrogen-gasoline-minibuses-of-1980. Accessed 10 February 2024
34. Roger Billings. *Dr. Roger Billings historical website: 1976—The hydrogen bus.* https://www.rogerebillings.com/1976-the-hydrogen-bus/. Accessed 21 April 2024
35. Wikipedia. *Church of Jesus Christ (Zion's Branch).* https://en.wikipedia.org/wiki/Church_of_Jesus_Christ_(Zion%27s_Branch). Accessed 1 May 2024
36. Medium, 9 October 2020. *A popular online learning platform was actually created by an underground religious 'Cult'.* https://onezero.medium.com/a-major-online-learning-platform-was-created-by-a-subterranean-religious-cult-whose-leader-has-cec99e7adcaf. Accessed 1 August 2024.
37. Fast Company, 14 September 2020. *The extremely weird story of a remote-learning company that's making parents livid* https://www.fastcompany.com/90549530/the-extremely-weird-story-of-a-remote-learning-company-thats-making-parents-livid. Accessed 10 September 2023
38. Green Car Reports. Historical account. https://www.greencarreports.com/news/1097672_more-olympics-green-cars-soviet-hydrogen-gasoline-minibuses-of-1980
39. C.A. MacCarley, *Final project report, development of a hydrogen fueled transit vehicle,* 1982. Online at https://digitalcommons.calpoly.edu/cgi/viewcontent.cgi?article=1312&context=eeng_fac
40. *Hydrogen fuel cars 1807–1986.* https://www.hydrogencarsnow.com/index.php/1807-1986/. Accessed 20 April 2024
41. *Hyster hydrogen fuel cells lift truck—Tech that performs.* https://www.hyster.com/en-us/north-america/technology/power-sources/hydrogen-fuel-cells/. Accessed 1 July 2023
42. *Everything you need to know about hydrogen fuel cell forklifts.* Forklift Inventory, 2020. https://www.forkliftinventory.com/blog/hydrogen-fuel-cell-forklifts/. Accessed 1 July 2023
43. B.R. Wigg, R.E. Coverdill, Chia-Fone, L. Chia-Fone, L. Dimitrios, C. Kyritsis Dimitrios, C. Kyritsis. *Emissions characteristics of neat butanol fuel using a port fuel-injected*, Spark-Ignition Engine April 2011 SAE Technical Papers, https://doi.org/10.4271/2011-01-0902
44. C.A. MacCarley. *Electronic fuel injection techniques for hydrogen powered I.C.* Engines, MS thesis in Engineering, University of California Los Angeles, 1978
45. C. Özyalcin, S. Sterlepper, S. Roiser, H. Eichlseder, S. Pischinger, Exhaust gas aftertreatment to minimize NOX emissions from hydrogen-fueled internal combustion engines. Appl Energy **353**, 122045 (2024). https://doi.org/10.1016/j.apenergy.2023.122045
46. C.A. MacCarley, *Electronic fuel injection for hydrogen fueled IC engines*, MS Thesis, UCLA, 1978. Accessible online at http://digitalcommons.calpoly.edu/cgi/viewcontent.cgi?article=1276&context=eeng_fac

47. J.E. Johnson, J.D. Naber, Atkinson cycle, Ch. 9, in *Alternative Fuels and Advanced Vehicle Technologies for Improved Environmental Performance*, 2nd edn., (2022) Available via Science Direct, https://www.sciencedirect.com/topics/engineering/atkinson-cycle
48. Robert Bosch GmbH. *Automotive Handbook*, 10th ed. 2018. Distributed by Society of Autmotive Engineers International, 2018. pp. 659-661.
49. C.A. MacCarley, A study of factors influencing thermally induced backfiring in hydrogen fueled engines, and methods for backfire control, in *Proceedings of the Intersociety Energy Conversion Engineering Conference*, (American Society of Mechanical Engineers, Atlanta, GA, 1981)
50. C.A. MacCarley, *Electronic Fuel Injection for Hydrogen Fueled IC Engines*, MS Thesis, UCLA, 1978. Accessible online at http://digitalcommons.calpoly.edu/cgi/viewcontent.cgi?article=1276&context=eeng_fac
51. C.A. MacCarley, Development of a high speed injection valve for electronic hydrogen fuel injection, in *Proceedings of the Third World Hydrogen Energy Conference, Tokyo, Japan, September 1980*, (International Association of Hydrogen Energy, Coral Gables, Florida, 1980)
52. C. A. MacCarley *A Study of Factors Influencing Thermally Induced Backfiring in Hydrogen Fueled Engines, and Methods for Backfire Control,* 1981
53. Richard Sachek, TopSpeed, 19 October 2023. *Toyota's Hydrogen Combustion Engine Has The Potential To Make EVs Obsolete.* https://www.topspeed.com/toyotas-hydrogen-combustion-engine-has-the-potential-to-make-evs-obsolete/. Accessed 10 January 2024.
54. AVL Press Release, 11 October 2023. *410 Hp from 2-Liter Displacement—Prototype of the H2-ICE Race Engine from AVL RACETECH Proves Itself on the Testbed* https://www.avl.com/en-us/press/press-release/410-hp-2-liter-displacement-prototype-h2-ice-race-engine-avl. Accessed 10 January 2024
55. Marry Bellis. *Hydrogen Fuel Cells Innovation for the 21st Century*, published online by ThoughtCo History and Culture, 2019, https://www.thoughtco.com/hydrogen-fuel-cells-1991799. Accessed 4 January 2024
56. Steve Barrett (ed.). *AeroVironment Flies World's First Hydrogen Powered Plane*. Fuel Cells Bulletin September 2005, Elsevier Advanced Technology, Oxford OX5 1AS, United Kingdom
57. Stellantis Corp. *What is a fuel cell?* https://www.stellantis.com/en/technology/hydrogen-fuel-cell-technology. Accessed 20 December 2023
58. Ballard Corp. https://www.ballard.com/fuel-cell-solutions. Accessed 1 Sept 2023
59. Michael Barnard. *The Odyssey of the Hydrogen Fleet: A Tragicomedy in Six Acts.* Clean Technica. 2023. https://cleantechnica.com/2023/11/24/the-odyssey-of-the-hydrogen-fleet-a-tragicomedy-in-six-acts/. Accessed 1 September 2024
60. Theunis Jansen van Rensburg. Popular Technology. *The first hydrogen-powered passenger train starts service.* https://bandwidthblog.co.za/2022/09/05/hydrogen-powered-trains-germany/. Accessed 1 September 2024
61. World Economic Forum. *Energy Transition: This Is how hydrogen trains could transform transport* 23 Sept, 2024. https://www.weforum.org/agenda/2024/09/hydrogen-trains-sustainable-transport/. Accessed 15 September 2024
62. Paul Ridden. New Atlas, 27 March 2024. *Fuel-cell train travels more than 1,700 miles on one tank of hydrogen.* https://newatlas.com/transport/flirt-h2-fuel-cell-train-guinness/. Accessed 15 September 2024.
63. Leigh Collins (ed), Hydrogen Insight for Accelerate Hydrogen, 3 October 2024. *'Fundamentally unreliable' | German authorities threaten to pull the plug on world's first hydrogen-only rail line* https://www.hydrogeninsight.com/transport/fundamentally-unreliable-german-authorities-threaten-to-pull-the-plug-on-worlds-first-hydrogen-only-rail-line/2-1-1719331. Accessed 4 October 2024
64. Q. Xu, Z. Guo, L. Xia, Q. He, Z. Li, I.T. Bello, K. Zheng, M. Ni, A comprehensive review of solid oxide fuel cells operating on various promising alternative fuels. Energy Convers Manage **253**, 115175 (2022)
65. Smithsonian Institute online tutorial. *Solid Oxide Fuel Cells.* https://americanhistory.si.edu/fuelcells/so/sofcmain.htm#history. Accessed 10 January 2023

66. *Bloom's Fuel Cells: Just How Green Is a Bloom Box?* Eric Weshoff for Green Tech Media, Sept 04, 2013. Online at https://www.greentechmedia.com/articles/read/Blooms-Fuel-Cells-Just-How-Green-is-a-Bloom-Box
67. Wikipedia. *Bloom Energy*. https://en.wikipedia.org/wiki/Bloom_Energy. Accessed 10 January 2024
68. Christopher Helman, Forbes Magazine, Feb 13, 2020, https://www.forbes.com/sites/christopherhelman/2020/02/13/the-forbes-investigation-how-bloom-energy-blew-through-billions-promising-cheap-green-tech-that-falls-short/?sh=38c1700c3e5f. Accessed 10 Jan 2024
69. Simply Wall St, March 30, 2022, *When Will Bloom Energy Corporation (NYSE:BE) Breakeven?* March 30, 2022 https://finance.yahoo.com/news/bloom-energy-corporation-nyse-breakeven-153255663.html. Accessed 10 January 2024
70. *Bloom Energy, Making Clean, Reliable Energy Affordable for All*. https://www.bloomenergy.com/company/. Accessed 10 January 2024
71. Evan Halper. Los Angeles Times, *Aug. 10, 2021 Is California's 'Hydrogen Highway' a road to nowhere?* Online at https://www.latimes.com/politics/story/2021-08-10/hydrogen-highway-or-highway-to-nowhere. Accessed 1 March 2024
72. Robert Bosch GmbH, *Automotive Handboook*, 10th edn. (Distributed by SAE International, Warrendale, PA, USA, 2018) pp. 345 and 346
73. M.R. Usman, Hydrogen storage methods: review and current status. Elsevier Renew Sustain Energy Rev **167**, 112743 (2022)
74. US Department of Energy. *Hydrogen Storage*. Hydrogen and Fuel Cell Technologies Office, https://www.energy.gov/eere/fuelcells/hydrogen-storage. Accessed 17 November 2023
75. *Energy requirements for hydrogen gas compression and liquefaction as related to vehicle storage needs*. DOE Hydrogen and Fuel Cells Program Record, Record #: 9013, July 7th, 2009. Available online https://www.hydrogen.energy.gov/docs/hydrogenprogramlibraries/pdfs/9013_energy_requirements_for_hydrogen_gas_compression.pdf?Status=Master. Accessed 20 January 2024
76. P.M. Perillo, Corrosion behavior of coatings of titanium nitride and titanium-titanium nitride on steel substrates. **62**(02) (J Corrosion, 2006). https://doi.org/10.5006/1.3278263. This observation is from personal experience with titanium nitride deposited as a protective layer on silicon or gallium arsenide integrated circuits
77. H.T. Hwang, A. Varma, Hydrogen storage for fuel cell vehicles. Curr Opin Chem Eng **5**, 42–48 (2014) https://www.sciencedirect.com/science/article/abs/pii/S2211339814000446
78. M.C. Galassi, D. Baraldi, B.A. Iborra, P. Moretto, CFD analysis of fast filling scenarios for 70 MPa hydrogen type IV tanks. Int J Hydrogen Energy., Elsevier 7 February (2012)
79. *Reason for negative Joule Thomson coefficient of Helium and Hydrogen at NTP conditions*. Stack Exchange Chemistry. https://chemistry.stackexchange.com/questions/61517/reason-for-negative-joule-thomson-coefficient-of-helium-and-hydrogen-at-ntp-cond. Accessed 24 Jan 2024
80. SAE International, Technical information report for fuel system in fuel cells and other hydrogen vehicles. SAE J **2579** (2009) revised January 2009
81. International Standard Organization. *Gaseous hydrogen and hydrogen blends—land vehicle fuel tanks*. ISO/TS 15869; 2009.
82. College Sidekick. *Non-Ideal gas behavior, 2024* https://www.collegesidekick.com/study-guides/sanjacinto-atdcoursereview-chemistry1-1/non-ideal-gas-behavior. Accessed 10 March 2024
83. R.C. Mathewson, Small liquid hydrogen dewars, design and use (1960), in *Advances in cryogenic engineering*, ed. by K.D. Timmerhaus, vol. 2, (Springer, Boston, MA, 2023). https://doi.org/10.1007/978-1-4684-3102-5_11. Accessed 15 December 2023
84. T. Zhang, J. Uratani, Y. Huang, L. Xu, S. Griffiths, Y. Ding, Hydrogen liquefaction and storage: recent progress and perspectives. Renew Sustain Energy Rev **176**, 113204 (2023) ISSN 1364-0321, https://www.sciencedirect.com/science/article/pii/S1364032123000606
85. Z.S. Saif et al., Hydrogen liquefaction: a review of the fundamental physics, engineering practice and future opportunities. Energy Environ Sci **15**(7), 2690–2731 (2022)

86. B. Sundén, Compressed cryogenic gas, in *Hydrogen, Batteries and Fuel Cells, 2019, Section 3.4.3.2*, (2024) Online at https://www.sciencedirect.com/topics/engineering/compressed-hydrogen. Accessed 12 April 2024
87. US Dept of Energy, Hydrogen and Fuel Cell Technologies Office. *Hydrogen Storage*. https://www.energy.gov/eere/fuelcells/hydrogen-storage. Accessed 15 December, 2023.
88. Lennie Klebanoff, Jay Keller, Sandia National Laboratory, DOE Metal Hydride Center of Excellence. *Final Report for the DOE Metal Hydride, 2012-0786*, 2012. Available online at https://www.energy.gov/eere/fuelcells/articles/final-report-doe-metal-hydride-center-excellence
89. B. Sakintuna, F. Lamari-Darkrim, M. Hirscher, Metal hydride materials for solid hydrogen storage: a review. Int J Hydrogen Energy **32**(9), 1121–1140 (2007). https://doi.org/10.1016/j.ijhydene.2006.11.022
90. US DOE *Metal Hydride Storage Materials*. Hydrogen and Fuel Cell Technologies Office. https://www.energy.gov/eere/fuelcells/metal-hydride-storage-materials
91. Wei Liu, Kondo-Francois Aguey-Zinsou. Low temperature synthesis of LaNi5 nanoparticles for hydrogen storage. Int J Hydrogen Energy, 41(3), 2016, Pgs 1679-1687. https://www.sciencedirect.com/science/article/abs/pii/S0360319915306443.
92. J.M. Pasini, C. Corgnale, B.A. van Hassel, T. Motyka, S. Kumar, K.L. Simmons, Metal hydride material requirements for automotive hydrogen storage systems. Int J Hydrogen Energy **38**(23), 9755–9765 (2013)
93. C.A. MacCarley, Development of a sodium borohydride hydrogen fuel storage system for vehicular applications, in *Symposium on Alternate Fuel Resources*, (American Association of Aeronautics and Astronautics), 1976
94. *Kipp's Apparatus*, invented in 1844 by Dutch pharmacist Petrius Kipp. https://digital.sciencehistory.org/works/7w4wlce. Accessed 20 February 2024
95. Bernard I. Robertson. *Sodium Borohydride—Fuel Cell Vehicle 'Natrium' uses clean, safe technology to provide hydrogen on demand*, June 10, 2002, Sindelfingen. Accessible online at https://media.stellantisnorthamerica.com/pdf.do?id=814
96. Wikipedia. *Chrysler Natrium* https://en.wikipedia.org/wiki/Chrysler_Natrium. Accessed 1 November 2024
97. Allpar.com *2001 fuel-cell minivan: Chrysler Town & Country Natrium*. https://www.allpar.com/d3/cars/concepts/natrium.html. Accessed 1 November 2024
98. W. Zhang, X. Zhang, Z. Huang, H.-W. Li, M. Gao, H. Pan, Y. Liu, Recent development of lithium borohydride-based materials for hydrogen storage. Adv Energy Sustain Res Open Access J **5** (2021). https://doi.org/10.1002/aesr.202100073. Accessed 1 March 2023
99. W.-S. Han, T.-J. Kim, S.-K. Kim, Y. Kim, Y. Kim, S.-W. Nam, S.O. Kang, Silane-based hydrogen storage materials for fuel cell application: hydrogen release via methanolysis and regeneration by hydride reduction from organosilanes. Int J Hydrogen Energy **36**(19), 12305–12312 (2011) https://www.sciencedirect.com/science/article/pii/S0360319911015904. Accessed 20 June 2024
100. Mahdi M. Abu-Omar, Guodong Du, Rex Corbin, Erin Smit *Silane Activation by Transition Metal Catalysts for Hydrogen Storage*. Report to the US Dept of Energy, Hydrogen Technology Office, 2008. https://www.hydrogen.energy.gov/docs/hydrogenprogramlibraries/pdfs/review07/besp_1_abu-omar.pdf
101. J.M. Brunel, *Polysilanes: The Grail for a Highly-Neglected Hydrogen Storage Source* (Hydrogen Energy Publications LLC., Elsevier Ltd., 2017), pp. 0360–3199. https://doi.org/10.1016/j.ijhydene
102. Wikipedia. Pentaborane. https://en.wikipedia.org/wiki/Pentaborane(9). Accessed 20 June 2024.
103. Takato Mitsudome, Teppei Urayama, Taizo Kiyohiro, Zen Maeno, Tomoo Mizugaki, Koichiro Jitsukawa, Kiyotomi Kaneda. *On-demand Hydrogen Production from Organosilanes at Ambient Temperature Using Heterogeneous Gold Catalysts*. http://www.nature.com/scientificreports/ November 2016.

104. K. Mori, M. Tano, T. Mizugaki, K. Ebitani, K. Kaneda, Efficient heterogeneous oxidation of organosilanes to silanols catalysed by a hydroxyapatite-bound Ru complex in the presence of water and molecular oxygen. New J Chem **26**, 1536–1538 (2002) Referenced in citation 107
105. Fraunhofer-Gesellschaft – Research News, 1 March 2022. *Obtaining hydrogen from methanol: Optimized reformers.* https://www.fraunhofer.de/en/press/research-news/2022/march-2022/obtaining-hydrogen-from-methanol.html. Accessed 1 May 2024
106. A.I. Osman, M. Nasr, E. Lichtfouse, et al., Hydrogen, ammonia and methanol for marine transportation. Environ Chem Lett **22**, 2151–2158 (2024). https://doi.org/10.1007/s10311-024-01757-9
107. G. Bozzano et al., Efficient methanol synthesis: perspectives, technologies and optimization strategies. Prog Energy Combust Sci **56**, 71–105 (2016)
108. Blue World Technologies. Fueling the Future with Methanol Fuel Cells I Hydrogen + Fuel Cells EUROPE. Presentation at H2FC Hannover, 2023. https://www.youtube.com/watch?v=ZQezOmL3Owo
109. J. Finegold, G. Glinski, G. Voecke *Disassociated Methanol Citation: Final Report.* US DOE, Solar Energy Research Institute, August 1984. SERI/TR-235-2083.
110. Fraunhofer Research News. 1 March 2022. Hydrogen technology. *Obtaining Hydrogen from Methanol: Optimized Reformers.* Accessed 1 Sept 2024
111. Stephen H. Crolius, Ammonia Energy Association. *NH3 vs. MCH: Energy Efficiency of Hydrogen Carriers Compared.* 9 May 2019. https://ammoniaenergy.org/articles/nh3-vs-mch-energy-efficiency-of-hydrogen-carriers-compared/. Accessed 10 May 2024
112. Kishore Kumar Sriramoju. Pressure exerted by ammonia at room temperature. Forum conversation 30 December 2021. https://www.researchgate.net/post/Pressure_exerted_by_ammonia_at_room_temperature. Accessed 1 May 2024
113. Hans Pasman. *Industrial Processing Systems, Their Products and Hazards.* Section 1.3 in Anhydrous Ammonia/Risk Analysis and Control for Industrial Processes—Gas, Oil and Chemicals, 2015. Also available online at https://www.sciencedirect.com/topics/engineering/anhydrous-ammonia
114. Leigh Collins, Hydrogen Insight, 22 May 2024. *Japanese rail operator eyes hydrogen trains fuelled by onboard liquid organic H_2 carrier.* https://www.hydrogeninsight.com/innovation/japanese-rail-operator-eyes-hydrogen-trains-fuelled-by-onboard-liquid-organic-h2-carrier/2-1-1647841. Accessed 23 March 2024
115. ENEOS Group, Japan. *Direct MCH® ~ one-step chemical hydride production using renewable energy.* https://www.eneos.co.jp/english/company/rd/intro/low_carbon/dmch.html. Accessed 12 May 2024.
116. P.J. Ansell, Review of sustainable energy carriers for aviation: benefits, challenges, and future viability. Prog Aerospace Sci **141**, 100919 (2023) ISSN 0376-0421
117. Airlines for America. *Deployment of sustainable aviation fuel in the United States* 25 September 2020. https://www.airlines.org/media/deployment-of-sustainable-aviation-fuel-in-the-united-states-2/. Accessed 10 April 2024
118. S. Tiwari, M.J. Pekris, J.J. Doherty, A review of liquid hydrogen aircraft and propulsion technologies. Int J Hydrogen Energy **57**, 1174–1196 (2024) https://www.sciencedirect.com/science/article/pii/S0360319923065631
119. Mark Piesing, BBC. *The epic attempts to power planes with hydrogen.* 21 March 2022 https://www.bbc.com/future/article/20220316-the-epic-attempts-to-power-planes-with-hydrogen. Accessed 24 March 2024
120. S. Tiwari, M.J. Pekris, J.J. Doherty, A review of liquid hydrogen aircraft and propulsion technologies. Int J Hydrogen Energy **57**, 1174–1196 (2024). https://doi.org/10.1016/j.ijhydene.2023.12.263
121. C. Bergero, G. Gosnell, D. Gielen, S. Kang, M. Bazilian, S.J. Davis, Pathways to net-zero emissions from aviation. Nat Sustain **6**(4), 404–414 (2023)
122. Katie Camero, Science 28 Jun 2019. *Aviation's dirty secret: Airplane contrails are a surprisingly potent cause of global warming. Warming effect of thin, white clouds will triple by*

2050. https://www.science.org/content/article/aviation-s-dirty-secret-airplane-contrails-are-surprisingly-potent-cause-global-warming. Accessed 12 April 2024.
123. Hannah Ritchie. Our World in Data, 2024. *What share of global CO2 emissions come from aviation? Non-CO2 climate impacts mean aviation accounts for around 4% of global warming to date* Online at https://ourworldindata.org/global-aviation-emissions. Accessed 10 April 2024.
124. M. Klöwer, M. Allen, D. Lee, S. Proud, L. Gallagher, A. Skowron, Quantifying aviation's contribution to global warming. Environ Res Lett (2021) as reported in citation[127]
125. US Dept of Energy, Bioenergy Technologies Office. *Sustainable Aviation Fuels.* https://www.energy.gov/eere/bioenergy/sustainable-aviation-fuels. Accessed 10 April 2024
126. US Dept of Energy, Bioenergy Technologies Office. *Sustainable aviation fuel grand challenge.* https://www.energy.gov/eere/bioenergy/sustainable-aviation-fuel-grand-challenge. Accessed 10 April 2024
127. Elan Head, The Air Current. *Understanding the path to 100% SAF.* 13 April 2022. https://theaircurrent.com/technology/path-to-100-saf-sustainable-aviation-fuel/. Accessed 10 April 2024
128. Airbus Commercial Aircraft. *First A380 powered by 100% sustainable aviation fuel takes to the skies,* 28 March 2022. https://www.airbus.com/en/newsroom/press-releases/2022-03-first-a380-powered-by-100-sustainable-aviation-fuel-takes-to-the. Accessed 10 April 2024
129. J.-K. Lin et al., The properties of sustainable aviation fuel. Energy **283**, 129125. Table 1. 15 November 2023, https://www.sciencedirect.com/science/article/pii/S0360544223025197. Accessed 10 April 2024
130. Properties of fuels from *Bosch Automotive Handbook*, 10th ed. 2018 Distributed by SAE International. pp. 345–6
131. Mahler AGS GmbH. *Hydrogen generation by steam reforming,* 2019. https://www.youtube.com/watch?v=xAjHJ49VOUM. Accessed 51 October 2024.
132. H. Song, S. Luo, H. Huang, B. Deng, J. Ye, Solar-driven hydrogen production: recent advances, challenges, and future perspectives. ACS Energy Lett **2022, 7**(3), 1043–1065. https://doi.org/10.1021/acsenergylett.1c02591. Accessed 1 March 2024
133. International Energy Agency. *Direct Air Capture, A key technology for net zero, updated October 2024.* May be obtained online at https://iea.blob.core.windows.net/assets/9766b4da-a5e3-4d76-874d-ea286e333956/DirectAirCapture_Akeytechnologyfornetzero.pdf
134. Justin Freeman, Chevron Oil Co., *Chevron natural gas carbon capture technology testing project*, Cooperative Agreement No. DE-FE0031944, 13 August 2021. https://netl.doe.gov/sites/default/files/netl-file/21CMOG_PSC_Freeman.pdf. Accessed 2 April 2024.
135. US Department of Energy, From Hydrogen Shot to Hydrogen Hubs—Hydrogen Program Annual Merit Review and Peer Evaluation Meeting, June 5, 2023. Meeting presentation online at https://www.hydrogen.energy.gov/docs/hydrogenprogramlibraries/pdfs/review23/plenary2_hydrogen_program_panel_2023_o.pdf?Status=Master. Accessed 2 April 2024
136. D.C. Rode, J.J. Anderson, H. Zhai, P.S. Fischbeck, Six principles to guide large-scale carbon capture and storage development. Energy Res Soc Sci **103**, 103214., ISSN 2214-6296 (2023). https://doi.org/10.1016/j.erss.2023.103214
137. (US) Congressional Research Service DOE's Carbon Capture and Storage (CCS) and Carbon *Removal Programs*, updated 19 January 2024. https://crsreports.congress.gov/product/pdf/IF/IF11861. Accessed 2 April 2024
138. Winset Farms, Santa Maria, California. *The Way We Grow* https://windsetfarms.com/the-way-we-grow/ Also observed during personal facility tour, 2017.
139. S. Taku Ide, S. Julio Friedmann, H.J. Herzog, *CO_2 leakage through existing wells: current technology and regulations* (Joint Publication of Stanford University, MIT and Lawrence Livermore National Laboratory. http://sequestration.mit.edu/pdf/GHGT8_Ide.pdf) Accessed 2 April 2024
140. US EPA, Office of Water (4606M) *Geologic Sequestration of Carbon Dioxide Draft Underground Injection Control (UIC) Program Guidance on Transitioning Class II Wells to Class VI Wells.* EPA 816-P-13-004 December 2013. Archive document available online at

https://archive.epa.gov/epa/sites/production/files/2020-10/documents/gs_transitioning_class-ii_clas_vi_epa816p13004_508corrected.pdf
141. Carbfix. *We turn CO2 into stone.* https://www.carbfix.com/ Accessed 1 October 2024.
142. David Schlissel, Dennis Wamsted, Seth Feaster, Suzanne Mattei, Omar Mawji and Tom Sanzillo, Institute for Energy Economics and Financial Analysis. *Blue Hydrogen Technology Challenges, Weak Commercial Prospects, and Not Green.* February 2022. https://ieefa.org/resources/blue-hydrogen-technology-challenges-weak-commercial-prospects-and-not-green Accessed 10 April 2024
143. First Element Fuels True Zero, *Environment.* (2024). https://www.truezero.com/videos/environment/. Accessed 30 March 2024
144. Hydrogen Insight, 1 Dec 2022. *'Climate penalty': High levels of blue hydrogen use could wipe out carbon reduction, or even make it worse.* https://www.hydrogeninsight.com/policy/climate-penalty-high-levels-of-blue-hydrogen-use-could-wipe-out-carbon-reduction-or-even-make-it-worse-study/2-1-1364843 Accessed 1 Feb 2024
145. First Element Fuels, True Zero. *How it works.* (2024) https://www.truezero.com/videos/how-it-works/ Accessed 30 March 2024
146. CSIRO, Australia's National Science Agency. *How the Hydrogen Refueling Station Works.* 20 December 2023. https://www.csiro.au/en/about/challenges-missions/Hydrogen/Hydrogen-Refuelling-Station. Accessed 3 March 2024.
147. Sierra Club International web page. *How Clean is Green Hydrogen.* https://www.sierraclub.org/sierra/2023-3-fall/feature/how-clean-is-green-hydrogen. Accessed 15 December 2023
148. David Schlissel, Dennis Wamsted, Suzanne Mattei, Omar Mawji *Reality Check on CO2 Emissions Capture at Hydrogen-From-Gas Plants*, February 2022 https://ieefa.org/wp-content/uploads/2022/02/Reality-Check-on-CO2-Emissions-Capture-at-Hydrogen-From-Gas-Plants_February-2022.pdf Accessed 5 April 2024
149. Jimmy Cloutier. *Burgeoning hydrogen industry draws $41 million in federal lobbying from fossil fuel companies.* Open Secrets—Following the Money in Politics, 12 December 2023. https://www.opensecrets.org/news/2023/12/hydrogen-industry-draws-millions-in-federal-lobbying-from-fossil-fuels. Accessed 1 May 2024
150. Julie McNamera, The Equation, 9 December 202. *What's the role of hydrogen in the clean energy transition?* Union of Concerned Scientists https://blog.ucsusa.org/julie-mcnamara/whats-the-role-of-hydrogen-in-the-clean-energy-transition/ Accessed 10 April 2024
151. R.W. Howarth, M.Z. Jacobson, How green is blue hydrogen? Energy Sci Eng. **9**, 1676–1687 (2021). https://doi.org/10.1002/ese3.956
152. An alternative definition of pink hydrogen, appearing in most web search results, e.g., *National Grid,* https://www.nationalgrid.com/stories/energy-explained/hydrogen-colour-spectrum Accessed 10 November 2023
153. California Public Utilities Commission, decision 21-06-005 June 4, 2021, COM/CR6/jnf *Order instituting rulemaking regarding policies, procedures and rules for the self-generation incentive program and related issues.* Rulemaking 20-05-012. pp. 34–36
154. Eric Hand, Science, 16 February 2023. *Hidden hydrogen: Does Earth hold vast stores of a renewable, carbon-free fuel?* https://www.science.org/content/article/hidden-hydrogen-earth-may-hold-vast-stores-renewable-carbon-free-fuel Accessed 1 May 2024
155. US Congress. *H.R.5376—inflation reduction act of 2022.* https://www.congress.gov/bill/117th-congress/house-bill/5376/text
156. US DOE, Energy Efficiency and Renewable Energy (EERE), Hydrogen and Fuel Cell Technologies Office. *Hydrogen Shot Program.* 7 June 2021. https://www.energy.gov/eere/fuelcells/hydrogen-shot. Accessed 2 April, 2024
157. Jijo Malayil, Interesting Engineering, Nov 02, 2023. *North America's largest green hydrogen plant now operational* https://interestingengineering.com/innovation/north-americas-largest-green-hydrogen-plant-now-operational Accessed 15 December 2023.
158. US EIA report, 2022. *Coal was the largest source of electricity generation for 15 states in 2021.* https://www.eia.gov/todayinenergy/detail.php?of=54919 Accessed 1 July 2023

159. US EIA. *California state profile and energy estimates* (2023). https://www.eia.gov/state/analysis.php?sid=CA Accessed 1 May 2024
160. United States Geological Survey (USGS). *The potential for geologic hydrogen for next generation energy.* https://www.usgs.gov/news/featured-story/potential-geologic-hydrogen-next-generation-energy Accessed 10 January 2024
161. US Geological Survey (USGS). *The potential for geologic hydrogen for next-generation energy.* https://www.usgs.gov/news/featured-story/potential-geologic-hydrogen-next-generation-energy Accessed 10 January 2024
162. US Geological Survey (USGS) online presentation, 21 December 2023. *Geologic hydrogen: global occurrences and case studies.* https://duckduckgo.com/?t=ffab&q=geologic+hydrogen&atb=v313-1&iax=videos&ia=videos&iai=https%3A%2F%2Fwww.youtube.com%2Fwatch%3Fv%3DHU1ExScjDrQ. Accessed 4 March 2024
163. Polly Martin, HydrogenInsight, 31 July 2023. *What level of greenhouse gas emissions can be expected when extracting natural hydrogen?* https://www.hydrogeninsight.com/production/what-level-of-greenhouse-gas-emissions-can-be-expected-when-extracting-natural-hydrogen-/2-1-1493298
164. Hydrogen Insight (online) *What Level of Greenhouse Gas Emissions Can be Expected from Natural Hydrogen,* 2023 https://www.hydrogeninsight.com/production/what-level-of-greenhouse-gas-emissions-can-be-expected-when-extracting-natural-hydrogen-/2-1-1493298 Accessed 1 March 2024
165. US Advanced Research Projects Agency-Energy (ARPA-E). *Geologic Hydrogen (abstracts from awardees of research grants on geologic hydrogen)* https://arpa-e.energy.gov/technologies/exploratory-topics/geologic-hydrogen Accessed 1 May 2024
166. *Hidden hydrogen: Does Earth hold vast stores of a renewable, carbon-free fuel?* Energy Today, published by the U.S. Energy Information Agency, 16 Feb 2023. https://www.eia.gov/todayinenergy/ Accessed 4 March 2024
167. Edgar Meza, Benjamin Wehrmann. *Natural hydrogen discoveries near French-German border fuel hopes for H2 abundance* Clean Energy Wire, translated report from Suddeutsche Zeitung, 16 January 2024. Accessed 5 March 2024
168. Leigh Collins, HydrogenInsight, 1 November 2023. *'Significant concentrations' of natural hydrogen detected at Australia's first exploration well—raising hopes of world-first commercial exploitation* https://www.hydrogeninsight.com/production/significant-concentrations-of-natural-hydrogen-detected-at-australias-first-exploration-well-raising-hopes-of-world-first-commercial-exploitation/2-1-1546424 Accessed 30 March 2024
169. James Dinneen, NewScientist *Huge deposit of natural hydrogen gas detected deep in Albanian mine* https://www.newscientist.com/article/2416060-huge-deposit-of-natural-hydrogen-gas-detected-deep-in-albanian-mine/. Accessed 24 April 2024
170. G.S. Ellis, S.E. Gelman, Model predictions of global geologic hydrogen resources. Sci Adv **10**(50) (2024). https://doi.org/10.1126/sciadv.ado0955
171. US Congressional Research Service. *Injection and geologic sequestration of carbon dioxide: Federal role and issues for congress.* 22 September 2022. Available online at https://crsreports.congress.gov/product/pdf/R/R46192
172. Wood MacKenzie Consulting. *Over the rainbow: Why understanding full value-chain carbon intensity is trumping the colour of hydrogen 2024.* https://www.woodmac.com/horizons/hydrogen-colours-value-chain-carbon-intensity/. Accessed 30 January 2024
173. Hydrogen Insight, 7 June 2023. *Hydrogen is a more potent greenhouse gas than previously reported, new study reveals.* https://www.hydrogeninsight.com/analysis/hydrogen-is-a-more-potent-greenhouse-gas-than-previously-reported-new-study-reveals/2-1-1463495 Accessed 1 Feb 2024
174. M. Sand, R.B. Skeie, M. Sandstad, S. Krishnan, G. Myhre, H. Bryant, R. Derwent, D. Hauglustaine, F. Paulot, M. Prather, D. Stevenson, A multi-model assessment of the global warming potential of hydrogen. Nat J Commun Earth Environ (2024). https://doi.org/10.1038/s43247-023-00857-8. Accessed 28 March 2024

175. I. Ocko, S. Hamburg, Climate consequences of hydrogen emissions. Eur Sci Union Atmos Chem Phys **22**(14), 20 (2022)
176. Cost of automotive fuel cells from National Renewable Energy Laboratory study using several different FCVs. Automotive fuel cell stack and system efficiency and fuel consumption based on vehicle testing on a chassis dynamometer at minus 18 °C to positive 35 °C temperatures. Int J Hydrogen Energy **45**(1), 861–872 (2020) Accessible online at https://www.sciencedirect.com/science/article/abs/pii/S0360319919339618
177. California Energy Commission, Clean Transportation Program. FINAL PROJECT REPORT Southern California Hydrogen Fill System, Delivery Trailers, and Fill System Automation ARV-10-048 and ARV-12-059. https://www.energy.ca.gov/sites/default/files/2021-04/CEC-600-2021-016.pdf
178. US Energy Info Admin (EIA) Today in Energy 21 July 2020. *More than 60% of energy used for electricity generation is lost in conversion* https://www.eia.gov/todayinenergy/detail.php?id=44436. Accessed 10 April 2024
179. Out World in Data (online). *Carbon intensity of electricity generation, 2022* https://ourworldindata.org/grapher/carbon-intensity-electricity?region=NorthAmerica#explore-the-data
180. Engineering Toolbox *Combustion of Fuels – Carbon Dioxide Emissions*. https://www.engineeringtoolbox.com/co2-emission-fuels-d_1085.html Accessed 12 May 2024.
181. Robert Bosch GmbH. *Automotive Handbook*, 10th ed. Distr by SAE International. p.346
182. T. Terlouw, C. Bauer, R. McKenna, M. Mazzotti, Large-scale hydrogen production via water electrolysis: a techno-economic and environmental assessment. Energy Environ Sci (Analysis) **15**, 3583–3602 (2022). https://doi.org/10.1039/D2EE01023B. Royal Society of Chemistry. https://pubs.rsc.org/en/content/articlehtml/2022/ee/d2ee01023b Accessed 10 April 2024
183. IEA (2023) *Comparison of the emissions intensity of different hydrogen production routes, 2021* https://www.iea.org/data-and-statistics/charts/comparison-of-the-emissions-intensity-of-different-hydrogen-production-routes-2021 Accessed 2 April 2024.
184. N. Muradov, Low carbon production of hydrogen: steam methane reforming with CCS, in *Compendium of Hydrogen Energy, V.1 Hydrogen Production and Purification*, (Elseiver Ltd., 2015). https://doi.org/10.1016/C2014-0-02671-8. https://www.sciencedirect.com/book/9781782423614/compendium-of-hydrogen-energy. Accessed 10 April 2024
185. I.I. Komarov et al., Comparative analysis of the efficiency of using hydrogen and steam methane reforming storage at combined cycle gas turbine for cogeneration. J Phys: Conf Ser **2053** (2021). https://doi.org/10.1088/1742-6596/2053/1/012007/pdf
186. N. Muradov, Low carbon production of hydrogen: steam methane reforming with CCS, in *Compendium of Hydrogen Energy, V.1 Hydrogen Production and Purification*, (Elseiver Ltd., 2015) https://www.sciencedirect.com/book/9781782423614/compendium-of-hydrogen-energy. Accessed 10 April 2024
187. MIT Climate Patrol. *How efficient is carbon capture and storage?* 23 February 2021 https://climate.mit.edu/ask-mit/how-efficient-carbon-capture-and-storage Accessed 2 April 2024
188. Climate Policy Watcher. *Efficiency parameters of CCS* 21 Mar 2023 https://www.climate-policy-watcher.org/energy-consumption/efficiency-parameters-of-ccs.html Accessed 2 April 2024
189. FossilFuel.com 1 Jan 2020. *How is natural gas refined?* https://fossilfuel.com/how-is-natural-gas-refined/ Accessed 10 January 2024
190. Bob Schildgen, Sierra Club. 8 January 2017 *Hey Mr. Green! How much CO2 is generated by producing and transporting a gallon of gas?* https://www.sierraclub.org/sierra/ask-mr-green/hey-mr-green-how-much-co2-generated-producing-and-transporting-gallon-gas Accessed 10 April 2024
191. Michael Levi, Council on Foreign Relations, Energy, Security, and Climate. *Do gasoline based cars really use more electricity than electric vehicles do*? 26 October 2011. https://www.cfr.org/blog/do-gasoline-based-cars-really-use-more-electricity-electric-vehicles-do Accessed 2 May 2024.

192. Jordan Wirfs-Brock, InsideEnergy, 6 November 2015. *Lost in transmission: how much electricity disappears between a power plant and your plug?* https://insideenergy.org/2015/11/06/lost-in-transmission-how-much-electricity-disappears-between-a-power-plant-and-your-plug/ Accessed 20 April 2024
193. Jiguparmar. *Total losses in Power Distribution and Transmission Lines*, 9 April 2013 https://electrical-engineering-portal.com/total-losses-in-power-distribution-and-transmission-lines-1 Accessed 10 April 2024
194. OECD/IEA *2018 United States – Electric power transmission and distribution losses.* http://www.iea.org/stats/index.asp
195. American Chemical Society, Limitations of ammonia as a hydrogen energy carrier for the transportation sector. ACS Energy Lett **6**(12), 4390–4394., November 15, 2021 (2021). https://doi.org/10.1021/acsenergylett.1c02189
196. Go-e 10 Jan 2023 *EV Charging Efficiency: Why Are There Energy Losses and How to Avoid Them?* https://go-e.com/en/magazine/ev-charging-losses Accessed 20 April 2024
197. T. Terlouw, C. Bauer, R. McKenna, M. Mazzotti, Large-scale hydrogen production via water electrolysis: a techno-economic and environmental assessment. Energy Environ Sci **15**(9), 2583–3602. https://doi.org/10.1039/D2EE01023B. https://pubs.rsc.org/en/content/articlehtml/2022/ee/d2ee01023b. Accessed 10 April 2024
198. PetroWiki, last edited 2 June 2015. *Compressed natural gas (CNG)* https://petrowiki.spe.org/Compressed_natural_gas_(CNG) Accessed 20 April 2024
199. F. Tong, I. Azevedo, P. Jaramillo, Economic viability of a natural gas refueling infrastructure for long-haul trucks. J Infrastruct Syst **25**(1) (2018). https://doi.org/10.1061/(ASCE)IS.1943-555X.0000460, https://doi.org/10.1061/%28ASCE%29IS.1943-555X.0000460
200. A.A. du Plessis, M.J. Booysen, Electric vehicle drivetrain efficiency and the multi-speed transmission question. World Electric Vehicle J **2023, 14**(12), 342. https://doi.org/10.3390/wevj14120342. https://www.mdpi.com/2032-6653/14/12/342, Accessed 20 April 2024
201. Mark Crittenden, IEEE Spectrum 19 Aug 2020, *With Ultralight Lithium-Sulfur Batteries, Electric Airplanes Could Finally Take Off.* https://spectrum.ieee.org/with-ultralight-lithiumsulfur-batteries-electric-airplanes-could-finally-take-off Print and online. Accessed 5 April 2024
202. Daniel Barba *Assessing the Efficiency Potential of Future Gasoline Engines* SAE 2018 High Efficiency IC Engine Symposium, https://www.epa.gov/sites/default/files/2018-10/documents/high-efficiency-ic-engine-sae-2018-04.pdf Accessed 1 March 2024.
203. US EPA, FuelEconomy.gov *Fuel Economy rating for 2015 Honda Civic NGV, 1.8L, 4 cyl, manual.* https://www.fueleconomy.gov/feg/Find.do?action=sbs&id=35886. Accessed 20 April 2024.
204. US EPA. Fuel Economy.gov *Fuel Economy rating for 2023 Honda Civic 4dr 2.0L, 4 cyl, automatic.* https://www.fueleconomy.gov/feg/Find.do?action=sbs&id=46506 Accessed 20 April 2024.
205. US EIA. *Carbon intensity of U.S. power generation continues to fall but varies widely by state, September 2022* https://www.eia.gov/todayinenergy/detail.php?id=53819. Accessed 1 May 2024
206. International Energy Agency (IEA). *Global EV outlook 2023.* https://www.iea.org/reports/global-ev-outlook-2023 Accessed 5 May 2023
207. Bret Williams. How safe are hydrogen fuel cell cars in a crash? Hydrogen Fuel News, May 22, 2023 https://www.hydrogenfuelnews.com/how-safe-are-hydrogen-fuel-cell-vehicles-in-a-crash/8539783/ Accessed 15 March 2024
208. Inside EVs *Watch How Euro NCAP Crashes Hydrogen Toyota Mirai.* September 2021. https://insideevs.com/news/531831/euroncap-crashes-hydrogen-toyota-mirai/ Accessed 1 November 2023
209. National Highway Transportation Safety Administration. (2020) *Hyundai Nexo*, Fire Risk Warning, NHTSA recall numbers 24V761000 and 22V258000, October 2024. https://www.nhtsa.gov/vehicle/2020/HYUNDAI/NEXO/SUV/FWD#recalls

210. Both accidents described in Agnete Klevstrand, Hydrogen Insight, *Explosion after explosion' | Three injured after truck towing 420kg of hydrogen crashes and explodes in Ohio.* 7 February 2023 https://www.hydrogeninsight.com/transport/explosion-after-explosion-three-injured-after-truck-towing-420kg-of-hydrogen-crashes-and-explodes-in-ohio/2-1-1400104 Accessed 1 June 2024
211. International Energy Agency (IEA). *World Energy Outlook, 2024.* https://www.iea.org/reports/world-energy-outlook-2024
212. Fred Lambert, Electrek. *Hydrogen station explodes, Toyota halts sales of fuel cell cars, is this the end? 11* June 2019. https://electrek.co/2019/06/11/hydrogen-station-explodes-toyota-halts-sales-fuel-cell-cars/ Accessed 15 July 2019
213. Historical data from *Hindenburg Design and Technology*, https://www.airships.net/hindenburg/hindenburg-design-technology/. Accessed 25 July 2022
214. Specifications from *Hindenburg Design and Technology* https://www.airships.net/hindenburg/size-speed/. Accessed 25 July 2022
215. Christopher Klein, *The Hindenburg Disaster: 9 surprising Facts*. History (online) 2 October, 2023. https://History.com/news/the-hindenburg-disaster-9-surprising-facts/. Accessed 1 March 2024
216. Trevin Casinader, Aviation Insider. Airbus A320 Guide and Specs: The World's Highest-Selling Narrowbody Airplane Brands. https://aviatorinsider.com/airplane-brands/airbus-a320/ Accessed 1 November 2024.
217. *Alternative Fueling Station Counts by State*, US Dept of Energy Alternative Fuels Data Center, https://afdc.energy.gov/stations/states accessed March 10, 2024
218. *Hydrogen Refueling Stations in California.* California Energy Commission, https://www.energy.ca.gov/data-reports/energy-almanac/zero-emission-vehicle-and-infrastructure-statistics/hydrogen-refueling. Accessed 15 Nov 2023
219. Al-Ahmed, S. Hossain, B. Mukhtar, S. U. Rahman, H. Abualhamayel,J. Zaidi, Hydrogen highway: an overview. in 2010 IEEE International Energy Conference, Manama, Bahrain, 2010, pp. 642-647, doi: https://doi.org/10.1109/ENERGYCON.2010.5771760.
220. M. Genovese, P. Fragiacomo, Hydrogen refueling station: Overview of the technological status and research enhancement. J Energy Storage **61**, 106758 (2023)
221. Anela Dokso, H_2 Energy News, 4 April 2024. *California's hydrogen highway difficulties: limited stations and $200 fill up costs.* https://energynews.biz/californias-hydrogen-highway-difficulties-limited-stations-and-200-fill-up-costs/?amp=1. Accessed 20 April 2024.
222. Evan Halper, Los Angeles Times, 10 August 2021. *Is California's 'Hydrogen Highway' a road to nowhere?* Online at https://www.latimes.com/politics/story/2021-08-10/hydrogen-highway-or-highway-to-nowhere Accessed 28 March 2024.
223. Hydrogen Central. *Powertap enters into an exclusive middle east distribution agreement deploying more than 100 hydrogen refuelling stations.* August 5, 2021 https://hydrogen-central.com/powertap-middle-east-distribution-100-hydrogen-refuelling-stations/ Accessed 2 April 2024
224. CSIRO, Australia's National Science Agency. *How the Hydrogen Refueling Station Works.* 20 December 2023. https://www.csiro.au/en/about/challenges-missions/Hydrogen/Hydrogen-Refuelling-Station Accessed 3 March 2024
225. Steve Hanley, CleanTechnica, February 2024. *Shell shuts down its us hydrogen filling stations* https://cleantechnica.com/2024/02/10/shell-shuts-down-its-us-hydrogen-filling-stations/ Accessed 4 April 2024
226. US Department of Energy, Office of Energy Efficiency and Renewable Energy, Hydrogen and Fuel Cell Technologies Office. *Hydrogen Production.* https://www.energy.gov/eere/fuelcells/hydrogen-production Accessed 18 April 2024
227. Alan Ohnsman, Forbes 16 April 2024. *Green hydrogen's hype hits some very expensive hurdles.* https://www.forbes.com/sites/alanohnsman/2024/04/16/green-hydrogens-hype-hits-some-very-expensive-hurdles/ Accessed 18 April 2024
228. Todd Woody, Bloomberg Green, 6 April 2024 *Fill'er up … for $200?* and 4 April 2024 *Few Stations and $200 to Fill Up: Life on California's 'Hydrogen Highway'* https://www.bloom-

berg.com/news/features/2024-04-04/california-s-hydrogen-fuel-cell-cars-lose-traction-against-battery-models?srnd=green. Accessed 6 April 2024
229. Polly Martin, Hydrogen Insight, 23 February 2024. *Toyota is now offering a 60% discount on this hydrogen-powered car in the US.* https://www.hydrogeninsight.com/transport/toyota-is-now-offering-a-60-discount-on-this-hydrogen-powered-car-in-the-us/2-1-1602906. Accessed 10 April 2024.
230. Data and comments from personal experience. Confirmed in Car and Driver Magazine, *2015 Hyundai Tucson Fuel Cell: Hydrogen-Powered Motoring for $499 a Month Fuel's included too (Well, until the lease is up.)* 20 Nov 2013. Also online at https://www.caranddriver.com/news/2015-hyundai-tucson-fuel-cell-photos-and-info-news
231. *Inside EVs*: *Official: Honda Discontinues Clarity PHEV And Clarity Fuel Cell. 16 June 2021.* https://insideevs.com/news/514412/honda-discontinues-clarity-phev-fcv/. Accessed 15 July 2021.
232. American Honda Motor Company, *The CR-V Meets Alternative Energy. 2024.* https://automobiles.honda.com/future-cars/cr-v-fcev. Accessed 10 February 2024.
233. Julie Campbell, Hydrogen News—Green Hydrogen Report. *What makes Toyota's hydrogen combustion engine special?* 5 January 2023. https://www.hydrogenfuelnews.com/hydrogen-combustion-engine-toyota-2/8556753/
234. Priyanka Dasgupta, E-Tech, 8 February 2024. *Balancing the grid with hydrogen storage.* https://etech.iec.ch/issue/2024-01/balancing-the-grid-with-hydrogen-storage Accessed 20 February 2024.
235. Hans Noordsij. *The actual efficiency of electrolysis* April 25, 2021 https://www.linkedin.com/pulse/actual-efficiency-electrolysis-hans-noordsij Accessed 15 Febrary 2024.
236. Philipp Lettenmeier. *Efficiency—Electrolysis* Siemens Energy AG, January 2019. https://assets.siemens-energy.com/siemens/assets/api/uuid:5342163d-2333-4c8d-ae85-2a0e8d45db56/white-paper-efficiency-en.pdf. Accessed 1 February 2024.
237. US Department of Energy. *Comparison of fuel cell technologies, 2018.* Also available online at https://www.energy.gov/eere/fuelcells/comparison-fuel-cell-technologies
238. M. Al-Shafi, O. Massarweh, A.S. Abushaikha, Y. Bicer, A review on underground gas storage systems: Natural gas, hydrogen and carbon sequestration. Energy Rep **9**., ISSN 2352-4847, 6251–6266 (2023)
239. Wei Liu, Zhixin Zhang, Jinyang Fan, Deyi Jiang, Zhengyi Li, Jie Chen. Research on gas leakage and collapse in the cavern roof of underground natural gas storage in thinly bedded salt rocks. J Energy Storage 2020., Elsevier.
240. Hydrogen Forward. *Understanding hydrogen leakage today and tomorrow.* 4 August 2022. https://www.hydrogenfwd.org/understanding-hydrogen-leakage-today-and-tomorrow/
241. K. Mongird, V. Viswanathan, J. Alam, C. Vartanian, V. Sprenkle, R. Baxter, *2020 Grid energy storage technology cost and performance assessment* (U.S. Department of Energy, 2020) https://www.energy.gov/energy-storage-grand-challenge/downloads/2020 grid-energy-storage-technology-cost-and-performance. Accessed 20 June 2023
242. Hayley Smith, Los Angeles Times, Climate California, 13 October 2023. *California to get up to $1.2 billion in federal funds to expand hydrogen energy projects.* https://www.latimes.com/environment/story/2023-10-13/california-gets-1-2-billion-for-hydrogen-power-development. Accessed 20 April 2024.
243. M. Yue, H. Lambert, E. Pahon, R. Roche, S. Jemei, D. Hissel, Hydrogen energy systems: a critical review of technologies, applications, trends and challenges. Renew Sustain Energy Rev **146**, 111180., ISSN 1364-0321 (2021)
244. Green Hydrogen Coalition. Green hydrogen: A game changer in the fight for our climate. 2024. https://www.ghcoalition.org/. Accessed 3 April 2024.
245. Cara Fogler, Sierra Club. Hydrogen: future of clean energy or a false solution? 4 January 2022 https://www.sierraclub.org/articles/2022/01/hydrogen-future-clean-energy-or-false-solution Accessed 1 July 2023

246. Earth Justice, Reclaiming hydrogen for a renewable future: distinguishing oil & gas industry spin from zero-emission solutions, 31 August 2021, https://earthjustice.org/feature/green-hydrogen-renewable-zero-emission. Accessed 1 July 2022
247. Environmental Defense Fund. *Hydrogen could have much bigger climate impact than most estimates, study shows.* February 21, 2024. https://www.edf.org/media/hydrogen-could-have-much-bigger-climate-impact-most-estimates-study-shows. Accessed 25 March 2024
248. Sarah Lutz, Friends of the earth. *The Hydrogen Hustle.* https://foe.org/wp-content/uploads/2024/03/HydrogenPollutionWIshlist_Climate_Report-1.pdf Accessed 2 April 2024
249. Julie McNamara, Union of Concerned Scientists. *The equation: UCS testimony on the clean hydrogen production Tas credit*, 24 March 2024. https://blog.ucsusa.org/julie-mcnamara/ucs-testimony-on-the-clean-hydrogen-production-tax-credit/. Accessed 18 April 2024
250. US DOE. Energy Efficiency and Renewable Energy (EERE), Hydrogen and Fuel Cell Technologies Office. *Hydrogen Shot Program.* 7 June 2021. https://www.energy.gov/eere/fuelcells/hydrogen-shot. Accessed 2 April 2024.
251. Nima Simon, Mike McCurdy and Heidi Marie Larson. *Examining the current and future economics of hydrogen energy.* Aug 13, 2021 https://www.icf.com/insights/energy/economics-hydrogen-energy. Accessed 4 March 2024
252. Dan McCarthy, Canary Media. *The clean hydrogen paradox* January 2024. https://www.canarymedia.com/articles/hydrogen/the-clean-hydrogen-paradox?_hsmi=291751420. Accessed 15 February 2024.
253. Amy Power. Hydrogen Technology Expo, Industry News. *MAN has created 200 hydrogen combustion-powered trucks* 12 Apr 2024. https://www.hydrogen-expo.com/hydrogen-industry-news. Accessed 12 April 2024
254. World Hydrogen Technology Expo North America 2024, NRG Center, Houston, TX, USA. https://www.hydrogen-expo.com/industry-news/innovative-compressor-designed-hydrogen-refuelling-launched-siad-mi?utm_source=spotler&utm_medium=email&utm_campaign=HT+World+News+USA+-+General+Newsletter+-+10%2f05+Engaged&utm_term=SIAD+MI+launches+a+compressor+specially+designed+for+hydrogen+refuelling&utm_content=158103&gator_td=jgtIYgSSyUK%2bIxy9fbQ0GQM8PO36g%2b1FcHJajZ5I2aM7TzzqwxCnY0QmvlxVhenNtOQ3Wx2TP%2fuFDsnewOo00nKxneeqPeGRgHMzgM2RTZgXfD2KnTfGy3RPM9MRhhIh929PEgYKd6DFFq9WKaCYlMsIiZ4PfFyEQi2be2k47Fx%2b9JcTKGIecK4ezEFTJvdQ6ZJCfVfdilRBDIff5mQoTw%3d%3d. Accessed 14 May 2024.
255. Ian Dexter Palmer, Forbes Business, 15 June 2023. *Here's how oil and gas companies are jumping on hydrogen, but is it over-rated?* https://www.forbes.com/sites/ianpalmer/2023/06/08/heres-how-oil-and-gas-companies-are-jumping-on-hydrogen-but-is-it-over-rated/. Accessed 1 September 2024
256. Aime Williams, Financial Times, 12 August 2022. *Why the US fossil fuel industry wants to get in on the hydrogen act.* https://www.ft.com/content/1dd75f4f-427e-41c2-9483-4550d96f9487. Accessed 15 August 2024
257. P. Brandl, M. Bui, J.P. Hallett, N. Mac Dowell, Beyond 90% capture: possible, but at what cost? Int J Greenhouse Gas Control **105** (2021). https://doi.org/10.1016/j.ijggc.2020.103239. Referenced in *Ask MIT Climate, How efficient is carbon capture and storage? February 23, 2021* https://climate.mit.edu/ask-mit/how-efficient-carbon-capture-and-storage. Accessed 20 March 2024
258. Yahoo Finance, Sep 13, 2023. United States Hydrogen Electrolyzer Market to Surpass US$ 790.8 Million by 2033 as Need for Sustainable Energy Sources Skyrockets | FMI. https://finance.yahoo.com/news/united-states-hydrogen-electrolyzer-market-143000813.html. Accessed 1 January 2024
259. Polly Martin, Accelerate hydrogen/hydrogen insight, *'Severe overcapacity' | The global supply of electrolysers far outstrips demand from green hydrogen projects:* BNEF 28 March 2024 https://www.hydrogeninsight.com/electrolysers/severe-overcapacity-the-global-supply-of-electrolysers-far-outstrips-demand-from-green-hydrogen-projects-bnef/2-1-1618327.

Original data source: *Electrolyser Manufacturing 2024: Too Many Fish in a Tiny Pond*, BloombergNEF. Accessed 28 March 2024

260. Nina Raffio, USC Today. I*s green hydrogen the fuel of the future?* 8 September 2023. https://today.usc.edu/is-green-hydrogen-the-fuel-of-the-future/. Accessed 6 May 2024
261. United Nations Climate Change. *Green hydrogen.* https://unfccc.int/technology/green-hydrogen. Accessed 6 May 2024
262. US Department of Energy, Office of Nuclear Energy. *3 Nuclear power plants gearing up for clean hydrogen production*, 9 November 2022 https://www.energy.gov/ne/articles/3-nuclear-power-plants-gearing-clean-hydrogen-production. Accessed 1 May 2024
263. West Coast Choppers, *Official web site.* https://westcoastchoppers.com/. Accessed 1 December 2023

Ammonia

10

Ammonia currently enjoys a resurgence of interest as a potential renewable fuel, with some claiming that its environmental impact is less egregious than electric vehicles [1].

Just to be clear, we are referring to *anhydrous ammonia* which is a gas at normal temperature and pressure (NTP, 20 °C, 1 atm), not the household liquid cleaning product referred to as ammonia, which is a very weak solution of ammonia gas in water.

Ammonia (NH_3) is possibly the most common synthesized chemical in the world, most often used for nitrogen fertilizer, either directly or as a precursor to ammonium nitrate. Its role in agricultural productivity cannot be underestimated. Agriculture in most developed countries relies on ammonia to replace depleted soil nitrogen after several seasons of monocrop cultivation. Approximately 95% of all hydrogen is produced from natural gas or coal solely for ammonia production, and this hydrogen production is believed to account for 2% of all anthropogenic greenhouse emissions.

Unlike its chemical cousins hydrogen and methane, there is less interest in ammonia as a direct combustion fuel than as a hydrogen carrier or chemical for synfuel production. Like hydrogen, it can be burned directly in an internal combustion engine. It is classified as a Class 2 flammable gas under the Harmonized System of Classification and Labelling of Chemicals (GHS). Its high hydrogen density makes it a favorable candidate for catalytic reformation into hydrogen. Along with hydrogen, it contains no carbon, so its combustion does not generate carbon dioxide. Consequently, ammonia could be referred to as a "zero-carbon" fuel, provided that the environmental impacts of its production are ignored.

The term *renewable ammonia* refers to ammonia that is made from *renewable hydrogen*, which is hydrogen made by electrolysis using *renewable electricity*, e.g., solar, wind, geothermal, and hydroelectric. The renewability of ammonia derives

C. A. MacCarley, *Non-Petroleum Automotive Transportation*,
https://doi.org/10.1007/978-3-031-85028-8_10

exclusively from the renewability of the electricity used to make it, optimistically assuming that "clean electricity" is available and that ammonia production would be the best use for it.

Physical and Biological Properties

Ammonia is a pungent, colorless, toxic, caustic (high pH), hazardous gas that occurs naturally from the breakdown of organic waste matter, most notably from high-density livestock operations [2]. Its manufacture in large quantities and liberal use in agriculture is a concern in the climate change puzzle. It is a reactive gas that combines with low-ppm acids in the atmosphere to form an estimated 30–50% of all PM2.5 particulate emissions worldwide [3]. It absorbs longwave infrared radiation, making it a potential greenhouse gas, but its high solubility in atmospheric water and broad reactivity gives it a short lifetime in the atmosphere, reducing its direct global warming impact to inconsequential levels. In fact, some have suggested (but not proven) that the aerosols and particulates created by ammonia could be a source of atmospheric cooling. Its direct toxicity and long-term health hazards are known causes of respiratory illnesses and early mortality. It is a powerful caustic that attacks flesh upon contact, requiring complete hazmat protection when handling. Concentrations as low as 20 ppm will cause eye irritation, and at 300 ppm (0.03%) it is considered immediately dangerous to health and can instantly cause permanent eye damage [4]. Even a tiny leak can be a hazmat incident. The prospect of automobile ammonia fuel tanks being filled by consumers would probably be a windfall for personal injury and environmental attorneys.

A Hydrogen Carrier

On a mass basis, ammonia is a dense hydrogen carrier, which is evident from its chemical formula (NH_3), with a hydrogen mass fraction of 0.177, superior to methanol (CH_3OH) at 0.126, but less than methane (CH_4) at 0.250. In liquid form, ammonia, with a density of 682 g/L, provides an equivalent hydrogen volume storage density of 292 g/L, higher than either liquid hydrogen (70.9 g/L) or compressed hydrogen (40.0 g/L). On a volumetric basis, the hydrogen content content in one liter of gaseous ammonia at NTP is 0.129 g/L, compared with gaseous hydrogen at NTP with a density of 0.83 g/L.

Because of its high hydrogen content, a high-value application of ammonia is long-distance hydrogen transport. It is this application that has received endorsements from researchers in the USA and China, including US national laboratories. In 2018, the US Department of Energy launched the program *H2@Scale* in an effort to take hydrogen from novelty into large scale usage. One of the workshop sessions was dedicated to ammonia as a tool to support that objective [5].

A Combustion Fuel

Most people do not think of ammonia as a fuel, since it has other more important uses. The combustion characteristics of ammonia are poor compared with most hydrocarbon fuels, including all fuels discussed in this book. Ammonia in any air/fuel ratio (AFR) has a slower flame propagation speed than that of other combustion fuels with the same AFR. This limits its utilization to low-speed engines with relatively low brake mean effective pressures (BMEPs), i.e., large low-RPM engines for tanker-scale marine applications. The addition of a small percentage of hydrogen to the ammonia–air blend is known to significantly increase the flame speed, although still not to the level of most other ICE fuels. The required supplemental hydrogen could potentially be produced onboard using a small ammonia-hydrogen reformer [6].

Ideally, combustion of ammonia in a homogeneous lean mixture results in only nitrogen and water exhaust. But like hydrogen, mixtures close to stoichiometric can produce unusually high levels of oxides of nitrogen, as well as nitrous oxide. Lean operation of an ammonia engine will reduce or eliminate NOx emissions, but has been found to exacerbate the already-poor combustion properties of ammonia, increasing the probability of unburned ammonia emissions.

While there have been experimental vehicles such as shown in Fig. 10.1 modified to run on ammonia with limited success, to date the only commercially successful applications of ammonia as an IC engine fuel have been large low-speed engines such as shown in Fig. 10.2 designed for marine transport vessels. In these applications, its low power limitations and potential NOx emissions are less consequential than in regulated automotive applications [7]. Ammonia especially makes sense as a fuel for supertankers that transport liquid ammonia since the fuel is already onboard, and the alternative is bunker oil which is environmentally the worst of any liquid fuel in common use. Starting in 2024, there has been considerable activity in formalizing standards and guidelines for the use of ammonia as a marine transport fuel, and two (possibly more) ammonia transport ships powered by ammonia are currently under construction [8].

Fig. 10.1 Ammonia-fueled ICE car built by Ammonia Research Group at the Korean Institute for Energy Research (KIER). https://nh3fuelassociation.org/2013/06/20/the-amveh-an-ammonia-fueled-car-from-south-korea/. Creative Commons Attribution 3.0 Unported License

Fig. 10.2 Ammonia engine for large marine use at MAN ES Research and Development Center in Copenhagen, 2023. © MAN Energy Solutions SE/Niels Busch https://www.man.eu/corporate/en/homepage.html. Used with permission

A press release in July 2023 by Chinese manufacturer GAC Group, which is minority owned by Toyota Motor Corporation, announced that they have "developed the world's first ammonia-powered engine for passenger vehicles", although the engine specifications revealed that it will actually "use an ammonia-gasoline blend to achieve acceptable combustion [9–11]." The announcement reignited media excitement about ammonia as the "automotive fuel of the future [12]," although there have not yet been any follow-up announcements of a production release date.

The required components of a possible fuel system for an ammonia-gasoline engine are shown in Fig. 10.3. Ammonia as a fuel for IC engines has been considered many times during the past century, each time starting with an epiphany that ammonia is a carbon-free, hydrogen-dense molecule, then quickly fading from the news cycle following closer analysis or actual experience. Results from reports on nine recent ammonia engine projects are summarized in citation [13]. None of the projects concluded that ammonia is suitable for use as a transportation fuel except in highly specialized applications, such as the marine application just discussed. This is not to say that ammonia does not have some desirable attributes as a fuel, only that it also has greater undesirable drawbacks compared with other alternatives.

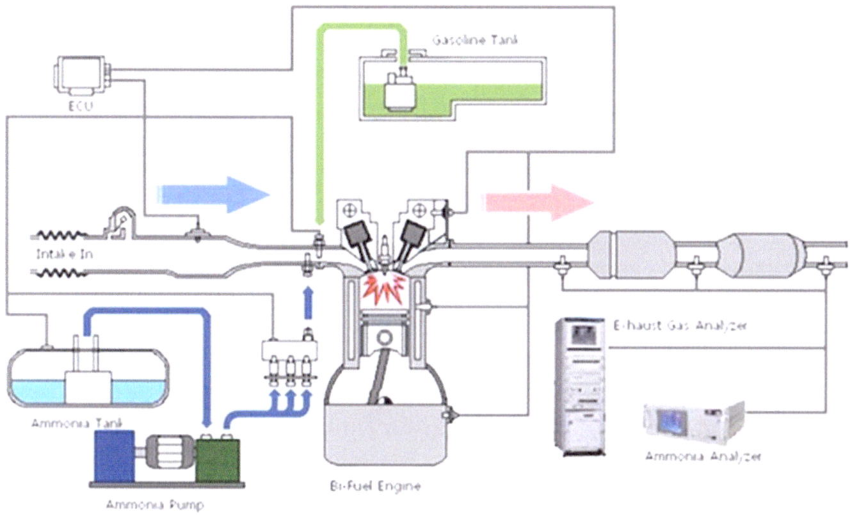

Fig. 10.3 Diagram of gasoline-supplemented ammonia engine test apparatus constructed by the Green Transportation Technology Group at KIER, South Korea. From NH3 Fuel Association web site, https://nh3fuelassociation.org/2013/06/20/the-amveh-an-ammonia-fueled-car-from-south-korea/. Creative Commons License, Attribution 3.0 Unported, CC BY 3.0

Ammonia Synthesis

Commercially, ammonia is produced from hydrogen by the Haber-Bosch process [14], using hydrogen made almost exclusively from natural gas or coal [15].

Steam methane reforming reaction

$$CH_4 + H_2O \rightarrow CO + 3H_2$$

Water–gas shift reaction

$$CO + H_2O \rightarrow CO_2 + H_2$$

Haber-Bosch ammonia synthesis

$$N_2 + 3H_2 \rightarrow 2NH_3$$

Overall process

$$CH_4 + 2H_2O + \frac{4}{3}N_2 \rightarrow \frac{8}{3}NH_3 + CO_2$$

Ammonia is an infrared absorber, therefore a greenhouse gas. It is more potent than carbon dioxide, but it is usually not classified as a GHG of concern because of its high solubility in atmospheric water, which gives it a half-life as short as one week before returning to the surface as precipitation or condensation.

Ammonia has a wide range of industrial applications [16], all of greater value than using it as a combustion fuel or as a carrier for hydrogen to be used as a fuel. These applications include the manufacture of synthetic fibers; bleaching of cotton; production of synthetic resins; petroleum refining; production of urea for diesel exhaust fluid (DEF); processing of raw latex for rubber production; surface hardening of steel by nitriding. It is the primary feedstock for the production of soda ash, nitric acid, and the high explosives trinitrotoluene [TNT], nitroglycerin, and nitrocellulose (guncotton). It is also the immediate precursor to hydrazine, the preferred high-impulse rocket propellant for the attitude control thrusters of spacecraft.

Critical to Agriculture, Harmful to the Environment

Ammonia is used as a nitrogen fertilizer for agriculture, either directly or as the precursor to urea and ammonium nitrate. Agriculture accounts for more than 95% of its use worldwide (Fig. 10.4).[1] In this capacity, ammonia is critical to agricultural productivity[2] worldwide, while at the same time, it is an air and water pollutant of growing concern. Agriculture is responsible for 81% of global ammonia emissions and 50% (EU) or 30% (US) of all PM2.5 particulate pollution [17]. The CO_2 released in the production of hydrogen for ammonia synthesis is reported to be responsible for approximately 2% of all anthropogenic greenhouse gas emissions worldwide [18].

Ammonia as a Refrigerant

Owing to its very high specific heat of vaporization and phase-transition characteristics near-ambient temperatures, ammonia has and remains today one of the most common industrial refrigerants (designated R717 in the USA) for absorption-type heat pumps. It is the oldest industrial refrigerant, in use for over a century. Ammonia absorption refrigeration is more efficient and less costly than consumer-type vapor-compression refrigeration cycles that use synthetic refrigerants. Synthetic refrigerants are various manufactured fluorocarbons or chlorofluorocarbons (CFCs) that are inert and non-toxic, but are among the most potent of all greenhouse gases if leaked into the atmosphere [19]. Ammonia is illegal in the USA and most Western countries for use in non-industrial refrigeration applications due to its flammability, and environmental and health hazards that justify its EPA, OSHA, and NIOSH classification as a hazardous substance [20].

As summarized in citation [21]

[1] Katie E. Wyer, et al. [3].

[2] *Monocrop* means the cultivation of the same crop every growing season, rather than crop rotation, which helps to preserve soil fertility. Examples of crops that are most often grown using this method are corn, soybeans, and wheat.

Fig. 10.4 Agricultural spraying. Photo from USDA voluntary guide to help farmers reduce ammonia emissions, July 27, 2018. Available online at https://www.openaccessgovernment.org/new-voluntary-guide-to-help-farmers-reduce-air-pollution-from-ammonia/4. Public domain

> When it comes to industrial usage R717 is the most suitable refrigerant for heat pump. The efficiency of Ammonia is high <and it> can be easily applied below 80 °C temperature. The most impressive feature of Ammonia is that it is a natural refrigerant and does not contribute to the greenhouse effect. It is inflammable and toxic. On the other hand, due to its odor, the leakage can be easily sensed.

While refrigeration is only a very small user of ammonia, it is known to be one of most common workplace hazards, where its toxicity and corrosive properties make it much more hazardous to workers than synthetic refrigerants, typically CFCs [22]. Its use in HVAC systems for habited structures is highly restricted in the USA by the EPA, and internationally by the Institute of Internal Auditors (IIA), despite its superior performance and lower cost compared with synthetic refrigerants. A typical industrial refrigeration system that uses ammonia refrigerant is shown in Fig. 10.5.

Ammonia Storage and Handling Safety

The production, transport, and use in agriculture and the chemical industry of ammonia have been responsible for some of the most devastating accidents (other than warfare) in recent world history. The public risk may be overstated because there is a lot of it in use, and the accident rate is relatively low compared with other industrial and agricultural hazards. Even considering this, the volume and severity

Fig. 10.5 Industrial food storage refrigeration apparatus. From 2023 Berg Chilling Systems Inc., https://berg-group.com/blog/food-and-beverage/refrigeration-equipment-food-beverage/. With permission

of casualties from ammonia handling would increase by a very large factor if ammonia was adopted as a common transportation fuel. The general public is poorly educated about its risks, most only aware of the weak hydrous ammonia/water solutions used for household cleaning or disinfection.[3] A selection of examples of incidents involving ammonia are reviewed below.

Roadway Accidents

11 May 1976, Houston, Texas, USA [23]—Characterized in media at that time as the worst non-fire roadway accident in Houston's history. A semitrailer carrying 7000 gallons of anhydrous ammonia ran off a freeway off-ramp, spilling its contents. A scenario not unlike roadway accidents involving gasoline tank trailers. Amazingly, the contents did not ignite. But the ruptured tank released extremely toxic and caustic anhydrous ammonia that left seven dead and nearly 200 severely injured from inhalation, skin or eye contact, some far from the accident site. It required the evacuation of a significant part of Houston.

Highly Toxic and Caustic

24 March 1992, Dakar, Senegal [24]—Considered the worst industrial ammonia release in history, despite the relatively small size of the facility. An ammonia tank truck at a peanut-oil-processing facility operated by Sonacos SA ruptured due to overpressure, releasing 22 metric tons of pressurized anhydrous ammonia. Ammonia vapor and liquid aerosol formed a dense vapor cloud that engulfed the oil mill, nearby offices, and adjacent residential and business areas. It was fortunate that the plume did not encounter an ignition source. The toxic and caustic properties alone killed 44 people immediately. Ultimately, the incident caused 129 fatalities and 1,150 injuries. Among the injured were emergency responders that, although

[3] For example, the highly effective window cleaner "Windex® with Ammonia-D".

equipped with hazmat gear, were ill-prepared to deal with a toxic chemical event of this magnitude.

Fire and Explosion of Ammonia

17 April 2013, near Waco Texas [25]—The West Fertilizer plant, located north of Waco Texas, USA, is located close to schools and residences. It had the capacity to store as much as 25 metric tons (54,000 lbs) of anhydrous ammonia, used for the production of various agricultural chemicals including ammonium nitrate. According to Texas regulators, the stored ammonia posed no fire or explosion risk, with their worst possible scenario being the slow release of ammonia gas that would harm no one. An ammonia storage tank exploded. 15 people died including 9 volunteer firefighters. 280 were treated for injuries, some severe. The blast was compared to a "nuclear bomb" by witnesses, completely destroying a four-block area and shaking the ground 50 miles away. The exact cause of the explosion was not determined. West Fertilizer was fined $118,300 for workplace violations including unsafe handling and storing dangerous chemicals [26].

Explosion of Ammonium Nitrate made from Ammonia

4 August 2020, Beiruit, Lebanon [27]—Recognized as the largest man-made non-nuclear explosion in history. An estimated 2,750 tons of ammonium nitrate was unsafely stored in an urban port area of the capital of Lebanon. Over 6000 injured, 200 fatalities, and over 100 missing casualties. It destroyed a major part of the city and did over $10 billion USD damage to infrastructure, leaving 300,000 displaced individuals. The cause of ignition was never definitively determined.

Harm to Farmworkers and Rural Populated Areas

Exposure to ammonia, either used as nitrogen fertilizer or consequential from intensive livestock farming, is a well-known cause of both acute and long-term injury, not only to agricultural workers but also to residents in areas proximate to the operations. Medical studies are plentiful [28], but since agricultural injuries are known to be underreported and poorly documented, accurate statistics are difficult to obtain [29].

15 April 2003, New Calamus, Iowa [30]. A typical incident documented by the NHTSA involved a "nurse tank" that partially ruptured after being filled with anhydrous ammonia. 1,300 gallons of the anhydrous ammonia gas escaped, seriously injuring two nurse tank loaders, one of whom died from his injuries 9 days after the accident. A typical injury from brief skin contact with anhydrous ammonia is shown in Fig. 10.6.

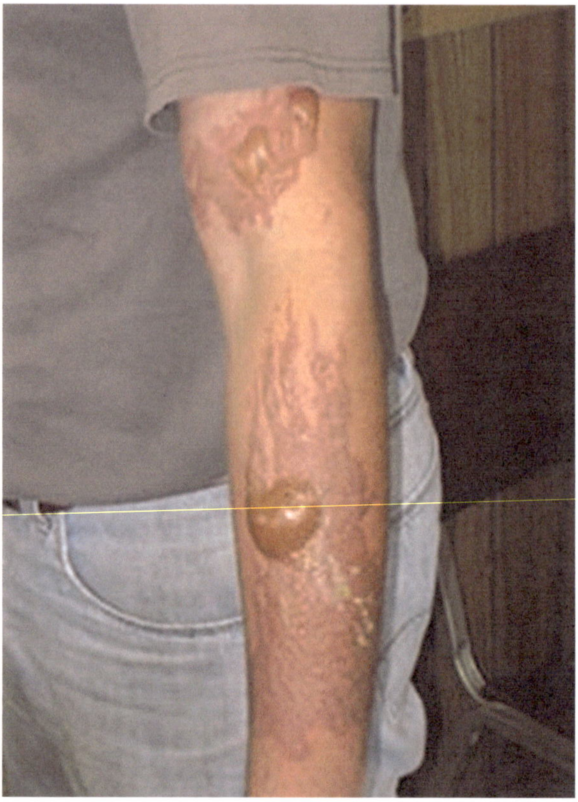

Fig. 10.6 Example of burns from brief exposure of flesh to anhydrous ammonia. From *Anhydrous Ammonia*, publication of Purdue University College of Agriculture, June 2021. https://ag.purdue.edu/department/extension/ppp/resources/ppp-publications/ppp-140.html (Educational fair use.)

Summary of Ammonia Fuel Characteristics

- High mass energy density, superior to most hydrocarbons except hydrogen and methane (natural gas).
- Physical properties similar to propane: liquifies 8.6 bar (125 psi) at 29 °C. International hazardous material class UN1005.
- Usually stored and transported at ambient temperature in 250 psi (18 bar) rated carbon steel tanks. Not compatible with materials attacked by caustics, e.g., aluminum.
- Ammonia is dangerous to handle. It is assigned an Occupational Safety and Health Administration (OSHA) hazardous health rating of 3, the highest level. It is flammable in gaseous form but difficult to ignite, assigned to Flammability Class H by the National Fire Protection Association (NFPA). It a poisonous caustic gas and potent lung and eye irritant even at very low concentrations. It rapidly attacks organic materials, including human flesh. Handling of anhydrous ammonia requires full-body personal protection equipment and respirator under 49 CFR §172.102.
- Air transport is not permitted on any commercial carrier internationally.
- Can be burned directly as a spark ignition (SI) or spark-assisted diesel engine fuel. Very poor combustion characteristics, leading to lower efficiency and low

power output, and unburned fuel in exhaust for both rich (stoichiometric) and lean (excess air) mixtures. Typically used only in large low-RPM engines for marine vessels that transport ammonia.

- Produces zero carbon dioxide when burned. However, it generates NOx emissions similar to or greater than gasoline at air/fuel ratios close to stoichiometric. NO and NO_2 are highly toxic and photochemically reactive in the atmosphere, making them the regulated pollutant of greatest concern for combustion engines. Unlike most other combustion fuels, it also produces N_2O, a non-toxic pollutant that is one of the most potent anthropogenic GHGs.
- Already has a distribution infrastructure since it is universally used in modern agriculture.
- Ammonia is made starting with hydrogen, which is produced from natural gas by SMR or coal by pyrolysis. These are highly carbon-intensive processes, to an extent that ammonia production is one of the largest contributors to global warming, even though ammonia itself is not usually classified as a greenhouse gas.
- So-called "green" ammonia is synthesized from green hydrogen, which is made from electricity, the majority of which is made from fossil fuels. The energy losses of each conversion step are high, so the efficiency of green ammonia production is extremely low (<10–20%). Far from being commercially viable without major subsidies, renewable ammonia has never actually been produced at commercial scale.
- Ammonia is one of two potential hydrogen energy carriers being considered as a means for international transport of hydrogen (the other is methanol).
- Bottom line: the environmental impact of ammonia production depends primarily on the "greenness" of the hydrogen from which it is made.
- 95+% of ammonia is produced for agricultural fertilizer using hydrogen that is made from natural gas or coal, with high CO_2 emissions.
- It has widespread uses as an industrial refrigerant, and as a working fluid for absorption heat pumps. Its popularity is because this type of system has the lowest operating cost of all refrigeration options, and lower temperatures can be reached compared with synthetic refrigerants. It is currently not permitted as a refrigerant for home HVAC systems.
- In the reverse of the ammonia production process, hydrogen can be made from ammonia in a reforming process that can potentially be made compact enough for use on a vehicle. This, at least theoretically, facilitates its potential use as a hydrogen carrier for fuel cell vehicles.
- Corrosive in contact with some metals (copper, brass, aluminum) and elastomer seals. OK with regular carbon or mild steel.
- Although ammonia has the infrared absorption characteristics of a greenhouse gas similar to methane, but it is not considered a contributor to global warming due to its high solubility in water, which removes it quickly from the atmosphere in precipitation or condensation.
- The largest current use for ammonia is nitrogen-bearing fertilizer for agriculture, either directly or in the manufacture of other nitrogen fertilizers such as ammonium nitrate. Ammonia dispersed in fields has been reported to be a substantial

contributor to fine particulate matter (PM2.5) emissions in the US and Europe, where it accounts for the formation of 30–50% of all PM2.5.[4]

Conclusions for and Against Ammonia as a Transportation Fuel

A Positive Assessment

From by S. Giddey et al in *Ammonia as a Renewable Energy Transportation Media, 2017* [31]:

> Ammonia synthesized using hydrogen from renewable sources offers a vast potential for the storage as well as transportation of renewable energy from regions with high intensity to regions lean in renewable sources. Ammonia can be used as an energy vector for an emissionless energy cycle in a variety of ways. Ammonia at the point of end use can be converted to hydrogen for fuel cell vehicles or alternatively utilized directly in solid oxide fuel cells, in an internal combustion engine or a gas turbine. One ton of ammonia production requires 9–15 MWh of energy. However, its conversion back to useful form or direct utilization can lead to substantial energy losses.

It is also worth noting that the agricultural productivity that provides food for the world also supports the production of renewable biofuels such as ethanol and biodiesel. Therefore, ammonia fertilizer already plays a critical role in the success of cultivated biofuels.

A Negative Assessment

Ammonia production is currently dominated by agriculture and industrial demand. Ammonia is a reactive flammable gas that is extremely dangerous to handle. It can potentially be used as a fuel for engines. Its carbon-free composition promises zero CO_2 emissions from combustion, although NO_x emissions greater than gasoline occur at higher power levels.

Ammonia is potentially a *green fuel*, but only if it is made exclusively from *green hydrogen* (made by electrolysis using carbon-free electricity). The overall efficiency of this multistep process is very low, which is why it accounts for only a trivial percentage of worldwide ammonia production. The Haber-Bosch process is used almost exclusively, which uses hydrogen as its input. As of 2023, there are no commercial ammonia producers that use green hydrogen, although such a substance is routinely referred to as critical to climate change remediation. Most hydrogen is made from natural gas or coal by steam reforming (SMR). This is gray hydrogen or brown/black hydrogen, respectively. The production of the hydrogen alone for making ammonia is (conservatively) reported to be responsible for an estimated 2% of all anthropogenic greenhouse gases contributing to global warming worldwide.[5]

---

[4] Katie E. Wyer, et al. [3].

[5] Sierra Club Journal (online), Fall 2023 issue. "…the life-cycle greenhouse gas emissions of blue hydrogen were more than 20 percent greater than those from burning straight methane gas. The use

Ammonia production uses 6% of all global natural gas and 2% of coal, generating massive carbon dioxide emissions as a by-product.

The high-yield monocrop agricultural practices worldwide depend on ammonia fertilizer to replace the nitrogen that is depleted from the soil after the first few growing cycles. It is often pointed out that much of the world population would starve without ammonia. This observation leads to two cautionary conclusions: (1) Agricultural food production is a more important use for ammonia than combustion as a fuel. (2) If truly green (electrolytic) hydrogen were ever to become a significant feedstock for ammonia production, it would be hugely energy wasteful and too costly for this application. Its supply chain weak link, "clean hydrogen", is the common denominator of many competing synthetic transportable alternative fuels, including the synthetic e-fuels that are considered the only viable replacements for petroleum aviation fuels.

Commentary

The mid-2023 announcement by Toyota and GAC Motors suggested that they may be planning to manufacture an ammonia-fueled passenger car in the indefinite future, but that news cycle came and went without action by late 2023 [6]. Even the use of reduced-carbon "blue hydrogen" relying on the maximum possible (90%) level of carbon capture is estimated to generate life-cycle greenhouse gas emissions estimated to be more than 20% greater than those produced by simply burning the natural gas from which it was made.[6] The only transportation application for which ammonia fuel earns a positive score for energy and economic sustainability seems to be large marine engines in ocean-going ammonia tankers which already carry ammonia as cargo and operate in unregulated waters. The intercontinental transport of hydrogen in the form of ammonia makes economic sense compared with liquid hydrogen transport.

As for environmental sustainability, the use of ammonia as a road vehicle fuel would not be beneficial for the environment. However, considering the enormous amount of ammonia produced and used for agriculture, any additional use of ammonia as a combustion fuel would probably result in only a minimal relative increase in atmospheric GHG levels.[7]

of blue hydrogen, appears difficult to justify on climate grounds." https://www.sierraclub.org/sierra/2023-3-fall/feature/how-clean-is-green-hydrogen.

[6] Sierra Club [5].

[7] Katie E. Wyer, et al. [3].

References

1. C. Tornatore, L. Marchitto, P. Sabia, M. De Joannon, Ammonia as green fuel in internal combustion engines: state-of-the-art and future perspectives. Front. Mech Eng., 8 (2022). https://doi.org/10.3389/fmech.2022.944201/full
2. PIN Petrochemical, Chemical and Energy, Jul 01 2021. *Where does ammonia come from?* https://www.petro-online.com/news/measurement-and-testing/14/breaking-news/where-does-ammonia-come-from/55630. Accessed 24 March 2024
3. K.E. Wyer, D.B. Kelleghan, V. Blanes-Vidal, G. Schauberger, T.P. Curran, Ammonia emissions from agriculture and their contribution to fine particulate matter: a review of implications for human health. J Environ Manage **323** (2022) https://www.sciencedirect.com/science/article/pii/S0301479722018588?via%3Dihub
4. US Center for Disease Control. ATSDR Portal. Medical management guidelines for ammonia. https://wwwn.cdc.gov/TSP/MMG/MMGDetails.aspx?mmgid=7&toxid=2. Accessed 1 July 2024
5. US DOE Hydrogen and Fuel Cells Technologies Office. H2@Scale: Ammonia for H2@Scale Virtual Panels. "Invitation-only virtual panels held May 6–7, 2021, focused on the growing opportunities for ammonia as a viable clean-hydrogen carrier serving diverse end uses, supporting important economic and environmental imperatives." https://www.energy.gov/eere/fuelcells/ammonia-h2scale-virtual-panels. Accessed 1 December 2021
6. B. Wang, H. Wang, B. Duan, C. Yang, H. Deng, Y. Wang, Effect of ammonia/hydrogen mixture ratio on engine combustion and emission performance at different inlet temperatures. Energy **272**, 127110 (2023). https://doi.org/10.1016/j.energy.2023.127110MANES. Moving Forward on Ammonia Engines, https://www.maritimepropulsion.com/news/man-moving-forward-ammonia-602261 accessed 29 Feb 2024
7. Julian Atchison, Ammonia Energy Association. *IMO moves forward with interim guidelines for ammonia fuel use*. https://ammoniaenergy.org/articles/ 2 October 2024. Accessed 2 October 2024
8. Engineering Mix, *Everything you need to know about the Toyota ammonia engine*, 24 October 2023. https://www.engineeringmix.com/toyota-ammonia-engine/. Accessed 24 March 2024
9. Tech Addicts on YouTube. *Toyota Just REVEALED An Ammonia-Powered 2.0-L Engine!* https://www.youtube.com/watch?v=PlJa4m3UfgQ&t=584s. Accessed May 1 2024
10. Bret Williams. *Hydrogen fuel news—green hydrogen report,* 25 October 2023. https://www.hydrogenfuelnews.com/ammonia-engine-toyota/8561347/. Accessed 5 September 2024
11. Micah Wright. *Corrosion of conformity: Toyota takes aim at EVs with ammonia engine*. Engine Labs, August 24, 2023 https://www.enginelabs.com/engine-tech/engine/corrosion-of-conformity-toyota-takes-aim-at-evs-with-ammonia-engine/. Accessed 24 March 2024
12. Stephen H. Crolius. *Ammonia-Powered Internal Combustion Engines. Ammonia Energy Association web page.* 2 November 2016. https://ammoniaenergy.org/articles/ammonia-powered-internal-combustion-engines/. Accessed 20 March 2024.
13. Jim Clark. The Haber Process. LibreTexts, Chemistry. Undated. https://chem.libretexts.org/Bookshelves/Physical_and_Theoretical_Chemistry_Textbook_Maps/Supplemental_Modules_(Physical_and_Theoretical_Chemistry)/Equilibria/Le_Chateliers_Principle/The_Haber_Process. Accessed 24 March 2024.
14. US Department of Energy, Office of Energy Efficiency & Renewable Energy, Hydrogen and Fuel Cell Technologies Office. *Hydrogen Production: Natural Gas Reforming* https://www.energy.gov/eere/fuelcells/hydrogen-production-natural-gas-reforming Undated. Accessed 24 March 2024.
15. Steven S. Zumdahl, Britannica Science and Tech. *Ammonia, chemical compound.* Mar 2, 2024 https://www.britannica.com/science/ammonia. Accessed 30 March 2024.
16. J.W. Erisman, How ammonia feeds and pollutes the world. Science **374**(6568), 685–686 (2021). https://doi.org/10.1126/science.abm3492

17. By Loz Blain, New Atlas. *World's largest green ammonia projects could clean up half the market* November 30, 2022 https://newatlas.com/energy/largest-green-ammonia-projects. Accessed 20 March 2024.
18. FrostPoint refrigeration products web site, *Everything you need to know about ammonia refrigeration.* Nov 24 2020 https://www.frostpoint.net/articles/everything-you-need-to-know-about-ammonia-refrigeration.html Accessed 1 March 2024.
19. Safety and Health. *Anhydrous ammonia: Know the dangers,* December 2017. https://www.safetyandhealthmagazine.com/articles/16468-anhydrous-ammonia-know-the-dangers. Accessed March 29, 2024.
20. Blog posts by Vijay. *Refrigerant used in heat pumps and how to choose it wisely,* Aspiration Cleantech Ventures, 29 March 2021. https://aspcv.com/refrigerant-used-in-heat-pumps-and-how-to-choose-it-wisely/. Accessed 20 March 2024.
21. US Dept of Labor., Occupational Health and Safety Administration (OSHA). *Ammonia Refrigeration, OSHA Stds 29 CFR 1910.* https://www.osha.gov/ammonia-refrigeration/standards. Accessed 10 March 2024.
22. Hazmat Nation. 11 May 2020. *On May 11, 1976, an ammonia truck disaster killed 7 Houstonians and injured nearly 200.* https://www.hazmatnation.com/news/on-may-11-1976-an-ammonia-truck-disaster-killed-7-houstonians-and-injured-nearly-200/. Accessed 1 June 2024.
23. Seshu Dharmavaram, Venkat Pattabathula . *Learning from the Worst Ammonia Accident.* AIChE Safety, July 2023. https://www.aiche.org/resources/publications/cep/2023/july/learning-worst-ammonia-accident
24. National Geographic, 20 April 2013. *Explosion highlights dangers of anhydrous ammonia* https://www.nationalgeographic.com/science/article/130418-west-texas-fertilizer-explosion-fire-anhydrous-ammonia-science. Accessed 1 June 2024.
25. BuzzFedd News, 10 October 2013. *Texas fertilizer plant where massive explosion happened fined $118,300 for workplace violations.* https://www.buzzfeednews.com/article/adriancarrasquillo/texas-fertilizer-plant-where-massive-explosion-happened-fine. Accessed 1 June 2024.
26. National Institute of Health (USA). Beirut Ammonium Nitrate Explosion: A Man-Made Disaster in Times of the COVID-19 Pandemic. https://pmc.ncbi.nlm.nih.gov/articles/PMC7985624/. Accessed 1 June 2024.
27. EPA Office or Agricultura and Rural Affairs (USA). *Official correspondence from Jennifer Tucker, Deputy Administrator, National Organic Program.* 16 May 2024. https://www.ams.usda.gov/sites/default/files/media/NOPEPAAmmoniaExtracts.pdf. Accessed 1 June 2024.
28. Ashley Van Winckel. The harmful risks from anhydrous ammonia use in agriculture, 29 May 2021. https://storymaps.arcgis.com/stories/a609acc6a4364ee78dd4b9015a9ed54c. Accessed 1 June 2024.
29. National Transportation Safety Board. *Nurse Tank Failure With Release of Hazardous Materials, April 15, 2003.* Hazardous Materials Accident Report, NTSB/HZM-04/01PB2004-917001 Online at https://www.ntsb.gov/investigations/AccidentReports/Reports/HZM0401.pdf
30. S. Giddey Orcid, S.P.S. Badwal, C. Munnings, M. Dolan, Ammonia as a renewable energy transportation media. ACS Sustain Chem Eng **5**(11), 10231–10239 (2017). https://doi.org/10.1021/acssuschemeng.7b02219. Accessed 5 January 2024

11 Synthetic and E-Fuels

What Are Synthetic Fuels?

Fuels classified as **synthetic fuels** or **synfuels** have been around since before refined gasoline. The definition is subject to some interpretation, e.g.,

> Probstein et al. [1]: "Gas or liquid fuel obtained through conversion of carbonaceous material to another form"

Both references below seek to narrow the definition:

> Wikipedia [2]: "Synthetic fuel or synfuel is a liquid fuel, or sometimes gaseous fuel, obtained from syngas, a mixture of carbon monoxide and hydrogen, in which the syngas was derived from gasification of solid feedstocks such as coal or biomass or by reforming of natural gas."

> Ruth and Stephanopoulos [3]: "…there is a good deal of confusion surrounding the term, synthetic fuels." "Synthetic fuels are fuels produced via chemical or biological reactions either (i) connecting reducing equivalents with a carrier molecule such as carbon dioxide or nitrogen, or (ii) mediating deconstruction of large molecular backbones, as in pyrolysis".

The objective of most manufactured (synthetic) fuels is the replacement of existing petroleum fuels with equivalent, usually superior substances. This makes synfuels the ultimate transition fuels, since no change is required to existing vehicles or the refueling infrastructure. While synthetic versions of petroleum fuels often have more desirable environmental characteristics, this may not be true in individual cases.

Most, but not all, synfuels start with hydrogen, and many involve methanol as an intermediary.

C. A. MacCarley, *Non-Petroleum Automotive Transportation*,
https://doi.org/10.1007/978-3-031-85028-8_11

Fig. 11.1 Wind turbines next to an unidentified processing facility, as a rendition of what a dedicated renewable e-Fuel facility might look like. Photo: Jonathan Billinger, CC BY-SA 2.0, https://commons.wikimedia.org/w/index.php?curid=114685132

Electrofuels (e-Fuels)

While there are exceptions to the definition,[1] e-fuels are synthetic fuels that are made using electricity. The usual first step involves making hydrogen by electrolysis, and then combining it with various sources of carbon to synthesize an intended hydrocarbon. All e-fuels are synfuels. Not all synfuels are e-fuels [4]. *Power-to-gas* (PtG), *power-to-liquid* (PtL), and *power-to-fuel* (PtF) are all references to combustion fuels made from electricity.

The source of the electricity is irrelevant to the definition. It is usually assumed that e-fuels are lower in carbon emissions than their fossil equivalents, but there is no requirement that an e-fuel be renewable, environmentally beneficial or even an efficient use of electricity. The motivation is usually the replacement of a common fossil fuel with a drop-in manufactured fuel that is less harmful to the environment. For a synfuel to produce lower carbon dioxide emissions, it must contain less carbon per unit of combustion energy than the fuel it replaces. Hydrogen is combined with carbon from various sources and processed to produce engineered molecules that have properties similar to petroleum fuels. Based on a review of recent practices, the electricity used to make the renewable hydrogen for synfuel production is usually not renewable; it comes from the utility grid, which is a mix of both fossil and carbon-free sources. The WTW (well-to-wheels) carbon emissions of an e-fuel can be no better than those of the electricity used to produce it. Any environmental advantage of an e-fuel over a fossil fuel relies on the use of renewable or carbon-free electricity from solar, wind, geothermal, or arguably, hydro or nuclear power. If synthesized using entirely zero-carbon electricity, the e-fuel can be considered *green*, free of CO_2 emissions at any point in its production and distribution. But this is rare. E-fuels are almost always more expensive and demanding of energy than an equivalent fossil fuel. Almost all are presently made from nonrenewable electricity sources.

Figure 11.2 depicts the basic steps for the manufacture of an e-fuel, starting with either renewable or fossil electricity.

[1] As described in the chapter *Fuels for Desperate Times*, acetylene generated by the reaction between calcium carbide and water fits the definition of an e-Fuel because the carbide is made in a high-temperature electric arc furnace.

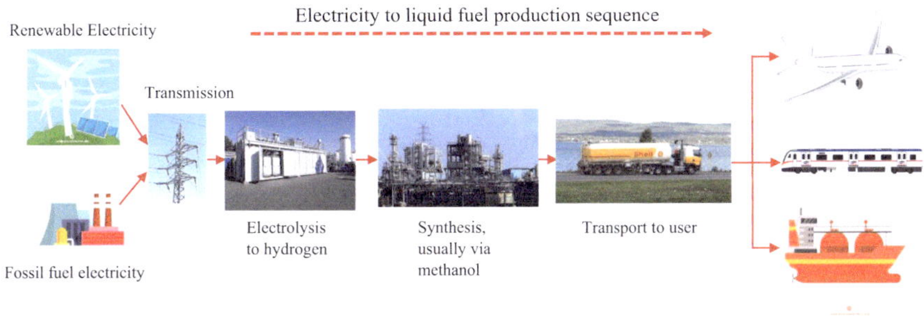

Fig. 11.2 Pictorial showing three or more conversion steps for production of liquid e-fuel from electricity. Diagram by author with clip art from https://publicdomainvectors.org/en/search/transportation/4 CC BY-SA or https://commons.wikimedia.org/w/index.php?search=power+lines&title=Special:MediaSearch&go=Go&type=image

Solar, wind, or other (truly) renewable electricity sources produce zero carbon per kWh produced. As discussed in Chapter 9 *Hydrogen*, electricity from the US grid incurs average CO_2 emissions of approximately 400 g CO_2/kwh, which after electrolysis translates to a carbon intensity for the produced hydrogen of approximately 10 kg CO_2/kg H_2. By comparison, hydrogen made by SMR (gray) produces hydrogen with typical CO_2 emissions of 19 kg CO_2/kg H_2. Blue hydrogen with 50% CCS could possibly reduce this figure by half (ignoring the energy required for the CCS), which in this simple comparison would make blue hydrogen lower in net carbon emissions than grid electrolytic (drab green) hydrogen.

At this time, only about 2% of hydrogen worldwide is made by electrolysis with a production (not retail) cost of \$3.00–\$8.00/kg. "Grey" hydrogen production costs between \$0.90 and \$1.78/kg, and "blue" (with at least some CCS) hydrogen ranges from \$1.20 to \$2.60/kg [5].

But the IRENA green hydrogen roadmap (see Chapter 9 *Hydrogen*) for Europe, North America, Asia, and Australia is predicated entirely upon electrolytic hydrogen. As more of the grid electricity is sourced by solar and wind energy, this situation can be expected to improve, but probably never to the point that production of hydrogen, and therefore e-fuels, will become carbon neutral. With the US DOE e-Fuel and SAF Grand Challenge objectives to utilize exclusively renewable or unusable off-peak electricity, this will hopefully become closer to reality. Meanwhile, most promotional references to e-fuels or *green synthetic fuels* continue to imply production from carbon-free energy. In reality, this is rarely the case.

Power-to-Gas (PtG)

Power in this case refers to electricity that is used to make hydrogen by electrolysis. The usual implied assumption is that the electricity is generated in some carbon-free way: solar, wind, hydro, or nuclear, but not necessarily. Even fully subsidized "renewable" e-fuel/PtG projects usually use renewable electricity "when it is available" but grid electricity at other times. Most PtG synfuels are currently made from grey or blue hydrogen, which is less costly.

Another spin on this idea is to use excess electricity during minimum demand periods for direct air carbon capture as well as for hydrogen electrolysis, and then use the hydrogen to synthesize methane or ammonia using the captured carbon. This would ideally be done at a co-located facility that has a need for e-methane or e-ammonia.

Hydrogen can itself be classified as the consummate e-Fuel, made directly from electricity by electrolysis. As mentioned in Chapter 9 *Hydrogen*, grid energy storage using hydrogen is one PtG application that could assist with grid demand variations, or serve as a means for leveling the output of intermittent solar and wind generation. In the latter case, it has even been given (yet another) name: *windgas* [6].

Power-to-Liquid (PtL)

G is replaced with L when the output is a liquid fuel. Ideally, renewable or excess grid electricity is used to make hydrogen by electrolysis. Then the hydrogen is used to synthesize a liquid fuel such as e-methanol, e-gasoline, e-diesel, or e-kerosene. The only distinction is the output product. The liquid form of the e-fuel makes it more likely to be used as a replacement vehicle fuel.

A high-profile example of PtL is the George Olah carbon dioxide-to-methanol facility in Iceland, operated by Carbon Recycling International that produces methanol fuel under the trade name *Vulcanol* (Fig. 11.3). Geothermal electricity is used to produce electrolytic hydrogen, and then methanol is reformed from the hydrogen

Fig. 11.3 *Vulcanol* is the trademarked name for e-methanol produced by Carbon Recycling International's CO_2-to-fuel process in Iceland. Uses geothermal electricity to produce electrolytic hydrogen, then methanol synthesis using CO_2 from atmospheric carbon capture. Vehicle is a 2020 Geely Emgrand methanol sedan produced for the Chinese market. Image copied from Damien Speight, Nature's recipe for a world run entirely on green energy: Green Carbon & Renewable Fuels. Original photo by Zhejiang Geely Holding Group Co Ltd. Public domain

and CO_2 recovered by DACC (direct air carbon capture). Here is a summary of a few of the best known e-fuels, starting with their building blocks, e-hydrogen, and/or e-methanol.

e-Hydrogen (aka green hydrogen)

E-fuels begin with hydrogen produced by the electrolysis of water using electric energy (or possibly geological or direct solar hydrogen in the future). While the definition of e-fuel does not constrain how the electricity was generated, for the e-fuel to be considered renewable, the source of the electricity must be renewable. This "green" constraint limits the original energy sources for the electricity to solar PV, wind, geothermal, or (arguably) hydroelectric energy. Electricity from nuclear energy also meets the carbon-free criteria, but it is usually not considered because of the perceived risk of nuclear power generation.

Production by SMR (gray) hydrogen from natural gas generates about twice the carbon dioxide of electrolytic hydrogen using grid electricity, which is itself carbon-intensive. The ratio is much worse for hydrogen produced from coal.

Green hydrogen as an end-product represents one of the most difficult clashes between the dual objectives of environmental sustainability and energy sustainability. Electrolytic production of hydrogen is wasteful compared to the energy value of the electricity, and even more wasteful when traced back to the original source of energy for the power generation. Yet, in my survey of published studies, the electrical energy required for the generation of the required e-hydrogen for e-fuel production was always assumed to be 'free' because it is "excess generation," most notably in Iceland, Hawaii, Norway, Sardinia (Italy), or other isolated nations having geothermal resources. Unfortunately, this assumption is not transferrable or scalable [7].

E-Methanol

E-Methanol is a PtL version of methanol, usually made from electrolytic hydrogen and some form of carbon, ideally carbon from CCS or waste carbon compounds from a co-located industrial or power generation facility. But in practice, the carbon comes from conventional sources, all involving combustion. Currently all except experimental quantities of methanol are made from fossil fuels [8, 9] as discussed in Chap. 6 *Methanol*. The GHG impact of making methanol from natural gas is not as egregious as making just hydrogen by SMR. Methanol is most commonly made from syngas, but the CO_2 is at least partially utilized in the methanol synthesis step that follows it, and heat can be partially reused in the process.

Methanol is often an intermediary in the production of other synthetic fuels. E-methanol (CH_3OH) starts with e-hydrogen made by electrolysis powered by some sources of renewable electricity. This is followed by synthesis of methanol by reaction of the hydrogen with carbon dioxide via one of several high-temperature variations of the Fischer–Tropsch process [10] as discussed in Chap. 6 *Methanol*.

The possibility of utilizing carbon dioxide captured by CCS to react with e-hydrogen to produce e-methanol is a big selling point, especially if the e-methanol facility is co-located with a large CO_2 emitter. Even more attractive (although far less practical) is carbon dioxide directly captured from the atmosphere by DCCS. This is the basis of much-hyped fuel-from-greenhouse-gas scenarios that are extremely attractive from an *environmental* sustainability point of view, but equally unattractive from an *energy* sustainability view: The required energy for making e-methanol from DCCS CO_2 greatly exceeds the energy that was generated by the combustion or SMR process that produced the CO_2 originally released into the atmosphere. DCCS e-methanol only pencils out if the required electric energy is free and would otherwise have been wasted [11].

Of all the e-fuels (other than e-hydrogen), e-methanol is the simplest and most energy-efficient to make and is the most ubiquitous intermediate feedstock for other e-fuels. For a remotely located e-fuel operation, the transportability and ease of storage of methanol are large advantages that presumably outweigh the low overall efficiency of the hydrogen-to-e-methanol synthesis [12]. And if the prima facie objective is the removal of carbon from a power plant stack or the atmosphere, CCS or DCCS methanol would be a desirable destination for the sequestered carbon, allowing it to be used again in the fuel chain.

At this time, worldwide, the production of e-methanol entirely from renewable energy is almost non-existent. The only such facility appears to be the previously mentioned George Olah CO_2 methanol plant near Reykjavik in Iceland which produces 5 million liters of renewable methanol per year from 5600 tons of CO_2 per year captured from a co-located (mostly) geothermal electric power generation facility. The methanol product is used locally as a motor fuel for the methanol fueled vehicles shown in Fig. 11.3. Without geothermal or some other near-zero-cost carbon-free source of electricity, such a facility could not be economically or environmentally sustainable.

But in the few situations in which the cost of renewable electricity is minimal, e-methanol as the product of DCCUS (direct air carbon capture, utilization, and storage) is almost surely the best of the possible destinations for the captured carbon, since methanol has high market value. E-methanol is easily storable and transportable and has a lower carbon-to-hydrogen ratio than any other liquid hydrocarbon, assuring that its use as a fuel is less carbon-intensive than other liquid fuels.

e-DME

Dimethyl ether (DME, CH_3OCH_3) is a synthetic fuel that burns very cleanly and has a cetane rating higher than petroleum diesel fuel, making it a viable replacement for diesel fuel. Its production process starts with natural gas, biomass or hydrogen and carbon dioxide and is almost identical to that of methanol. As discussed in Chapter 13 *Dimethyl Ether*, DME has physical properties similar to butane (cigarette lighter fluid) or LPG. Like methanol, most commercial production of DME involves the

Fischer–Tropsch process, with the hydrogen, carbon dioxide, and carbon monoxide of the resulting syngas serving as the feedstock for the catalyzed synthesis reaction.

E-DME differs from regular synthetic DME only in that it uses e-hydrogen generated by electrolysis of water using electricity rather than from reformed methane or syngas from biomass. Ideally, but unlikely, the required carbon comes from CO_2 captured from flue gases or from the atmosphere by DCCS. Two paths are most common—dehydration of intermediate methanol, or a direct catalytic process under temperature and pressure conditions slightly higher than those used for methanol synthesis. It is almost surely the cleanest fuel for compression ignition (diesel) engines, with lower emissions of CO_2 as well as particulates and NO_x. It is not a drop-in replacement for diesel fuel, but the fuel storage and injection systems of diesel vehicles can be re-engineered to handle DME.

The production energy efficiencies for e-DME are very similar to those for methanol. Like all e-fuels, the justification for use of renewable electricity to generate the requisite hydrogen is a concern, especially since DME can be produced from so many feedstocks, including most forms of biomass. Nevertheless, the potential to replace diesel, biodiesel, or bunker oil in transcontinental marine transport could have a large impact on the GHG emissions of these classes of transportation, which are among the worst polluting.

e-Ammonia

E-ammonia (NH_3) is made by a process that starts with e-hydrogen, followed by catalytic reaction of the hydrogen with nitrogen, usually from the atmosphere. The allure of ammonia as a combustion fuel or energy carrier is of course due to its lack of carbon, resulting in zero CO_2 combustion emissions (although oxides of nitrogen remain a serious problem). It is a potential hydrogen carrier for mobile applications—it can be easily reformed into hydrogen and nitrogen onboard a vehicle. It is discussed in greater detail in Chapter 10 *Ammonia*.

Ammonia has never gained traction beyond demonstration projects or speculative product announcements.[2] Its physical properties and safety issues make its handling as a fuel intrinsically unsafe. However, there are some applications in which it is an acceptable or preferred alternative compared with traditional fossil fuels, in particular, transcontinental marine shipping of ammonia since the ammonia is already the cargo.

E-ammonia made from renewable electricity via hydrogen electrolysis will always be noncompetitive compared with the production of hydrogen from natural gas or coal, justified only if the source electricity has low or zero incremental cost. Three such facilities are currently under construction, all located in areas in which

[2]Toyota announcement about an ammonia powered ICE, reported in *Engine Labs*, August 24, 2023, *Corrosion Of Conformity: Toyota Takes Aim At EVs With Ammonia Engine* by Micah Wright. https://www.enginelabs.com/engine-tech/engine/corrosion-of conformity-toyota-takes-aim-at-evs-with-ammonia-engine/.

Fig. 11.4 Photo representative of the announced "World's Largest Green Ammonia Plant" under construction in South Africa. Not actually an e-ammonia facility, since hydrogen will be (at least initially) from blue hydrogen. From CCE, 18 January 2022. https://cceonlinenews.com/2022/01/18/worlds-largest-green-ammonia-plant-for-nelson-mandela-bay-south-africa/. Public domain

these criteria are met because of an excess of hydroelectricity and wind power: The REDDAP (Renewable Dynamic Distributed Ammonia Plant) located in Ramme, Jutland, Denmark [13]; also the ATOME Energy Project in Villeta, Paraguay [14]; and the Hive Hydrogen project near Nelson Mandela Bay in South Africa [15]. All facilities are intended for production of ammonia or ammonium nitrate for fertilizer or chemical production (Fig. 11.4).

E-Hydrocarbon Liquid Fuels

e-Gasoline, e-Diesel, e-Kerosene, e-Synthetic/Sustainable Aviation Fuel (e-SAF).
Synthetic replacements for common transportation fuels are almost always derived from natural gas, with a small minority from coal or biomass. They are not e-fuels, but they *could* be e-fuels if made starting with electricity. The distinction is shared by all synthetic fuels—the cost of a truly "green" version of any synfuel is low energy efficiency and higher cost.

As with the simpler e-fuels listed above, all e-hydrocarbon fuels start with electrolytic hydrogen, often with methanol as an intermediary, but followed by extensive processing involving heat and selective catalysts to produce liquid fuels that can approximate gasoline or diesel fuel and thus are drop-in replacements. It has been said that with enough energy and processing steps, it is theoretically possible to manufacture almost any hydrocarbon starting with hydrogen and some source of carbon [16]. This is, in fact, the foundation of much of the petrochemical, plastic, pharmaceutical, and processed food product industries.

The obvious advantage of e-gasoline, e-diesel, or e-SAF is that these fuels can be used as drop-in replacements for the fossil fuel versions of these fuels and the existing automotive or aero fuel distribution infrastructures, with no or minimal modifications required for the vehicles. The best-justified application is e-SAF for aviation, since there are currently no other acceptable alternatives to petroleum jet fuels.

Synthetic (or Sustainable) Aviation Fuel (SAF)

There is strong motivation for sustainable replacements for petroleum jet fuel and other aviation fuels, considering that aviation is responsible for 3.5% of climate change, according to the US National Oceanic and Atmospheric Administration (NOAA) (Fig. 11.5) [17].

There are many pathways for producing SAF from both renewable and nonrenewable feedstocks. Of greatest interest are SAFs that can be produced from non-petroleum-based renewable feedstocks including cultivated fuel crops, woody (cellulosic) biomass, and waste cooking oils and fats. Fuels for commercial jet or turboprop aviation must conform to ASTM Standard D7566—Specification for Aviation Turbine Fuel Containing Synthesized Hydrocarbons. The maximum blending ratios for the various SAFs with petroleum Jet-A allowed by the FAA are between 5% and 50% depending on the SAF and the engine. Use of SAF in US commercial aviation is growing at a rapid pace, with approximately 5 million gallons (19 million liters) in 2021 increasing to 24.5 million gallons (93 million liters) in 2023, according to the EPA. An excellent summary of the various synthesis processes can be found at citation [18].

In 2020 the US Dept of Energy Office or Energy Efficiency and Bioenergy published the technology review and roadmap document, *Sustainable Aviation Fuel Review of Technical Pathways* [19]. Since commercial aviation fuels must be standardized worldwide, the United Nations established the International Civil Aviation Organization (ICAO) to oversee fuel standards [20] and has published the ICAO Global Framework for Aviation Alternative Fuels international roadmap.

Among the ASTM-conforming SAFs considered of greatest promise is Fischer–Tropsch (FT) synthetic paraffinic kerosene (SPK) that has been allowed the maximum blend ratio 50% since 2009. Like most renewable synthetic fuels, its production uses woody biomass to produce low-grade syngas using gasification, which is

Fig. 11.5 Rolls Royce Trent 1000 engine modified to run on 100% sustainable aviation fuel. Image copyright Rolls Royce Ltd., https://www.adsadvance.co.uk/rolls-royce-joins-boeing-and-world-energy-for-successful-100-saf-flight.html. With permission

Fig 11.6 Reported to be the world's only natural gas-to-gasoline complex, located in Turkmenistan, in operation since June 2019. The subject of multiple publications and announcements by Haldor Topsoe A/S of Denmark. https://www.topsoe.com/blog/worlds-largest-atr-based-methanol-plant-has-been-put-into-successful-operation. Image public domain

supplemented with hydrogen from either a renewable or conventional process. The enhanced syngas then undergoes the usual Fischer–Tropsch reaction as the first of several synthesis steps to convert the syngas ultimately into jet fuel. Potential feedstocks include various sources of renewable biomass, primarily biomass such as municipal solid waste, agricultural wastes, forest wastes, wood, and energy crops.

The production of FT-SPK generates large amounts of carbon dioxide, but because of the biomass feedstock, the US DOE classifies it as a renewable fuel that could reduce emissions compared with conventional jet fuel by up to 94% [15]. The substitution of electrolytic hydrogen and CCS carbon for the hydrogen and carbon monoxide in syngas improves its carbon footprint further, allowing the e-SAF designation.

In the USA, sustainable aviation fuels are subsidized under the 2005 Renewable Fuel Standard Renewable Fuel Standard (RFS), subject to the requirements of the Energy Independence and Security Act of 2007 (EISA). This regulation required that renewable fuel (SAF or e-SAF) be blended into commercial aviation fuels in increasing amounts each year, progressively lowering levels of greenhouse gases (GHGs). The US EPA tracks compliance through the Renewable Identification Number (RIN) system, which assigns a RIN to each gallon of renewable fuel. Entities regulated by the RFS can meet their renewable volume obligations (RVOs) by either selling the required biofuels volumes or purchasing RINs from parties that exceed their renewable fuel production requirements (Fig. 11.6) [21].

The Fuel vs. Electricity Dilemma

Recalling the second law of thermodynamics, an obvious question is why go to all the trouble of taking electricity, the highest *quality* form of energy, and performing additional energy-losing conversion steps to produce a liquid or gaseous fuel, the use of which is much less efficient than the direct use of the electricity? There seems to be three legitimate justifications:

1. *A drop-in replacement fuel is required for an application that requires a narrowly specified fuel.* Aviation fuel is the best and probably the only legitimate example here. For example, nothing else can provide the same energy density and combustion characteristics of Jet A-1. The fuel is needed even if its production is energy-wasteful.
2. *Fuel energy storage density and transfer rate.* This mainly applies to the use of an e-fuel as a storage medium for electric grid energy storage. Electricity cannot be stored in batteries at anything near the energy density of a liquid or pressurized gaseous e-fuel.

   This also applies to mobile uses: Much more energy can be stored onboard a vehicle in the form of the e-fuel made from the electricity, even if the fuel will be burned to produce the motive power. Prior to the development of advanced batteries, this was the main argument in favor of e-hydrogen for fuel cell vehicles, although this advantage is now diminished by advanced lithium batteries. Another example: a large transport truck, plane, or ship can transport a much greater amount of energy in the form of a liquid e-fuel than it could using battery-stored electricity.
3. *An ancillary objective.* The most compelling environmental reason is atmospheric carbon removal and sequestration in the form of e-methanol, e-DME, or e-ammonia. In this case, the primary objective is *not to make the fuel but to capture* CO_2. The e-fuel resulting from the process is simply a sequestered form of the carbon (CO2) that would have ended up in the atmosphere.

As always, the larger the number of energy conversion steps, the greater the energy losses. In grid storage applications, power-to-gas (PtG) would suffer from the original generation efficiency of the electricity, the electrolysis to generate hydrogen, the compression energy if stored in tanks or the gas losses from underground storage, the fuel cell reconversion to electricity, and the electrical power conversion losses in AC-DC and then DC-AC conversion at the grid interface. State-of-the-art regenerative hydrogen fuel cells (RHFCs) have a round-trip efficiency RTE of only 30%. The combined losses could take this as low as 20%. Since there is no apparent advantage of PtG over lithium batteries for grid energy storage, such a low RTE would not compare favorably with the RTE of grid battery storage, typically 75–85% for pumped hydrostorage (PSH).

PtG and PtL both face the question: Are the energy losses from making the e-fuel justified for what might be only a small improvement in environmental impact compared with the fossil fuel it replaces? And how does the CO_2 that would have been released to the atmosphere by combustion of a fossil fuel compare with that released by the production and combustion of the e-fuel?

Commentary

There are applications that can only be served by combustion fuels. Therefore, there are legitimate needs for alternatives to incumbent fossil fuels. Fuels made from electricity are solutions for applications without low-carbon alternatives that meet requirements for high energy density and fuel properties similar to the fossil fuels they are intended to replace. Synthetic aviation fuel made from electrolytic hydrogen, aka e-SAF, is such an application. However, the only situation in which e-gasoline or e-diesel makes environmental or energy sense is if excess renewable electric energy is available while petroleum fuels are scarce or deemed too harmful for fuel use. Or, possibly, if the primary objective is CCUS or DCCS rather than production of the e-fuel for its own sake. In this case, poor net energy efficiency can be legitimately overlooked, at least until the environmental impact from the production exceeds that of the fossil fuel that is being replaced [22].

References

1. R.F. Probstein, E.R. Hicks, *Synthetic Fuels* (Dover Publications, 2006) Referenced in citation 3
2. Wikipedia. *Synthetic Fuel.* Last edited on 28 March 2024 https://en.wikipedia.org/wiki/Synthetic_fuel. Accessed 30 May 2024
3. J.C. Ruth, G. Stephanopoulos, Sythetic fuels: what they are and where do they come from? Curr Opin Biotechnol **81**, 102919. Elsevier (2023). https://doi.org/10.1016/j.copbio.2023.102919
4. Automotive Daily 6 March 2023. *A complete guide to eFuels.* https://www.autodaily.com.au/complete-guide-to-efuels/. Accessed 30 May 2024
5. Emeka Ochu, Sarah Braverman, Griffin Smith. Hydrogen Fact Sheet: Production of Low-Carbon Hydrogen. Center on Global Energy Policy at Columbia University, School of International and Public Affairs. 17 June 2021. Accessible from https://www.energypolicy.columbia.edu/publications/hydrogen-fact-sheet-production-of-low-carbon-hydrogen/
6. Sonal Patel, Power Magazine, 1 September 2020 *WindGas Falkenhagen: Pioneering Green Gas Production.* https://www.powermag.com/windgas-falkenhagen-pioneering-green-gas-production/. Accessed 1 December 2024
7. S. Sollai, A. Porcu, V. Tola, F. Ferrara, A. Pettinau, Renewable methanol production from green hydrogen and captured CO_2: a techno-economic assessment. Elsevier Sci J CO2 Util **68**, 102345 (2023)
8. International Renewable Energy Agency (IRENA). Innovation outlook: renewable methanol, 2021. https://www.irena.org/-/media/Files/IRENA/Agency/Publication/2021/Jan/IRENA_Innovation_Renewable_Methanol_2021.pdf. Accessed 4 November 2023
9. D. Wang, W. Meng, H. Zhou, Y. Yang, J. Xie, S. Yang, G. Li, Novel coal-to-methanol process with near-zero carbon emission: pulverized coal gasification-integrated green hydrogen process. J. Clean. Prod. **339**, 130500 (2022). https://doi.org/10.1016/j.jclepro.2022.130500
10. Probably the most widely deployed is the Haldor Topsøe process, see *Methanol Synthesis* https://www.topsoe.com/processes/methanol/synthesis. Accessed 17 December 2023
11. ClimeWorks. *Act now—Remove to zero.* https://climeworks.com/. Accessed 30 May 2024
12. Haldor Topsøe presentation and web page on e-fuel production facility under construction in Denmark. *2022–006-denmark-haldor-topsoe-signs-eur45-million-funding-deal-with-eib-to-drive-green-energy-transition.* https://www.eib.org. Accessed 2 Dec, 2023
13. Skovgaard Energy (press release). *REDDAP: Power-to-Ammonia Ramme* https://www.reddap.dk/en Accessed 5 July 2024
14. World Fertilizer News 01 Nov 23. *ATOME Energy provides update on green ammonia facility in Paraguay* https://www.worldfertilizer.com/project-news/01112023/atome-energy-has-provided-on-green-ammonia-facility-in-paraguay/ Accessed 5 July 2024
15. CCE News, 18 January 2022. World's Largest Green Ammonia Plant set for construction in South Africa https://cceonlinenews.com/news/worlds-largest-green-ammonia-plant-for-nelson-mandela-bay-south-africa/ Accessed 5 July 2024
16. eFuel-Today. Efficiency of e-fuels—How effective are synthetic fuels, 5 March 2023, https://efuel-today.com/en/efficiency-of-e-fuels/

17. US Dept. of Commerce, National Oceanic and Atmospheric Administration. Aviation is responsible for 3.5 percent of climate change, study finds. *3* September 2020. https://research.noaa.gov/2020/09/03/aviation-is-responsible-for-35-percent-of-climate-change-study-finds/ Accessed 6 July 2024
18. US Dept of Energy, Alternative Fuels Data Center. *Sustainable Aviation Fuels* https://afdc.energy.gov/fuels/sustainable-aviation-fuel Accessed 5 July 2024
19. US Dept of Energy, Office of Energy Efficiency and Renewable Energy. Sustainable aviation fuel review of technical pathways https://www.energy.gov/sites/prod/files/2020/09/f78/beto-sust-aviation-fuel-sep-2020.pdf. Accessed 5 July 2024
20. United Nations International Civil Aviation Organization (ICAO), Sustainable Aviation Fuel (SAF). https://www.icao.int/environmental-protection/pages/SAF.aspx Accessed 5 July 2024
21. US Dept of Energy, Alternative Fuels Data Center. Renewable fuel standard. https://afdc.energy.gov/laws/RFS. Accessed 5 July 2024
22. Reuters (online). *Explainer: What are e-fuels, and can they help make cars CO2-free?* https://www.reuters.com/business/autos-transportation/what-are-e-fuels-can-they-help-make-cars-co2-free-2023-03-07/. Accessed 4 October 2024

Plant Oils, Biofuels, and Renewable Diesel

12

Many plant and seed oils can be used to make compression ignition (diesel) fuels with only minimal processing. In the last 25+ years, *biodiesel*, a plant-derived renewable fuel, has been popular, and in many countries, government-subsidized as an alternative fuel. It can be derived from a wide range of agricultural crops and even recycled cooking oil.

Alterantive Diesel Fuel Facts

- *Vegetable oils* (straight vegetable oil, SVO) will work as fuels in a diesel engine but are not recommended due to their physical properties. Not allowed under EPA regulations, although not uncommon to be blended by users with mineral diesel in non-emission-regulated off-road equipment.
- *Biodiesel fuel* (B-diesel) is made from plant oils, methanol, and a caustic chemical such as sodium hydroxide. It is heavily subsidized in the USA, usually blended with mineral diesel fuel to a maximum percentage of 20% (B20), shown as an alternative to diesel No.2 fuel, at a fuel station in Fig. 12.1.
- *Renewable diesel fuel* (R-diesel) is not the same as biodiesel. It is a direct replacement for mineral diesel fuel that has nearly identical properties. It is more expensive than biodiesel or mineral diesel.
- Biodiesel, as currently produced from crops, has WTW (well-to-wheel) GHG emissions usually lower than Diesel No. 2 (D2). The exact difference is the subject of extensive debate, with wide variations reported based upon the analysis design and assumptions.
- Biodiesel cetane ratings are lower than mineral diesel or renewable diesel, but usually higher than raw vegetable (plant seed) oils, which vary greatly.
- All biofuels have intrinsically low sulfur content. They are exempted from California ULSD (Ultra-Low-Sulfur) requirements.

C. A. MacCarley, *Non-Petroleum Automotive Transportation*,
https://doi.org/10.1007/978-3-031-85028-8_12

Fig. 12.1 Biodiesel fuel dispenser. Image from United Soybean Board, USA. https://www.flickr.com/photos/unitedsoybean/10479076844/in/photostream/ Attribution 2.0 Generic, CC BY 2.0

- Biodiesel usually produces higher NOx emissions compared with D2 or renewable diesel, although not beyond the range of existing DEF-based engine NO_x control systems.
- Biodiesel particulate emissions are generally lower than D2, although there is some evidence that the smaller particulate size poses a greater health risk.
- All can use existing diesel fuel distribution system.
- Biodiesel and renewable diesel are classified as renewable fuels, though energy and carbon balances are questionable. Probably the only legitimately renewable fuels are unprocessed vegetable oils.
- Water requirements and land use conversion are major concerns for all bio-derived diesel fuels.
- Energy yield per cultivated acre is poor. Land requirements for biodiesel are higher than corn ethanol for equivalent energy.
- All bio-derived diesel fuels are hard to justify when compared with the use of the same plant feedstocks for food.
- All are more expensive than D2 for a given amount of energy or travel distance.
- The largest fraction of biodiesel production in the USA is from soybean (soya) oil. Recycled cooking oil second, used when available.
- Diesel engine manufacturers warn against using straight vegetable oil in modern diesel engines.
- They also disallows the use of any biodiesel blend that contains more than 20% biodiesel in on-road vehicles
- Biodiesel has lower energy density than D2, renewable diesel, *or even the vegetable oil used to make it*.
- Biodiesel and renewable diesel fuels were developed to replace mineral (petroleum distillate) diesel fuel for automobiles and trucks. Both qualify for incen-

tives under the US Renewable Fuel Standard (RFS) Program and may also qualify for tax incentives in some states.

Plant (Vegetable) Oils

Unprocessed plant oils (aka vegetable oils) have been used as fuels for compression ignition engines since the invention of the diesel engine by Rudolf Diesel 1897. At the demonstration of his compression ignition engine at the 1900 Paris Exhibition, the engine ran on peanut oil. He experimented with various plant oils and animal fats [1]. Circa 1930, plant-based oils were forgotten when petroleum (mineral) oils became common and inexpensive. A diesel engine can actually run on most raw vegetable oils, although it violates the warranties of all major diesel engine manufacturers. The cetane ratings of vegetable oils are highly variable, and their high viscosity, fat, and fatty acid content cause problems due to their tendency to solidify at temperatures below about 20 °C as well as excess wear from corrosion and generally lower lubricity compared with petroleum diesel. The DOE strongly discourages the use of unprocessed vegetable oils as fuels in modern *emission-controlled diesel* vehicles [2].

Renewable Diesel

Both renewable and biodiesel fuels are intended for the same applications as petroleum distillate (aka mineral) diesel. Along with fuel ethanol, they qualify for the US Renewable Fuel Standard (RFS) Program and may also qualify for state government fuel standards and programs. Both biodiesel and renewable diesel are produced from plant (veg) oils and animal fats. However, the production process, blending limits, and cost of these two bio-derived fuels are very different. The chemical and physical properties of renewable diesel are closer to those of petroleum (mineral) diesel. Renewable diesel meets the federal registration requirements for mineral diesel fuels and fuel additives, and the ASTM D975-21 Standard Specification for Diesel Fuel [3]. Renewable diesel is acceptable by all engine manufacturers as a direct substitute for mineral diesel. Renewable diesel is labeled with an R followed by the percentage (by volume) of the renewable diesel content, e.g., 100% renewable diesel fuel is designated R100. A 50% blend would be designated R50.

There are actually several different processes for converting plant oils to renewable diesel fuel. These are summarized on the DOE Alternative Fuels Data Center (FDC) website [4]. The most common method used in the USA is hydrogenation-derived renewable diesel (HDRD) or hydroprocessed esters and fatty acids (HEFA) produced by hydrogenation of triglycerides, a similar process used for desulfurizing petroleum diesel.[1] This hydrotreating process involves reacting the feedstock (lipids) with

[1] US Dept of Energy [4].

hydrogen under elevated temperatures and pressures in the presence of a catalyst. Hydrotreating is already part of the petroleum refining process, so existing refineries can be converted to renewable diesel production with minimal changes.

The various processes all involve large quantities of hydrogen, making renewable diesel more expensive than biodiesel production. And the renewability of the hydrogen source factors into whether the renewable diesel can meet national or state standards for biofuels. If made from entirely renewable (i.e., electrolytic or naturally occurring) hydrogen, the resulting renewable diesel would be unreasonably costly to produce, regardless of incentives. In practice, the required hydrogen for hydrotreating both renewable diesel and mineral diesel is made by conventional reforming of natural gas or feedstock petroleum, which calls into question the *renewable diesel* name.

Renewable diesel usually has a slightly higher H:C ratio than mineral diesel, so it produces lower CO_2 per unit of fuel energy. Its cetane rating—a measure of ignition lag and therefore compression ignition fuel quality—is typically above 70, about the same as mineral diesel, and higher than biodiesel.

Biodiesel

Biodiesel is produced from plant-based oils or animal fats. Soybeans are the dominant feedstock for US biodiesel production (Fig. 12.2). Recycled cooking oil has also become a major feedstock for biodiesel producers. Rapeseed oil, sunflower oil, and palm oil are also used as feedstocks for biodiesel production, but are more common in other countries (Fig. 12.3). Algae has long been believed to have potential for biofuel production, but currently is not used for production. The feedstocks used for biodiesel production can affect the physical properties and possible uses of biodiesel [5].

Biodiesel is a legacy alternative fuel. It was first commercially manufactured in the USA in Kansas City, Missouri, in 1991 [6]. By 2009, the National Biodiesel Board listed over 200 US producers. Biodiesel is usually sold as a blend with mineral diesel fuel, most commonly a volumetric blend of 20% biodiesel and 80% D2. Its production involves a process called transesterification which uses methanol to convert the oil and fats into fatty acid methyl esters (FAME) and a coproduct, glycerol. The process can be catalyzed by either an acid or base, with sodium hydroxide most commonly used in the USA.

A typical "recipe" for small-scale production of one gallon (3.8 L) of biodiesel fuel from rapeseed is [7]:

1 gal (3.8 L) rapeseed (canola) oil, extracted from 1/3 bushel (9.3 gal US, 9.0 kg), boiled and pressed dry rapeseed.

0.20 gal (760 ml) methanol (20% of oil volume)

0.046 lb (21 g) sodium hydroxide (lye, caustic soda)

The home-brew process is described in many publications and online instructional videos, but beware of the inadequate safety provisions and lack of scientific understanding in many of the DIY productions. One article with some level of

Fig. 12.2 Harvested soybeans, ready for processing into soy oil and ultimately biodiesel fuel. Wikipedia commons, File:CSIRO Science Image 3272 Soybeans.jpg, Creative Commons Attribution 3.0 Unported license

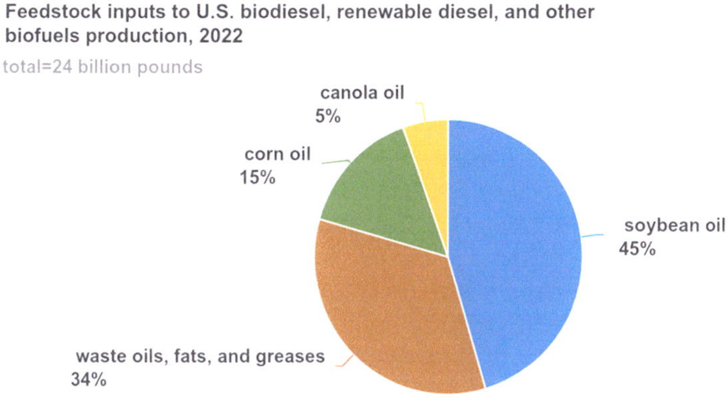

Fig. 12.3 Feedstocks for biodiesel production in USA. Monthly Biofuels Capacity and Feedstocks Update, US EIA, December 2023. https://www.eia.gov/energyexplained/biofuels/biodiesel-rd-other-basics.php. Public domain

accountability is citation [8]. The transesterification reaction requires a temperature of 70 °C, which is usually provided by external combustion heat, typically from natural gas or waste biomass from grain crops. It is possible to make DIY biodiesel fuel in a batch process, but is dangerous due to the risk of fire or explosion from the combination of combustion heating and large quantities of oil and methanol, in addition to possible chemical burns from sodium hydroxide. Amateur biodiesel production is a known cause of serious accidents and even fatalities [9].

The output should be 1 gallon of biodiesel fuel and variable amounts of glycerin (liquid) and methanol. Fractional distillation is used to separate the glycerin/methanol co-product solution, allowing recovery of methanol for reuse and the possible sale of the glycerin for chemical production and consumer products including soaps and cosmetics. The production of biodiesel has led to a large surplus of glycerin in the USA, where 63% of worldwide biodiesel production occurs [8]. The low market value of the waste glycerin and low cost of methanol may not economically justify recycling [10].

In the USA, commercial biodiesel must meet American Society for Testing and Materials (ASTM) specification D6751 [11] for blending with petroleum diesel/distillate. Production of biodiesel, both commercial and DIY, uses the transesterification and refining process illustrated in Fig. 12.4, from the US Dept of Energy.

Biodiesel has a higher cloud point (or wax point), the temperature at which it starts to solidify. It has a higher viscosity (thickness) compared with petroleum diesel. Consequently, it can easily encounter flow or solidification problems in cold weather—one of the main reasons why it is blended with mineral diesel. Unlike petroleum diesel, biodiesel has a finite shelf life and is more hygroscopic, attracting water that can lead to potential algae and bacteria growth. This is a major problem for recycled waste vegetable oil (WVO) that may have high levels of nutrients that encourage bacteria or algae growth. (WVO is discussed in Chapter 15 *Fuels for desperate times.*) Professional truckers may be familiar with finding algae jungles growing in their fuel tanks after sitting out a season [12]. Unblended biodiesel (B100) is not approved by any current manufacturers of on-road vehicles. The cetane number, a measure of combustion quality in a diesel engine, is 45–67 for biodiesel, lower than that of petroleum diesel or renewable diesel which is typically 70. Biodiesel also has a lower fuel energy value (specific calorific value) than mineral diesel, renewable diesel, and even the vegetable oil that it was made from [13]. For these reasons, the biodiesel content of blended Bxx fuels is limited by all manufacturers of modern emission-controlled diesel engines to a maximum of 20% biodiesel (B20).

Biodiesel typically has a higher H:C ratio than petroleum diesel and therefore produces somewhat lower CO_2 emissions, similar to renewable diesel. But biodiesel is known to increase oxides of nitrogen (NOx) emissions, especially in higher compression open-chamber diesel engines, another characteristic that limits its blend

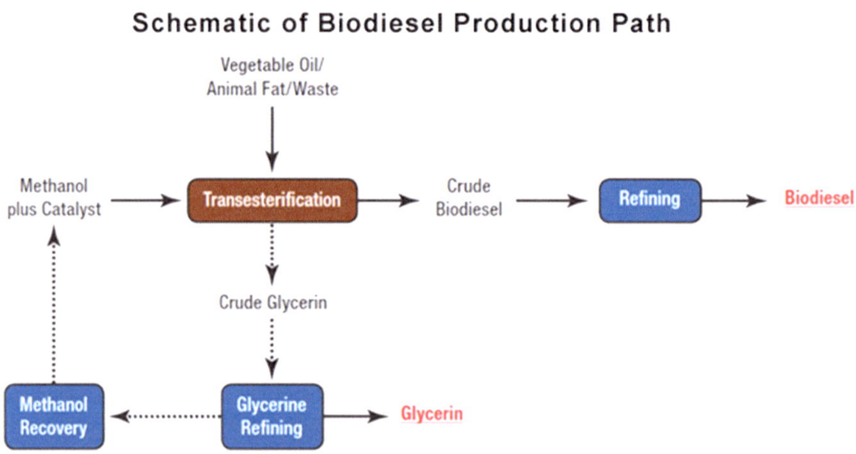

Fig. 12.4 Commercial production of biodiesel from plant oil feedstocks. Some of the methanol used for transesterification is recovered. A large byproduct of glycerin is also produced, which has commercial value but is often discarded. Diagram from https://afdc.energy.gov/fuels/biodiesel-production. Public domain

level with diesel fuel. In California, CARB has established separate regulations for biodiesel-fueled vehicles to mitigate the potentially higher NO_x compared with ultra-low-sulfur petroleum diesel fuel (ULSDF).

GHG Impacts of Bio-derived Diesel Fuels

The environmental impacts of the main biofuels, biodiesel and corn ethanol, have often been subjects driven more by politics than objective science. Energy independence and job arguments have dominated the biodiesel conversation in the USA since the late 1980s, while objective evaluations of the complete (including land use conversion) life cycle GHG emissions have been rare. Consequently, available data is highly variable.

All currently produced biofuels, especially biofuels for diesel engines, originate from agricultural crops that could be used as food or animal feed. Soybeans (soya), corn oil, and rapeseed (canola) oil provide 64% of the biodiesel feedstocks in the USA, while rapeseed is most common in Europe and palm oil dominates in southeastern Asia. These are classified as *first-generation biofuels*. Inasmuch as these fuels must come from cultivated crops that could be used as foodstuffs, the environmental calculus is murky, with a plethora of ways to frame arguments both in favor and opposed to this diversion of food to fuel in agriculture [12].

Second-generation biofuels are derived from non-food feedstocks: lignocellulosic plants cultivated specifically for fuel production, but also agricultural residues, forest residues, solid waste, etc. Second-generation biofuels can be much less environmentally consequential if their cultivation does not require land use conversion or their feedstocks are existing waste products that would otherwise decay, be burned, or end up in landfills. Unfortunately, the technology to produce biodiesel from second generation sources is even more complex and costly than for production of cellulosic ethanol, which has been studied for decades [12] but remains unfeasible.

Recyclable vegetable oil first used for cooking is ambiguously classified between first- and second-generation feedstocks, having attributes of both, depending on the degree to which its original carbon footprint can be ignored by allowing it to be considered a waste product of no additional consequence to the environment [14].

While the classification is potentially very broad, *third-generation biofuels* are usually synonymous with oil or ethanol feedstocks produced from microalgae. There have been compelling arguments in favor of algal biofuels, citing its exceptional growth rate and solar conversion efficiency. Still in research deployments only, this feedstock may have future potential for GHG reduction, but the majority of unbiased studies report net GHG impacts as possibly worse than fossil fuels. At the present state of the science, the consensus is that third-generation biodiesel is neither economically nor energy sustainable [15]. Second- and third-generation biofuels are sometimes referred to in the USA and Europe as *advanced biofuels*.

Among the environmental impacts of greatest concern for first-generation biodiesel are direct and indirect land-use change (LUC) and the degradation of uncultivated land, forests, water resources, and ecosystems [16]. First-generation feedstocks such as soy or corn that have value as food or livestock feed are subject to the well-known food vs fuel conundrum, especially with regard to land and water use, as well as the climate and health impacts of agricultural chemicals. Climate and health compromises are justified for food production, but are they justified for the production of combustion fuels? Both end products are heavily subsidized, the first for *food security*, the second for *energy security*.

Reported values of GHG emissions for first-generation biodiesel show large variations between the many published Life Cycle Analyses, with a range of 3–111 g CO_2 per MJ of fuel energy reported. The average GWP of biodiesel production from first-generation feedstocks is generally lower than that of mineral diesel. This conclusion and a detailed analysis of the environmental impacts of bio-derived diesel fuels can be found in citation [17].

Cost

US government subsidies for soybean (and corn) production apply regardless of whether the crops are used for food or fuel production. The Commodity Credit Corporation (CCC) and the Federal Crop Insurance Corporation (FCIC) aid farms and implements measures to stabilize the prices of farm commodities. These subsidies have endured since the US Great Depression of the 1930s as provisions of US agricultural policy to support critical food production.

In addition to the core agricultural subsidies are credits to encourage the cultivation and processing of fuel crops, including the Volumetric Biodiesel Excise Tax Credit and the Renewable Biodiesel Tax Credit. According to the Iowa Soybean Association [18],

> There are two tax credits that soybean oil-based biofuel producers are eligible for. The long-standing Blenders Tax Credit provides $1 per gallon of biodiesel, renewable diesel and sustainable aviation fuel (SAF) that was blended into the US fuel pool, even if the fuel is imported. In August 2022, the Inflation Reduction Act (IRA) extended the blenders tax credit through the end of 2024. The IRA created two additional biofuel tax credits; the Sustainable Aviation Fuel Tax Credit (40B) which will expire at the end of 2024, and the Clean Fuels Production Tax Credit (45Z) starting in 2025.

According to the Institute for Energy Research [19],

> U.S. renewable diesel production capacity nearly quadrupled from 791 million gallons a year in 2021 to 3 billion gallons by 2023. Combined with biodiesel, total U.S. output capacity for biomass-based diesel surpassed 5 billion gallons by 2023. Renewable diesel is a substitute for oil-based diesel, whereas biodiesel can only be used as a blend, making the former more attractive for producers. Renewable diesel can be produced in existing refinery equipment, but the yields are lower than with conventional diesel.

Fuel-specific credits and direct subsidies create different profitability models which favor fuel over food, but also place the fate of the biofuel industry in the ever-changing hands of political leadership.

While both ***renewable diesel*** and ***biodiesel*** fuel are derived from similar feedstocks, renewable diesel is a higher-quality fuel, justifying 100% R100 rather than the 20% B20 limit for biodiesel. But it is more expensive and more energy intensive to produce compared with petroleum diesel or (subsidized) biodiesel fuel: from $.50 more to twice the cost of biodiesel depending on location and market. Nevertheless, there is a trend (at least in California, late 2024)[2] for fuel stations to start offering renewable diesel R100 instead of or in addition to biodiesel B20, despite the higher cost at the pump.

Both straight vegetable oil (SVO) and waste vegetable oil (WVO) are not approved by any engine manufacturer, and their use is illegal under EPA and CARB anti-tampering regulations. Nevertheless, older off-road diesel engines continue to be fueled by locally grown SVO feedstocks. The use of WVO, once given away free by fast food restaurants, has almost completely disappeared because it is now purchased by biodiesel producers. No more aroma of French fries when following a *greasel* pickup.

B20 biodiesel, with the help of government subsidies, has a regulated price that is set slightly below the volumetric cost of Diesel No. 2 (D2). But due to its relatively lower fuel energy value, it is actually slightly more expensive per unit of energy, i.e., vehicle range. Current US national average prices for petroleum and biodiesel fuels can be checked monthly at citation [20], an example from summer 2024 is shown in Table 12.1. Up until just recently (late 2024), *renewable diesel* had been sold almost exclusively to institutional users and government agencies rather than consumer retail outlets, due to its higher cost. R99 renewable diesel can replaces D2, since their properties are nearly identical.

Table 12.1 Comparative retail prices for standard and alternative fuels, summer 2024. From [20]

Fuel	Price
Biodiesel (B20)	$3.94/gallon
Biodiesel (B99–B100)	$4.57/gallon
Ethanol (E85)	$2.96/gallon
Natural Gas (CNG)	$2.90/GGE
Liquefied natural gas	$3.85/GGE
Propane	$3.45/gallon
Gasoline	$3.65/gallon
Diesel	$4.07/gallon

Data from https://afdc.energy.gov/fuels/prices.html

[2] Based on personal survey of 20 diesel fuel stations in Central and Coastal California.

References

1. A.M. Brownstein, Diesel fuel (Chapter 7), in *Renewable Motor Fuels The Past, the Present and the Uncertain Future (Book)*, (2015) https://www.sciencedirect.com/book/9780128009703/renewable-motor-fuels. Accessed 1 November 2024
2. US Dept. of Energy AFDC. The Myth of Using Straight Vegetable Oil as a Diesel Fuel Fact Sheet (2024). https://www.energy.gov/energysaver/myth-using-straight-vegetable-oil-diesel-fuel-fact-sheet. Accessed 5 July 2024
3. California Air Resources Board. https://ww2.arb.ca.gov/resources/fact-sheets/fact-sheet-renewable-diesel-fuel-requirements. Accessed 2 June 2024.
4. US Dept of Energy, Alternative Fuels Data Center. *Renewable Diesel*. https://afdc.energy.gov/fuels/renewable-diesel. Accessed 1 July 2024.
5. US Energy Information Agency. Today in Energy, 3 May 2018. Biodiesels produced from certain feedstocks have distinct properties from petroleum diesel. https://www.eia.gov/todayinenergy/detail.php?id=36052
6. Farm Energy. History of Biodiesel 3 April 2019. https://farm-energy.extension.org/history-of-biodiesel/. Accessed 1 December 2024.
7. Average of multiple online sources. Methanol and lye requirements from ThoughtCo. How to Make Biodiesel from Vegetable Oil. https://www.thoughtco.com/make-biodiesel-from-vegetable-oil-605975. Accessed 1 May 2024
8. Mother Earth News, 2024. How to Make Biodiesel: DIY Home Biodiesel Production. https://www.motherearthnews.com/sustainable-living/green-transportation/home-biodiesel-production-zm0z15aszmar/ Accessed 5 July 2024
9. Marmo L, et al. Fire and explosion risk in biodiesel production plants: a case study. Hazards 27, IChemE, 2017.
10. B.R. Margarida, L.F. Luz Jr., Reutilization of crude glycerol in a circular biodiesel production process. Chem Eng Trans **92**, 445–450 (2022) Published by AIDIC, Italy
11. ASTM International. Standard Specification for Biodiesel Fuel Blend Stock (B100) for Middle Distillate Fuels. Current as of 22 March 2023. https://www.astm.org/d6751-20a.html Accessed 5 July 2022
12. Diesel Fuel HQ How to Prevent Diesel Fuel Algae? (10 Proven Ways). https://dieselfuelhq.com/how-to-prevent-diesel-fuel-algae/ Copyright © 2024 Diesel Fuel HQ. Accessed 1 December 2024
13. Robert Bosch GmbH, Properties of liquid fuels and hydrocarbons, in *Automotive Handbook*, 10th edn., (2022), p. 346
14. Wikipedia. *Second Generation Biofuels*. https://en.wikipedia.org/wiki/Second-generation_iofuels. Accessed 1 November 2024
15. H. Xu, L. Ou, Y. Li, T.R. Hawkins, M. Wang, Life cycle greenhouse gas emissions of biodiesel and renewable diesel production in the United States. Environ Sci Technol **56**(12) (2022) ACS Publications. Policy Analysis, 16 May 2022. https://pubs.acs.org/doi/10.1021/acs.est.2c00289?__cf_chl_tk=k91jd7j3_rVDYBoLWxQsGcfnkNU5gXLYs3vt6fyRzyU-1732743962-1.0.1.1-9WKErIydFuxmg3ZHXFdfWMo0IAAnse0C6w.AS2UM.NI
16. K.G. Austin, J.P.H. Jones, C.M. Clark, A review of domestic land use change attributable to U.S. biofuel policy. Renew Sustain Energy Rev **184** (2023) Accessible at https://www.sciencedirect.com/science/article/pii/S136403212200106X Accessed 1 November 2024
17. H.K. Jeswani, A. Chilvers, A. Azapagic, Environmental sustainability of biofuels: a review. Proce Royal Soc A **476**, 20200351 (2020). https://doi.org/10.1098/rspa.2020.0351. Accessed 1 November 2024
18. Iowa Soybean Association. What are the current tax credits in place for fuel made from soybeans? https://www.iasoybeans.com/newsroom/article/isr-october-2024-issue-update-proposed-tax-credits. Accessed 1 November 2024.
19. Institute for Energy Research. Renewable diesel and biodiesel get a setback. June 4, 2024 https://www.instituteforenergyresearch.org/fossil-fuels/renewable-diesel-and-biodiesel-get-a-setback/ Accessed 1 November 2024.
20. US Dept. of Energy. Alternative fuel price report, April 2024. https://afdc.energy.gov/files/u/publication/alternative_fuel_price_report_april_2024.pdf. Accessed 10 July 2024.

13 Dimethyl Ether

Dimethyl ether (DME) is probably the least known of all available alternative fuels, but it has the potential to provide the largest positive energy and environmental impact.

It can be used as a substitute or blending agent with LPG for domestic cooking or heating, but its greatest value is as a replacement for petroleum diesel fuel in compression ignition (CI) engines. It has also shown potential as a fuel of gas turbine engines, either alone or as a supplement to jet fuel. In a CI engine, it burns extremely clean compared with diesel fuel, producing no particulates and less than half the CO_2 per unit of energy (i.e., travel distance). It is a gas at NTP that liquifies at less than 5 bar pressure at room temperature, and has physical properties similar to butane or propane (LPG). It requires different fuel injection components and low-pressure compressed storage tanks on a vehicle. Worldwide, DME is usually made from natural gas, or from coal in China and India. Like methanol, DME is made by reformation of syngas, which can be produced from almost any carbonaceous material, in particular, cellulosic biomass. If produced from waste biomass, it is a renewable fuel with a relatively small carbon footprint [1, 2].

DME is often added to liquid and gaseous fossil fuels and biofuels to improve their combustion quality and reduce emissions. In China, it is commonly used in place of or blended with LPG, which is a more expensive commodity.

As a vehicle fuel, an important consideration is that its volumetric energy density is about half that of diesel fuel, requiring larger fuel tanks for equivalent range. Figure 13.1 shows a prototype Volvo box truck specifically designed to run on DME, with larger fuel storage tanks on either side covered by white access panels.

C. A. MacCarley, *Non-Petroleum Automotive Transportation*,
https://doi.org/10.1007/978-3-031-85028-8_13

Fig. 13.1 Volvo Box Truck designed to run on DME, 2015. Image unattributed, believed to be Volvo AG Truck Group. Copy from https://cordis.europa.eu/article/id/84358-a-clean-alternative-truck

Production

DME is a replacement diesel engine fuel that can be thought of as the CI counterpart to methanol. Chemically, it has the same molecular composition as ethanol, but has the symmetric molecular structure of an ether rather than an alcohol. In its production, DME is closely related to methanol. In fact, it is commonly made from methanol in a process called *dehydration of methanol*.

This *indirect* process converts two methanol molecules into one molecule of DME and one molecule of water, in the presence of an aluminum oxide or aluminosilicate (zeolite) catalyst.

$$2CH_3OH \rightarrow CH_3OCH_3 + H_2O$$

$$2 \times \text{methanol} \rightarrow \text{dimethyl ether} + \text{water}$$

It is exothermic, releasing excess heat which reduces the net efficiency.

For *direct* synthesis, DME is made directly from syngas: the methanol synthesis and dehydration occur concurrently in a single reactor using a bifunctional catalyst [3]. As the products of the reformation are generated, they are consumed by the synthesis reaction. The reactions are endothermic and exothermic respectively, with the heat released by the synthesis utilized by the reforming process.

The overall reaction for direct DME synthesis from syngas is net exothermic:

$$3CO + 3H_2 \rightarrow CH_3OCH_3 + CO_2 + \text{heat}$$

There are several variations of this process for different production scales. A detailed study [4] comparing indirect vs. direct DME synthesis found an optimized conversion efficiency of approximately 65% for both methods [5] each depicted in the process diagrams of (Fig. 13.2) for various fossil and renewable feedstocks.

As reported in reference,[1] the reactor for *direct* synthesis is more complex, and optimization of each process is a compromise. Catalysts with higher selectivity are required.

[1] Cyril Fortin et al. [4].

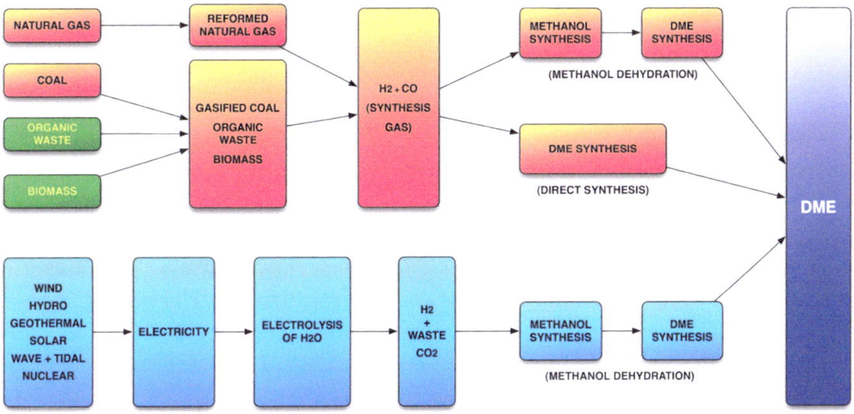

Fig. 13.2 Paths for production of DME from various energy feedstocks. Diagram from https://upload.wikimedia.org/wikipedia/commons/0/03/Dme_production_dialg_090423.jpg. Creative Commons CC BY 2.0

In the *indirect* method, the separation of DME from the byproducts (methanol, CO_2, ethylene, hydrogen, and water) is more difficult and energy-intensive, especially if the feedstock for the syngas is broadly specified biomass.

However, if the objective is fuel-grade DME, contaminants in small amounts may be tolerable; therefore, fuel-grade DME can be considerably less costly.

Physical and Fuel Properties

- A potential CI engine fuel, replacing diesel fuel.
- Naturally low-carbon, no-sulfur if made from natural gas. Biomass DME may contain contaminants such as sulfur, but selectivity of candidate biomass can avoid this.
- DME has a higher cetane rating than diesel or biodiesel making it a better CI engine fuel than diesel or biodiesel fuel.
- A gas at NTP. Liquefies under very little pressure, 5 bar (73 psia) at 22 °C. This is convenient for fuel storage, but it makes challenging compared with conventional liquid diesel fuels.
- Easily stored and handled, similar to butane or LPG.
- Existing infrastructure for LPG distribution and retail sales can be utilized with no changes.
- Commonly used as a consumer spray-can propellant, e.g., hair spray, paint, spray lubricants, and disinfectants.
- Also used as a commercial or specialty refrigerant. Well-established handling procedures.
- Almost zero lubricity, so a lubrication additive is required to prevent wear of fuel injection components and cylinder walls. Castor or Lesquerella oil have been used successfully.

- More efficiently made from natural gas or biomass than any synthetic liquid fuel except methanol.
- Very low well-to-tank and well-to-wheel carbon intensity for any feedstock, similar to methanol or ethanol.
- Combustion produces zero particulates or soot emissions, even with excessively rich AFRs.
- Efficiency of production from biomass gives it the potential to be a low cost, renewable fuel for CI engines.
- If leaked or released into atmosphere, DME has a global warming potential of 1, and an atmospheric lifetime of less than one week. Trivial GHG impact.
- Similar safety profile to propane except that higher upper flammability limit makes it easier to ignite in the event of a leak. High flame propagation velocity leads to potentially explosive mixtures in enclosed spaces.
- In vapor form, heavier than air, similar to LPG. Similar area ventilation requirements.
- Fuel tank similar to LPG, with 250 psi maximum pressure rating.
- Distinct "sweet" odor makes it easier to detect leaks without added odorant.
- Transport market is emerging with truck manufacturer partners like AB Volvo Group, Mack (USA), and Nissan. Pilots in EU, Japan, and China.
- DME has also been used experimentally as a fuel for gas turbine engines, replacing jet fuel. This was discussed as a possible SAF option in Chap. 11 *Synthetic and e-Fuels* [6].
- DME can also substitute for or be blended with LPG for domestic cooking. It is a favorable option where petroleum is scarce but coal is plentiful.
- Stoichiometric mass AFR = 9.0.
- Mass density = 2.05 g/L in liquid form.
- Mass energy density, LHV = 28.8 kJ/g.
- Volume energy density in liquid form = 0.669 g/ml.

Limitations

- Volumetric energy density about half that of D2. Twice the volume for each fuel injection for same power output. Twice the fuel tank volume required for the same range.
- DME is a challenging fuel to inject in a diesel engine since it is difficult to keep it in liquid form in fuel lines and injectors exposed to engine heat. Low-pressure systems prone to vapor lock.
- Nontoxic, but NIOSH Class 2 health hazard. An asphyxiant like carbon dioxide. Slightly inebriating or nauseating if inhaled.
- Light blue low-luminosity flame. Flame is difficult to observe in daylight.
- The boiling point of DME at 1 atm is −24 °C. This means that if the outdoor (tank) temperature is below this, the fuel will remain a liquid with the tank pressure at or below 1 atm.

- DME is an excellent solvent and therefore has compatibility problems with many elastomers used in fuel systems designed for petroleum products. Like methanol, it is not compatible with synthetic rubbers designed exclusively for use with gasoline or diesel fuel, but it is compatible with less expensive and more common Buna-N or similar nitrile rubbers.

Current Commercial Uses for DME

- DME is commonly blended with LPG (propane + butane) for domestic cooking in locations where LPG or natural gas is scarce or expensive, e.g., Western China. It is nearly a drop-in replacement for LPG requiring only a slightly richer AFR. Outdoor gas BBQ grills such as shown in Fig. 13.3 can work well with DME replacing LPG, requiring only slight enlargement of the burner jets.
- Sold under the DuPont trade name Dymel®, DME has long been considered an ideal propellant for consumer spray cans, e.g., hair spray, paint, spray lubricants, and disinfectants. Although it is not toxic, it is not used for pressure-dispensed food products, since it is soluble in many liquids, including cooking oils. Pressure in an aerosol spray can is maintained at the boiling point of the DME propellant, about 63 psig (77 psia, 5.3 bar) at 22 °C [7], but varies with temperature. The pressure drops to atmospheric at −25 °C, which limits use of the spray can at very cold temperatures.
- As a phase-change refrigerant, DME has an ideal boiling point (−25 °C at 1 atm), and a high heat of vaporization (21.5 kJ/mole) compared with 15.5 kJ/

Fig. 13.3 Sustainable barbecue: Generic photo of a gas grill for outdoor cooking. Conversion to DME requires slightly increasing the orifice area of the burner jets. Not attributed

mole for R134a. However, DME is flammable, similar to R290 (propane) with 18.8 kJ/mole energy content.

GHG Emissions

How do the CO_2 emissions per unit of distance of a DME-fueled vehicle compare with those of an equivalent diesel or electric vehicle? Here is a simplified comparison based on a 2024 light truck that is available in either diesel or electric versions. The DME engine is the same diesel engine, but converted to operate on DME.

A 2024 Dodge (Stellantis) RAM 1500 with the 3.0 L turbo diesel option is fuel economy rated at 47.7 km/gal US (29 mpg) of diesel fuel having an LHV of 43.0 MJ/kg, equivalent to an energy use rate of 2.80 MJ/km:

$$\text{Diesel}: \left(\frac{3.15\,\text{g CO}_2}{1.0\,\text{g D2}}\right)\left(\frac{1\,\text{g D2}}{43.0\,\text{kJ}}\right)\left(\frac{2800\,\text{kJ}}{\text{km}}\right) = 205\,\text{g CO}_2\,/\,\text{km}$$

If the diesel engine is converted to run on DME,

$$\text{DME}: \left(\frac{88\,\text{g CO}_2}{46\,\text{g DME}}\right)\left(\frac{1\text{g DME}}{28.9\,\text{kJ}}\right)\left(\frac{2800\,\text{kJ}}{\text{km}}\right) = 185\,\text{g CO}_2\,/\,\text{km}$$

For perspective, a 2024/2025 RAM 1500 Electric has a rated energy use of 0.31 kWh/km. The CO_2 emissions per kWh from the US electric grid, ignoring distribution losses, are 400 g CO_2/kWh:

$$\text{Electric}: \left(\frac{400\text{g CO}_2}{\text{kWh}}\text{for grid electricity}\right)\left(\frac{0.35\,\text{kWh used by truck}}{\text{km traveled}}\right) = 140\,\text{g CO}_2\,/\,\text{km}$$

In this simple comparison, DME generates 10% less CO_2 per distance than diesel fuel in the same engine/vehicle, but 32% more net CO_2 per distance than the electric model of this truck.

Power Output (at Stoichiometric Limit)

Assume 2.0 L diesel engine, naturally aspirated, 4000 RPM max power speed.

At 22 °C, 1 atm, 1 L air weighs 1.20 g.

DME is injected after intake valve closed, so it does not displace any intake air.

At 4000 RPM, intake air mass flow: g

$$4000\,\text{RPM}\left(\frac{1\,\text{min}}{60\,\text{sec}}\right)\left(\frac{1\,\text{intake}}{2\,\text{revolutions}}\right)\left(\frac{2.0\,\text{L air}}{1\,\text{intake}}\right)\left(\frac{1.20\text{g air}}{\text{L air}}\right) = 80\text{g air}\,/\,\text{sec}$$

DME:

$$\left(\frac{80\,\text{g air}}{\text{s}}\right)\left(\frac{1\text{g DME}}{9.0\text{g air}}\right)\left(\frac{28.8\text{ kJ}}{1\text{g DME}}\right)\left(\frac{\text{kW}}{\text{kJ/s}}\right)(0.30\text{ efficiency}) = 76.8\,\text{kW} = 103\,\text{HP}$$

$$\text{Diesel}: \left(\frac{80\text{g air}}{\text{s}}\right)\left(\frac{1\text{g D2}}{14.5\text{g air}}\right)\left(\frac{43.0\text{ kJ}}{1\text{g D2}}\right)\left(\frac{\text{kW}}{\text{kJ/s}}\right)(0.30\text{efficiency}) = 72.2\text{kW}$$
$$= 96.2\text{ HP}$$

DME will produce 7.1% greater power output than diesel fuel in the same engine.

DME Engines

As reported by Fleet Equipment Magazine in 2014 [8],

> Volvo recently tested eight different fuels along criteria including energy efficiency, cost, infrastructure and climate impact. DME came out on top in their testing, beating out CNG, LNG, ethanol and biodiesel.

In theory, any CI (diesel) engine can run on DME. But it far from a drop-in replacement. Its physical properties are more like propane or butane than diesel fuel. Its low viscosity in liquid form and nearly zero lubricity make it difficult to inject, and internal pump leakage can be as high as 50%. Approximately twice the fuel volume must be injected to provide power output equivalent to diesel fuel. Other than these fuel system challenges, the diesel engine itself needs no significant modification. The difference is entirely in the fuel injection system. The core challenge is getting a metered quantity of DME fuel into the cylinders with each engine cycle. This requires that the fuel remain in liquid phase at all points in the system prior to being injected at high pressure into the cylinders. Two approaches are presented below.

Positive Displacement Fuel Injection

Positive displacement injection systems are mechanical or electromechanical fuel injection systems for CI engines. All automotive diesel fuel injection systems made until the early 1990s were of this type, and unregulated diesel engines continue to use them today. Other than a fuel shut-off solenoid, traditional distributor or rotary diesel injection pumps need no electrical components or controls.

Figure 13.4 shows a 4-cylinder diesel engine on a test stand, converted to positive displacement DME fuel injection, using a single distributor pump that provides fuel to the four injection nozzles (or injectors) of the engine.

Referring to Fig. 13.5, a distributor injection pump meters fuel at high pressure to the injector for each cylinder, once every two engine rotations. This is done by a very

Fig. 13.4 Isuzu 4-cyl 1.9L diesel engine modified for operation on dimethyl ether undergoing tests on dynamometer, at California Polytechnic State University. Photo by author

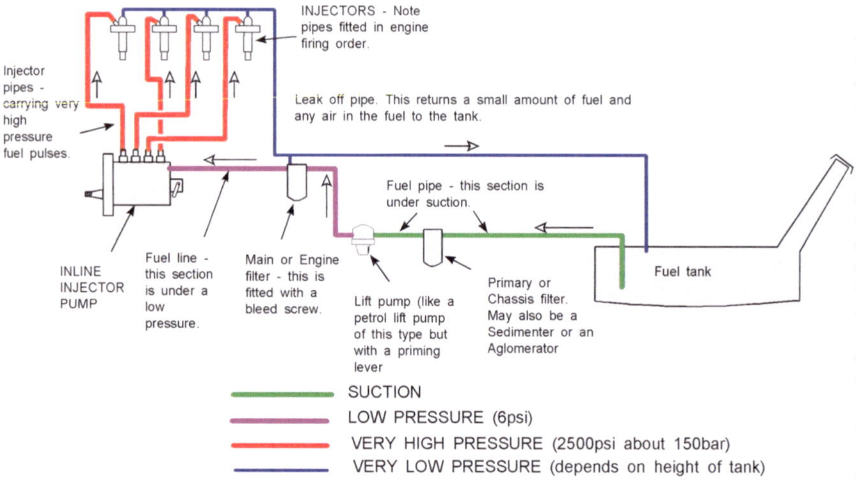

Fig. 13.5 Diagram of a conventional low-pressure diesel injection system, almost identical to the topology of a low-pressure DME injection system. Figure from *The CI fuel system*. http://www.tb-training.co.uk/CIsys.htm. With permission

small piston in a cylinder inside the injection pump that, for every injection, compresses a few microliters of fuel, forcing it at high pressure through a steel tube (line), into an injection nozzle (injector), and then into the engine cylinder. The injection pulse occurs during the late compression and early power strokes, when the cylinder combustion pressure is 250–500 bar. For this tiny-displacement (50–1000 μl) pump to work, the fluid it acts upon must be incompressible and of sufficiently high viscosity to avoid excessive leakage past the tiny piston of the pumping chamber. This is not a problem for mineral diesel fuel, but it is for DME due to its very low liquid viscosity, and the possibility of *vapor lock* at engine temperatures.

DME remains a liquid only if under moderate pressure, like propane or butane. As the temperature increases, the pressure required to keep it in liquid form increases exponentially, as shown in Fig. 13.6 [9, 10]. The plot shows its *vapor pressure* at any given temperature. The distributor pump can be kept cool by locating it away from the

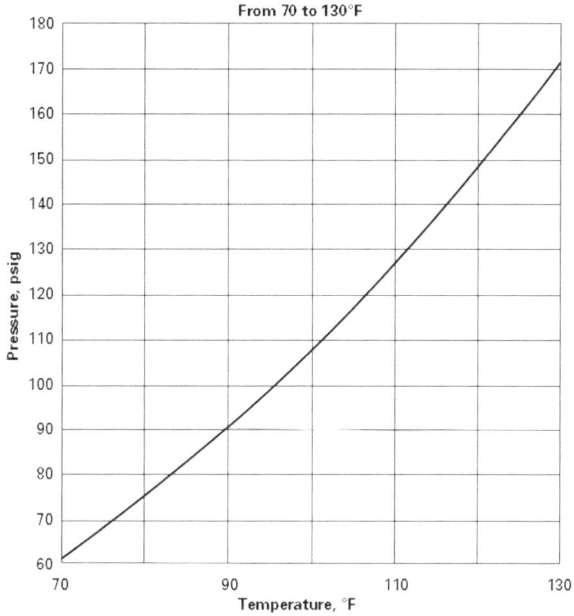

Fig. 13.6 Vapor pressure of DME as a function of temperature (1 bar = 14.5 psia ≈ 0 psig). Plot from Chemours HP DME technical reference manual. https://www.chemours.com/en/-/media/files/propellants/hp-dme-technical-information.pdf?rev=094d8ed791f84854a37f75e85d1672f7. Used under STM guidelines

engine heat, but that isn't possible for the high pressure steel fuel lines or the injection nozzles, that are in direct thermal contact with cylinder head. This remains a problem for this fuel injection topology.

The *vapor pressure* of DME at 22°C (72°F) is only 5 bar (72 psi), but at the 100+ deg C temperature (depending on the engine) near the cylinder head, its vapor pressure can be 33 bar (479 psi) or higher. To assure that the DME always remains in liquid phase, even at this increased temperature, the entire fuel system pressure must be maintained above about 35 bar (500 psi).

In a laboratory environment, it's no problem to maintain DME above 35 bar: just store the fuel in a conventional gas cylinder, and pressurize it to 35+ bar with some nonreactive gas that has a low solubility in DME. Nitrogen is the usual choice. But this is not convenient in a vehicle, since it would require reducing the tank pressure below 5 bar to transfer DME into the tank when refueling, and repressurizing after refilling from a supply of pressurized nitrogen. (Obviously, compressed air would be a risky replacement because it forms a flammable mixture with the DME.)

The more practical solution is to pressurize the liquid DME using a low-volume pump that is capable of continuously compressing DME to a pressure above 35 bar. This is beyond the capability of current liquid or dual-phase pumps of reasonable cost, although it is a solvable engineering challenge for injection pump design.

Alternatives to the high primary-side DME pressure all involve ways to reduce heat transfer into the DME, especially in the fuel injection nozzles or fuel lines. Insulation of fuel lines is helpful. Liquid cooling of the fuel injection nozzles is another possibility, or an azeotropic blend of a high-latent heat fuel such as methanol (37.6 kJ/mole) with the pressurized liquid DME. There are opportunities for innovation here.

The lubricity issue can be reduced but not completely overcome by the use of a lubricating fuel additive. Most organic and to a lesser degree, mineral lubricants are reasonably soluble in DME. Castor oil in a 1–5% volumetric liquid blend seems to work well, borrowed from experience with methanol engines. The additive serves two purposes: lubrication of sliding surfaces, and improvement of the fuel viscosity to reduce leakage past the piston, gear, or nutator assembly in a fuel transfer or injection pump.

The positive displacement injection approach can use a conventional diesel fuel injection rotary distributor or inline pump to work with DME. It is not an optimum solution, but has been used for experimental engine conversions. The extremely low viscosity of DME in a distributor pump will cause significant internal leakage; as much as 50% has been reported [11]. Exacerbating the problem is that DME in liquid phase has 5–6 times the compressibility of D2. This further reduces the volumetric efficiency of the pump. The low pumping efficiency combined with the requirement for twice the fuel delivery volume per injection will likely require a higher delivery injection pump than the original diesel pump. It is imperative that the rotary distributor or inline pump be equipped with delivery valves on each injection line to prevent depressurization between each injection. (Delivery valves are check valves on each output port of the pump.) A simplified diagram of a positive displacement fuel injection system is shown in Fig. 13.8.

Regarding the routing of the fuel return lines from the pump and injectors into the intake manifold: This is an acceptable alternative to returning the internal leakage DME to the fuel tank, but only if the DME-air mixture in the manifold does not exceed the lower flammability limit of DME in air, 3.4% [12] by volume. This was the approach used successfully in the DME-converted single cylinder 650 cc diesel engine shown in the student project vehicle of Fig. 13.7.

Fig. 13.7 Single-cylinder 650cc DME-converted diesel engine. Demonstrated in a series hybrid electric service cart. A class project in "Automotive Engineering for a Sustainable Future", Olin College, 2011. Photo by author

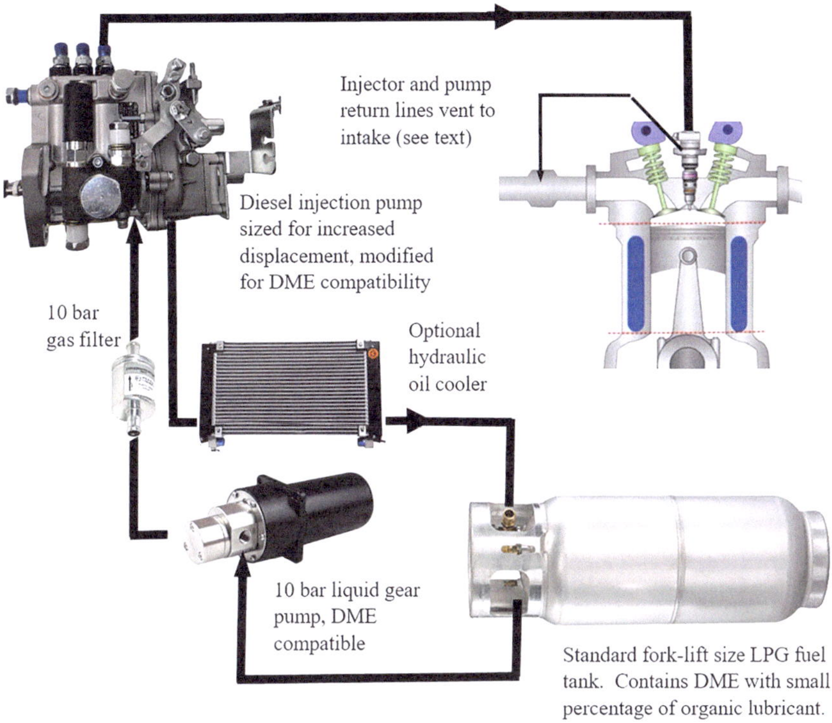

Fig. 13.8 Diagram of main components of simple positive displacement DME fuel injection system. Diagram by author using public domain component graphics

High-Pressure Common-Rail (HPCR) Fuel Injection

A superior injection option is an HPCR injection system of the type that is standard on almost all modern emission-controlled diesel engines. The configuration is also used for gasoline direct cylinder injection (DISI) engines increasingly common on late-model cars. For electronically controlled diesel HPCR injection systems, the option to inject more than once during the compression and expansion stroke is an advantage for efficiency and emissions.

In a DME HPCR system, shown conceptually in Fig. 13.9, the fuel is continuously pressurized by a high-pressure fuel pump to at least 250 bar, to facilitate injection into the high cylinder pressure during the compression stroke and early power stroke. Diesel HPCR systems use much higher pressures, typically 1500–2500 bar (22,000–36,000 psi), to assure ultra-fine fuel atomization in the injector spray. DME does not require such high injection pressures since it vaporizes instantly upon injection into the hot combustion chamber. This reduces the challenge of

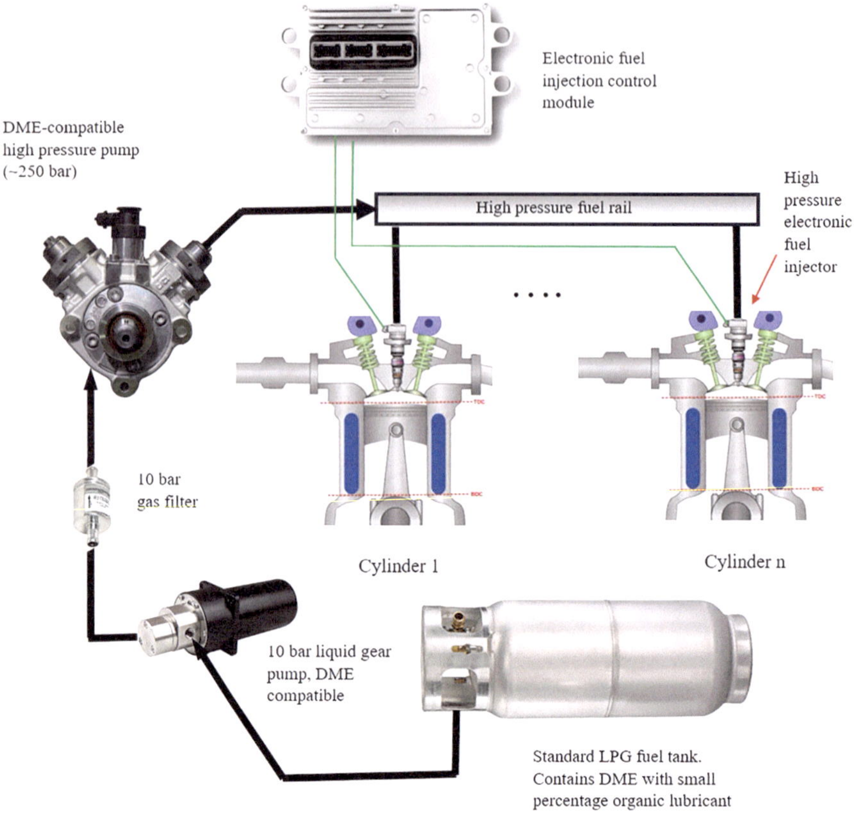

Fig. 13.9 Components of a high-pressure common-rail DME fuel injection system for an n-cylinder diesel engine. All are off-the-shelf except for the DME-compatible high-pressure common-rail fuel pump. Diagram by author using public domain clip art

common-rail DME injection somewhat, but not sufficiently to allow the use of off-the-shelf HPCR pumps. Continuously pressurizing DME to the pressures required for an electronically controlled common-rail direct cylinder injection system is not possible with existing gasoline or diesel common-rail pumps.

Modern electronically controlled common-rail diesel systems are basically the same as gasoline direct cylinder injection systems. And both are identical in topology to gasoline port injection systems dating back to the late 1960s, except that they operate at much higher fuel pressure. (Conventional gasoline port injection systems operate at only about 3 bar rail pressure.)

An HPCR injection system is simpler than a displacement-metered system. The fuel in the rail (a manifold feeding all injectors) is dispensed to each injector, which has high-pressure solenoid valves that spray into each combustion chamber when actuated by the injection control computer. Fuel injection timing is referenced to a camshaft position sensor, and fuel quantity is controlled by the pulse duration of the injectors rather than the variable displacement of a diesel distributor pump.

But the requirement for a high-pressure DME pump that is capable of maintaining pressures in excess of 250 bar is extremely challenging due to the very low viscosity and lack of lubricity of DME. Also, about twice the volume of DME must be injected compared with diesel fuel. A gasoline direct injection common-rail pump is the closest candidate for modification but is still inadequate due to DME's combination of low viscosity, zero lubricity, and material compatibility challenges. DME is an exceptional solvent—very good at stripping the oil film from the injection pump chamber surfaces as well as the engine cylinder walls. A summary of the state of the art for DME fuel system components can be found in citation [13].

The HPCR configuration eliminates concern about fuel vaporization, even at the temperatures near the cylinder head, because the fuel rail pressure is well above the boiling point of DME. A number of clever solutions have been cobbled together from existing components. One worth mentioning was a laboratory system that used a plunger-type laboratory dual-phase pump to pressurize the fuel rail. This type of pump generates a highly pulsatile output, so a 300 ml accumulator was used between the pump and the fuel rail, but even this was found to be inadequate to stabilize the fuel rail pressure sufficiently for engine operation other than for single-injection spray tests [14].

Fuel System Modifications Common to Both Fuel Injection Methods

For either fuel injection system topology, the fuel system components for containment and transfer of fuel from the tank to the fuel injection pump engine are the same.

A vehicular DME tank such as shown in Fig. 13.10 can be a standard 250-psi (1.7 MPa) LPG (propane) tank. The DME tank pressure at ambient temperatures is

Fig. 13.10 DME fuel tank on Volvo Class 7 truck. Same specifications as an LPG tank, with 125 psig working pressure, 250 psi maximum pressure. https://www.truckinginfo.com/152918/volvo-thinks-the-fuel-of-the-future-is-dme-and-its-almost-here. With permission

Fig. 13.11 Volvo DME truck engine, 2013. Uses a relatively low-pressure common-rail injection system as described in text. From https://www.truckinginfo.com/152918/volvo-thinks-the-fuel-of-the-future-is-dme-and-its-almost-here. With permission

well within the allowable tank pressure. The tank pressure will remain above atmospheric for ambient temperatures above −25°C.

DME is drawn from the bottom of the tank as a liquid by a nonpulsatile transfer pump. 10 bar minimum outlet pressure will assure against vapor lock in this leg of the fuel flow as long as components are not directly exposed to engine heat. The moderately pressurized DME output of this pump is then fed to the input of an oversized conventional diesel injection pump in a positive displacement system, or to the inlet of an HPCR pump. The preferred type of pump for this moderate-pressure leg is a "dual-phase" pump designed for compressing thin solvents or multiphase fluids prone to vapor lock or pump cavitation.

The 2013 Volvo VNL truck with converted D13 DME engine shown in Fig. 13.11 uses an experimental HPCR DME fuel injection system as described above.

Commentary

DME is a combustion fuel that may be considered a compromise solution for meeting zero-carbon objectives. But it is surprising that DME has not already seen greater adoption as a clean replacement for diesel fuel, for which sustainable options are limited. Its well-to-wheel efficiency and carbon footprint are superior to biodiesel and renewable diesel fuel, the only currently viable substitutes for CI (diesel) engines. Due to its lower energy density and low boiling temperature/pressure, handling procedures for DME are less convenient than conventional diesel or biodiesel options. But these limitations are about the same as LPG (propane), for which there is extensive experience and an existing fuel distribution network. DME is certainly safer than many other alternatives due to its non-toxicity.

The major obstacle to its use in existing diesel engines seems to be the lack of robust DME-compatible fuel injection and engine material technologies. The solution seems only a matter of incremental engineering, within the capability of existing diesel engine and fuel system manufacturers.

DME is currently produced in the USA from natural gas for the same reason as almost every other synthetic renewable fuel: cost. This energy origin ties all synthetic fuels, including DME, to the environmental consequences of natural gas. Unfortunately for the environment, natural gas (methane) is abundant and the cheapest source of portable energy in existence. The technology and cost required to produce DME from waste biomass are greater than from natural gas, but almost identical to those for production of methanol from biomass. Extraction of natural gas will not last forever, and when environmental concerns reach a high enough level, natural gas will become less financially competitive. DME checks the boxes for energy, environmental, and economic sustainability, making it arguably the best transition fuel for CI engines, replacing petroleum with minimal disruption of the supply chain.

References

1. Project BioDME a consortium of European auto manufacturers and bio-DME producers). http://www.biodme.eu/. Accessed June 10, 2024
2. K. Im-orb, P. Piroonlerkgul, Sustainability analysis of the bio-dimethyl ether (bio-DME) production via integrated biomass gasification and direct DME synthesis process. J Renew Energy **208**, 324–330 (2023). https://doi.org/10.1016/j.renene.2023.03.092
3. EPCM. Direct DME Synthesis from natural gas. https://epcmholdings.com/direct-dme-synthesis-from-natural-gas/. Accessed June 20, 2024
4. C. Fortin, N. Gianfolcaro, R. Gonzalez, J. Lohest, A. Lonneux, A. Kesnelle, N. Siliki, T. Peiffer, R. Renson, C. Schmitz, Dimethyl ether, a review of production processes and a modeling of the indirect route. Liege Univ, 2019–2020 (2020) Liege University (Belgium) report PROJ0012-1, 2020 https://www.chemeng.uliege.be/upload/docs/application/pdf/2020-05/projetdme_finalreport.pdf
5. Z. Azizi, M. Rezaeimanesh, T. Tohidian, M.R. Rahimpour, Dimethyl ether: a review of technologies and production challenges. Chem Eng Process: Process Intensification **82**, 150–172 (2014)
6. P.A. Glaude, R. Fournet, R. Bounaceur, M. Moliere, *DME as a potential alternative fuel for gas turbines: A numerical approach to combustion and oxidation kinetics*. Proceedings of ASME Turbo Expo 2011, GT2011, Vancouver, British Columbia, Canada, June 6-10, 2011
7. Chemours, HP DME. Figure 2. HP DME vapor pressure. https://www.chemours.com/en/-/media/files/propellants/hp-dme-technical-information.pdf?rev=094d8ed791f84854a37f75c85d1672f7. Accessed July 1, 2024
8. Carol Birkland, Fleet equipment, March 18, 2014. *Dimethyl ether: Fuel of the future?* https://www.fleetequipmentmag.com/dme-fuel-future/. Accessed September 1, 2024
9. Diversified CPC. Vapor pressure dimethyl ether. https://diversifiedcpc.com/wp-content/uploads/2021/12/vpdme.pdf. Accessed June 24, 2024
10. J. Wu, Z. Liu, J. Pan, X. Zhao, Vapor pressure measurements of dimethyl ether from (233–399) K. J Chem Eng Data **49**(1), 32–34 (2004). https://doi.org/10.1021/je0340046
11. A.K. Agarwal, S. Mehra, H. Valera, N.K. Mukherjee, V. Kumar, D. Nene, Dimethyl ether fuel injection system development for a compression ignition engine for increasing the thermal efficiency and reducing emissions. Energy Convers Manage **287**, 117067 (2023) April 2023. https://www.sciencedirect.com/science/article/abs/pii/S0196890423004132?via%3Dihub. Accessed July 15, 2023
12. Robert Bosch GmbH, *Automotive Handbook*, 10th edn. (Distributed by SAE International, 2018), p. 346

13. Transport Canada. *Dimethyl ether fuel literature review*. https://tc.canada.ca/en/programs/non-funding-programs/ecotechnology-vehicles-program/dimethyl-ether-fuel-literature-review. Accessed July 15, 2024.
14. S. LeBlanc, L. Jin, N.S. Sandhu, et al., *Combustion characterization of DME-fueled dual fuel combustion with premixed ethanol* (2022). https://doi.org/10.4271/2022-01-0461. SAE Technical Paper 2022, No. 2022-01-04 61. Cited in Dimethyl Ether to Power Next-Generation Road Transportation, online at https://www.sciltp.com/journals/ijamm/article/view/245. Accessed July 15, 2024

Electric Vehicles 14

Synopsis

This chapter is about battery electric vehicles (BEVs) and their technologies, safety and environmental and energy impacts, infrastructure requirements, and the effects of transportation electrification on personal transportation as we know it. I will briefly review the history of electric cars beginning in 1881. Then, the technology of electric vehicles: batteries, motors, controls, charging systems, regenerative braking, and recharging infrastructure requirements. BEVs and internal combustion vehicles (ICVs) will be compared for their net environmental impact. Electrical generation and distribution issues associated with EVs will be discussed, including compatibility with renewable electricity sources. Also, safety and roadway issues related to the increased weight of EVs. The broader ramifications of transferring the automotive energy sector from petroleum to electricity will be summarized.

Even some of the most out-of-the-way outposts have embraced electric vehicles, as depicted in the photograph of Fig. 14.1, taken at a former general store in a backwoods town. As of 2024, the future of automobiles seems ordained to be battery electric. Not exclusively, but certainly to a majority extent. Therefore, more attention will be devoted to electric vehicles than other alternative automotive propulsion technologies.

Supplementary Information The online version contains supplementary material available at https://doi.org/10.1007/978-3-031-85028-8_14.

C. A. MacCarley, *Non-Petroleum Automotive Transportation*, https://doi.org/10.1007/978-3-031-85028-8_14

Fig. 14.1 Electric cars are everywhere. Tesla public charger outside a country store in Mitchell, Oregon. Population 136. Photo: Author, 2023

Some Definitions

A vehicle classified as a ***battery electric vehicle*** **(BEV),** or commonly just **EV,** has two distinguishing attributes:

1. Motive power is provided solely by an electric traction motor rather than an ICE.
2. Energy is stored onboard the vehicle in a battery of rechargeable electrochemical cells.

Technically, any form of onboard electric energy storage should qualify a vehicle to be classified as an EV, but distinct categories have been assigned based on the energy storage method, e.g., battery electric vehicles (BEVs), hydrogen FCVs, batteries plus supercapacitors, flywheel kinetic energy storage, solar photovoltaic trickle charged batteries, etc. It is worth noting, however, that all electric motor-driven vehicles still require a few kWh of batteries as an energy buffer to provide continuous power to the drive wheels. This includes FCVs which were covered in Chap. 9 *Hydrogen*, even though electricity is generated onboard by hydrogen fuel cells.

A variation of the electric vehicle is a *Hybrid Electric Vehicle* that uses both an electric motor and an ICE as motive power sources. Gasoline hybrid electric vehicles will be covered in the Chap. 15, *Hybrid Vehicles*, although many of the components are the same as those used in BEVs, such as the motor, battery, and the power electronics.

Before the development of modern lithium-based batteries in the late 1990s, many clever ways to overcome the range limitations of electric vehicles were attempted with various degrees of success. Common were external *Range Extenders* (REx) in the form of self-contained ICE generators, either onboard or on towable trailers, that functionally transform the vehicle into a series hybrid with a weaker ICE than the electric motor. Other innovative solutions, such as mechanized exchange of the vehicle battery have been deployed since the late 1800s, to be discussed later in this chapter.

The most beneficial attribute of an EV is the lack of a combustion engine that generates exhaust gas. This benefit is huge considering that there are approximately 300 million registered motor vehicles in the USA alone, almost more cars than people. In the late 1960s, toxic emissions (HCs, CO, NO_x, and particulates) from ICVs had led to dangerously poor air quality (aka health hazard) that required the declaration of some days as "smog alert days" in urban areas such as Los Angeles. These tailpipe emissions were directly toxic, responsible for causing or aggravating respiratory diseases, and carcinogenic to a degree greater than cigarette smoking. People were advised to avoid outdoor activities or in some cases, curtail the use of their cars on certain days. As discussed in Chap. 3 Automobiles and the Environment, while the main culprit was petroleum fuels—gasoline and diesel—there were obvious technical reasons within our control: e.g., the use of carburetors, which are incapable of maintaining an exact gasoline-air mixture, lack of effective exhaust aftertreatment controls, e.g., catalytic converters, and poor engine and vehicle design in general. Innovation driven by public pressure eventually led to radical reductions in tailpipe emissions: the introduction of electronic fuel injection in the late 1960s, and the invention in the mid-1970s of oxygen-sensing closed-loop feedback fuel injection that enabled three-way catalytic converters. These systems greatly reduced regulated emissions, but had no effect on GHG emissions. At that time, no thought was given to carbon dioxide emissions. The greenhouse effect was not recognized as an automotive emission concern, and the connection between anthropogenic CO_2 and global warming were not topics of conversation or scientific investigation. Global warming trends were measurable, but no one took the threat seriously until the 1990s.

Carbon dioxide is not a pollutant in the contemporary sense since it is neither toxic nor carcinogenic, and it is a natural component in the life-cycle processes of the planet. Our awareness has greatly changed since then. Anthropogenic GHGs are not just recognized as harmful, but more harmful in the long term than directly toxic emissions from ICVs. We now search for ways to mitigate GHG emissions, especially CO_2 generated by automobiles. But CO_2 cannot be removed by exhaust post-treatment. Carbon dioxide emission from cars are directly proportional to the amount of gasoline or diesel fuel used by an ICE. It can only be reduced by either (1) reducing the amount of fuel used, or (2) changing the fuel. It is solution (2) that is addressed by the electrification of automobiles, and if we consider only vehicle tailpipe emissions, it is the ideal solution. No combustion, no CO_2. The moniker ***Zero-Emission Vehicle*** (ZEVs) [1] was coined originally by the California Air Resources Board (CARB) and later adopted by the US EPA, originally referring to regulated emissions but later included GHGs as well. Only battery electric and hydrogen-fueled (fuel cell or combustion) vehicles are allowed this legal designation.

EVs *were* an effective solution to *tailpipe emissions*, and CO_2 emissions were of far less concern than regulated emissions. But anyone with a modest understanding of automotive technology recognized from the beginning that zero exhaust emissions did not necessarily mean that the operation of the car resulted in no harmful atmospheric emissions, since it only considered tailpipe emissions, not the generation of the electric power used to charge EV batteries. With the limitations of batteries at the time, owning a BEV required a commitment by owners to accept range limits and long refueling time compared with ICVs. The hope was and remains that electricity will continue to become less carbon intensive, as renewable sources eventually replace the fossil fuels (coal or natural gas) that still dominate electricity generation.

The electric vehicle expansion over the past 20 years followed a century-long quest to develop the elusive "better battery" prophesized since before 1900. It arrived in the form of lithium-ion batteries, which became acceptable for EVs only after more than a decade of additional research and development to minimize the intrinsic instability and packaging challenges of this new chemistry. While still nowhere near the energy density of gasoline, lithium-based batteries could store more electric energy than any prior rechargeable battery chemistry. Of importance here is that lithium batteries could provide adequate EV range to meet the demands of automobile drivers who were accustomed to the several hundred km or miles of range provided by gasoline or diesel fuel.

Prior to the new electric vehicle revolution, electric propulsion had already been used for over a century for railway locomotives, trams, trollies or subway trains, industrial vehicles and submarines. Even the NASA Apollo Moon Rovers of the early 1970s. Electric rail applications were powered by overhead catenary wires or a "third rail" for trains such as subways with safely isolated tracks. The submarines first used to great effect by the German navy in World Wars One and Two were diesel-battery electric hybrids. For road-going vehicles (automobiles, trucks, buses, offroad equipment), electric propulsion with onboard energy storage enjoyed a short-lived dominant era from 1890 through 1910, before becoming uncompetitive with ICE vehicles.

The Evolution of Electric Vehicles

According to one historical account [2], the first road-going electric sort-of-vehicle was demonstrated in 1828 by Hungarian Ányos Jedlik: a small-scale model car powered by a crude electric motor that he designed. Other instances followed: Circa 1832 Robert Anderson of Scotland was reported to have invented an electric-powered carriage, although little is known of the details. In 1835, another scale model electric car was designed by Professor Stratingh of Groningen, Holland, at the same time that Thomas Davenport, a blacksmith from Brandon, Vermont, USA, built a scale model electric car. Aside from the lack of utility of these early attempts, they all predated the development of the lead-acid cell developed by Gaston Plante' in 1859, so they had no practical means to store adequate electric energy onboard, and would therefore not qualify as true electric vehicles by the modern definition. Fast forward to the 1880s, a time of prolific invention and imagination in Europe and the USA. The age of steam power was still at its peak but waning. Electricity was a magical phenomenon that seemed to have unlimited potential applications.

American inventor Thomas Edison's patent for the electric light bulb had just been granted on 27 January 1880. Electricity was a frequent topic of science fiction, and creative minds such as French author Jules Verne portrayed visions of an amazing electric future in print media [3].

Depending on the historical account, the first practical electric vehicle as defined today was attributed to two independent inventors in 1881: the electric motor-assisted pedal tricycle of French inventor Gustave Trouvé (Fig. 14.2) [4, 5], and the battery electric-only tricycle constructed by Professors William Ayrton and John Perry in England (Fig. 14.3) [6]. Both were built starting with English pedal-tricycles, a common form of personal transportation at the time. Both used Plante's lead-acid electrochemical cells and brush-commutated DC motors of the type first demonstrated by German Physicist Moritz von Jacobi in 1834 [7]. Both were publicly demonstrated in 1881 approximately 2 months apart, with Trouvé's demonstration at the Paris Electric Exhibition in April and Ayrton's at the Royal Academy of Science in London in June. What is clear from a composite of several historical accounts is that the Trouve trike was more of a static display at the Paris Electric Exhibition that, when demonstrated, achieved a speed of 3.6 km/h (2.25 mph) over just a few meters distance. The Ayrton and Perry trike, with its 48V battery, was reported in media at the time (surely optimistically) as achieving a range of 40 km and a maximum speed of 14 km/h (8.75 mph). Neither of these accomplishments were actually novel since the combination of an electric motor and a lead-acid battery had already been demonstrated for powering boats and factory machinery. Applying this method to road-vehicle propulsion would have been the obvious next application. The years immediately following these demonstrations saw unprescidented invention related to electric propulsion, including practical four-wheel versions of the Trouve/Ayrton/Perry electric trikes. The first 'production' electric car was probably built by English

Fig. 14.2 Electric-assisted pedal trike built by inventor Gustave Pierre Trouvé, demonstrated at the Paris Electric Exhibition April 1881. Novel but dangerous power control involved raising and lowering lead plates in the sulfuric acid battery electrolyte (seen behind the driver). Image from Scientific American, pre-1800, public domain, as included in History of the Electric Automobile, Ernest Wakefield, SAE Publication, 1994

Fig. 14.3 The Ayrton and Perry battery-electric trike, demonstrated for the Royal Academy of Science in London, June 1881. It appears to have included the first electric lights on any automobile, the electric arc lamps seen above the drive wheels, likely a contribution of his former student and wife Hertha, a physicist and expert on arc lighting [9–11]. Image from Scientific American, pre-1800, public domain

inventor Thomas Parker in 1884, who formed what was arguably the first EV manufacturing enterprise, the Elwell-Parker Company. Parker is also credited with electrifying the London underground tramways [8].

Early Electric Rail

The history of electric cars would not be complete without looking to the early electric railway and tram locomotives, with electricity supplied by overhead lines, third rails, or onboard batteries charged by diesel-electric generators (see Chap. 15, *Hybrid Vehicles*).

According to citation [12] the first operational battery-electric locomotive was built in 1837 in Scotland by chemist Robert Davidson of Aberdeen, using copper-zinc galvanic cells and two crude reluctance motors directly coupled to drive wheels. In 1842, his second larger, electric locomotive, named the *Galvani*, was tested successfully on the Edinburgh and Glasgow Railway but was destroyed by (coal) railway workers fearful that it might make their jobs obsolete.

Fig. 14.4 EL-1 Electric Switch Engine used in routes through the tunnels of the Baltimore Belt Line, USA, 1895. A steam locomotive was towed behind the EL-1 for use outside tunnels in non-electrified parts of the route. From Scientific American, 10 August 1895. Public domain

The most compelling motivation for electric vs coal locomotives was the problem of black coal exhaust in urban areas and railway tunnels (Fig. 14.4). Despite the fact that almost all of the electricity to power electric rail came from coal-fired power plants (a small amount was hydroelectric), several overriding advantages of electric rail soon became apparent:

- The relocation of the air pollution away from the rail right-of-way made rail stations and track sidings instantly cleaner (but not necessarily healthier since coal was still the dominant home and industrial heating fuel). Visibility and the ability to breathe in long tunnels were greatly improved. Hydroelectric generation, where available, could also provide this energy, a huge improvement over coal.
- The overhead wires (catenaries) or third rail electric distribution systems allowed regenerative braking and system-wide power recovery. Simply stated, a train going downhill pushed power back to the distribution system which could be concurrently used by another train going uphill. At a time when energy efficiency was not really a concern, this feature reduced operating costs, which have always been important.
- Electric trains and trams were much quieter than steam—a big advantage in cities, encountering less local resistance to rail right-of-way.
- Electric trains could be potentially much faster than steam locomotives. All high-speed passenger trains in the world today are electric.
- At a time when paved roads were nonexistent except in affluent urban areas, rail tracks were ubiquitous. Rail was the way people and materials were moved on land for longer distances and heavier loads.

The decision to electrify urban rail/tram lines was obvious; the advantages over horse-drawn trolleys and coal locomotives outweighed the cost of overhead catenaries and local electric power plants. This practice was soon adopted on almost all rail and tram lines throughout Europe, with electricity cost-effective because of the relatively short distances between cities. But in the USA, with longer distances, the railway electrification costs were considered unwarranted, so steam trains continued to dominate until they were eventually replaced by the diesel-electric hybrid trains ubiquitous in US rail freight and inter-city passenger service today. According to citation [13], this situation is unlikely to change because almost all railroad right-of-way and the freight hauling infrastructure in the USA are privately owned, and railroads are unwilling to invest in a complete transition to electrification that would be very costly. In Europe and most developed countries, railway networks are considered part of the national transportation infrastructure, just like roads, highways, and waterways, and they are financed and/or managed by government agencies and partnerships.

The Golden Age of Electric Cars

A tsunami of innovation followed the first EV demonstrations in the 1890s. Electric cars and delivery vehicles were *the* hot technical topic, even as internal combustion and steam engines were also increasingly popular for automotive propulsion. In fact, electric vehicle registrations in the US exceeded those of gasoline cars for nearly 20 years before loud, smelly, and dangerous internal combustion engine vehicles eventually exceeded their popularity. By 1900, 38% of road vehicles were electric, while only 22% were gasoline powered. (The remaining 40% were mostly steam powered, with coal or various flammable liquids as the fuel source.) Electrified wagons, carriages, coaches, and tricycles had arrived, and possibly 50+ electric conveyance manufacturers had sprouted in England, France, Germany, and the United States. The period from 1890 through 1910 is unofficially referred to as the *Golden Age of Electric Cars*, a time when gasoline ICVs were anything but a certain proposition.

Figure 14.5 is a reprinted photograph of the 1896 Morris and Salom *Electrobat*, considered the first *production* electric car made in the USA. Like most other pioneering electric vehicles, it stored energy in lead-acid batteries [14].

It is difficult to do justice to the innovation and entrepreneurial adventures during the 20-year Golden Age. An entire industry rose from infancy to commercial prominence and then faded away just as quickly in the face of competing combustion engine technology. Had a better battery actually been "just around the corner" [15], as Thomas Edison is often incorrectly credited (or blamed) for saying, the automotive world today would be much different. A comprehensive chronicle of these early years can be found in Wakefield [14] and a compact historical account can be found in Car and Driver Magazine [16]. An early EV pioneer in the USA, Andrew Riker of Brooklyn New York, constructed his first electric tricycle in 1890, and by 1900 had manufactured and deployed a large fleet of electric taxicabs in New York City, USA.

Electric passenger cars of the era were expensive, owned exclusively by the affluent in cities where range requirements were defined by the previous mode of

Fig. 14.5 The 1896 Morris and Salom Electrobat, the first US-manufactured electric car. Scientific American, 1896. Public domain

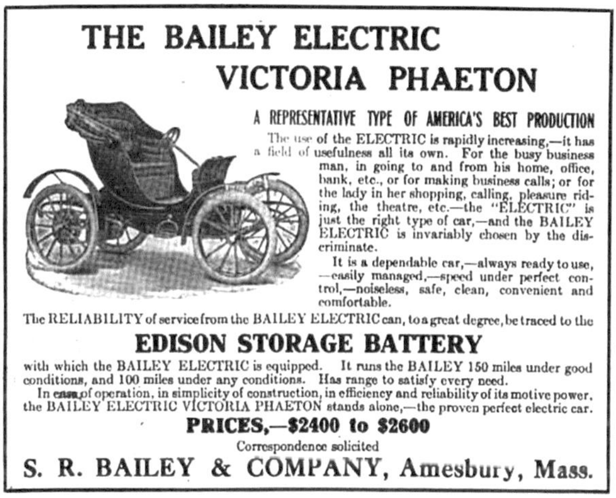

Fig. 14.6 The 1913 Bailey Electric Victoria Phaeton used Edison Nickel-Iron Batteries. The $2400 USD price would be equivalent to $78,000 in 2024, approximately the cost of a 2024 Base Tesla S. https://de.m.wikipedia.org/wiki/Datei:Bailey_Electric_ad_1911.jpg CC-BY-2.0

transportation, horse-drawn carriages or buggies, and home charging was possible because the owners were among the few privileged to have either AC (Westinghouse) or DC (Edison) utility electric service. They were heavier than their fledgling ICE contemporaries due to the mass of the lead-acid Plante' cells (or Edison nickel-iron batteries, e.g. Fig. 14.6) that they carried. At a time when few roads were paved, getting stuck in mud or soft dirt was more likely than horse-drawn conveyances and lighter ICE vehicles. These considerations limited interest in electric cars outside of major metropolitan areas.

By 1910, EVs were being gradually displaced by IC engine cars powered by alcohol, benzine, or gasoline such as the low-cost 1908 Ford Model T. The Detroit Electric Car Company continued to profitably manufacture battery electric luxury

Fig. 14.7 Advertisement for the Detroit Electric, circa 1910, a luxury car designed for wealthy urban commuters, 1910. https://archive.org/details/DetroitElectricCars1912A Public domain

cars until 1939, owing to their clean and quiet characteristics, but to an increasingly elite customer base. Fig. 14.7 is a copy of an advertisement for the Detroit Electric, circa 1910, a luxury (quiet and non-smelly) car for wealthy urban commuters. Detroit Electric built 13,000 electric cars from 1907 to 1939 [17].

A 1908 Baker Electric Model V Victoria claimed a 40–50 mile range, using a 48-volt battery and motor with shaft drive (an innovation at the time) to the rear axle [18]. Baker produced electric cars from 1899 to 1921, and in 1906 was the world's largest manufacturer of electric (possibly all) cars, pre-dating the first Ford Model T. In 1902, Walter Baker established a long-standing land speed record for electric vehicles, as discussed later.

As early as 1890, American inventor Thomas Edison saw the future of automobiles as electric [19]. He immediately realized that the primary limitation was the battery, so he tasked his Menlo Park laboratory with finding alternatives to heavy lead-acid Plante batteries. His 1901 patented nickel-iron *Edison Battery* provided higher energy and power density but had a high self-discharge rate and poor round trip efficiency (RTE). It was much more expensive compared with the usual lead-acid Plante batteries of the time [20]. Figure 14.8 shows Mr. Edison with a Bailey

Fig. 14.8 Bailey Electric Runabout equipped with Edison batteries after 1000 mile trip across England, 1911. Thomas Edison NHP archive. Public domain

Fig. 14.9 Retro Nickel-iron batteries briefly produced 1972-75 by Exide Technologies, last owner of Edison Storage Battery Co., CC-BY SA Image by "Z22", Wikipedia Free Repository

Electric Runabout equipped with Edison batteries, after it completed a 1000-mile demonstration trip around England in 1911, allegedly covering the final 170-mile leg without recharging, an EV range record that (if true) would today be competitive with cars using lithium batteries [17].

Figure 14.9 shows nickel-iron batteries produced briefly by Exide Technologies (formerly the Electric Storage Battery Company) following its acquisition of the Edison Storage Battery Co. in 1972 [21]. Iron-based battery chemistries have seen a recent resurgence of interest (2010–) due to concerns over scarce lithium, cobalt, and nickel resources required for mass production of lithium-ion batteries. Edison, legendary for his persistence, had been working on improved batteries for electric vehicles since 1890, and firmly believed that improved batteries would soon be

Fig. 14.10 1914 Baker Electric Victoria being home-charged. "Girl charging a Baker electric automobile", c 1914, National Automotive History Collection, Detroit Public Library. Public domain

forthcoming. It is ironic that in 2023, nickel-iron batteries are still considered an advanced battery chemistry worthy of continued refinement [22, 23].

While some homes and businesses were equipped with DC electric service as a result of Edison's determination to establish DC as the standard electric distribution in the USA, the 1914 Baker Electric Victoria Model shown in Fig. 14.10 is shown being home-charged from Edison's arch-rival George Westinghouse's AC power distribution. The large cylindrical device against the wall in the photo is a motor-generator (MG) unit for converting AC power to DC of the appropriate voltage for charging the car [24]. This was the usual method for AC-DC and DC-AC power conversion through the 1950's, because the vacuum tube rectifier diode that had been invented in 1904 by British scientist John Fleming, would not be capable of high-power rectification (AC-to-DC conversion) for another 30 years.

Along with the early explosion of interest in electric vehicles prior to the turn of the century came the first dedicated trade periodical, *Horseless Age* [25] 1895–1918, the front page of the inaugural edition shown in Fig. 14.11, which chronicled the innovations and business of the new EV industry.

For the duration of the EV Golden Age, electric cars were about as unaffordable for the proletariat as they are now. Meanwhile, the Ford Model T was introduced in 1908, and by 1923, could be purchased for under $300 USD, while most electric cars of the time cost over $3000 [26]. Electric cars had always been marketed to women, or more specifically, to the affluent husbands for whom the indulgence could be justified. They were vastly more civilized than the crude, smoke-spewing, noisy gasoline cars that had to be hand-cranked to start until 1911, when the Kettering electric starter was patented and used on the 1912 Cadillac Model 30 [27].[1] As the EV market started to decline after this (and a few other ICE) innovations, EV marketing efforts increasingly focused on women as the targets for sales,

[1] The Kettering electric starter was first sold on a 1912 Cadillac Model 30. Its adoption in the years that followed radically changed the acceptability of gasoline cars, and presaged the decline of electric cars that had never needed a starter to operate.

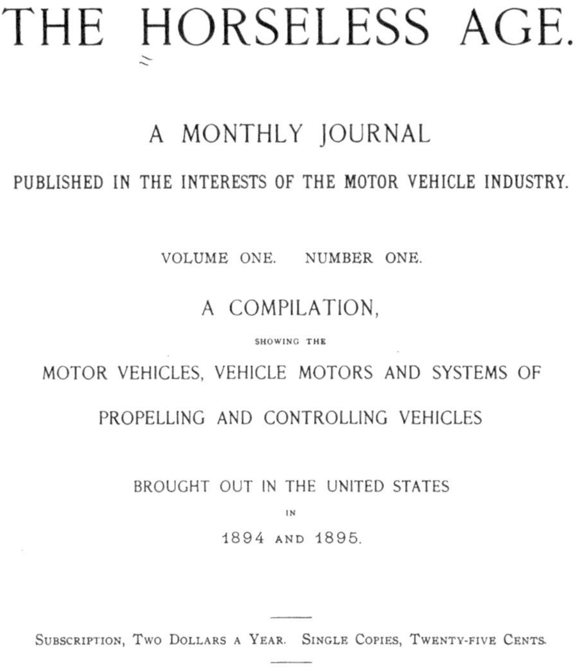

THE HORSELESS AGE.

A MONTHLY JOURNAL

PUBLISHED IN THE INTERESTS OF THE MOTOR VEHICLE INDUSTRY.

VOLUME ONE. NUMBER ONE.

A COMPILATION,

SHOWING THE

MOTOR VEHICLES, VEHICLE MOTORS AND SYSTEMS OF

PROPELLING AND CONTROLLING VEHICLES

BROUGHT OUT IN THE UNITED STATES

IN

1894 AND 1895.

SUBSCRIPTION, TWO DOLLARS A YEAR. SINGLE COPIES, TWENTY-FIVE CENTS.

Fig. 14.11 Title page of Horseless Age, Vol 1 No. 1, 1894. From Hathi Trust, US Library of Congress, https://catalog.hathitrust.org/Record/000543204

Fig. 14.12 Marketing the luxurious and easy-to-drive Detroit Electric to women, circa 1917. Patriotic reference is to World War I. Wikipedia Commons. Detroit Electric Automobiles (1917). https://commons.wikimedia.org/wiki/File:Detroit_Electric_Automobiles_%281917%29_%28ADVERT_115%29.jpeg CC-BY-SA

as exemplified by the print ad in Fig. 14.12. The message was "*Ladies be patriotic and drive your own*" electric car, since they are simpler to operate, clean and quiet, and neither gasoline nor men were needed to operate them, both having greater value in the WWI effort. The era of stable employment for *motormen* was ending. By 1930, electric cars represented less than 1% of the US automobile market.

EVs Held All the Early Land Speed Records

Not withstanding their range limitations, the first electric cars were (much) faster than early IC engine cars, and they held all automobile speed records during the Golden Era.

Electric passenger cars such as the Detroit Electric and Baker Electric were designed for low-speed use, typically 20 mph, which was considered suitable or even fast for urban use since the competition was literally horse power. 20 mph (32 kph) would have been considered a maximum safe speed on rough unpaved roads built for horse-drawn conveyances. There were no traffic or pedestrian statutes. The general incompatibility of motor vehicles with vehicles powered by horses lead to ridiculously restrictive laws regulating "horseless carriages" enacted out of public fear of the new technology, e.g., the short-lived "Locomotive Act of 1864" in England and the "Red Flag Law of 1894" in the USA, both repealed in 1896.

However, even with the crude motor and battery technology at the time, it was soon realized that electric propulsion had the potential for much higher speeds and faster acceleration if the vehicle structure, aerodynamics, and roadways could handle it.

Figure 14.13 is a photograph of Belgian Count Camille Jenatzy's 1898 *La Jamais Contente*, credited as the first vehicle of any type to exceed 100 km/h (62 mph). It used high-drain single-use (probably excessive sulfuric acid)[2] lead-acid batteries and an aluminum torpedo-shaped body shell that looked fast but actually gained nothing since the unshielded position of the driver and the fully exposed undercarriage negated any aerodynamic advantage of the body [28]. The small stout wheels were from an artillery cannon, necessary because of the heavy battery mass. Jenatzy's self-congratulatory parade in Paris shown in Fig. 14.13 introduced another enduring tradition of automotive marketing, the obligatory attractive woman with the car.

In the USA, Walter C. Baker, a 34-year-old engineer and his father were already building electric vehicles at their Baker Electric Automobile Company in Cleveland 1899–1915 [29]. On May 31, 1902, he established an official world land speed record of 104 mph (167 kph) at Ormond Beach, Florida. A local newspaper photo of the vehicle is shown in Fig. 14.14 [14]. In a subsequent run, he reached either 127 mph or 136 mph[3] (according to different press accounts) [30] and may have been able to travel even faster had not, according to the reports, all four wheels left the ground after encountering street car tracks following the official end of the speed run [14, pp. 244–246] (the car had no suspension) leading to the loss of one of the wheels and control of the heavy lead-acid racer, killing one spectator and injuring several others. Baker and his brakeman inside the vehicle were uninjured because of

[2]Veterans of electric car racing in the 1990's were aware that if they wanted to break a speed record, at the expense of destroying their lead-acid batteries, they refilled them using battery acid rather than water.

[3]The Baker speed record was unofficially reported as 136 mph.

Fig. 14.13 In 1898 the "La Jamais Contente" (never content) exceeded 100 km/h (62 mph) after multiple attempts over 5 years. From Scientific American, 1898. Public domain

Fig. 14.14 Walter Baker of start-up Baker Electric Company of Cleveland officially achieved over 104 mph on a packed sand beach in Florida in 1902. A record for EVs that stood for many decades. Original image author unknown. Image copy from https://www.macsmotorcitygarage.com/this-car-went-100-mph-in-1902-and-it-was-electric/. Public domain

another innovation: a fully enclosed aerodynamic aluminum body. Possibly because of the accident, his 127 or 136 mph record did not enter official record books. In fact, he was arrested for manslaughter, but the charges were dropped when it was found that the victim illegally crossed the safety barriers into the racer's path. He received no Jenatzy-type parade, but his 104 mph record in 1902 is recognized as the first road vehicle of any kind to exceed 100 mph [31].

Confronted for the first time with the potential carnage that could be caused by high-speed automobiles (gasoline or electric), the recently founded Automobile Club of America (now AAA) immediately suspended any involvement in motor racing, a ban that lasted until 2005 [14, pp. 236–237].

As a historic note, the first *gasoline-powered* vehicle to exceed 100 mph and (just barely) exceed Baker's 104 mph record was the Napier Sampson L48 with a 15 L (later 20 L) six-cylinder engine, achieving 104.651 mph 3 years later in 1905 [32]. The world land speed record was again broken in 1906, this time by a *Stanley Steamer* at 127.7 mph. Baker's 104 mph speed record for battery electric vehicles remained unsurpassed until 1968 when the *Autolite Lead Wedge* achieved a "flying mile" record of 128.9 mph [14].[4]

That Walter Baker and his father were ahead of their time would be an understatement. His aero vehicle body and low-drag wheel designs were the first departure from the carriage-descended planform of all vehicles up to that time, and the forerunner of all speed record attempt cars to follow. This was entirely an intuitive exercise for the Bakers since nothing like it had ever been built or even proposed before.

Another pair of interesting footnotes in electric automotive history are the first two documented formal automobile races in the USA. These admitted any type of motor vehicle, including electric, steam, and ICVs running on any fuel. These early events were well-publicized demonstrations of both the advantages and the limitations of each motive power type. The first race in 1895 was the *Chicago-Evanston Road Race* promoted by the Chicago Times-Herald, that was held on a 53-mile unpaved public road that exceeded the battery range of the electric cars entered. All of the electric entries were leading at the half-way point but were forced to drop out before the finish line as their batteries became depleted. The second race, also sponsored by the Chicago Times-Herald in 1896, was held on a closed oval course at the Narragansett Park (horse) race track outside of Chicago. Entered were six ICE cars and two electric cars. The race was won by a Riker Electric with an average speed of 24 mph.

A list of all 2024 electric land speed record holders and challengers can be found in reference [33].

First EV Renaissance

The years from 1920 through approximately 1980 saw the dominance of the piston ICE for powering land and air vehicles, with steam retaining its position only for powering ships or for stationary applications. Despite the noise, grease, oil leaks, smoke and toxic liquid fuels, the high energy densities of liquid fuels (gasoline, diesel fuel, benzene, or alcohols) provided range limited only by the size of the fuel tank, as well as fast refueling times. To a major extent, internal combustion automobiles defined the progress and culture of the twentieth and early twenty-first centuries, as electric vehicles faded from public memory, relegated to specialty applications only. The low energy density of electrochemical batteries simply could not compete. The lead-acid battery remained the most viable electric energy storage device for vehicles for the next 100 years. EVs enjoyed a resurgence of interest (but little commercial success) between the early 1960s and the late 1990s, a period that could be referred to as the First EV Renaissance. This period was characterized by numerous hopeful startup companies and a few electric "concept cars" from major manufacturers, but all were limited by the lack of a battery that could provide an adequate range for consumer acceptance.

[4] Electric Vehicle News, Vol 1. No. 2, March 1969, as referenced in Wakefield.

Battery electric propulsion remained in use for indoor-operated forklifts, mining vehicles, golf carts, and airport 'tugs'. But their limited range and recharge time limitations made them unpopular for roadway transportation, despite the considerably lower cost per mile of electric energy in those years.

One example of a less unsuccessful road-going EV of the era was the CitiCar produced in 1974-1977 by Sebring Vanguard of Florida, USA [34]. It was basically an enclosed golf cart, but it differed from previous electrified vehicles of the 1960s and 1970s that were converted gasoline vehicles. It was a 567 kg purpose-built micro-car with a useful range of only 40 miles and a top speed of 36 mph, specifically for urban commuters in need of a low-cost low-speed ride for short distances. It predated the US DOT vehicle classification *Neighborhood Electric Vehicle* (NEV) [35] and was registered as a low-speed vehicle (LSV), the same as road-operated farm machinery.

One ingenious but goofy-looking feature of the CityCar was the placement of its six (36V model) or eight (48V model) lead-acid batteries in two groups, inside the front and rear bumpers, allowing them to serve as crash energy absorbers (aka 'battery rams') to improve survivability in front or rear impacts with larger vehicles. It was never formally established that the battery-laden bumpers actually did save lives because so few CitiCars were involved in collisions. But (from personal experience)[5] it was uniquely satisfying to know that if struck from the rear by a tailgater frustrated by your slow speed, the offending vehicle would be covered in sulfuric acid and lead debris. Sebring Vanguard was acquired by Commuter Cars Inc. in 1977, and the CityCar's unique battery placement was eliminated from the new version called the Commutacar, which never enjoyed the success of the funky CityCar and ended production in 1982.

The battery bumper concept of the CityCar could never use modern lithium batteries which are prone to ignition and extreme fires if ruptured. They must be well protected from any impact or fire, even more so than gasoline tanks. But lead-acid batteries are relatively benign when ruptured. The worst result is a toxic but manageable mess of sulfuric acid and lead, but nothing flammable other than a small amount of hydrogen accumulated inside the battery case.

The Aquarian Age[6] [36] dream of a future with small, ultra-efficient, nonpolluting electric cars was never realized.[7] The vehicle size relationship between ICVs and EVs is now actually reversed, with oversized *electric SUVs and trucks* dominating the US electric vehicle market, leading to much greater collision damage and 50% (IIHS data) more fatalities in collisions. The increasing size/weight trend is certainly not unique to electric vehicles. Vehicle size bloat in the USA has gotten

[5] Personal experience driving a Sebring Vanguard CitiCar in Denver CO, 1979.

[6] *Age of Aquarius* as used here refers to the American counterculture period, generally between 1965 and 1975, known for hope and somewhat naive belief in a better future for all personkind. The name was adopted by members of that generation even though it actually had nothing to do with the astrological Age of Aquarius, spanning (arguably) 1443 CE to some date between 1447 and 3597 CE depending on the authority.

[7] As a believer in the potential of technology to do good for humanity, the bloatification of electric vehicles is a great disappointment.

Fig. 14.15 1974-1977 Sebring-Vanguard CitiCar. (left) compared with a 2020 Tesla Model X (right). Photographed 2016 in San Francisco by Steve Jurvetson. Creative Commons Attribution 2.0. https://commons.wikimedia.org

continuously worse since the 1980s, contributing to the extinction of subcompact cars in the US market. According to 2022 data from the U.S. Environmental Protection Agency, "three-quarters of new vehicles produced in the U.S. are light trucks" [37]. Apparently, every suburban cowperson needs a truck to move their garden mulch.

Not of any particular relevance, but the weight of the popular Tesla Model X standard version is 2352 kg, over four times as much as the 567 kg CitiCar shown next to it in Fig. 14.15.

The California Zero-Emission Vehicle Mandate

The US State of California has had a long history of being on the leading edge of environmental regulation. Empowered by previous attempts to drive automotive emission technology by political means, the state legislature squared off against the major automakers who had a long history of fabricating reasons why ambitious air quality objectives could not be met. There was little trust between increasingly polarized opponents in the clean air battle.

In 1988, the California Clean Air Act was passed by the legislature. It authorized the implementation of any "technologically feasible" standards on vehicles and fuels, with the stated objective of "the maximum degree of emission reduction possible."

This led in December 1989 to the first incarnation of the California LEV (Low Emission Vehicle) Program. It empowered the California Air Resources Board to specify the first three "Tiers" of a new regulatory structure. The tiers were based on automotive HC emissions per mile: Transition Low Emissions Vehicles (**TLEV**, HC < 0.125 g/mi), Low Emission Vehicles (**LEV**, HC < 0.075 g/mile), and Ultra Low Emission Vehicles (**ULEV**, HC < 0.04 g/mi).

As expected, the three major US automakers strenuously objected.

Nevertheless, in 1990 General Motors (USA), seeing an opportunity to dominate a nascent market with unlimited growth potential, introduced the *Impact*, a two-passenger battery electric vehicle, later renamed the EV-1 (Fig. 14.16). It was a

Fig. 14.16 The 1997-99 GM EV-1 battery electric vehicle. https://commons.wikimedia.org/wiki/File:GM_EV1_2017.jpg Wikipedia Commons. International CC BY-SA 4.0

technical tour-de-force among not just electric but all passenger cars of the time. Its aerodynamic shape and "boat-tail" helped it achieve an aerodynamic drag coefficient C_d under 0.2, possibly the lowest of any production vehicle up to that time. Its 26 maintenance-free lead-acid batteries stored 16.5 kWh and its 102kW (137 HP) induction motor provided 60 miles range and a top speed electronically limited to 80 mph. On March 11, 1994, the Impact established a new world land speed record for electric vehicles at 183.8 mph.

The events that followed this remarkable accomplishment were apparently not anticipated by GM marketing (the same creative folks that named the car "Impact" because it was supposed to make a big technical impact[8]). In their enthusiasm to showcase their technical prowess, GM had inadvertently given the world a glimpse of a future in which cool-looking, quiet vehicles with no tailpipes displaced the gas-guzzling, pollution-spewing products *upon which they depended for profitability*. Edisonesque confidence in their technology and new aggressive EPA/CARB emission targets inspired GM and other automakers to attempt to produce and (at least pretend to try to) sell an affordable electric car.

There was *just one problem:*[9] battery technology was not much further along in 1996 than it was 100 years prior, even in the hands of Edison himself. The tiny (by today's standards) 16.5 kWh maintenance-free lead-acid battery in the GM Impact weighed 1310 lbs almost half the 2908 lb (1,319 kg) vehicle weight [38].

Meanwhile in California, empowered by the prevailing belief that scientific breakthroughs can be legislated, in 1990, a new vehicle classification was created: the *ZEV*, reserved exclusively for battery electric vehicles and (yet to be on anyone's radar) hydrogen fuel cell vehicles. Along with this came the infamous policy now posthumously referred to as the *California Zero Emission Vehicle Mandate*. Three planned phases were specified with the force of state law [39]:

[8] There have certainly been worse product names, such as https://www.watchmojo.com/articles/top-20-worst-product-names-ever.

[9] In case you missed the connection, "There's just one problem ..." is a phrase often used by physicist and educator Dr. Sabine Hossenfelder in her popular podcast Science with Sabine. https://en.wikipedia.org/wiki/Sabine_Hossenfelder.

While meeting the fleet average requirement, each manufacturer's sales fleet of passenger cars and light-duty trucks from 0-3750 lbs, LVW[10] shall be composed of at least **2% ZEVs** each model year from 1998 through 2000, **5% ZEVs** in 2001, and **10% ZEVs** in 2003 and subsequent model years.

(a) A manufacturer may meet the ZEV requirements by submitting to the Executive Officer a commensurate amount of g/mi NMOG emission credits earned exclusively from the sale of ZEVs. These credits may be earned previously by the manufacturer or acquired from another manufacturer.
(b) Manufacturers which sell fewer ZEVs than required in a given model year shall make up the deficit by the end of the next model year, by selling an additional number of ZEVs equal to their deficit or by submitting to the Executive Officer a commensurate amount of g/mi NMOG credits earned exclusively from the sale of ZEVs.
(c) Small volume manufacturers shall not be required to meet the percentage ZEV requirements. However, small-volume manufacturers may earn and market credits for ZEVs they produce and sell.
(d) Intermediate volume manufacturers shall not be required to meet the percentage ZEV requirements before the 2003 model year.

The terrified auto industry (*worldwide,* not just the USA) responded earnestly with massive investment and R&D on batteries and charging equipment, with abiding hope that these goals might somehow be doable. GM acquired several small companies that had technologies they might need. But unlike the incremental engineering developments that resulted from other optimistic science-free regulations, the obstacle here was electrochemistry: finding battery chemistries and configurations that had a chance of providing at least four times the mass energy density of state-of-the-art lead acid batteries, allowing the range necessary to make EVs palatable to the general public without exposing them to nightmare-level risks of exotic chemicals. It soon became clear that only a limited number of affluent environmentalists, or government agencies that had no choice, would buy the best EV they could produce at that time. The California mandate legislation specifically said, "manufacturer's sales fleet," which meant that they had to build and offer the required percentage of electric cars for sale, even if no one bought them.

Also, whether intentionally or not, a subtle back door had been inserted in the legislation, the long term effects of which are still being felt today. The mandate was restricted to *"... passenger cars and light-duty trucks from 0-3,750 lbs, laden or loaded vehicle weight (LVW)."* Note that the 1990 average "curb" or *unladen vehicle weight or UVW* of all cars and light trucks sold in the USA was already nearly 3750 lbs. Given the state of battery technology in 1996, any electric vehicle that could provide a range acceptable to consumers would have to weigh more than the mandate's maximum regulated weight limit. Even with much better batteries today (2024), very few EVs in the US would count toward the mandate's requirements ... The average curb weight of an electric automobile, light truck, or SUV in the USA today is 4800 lbs (2177 kg) [40].

Conversely, the mandate had unintentionally incentivized increasing the weight of all non-electric cars to above 3,750 lbs LVW to keep them out of the targeted

[10] LVW stands for *Loaded Vehicle Weight*, which is the unladen or curb weight plus the maximum allowable weight of passengers and cargo.

weight class. It also actually discouraged sales of electric cars and trucks because, unless there was a major battery technology breakthrough, few people would buy an electric vehicle with a range of only 50 miles, limited by the maximum qualifying vehicle weight.

What followed was one of the largest lobbying efforts in California's history to repeal, reduce, or dilute the mandate and its aggressive adoption timetable. Three successive compromises were brokered as each deadline approached: in 1996, the 1998 target was renamed as a "demonstration program." In Round Two, the 2001 requirement was modified by adding another new classification, the *PZEV (Partial Zero-Emission Vehicle)* and the *ATPZEV (Advanced Technology Partial Zero-Emission Vehicle)*. Most (but not all) PZEVs are gasoline *hybrids*, and most ATPZEVs are gasoline-electric *plug hybrids* [41].

The most important provision of the first and second compromises was a *2-for-1 hybrid/electric vehicle swap*. This allowed recently-introduced (from Japan) *hybrid electric vehicles* (HEVs classified as PZEVs or ATPZEVs) *to each be counted as "half of an electric vehicle."* The idea was that to meet the percentage of BEVs mandated, automakers needed only build twice the number of hybrid ICVs instead of BEVs. Another important provision classified the newly available *hydrogen fuel cell vehicles* (HFCVs or FCVs) as ZEVs.

Since (non-plug) hybrids are gasoline-only vehicles, the compromise kept everyone happy. Oil interests would still control the entire vehicular energy market, automakers would not need to retool and invest in EV manufacturing, gas station operators could continue to sell gasoline, dealers could sell products they understood and could promote, and the deep roots of gasoline culture would remain intact. At least for a few years. Japanese manufacturers must have been particularly happy because it would not be until 2004 that an HEV would be made by an American automaker, and no fuel cell hydrogen vehicle has ever been sold by a US automaker.

This bait-and-switch allowed the California Legislature to save face. And from the perspective of nearly 3 decades, this compromise was the best thing that came out of the irrational exuberance of the EV Mandate. Hybrid technology leaped forward in Japan (while US manufacturers continued to complain about cost). Hybrid sales worldwide exploded. They did not need to wait for lithium batteries. Hybrids needed only a tiny (e.g., 1.0 kWh) "ballast" battery to store energy recovered by regenerative braking and allow start-stop IC engine control. NiMH batteries that were available years before lithium batteries would suffice. And at long last, the first US-made hybrid (the 2004 Ford Escape mini-SUV) entered the market to great acclaim, incorporating power-split hybrid drive technology licensed from Toyota [42].[11]

By the third dilution in 2003, the mandate was reduced to a bizarre collection of formula-driven requirements involving percentages of 6% PZEVs, 2% ATPZEVs, and 2% ZEVs [39] .

[11] It is ironic that GM licensed power split hybrid drive technology from Toyota, since the concept was developed and patented in the USA, originally by TRW in the late 1960s, and in the 1990s by Paice LLC, a small investor-backed company that successfully sued Toyota for infringement.

In 1996, GM made an earnest attempt to comply with the ZEV mandate by manufacturing a production version of the *Impact*, renamed the *EV-1* (good decision), and offering it for lease (not sale) to the public.

In the 1997 and '98 EV-1 model years, energy was stored in advanced maintenance-free lead-acid batteries. But for 1999 (the final EV-1 model year), following GM's acquisition of a controlling interest in battery manufacturer Ovonics [43],[12] NiMH batteries were adopted for the EV-1, the first production car to use anything other than lead acid. This new battery chemistry provided a superior range compared with the best lead-acid or nickel-cadmium (NiCd) batteries of the time, but still inadequate for the range demands of the mass market. A total of 1117 EV-1s were produced and leased to the public 1997–1999. Leasees were selected from a large pool of *applicants* based on their enthusiasm for EVs and demographics: an eclectic cohort, all in Southern California.

The leasees adored their EV-1s despite the 50–60 mile range [44]. For many, the car had become a part of their identity. This became a problem that GM had not anticipated: lessees did not want to give their cars back at the end of their 3-year lease.

Now relegated to automotive folklore, the EV-1 ended up a public relations disaster for General Motors when the company attempted to recover their cars from the leasees in 2000. The drama of 1000 sobbing drivers, each saying passionate goodbyes to their beloved cars as a cold-hearted corporation towed the cars out of their driveways was heavily documented by sympathetic news media. Conspiracy theories blossomed. A cult movie was made, "Who Killed the Electric Vehicle?" [45] The legal but mismanaged corporate behavior is now the subject of case studies in graduate business courses and law schools. The official corporate sound bite was that the EV-1 was never actually intended for production (NOT what they promised the leasees) and that GM could not continue to service 1000 or so EV-1s in public hands. Indeed, GM's Saturn division dealers that leased the EV-1s to the selected leasees were not qualified (equipment or staff) to service these space-age vehicles. Repairs during the lease period had been almost exclusively handled by dealer mechanics following Ikea-like remove-n-replace instructions for entire assemblies, with no troubleshooting needed or allowed. The secrecy of proprietary technology was a likely concern. Possibly most important was that, unknown to the sobbing leasees, the lead acid (1996-1997) or NiMH (1998) batteries were nearing the end of their expected 3 years lives, with replacement costs that neither GM nor the future owners would be willing to accept.

But no one understood why it was necessary to destroy all evidence of the car's existence. The visuals were the height of bad public relations, e.g., Fig. 14.17.

A more conspiratorial (but probably true) explanation is that GM's most popular and profitable vehicles at the time happened to be super-sized gas/diesel trucks and SUVs, cash cows that would be threatened if too many consumers started prioritizing environmental impact or energy security over having the biggest baddest four wheels on the block. In 1999, AM General, the manufacturer of the military HMMV, aka *Humvee*, had morphed the military truck into a commercial monster truck called

[12] Founded in 1985 by Stanford R. Ovshinsky, inventor of the Nickel Metal Hydride battery.

Fig. 14.17 Crushed EV-1s, 2003. Wikipedia Commons File:Ev1 crush5.jpg, Creative Commons Attribution-Share Alike 2.0 Generic license. CC-BY-SA-2.0

the *Hummer*, and then sold the right to manufacture and distribute the vehicle in the civilian market to GM. In late 1999, GM introduced it as the Hummer H1, and soon afterward the Hummer H2 was born, nothing more than a re-bodied Chevy Suburban that consumers were eager to pay $10,000 USD more for because it looked so menacing. These two passenger/utility vehicles were the largest and heaviest in the GM model lineup. Both were hugely profitable for GM, and intense buyer demand countered concerns about fuel economy or emissions. Gasoline was still relatively cheap, and buyers' tastes had been successfully cultivated to favor large off-road-capable vehicles, a fashion statement that continues today.

GM also pulled the plug on their heavily-invested attempt to establish their Delco/Hughes Electronics proprietary Magne-Charge[13] [46] inductive charging coupler shown in Fig. 14.18 [47]) as the EV charging standard in the USA and Europe [48].[14] The Magne-Charge (SAE J1773) coupler was indeed the ultimate in consumer safety since no electrical contacts were exposed. It was the only option for charging the last MY 1992–93 of GM/Conceptor electric G-Vans and all of the EV-1s, which lead to speculation that an ulterior motive for the EV-1 was to establish an early dominant market position for the charging standard. But there were technical limitations. A Magne-Charge paddle was essentially one side of an air-coupled transformer, which is electrically lossy and subject to a rather low power transfer limit. None of the up-and-coming EV manufacturers except Toyota were willing to adopt it. Conductive charging ruled, despite greater safety concerns. Concurrent with the Magne-Charge standard, the SAE established a conductive coupler and control standard, SAE J1772. By 1996, this had become the defacto world standard, later adopted by every manufacturer except Tesla that insisted on

[13] Inductive coupler assembly having its primary winding formed in a printed wiring board.

[14] EV1 Club Home Page, EV1 Club. Retrieved 2007-08-23. GM Pulls the Plug on Inductive Charging. Letter dated 2002-03-15 from General Motors Advanced Technology Vehicles Division, as cited in Magne Charge.

Fig. 14.18 GM/Delco Magne-Charge inductively coupled EV charge paddle. Shown in front of a 1998 GM EV-1 that used this charging method exclusively. Image from Renewable and Sustainable Energy Reviews. Publisher: Elsevier. https://www.researchgate.net/publication/309666212_00005361_125788 Used under STM guidelines

using their own (misleadingly named) North American Conductive Charging Standard. By 2022, Tesla strongarmed all manufacturers into adopting the NACC (and later NAC) standard by virtue of the fact that they had deployed nearly half of all public or home EV chargers.

Modern EVs Began as Converted ICVs

Prior to the mid-1990's, almost all road-going EVs were aftermarket conversions of existing gasoline vehicles that were allowed to retain the original vehicles' Federal Motor Vehicle Safety Standards (FMVSS) safety certifications. This meant that their design requirements remained based on the weight, structure, and control systems of the original ICE vehicle, including the suspension, brakes, steering, and structural integrity of the vehicle. Batteries used in conversions prior to the 2010s were almost exclusively lead acid, and the mass of these batteries added 50–100% to the original vehicle weight. Under US law, any "new" EVs were subject to the same safety and road worthiness standards as ICE vehicles, regardless of weight. This was mostly pragmatic considering the miniscule number of electric conversions. The cost implications of standardized crash testing would be severe for a fledgling startup company with limited production and financial resources. Custom low-production EVs were allowed to be registered under an *NHTSA variance* reserved for production runs usually less than 1000 vehicles. This avoided the cost burden of safety certification, including destructive testing of several "validation prototypes."

Fig. 14.19 Example of a DIY conversion of a 1994 Honda del Sol into an EV. It carried 18 eight-volt lead-acid batteries that nearly doubled the vehicle mass, compromising braking and steering. Photo by author

This was not unusual in the auto industry. The safety testing variances were common practice for most exotic vehicles, including the original Lotus Elise from which the Tesla Roadster was derived, and most rare performance cars including Ferrari, McLaren, Lamborghini, Maserati, Bugatti, and some Porsche, BMW, and Mercedes. Owners of these vehicles are (if they read the contractual documents) aware that they are their own crash dummies.

Home-Built EVs

The conversion of existing ICE vehicles to electric propulsion has for decades been a fairly widespread advocation of automotive enthusiasts that also cared about the environment, or just wanted to build a cool project to show off (like the author). Since home-built or limited-production electric conversions almost exclusively used heavy lead-acid batteries, these projects earned the derogatory label "lead sleds" because of the increased vehicle weight—sometimes well beyond the original vehicle's Gross Vehicle Weight Rating (GVWR). Many, like the 1994 Honda Del Sol conversion in Fig. 14.19, were borderline dangerous, with excessive braking distances, heavy steering, and suspensions inadequate for the increased vehicle weight. Nevertheless, such conversions only put their owner/builders at risk, not mass-market buyers. And much of the technical knowledge and enthusiasm for EVs today began with these amateur electric conversions.

AC Propulsion tZero

The first of the new generation of high-performance electric cars was almost surely the *tZero*, built 1997–2003 by AC Propulsion of Monrovia, California.

The tZero shown in Fig. 14.20 was a remarkable merger of state-of-the-art power electronics, battery technology, and automotive artistry. The tZero was a modified Piontek *Sportech* kit car to which AC Propulsion added their revolutionary AC-150 (150 kW, 201 HP) [49] induction motor drive system. Of the three operational prototypes built, the first two stored energy in 28 Johnson Controls 12V D31T *Optima*

Fig. 14.20 One of three 1997 AC Propulsion tZero electric roadsters, the predecessor of the 2006 Tesla Roadster. At the time, the T-Zero was the fasted EV for sale to the public. https://en.wikipedia.org/wiki/AC_Propulsion_tzero CC BY-SA 4.0

Yellow Top batteries to provide the nominal 336 volts for the AC-150 power unit. Based on the Optima battery specs, this would have amounted to a 25.2 kWh battery pack weighing 1680 lbs, probably resulting in an unladen weight about 1000 lbs heavier than a Sportech kit car powered by an IC engine. In the late 90s, before lithium batteries were available for EVs, the Clarios LLC (formerly Johnson Controls) *Optima* battery with its tubular maintenance-free cells was revered by Do-It-Yourself (DIY) builders as the best available deep-cycle battery for conversions due to its durability and reasonable cost. The third tZero was equipped with a hand-assembled battery pack consisting of an undocumented large number of lithium-ion cells intended for or actually scavenged from laptop computers. Unofficially, this would make it the first EV to use lithium-ion batteries, which at that time, were still prone to spontaneous ignition.

The lithium version of the tZero was 500 lbs lighter than its two lead-acid predecessors and was the fastest licensed electric car of its time with a potential maximum speed of 140 mph, a 0–60 mph time of 3.6 s, and a claimed range of 300 miles. But as a hand-built exotic car, its selling price was $220,000 USD, far above the cost tolerance of hobbyists and even wealthy enviro-nuts[15] [50]. In 2003, current Tesla Inc. president Elon Musk attempted to purchase AC Propulsion, but his offer was not accepted. The tZero never entered production and commercialization plans were abandoned later that year for lack of the financial resources needed to launch, manufacture, sell and support a new automobile in a ruthlessly competitive industry.

Tesla Roadster

Flush with cash from his share of the sale of PayPal in 2002, Mr, Musk purchased a controlling interest in another small startup, Tesla Motors Inc. in 2004. After discharging the two company founders shortly after gaining control, in 2006 Mr. Musk

[15]"Enviro-nut" is a US slang term from the 1970's meaning *an environmental action extremist,* according to *Slang Define*.

Fig. 14.21 2010 Tesla Roadster. https://creativecommons.org/licenses/by/2.0/ Attribution 2.0 Generic. CC BY 2.0 International

personally introduced the Tesla Roadster [51] that had been nearly completed by the original Tesla ownership. All persons involved that could still be interviewed in 2023 confirmed that the Roadster was an attempt to copy the short-lived *AC Propulsion tZero*[16] that Mr. Musk had failed to acquire [52] shown in Fig. 14.20, that had been demonstrated in early form in 1997 and test-driven by Mr. Musk in 2003 [53]. Lithium-ion batteries were used in all of the Tesla Roadsters, 2008–2012. This feature alone was a historic advance since it gave these high-performance sports cars a much-improved range, shedding the popular *lead sled* legacy. Heavily promoted in media, with its attractive Lotus-manufactured fiberglass body, Tesla Roadsters soon became objects of mid-life crisis dreams (Fig. 14.21).

Like the tZero, the Tesla Roadster was a modified version of an existing fiberglass-bodied car, in this case, the Lotus Elise, manufactured in Hethel, England, by Lotus Cars Ltd. 2008–2012. A wise choice since the Elise was an exceptionally attractive niche sports car by a legendary limited-production manufacturer. 2450 Roadsters were build, with 1000 (possibly a few more) legally sold and registered in the USA under an FMVSS low-production safety exemption, not including the one that was launched into orbit as a marketing "opportunity" by SpaceX, the aerospace company owned by Mr. Musk. The Tesla Roadster initially used the same AC-150 drivetrain as the tZero, purchased from AC Propulsion, and a hand-assembled liquid-cooled 53 kWh battery pack consisting of 6,831 size "18650" (18 mm diameter, 650 mm length) lithium-ion cells, stated to have been purchased from "reputable Fortune 500 battery suppliers" [54], located behind the seats [55]. The battery pack alone justified the $109,000 MSRP of the Gen 1 Roadster. The 150 kW AC-150 drive motor was soon replaced by a more powerful Tesla-engineered 185 kW induction motor, differing sufficiently from the AC-150 to circumvent AC Propulsion patents. One of the compromises was that the Tesla motor used air cooling rather than the AC-150's superior liquid cooling, which was particularly ironic since the battery pack of the Roadster was liquid cooled. Four years later, when the Tesla Model S was introduced for sale in 2012, its induction motor adopted liquid cooling.

[16] AC-Propulsion of Monrovia, California only produced three running tZeros, lacking the financial clout of Tesla to weather years of production at a loss made possible by Mr. Muck's investment.

Despite its higher power motor, the Roadster's performance was slightly less than the tZero, probably because of its larger and heavier battery pack. But even with only a 53 kWh battery (tiny by today's standards) and 150 kW motor, it was greatly overpowered compared with the original Elise. It was typical of the level and finish of high-end custom EV conversions at the time, each car being hand-made in a small facility in Palo Alto, California. Its MSRP in 2011 of $109,000 made it accessible only to buyers of exotic cars, but even this higher-than-market cost was below its production and development costs. It is rumored to have never generated a profit for Tesla, which survived borderline insolvency for more than 5 years only because of the private resources of Mr. Musk and several unprecedented government-backed loans. Understandably for any new technology, it had "early opter" problems, but the public's perception was well managed, and affluent buyers queued up to buy the exotic "it-car" of the late 2000s. Credit is due that, more than any other previous EV, the Roadster changed the popular image of electric cars from that of oversized golf carts and indoor forklifts into the ultimate in cool fashion accessories. One memorable example: a 2010 Tesla Roadster was showcased as the automotive object of lust by lead character Harvey Spector in Season 1, Episode 3 of the 2011 Hulu/Netflix legal drama series "Suits" (also featuring aspiring actress Meghan Markle) [56].

In January 2008, the NHTSA announced that it would grant a limited-production waiver of FMVSS crash safety testing and advanced airbag requirements to Tesla Motors for the Roadster. 1000 units were legally registered in the USA, while 1418 units were sold worldwide between September 2008 and September 2012 using Lotus Elise "gliders" (cars without engine and transmission) imported from England. The FMVSS exemption expired for cars made after 2011, so the last Roadsters could not be sold in the American market. Fifteen *Final Edition Roadsters* were produced and all exported to close out the manufacturing cycle of Tesla's first electric car [57]. The rest is history.

The Roadster was a great success despite its technical issues and high cost. Its straight-line acceleration was stellar. Not generally recognized by the public or media at the time was the same overweight problem that plagued all DIY EV conversions. The 1996 Lotus Elise weighed 725 kg (1598 lb). Maximally accessorized, the last-produced 2010-12 Lotus Elise had a maximum curb weight of 900 kg (1984 lb) [58]. During the 1960s, Lotus Cars Ltd. had achieved great success in Formula 1 with their ultra-light race cars and similar passenger cars. The Elise had a composite body and the traditional Lotus just-enough-structure-to-win-the-race, i.e., it was designed to meet but not exceed requirements. The 2010 Tesla Roadster base model had an advertised unladen weight of 1305 kg (2877 lbs) [59], a 45% weight increase over the Elise. Its battery pack alone added 450 kg (992 lbs) to the car [60]. This increased the weight of the Roadster beyond the full-loaded GVWR of the Elise chassis, 1055 kg (2326 lbs) [61] (GVWR = the maximum safe fully-loaded vehicle mass). With a normal passenger and luggage load, the Roadster's weight was as much as 70% more than the original Elise. No published information was available regarding structural failures, but the overloaded suspension and steering were known issues. I could find no accounts of problems with the 4-wheel disk brakes, probably because of the brake-pad-sparing benefits of EV regenerative braking, and the oversized brakes from the racing lineage of the Elise. But at least one member of the Tesla Motor Club reported in 2011 that his 2010 Roadster had a "big tendency to understeer" during a track-day outing [62]. Understeer is a well-known characteristic of excessively heavy cars with poor weight distribution. It was a staple of front-heavy iron V8 American "muscle cars" of the 1960s and 1970s. The very quick 0–60 mph time of 4.0 s of the Roadster only exacerbated the safety issues. The Roadster was probably more a story of image than of objectively evaluated performance. This was the anthem of many of the cool cars of that era.[17]

Several years later, Road and Track Magazine in its June 2016 issue quoted from a presentation by Mr. Musk to Tesla investors:

> Musk admitted that the early Roadster was completely unsafe," that it "broke down all the time," and it "didn't really work.

But in fairness, problems are always expected with infant technologies, and despite its issues, the Roadster remains one of the most historically significant EV developments of all time. In 2024, it is a much-sought-after collector's car.

EVs Come of Age: FMVSS

Prior to 1989, no electric vehicles had been subjected to the rigorous safety testing requirements of the US Federal Motor Vehicle Safety Standards (FMVSS), enforced by the National High Transportation Administration (NHTSA). However, with increasing interest in electric cars motivated by the gasoline shortages of 1973 and 1979 in the USA, a few limited offerings of production electric vehicles were made by major auto manufacturers, including the General Motors/Conceptor e;ectric G-Van (USA), Griffon (England) and the Chrysler TEVan (USA). The first production electric vehicle to be fully certified as FMVSS compliant was the General Motors/Conceptor electric G-Van, in limited production 1989–1992 (Fig. 14.22). Almost absurdly expensive ($72,000 USD), heavy (8500 lbs), and slow (60 mph max), it was purchased only by electric utilities and government agencies. But government and utility purchasers required full compliance with all US Federal safety regulations.

FMVSS vehicle safety certification required the destructive collision testing of sample vehicles to evaluate passenger survivability. Details of its testing were not well documented. The tests involved staged impacts at various speeds and angles [63]. Since fewer than 100 Electric G-Vans were built, it is reasonable to assume that the high cost of the electric G-Van was at least partially due to the amortized cost of the destroyed certification vehicles.

[17] Interview with Nicholas Zart, automotive and EV journalist covering Tesla's history 2003 present, 20 November 2023.

Fig. 14.22 1991 GM/Conceptor Electric G-Van, originally owned by Pacific Gas and Electric Co. Photo: author

The Second EV Renaissance: EVs Enter the Mass Market

The 2012 Nissan Leaf was the first mass-market BEV that was available for sale to the public outside of Japan (Fig. 14.23). Despite its limited 116 km (75 miles) EPA-rated range, it was immediately popular and set the standard for the tidal wave of BEVs that followed only a few months later. But like so many innovative vehicles, it probably never made money for Nissan.

However, the iconic car that almost single-handedly wrote the chapter of the *second EV renaissance* was the Tesla Model S, shown in Fig. 14.24. After accepting $5000 USD deposits as early as 2007, the first production model was delivered in June of 2012. Despite its high cost (over $100,000 USD for the fully-optioned early deliveries), it eventually became a high-demand item at a cost as low as $57,400 USD for the base model, a new symbol for moderately affluent owners, both environmentalists and affluent status seekers. Its relative affordability and high profit margin for Tesla benefited from the newly enacted $7500 USD Federal tax credit for electric vehicles, and programs to assist in the installation costs of Level II home charging equipment capable of refueling the vehicle overnight.

The Model S was followed in a few years by the lower priced Model 3, which quickly became the most popular BEV in the world, and in 2024 is still among the top three best sellers. Figure 14.25 shows a 2023 Tesla Model 3 Long Range. The Model 3 was a technical tour de force, with features exceeding those of most contemporary ICVs. It was rated as the highest efficiency EV on the market through 2020. Particular attention was paid to operation in cold climates, in which the battery output and therefore the range of the emerging EVs were known to be greatly reduced.

From 2012 through the early 2020s, the EV and PHEV market grew exponentially, along with the trend toward ever-larger batteries to improve vehicle range. Small EVs were eventually discontinued in favor of larger and heavier electric SUVs and trucks that could carry the enormous batteries and commanded higher profit margins. In China, EV manufacturing grew from nothing to world

Fig. 14.23 2012 Nissan Leaf, sold in late 2011 in the USA. First mass-production electric car sold in the USA. Photo from https://c.pxhere.com/photos/80/2f/car_nissan_leaf_electric-1186974.jpg!d Public domain

Fig. 14.24 2012 (first MY) Tesla Model S at NAIAS 2012. Photographed by Autoviva.com. https://commons.wikimedia.org/wiki/File:Tesla_Model_S_at_NAIAS_2012_%2866 72639485%29.jpg Creative Commons Attribution 2.0 Generic license

Fig. 14.25 A customized 2024 Tesla Model 3, Long Range, Version 9, the largest-selling EV 2019-22, also rated in 2020 as the highest efficiency EV on the market. Image from "Wheel Front, Gallery". https://wheelfront.com/. Public domain

dominance in only a few years, led by Geely, SAIC and BYD, the last founded only about 8 years ago and by 2024, the world's largest automaker. Startup companies sprouted worldwide but especially in the USA, hoping to capitalize on the nascent EV market with premium vehicle prices, empowered by free-flowing government subsidies and easy investor money. The market quickly became flooded with large electric trucks and SUVs. By 2024, about half of these optimistic startups had become insolvent, as major manufacturers displaced them in the USA and the EU, and low-cost competition from China loomed.

At the same time as the EV tidal wave, the public charging infrastructure was built out at a record pace (compared with public works projects). By the end of 2022, there were 2.7 million public charging points worldwide, nearly half in China [64]. Tesla's high-rate "Superchargers" soon became ubiquitous, numbering more than 40,000 worldwide [65]. Countries, states, provinces, and automakers made bold promises to be 50% electric by 2025 and all-electric as soon as 2030 or 2035 at the latest. All of these promises were dialed back in late 2024 as the market normalized, and forgotten in 2025 in the USA with the change of Federal administrations. Collateral requirements such as the electric power infrastructure needed to support these massive new demands continue to be downplayed, even though any electric power engineer can see a potential train wreck ahead if a multi-trillion dollar public investment in the electric grid does not happen concurrently. Regardless, EVs are here to stay.

Third-Generation Electric Vehicles

According to the International Energy Agency electric cars accounted for around 18% of all cars sold worldwide in 2023 [66]. There were more than 500 different electric vehicle models, but only about 80 of these models are passenger cars; the large majority are classified as light trucks and SUVs. Official predictions are for over 1000 different EV models by 2028. As of early 2024, there are 40 world manufacturers of EVs that have achieved reasonable sales, but 90% of global sales were from only 20 of these manufacturers. Boutique manufacturers of high-end limited-production vehicles are excluded from this number. Existing models of ICVs have often been replaced by new BEVs or starting in 2025, plug hybrids.

It is arguable that a *third generation* of EVs can be designated, but it seems appropriate because of the maturity of the technology and market. While electrics still have lower automotive market penetration than ICVs, for countries in which major public investments in charging infrastructure have been made, EVs are no longer niche-market items. They are the new norm, exemplified by Norway, with 88% of all vehicles sales being electric in 2024. Almost anyone interested in buying a new car has probably considered a BEV or PHEV, at least until comparing costs.

Led by apparent vehicle preferences in the USA, the EV market is now dominated by trucks and SUVs. This trend is not unique to EVs, but has become a major issue in the EV industry: huge vehicles help to disguise the huge size and mass of the battery. With larger batteries come the potential for high-power outputs, another selling point for EVs compared with gasoline vehicles. The last EV *compact* car made in the USA, the Chevrolet Bolt, was reintroduced in 2024 as a larger and more expensive Electric Utility Vehicle (EUV). No production *subcompact* EV from a US manufacturer has been produced in many years. The Tesla Model 3 is classified as a *midsize car*, not a compact). The financial forces behind this trend will be discussed later in this chapter.

The EV technology landscape is truly international, involving almost every industrialized nation. Over half of world EV manufacturers are Chinese. Revered legacy US and European nameplates are now owned by China and India, including Volvo (Geely/China), MG (SAIC/China), Lucid (PIF/Saudi Arabia), Jaguar and Rover (Tata/Indian), Chrysler/Dodge/Jeep (Stellantis/Italy), and former Swedish company Polestar (Geely/China).

Legacy automotive firms Mercedes, Porsche, and BMW are the EV leaders in Germany. Toyota, Honda, and Nissan/Mitsubishi dominate Japanese EV production. South Korean Hyundai/Kia EVs became the top sellers in the USA in 2024. Even Ferrari and Lamborghini of Italy now offer plug hybrid electric models. In 2023 Tesla alone sold 1.8 million EVs worldwide. Also in 2023, BYD (China) sold 1.6 million BEVs and 1.4 million plug hybrids, none of these in the USA. EV models from Geely of China dominate European, Asian, and Subcontinent markets.

Faster *and* Longer Range?

The intrinsic ability of electric motors to provide maximum torque at low or zero speed, and high power in a small volume has been known for decades. These features have been exploited maximally by EV manufacturers, with even modestly priced EVs out-accelerating gasoline performance cars. The principles behind this are discussed later in this chapter.

On October 1, 2021, driver Eric Ritter and team Vesco, with a streamliner built by reVolt Systems (Fig. 14.26) reached an official speed of 568 km/h (353 mph), a US (probably world) record. The "Little Giant" streamliner used two "heavily modified" Tesla motors and 1152 prismatic lithium-ion batteries [67]. It is likely that this record will already have been exceeded by the time of this publication.

One of the distinguishing attributes of third-generation EVs is the degree to which the physics of electric propulsion has been used as a consumer marketing tool. Electric vehicles may still elicit some small concern for environmental and energy sustainability, but now the power output of production EVs has become their second most compelling sales attribute, just behind range. The ever-more-massive battery packs required for increased range endow these EVs with previously unheard of acceleration, giving daily EV drivers the ability to exceed legal speed limits by a factor of three.

The well-publicized leader of this product performance competition until 2023 was the Tesla S Plaid. According to Car and Driver Magazine, the Plaid achieved a 2.1-s 0–60 mph, 4.3 s 0–100 mph time, and a 9.4-s quarter-mile, which in 2022 made it the fastest production car they had ever tested [68].

Figure 14.27 shows a concept rendering of the new version of the Tesla Roadster, originally announced in 2017 for production in 2022. According to CEO Elon

Fig. 14.26 Team Vesco 444 "Little Giant" Team Vesco 444, US EV Speed Record holder 2021, at 353 mph powered by 1152 prismatic lithium-ion batteries and heavily modified Tesla motors, 2021. Image from https://www.carscoops.com/2021/10/team-vesco-444-little-giant-broke-the-national-ev-record-with-353-mph/ Public domain

Fig. 14.27 Tesla Marketing rendering of the "New"Tesla Roadster, announced in 2017. Claimed in 2023 to be able to achieve 0-60 mph acceleration under 1 s, maximum speed of 250 mph. Image from Tesla Gallery, open image distribution for press/publication, https://www.tesla.com/tesla-gallery

Musk, the new Roadster will be "the greatest supercar of all time," with a quoted 250 mph top speed, a sub-1 second 0–60 mph time, and a 625-mile range [69] using three motors and a 200 kWh battery.

The top dog in the ongoing speed and range competition now changes almost monthly, but the current (April 2024) leader among production EVs in both speed and range is the $250,000 USD Lucid Air Sapphire, shown in Fig. 14.28. This is not surprising considering that it has a 118 kWh battery that is the largest available in any electric passenger sedan, and three motors totaling 927 kW (1234 hp). It weighs 2430 kg (5345 lbs) which makes it the heaviest of all current electric sedans.

An electric full-size sedan that weighs near twice an equivalent ICV, can accelerate with force exceeding gravity, and travel silently at more than three times the maximum speed limit in any country. What could possibly go wrong?

Fig. 14.28 2024 Lucid Air Touring. The Sapphire version of the Lucid Air is currently the fastest production electric sedan, at least for now. Image from https://commons.wikimedia.org/wiki/File:Lucid_Air_Grand_Touring_-_front.jpg Attribution-ShareAlike 4.0 International. CC BY-SA 4.0

Fig. 14.29 The 2022 Production Land Speed Record holder, the Rimac Nevara. 1914 HP, 256 mph (412 kph) maximum speed, 0–300 kph acceleration in 9.22 s, 120 kWh LiMgNiCo battery, undisclosed weight. Capable of speeds four times the US freeway speed limit and more expensive than some luxury yachts.

As of 2024, the fastest electric vehicle for sale to the public was the 2022 Rimac Nevara [70] (Croatia), capable of 0–97 kph (60 mph) acceleration time of 1.74 s,[18] 0–300 kph (186 mph) in 9.22 s, and a top speed of 412 kph (256 mph) (Fig. 14.29). It uses four motors driving all wheels, with a total of 1,427 kW (1914 HP), and a 120 kWh LiMgNiCo battery. The preproduction advertised price $300,000 USD has been most recently restated as $2.4 million USD, and it is (April 2025) subject to a 25% US import tariff. [71].

[18] For comparison, the acceleration of gravity is equivalent to a 0–60 mph elapsed time of 2.75 seconds. To achieve a 0–60 time of 1.77 seconds, requires a coefficient of friction between the tires and the road surface that greatly exceeds 1.0, which can be thought of as requiring tires that are sticky enough to literally glue the car to the ground.

Even outside of the exotic performance car class, absurdly powerful electric cars of all configurations have become the norm, enabled by the massive batteries that give them long range, and that justify even more massive prices. The consequences for traffic, pedestrian, and cyclist safety are obvious but rarely discussed, shielded by their classification as environmentally friendly. This dichotomy will be discussed later in this chapter.

For those old enough to have experienced it, or who may have listened to 60's music by the Beach Boys or Jan and Dean, the current electric vehicle horsepower and range war is eerily reminiscent of the horsepower race of the 1960s in the USA that begat an entire generation of large, powerful, grossly polluting "muscle cars" with fuel economies in the single digits. Except the bragging point of engine displacement has been replaced by battery capacity.

Not Exactly the Electric Future some Had Hoped for

What we envisioned in the 1970s
2019 Smart EQ Electric (Fig. 14.30)
17.6 kWh battery
58 miles (94 km) range
2383 lbs (1083 kg)
80 HP (60 kW)
108 MPGe
$24,550 USD
Discontinued in the USA in 2019 but still sold in Europe.

Fig. 14.30 2019 SmartForTwo minicar, electric version. Photo by David Villarreal Fernández. https://commons.wikimedia.org/wiki/File:Smart-electric-drive.jpg CC Attribution-Share Alike 2.0 Generic

Fig. 14.31 2022 GM Hummer Electric. Photo by 42-BRT. https://commons.wikimedia.org/wiki/File:2022_GMC_Hummer_EV_Truck_%28Black%29_%28cropped%29.jpg Creative Commons Attribution-Share Alike 4.0 International

What we got in the 2020s
2022- GMC Hummer EV (Fig. 14.31)
205 kWh battery base, 247 kWh battery optional
329 miles (531 km) range
9700 lbs (4400 kg) w 205 kWh battery
1000 HP (746 kW)
47 MPGe
$104,000–$110,295
6-month waiting list to purchase

Lacking a radical improvement in battery energy density, it is inevitable that the EV range and power competition will continue to drive increased battery and vehicle mass, as well as increased acquisition and operational costs. This trend comes at a cost to society in the form of increased traffic fatalities, infrastructure costs, and environmental harm from the manufacturing of ever-larger batteries. The electrification of automobiles is a long-overdue improvement over fossil fuels, but there is a clear need for renewed focus on the greater good to avoid a calamity potentially worse than fossil fuels.

Electric Vehicle Components

While battery electric vehicles (BEVs) share most of the same body, interior, suspension, braking, steering, and active safety components as a gasoline or diesel vehicle, the main difference is in the drivetrain, which must include at a minimum:

1. One or more electric motors coupled to the drive wheels of the vehicle.
2. A battery consisting of the series and parallel connection of many electrochemical cells.
3. Motor/drivetrain control electronics that power the motor and implement regenerative braking, as commanded by the vehicle operator.

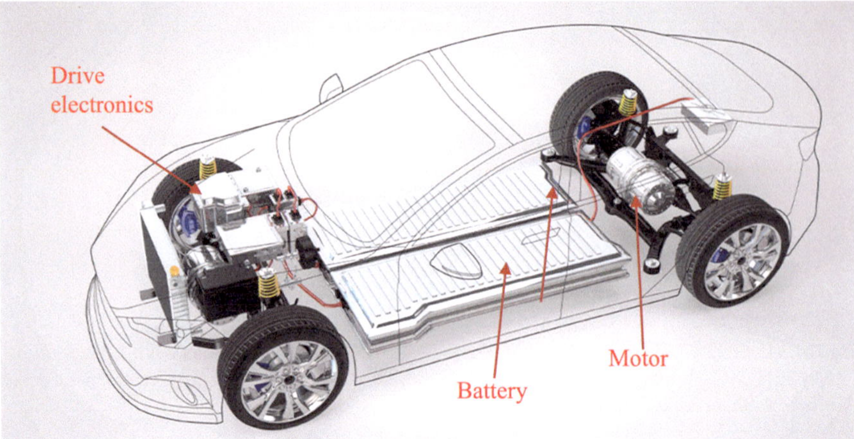

Fig. 14.32 Diagram showing electric vehicle drivetrain components of Tesla Model 3. Background image of electric vehicle drivetrain system from Infineon product web page, unattributed. https://www.infineon.com/cms/en/applications/automotive/electric-drive-train/)

A phantom view of the Tesla Model 3 appears in Fig. 14.32, showing these components [72].

The Electric Drive (Traction) Motor

While not the first electric motor, the first motor suitable for automotive propulsion was patented in the USA in 1887 by inventor Frank Sprague [73], probably following the work of Werner Siemens in Germany, who demonstrated an electric dynamo in 1856. This was a time of prolific invention, so credit must be shared with other innovators of the time including Nikola Tesla, James Maxwell, Oliver Heaviside, and Thomas Davenport.

The first widespread applications of electric motors in transportation were railroad and tram electric drive systems. These were (and still are) powered from overhead DC electrical caternaries. This application caught on quickly throughout the world. The technology of the electric motor drive for rail applications was so popular that by 1890, according to British historian Michael Robbins [74]:

> By 1 July 1890 there were 914 miles [of street railway] electrified in the United States—twice as many as were steam-operated and more than three times the length of cable lines. Within three years 200 streetcar systems had been or were in the process of being converted, 50 per cent by Sprague's company, more than 90 per cent based on his patents. Sixty per cent of all mileage was electrified by the end of 1893, 98 per cent ten years later.

Many different types of electric motors (aka rotating machines) have been used for EV propulsion. Batteries and motors with voltage ratings in the range of 48–144 VDC were most commonly used in first renaissance EVs and conversions. But electrical efficiency favors higher voltages, so since the early 2000s, all production EVs have used voltages in the range of 202V (Gen 2 Prius) to 800V (Lucid Air).

Because of the availability of maximum torque at zero speed, and a broad power curve, most EVs use a single-speed transmission, having no need for a clutch or fluid torque converter of the types that ICEs require. From a cost point of view, the lack of these two costly components is a huge commercial advantage for EVs, in addition to the low complexity of the electric motor compared with an ICE.

We will focus here only on the motor types that have been used in EVs during the past approximately 25 years.

Series or Parallel-Wound Brushed DC Motor

A brush-commutated DC motor is a type of *reluctance motor* that produces torque by the attraction or repulsion between the electromagnet poles in the stator (the outside part) with the electromagnet poles of the rotor (rotating part) that change polarity as it rotates by electrical switching (commutation) of current to the rotor via contact between graphite brushes and copper alloy contacts on the end of the rotor shaft. As the shaft rotates the carbon brushes make contact with successively different sections of the cylindrical copper commutator, changing the polarity of the rotor windings to cause repulsion or attraction between the rotor and the fixed stator electromagnets as it rotates.

The field may be wired with the rotor either in series or parallel (shunt) with the stator. In either configuration, it can theoretically operate on either AC or DC power since both the stator (field) and the rotor are electromagnets, so their electromagnet attraction or repulsion relationships are unchanged during both the positive and negative half-phases of Alternating Current (AC) power. Brushed DC motors are sometimes referred to as *AC/DC* or *Universal* motors, the type of motors used in almost all corded power tools, which are powered by AC and use AC phase chopper control for variable speed control. Its principle of operation means that its direction of rotation does not reverse with changes in the polarity of the power source. To reverse this type of motor requires reversing the polarity of either the rotor brushes or the stator electromagnet windings, which essentially changes the intervals of magnetic repulsion into magnetic attraction. This is true for either AC or DC power. The reversing switch on a power drill, either corded or cordless, accomplishes this.

This was the type of motor used in the very first EVs.[359] It remained the most common traction motor type, used in probably all electric vehicles for over 100 years until the advent of electronic commutation and induction motor drives in the late-1980s mid-1990s.

Series-connected (rather than shunt) motors are preferred in traction applications since the sum of the voltage across the field and the rotor is the total motor input voltage. This arrangement reduces the current in both the rotor and stator as the motor speed increases due to the phenomena of back-EMF (discussed below), assuring upper limit speed regulation at high speeds while allowing maximum current at zero speed. Series connections were nearly standard for electric vehicles prior to the mid-1990s, before variable frequency induction motor drives and high-coercivity magnetic materials became available for permanent magnet motors. They continue to be used in simple applications such as golf carts, small utility vehicles, and DIY electric vehicle conversions.

One example of an older series-wound DC motor that was used in many electric shuttle buses built between the late 1980s and about 2000, was the classic English-manufactured Nelco N200L, shown in Fig. 14.33. Its nominal continuous output was 35 kW (47HP), with a massive 400Nm torque at zero RPM, greater than the maximum torque of many 300+ HP muscle car engines of the 1960's and '70s. This characteristic easily met the starting torque requirements of electric buses, forklift trucks, and service vehicles without the need for a transmission.

Fig. 14.33 The Nelco N200L, a durable series-wound brush-commutated motor. Prior to approximately 1995 it was the most widely used motor for small-to-medium capacity electric buses and shuttles. The motor in this photos is a NOS replacement part for a 1989-1992 GM/Conceptor electric G-Van. Photo:author

Like most of the conservatively rated industrial brushed DC motors, DIY'ers often powered this type of motor at voltages well above its nominal nameplate rating of 170 VDC without harm. This is allowable because, for most vehicles, maximum motor power is only needed for brief periods, whereas the NEMA motor rating is based on thermal limits during continuous operation. The previously mentioned 1989-92 GM Electric G-Vans used this motor with a 216VDC nominal (234V actual) battery pack. Motor specifications are shown below, and (Fig. 14.34) is a line-drawing of the motor from the G-Van operator's manual.

Diameter	335/355mm Across Flats
Length	580mm
Voltage	216 VDC
Speed	Up to 5000 rpm
Rotation	Reversible
Enclosure	Ventilated with internal fan (IP20)
Insulation	Class H
Field Connection	Separately Excited with Interpoles
Termination	Flying Leads (not shown)
Weight	182 kg (400 lbs)

A lighter weight series-wound DC motor used in many DIY vehicle conversions is shown in Fig. 14.35.

A small increase in the motor voltage will produce a disproportionately greater increase in the torque and power output of this motor. At a given speed, the applied voltage for this type of motor has an approximately square-law relationship with motor torque because of the multiplicative effect between the stator and rotor magnetic fields. Both are proportional (up to a magnetic saturation limit) to their current, which is the same in each because of the series connection. At a given speed the

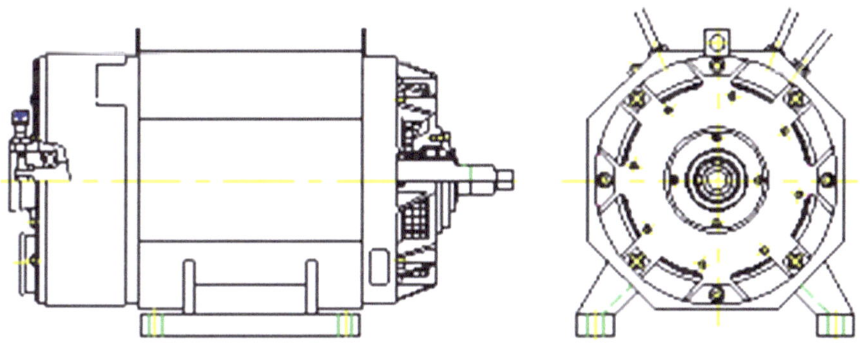

Fig. 14.34 Nelco 35 kW brushed series-wound DC motor, popular in electric truck and bus conversions in the 1990's. Specifications at left. From GM/Conceptor Electric G-Van service manual. Public domain.

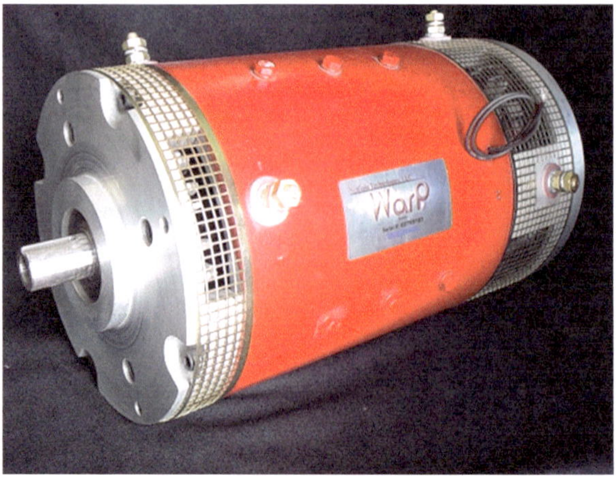

Fig. 14.35 The Warp 9.25" series-wound DC motor, rated at 32.5 HP continuous, but as much as twice for brief periods. Among the most popular motors for DIY EV conversions due to its lower weight compared with industrial electric motors. Photo found on http://rebirthauto.com/nmwarp9.aspx DIY EV parts exchange. Public domain.

Fig. 14.36 Cal Poly San Luis Obispo electric race car, second place finisher in 1998 EVTC Electrics race, Phoenix International Raceway. Photo: author

current increases linearly with motor voltage so the electromagnetic repulsion or attraction increases roughly as the product of the two.

As mentioned above, heavy-duty industrial motors such as the Nelco N200L have large heat capacities due to their massive steel bodies. They can be pushed for short periods to power outputs significantly higher than their continuous nameplate rating by using higher-than-rated voltages. *Over-voltaging* was a common trick used in the crude electric race cars that were popular during the 1990s, exemplified by the *EVTC Electrics* races sponsored by the Electric Vehicle Technology Corp. and Arizona Public Service in Arizona, USA, 1995–1999. For example, the second-place finisher in the lead-acid battery class in 1998 was the university entry shown in Fig. 14.36 which used an industrial forklift motor continuously rated at 48V, but with a 144V battery pack. Of course, taking this too far can and often did result in the explosion of motor brushes, a common attention-grabbing failure in these races. This tolerance for abuse is unique to large heavy industrial motors used in converted EVs and is not applicable to the induction or electronically commutated permanent magnet (ECPM) motors discussed later in this chapter.

Regarding the Nelco motor, the motor performance plot of Fig. 14.37 is copied from the 1989 GM/Conceptor *Electric G-Van Service Manual* [75]. Stated in the manual is:

> The N200L was originally developed for a large electric vehicle application. The motor has a rated power of 35kW at 216VDC (60 min continuous rating) and a peak torque of almost 400Nm. The motor can be externally forced-air cooled to improve the rated power.

Test data for this motor shown in Fig. 14.37 show the difference between the motor's 35 kW continuous nameplate rating, and its capability to produce 80 kW (107 HP) at 400 Amps for a maximum of two minutes. Power output in kW is calculated from the formula:

$$\text{Power}(\text{kW}) = \text{Torque}(\text{N·m}) \times \text{Speed}(\text{rpm}) / 9548.8$$

The large frames of industrial rotating machines are not just for mechanical strength, but to increase the thermal mass and therefore the maximum short-duration power capability. It should be noted that the power output ratings of a liquid-cooled ECPM (to be explained below) and induction motors in modern EVs are not tested against industrial standards, but rather a less conservative schedule appropriate to automotive propulsion in which maximum output power is of short duration.

Brushed Permanent Magnet DC Motor

Essentially the same type of motor as a series or parallel-wound electric motor, but the fixed electromagnets of the stator/field are replaced by high-strength permanent

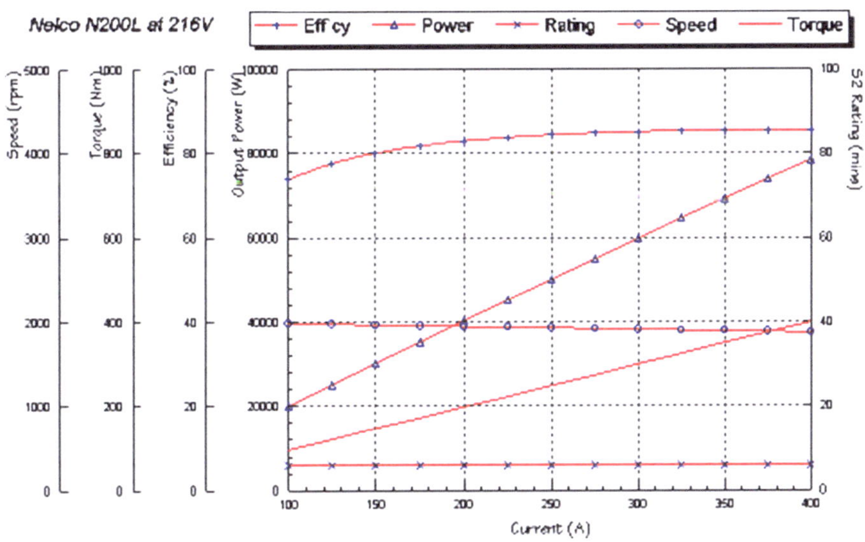

Fig. 14.37 Torque, power and efficiency vs. current of the Nelco N200L motor. From The GM/Conceptor Electric G-Van Service Manual, 1989. Out of copyright

magnets. It is among the most basic electric motor configurations, typically used in toys and low-power servos, such as the residential gate opener motor shown in Fig. 14.38. Until the 1980s affordable permanent magnets were not capable of the high flux levels of electromagnets. The introduction of extremely strong rare-earth magnets changed this, making possible lightweight motors with high power ratings—often used in hobby and surveillance drones. Unlike wound-field DC motors, a PM motor reverses direction if the voltage polarity is reversed, the reason that it is the preferred motor type for bidirectional servocontrol applications, and the reason that it cannot be powered by AC electric current.

Figure 14.39 shows a simplified diagram of a hypothetical 2-pole brushed permanent magnet DC motor. As the shaft rotates, the polarity of the electromagnets of the rotor is reversed every 180 degrees of rotation, assuring magnetic repulsion and magnetic attraction continuously to rotate it in the desired direction.

Electronically Commutated Permanent Magnet (ECPM) Motor

Mechanically, an ECPM motor (aka Brushless DC or BLDC motor) is the same as a brush-commutated PM motor, except that the electrical commutation is done by power switching semiconductors, eliminating the need for graphite brushes and the copper commutator of brushed DC motors. A precise rotary shaft position sensor is used to control the switching transistors as a function of the rotor position as it rotates. In most EV applications, the rotor has the permanent magnets embedded around its perimeter, and the wound-field electromagnets of the stator are commutated by the controller. This is the opposite of the configuration used for brush-commutated DC motors. Therefore, the stator is commutated rather than the rotor, but the magnetic interactions are the same.

This type of motor has been found to have the highest efficiency of all EV motor types. Its peak output is ultimately limited by the strength of the permanent magnets. Exotic magnetic materials are required, usually alloys of iron with nickel,

Fig. 14.38 Mabuchi 3T brush-commutated permanent magnet DC actuator motor. Permanent magnet design makes motor reversible by change in polarity. Photo: author

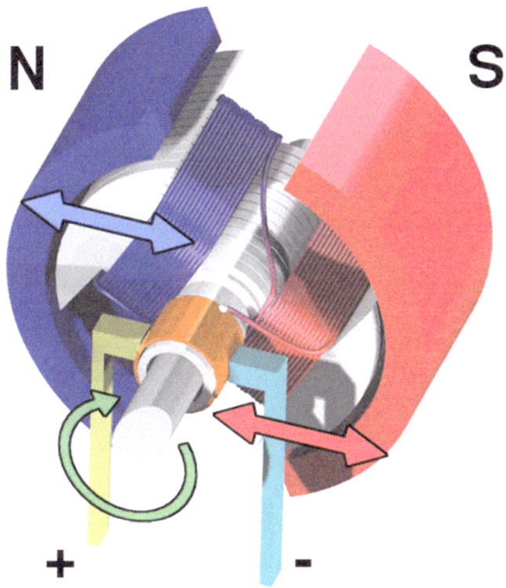

Fig. 14.39 Diagram of a simple 2-pole brush-commutated permanent magnet DC motor. From Wikipedia, "Permanent Magnet Motor", 2024. Creative commons license CC BY-SA 4.0. https://en.wikipedia.org/wiki/Permanent_magnet_motor

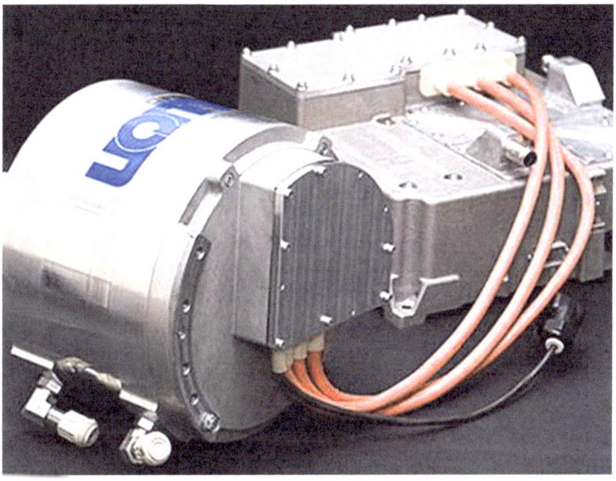

Fig. 14.40 UQM (formerly Unique Mobility) electronically commutated permanent magnet EV motor and drive electronics. UQM HDPP100 100kW Drive System. Image from Unique Mobility, early 1990s, https://www.thunderstruck-ev.com/uqm-pp100-coda.html DIY EV parts exchange. Public domain

neodymium, boron, cobalt, or samarium. The use of permanent magnets improves the motor efficiency since no power is required to excite the field windings as required in a powered-stator series or parallel-wound motor.

Figure 14.40 shows one of the first high-power ECPM motors specifically for electric vehicles, the UQM 100 kW liquid-cooled motor and integrated controller. ECPM motors are usually co-designed with their drive electronics, as in this case.

Axial Flux ECPM Motor

There have been several alternative mechanical ECPM motor configurations, each with advantages and disadvantages. An *Axial Flux* DC motor is a type of ECPM motor with an alternative geometry that dates back to Faraday's original "disk generator" in 1831 [76] (Fig. 14.41). The permanent and powered magnets are located

Fig. 14.41 CAD rendering of an axial flux EV motor from Magnax BV of Belgium, 2023. Larger in diameter but shorter in length than a conventional ECPM DC motor with similar ratings. https://spectrum.ieee.org/media-library/eyJhbGciOiJIUzI1NiIsInR5cCI6IkpXVCJ9.eyJpbWFnZSI6Imh0dHBzOi8vYXNzZXRzLnJibC5tcy8yNjc4NTM0MC9vcmlnaW4uanBnIiwiZXhwaXJlc19hdCI6MTY5MTI2NzA0MH0.MoS74fODyjnc5TcpjnnHglC3DJO8hqxrb5uvmGpIwJE/image.jpg.

opposite each other on both sides of a rotor disk rather than along its periphery. Axial flux motors have efficiencies similar to the more common radial flux motors, but they excel in their power density and maximum operating speed, which means greater power output for a given motor size or mass. The relatively flat packaging makes them considerably more compact. This has made them excellent candidates for electric aircraft, both at the hobby scale and actual scale.

Axial flux motors are capable of higher maximum speeds than radial flux motors for an unusual reason: at very high speeds, the diameter of the rotor increases slightly due to centrifugal force. The gap between the field electromagnets and the rotor permanent magnets (or vice versa) is critical to the output torque; the smaller the gap the higher the efficiency and peak torque. By placing the opposing magnets along the sides of the rotor rather than along the perimeter, the rotor diameter can expand without risk of contact between the spinning rotor and the surrounding stator, allowing smaller rotor/stator gaps and higher motor speeds, both of which contribute to higher power density.

AC Induction Motors and Variable Frequency Drives

Also called a *rotating transformer* or *variable frequency* AC motor, since there is no electrical coupling of power to the rotor. No rotor windings. Not commutated. The induction motor was originally patented by Nikola Tesla in 1888, the main patent drawing shown in Fig. 14.42.

The rotor is constructed of an electrically conductive but non-magnetic material, usually copper or aluminum alloy, and in traditional designs it can look like an exercise wheel for a pet rodent, giving this particular configuration the common name *squirrel cage motor*. Electrical current and the resulting electromagnetic force acting on the rotor are "induced" by its rotational speed relative to the frequency of the

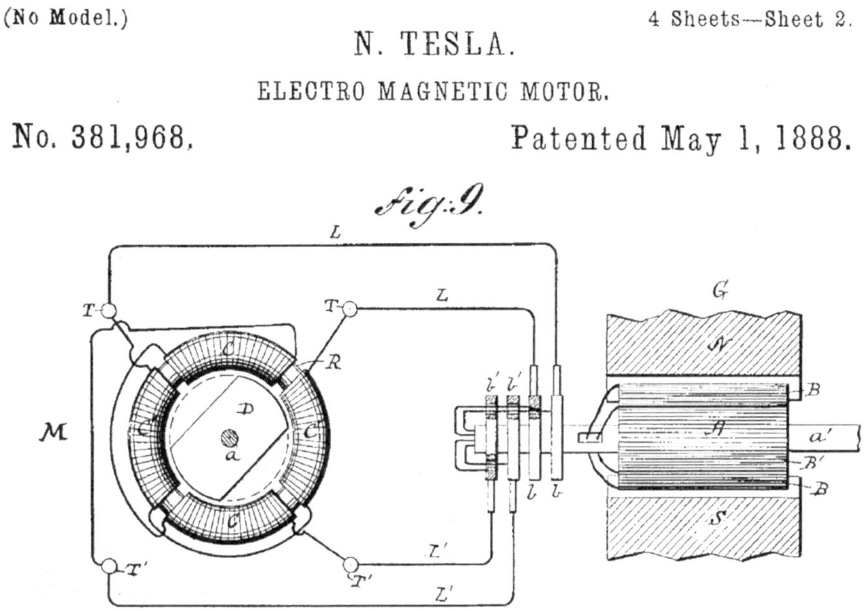

Fig. 14.42 Nikola Tesla's patent for the Induction Motor, 1888. This type of AC motor was a revolutionary change in motor technology, providing high power in a lightweight package. Induction motors are used in most industrial applications powered directly by 50 or 60 Hz electric power, or in variable frequency drive systems that allow speed control USPTO, public domain

alternating stator current, usually a 3-phase synthetic sine wave. The rotor is said to "slip" as its speed must be slower than the rotating electromagnetic field of the stator. The greater the speed difference, the greater the motor torque. This topology allows two mechanisms for speed and torque control: the frequency and the current applied to the stator windings that produce the rotating field. The variable frequency three-phase sine wave field is synthesized using a high-power solid-state controller. These two control mechanisms implement a variable frequency / variable torque drive. The application of induction motors with solid state power electronics to electric vehicle propulsion is attributed to Alan Cocconi of AC propulsion in 1992 [77]. Owing to the strength of the non-wound rotor which allows very high-speed operation, induction motors are capable of the highest specific power density among EV motor types, although they are not as energy efficient as ECPM motors.

While AC induction motors originally dominated the new EV market in the 2000s and early 2010s, rapid improvements in ECPM motors have largely displaced them except in the highest performance EVs. For example, the Tesla Model S uses an AC induction motor for its primary drive motor. A 185 kW Tesla induction motor is shown in Fig. 14.43. Induction motors have one important supply chain advantage over permanent magnet motors: the lack of need for rare-earth magnetic materials. Starting in 2025 these have become chips in a high-stakes trade war between the USA and China.

Fig. 14.43 185 kW air-cooled induction motor used in the Tesla Roadster. Image by Tesla Inc., 31 December 2015, copied from https://fuel-efficient-vehicles.org/energy-news/?page_id=1065 Public domain.

The Torque/Speed Characteristics of EV Motors

Most EV owners are aware that their EVs have different torque characteristics than ICVs, allowing them to accelerate quickly and to use motor reverse torque for regenerative braking.

While there are many types of electric motors, only two fundamental principles of operation are common in EV motors: (1) AC Induction, e.g., the AC-150 and the Tesla S motor, and (2) ECPM motors are used in almost all other EVs and hybrids, with the exception of a few high-end vehicles such as the Audi e-tron SUV and the Mercedes-Benz EQC. While induction motors have fundamentally different operating principles than commutated (brushed or brushless, permanent magnet or electromagnet) DC motors, both are capable of functioning as both a motor and a generator.

To explain this phenomena, let's focus on the commutated permanent magnet motors that dominate EV and hybrid propulsion at this time. Figure 14.44 shows a simplified electrical model to demonstrate how the motor can be used to either produce drive torque (as a motor) or braking torque (as a generator).

A slightly more complete model would include an inductor in series with the resistor R_a but it is not necessary for an understanding of the basic operation of the motor.

The relationship between the motor current and the motor voltage applied to the motor is

$$i_m = \frac{v_m - K_b N}{R_a}$$

The motor torque is proportional to the motor current:

$$T_m = K_c i_m \text{ where } K_c \text{ is the motor torque constant}$$

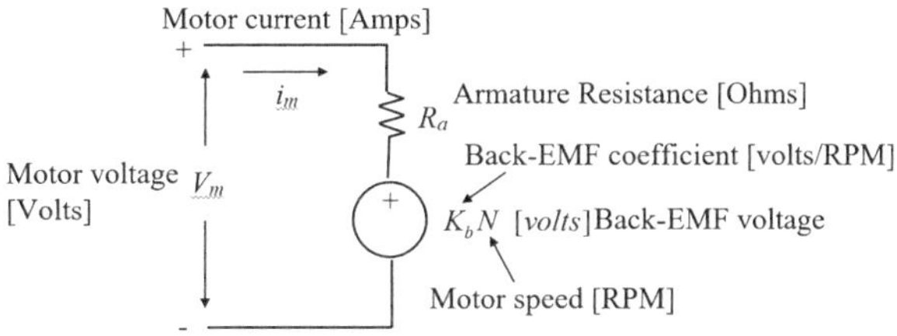

Fig. 14.44 Simplified DC motor electrical model. When the Back-EMF voltage exceeds the applied motor voltage, the motor current i_m reverses direction and it becomes a generator

The torque-to-voltage relationship is

$$T_m = K_c i_m = \left(\frac{K_{cm}}{R_a}\right)\left(v_m - K_b N\right) N - m$$

In this electrical model, the controlled voltage source labeled K_bN represents the "back-EMF" voltage of the motor, where N is the motor speed in RPM, and K_b is the "back-EMF coefficient" with units of volts/RPM. It is not a physical component but rather a voltage generated by the motor itself that opposes the input voltage V_m. As the motor rotates faster, the back-EMF voltage increases. This opposes the externally applied motor voltage, reducing the current flow in the commutated stator or rotor (aka armature) electromagnets. At zero speed the back-EMF voltage is zero, thus the current and therefore the motor torque is maximum. At high speed, when the back-EMF increases enough to match the applied motor voltage, no current flows and the torque produced by the motor is zero. And if the motor speed is externally driven to an even higher speed by the momentum of the vehicle, the back-EMF voltage exceeds the applied motor voltage, and the motor functions as a generator, producing current in the reverse direction and creating braking torque. If the generated reverse current is returned to the battery, it is called *regenerative braking*, discussed later in this chapter, and in Chap. 15, *Hybrid Vehicles*.

In practice, the motor stall current at or near zero RPM is limited electronically to prevent excessively high current if the full motor voltage V_m is applied while the motor is stopped. But this safety feature may be overridden for short periods to produce maximum take-off torque for exhibition purposes, e.g., in the Tesla Plaid "Ludicrous Mode."

For this idealized model, at any given input motor voltage V_m the motor torque T_m is maximum at zero RPM, and declines linearly with the motor speed, as shown in Fig. 14.45. This is almost the opposite of an ICE torque curve which produces zero torque at zero RPM (when the engine isn't turning). An ICE must start from idle speed. Were it not for a clutch or fluid coupling mechanism, an ICV could not start from a standing stop.

As the EV motor speed increases, the torque declines, eventually dropping to a level just sufficient to overcome the vehicle's air and mechanical frictional losses, which determines the maximum vehicle speed at a particular motor terminal voltage. This would correspond to the cruising speed of the vehicle on level ground at a given pedal position. While most EVs do not have a multi-ratio transmission, high-performance EVs capable of reaching very high speeds often incorporate a

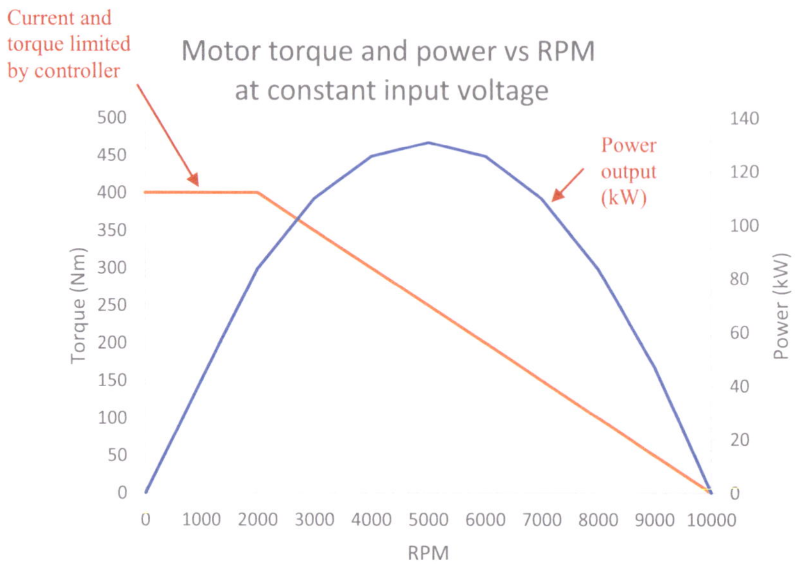

Fig. 14.45 Idealized 125 kW ECPM EV motor torque and power vs speed at maximum input voltage. Source: author

two-speed transmission, with the higher gear allowing the vehicle to overcome this speed-torque limitation of the motor. As perceived by the EV driver, if the control electronics did not tailor the torque curve, the driver experience might be a bit disconcerting, more like a slingshot than a motor vehicle.

Figure 14.45 shows a plot of the ideal torque and power curve for the generic DC motor model of Fig. 14.44. Torque declines with RPM, but since the power output (kW or HP) is the product of the speed and torque, it peaks midrange at 5000 RPM. The controller's current limiting at low speed can be seen from the shape of the torque curve below 2000 RPM. Actual data from a Delco Remy HVH410 traction motor are shown in Fig. 14.46. Note how the controller limits the current flow below 1000 RPM.

The high torque at low speed of electric motors has been showcased in amateur drag race videos found on YouTube circa 2006-7, some with home-built EVs powered by lead-acid batteries accelerating from a standing start quicker than well-known muscle cars [78]. Off the line, an EV has an advantage over an ICV, although at higher speeds the advantage is lost [79], as can be seen in online video clips of acceleration comparisons, e.g., a Tesla S Plaid will beat a Bugatti Veyron in a 0–60 mph (1–100 kph) contest, but the Veyron will pass it in a longer contest, and reach 253 mph (407 kph) while the Tesla Plaid maxes out at a mere 200 mph (322 kph).[19]

[19] A Google search with the phrase *Tesla S plaid vs. Bugatti* will produce at least 100 videos of drag races between the Tesla S Plaid and the Bugatti Chiron or Veyron.

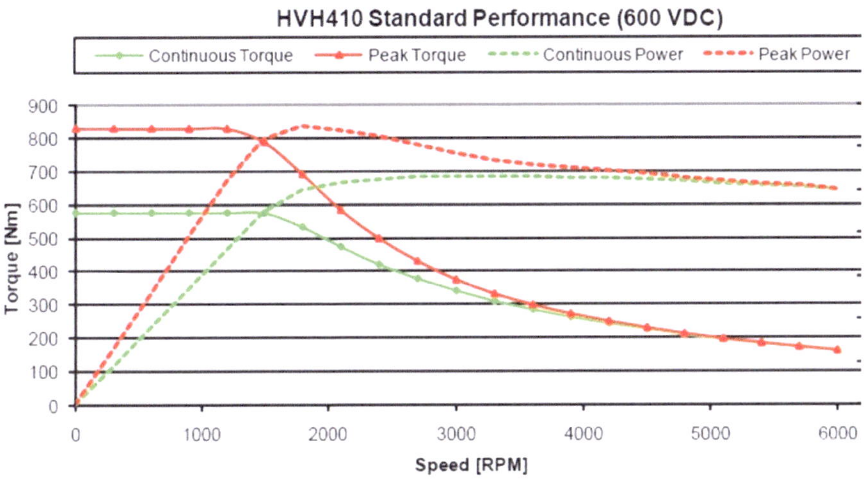

Fig. 14.46 Maximum torque and power curves for a Remy HVH410 high voltage 112 kW (150 HP) ECPM motor. Current, and therefore torque, is limited below 1500 RPM by the controller. From New Eagle Products Wiki, July 2024. https://wiki.neweagle.net/index.php?title=Remy CC BY Attribution 3.0

The Challenge of Electric Vehicle Motor Control

While the greatest technical challenges for electric propulsion were and continue to be the energy storage mechanisms, the lesser known problem of controlling the torque and speed of an electric traction motor lurked in the background for almost 100 years, with many novel electromechanical solutions prior to the invention of modern solid-state power electronics. This lesser known technology hurdle is at least worth mentioning when chronicling the early progress of electric vehicles.

The previously-mentioned 1881 Trouve trike employed an ingenious but impractical means for speed control: the lead plates of the Plante cells (lead-acid battery) were raised and lowered mechanically into the battery's sulfuric acid electrolyte to adjust current to the motor (!) Trouve did not attempt to patent the invention, either believing it was too obvious or realizing that it was impractical [80].

The method for speed control of the Ayrton and Perry trike was not published, but from the graphic of Fig. 14.3 it appears that the only mechanism for vehicle speed control was application of the friction brake on the single drive wheel. In other words, it was a one-speed, max power drive system that could only be slowed by application of the brake. With a maximum speed about the same as a walking pace, this hardly mattered. To its novelty credit, however, the trike was equipped with a hand-made voltmeter and ammeter [9–11].

Some early EVs used a rheostat (variable resistor or potentiometer) in series between the battery and the motor as a "current limiting" device similar to the old-school rheostats formerly used for theater stage light dimming. But the use of a series resistor for current control is very wasteful of energy at part loads since as much power would be dissipated as heat in the rheostat as in the motor. However, although still inefficient, the motor field current could be controlled by the rheostat at a lower current to control motor torque.

A slightly more efficient but equally impractical method for controlling power to the motor was battery switching (aka voltage switching) in which successive individual batteries were added or removed from a series string via a selector switch. This uneven use of the batteries resulted in different states of charge for different batteries in the string, which wasted battery capacity and range, and necessitated shuffling individual batteries as periodic maintenance. Soon, there followed many other EV speed control inventions, most unworkable. Apparently, 1899 was a watershed year for US patents for automotive basic mechanisms, including electric motor control. One solution, shown in the patent drawing of Fig. 14.47, used a mechanical selector switch that electrically rotated the batteries in the string to equalize the discharge of individual cells. It also included a mechanical motor speed governor, as if one was needed for a 10 mph vehicle [81]. Battery switching with battery rotation would remain in use for speed control of electric vehicles for the next 50 years, and still can be found in some electronics-free legacy EVs in operation today.

Some interesting applications unrelated to electric vehicles evolved from these early electric vehicle speed control innovations. A sophisticated electro-hydraulic system was patented in 1899 by W. B. Potter that used a hydraulic pump to decouple the motor from the drive wheels [82].

Also in 1899, the same approach was taken by Henry Dey of the Dey-Griswold Company, who developed and patented a variable speed *fluid drive* that allowed speed control with a constant speed motor input, the precursor to the hydrostatic drives that are common today on almost all agricultural and construction heavy equipment [14].[20] While I could find no documented examples of the use of these power and speed control ideas in electric vehicles, an opportunity may have been missed years ago: The ability to continuously vary the gear ratio between the electric motor and the drive wheels made possible a mechanical form of regenerative braking without electronics [83]. This concept would, in the late 1990s, at least partially underlie the principle of operation of the Toyota Hybrid Synergy Drive (HSD) that revolutionized hybrid vehicle technology. It would be nearly 70 years before solid-state power regulation by pulse duration modulation, the principle used in modern EV motor controllers, was made possible by the development of high-power switching semiconductor devices in the 1970s.

Types of EV Motor Controls

EV drive electronics, aka electric motor controllers, fall into three classes, each with an increasing level of sophistication:

[20] Henry Dey of the Dey-Griswold Company, reported in Wakefield.

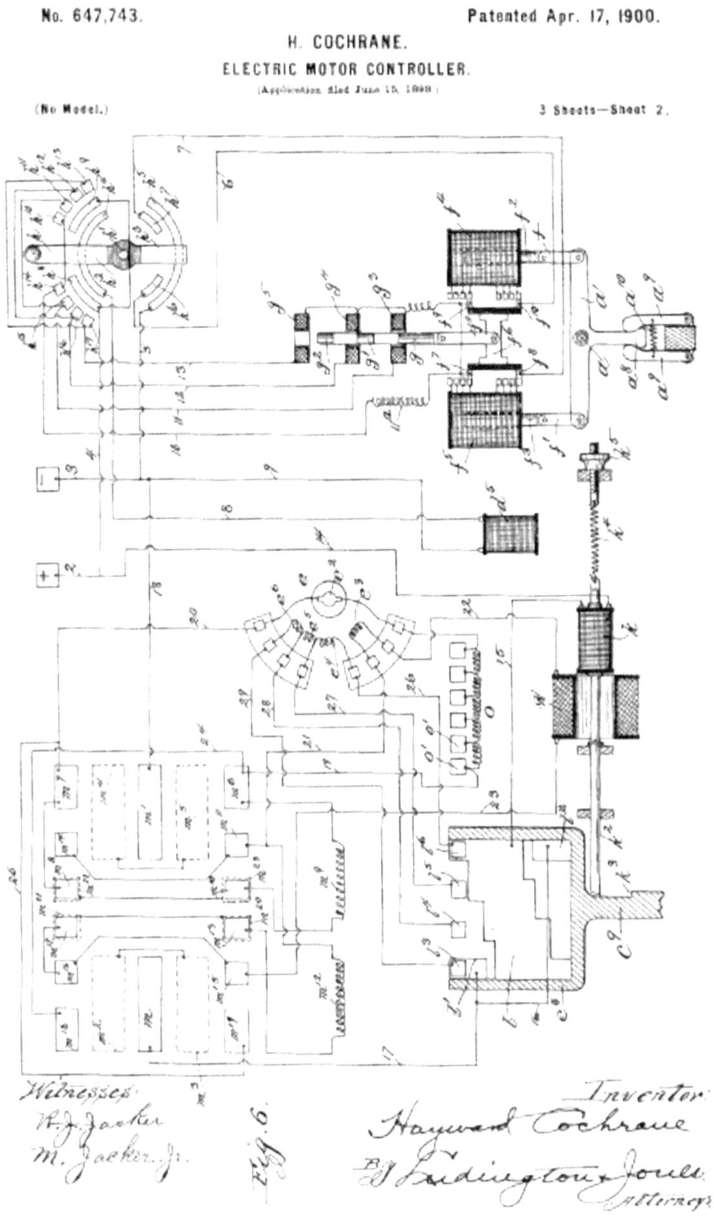

Fig. 14.47 One of many US patents circa 1900 on electric vehicle speed control based on mechanical voltage switching. Image from USPTO Patent 647,743 issued 17 April 1900. Public domain

PWM "Chopper" Controllers

The most basic DC motor controllers are simple pulse width modulation (PWM, chopper) controllers that rapidly turn on and off the full battery voltage applied to the motor. Pulse duration modulation is an efficient method for electric power control, because no power is wasted when it is only switched ON or OFF, only when it is linearly controlled, e.g., using a rheostat. It efficiently implements motor torque/speed control for series or parallel-wound DC motors and brushed permanent magnet motors. Golf carts and small electric utility vehicles are the dominant application. Pulse duration modulation (usually between 5 and 15 kHz) controls the time-averaged motor voltage and current, and therefore the motor torque. Low-cost chopper controllers like the one pictured in Fig. 14.48 do not usually incorporate regenerative braking. They are inexpensive and adequate for golf carts and other low-speed specialty vehicles.

Controllers for Brushless (Electronically Commutated) Motors

This classification includes both permanent magnet (ECPM) and common electromagnet field (series or parallel-wound) motors. For brushless motors, the controller is usually designed to work with a specific rotating machine, since it requires a shaft position sensor to precisely control its commutation of the (usually) three groups of electromagnets in the stator. The electronically commutated motor and controller replace mechanical commutation with electronic commutation avoiding the inevitable wear and ohmic losses of graphite brushes on a rotating copper commutator. The required shaft position sensor may be a quadrature encoder, a linear resolver, a Hall Effect sensor, or a radial optical position sensor. Alternatively, with some loss of precision, the instantaneous motor shaft position may be inferred by the controller from the stator voltage and current.

Figure 14.49 shows a photo of an ECPM motor controller removed from a 2012 Nissan Leaf.

Induction Motor Drive Electronics

For an induction motor, the frequency applied to the electromagnets of the field is not synchronized to the angular position of the rotor. In fact, the motor would

Fig. 14.48 Curtis 1204M PMC motor controller for golf cars and low-power electric vehicles. Photo: author

Fig. 14.49 Motor controller electronic assembly removed from a 2012 Nissan Leaf. Source: E-Bay "2011-2012 NISSAN LEAF EV INVERTER ASSEMBLY POWER HEAD" Accessed 1 December 2024

produce no torque if these were in synch. The torque is an increasing nonlinear function of the difference between the rotating field frequency and the rotor speed.

- The motor drive electronics are custom designed and programmed for a specific motor, and if not build by the vehicle manufacturer, are usually sold as a package with the motor.
- High-power semiconductors (PFETs or IGBTs) provide bidirectional current control to each phase.
- Always includes regenerative braking.
- Usually integrates battery charging using existing electronics.
- Highest power density of any motor/controller combination.
- Less efficient than ECPM motor and drive systems.
- Usually incorporates forced-air or liquid cooling of both the electronics and the motor.
- Alan Cocconi, inventor of modern EV induction motor drive, once quipped that the controller of his AC150 drive system is basically "a 150,000 Watt audio amplifier."

Figure 14.50 shows an AC 150 (200 HP) Power Controller made by AC Propulsion installed in the prototype "eBox," a modified Toyota Scion xB minivan. The AC-100 and AC-150 were recognized as the first high-power induction motor drive systems specifically for EVs. It was a triumph of power electronics design, but at a time before the US Federal and State subsidies that benefited later EV manufacturers, AC Propulsion's efforts to profitably manufacture any of its revolutionary products were unsuccessful.

Worthy of special mention is the short-lived Hughes Electronics 50kW "Dolphin" induction motor controller, shown in Figure 14.51. Its was used by US Electricar in its converted Geo Metros, Chevrolet S10 trucks, and various small electric vans circa 1994-08. This solid-state 3-phase motor controller represented the state of the art in drive electronics at the time, preceded only by the AC Propulsion AC-100 (100kW) induction motor drive system in 1992.

Fig. 14.50 AC-150 (200 HP) Power Controller Unit by AC Propulsion of San Dimas, CA. shown installed in a 2007 AC Propulsion "e-Box", a modified Toyota Scion xB minivan. Photo from http://www.acpropulsion.com/products-tzero.html. Photo taken at Cal Poly EV exhibition in 2007

Fig. 14.51 Hughes Electronics 50 kW 312V Dolphin Controller, used in the U.S. Electricar electric S-10 light truck, 1994. Also used in upgraded 100kW form for the GM EV-1. Photo by author

The Hughes Electronics (division of General Motors) Dolphin Controller was produced in limited quantities in 1994. (The Hughes Aircraft Company was broken up in 1985, and the electronics division purchased by GM). It incorporated an internal 3kW charger that shared the power electronics with the motor controller, and was compatible with the Delco Magne Charge inductive coupled charger, both firsts in the industry. But like all other attempts (except possibly DirecTV) by Hughes under GM to produce competitive commercial products, the Hughes Aircraft Company culture as a preeminent aerospace company was incompatible with manufacturing a mass-market product. The high-tech controller would never be affordable in a consumer EV market.

Regenerative Braking

As mentioned previously, the ability to partially recover braking energy is one of the most important efficiency-enhancing attributes of a BEV or hybrid vehicle. It is logical to assume that the high MPGe numbers achieved by BEVs and hybrids are due to the greater efficiency of an electric motor compared with an ICE. And that the greater vehicle mass would lead to increased wear of the vehicle brakes. But neither are the case. The reason is *regenerative braking*, which uses the electric motor as a generator to slow the vehicle rather than simply dissipating the vehicle's kinetic energy as heat in friction brakes.

When a vehicle is accelerating, fuel/electric energy is invested to increase the speed. Ignoring wind resistance and mechanical frictional losses for the moment, the energy E (kWh, Joules or HP-hours) required to accelerate a vehicle with mass m (kg, lbs) to a velocity v (kph, mph) is proportional to the increase in the vehicle's kinetic energy $\frac{mv^2}{2}$. This is true regardless of the energy source.

During braking in a non-hybrid ICV, this energy is lost to heating the vehicle disk brake rotors and pads. The kinetic energy of the vehicle's motion is completely wasted. But all modern hybrids, plug hybrids, and BEVs incorporate the ability to use at least some of the braking energy to recharge the battery. The recovery of wasted braking energy increases the overall vehicle efficiency (measured as MPGe or Miles per kWh), especially in stop-and-go driving such as in urban environments. If regenerative braking and the battery allowed 100% energy recovery, and there were no energy losses other than braking (e.g., increase in elevation, air friction, tire losses, drivetrain friction, etc.), it would be theoretically possible for the vehicle to run forever with no added energy after the first acceleration. Of course this is impossible, but that realization illuminates this unique capability of electric drive systems used in hybrid and battery electric vehicles—they can recover braking energy. An ICV cannot. Rated mileage increases. Brake components do not wear as much. This is the reason that all modern vehicles with hybrid or all-electric drivetrains achieve higher fuel economy ratings in the EPA *city* (urban) drive cycle compared with the *highway* drive cycle which involves much less braking. This trend is the opposite for conventional ICVs, which have always achieved higher highway mileage ratings than city mileage ratings.

The size of the vehicle battery establishes a limit on the rate at which braking energy can be absorbed, so full BEVs typically have a bit more capable regenerative braking ability than hybrid or plug hybrid vehicles.

This characteristic favors stop-and-go driving. The energy investment E to accelerate a vehicle of mass m to speed v increases *as the square of the terminal speed.* During braking from that speed down to a full stop, the vehicle's kinetic energy must be dissipated in some form. Conversion to electricity by the drive motor acting as a generator and returning it to the battery accounts for a large energy savings every time the brakes are applied.

At the same time, air resistance to the motion of the vehicle produces an opposing force that increases as approximately the square of the speed. While mechanical and tire friction also contribute to the *road load*, air resistance dominates the load at speeds typically above 30 mph (or higher speed thresholds for very aerodynamic cars). To maintain a given speed, the engine torque (forward force) must also increase as the square of the speed. And since the vehicle is traveling an incrementally greater distance per unit of time at a higher speed, the rate of energy transfer (aka power in HP or kW) *increases as the cube of the vehicle speed.* (Although the energy used *per unit of distance* only increases as the square of the speed.) This is why vehicles capable of very high speeds must have power plants with outputs much greater than one would assume if the power requirement was proportional to speed (a common misconception). Since air resistance (that equates to drive torque) increases with the square of the speed, the power is proportional to torque times the speed., a vehicle that requires 100 hp US (75 kW) to sustain 100 mph (161 km/h) on flat ground would theoretically need 800 hp to reach 200 mph:

$$\text{HP at 200 mph} = 100\ \text{HP} \times (200\ \text{mph}/100\ \text{mph})^3 = 100 \times (2)^3 = 100 \times 8 = 800\ \text{HP}$$

It is only in this regime that ICEs have some advantage over electric cars that have single-speed transmissions due to maximum power near the top of the power band, rather than in the low/mid speed range of electric motors.

Considering again the simplified motor model previously discussed, the idea of using the traction motor as a generator to slow the vehicle and recover braking energy seems straightforward. Motor braking occurs when the back-EMF voltage of the motor exceeds the input voltage. The motor then acts as a generator, with the power input coming from the vehicle inertia or downhill force. But how to get the motor back-EMF voltage to exceed the battery voltage when it is minimal as the vehicle is slowing to a stop?

For the motor to produce forward torque, the voltage from the battery to the motor must exceed the back-EMF voltage generated by the motor itself that increases with the motor speed. Ohm's law says that current will flow from a higher voltage to a lower voltage at the motor terminals. But for regenerative braking to work, the opposite is needed. For current to flow *into* the battery, the voltage of the motor while acting as a generator must exceed that of the battery. This motor/generator voltage is the back-EMF voltage of the motor. Power management electronics must have the ability to convert this lower voltage from the motor acting as a generator to a voltage above that of the battery (Fig. 14.52). This charges the battery while placing a load on the motor/generator which causes braking force. And varying the braking forces requires the ability to continuously control the voltage conversion ratio to adjust the power flow from the motor/generator into the battery. A circuit that is capable of this magic is a *bidirectional buck-boost converter* (BBBC) which is the key to implementing regenerative braking in almost all-electric vehicles.

For those interested in the power electronics, Figs. 14.53 and 14.54 show the schematic diagram of a BBBC of the type found in most electric and hybrid cars. The electrical specifications are approximately those of a Gen 1 Nissan Leaf BEV or a Gen 2 Prius HEV. The motor in this circuit is an ECPM Motor, but the electrical principles apply to any motor type. The only way to implement regenerative braking is to boost the motor terminal voltage when it is acting as a generator to a level higher than the battery voltage. The simplified electrical model of a permanent magnet DC motor appears at the far right in each schematic. The buck-boost converter rapidly alternates between charging and discharging inductor L1. Even for this relatively low-power EV BBBC, the pulsed current carried by the power transistors U1 and U2 can be as high as 1000 Amps, so each is implemented as the parallel

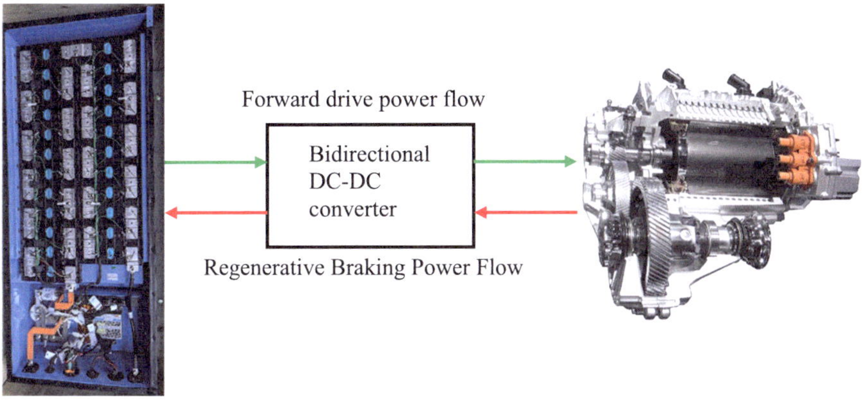

Fig. 14.52 A simplified diagram showing forward power transfer in green, reverse power transfer for regenerative braking in red. Diagram and EV battery photo by author. Volkswagen Golf electric motor from https://commons.wikimedia.org/wiki/File:E-golf-engine.jpg Attribution-Share Alike 4.0 International, CC BY-SA 4.0 Deed Motor/wheel graphic from http://www.buggies.builtforfun.co.uk/Howtoos/index.html

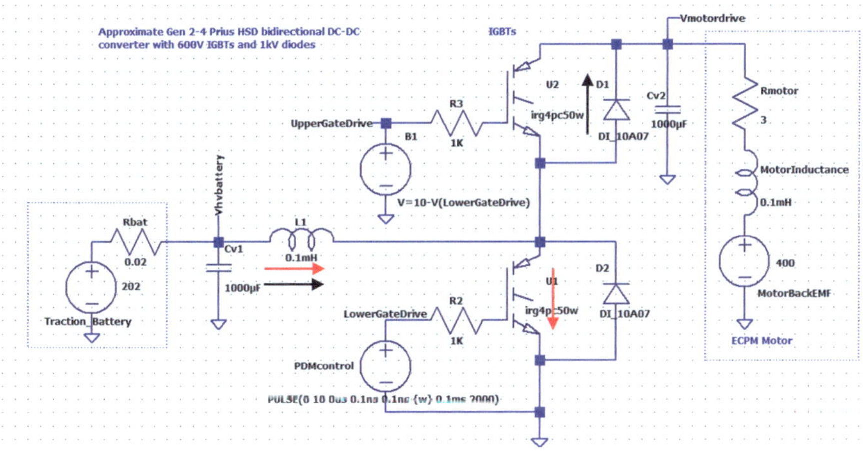

Fig. 14.53 Motor control circuit with regenerative braking capability, shown during forward power. Red arrows show current flow during charging of inductor L1, and black arrows show current flow when inductor discharges into motor.

combination of several Insulated Gate Bipolar Transistors (IGBTs), represented as a single device in the schematic. Figure 14.55 shows a photograph of the actual parallel-connected IGBTs in the controller.

The power flow direction shown in Figures 14.53 and 14.54 are controlled by the timing and relative ON/OFF times of the two IGBTs that act as switches to control current flow, alternately charging and discharging the inductor at a very high rate, typically 100 kHz. The red and black arrows in Fig. 14.53 show the direction

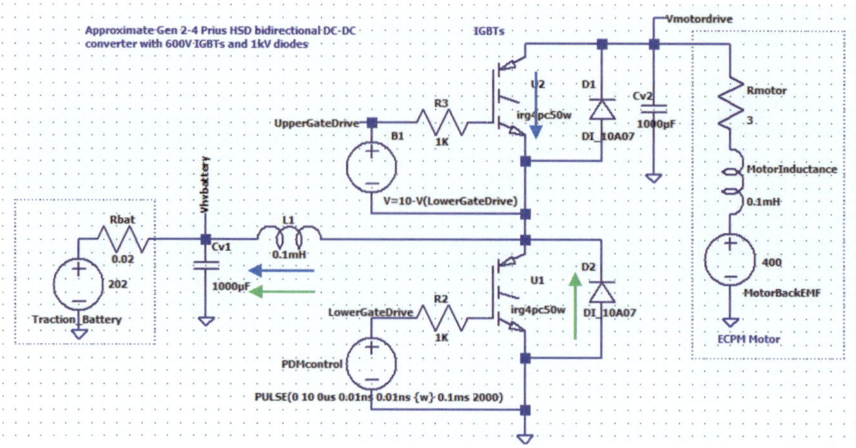

Fig. 14.54 Current flow directions during regenerative braking shown: green arrows during charging of inductor L1, and blue arrows during power transfer to the motor

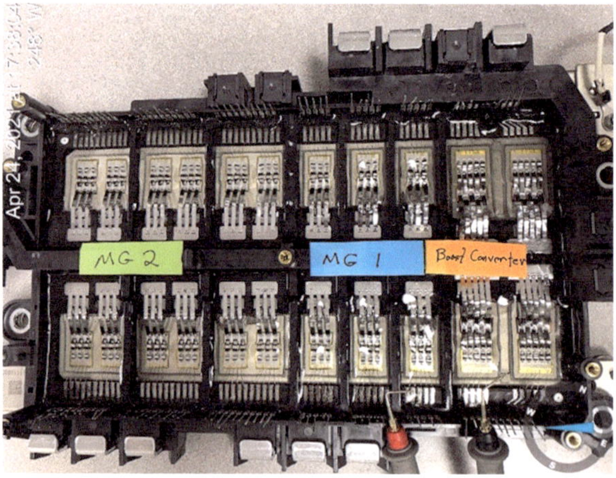

Fig. 14.55 Liquid-cooled power transistors from a 2016 Gen 3 Toyota Prius Prime plug hybrid. Transistors are mounted in thermal contact with a heavy aluminum heatsink through which liquid coolant flows. Photo by author

of time-averaged current flow during forward power. The green and black arrows in Fig. 14.54 show the average current flow direction during regenerative braking. Therefore, by precisely controlling the timing and ON/OFF intervals for each IGBT, the circuit can either *step up* (boost) or *step down* (buck) the output voltage compared with the input battery voltage, forcing current to flow on command in either the forward or reverse direction between the battery and the motor/generator. The transition from forward acceleration to braking can occur almost instantaneously.

The BBBC also makes possible electric drives that utilize *supercapacitors* (supercaps) to provide very high current but not as much energy as a battery. Supercaps have an advantage over batteries in that they can charge and discharge at very high rates to provide or absorb bursts of power for acceleration or braking or high-braking torque. But unlike batteries, their voltage is a nonlinear function of their state of charge. Charging or discharging supercaps requires the ability to continuously adjust the voltage to match the state of charge.

The power-handling requirements of the switching transistors and bypass diodes are extremely high, necessitating liquid cooling to prevent excessive temperatures that would destroy the electronics. The massive liquid-cooled heat sink and IGBTs of the bidirectional power controller from a 2016 Toyota Prius Prime (plug hybrid) are shown in Fig. 14.55. The size of the power switching transistors increases with the maximum motor power rating, so high-performance EVs require much larger power arrays and cooling systems compared with the relatively low-performance requirements of the Prius. The recent development of Silicon Carbide and Gallium Nitride power semiconductors has, since about 2022, started to replace Silicon IGBTs in EV power electronics, owing to their ability to operate at higher temperatures and switch faster than silicon IGBTs or Power Field Effect Transistors (PFETs).

Other EV Drive System Functions

Intelligent Control of Motor/Generators for Power-Split HEVs

In the Chap. 15 *Hybrid Vehicles*, we will look at the Toyota Hybrid Synergy Drive (HSD), the first power-spilt hybrid drive system, as currently used in most hybrid cars. It uses two electric motors/generators: MG2 is the main drive motor/generator with an output of 53 kW (71 HP). MG1 is the control motor/generator with an output of 7.3 kW 7.1 HP. These are coupled to a planetary gearbox along with the input of the gasoline ICE. MG1 primarily serves as control for the continuous gear ratio between the main motor MG2 that is connected directly to the drive axle, and the ICE that can serve as a power source either to supplement the torque from MG2 or to drive MG1 to charge the hybrid battery.

The additional control input provided by the MG1 motor/generator allows even greater flexibility in optimizing torque delivery and implementing regenerative braking. The controller for both motors is more complex since it implements a torque-balancing algorithm between the ICE and both motor/generators. In pure BEVs, there is no need for a power-split drive system of this type or even a transmission.

Multi-Motor Configurations

Most high-performance EVs use multiple motors (e.g., dual motors or tri-motors) to optimize torque delivery. With the ability to selectively adjust power to individual motors at each axle or wheel, multi-motor EVs can provide advanced traction control, stability enhancement, and torque vectoring, for enhanced braking, stability, and handling. Individual wheel torque control is almost impossible to implement in ICVs.

Autonomous Driving and Driver Assistance Control Inputs

It is much simpler to electronically control an electric motor than an ICE. While some active safety features such as automatic braking and lane departure warning/prevention are unrelated to the vehicle's motive power source, they are easier to implement with an electric drive system. Also, electric vehicles are much more amenable to fully or partially autonomous control of driving. Possibly the best known and most controversial such system at this time is the Full Self-Driving (FSD) option on Tesla electric vehicles. Other than early experiments with ICE-powered autonomous vehicles, the entire focus now is on autonomous electric vehicles. This is a deep topic outside the scope of this book. But it is almost certain that all future *safe* autonomous vehicles will be EVs.

EV Batteries

While electrochemical methods for storing electric energy were known as early as 250 BC in the Parthian Empire (present-day Iraq) [84], the first series connection of multiple cells to form a usable battery probably dates to the work of Alessandro Volta in 1800. It consisted of silver and zinc plates submerged in salt water or a water solution of sodium hydroxide serving as an electrolyte. A copy of a drawing from his 1800 paper presented to the Philosophical Transactions of the Royal Society, 1800 appears in Fig. 14.56. The series connection of multiple cells became known as the Voltaic Pile or Voltaic Column [85]. The basic construction of battery cells remains unchanged today: metals or metallic compounds separated by an ion-exchange medium known as an electrolyte.

Practical uses for the new electricity storage device were few until applications such as the Morse telegraph or Edison electric light bulb were invented many years later. Probably the first high-power application was the electric vehicle, which used a Voltaic Pile (battery) of lead plates in sulfuring acid invented by Gaston Planté in 1859 [86]. The lead-acid battery would remain the most common battery chemistry for electric propulsion until Oveshevski's nickel metal hydride battery in 1992 [87].

The quest for rechargeable batteries with higher energy densities and safer operational characteristics has been in progress worldwide since the Golden Age of electric vehicles and continues today. Hopes have soared and ebbed as hundreds of alternatives to the lead-acid battery have been proposed, tested, demonstrated, invested in, and ultimately proven unsuccessful in achieving these simultaneous goals. While many chemistries with higher specific energy densities have been identified and demonstrated at an experimental scale, all have had at least one critical limitation, e.g., high-temperature (molten sodium or sulfur) operation, highly toxic materials (bromine), safety concerns (almost all), dangerous manufacturing, and ultimately, cost.

The term "battery" in this context refers to one or more electrochemical cells connected in series in the same package to provide a desired voltage. The battery stores electric energy in chemical form, making its function in an automobile similar to the fuel tank of an ICV.

Fig. 14.56 Illustration from Alessandro Volta, "On the Electricity Excited by the Mere Contact of Conducting Substances of Different Kinds" Philosophical Transactions of the Royal Society, 1800. Public domain

The three most common batteries that have been used in battery electric and hybrid vehicles are Lead Acid, Nickel Metal Hydride (NiMH), and lithium-ion (Li-x) where x = combinations of iron, phosphorous, manganese, nickel, cobalt, aluminum, titanium, sulfur, etc.)

EV Battery Design Factors:

- Energy mass density
- Energy volume density
- Cost/kWh storage capacity
- Power density—maximum discharge and recharge rates, with and without external cooling
- Usable life, defined in terms of total charge-discharge cycles and absolute lifetime in years
- Safety: Stability, ie., resistance to thermal runaway, internal short circuits, dimensional change with state of charge – factors that affect susceptibility to spontaneous ignition. Toxicity and flammability in the event of rupture. Tolerance to overcharging or discharging.

Since the early days of the EV Golden Age, but especially in the last 30 years, research has sought improved alternatives for electric energy storage, especially those that are not as harmful to the environment or sensitive to future supply chain issues. A safe, sustainably manufactured battery that could achieve even higher energy densities and shorter recharge times would be the next EV game changer. In view of the potential value, the best and worst of society have been associated with better battery promises, usually as startup companies seeking investors. An interview of Thomas Edison published in The Electrician (London), February 17, 1883, accurately summed up both the past and present states of battery development:

> The storage battery is, in my opinion, a catchpenny, a sensation, a mechanism for swindling the public by stock companies. The storage battery is one of those peculiar things which appeals to the imagination, and no more perfect thing could be desired by stock swindlers than that very selfsame thing.... Just as soon as a man gets working on the secondary battery it brings out his latent capacity for lying.

Edison had the consummate qualifications, both as an engineer and a showman, to make such an accusation about the seedy side of innovation. Indeed, he was among the worst in this advocation. In one of his darker less-publicized events, Edison attempted to mislead the public that competitor Westinghouse's AC electrical power distribution was more dangerous than his DC distribution by publicly electrocuting a dog with AC power, to the horror of a gathered crowd. No mention of the fact that DC power would have been equally fatal.

Lead-Acid (Planté Cell) Batteries

The low cost, simplicity, and recyclability of batteries that use metallic lead or lead-calcium-antimony plates in a sulfuric acid solution have established lead-acid (Pb-H_2SO_4) batteries as the dominant battery chemistry for well over 100 years. The energy density of a typical deep-cycle lead-acid battery (Fig. 14.57) is between 10

Fig. 14.57 Typical sealed lead-acid 12 V automotive SLI battery. Creative Commons license CC BY-NC-SA 4.0

and 20 Wh/lb (22- 44 Wh/kg). Considering that EVs have a typical efficiency of 3 miles/kWh (5 km/kWh). a range of 60 miles (100 km) requires a battery mass of 1,000–2,000 lbs (450-900 kg). They also require a minimum of 4 h to 80% recharge, and their maximum useful life is typically 300 80% charge cycles.

While the chemistry has remained unchanged, there have been numerous engineering improvements that have improved lead-acid batteries. But developments such as maintenance-free valve-regulated lead acid (VRLA) or recombinant gas batteries, cylindrical cells, and lead-calcium alloy plates have not significantly increased the low energy density of lead-acid batteries. They have been used so long because we had nothing better.

Flooded Lead acid

The least expensive per kWh and most common automotive starting battery. Each of the six 2-volt cells in a 12 Volt battery can be accessed via fill plug on the top of the battery to check sulfuric acid levels and allow topping (refilling) cells with distilled water. Cells are vented to the atmosphere to relieve pressure created by the generation of hydrogen and oxygen due to electrolysis of the electrolyte.

Absorbed Glass Mat (AGM)

A sealed or maintenance-free battery: The first of two types of sealed (aka maintenance-free) lead-acid batteries. Most include a pressure relief valve that allows some pressure to build up in the battery (typically 1-4 psi), up to the limit of a pressure relief valve, but not allow air to enter the battery. AGM batteries are essentially the same as flooded batteries, except that the electrolyte is held as a paste in fiberglass mats in contact with the lead or lead-alloy plates. Sometimes classified as valve-regulated recombinant gas (VRLA) batteries, meaning that any hydrogen and oxygen generated due to slight overcharging are recombined in the electrolyte paste, forming water again. The big advantage is no need to periodically water the batteries. Energy storage capacity is usually less than flooded batteries due to the space required for the additional features. Cost is higher than flooded batteries, and AGM batteries cannot tolerate overcharging. Despite the higher cost and slightly lower energy density, VRLA batteries were preferred in Generation 2 EVs and conversions to avoid the monthly maintenance task of watering a large number of lead acid batteries. The well-known Optima "Tubular Cell" battery is a type of AGM battery.

Gelled Electrolyte (or Gel Cell)

Another type of recombinant gas VRLA battery, but using a gel electrolyte, typically sulfuric acid mixed with silica fume (fine-grain Silicon Dioxide), which fills the space between the lead or lead-alloy (usually sponge-type) plates. Attributes are similar to AGM batteries, except that gel batteries do not need to be kept upright due the high viscosity of the gel. Standard battery for UPS backup power systems and sealed consumer product batteries.

Three automotive *battery purpose designations* have become standard. These apply to all battery types, but they are only important for lead-acid batteries:

Starting Lighting and Ignition (SLI)

An automotive starting battery. Designed to produce high current for short periods by maximizing active plate surface area and minimizing plate separation. Will not last long if deep-cycled, since its thin fragile plates will be quickly eroded.

Deep Cycle

Thicker, solid plates with less surface area and greater separation allow these batteries to be almost fully (down to 20%) discharged without permanent damage. Less power, but longer life than SLI batteries. These are the batteries used in golf carts, utility service vehicles, and any EVs which use lead-acid batteries.

Marine Deep Cycle

Marine deep-cycle batteries are half-way between a true deep cycle and an SLI battery, intended to provide continuous power for electrical loads on watercraft, but also short bursts of very high current to start a marine (usually a diesel) engine.

The most common battery type used for pre-lithium electric vehicle conversions was the *flooded deep-cycle battery*. This is because the cost was low, the energy density (kWh/kg) was adequate, and the battery electrolyte could be serviced to maximize the life of the battery.

NiCad and NiMH

Prior to lithium, the only widely-adopted improvements over lead-acid used in EVs were nickel cadmium (NiCad) and nickel metal hydride (NiMH). These batteries remain popular for small low-cost electronics such as rechargeable flashlights and toys. And despite its lower power density than lithium, NiMH continues to be used in the non-plug versions of some hybrids including the Prius (Fig. 14.58). NiCad, which was the only rechargeable alternative to lead acid for at least five decades, no longer has any vehicle applications since practical lithium batteries became available, and because of end-of-life cadmium toxicity issues.

Fig. 14.58 Ovonics Nickel Metal Hydride battery, 2001. Still used in non-plug versions of the Toyota Prius. https://www1.eere.energy.gov/vehiclesandfuels/pdfs/success/nimh_batteries_mar_2001.pdf Public domain

It is understandable that for over 100 years, battery limitations have relegated electric vehicles to specialty applications only, rather than replacements for ICVs.

Lithium-Ion

It was not until lithium-based battery chemistries were (re)discovered in the 1970s, followed by intensive worldwide engineering efforts starting in the 1990s, that lithium-based chemistries found their way into small electronic devices that required longer run times. Circa 2006, they appeared in the first production electric vehicles, notably the AC Propulsion tZero, followed by the Tesla Roadster in 2008. The effects were felt immediately in the industry. The average cost of lithium-based batteries ~$1,100 USD/kWh in 2010, has fallen to $139 USD/kWh in late 2023. With continuous innovations in cell design and many possible combinations of lithium and proximate elements on the periodic chart, lithium batteries were indeed the Holy Grail sought for more than 100 years. They are now used in every full electric or plug hybrid car in production today. As of March 2023, there were 50 BEVs and 33 PHEVs for sale to the public in the USA, with even larger numbers in China and the EU. All use various lithium metal oxides as the cathode material.

Lithium-based batteries are not without issues, with a history of spontaneous vehicle fires during charging or storage, concerns over the supply chain for critical battery materials, and the lack of a standardized end-of-life recycling strategy. But lithium chemistries are the basis of almost all energy-dense (practical) battery options available for EVs today. And new developments are being announced almost every day.

Lithium batteries still have further to go, with two particular variations showing possibly the greatest potential for greater energy density, power density, and/or cycle life at least at this time.

Metallic Lithium and Solid-State Electrolytes

Batteries that use metallic lithium anodes (instead of graphite) and ceramic solid-state electrolytes are probably the most promising development path at this time.

Metallic lithium batteries contain more than twice the lithium metal of existing lithium-ion chemistries and can have up to twice the energy density. But along with higher energy and power density, they are more prone to dendrite growth between the anode and cathode in each cell, a mechanical problem that internally short circuits cells and is the most common cause of premature cell failure. Metallic lithium anodes are also more prone to expansion and contraction with temperature and state of charge, which could lead to the mechanical failure of a cell. In addition to degradation of performance, the heat generated by dendrite and expansion short circuits are well-known causes of spontaneous battery fires. And with the higher energy density, the consequences of a battery fire are greater. But problems of this type usually fall into the category of incremental engineering challenges, not fundamental chemistry or physics limitations. Breakthrough solutions with this technology could be a huge plus for EVs. Despite the greater lithium requirement, metallic lithium batteries are expected to provide greater vehicle range, extended battery life, tolerance for high charge rates, and reduced need for the other strategic battery metals

Ni, Co, and Mg. But as with other frequently announced "breakthroughs," expectations must be tempered. As of January 2025, no production EVs use lithium metal batteries.

Another synergistic technology, solid-state (usually ceramic) electrolytes, promises ion transport properties nearly as good as current polymer or saturated liquid electrolytes while serving as superior mechanical barriers to the formation of dendrites between the anode and cathode, described above. Dendrite growth is actually a problem with almost all batteries, lithium or otherwise. Growth occurs simultaneously with the number of charge-discharge cycles, eventually penetrating the polymer, saturated liquid, or ceramic electrolyte barrier separating the anode and cathode, causing an internal short circuit. With greater resistance to dendrite penetration as well as the ability to tolerate anode dimensional changes during charge and discharge, solid-state electrolytes are viewed as a possible enabling technology for metallic lithium batteries.

According to Greg Hitz of ION Storage Systems [88],

> No solid state battery manufacturer has yet to offer a 100% solution. There are technologies that support high charge/discharge rates, have high energy density, stable chemistries, and are scalable for manufacturing and integration, but no single product offers all of that without significantly compromising one or more of the other aspects.

For perspective, it is worth noting that engineering improvements in lithium-ion batteries moved the technology from dangerous (laptop computers and cell phones were not allowed on commercial aircraft), to safe for the general consumer and the EV market in only ten years.

New electrochemical materials technologies seem to make the news every day; some have potential, but most are missing some critical requirement that the inventors or promoters deprecate with phrases such as "with further development we expect …"

Game-changing improvements are certainly possible, but it is hard to beat the mass energy density and electrochemical properties of lithium, which is the lightest metal. There are still undiscovered combinations of materials and structures that justify hope, but no slam-dunks on the horizon.

Sodium-Ion Batteries

One of many incremental developments worthy of attention is the substitution for lithium of the next two elements in the alkali metals column of the periodic chart, sodium or potassium, as the active anode material (Fig. 14.59). Both are highly reactive metals like lithium, although not as light.

The development of sodium-ion (Na-ion) and to a lesser degree, potassium-ion batteries, occurred in parallel with lithium-ion batteries starting in the 1970s. But by the mid-1990s commercial interest had waned following improvements in lithium technologies which provided the highest energy and power density. The early thermal stability problems and the cost/availability of the other metals required for

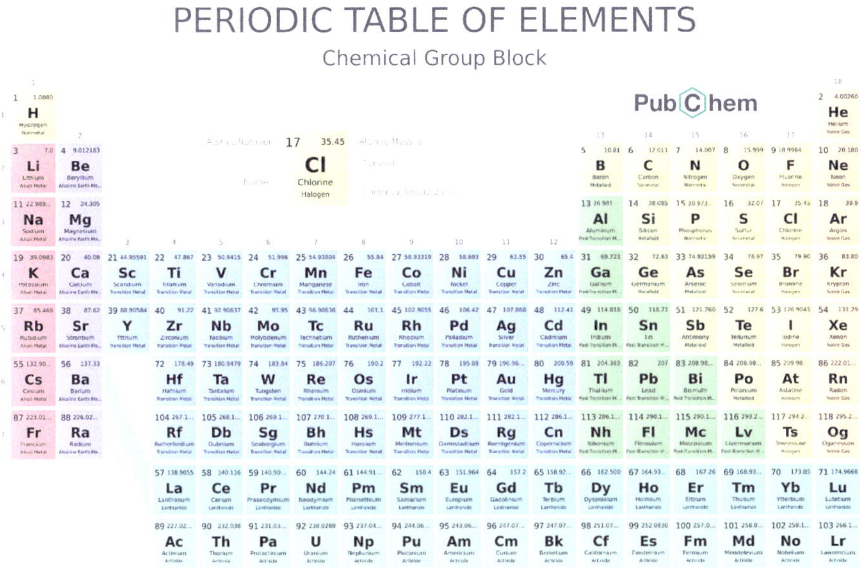

Fig. 14.59 Periodic Table of Elements. Lithium, sodium and potassium in the left-most (alkali metals) column, have desireable properties for advanced battery materials. Image from US National Institute of Health. https://pubchem.ncbi.nlm.nih.gov/periodic-table/png/Periodic_Table_of_Elements_w_Chemical_Group_Block_PubChem.png. Public domain

lithium chemistries were not yet seen as an impediment to widespread adoption. But with the scale driven by the EV revolution this has changed. The late 2010s brought renewed interest in Na-ion batteries almost entirely due to the exponentially increasing cost and geopolitical sensitivity of the lithium/cobalt/nickel/manganese supply chain on which they are based.

Compared with lithium, sodium-ion batteries are intrinsically more stable and provide energy storage densities much greater than lead acid, although not as high as Li-ion cells. Na-ion batteries of the 1980s were also known for limited cycle life but were no worse than lithium chemistries of that time. Longevity remains a challenge but has dramatically improved since then. At this time, the remaining limitation seems to be power density, that is, the ability to source or accept the very high discharge and charge rates that EV owners now take for granted with advanced lithium chemistries. Sodium remains an interesting alternative for stationary applications because, despite the lower performance. Na-ion batteries have none of the material supply limitations of lithium batteries, are easily scalable, intrinsically safe, and increasingly competitive with lithium in applications for which high current charging and discharging is not a requirement [89]. The current consensus (2024) seems to be that they are better suited to applications such as solar or wind electricity storage as replacements for the $LiFePO_4$ batteries that now dominate this application.

Lithium Sulfur Batteries

Finally, worthy of mention is the pairing of a lithium metal anode with a sulfur cathode. Lithium sulfur (Li-S) batteries have been around since the 1990s, and experimentally, this chemistry is capable of more than twice the energy density of lithium-ion batteries [90, 91]. The reduced requirement for costly or politically sensitive metals compared with contemporary lithium chemistries is its major advantage. The high energy and power density have supported their use in electric aircraft applications [92]. Cycle life is less advantageous. While Li–S batteries with up to 1500 charge and discharge cycles were demonstrated in 2017, their cycle life at commercial scale remains an impediment, an area of continued development. As of 2023, Li-S batteries were not yet commercially available. Time will tell if it could be the next increment in battery technology or another hopeful attempt.

Metal-Air Batteries

A different and older approach that still holds future promise is metal-air batteries, which have attributes that are mid-way between rechargeable storage batteries and fuel cells. Metal-air batteries, especially zinc-air, have been known and used for specialty applications since the 1990s. Lithium-air, zinc-air, aluminum-air, or magnesium-air "fuel cells" promise greater energy densities than non-air-breathing chemistries. These all have the theoretical potential for energy densities several times that of any other battery chemistries, including modern lithium-ion batteries. Yet despite the energy density (kWh/kg), they are typically limited to lower power (kW/kg) than other batteries. To date, metal-air batteries have only been used as primary (non-rechargeable) batteries, although they can accept some degree of electrical recharging. For vehicle use, they must be refueled by replacement of the active metal and liquid electrolyte rather than be electrically recharged [93].

One of the only attempts at practical use in tranportation was demonstrated in 2011 when an existing electric shuttle bus was modified by the Santa Barbara (California) Municipal Transit District (MTD) to use prototype zinc-air batteries (Fig. 14.60). The batteries were "recharged" by continuously feeding the battery with zinc pellets and recirculating the potassium hydroxide electrolyte solution. Other technical complexities (but not insurmountable obstacles) facing the metal-air batteries included the need to scrub carbon dioxide from the intake air since it will form carbonate deposits on the air cathode. This requirement is not unusual - it is shared with all hydrogen fuel cells. The effective RTE of a metal-air battery is usually inferior to other chemistries because the production or recycling/regeneration of the spent metal (zinc, aluminum, lithium) electrodes requires greater electric energy than that required to electrically recharge a regular lithium battery.

Refueling vehicles equipped with metal-air batteries requires a completely different infrastructure than the usual electric recharging network. Home refueling would probably be impossible unless the material exchange process could be fully automated at a reasonable cost. But for institutional applications such as public transit or freight transport, it is not any more exotic than refueling hydrogen fuel cell vehicles with ultra-high-pressure or cryogenic liquid hydrogen. Refueling would be more of an engineering than a chemistry challenge, with a lot of room for

Fig. 14.60 Zinc-air batteries were used experimentally in one of the Santa Barbara Downtown shuttle buses in 2011. From Marta Baginska "Metal-Air Batteries: Types, Applications, and Challenges," NPRE 498 Energy Storage Systems 12.07.2011. Mar 13, 2019. Public domain

innovation. Could metal-air batteries be an alternative to Li-based EV batteries? Even if battery reactivation was fully automated or incorporated into a battery exchange scheme, the likelihood of yet another multi-billion dollar public investment for EV refueling infrastructure is probably remote.

Many metal-air battery companies have come and gone in the past 30+ years, all promising but not delivering batteries that could greatly extend the range of EVs while meeting safety, lifetime and power requirements. With the support of subsidies and hopeful investors, research groups and startup companies working on metal-air batteries continue to appear [94] albeit below the usual media attention.

Higher Battery Voltage Improves Efficiency But Increases Hazards

Batteries in electric and hybrid cars and trucks usually have voltages between 202V (2005 Gen 2 Prius) and 900V (2022 Lucid Air Sapphire), with 500–800V being most common in 2024. The trend to higher voltage battery packs continues, with industry objectives of 1000 VDC [95]. Since power increases with the square of voltage, this is a logical development which allows the reduction of the gauge of electrical conductors to carry a given amount of power.

What Voltage Is Dangerous?

Any voltage over 50 VDC is considered hazardous, therefore all existing automotive EV voltages pose a potentially lethal risk if the electrical system of an EV is damaged in a collision or if the vehicle is serviced by inadequately trained personnel.

As discussed in the citation [96] the (US) National Electric Code (NEC) NFPA 7OE, Par. 1 10.7(F) sets the DC threshold of safety at 50 volts [97]. Since a number of safety regulations are applicable for voltages above 50V, it is not surprising that golf carts, neighborhood electric vehicles, and utility service carts usually have battery packs limited to 48V. However, IEEE Specification TS-60479-1 suggests that these long-standing limits are actually too high. The US Occupational Safety and Health Agency (OSHA) [98] now sets the safe "touch voltage" limit as 35 Volts in the USA. The Canadian Electrical Code defines "extra low voltage" not requiring additional safety measures as "up to and including 30 volts" and Canadian National Health and Safety regulations consider 30V as the worker safety threshold. International Electrotechnical Commission Specification IEC 60479-5 (the IEC European Standard) states: "body contact between 36 and 49 volts could cause ventricular fibrillation; recommends safe limit between 25 and 30V."

What Would Happen If a Person Was Exposed to an EV Battery Voltage?

You could expect severe electrical burns and electric shock, likely loss of a limb or death. This is not an exaggeration.

The amount of instantaneous power available from an EV battery pack is almost too large to comprehend. An illustration: the 1981 GM/Conceptor Electric G-Van was powered by lead-acid batteries of much lower energy and power than the lithium batteries in newer EVs [95]. The G-Van had a large heavy under-slung battery pack consisting of 36 6V deep-cycle industrial batteries, with a nominal series voltage of 216V, a much lower battery capacity and voltage than any present EVs. But fully charged, each 6V monoblock battery was capable of supplying momentary current bursts as high as 1700 amps, more than eight times the peak capacity of a standard residential electrical service panel at a lethal voltage. This means that the maximum momentary power that the battery pack could deliver in the case of a short circuit was 216V × 1700A = 367kW. For the benefit of the horsepower generation (like me), that's approximately 500 metric horsepower, where 1 metric horsepower = 0.986 US SAE horsepower = 0.736 kW. This is much more than the battery's continuous power rating and enough to instantly vaporize a metal wrench or body part that bridges battery contacts. Colloquial evidence from years of DIY experience is that most EV electrical injuries have been burns of a hand holding a wrench that accidentally bridged battery voltage contacts and became white-hot in milliseconds. Handling even lead-acid EV batteries requires great caution and protective measures such as insulated gloves, goggles, and most important, a partner to assist if necessary. The increasingly high voltages of recent EVs (e.g., 800V) exacerbate the importance of formal procedures and precautions during servicing, as well as in the aftermath of collisions that compromise an EV electrical system.

It is almost surprising that there have been few reported EV injuries due to high-voltage exposure. But this data is likely underreported, since non-serious battery handling errors are rarely reported. The severity of the risk increases with the number of EVs on the road. It's of particular concern for first responders at the site of an EV crash involving a damaged battery or electrical system [99]. Nevertheless, electric injuries are not common among mechanics and are nonexistent for consumers who heed warnings to not tamper with any part of the electrical system of their EVs.

Lithium battery fires and explosions are considered a much greater risk for first responders, outweighing the risk of electrocution due to high-voltage contact.

Still, with the growing deployment of EVs, dealer-only service policies have become the growing focus of disputes between manufacturers and independent service facilities, with safety usually the justification. As the industry matures, these issues should inevitably be resolved. Leading the mission to better train service personnel and gain access to service and repair business have been the US-based ASE (National Institute for Automotive Service Excellence), the ASA (Automotive Service Association) [100], and the International Hybrid and Electric Vehicle Repair Alliance (HEVRA) [101]. As of 2024, there are only two HERVA-certified EV repair facilities in the USA, but considerably more in the EU.

As reported by the US marketing firm IMR Inc. in 2024 [102]:

> Three out of every 100 vehicles (3.1%) serviced in the USA at independent repair shops are BEVs, while HEVs account for 6.2% of the vehicles serviced. Of shops that service BEVs/HEVs, almost 70% do not specifically market or advertise the service.

The Environmental and Social Impact of EV Batteries

The manufacturing of advanced batteries of all types carries significant environmental impacts as well as huge energy costs. This is especially true for lithium battery chemistries, arguably the most consequential of all production battery technologies [103].

The Peoples' Republic of China (PRC) produces the large majority of the lithium cells used in electric vehicles, including those that are assembled into battery packs by other manufacturers worldwide. According to current information, the PRC manufactures the large majority of lithium batteries worldwide, and has the fourth largest domestic lithium reserves in the world (Chile, Australia, and Argentina are numbers 1–3) (Fig. 14.61).

Of possibly greater environmental and social consequence than lithium metal extraction is the mining of the cobalt, nickel, manganese, and copper required for the production of EV batteries. For example, more than 70% of world cobalt reserves and mining operations are located in the Democratic Republic of Congo. In an area where human survival is often in the balance, it is reported (conservatively) that 15–30% of the cobalt ore is extracted by *artisanal mining*, a greenwashed term that means mining solely by human labor without heavy equipment. The social injustice and ultimate human toll is abhorrent [104] (Fig. 14.62).

The manufacture of batteries is hugely energy intensive, with most of the energy in the form of electricity which must be generated from other energy sources. Total CO_2 emissions from the production of lithium-based batteries vary depending on the materials (lithium, cobalt, nickel, manganese, copper, iron, etc.) for which the mining/extraction-related emissions are highly variable. The vast majority of lithium-ion batteries—about 79% of the world's supply—are manufactured in China where coal is the primary energy source. Coal is by far the dirtiest electric energy

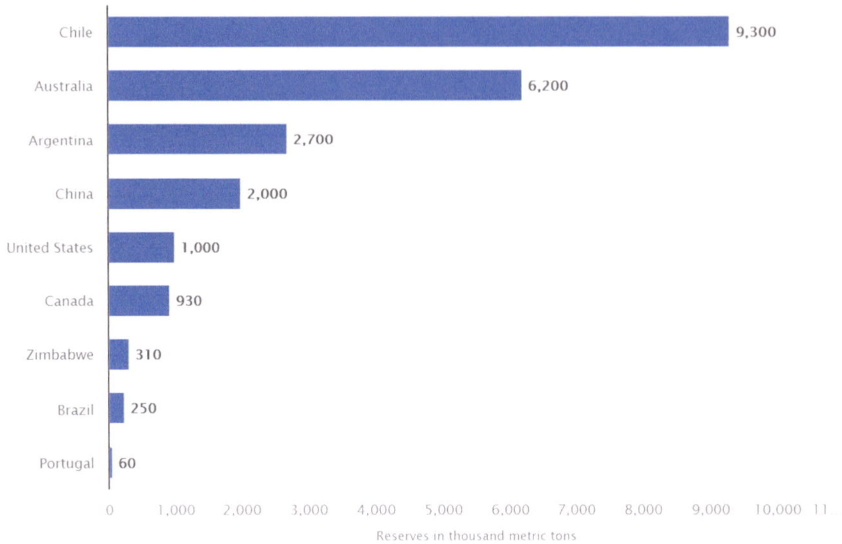

Fig. 14.61 Proven lithium reserves by country, 2022. (From Statistica. https://www.statista.com/statistics/268790/countries-with-the-largest-lithium-reserves-worldwide/ Published August 2023. Statistica university license

Fig. 14.62 Artisanal cobalt mining in the Democratic Republic of Congo, 2018. https://kosciol.wiara.pl/doc/4521326.Krwawy-konflikt-bez-konca MONUSCO / CC-SA 2.5

source. The dominance of China in EV battery production using the dirtiest form of electric generation is a factor when assessing the overall environmental sustainability of EVs.

A 2018 meta-study of published world data on lithium battery manufacturing 2015–2017 reported a range of values from 56 to 494 kg/CO_2e/kWh, depending on the country and CO_2 sources included in the analysis [105]. The unweighted average of 16 conclusions was 152 kg CO_2/kWh of maximum battery capacity. The wide variability is consistent with an example found at the *MIT Climate Portal*, updated July 2022:

> For illustration, the Tesla Model 3 holds an 80 kWh lithium-ion battery. CO2 emissions for manufacturing that battery would range between 2400 kg (almost two and a half metric tons) and 16,000 kg (16 metric tons). [106]

> Battery manufacturing CO_2 contributions are in addition to the manufacturing costs of the vehicle itself, which on a kg CO_2 per kg of vehicle mass basis, differs little between EVs and ICVs within the same vehicle class.

End-of-Life Issues for Lithium Batteries

With the proliferation of electric vehicles during the past 10 years, one critically important environmental issue has garnered relatively little attention: the disposal or recycling of lithium batteries. The battery is the most important and most costly component of an EV or plug hybrid. The active materials in the expired battery have considerable value if they could be recycled at a reasonable cost compared with new materials. The lack of standardization of batteries is a large factor that prevents cost-effective recovery of high-value materials. And as of 2023, the replacement of the battery in an EV at the end of its useful life is likely to be more costly than the residual value of the vehicle. According to the used car valuation company Edmunds (USA),

> On average, you can expect the replacement cost of an electric car's battery to run from $5,000 to upward of $15,000 [107]

One franchised EV battery service business in the USA quotes "over $35,000" for a 100kW+ battery pack in a typical high-end electric SUV or truck. The situation is evolving, but to date, there is little evidence that EV owners replace batteries; they just get rid of the car due to the cost. Major EV manufacturers are now working to address end-of-life issues with EV batteries, but the environmental impacts of this built-in obsolescence have not yet been fully embraced. Worldwide, many thousands of early model EVs have been abandoned with relatively low accumulated mileage. Worldwide, economics dictate decisions to recycle or discard. [108]

To address the fears of potential EV buyers about the future cost liability of a failed battery, in the USA, federal regulations require EV manufacturers to warranty vehicle batteries for 8 years or 100,000 miles. As of 2023, California rules are even more stringent: the lesser of 8 years or 150,000 miles. While the lifetime of an EV battery depends most on the number of complete charge-discharge cycles and the rate of charging, a mean lifetime range of 10–15 years seems to be the typical prediction for batteries manufactured in the past 5 years. This means that the first large wave of spent lithium vehicle batteries will probably hit between 2025 and 2030. The reality is that there are not yet any environmentally benign methods or regulatory guidelines for lithium battery disposal or recycling. Some argue that the lithium, cobalt, nickel, and manganese in this type of battery do not pose a direct health or environmental risk if simply buried in a landfill. This is certainly arguable considering the interaction of the battery metals and electrolytes with groundwater, and the tendency of these high energy density batteries to spontaneously ignite when damaged, burning even in the absence of air. But the pivotal issue seems to be economic sustainability: at this time, the cost to recover and reuse the metals and minerals is equal to or greater than their market value.

Perhaps the greatest incentive is the tenuous supply chain and increasing scarcity of these materials. Unless there is an unlikely innovation that allows superior battery performance than lithium-ion, recycling of these materials will become a necessity for the environmental and economic sustainability of EVs. With greater and greater range being the current trend for EVs, there is no reason to assume that demand for ultra-high energy density batteries will decline or even stabilize. Recycling is a necessary evil in service to the greater good. There seem to be three end-of-life strategies, although none have been deployed at the scale needed to handle the anticipated "dead battery wave":

Reuse/Repurposing

This method of disposal has seemed to receive the greatest interest recently. The idea is that an EV battery may be depleted to the point that it cannot provide adequate vehicle range, but it still may have as much as 75% of its original energy storage capacity. So continuing to use end-of-life EV batteries in stationary electric energy storage applications for which ultra-high energy density is not needed is an alternative to recycling, at least until the battery is exhausted beyond usability. Among the advocates of this temporary solution is Consumer Reports (USA) [109]. At this time, I am not aware of any grid-scale battery energy storage projects that formally use exhausted EV batteries, but it is a well-known battery storage method used for individual residential storage. And as more reach the limits of their useful life in an EV, this situation will surely evolve. Estimates of how much more use can be obtained from repurposed EV batteries are highly dependent upon the state of the exhausted batteries and the reuse application. The broad assumption that all or even most lithium-based batteries will remain in EV service for 10–15 years and then be usable for extended reuse is optimistic in view of the irreversible failure mechanisms most common in lithium batteries late in their usable life. The failure of individual cells in a large battery pack is much more likely than an equal and gradual degradation of all cells simultaneously. The well-known problem of anode-cathode dendrite growth is that internally short circuited cells are among the most common failure mechanisms, and it is exacerbated by the high-rate charging that EV owners have been encouraged to rely upon for convenience. Continued use of lithium battery packs having even a few cells that are internally shorted greatly increases the possibility of a spontaneous battery fire, especially during charging. The most authentic experience in this area seems to be coming from small businesses and franchises that "rebuild" or "recondition" EV batteries by testing and replacing individual cells, then reselling the battery packs for continued EV use, for example, reference [110].

Reuse/repurposing is an effective way to amortize the original environmental and energy impacts of the battery manufacturing over a longer period of time, but it does not address the ultimate issue: what to do at the actual end of battery life.

Recycling of Battery Materials

Estimates from multiple sources in 2023 place the current worldwide rate of recycling of lithium batteries at 5–10%. And this number may be overoptimistic, motivated by questionable means to claim large subsidies in the USA and EU, and the practice of counting each of the multiple recycling steps as a complete operation. For example, grinding up old batteries prior to shipping to another entity that extracts metals from the resulting "Black Mass" (Fig. 14.63), which is then shipped to refiners that may or may not produce new battery materials from the recovered metals.

Why such a low recovery rate compared with lead-acid batteries that recover 98% of their active materials for direct use in manufacturing new batteries? Lithium battery recycling is far more challenging, hazardous, costly, and environmentally consequential. Lithium battery chemistries are highly variable rather than the simple lead and sulfuric acid recovery for lead-acid batteries. Only since approximately 2020 has there been serious interest and government support for research, development, and deployment of improved battery recycling methods. This is understandable because even at this time, almost a decade into mass EV adoption, batteries at end of life are scarce, with most of the recyclable dead battery stream coming from small electronics, or EV batteries that were the result of large-scale safety recalls or premature failure due to defects. The USA is well behind the curve in failing to plan for the large number of end-of-life batteries less than 10 years away. Since 2006 in the EU, 50% of EV batteries are supposed to be recycled, but implementation has not necessarily followed this aggressive schedule. While grinding up old batteries seems to be a universal and necessary first step, many methods for battery metal recovery have been optimistically referred to as "recycling." But in the USA and the EU, only two such technologies have achieved a significant level of deployment at this time, and these only due to government subsidies [111].

Fig. 14.63 "Black mass" from shredded lithium-ion batteries. Source: Argonne National Laboratory. Public domain

Smelting

The most common method for recycling the battery materials at this time is pyrometallurgical processing (aka smelting), which involves first disassembling and removing copper connectors, steel and aluminum enclosures, and plastics from battery assemblies to isolate the individual battery cells. The cells are usually mechanically shredded resulting in "black mass," which is then incinerated in a high-temperature furnace, leaving a molten mass from which the valuable metals are separated: cobalt, nickel, manganese, and of course, lithium. The process is very energy intensive, and the environmental consequences include but are not limited to huge CO_2 emissions at least partially due to the natural gas or electricity used for incineration. Waste with little market value must still be disposed of in some way that has an acceptable environmental impact, a challenge not yet adequately addressed.

Chemical Reduction

Hydrometallurgical (chemical) processing uses liquid reactants and solvents to extract recoverable metals. Less energy intensive than smelting but involves large quantities of water and hazardous materials that cannot themselves be recycled. Lower CO_2 direct emissions, but a larger and more hazardous waste stream. More costly than smelting.

One recent variation vying for government funding is the electrochemical extraction of valuable metals. This process is probably the most complex of all the possible chemical processes and is claimed to be less environmentally inconsequential than conventional chemical extraction processes, although at much greater electric energy costs. As with smelting, hydrometallurgical processes result in unrecyclable waste after the metal extraction that must still be disposed of, hopefully not in environmentally consequential ways [112].

Progress and Complicating Factors

- In the USA, battery recycling is a significant component of the 2022 (USA) Inflation Reduction Act, setting aside incentives for improved methods and accelerated deployment despite less-than-ideal methods.
- Major EV manufacturers have formed alliances with battery recycling partners, subsidizing some of their capital and operating costs. The current leader in this endeavor is Toyota, followed by Tesla, VW, GM, Ford, and BMW.
- The wide range of battery chemistries prevents standardizing any one process. Accommodation of all EV batteries is much more costly than the more familiar recycling of lead-acid SLI batteries. This is further complicated by the highly specialized sizes, shapes, and mechanical construction of batteries, each of which is purpose-built for a single manufacturer, and usually for a single vehicle model.
- The cost of recovering these metals is currently (2024) greater than the cost of mining new materials, despite the high cost of mining and refining.
- Recovery rates of cobalt, nickel, and lithium remain poor for all recycling processes and have never reached a level of economic sustainability.

- The rapid pace of worldwide electric vehicle deployment greatly exceeds provisions for managing the anticipated "dead battery wave." Like other EV-related infrastructure issues such as electric distribution and road design, sustainable battery recycling lacks adequate national and international emphasis.

Advanced Battery Chemistries Currently in Production

Table 14.1 lists the major differences between battery chemistries that have seen significant applications in electric vehicles.

The State of the Art in EV Batteries, 2023–2024

At the time of writing, the trend among EV manufacturers, notably Tesla (USA) and BYD (China), is to transition to Lithium Iron (Ferrous) Phosphate (LFP) batteries in their lower cost vehicles with shorter range specifications, while continuing to use NMC or NMCA (GM's Ultrium battery chemistry containing aluminum) batteries in more expensive models with longer range and higher performance. The difference is the lower cost but lower energy density of LFP compared with NMC, meaning less range. Broader adoption of LFP in the highly competitive race to provide greater range is unlikely, since it would be at the cost of greater vehicle mass for EVs which are already 30–60% heavier than ICE models equivalent to the same vehicle (Fig. 14.64).

Possibilities remain, however, for the previously mentioned sodium-ion batteries in lower cost, low-range vehicle applications for which the lower energy density is a tolerable tradeoff with lower cost and availability. Potassium is a less likely candidate due to fundamental issues that limit its cycle life, and higher risk of spontaneous combustion if the battery is damaged.

While cusp battery technologies hold promise for the future, at this time there appear to be only two technologies in widespread use by vehicle manufacturers, both lithium-based.

1. First, lithium oxide in conjunction with cobalt, nickel, and manganese. For example, the most popular lithium MNC batteries used in most EVs with longer ranges. These provide the highest energy storage densities, as high as 300 wH/kg., but are expense, environmentally harmful and entail some risk of battery instability leading to fires.
2. The second option is the lower performance but intrinsically stable lithium iron phosphate (aka LiFePO4, LIPO, or LFP). This battery chemistry is not subject to thermal runaway if overcharged, tolerates high charge and discharge rates without excessive degradation, and most importantly is not dependent upon increasingly scarce metals. But the energy density of LFP batteries is at most 150 Wh/kg, approximately half that of any variation of lithium nickel manganese cobalt batteries.

Table 14.1 Comparison of EV battery chemistries as of 2023. From [113]

				Li-ion		
Specifications	Lead acid	NiCd	NiMH	Cobalt	Manganese	Phosphate
Specific energy density (Wh/kg)	30–50	45–80	60–120	150–190	100–135	90–120
Internal resistance (mO)	<100 12V pack	100–200 6V pack	200–300 6V pack	150–300 7.2V	25–75 per cell	25–50 per cell
Life cycle (80% discharge)	200–300	1000	300–500	500–1000	500–1000	1000–2000
Fast-charge time	8–16 h	1 h typical	2–4 h	2–4 h	1 h or less	1 h or less
Overcharge tolerance	High	Moderate	Low	Low. Cannot tolerate trickle charge		
Self-discharge/month (room temp)	5%	20%	30%	<10%		
Cell voltage (nominal)	2V	1.2V	1.2V	3.6V	3.8V	3.3V
Charge cutoff voltage (V/cell)	2.40 Float 2.25	Full charge detection by voltage signature		4.20		3.60
Discharge cutoff voltage (V/cell, 1C)	1.75	1.00		2.50–3.00		2.80
Peak load current best result	5C 0.2C	20C 1C	5C 0.5C	>3C <1C	>30C <10C	>30C <10C
Charge temperature	-20 to 50 °C -4 to 122 °F	0–45 °C 32–113 °F		0–45 °C 32–113 °F		
Discharge temperature	-20 to 50 °C -4 to 122 °F	-20 to 65 °C -4 to 149 °F		-20 to 60 °C-4 to 140 °F		
Maintenance requirement	3–6 months (topping charge)	30–60 days (discharge)	60–90 days (discharge)	Not required		
Safety requirements	Thermally stable	Thermally stable, fuse protection common		Protection circuit mandatory		
In use since	Late 1800s	1950	1990	1991	1996	1999
Toxicity	High	Very High	Low	Low		

Data source NASA, citation [113]

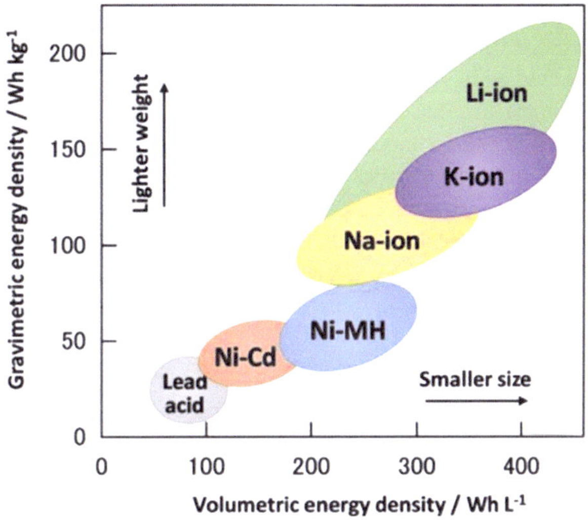

Fig. 14.64 Comparison of battery chemistries: volume energy density on abscissa and mass energy density on ordinate. Graphic from *Perspectives on Nickel Hydroxide Electrodes Suitable for Rechargeable Batteries: Electrolytic vs. Chemical Synthesis Routes*, MDPI Nanomaterials, Sept 2010. License CC BY 4.0

There are other lithium variations, but these seem to be the main ones in the near term [114]. From a vehicle point of view, despite the high energy density of NMC batteries, EVs that are already heavy will get even heavier to achieve the range demanded by consumers. So the incentive for innovation with EV batteries remains high. A comprehensive comparison of the pros and cons of various lithium battery chemistries can be found in citation [115].

EV Battery Risks

The fire risk of liquid fuel storage in ICVs is well understood and the risk has been accepted for over a century. The consequences of a vehicle fire can be horrendous, yet these accidents happen every day.

The equivalent situation for EVs is the fire risk associated with battery loss of integrity and damage to high-voltage conductors and components in a collision. These include battery fires while driving or as the result of a collision, or spontaneous battery fires while charging unattended. These risks are real and cannot be ignored, but in the context of all motor vehicle fires, may be overstated in news and social media due to the shocking visuals that accompany reports of such incidents. Considering that automotive fuel tanks are ruptured possibly even more frequently than EV batteries are seriously damaged [116], the only reasonable accommodation is enhanced equipment and training for first responders and hazmat teams for handling battery damage resulting in fires. A lithium battery fires are considerably harder to extinguish compared with gasoline or diesel fires.

Figure 14.65 is a television news photo of a Tesla Model 3 battery fire following a collision with a signpost in Baltimore, Maryland, on July 22, 2021. The collision of the car with a lightweight signpost should not have been sufficient to cause the catastrophic fire that immediately broke out. As reported by the responding fire department [117], extinguishing the EV battery fire required several hours and a huge amount of water, over 10,000 gallons. Lithium battery fires cannot be extinguished by simply depriving them of oxygen like other combustion sources, and the vehicle can and probably will spontaneously reignite as long as 48 h after the original fire. This has led to the refusal by some police impound and salvage facilities to accept damaged electric or plug HEVs [118, 119]. Also, some cautious parking garage owners discourage EVs by refusing to install EV chargers, and some have restricted EVs to specific parking spaces on the lowest level (Fig. 14.66).

Fig. 14.65 Tesla Model 3 fire after single car collision with signpost, 22 July 2021. Image from https://foxbaltimore.com/news/local/tesla-on-fire-in-towson-requires-large-emergency-response. Public domain

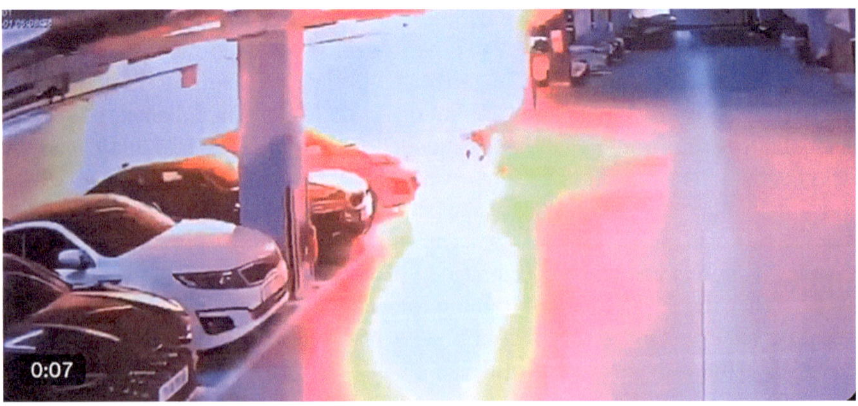

Fig. 14.66 Still frame from surveillance video posted by @btctslakeepgo on X of an unidentified electric SUV battery catching fire and exploding in parking garage in South Korea, 31 Jul 2024. https://x.com/btc_tsla_keepgo/status/1818858174423933183 Unrestricted use

Road Safety and Infrastructure Impacts of EVs

All vehicles (worldwide) are subject to at least some safety and emission requirements, although this is highly variable. In the USA, all federal safety regulations are enshrined in the FMVSS requirements, which are published annually in the US Federal Register [120].

Electric vehicles are similar to conventional ICE vehicles, but they have a few unique safety issues that ICE vehicles do not have as a result of their extra mass, quiet operation, and the lithium battery fire risk. While the nature of the hazards are different, the EV safety benefit of having no gasoline tank is offset by the hazards associated with the battery and high-voltage electrical systems.

Heavier Vehicles Cause Greater Damage in a Collision

Aside from the common belief that larger vehicles are safer (tougher, more intimidating, etc.), the motivation for massive EVs is driven mostly by consumers' range anxiety, with manufacturers competing to increase EV range to near gasoline levels.

Lithium batteries are superior in energy density compared with legacy lead-acid batteries, but they still provide vastly less energy than the equivalent mass of gasoline or diesel fuel. The ongoing EV range war has led to a return of the *lead sleds* of the lead-acid EV generation. Roughly speaking, every additional kWh of NMC battery capacity, including containment and cooling, adds approximately 4.5 kg (10 lbs) to the vehicle. Thus it is not surprising that a base model 2023 GM Electric Hummer with its state-of-the-art 200 kWh *Ultrium* Nickel Cobalt Manganese Aluminum (NCMA) battery and an advertised range of 250 miles weighs 4354 kg (9600 lbs), more than twice that of the similarly sized 2087 kg (4600 lb) gasoline 2023 Chevrolet Silverado 1500 Custom Crew Cab truck that has a much larger payload capacity.

A more fair comparison might be between two versions of the same car or truck from the same manufacturer, one an ICE and the other an EV: the base models of the gasoline 2023 Ford F150 XL gasoline has a curb weight of 1824 kg (4021 lbs) [121]. The base model 2023 Ford Lightning electric weighs 2728 kg (6015 lbs) with a standard battery or 2885 kg (6361 lbs) with an extended range battery [122]. Approximately a 50% weight penalty for the electric, which is typical when comparing EVs vs ICVs of the same type and with similar options.

Comparing passenger cars with similar capacities and classifications using US EPA data, a 2022 Tesla S Long Range has a curb weight of 2068 kg (4561 lbs). The average weight for all automobiles and light trucks (ICE and EV) sold in the USA in MY 2021 was 1885 kg (4156 lbs) [123, 124]. For MY 2022, the differential weights for similar electric vs gasoline or diesel vehicles as reported by multiple US agencies and independent reviews varied from 33% to 50%. Manufacturer websites and published specs for 2022-24 EVs now usually omit the vehicle weight, even while listing specifications such as upholstery materials. Table 14.2 provides a breakdown of weights by vehicle class in 2022 assembled from multiple sources: Carspec.org, Mechanic Base, Motor Trend, and ultimatespecs.com.

Table 14.2 Average weight of equivalent MY 2022 ICVs and BEVs

Body style	ICV Examples	Curb Wt (kg)	Equivalent EV	Curb Wt (kg)
Subcompact	BMW Mini	1232	BMW Mini electric	1429
Compact car	Toyota Corolla Hybrid LE CVT	1295	Tesla Model 3, base	1658
Midsize car	Toyota Camry SE Auto	1518	Tesla Model S, base	2073
Large car	Audi A8	2169	Lucid Air	2.350
Sports Car	Chevrolet Corvette	1529	Porsche Taycan	2380
Crossover SUV	Volkswagen Tiguan SE 2.0T	1711	Volkswagen ID.4	2239
Standard SUV	Honda Pilot	1836	Volkswagen ID. Buzz	2352
Half-ton Truck	Ford F-150 XL 2WD Reg Cab	1827	Ford F-150 Lightning, 2WD Reg Cab	2948

Vehicle weight and efficiency have always taken a back seat to advertised capabilities for trucks and large SUVs, a market segment that is more insulated from environmental or energy sustainability concerns. Product promotions now stress towing capacity in an effort to portray the increased mass in a positive light. For towing, heavier is better. It is simply a matter of physics that a heavier vehicle is needed to safely tow a heavy trailer. But how many trucks are actually used for towing? According to Motorbiscuit [125], about 7% are used to tow with any frequency, while 63% have never towed anything. The mismatch between the towing capacity and the payload capacity is incongruous to experienced truck owners: The 2024 Hummer Electric has a towing capacity of 3856–5443 kg (8500–12,000 lbs) but a cargo capacity of only 590 kg (1300 lbs) [126]. This is because the battery of even the shortest-range version weighs 1326 kg (2923 lbs), taking up most of the weight-carrying capacity (Fig. 14.67). Starting in MY 2024, a fully enclosed SUV version is available weighing even more than the pickup version, making it the heaviest non-armored SUV ever made [127].

From a study of battery electric automobiles (not trucks) for MY 2021 published by InsideEVs [128]:

> The heaviest car on the list of almost 70 EVs is the Mercedes-Benz EQV luxury passenger van, which is not far from 3,000 kg (6,612 lbs). The Audi e-tron 55 SUV is at 2,720 kg.
>
> Overall, well over a third of the EVs tested weigh more than 2,000 kg, and thus the average was inflated to almost 1,940 kg. There are only seven models below 1,500 kg, while more than half of the models are between 1,500 and 2,000 kg.

Since sales data for all EVs were not publicly available, it was not possible to determine an appropriately weighted mean mass for all battery electric automobiles in a given model year. Also, the exclusion of electric trucks by InsideEVs, which dominate the US market, would skew such an average to an unrealistic value. But it

Fig. 14.67 2023 GM Hummer Electric pickup truck. At approximately 4350 kg (9600 lbs) curb weight, the heaviest and least efficient EV ever sold to the general public. The lowest range battery alone weighs 1326 kg (2923 lbs), which is more than the weight of most gasoline cars. MSRP from $87,000–$110,000 USD depending on trim and battery capacity, but since its release in December 2021, dealer markups have increased the selling price by an additional $50,000–$140,000 USD. https://www.gmc.com/electric/hummer-ev/pickup-truck

was reported that starting with MY2024, every EV manufactured in the USA weighed over 2700 kg (6000 lbs). The growth in the number of excessively large EVs and near-extinction of small EVs has been referred to as the "EV Obesity Epidemic" in business media [129].

Among similar 2023 SUVs, the *lowest* EV/ICV weight differential (21%) appears to be the Volvo XC40 crossover which is available as both an ICV (the B5 model) and a BEV (Recharge model), with identical trim levels [130] and a difference of only 371 kg (818 lbs).

- XC40 B5 AWD (ICE) 1750 kg (3861 lbs)
- XC40 Recharge Twin AWD (223-mile range EV) 2122 kg (4679 lbs)

From a historical perspective, the heaviest non-commercial truck sold in 1988 was the Ford F350 Lariat, extended cab dually which had a curb weight of 2063 kg (4548 lbs) and a 2500 kg (5,500 lbs) payload capacity [131].

EV Braking: Good and Bad Characteristics

Unique to electric and HEVs are two different mechanisms to slow and stop the vehicle: conventional hydraulic disk brakes, and regenerative braking, previously discussed. Regen braking provides a huge increase in efficiency for driving that involves mostly braking and accelerating. But it has no effect on the vehicle's braking distance. Other than the influence of a more effective automatic braking system (ABS) for either vehicle type, regeneration does not provide any improvement in actual stopping ability. While Newton's law theoretically says that vehicles with different masses but the same coefficient of tire friction should have the same stopping distance, vehicles with greater mass almost always have longer braking distances compared to lighter vehicles. This has a lot to do with the non-linear nature of the coefficient of friction between the tires and the road surface, and the vehicle weight distribution, height, and tire contact area. Assuming the theoretical case of a fixed coefficient of friction between the tires and the road surface, the braking distance should not be affected by the vehicle mass, since both the downward force and the braking force are proportional to the vehicle mass. But in most cases, the larger contact-area-to-load that allows the use of softer (stickier) tread components on lighter vehicles helps to reduce the braking distance. EV-specific tires, in the interest of improved wear life under greater tire loads, use harder, more wear-resistance tires, the opposite of tires designed for handling and effective braking.

The only design advantage of a BEV that affects braking distance is the placement of the batteries low in the chassis, which reduces the forward pitching of the vehicle during hard braking compared with vehicles that have higher centers of gravity. Although with the trend toward increased ground clearance for off-road-capable electric trucks and SUVs, the increased height defeats much of the advantage of the low placement of the battery in the chassis.

This observation is not exclusive to EVs.

A list of the top 100 cars of all types with the shortest braking distances was compiled in 2023 by the racing publication Fastlaps [132]. The top six cars are Formula 1 or Le Mans LMP1 race cars. Excluding these to reduce the list to cars that are actually sold to the public, the shortest 60-0 mph braking distance (27 m, 72 ft) was a six-way tie between the Dodge Viper Mk V, Ford Mustang S650, Porsche 718 Cayman, and all three versions of the Chevrolet Corvette. It is noteworthy that nowhere on the list of 100 cars is a battery electric car, even though most high-end electric cars can accelerate from a standing stop faster than any of these cars.

Specific to EVs, in 2022 the Insurance Institute for Highway Safety tested several (not all) of the most popular battery electric cars. Among the cars they tested, the braking distance winner was the Tesla Model 3 Long Range, with a 70–0 mph (not 60-0) braking distance of 54m (176 ft). (This number was disputed by Tesla, stating that it should have been 33m (108ft) [133] based on their own tests). This claim is questionable, considering that the best braking car ever tested by Car and Driver magazine (previously referenced) was a 2017 Chevrolet Camaro ZL1, with a 70 to zero mph stopping distance of 39 m (129 feet). A bar chart illustrating the braking distances of several popular EVs is shown in Fig. 14.68, from the IIHS.

EVs vs. Pedestrians and Bicycles

In 2011, an NHTSA study [134] was published on hybrid electric vehicle (HEV) safety. 2011 was the first year that any mass-market EV was available (2012 Nissan Leaf). But hybrids had been sold in the USA since 1997 (Gen 1 Honda Insight). Among the conclusions:

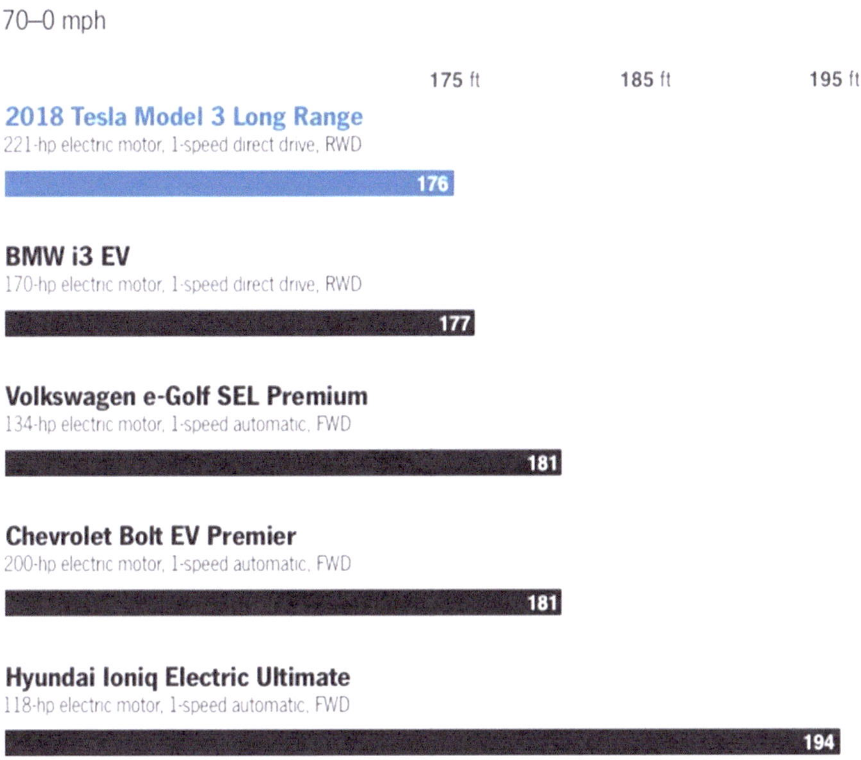

Fig. 14.68 Emergency braking distances in feet to stop from 70 mph for 2018 MY EVs. Graphic from the IIHS (Insurance Institute for Highway Safety) via Road and Track Magazine

> … an HE (hybrid electric) vehicle was two times more likely to be involved in a pedestrian crash than an ICE vehicle in situations involving low-speed maneuvers (Hanna, 2009)." "Overall, the odds ratios indicate that the odds of an HE vehicle being in either a pedestrian or bicycle crash are greater, 35 percent and 57 percent respectively, than the odds of an ICE vehicle being in a similar crash.

The primary causality in pedestrian collisions was found to be conspicuity due to the quiet operation, a characteristic that for fully electric vehicles is particularly bad at slow speeds on noisy city streets. FMVSS guidelines were subsequently supplemented to require the addition of audible alert devices for HEVs and EVs when the directional signals are on, warning pedestrians and cyclists of an impending turn. The study also reported that in the event of an accident, greater mass and increased stopping distance contributed to accident severity.

Managing the Public Infrastructure Costs of EVs

In the USA, the majority of road maintenance and infrastructure tax revenues are tied to gasoline and diesel fuel sales—a hidden tax built into the fuel cost at the pump. Vehicles larger than a certain size (variable with state) are classified as "commercial" regardless of whether they are used in commerce or not. The annual vehicle registration fees for commercial vehicles are much higher than those for non-commercial vehicles since they are tied to the vehicle weight, reflecting their greater impact on the roadway infrastructure.

In California, a small portion of the annual registration fees for EVs is designated for support of infrastructure. But considering the consistently greater mass of EVs, a fair assessment of the incremental cost of operating an EV on public roadways is lacking. This is a policy area that is still evolving. As EVs continue to significantly displace ICVs in the comming years, this imbalance must soon be addressed in the interest of safety and financial sustainability.

Road Surface and Tire Wear

Most people give little thought to the cumulative wear of vehicles on road surfaces, despite the fact that road maintenance is usually the largest expense of local, state, and federal highway agencies (Fig. 14.69). The problem of increased vehicle mass is an even greater issue for elevated roadway structures and bridges. With decades of experience, the US Transportation Research Board (TRB) and American Association of State Highway Officials (AASHO) have reduced to science the correlation between vehicle tire load (weight per tire contact area) and cumulative roadway damage, leading to a ratiometric formula [135]:

$$\text{Relative road } \textit{damage} \propto \left(\frac{W_2}{W_1} \right)^4$$

Fig. 14.69 Road surface wear due to tire loading. Pixabay free stock image

where W_1 and W_2 are the ratio of weights between two vehicles, and $\propto$ means "is proportional to." Relative road damage is synonymous with relative cost to maintain and repair. The damage metric is a ratio because different road surfaces are more or less durable. But note the 4th power exponent. Heavier vehicles do *much* more damage to roadways.

For example, the roadway maintenance cost attributed to the aforementioned 4561 lb Tesla S is about 4 times greater than that of a 3239 lb 2023 Honda Accord with approximately the same passenger and cargo capacity:

$$\text{Relative road } \textit{damage} \propto \left(\frac{4561}{3239}\right)^4 = 3.93$$

The rapidly growing transition to heavier trucks and SUVs (ICV or EV) has been a growing concern over the past decade. In the unlikely case that there were suddenly many more GM Hummer Electric "light trucks" driven by former Honda Prelude drivers, the cost to taxpayers would increase by

$$\text{Relative road } \textit{damage} \propto \left(\frac{9500}{3239}\right)^4 = 74 \textit{ times the road maintenance cost}$$

Road wear and damage translate to increased maintenance, energy, and environmental harm, although I am not aware of research that has quantified the latter two of these ramifications.

As might be assumed from the increased road surface damage done by heavier EVs, wear on tires is also accelerated. Harder compounds in tires specifically made for EVs reduces the severity of the increased tire wear, but at the expense of further road surface wear.

Tire Particulates

Only recently have the increased tire residues from heavier vehicles been examined. The same exponential formula relating road surface wear to vehicle mass applies to tire wear rate [136]. As tires wear, they shed carbon black and synthetic polymer materials. The particulates are centered around the 2.5 μm diameter that are very damaging to the environment, and carcinogenic for humans.

From Helixx [137]:

> The particles released by tires can contribute to the formation of fine particulate matter (PM2.5), which has been linked to a range of health issues, including respiratory and cardiovascular problems. This can be particularly concerning in urban areas where air pollution is already a major issue.
>
> In addition to the health impacts, tire particulate pollution can also have ecological effects. The particles can accumulate in soil and waterways, potentially harming plants and aquatic life. They can also contribute to the degradation of infrastructure, such as roads and buildings, by causing corrosion.

The high performance of late model EVs further increases the wear rate [138].

The problem may be further exacerbated by efforts of tire manufacturers to increase the wear life of EV tires by using harder tread compounds that may shed a lower mass but a numerically greater number of particles of smaller diameter per mile, which may pose even greater environmental risks.

Parking Structures and Garages

Another often ignored area of infrastructure affected by the increase in mean vehicle weight is parking structures that were designed, often many years ago, with the expectation that passenger car weights would either not change or gradually get lighter rather than heavier—the trend in the 1970–1980s. Design safety factors in structural handbooks were based on an assumed average vehicle weight. With the unexpected increase in weight of vehicles of all types, but especially the 33–50% premium of EVs, some structures have been found to be at risk of collapse [139, 140] (Fig. 14.70).

The remedial options are limited for existing older structures: reinforce the structure, reduce the total number of parking spaces in the structure, or prohibit EVs on any except the lowest level. Most problem cases will go unidentified for years, since the costs of upgrading parking structures to restore design load safety margins are considerable. Inevitably, the cost of parking a car must increase as an indirect result of the increased average vehicle mass.

Fig. 14.70 Sensational headline from a newscast following the collapse of a Manhattan NY parking structure on 18 April 2023. 48+ cars destroyed, 1 fatality. Few of the vehicles involved were actually EVs, but almost all were large SUVs. https://www.youtube.com/watch?v=n_Hxx5nZzRQ. Unrestricted use

Technology Changes, People Do Not

In just the last decade, the response of the automotive industry to EV range anxiety and lust for the fastest car has been driving the increasing mass of batteries and larger motors, responsible for the increasing mass of EVs. According to Tesla marketing in their 2022 Impact Report [141]:

> Consumers do not buy a vehicle that can meet most of their driving needs; they buy a vehicle that meets all of their driving needs. Since its introduction in 2012, we have increased the range of Model S by over 50% from 265 miles to 405 miles of range for the long-range version.

Not stated in the Tesla report was that the mean curb weight of Tesla S models rose by nearly the same percentage during this period to 2200 kg (5000 lbs). It is well known that the Tesla Model S or X Plaid models were the quickest-accelerating four-door cars available to consumers through 2023 (surpassed in 2024 by the Lucid Air Sapphire, weighing approximately 2500 kg (5500 lbs)). This was a major selling point for either car.

Figure 14.71 shows a photo of a crash test conducted by Swiss insurance company AXA. Two otherwise identical Volkswagen Golf sedans, one an ICV and the other an EV, underwent a head-on collision. Following a series of such tests, the insurer issued a statement that included: [142]

> Electric cars cause more damage in collisions than conventional cars—partly due to their incredible acceleration—and the increased weight of e-cars and trucks is a huge concern for occupants of lighter cars and will lead to increased pedestrian deaths, a Swiss insurance company announced after crash tests last month. … A look at the accident statistics of AXA Switzerland shows that drivers of electric cars cause 50 percent more collisions with damage to their own vehicles than those of conventional combustion engines

Fig. 14.71 Frontal crash test conducted by the Swiss insurance company AXA, between two nearly identical 2022 Volkswagen Golfs, one an ICE model, the other an EV model. Guess which one is the EV? (the yellow one on the left). Photo by Michael Buholzer/ AXA. Unrestricted use

The trend toward larger, heavier vehicles is an epidemic that is certainly not limited to EVs. As reported by Evercore ISI analysts on the Axios website *EVs-weight-safety-problems* [143]:

> The average weight of U.S. vehicles has already increased from about 3,400 pounds to 4,300 pounds over the last 30 years as Americans have ditched passenger cars for pickups.

In few discussions of safety standards and regulations during this time have the consequences of vehicle bloat been seriously considered. An early alarm was raised by the *Pedestrians, Bicycles, Human Factors Committee* (ACH00) [144] of the National Academy of Engineering Transportation Research Board, which maintains statistics on accidents involving bicycles, motorcycles, and pedestrians. They observed that a large increase in fatalities was due to *high-profile vehicles*, i.e., SUVs and trucks. Unlike passenger cars of limited height, the impact of a bicycle, motorcycle, or pedestrian with a high-profile vehicle results in total deceleration rather than the rider or pedestrian flying over the top of the vehicle, which reduces impact severity and increases surivability. Another self-defeating safety trend is the continued focus on the safety of the vehicle occupants, without concern for the harm to the occupants of other vehicles, pedestrians, or cyclists. This interpretation of vehicle safety further favors larger, heavier vehicles.

According to the NHTSA, *three-quarters of the vehicles manufactured in the USA in 2024 are trucks*, with the remaining one-quarter almost entirely SUVs and a small number of performance cars. As of 2024, *no compact or subcompact cars are produced in the USA*. Even in environmentally conscious California, 48% of registered vehicles are light trucks (over 85% in Texas and Wyoming) [145]. According to the National Automobile Dealers Association (NADA), in 2022, 80% of all vehicles sold in the USA were trucks or SUVs [146]. The Ford F250 truck is the largest-selling vehicle of any type in the USA. The cost of the increased harm to smaller vehicles, pedestrians, and cyclists is not included in the vehicle purchase price.

According to Bloomberg Hyperdrive, EV range has been increasing an average of 10% per year since 2018, leading to a commensurate increase in their average weight:

> From 2018 to 2022, the average range of fully electric vehicle models globally jumped from 143 miles (230 kilometers) to 210 miles (337 km). US figures are even higher due to the combination of larger vehicles, longer driving distances and the dominance of Tesla, which sells higher-range models. [147]

The most direct consequence of the EV “Range Race” is safety—especially for everyone other than the EV occupants. The June 2023 edition of the Atlantic Magazine summarized the situation, saying:

> … the relentless enlargement of American EVs is an ominous development for road safety, because added weight and height make cars more dangerous for anyone walking, biking, or inside smaller vehicles. Deaths among both pedestrians and cyclists recently reached 40-year highs in the U.S., and researchers have found vehicle size to be a cause. [148]

From National Public Radio Jan 11, 2023 [149]:

NTSB head warns of risks posed by heavy electric vehicles colliding with lighter cars

Jennifer Homendy, chairwoman of the National Transportation Safety Board, issued a statement on Oct. 3, 2019 warning about the safety ramifications of EVs, due to their mass alone. Other hazards include the lack of sound emissions, and high profiles of these vehicles which block vision of following drivers and make collision with a motorcycle or bicycle much more deadly.

Quoting NTSB Director Homendy:

The extra weight that EVs typically carry stems from the outsize mass of their batteries. To achieve 300 or more miles (480 or more kilometers) of range per charge from an EV, batteries have to weigh thousands of pounds.

"We have to be careful that we aren't also creating unintended consequences: More death on our roads," she said. "Safety, especially when it comes to new transportation policies and new technologies, cannot be overlooked."

Homendy noted that Ford's F-150 Lightning EV pickup is 2000–3000 lbs (900–1350 kg) heavier than the same model's combustion version.

Even apart from EVs, the nation's roads are crowded with heavy vehicles, thanks to a decade-long boom in sales of larger cars, trucks and SUVs that's led to extreme mismatches in collisions with smaller vehicles. But electric vehicles are typically much heavier than even the largest trucks and SUVs that are powered by gasoline or diesel.

Michael Brooks, executive director of the non-profit Center for Auto Safety, raised concerns about the weight of EVs because buyers seem to be demanding a range of 300 or more miles per charge, requiring heavy batteries:

Setting up a charging network to accommodate that may be a mistake from a safety perspective, Brooks said.

These bigger, heavier batteries are going to cause more damage," he said. "It's a simple matter of mass and speed."

Brooks said he knows of little research done on the safety risks of increasing vehicle weights. In 2011, the National Bureau of Economic Research published a paper that said being hit by a vehicle with ***an added 1,000 pounds increases by 47% the probability of being killed in a crash***.

Adrian Lund, director of the IIHS confirmed the obvious after extensive crash testing when he stated:

All things being equal, people in larger heavier cars will fare better in their crashes than people in smaller lighter cars [150].

His comment was not specific to EVs, but since an EV version of any car is heavier than the gasoline model, it is apropos. Unsaid but possibly more important to public health is that that buyers are heavily persuaded by the advertised “IIHS Crash Safety Rating” which as of 2022 were solely based upon the protection of the vehicle occupants, not the pedestrian, cyclist, or occupant of the other vehicle in a crash. Moms feel safer driving their kids to school in vehicles as massive as armored personnel carriers.

As an entirely new generation of electric vehicles replace existing ICVs, the weight increase will make the current safety situation even worse [151, 152].

Future EV Trend: Bigger and Badder

Current (2023) trends in EVs continue toward ever-greater weight, size, and power. Announced by Mr. Musk in 2020, and finally introduced in late 2024, the Tesla Cybertruck (Fig. 14.72) has quickly become a flashpoint for everything both good and bad about the new generation of electric vehicles.

2024 Tesla Cybertruck Official specifications from [153]

Weight	7000 lbs (3200 kg) reported by Motor Trend
Cargo	120.9 cu ft
Wheels	20”
Seating	5 Adults
Ground Clearance	17.44” in Extract Mode
Overall Width	Folded mirrors: 86.6”, Extended mirrors: 95”
Overall Height	70.5”
Overall Length	223.7”
Range (est.)	320 miles
Acceleration	2.6 s 0–60 mph
Drive	All-Wheel Drive
Top Speed	130 mph
Towing	11,000 lbs

Through the 1980s, light-duty (aka pickup) trucks were rated and named by their payload capacity, e.g., a Chevy 1500 had a 1500 lb (680 kg) safe load capacity, mostly determined by the rear axle rating. Traditionally, the tare (zero load) weight of an American pickup truck would usually be about the same as its payload

Fig. 14.72 3200 kg 2024 Tesla Cybertruck. Tesla Lineup of Vehicles, Cybertruck, Photo by Steve Jurvetson, CC BY 4.0 Deed Attribution 4.0 International

capacity, e.g., a Ford F-350 that weighed 4,700 lbs (2,132 kg) could safely carry up to 5,500 lbs (2,500 kg) of cargo. The GM electric Hummer weighs 9,700 lbs (4,400 kg) but has a strictly limited payload capacity of 1300 lbs. (590 kg). That is about the same as the passengers plus luggage of a typical midsize sedan. Are electric trucks cargo movers or actually overweight passenger cars to which a cosmetic cargo bed has been added just to merit the classification "truck"?

On the positive side of the extra weight of an EV is that the increased truck mass warrants a nearly proportional increase in towing capacity. It is not surprising that the 8000 lb Tesla Cybertruck has an 11,000 lb towing capacity.

Tesla has received a record-breaking number of customer orders for Cybertrucks, despite a record number of early-release criticisms in media about design and manufacturing issues [154]. Not exactly a truck for the working person, but that's not the reason people buy it.

The growth in the size and mass of *all* vehicles (EV and ICV) sold in the USA, China, and Western Europe affects more than safety and infrastructure requirements; it accomplishes the *opposite* of all environmental and energy sustainability objectives. The correlation between vehicle mass (inertial weight) and CO_2 emissions (proportional to fuel consumption) is shown in Fig. 14.73 for vehicles produced in 1978 vs 2020 [155]. It is clear that vehicle weight correlates with CO_2 emissions, which are proportional to MPGe regardless of the vehicle type or age. A traditional automotive "rule of thumb" [156] is

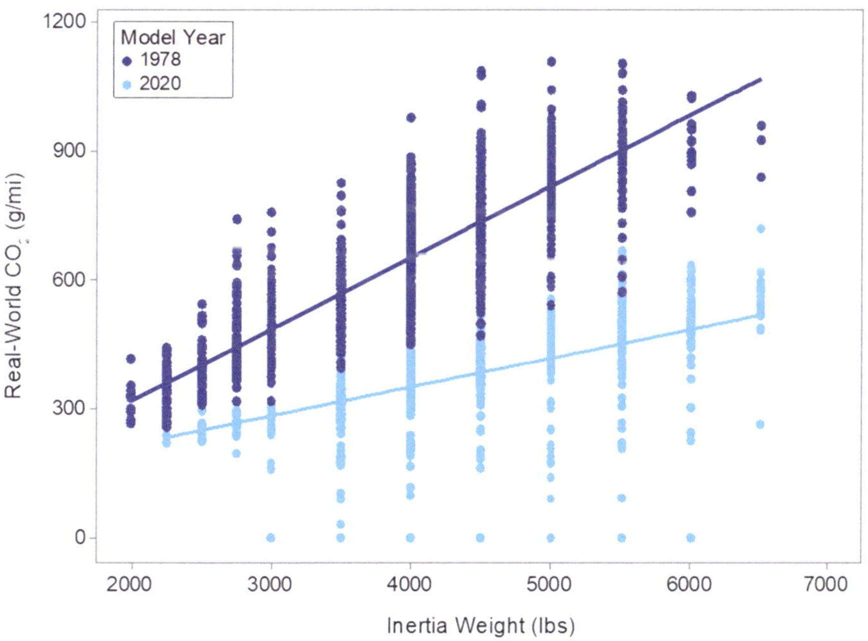

Fig. 14.73 CO_2 emission as a function of vehicle weight. US EPA. Public domain

for every additional 100 pounds, fuel economy typically decreases by 1-2%. So a vehicle that is 1000 pounds heavier than another similar model could see its MPG reduced by 10-20% due to the weight differential alone.

One partially redeeming attribute of EVs is that due to regenerative braking, the city (urban EPA test) mileage is less dependent upon vehicle mass, since regardless of mass, about the same percentage of the braking energy is recovered. The same is true of ICE hybrids.

Not a Matter of Consumer Need

Jessica Caldwell, executive director of insights at Edmunds Car Reviews [157] asked on their website:

> Why Is Everyone Building an Electric Pickup Truck? "It's not like people have been asking for this."
>
> For automakers, pickups are a great opportunity: They have high margins and are more profitable than most other passenger vehicles. The rash of startups making electric pickups and SUVs is not an accident. Their higher prices make it easier to "hide" the up-front costs of research, development, and batteries than with a cheaper sedan or compact. [158]

The skewed production of more massive EVs compared with ICVs (available in the US market), is illustrated by (Fig. 14.74) copied with permission from [159]. The "Green Score" is the ACEEE *Greener Cars Rating*, which factors in both operational and production energy use and CO_{2e} emissions.

On May 4, 2023 an article by Oliver Milman of the Guardian USA observed:

> … the ballooning size of electric vehicles, crowding out smaller, more affordable models that strip fewer resources from the environment …
>
> … General Motors, which aims to sell 1m EVs in the US by 2025, said that the Michigan plant currently churning out Bolts will switch to new electric models of the Silverado and the GMC Sierra – hulking, and more expensive, alternatives that will probably provide the auto company a greater financial return than the modest Bolt.
>
> The EV market is now almost entirely dominated by large, luxury, expensive vehicles," she said. "That isn't really helping low-income people and those on the frontlines of polluting facilities and climate change.

According to J.D. Power [160] in December 2023, trucks/SUVs accounted for 81% of new-vehicle retail sales in the USA. In 2013, this number was 52.1% [161]. This represents an increase over just 10 years of 55%, with profit margins higher for larger vehicles (Fig. 14.74).

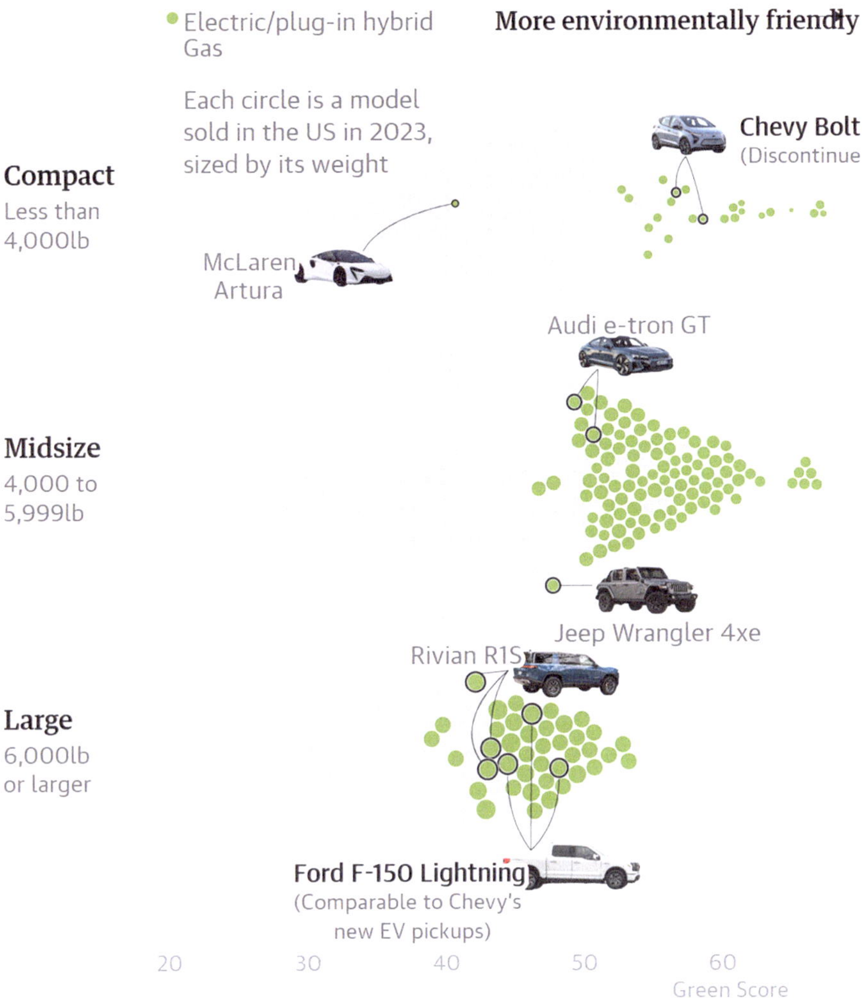

Fig. 14.74 Illustration from "Large electric vehicles are worse for the environment than some smaller, gas-fueled cars" Graphic from Car and Drive Magazine, December 2023 with permission. Data from https://www.theguardian.com/business/2023/may/04/electric-vehicles-suvs-us-vehicle-fleet

E-Bikes and E-Scooters

Less than 10 years ago, electric motor-assisted bicycles (e-bikes) and scooters (e-scooters) were novel and used mostly for recreation rather than for practical transportation. As of 2023, that situation has radically changed, with e-bikes and scooters ubiquitous in the daily urban and suburban vehicle mix, representing the largest growing transportation sector. These low-cost unlicensed vehicles, which can be ridden by anyone without a driver's license, already exceed the number of

Fig. 14.75 A new canvas for innovation and self-expression. The nCycle e-bike, designed in 2014 by Hussain Almossawi & Marin Myftiu. https://upload.wikimedia.org/wikipedia/commons/3/3e/NCycle_e--bike_%282014%29.jpg CC-BY-SA-4.0

motorcycles on the road. They have and will continue to serve as an environmentally benign alternative to traditional fully powered transportation, and are incentivized to a greater percentage of the purchase cost than any other vehicle type (Fig. 14.75). According to Forbes (4 Oct 2023):

> With a boost from a growing number of cities and states, the souped-up two-wheelers are increasingly displacing cars for short trips—and outselling EVs.

This new entry in the urban transportation cloud is not without growing pains. Not as fast as motorcycles, but potentially much faster than pedal-powered bicycles, their abrupt arrival has confounded local and state traffic laws and infrastructure demands. But since they displace ICVs and 4-wheel EVs while encouraging healthier lifestyles among formerly sedentary drivers, their benefits are undeniable. The reduction in the need for 4-wheel vehicle parking is also a huge benefit from an urban planning point of view.

This class of vehicles is considered in most municipalities as a subset of human-powered bikes, and subject to laws regulating bicycles. If operated at low speeds in designated bike lanes under mostly human power, this designation is justified. But that is not how most are operated; they are often ridden at much higher speeds than bicycles on both roadways and sidewalks and parked in both bike racks and motorcycle parking areas. They are ridden aggressively on dirt trails previously the exclusive realm of hikers and human-powered mountain bikes [162].

According to CBS News, referring just to the USA [163]:

> Viewed as eco-friendly (studies have found that's not exactly the case) and cost-effective modes of transportation for short distances, the battery-powered products were part of the equation in more than 190,000 E.R. visits and at least 71 deaths from 2017 through 2020, according to the federal agency.

The lack of restrictions and speed difference, combined with rapidly growing numbers of usually less-experienced riders has created a much greater risk of collisions with both motor vehicles and pedestrians. In a reaction to subscription e-scooters, e.g., the Bird Scooter franchise launched in San Francisco and expanded to several other major cities, local ordinances have been hastily drafted to restrict or even ban their use, following a near-epidemic of scooter-pedestrian collisions. Some universities and commercial centers have restricted parking to designated motorcycle lots. And recently there have been increasing cases of battery fires during battery charging, which can potentially be more dangerous than EV battery fires since e-bike and e-scooter batteries are often charged indoors [164].

US industry guidelines, enforced in their specifications only, limit the motor-assisted speed to 28 mph, but with the combined input of the rider and the motor, speeds much greater than this are possible, especially on downhill grades. Consider the extreme case of the SWIND EB-01, sold since 2018, a British manufactured e-bike billed as a "hyper bicycle" with a 15kW (20 HP) motor and an official top speed of 60 mph, but potentially as high as 80 mph with simple electrical modifications. Yet it is classified as an e-assist mountain bike, not even a "moped," the traditional designation for ICE/human-powered hybrid two-wheeled vehicles [165].

Even a casual observation of current transportation trends will lead to the conclusion that safety and parking issues alone will dictate significant changes to surface streets and roads, which were traditionally designed to accommodate only motor vehicles and pedestrians, with bicycles as an afterthought [166]. This transition was inevitable and most cyclists agree that it is long overdue. In the USA, where automobiles are by far the most dominant form of personal transportation, 49,000 bicyclists were injured in crashes involving automobiles in 2019 [167].

The challenge is obvious: there is only so much road right-of-way, and this infrastructure has been in place for many decades. Electric motor-assisted bicycles and scooters represent a fifth vehicle speed class of roadway users: automobiles, trucks and motorcycles (25–75 mph), pedestrians (0–3 mph), human-powered bicycles (3–20 mph), e-bikes/scooters (5–40 mph), and skateboards and skates (0–10 mph).

The usual solution, forcing e-assisted human-powered vehicles onto the vehicular motorway or into bike lanes puts their riders at great risk since they are quick, numerous, unpredictable, and difficult for motor vehicle drivers to see. Yet sidewalks in some urban areas have become danger zones for pedestrians due to their use by e-assist riders, mostly e-scooters (Fig. 14.76). According to the IIHS [168]:

> Broadly speaking, the researchers found that e-scooter riders suffered injuries more frequently per mile traveled than bicyclists, but bicyclists were 3 times as likely as scooter riders to be hit by motor vehicles. In contrast, e-scooter riders were twice as likely as bicyclists to get injured because of a pothole or crack in the pavement or other infrastructure like a signpost or curb. … Cities like Denver and San Antonio have banned e-scooters from the sidewalks altogether. … The picture is still not clear when it comes to where scooters should be ridden.
>
> Our results suggest that moving scooters off the sidewalk could put riders at risk of more severe injuries, but as things stand they might be suffering these lesser injuries more often.

The e-bike/e-scooter issue is complex, with no ideal solutions for safe accommodation on roadways. This situation serves as one of many common examples of the impacts of technical innovations on society and the legal system. Without

Fig. 14.76 Two quiet vehicles: Stock photo (possibly staged) of collision between a Nissan Leaf EV and e-scooter, 2020. Image found at https://fundcapitalamerica.com/wp-/uploads/2022/06/Electric-Scooter-Accident.jpg. Public image upload, unknown origin

broad-scale changes in roadway design and traffic laws, this will continue to be an area of inconsistency between environmental benefit and risk of injury or death. The worst possible outcome is starting to play out in some early-opting cities: restrictions on e-assist vehicles that have become so egregious as to discourage the use of these otherwise positive alternatives to traditional automobiles.

Electric Vehicles vs. the Electric Power Grid

The proliferation of EVs as replacements for gasoline and diesel vehicles obviously depends on access to reliable utility electricity at adequate power levels, either continuously or at least with a predictable availability schedule. Even in relatively affluent areas, aging or poorly maintained power distribution infrastructure remains a major concern. In many areas in the USA, the electrical distribution is already overloaded.

A simplistic estimate to frame the challenge of powering EVs with grid electricity: Most published analyses assume that in the USA, the average EV will accrue 13,500 miles/year with an efficiency of 3-4 miles/kWh. This totals 3.9 MWh per year. An average US home uses 10 MWh/year. If on average, every household had one EV charged at home, the annual electric energy use for the home and the residential area would increase by 39%. A typical utility can handle an annual load increase of 1–3% using existing infrastructure and reserve margins [169]. More than that will require new facilities.

But the problem is not this simple. This is just annualized average energy use. Its impact on the grid depends on the power delivered at any given moment. We examine this reality below.

In the USA, the electric grid infrastructure is woefully out of date, most of it constructed nearly 75 years ago. In 2023, more than 1.2 million EVs we sold in the US, while there has been only minimal public investment in infrastructure. Fires in the Western USA and Canada have created an electric transmission and distribution crisis even at current load levels, with few EVs compared with cars. The infrastructure investment of the 2022 Inflation Reduction Act has only barely helped to reduce existing risks and inadequacies, not yet at the point of addressing the whole shift of transportation and domestic energy use from fossil fuels to electricity. Possibly worse is the public's lack of awareness of the severity of the problem and the extent to which it is compounded by the incremental load of every new electric vehicle and every electric heat pump that replaces natural gas heating [170]. Since about 2015, electric power grid failures and planned outages have become the new norm, not seen in the USA since the mid-twentieth century.

In the power industry, the triage has prioritized generation capacity and transmission, both with the broadest impact. An independent analysis by the energy industry watchdog *EnergyHub* [169] study concluded that at the moment,

> the bulk grid is in relatively good shape when it comes to meeting rising electricity demand, while the country's aging distribution infrastructure provides more cause for concern.

The electric grid can be subdivided into three infrastructure components [169, 171]:

Bulk electric system: the centralized generation facilities, grid management and electric energy storage facilities.

Transmission networks: high-voltage long-distance electricity exchange over six regional interconnect network operators in the USA, Canada and Mexico.

Distribution networks: The power poles, transmission wires, substations, and secondary transformers that provide power to individual users.

Numbers vary for the impact of a new EV on neighborhood electric distribution [172]. But a crude calculation might assume 13,500 miles per year traveled, all on weekdays, in an EV that achieves an average of 3.5 mile/kWh. This equilibrates to approximately 15 kWh to charge the EV on each weekday, ignoring efficiency losses. For comparison, the average daily electric energy use per household in California, where the majority of US EVs are located, is 18 kWh/day. So as a back-of-envelope estimate, the EV would approximately double the household energy use on weekdays. If all homes had an EV with s similar use and charging schedule, the EVs would have nearly the same effect as doubling the number of residences in a neighborhood. In newer suburbs, this may not be an issue. In areas of older electric infrastructure, it could create an overload crisis.

What matters in electric distribution more than the average energy (kWh) is the maximum (peak) power or load (kW), which may only be needed for 1 or 2 h each day. The need must be met by the utility regardless of its energy resources at that time. The time of day that the vehicle is charged matters greatly (Fig. 14.77). Prior to the proliferation of home solar PV installations, transferring as much load to late night was the best load-balancing strategy. This was reflected in optional time-of-use power plans offered to power customers. For homes that have grid-tie solar PV systems and an EV to charge, the scheduling becomes a bit more complex, since the solar system produces peak power mid-day, *the time when a commuter vehicle is not at home.* Widespread adoption of electric vehicles can either enhance or harm grid stability, entirely depending on when they are charged. Any time of day other than

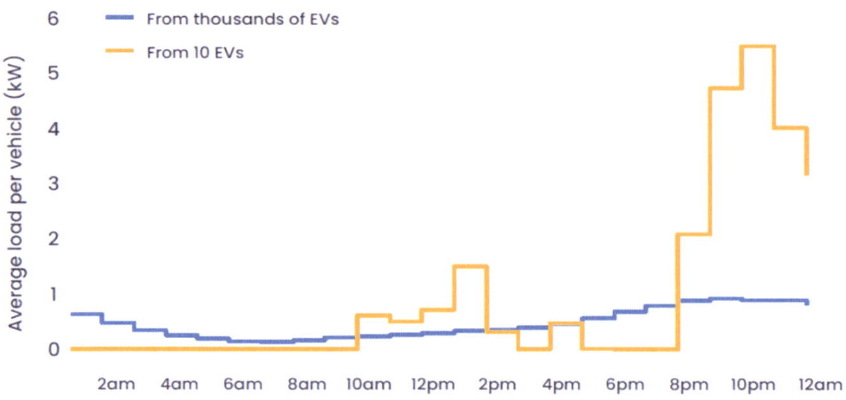

Fig. 14.77 Results of study of EV charging load vs. time of day in the USA. From citation [185], Academic fair use

mid-morning to mid-afternoon, solar is unavailable and wind is uncertain. The result is the energy demand *duck curve*, an example appearing in Fig. 14.92 of a later section, for a typical weekday of electrical demand in Hawaii, provided by the *Hawaii Electric Company* (HECO). During sunlight hours, demand on the utility drops, sometimes even becoming negative. Peak demand occurs 4–8 PM, when intermittent generators are offline. 5:30 PM is the worst time of day for utility grid load. But an EV commuter returning home may have a need to charge the car for use later in the evening, or even just maintain a charging routine that works for all household users of the EV. In this situation, the EV contributes to the overload of distribution and generation.

Figure 14.78 shows the result of a study of electric vehicle charging loads over a two-day (weekday + weekend) period, performed by the Northwest Power and Conservation Council in 2021 [173]. It makes clear the peak load problem, the largest charging load occurring between 5:00 PM and midnight on weekdays.

Secondary Transformers: Low-Hanging Fruit in the Grid Failure Tree

Another widely held assumption is that the lack of transmission ampacity will be the major limiting factor for home or local public EV charging. This is indeed a limiting factor. But it has been known for the past 25 years that the most vulnerable areas of the grid are neighborhood distribution. There are over 50 million distribution transformers in the USA alone. Over two-thirds of these substation and secondary transformers are more than 25 years old [171]. Of special concern are the *secondary transformers,* each providing power to between 1 and 16 homes. In the USA, older transformers were almost always pole-mounted as shown in Fig. 14.79, while newer developments with underground utilities have ground-level transformer cabinets. Many of the transformers have been in service for decades, some past their

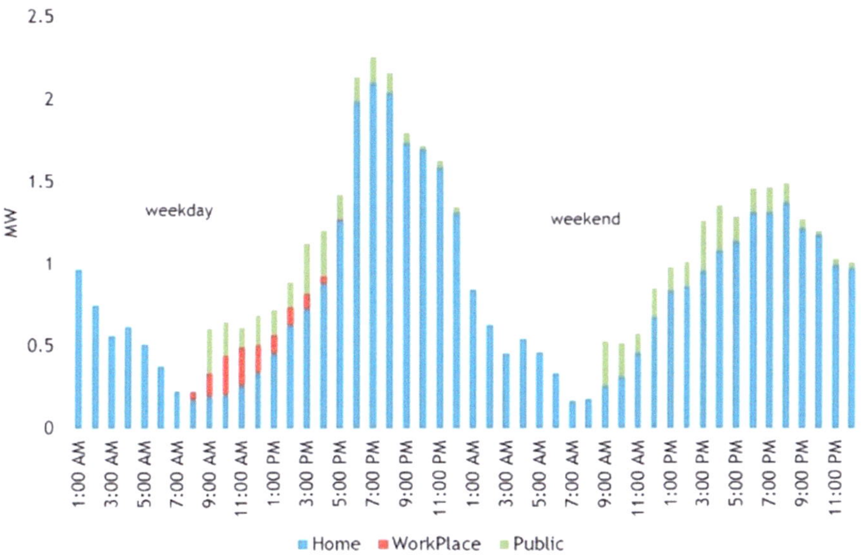

Fig. 14.78 Electric vehicle charging loads over a 2-day period. Peak charging load on either day occurs at 7:00 PM, the worst possible time for loading the electric grid. From Northwest Power and Conservation Council https://www.nwcouncil.org/2021Power and Conservation Council https://www.nwcouncil.org/2021powerplan_plug-electric-load-profiles/ Public domain

Fig. 14.79 Secondary distribution transformers are at risk of overload when an EV is charged at home. This transformer shows discoloration that may have been caused by overheating or leakage of transformer oil. Photo: author

design lifetime of 20–30 years. Increasing suburban housing density places additional loads on the transformers. These critical components were often neglected even before the first EV showed up. The engineering philosophy of electric utilities prior to the introduction of grid-tie solar PV and EVs was to allow limited overload during the day, assuming that it will cool off during the night. But if an EV is charged during the night, the nighttime load will likely exceed even the daytime demand, doubling the daily demand and invalidating this assumption.

The overwhelming majority of home solar PV systems worldwide are grid-connected and do not include battery energy storage. During peak sun hours mid-day, they push power to the grid in the opposite direction for which the secondary transformers were designed. In the USA, most utilities deliver power to the secondary transformer at between 4.4 kV and 44 kV. The transformer reduces this high voltage to the US standard 115/230 VAC AC delivered to each user. When excess local solar generation is being sourced by a grid-tie solar inverter to the grid, instead of stepping down this high voltage, the transformer must step up 230 VAC to the high distribution voltage. Theoretically, a transformer should work as well in one direction as the other. But secondary transformers are optimized to work in step-down mode. Operation in the reverse direction may be only slightly more lossy, but it can have an impact on efficiency and equipment lifetime [174].

And one other subtle issue for grid-tie systems: For the excess solar power stored in the grid, the energy transferred through the transformer can theoretically double as a result of solar feed to the grid and use later: power must flow through it first in one direction while a solar system is pushing energy to the grid, and then back when the energy is used later. This is not much of a problem if there are other households without solar energy connected to the transformer, since they will use the power fed to the transformer, but every watt that is pushed backwards through the transformer is an additional source of transfomer heating that does not show up when net power use is calculated. Local battery storage is the best solution to this problem, but solar batteries rarely have the capacity to even partially charge an EV during peak demand hours in addition to supplementing the peak household load during that time.

One other well-intended but problematic trend that will affect local distribution and distribution transformers is the growing number of municipalities that are banning or restricting the use of natural gas in homes, based on the well-intended but questionable assumption that if they stop using it for home heating and cooking by adopting electric heat pumps and induction cook tops, it will reduce natural gas production and therefore reduce global warming. There are valid arguments for and against this movement, but most involve considerations unrelated to environmental or energy sustainability. The impact of transferring the entirety of the energy supplied to homes to electric power will be profound, since on average, more than half of the energy used in a home that has gas service is from the natural gas, a choice that has always been favored because of its lower cost compared with electricity. Cutting off natural gas service can as much as double the electrical demand of a home. The combination of one or more home-charged EVs, an all-electric home, and a grid-tie solar PV system without storage can increase the home's stress on the grid by a large factor, compounding existing problems with aging distribution infrastructure starting with the home's local distribution transformer.

It is estimated that in California alone, which generally has newer electrical infrastructure than eastern US states, the cost to update the electrical distribution to accommodate EVs may be up to $20 billion USD. The metastudy report at citation [175] concludes:

> Charging electric vehicles at home will exceed most power lines' capacity.

If this incremental increase in electrification were to play out gradually over at least a decade, a long-term plan could ease the transition. But with recently announced targets for electric vehicle market dominance as well the goal of full home electrification by 2030 (California), the abrupt transition is a potential crisis in the making. Infrastructure improvements usually have long time constants, e.g., roadway traffic hazards are usually fixed or upgraded only after they cause a serious accident. The "hazard" in this case is far more serious.

Worldwide, it is estimated that at least 10% of the population has no electric service at all, and a considerably higher percentage lacks reliable electricity of adequate capacity [176]. Their ability to support the electrification of automobiles is seriously lacking. The obvious scenarios of impoverished urban citizens in need of inexpensive transportation in Latin America, Africa, or Southeast Asia usually come to mind. But the affected population is actually much larger, including many low-income populations in North and South America, Europe, Asia, and Australia.

In remote areas worldwide with weak or no utility electricity, EVs are sometimes charged by diesel generators. And vacationers visiting off-highway or wilderness areas often bring a portable gasoline generator in case they have an unexpected need to charge their "off-road" electric SUVs. This is an interesting dichotomy in the pristine wilderness.

This rather negative assessment of our readiness for full electrification is not intended to discourage the transition of transportation to electricity, but to serve as a reminder of the magnitude of the infrastructure development task necessary to fulfill easily-made political or corporate promises such as "all-electric by 2030."

EV Charging Standards and Infrastructure Requirements

Up until recently (2023), the majority of the charging energy dispensed to personally owned BEVs and PHEVs came from SAE J1772 Level 2 (230 VAC, 50A max) home chargers, usually installed at the time of purchase of the vehicle. 230 VAC, 200 Amp single phase power is the electrical service found in almost all newer US homes for high-power loads such as electric heating and cooling, electric dryers, water heaters, cooktops, and ranges. Or in most of the rest of the world, it is the standard outlet voltage. The majority of EV charging occurs at night, with the vehicle parked in a garage or carport. The Level 2 battery charger is onboard the vehicle so that the wall-mount "charger" is really just a safety interlock for switching on or off the 230VAC charging connector following successful communications with the vehicle's charger.

But the vehicle range limitations combined with the unpredictability needs on a given day necessitate public charging facilities. Most public chargers (charging stations) are also Level 2, which is adequate for charging 60–70% of the battery capacity in 4–6 hours. For public chargers, the user is billed, usually on a kWh energy use basis, for the charging energy used, although time limitations and time-on-charger are often incorporated in the charging port's algorithm. Payment is usually made by credit card, or it is covered by some prepaid or free-with-vehicle-purchase plan.

But with increasingly large batteries and the expectation of longer travel distances between charging stops, higher power charging is now a necessity. A minority but increasing number of public chargers are Level 3 "fast chargers" which use a much larger battery charger that are located outside the vehicle. For these, power is transferred to the vehicle in the form of DC at high currents. Charging at high rates, typically 100–200 kW, can greatly reduce the battery charging time, which may be critical to consumer acceptance when electric vehicles are used for trips longer than the single-charge round-trip vehicle range. The largest cost and greatest limitation for Level 3 chargers is the need for grid connections capable of sourcing high power. Each charging connection can easily exceed the existing load of an entire neighborhood. This almost always requires the installation of a dedicated higher capacity grid secondary (local) transformer and sometimes upgrades to the distribution lines and switchgear. This does not even include the inevitable need to increase electric generation capacity as more electric vehicles must be charged. Power must be available at any time needed for charging, which usually does not align with the output periods of intermittent generators such as solar or wind. This means either increased dispatchable power generation using fossil fuels or nuclear, or the addition of local or grid-scale battery or pumped hydro energy storage.

The notion of "fast charging" is highly variable, usually referring to any charge rate above that of Level 2, although it isn't uncommon to hear Level 2 chargers referred to as fast chargers simply because they are faster than Level 1 chargers that use 115VAC power. At the high end, Level 3 includes ultra-high-rate chargers such as Tesla Superchargers (V3 or V4), which dispense DC power at high voltage (400–800 VDC) at rates up to 250 kW, to theoretically charge to 80% a 100 kWh battery in as little as 20 min, or a 40 kWh partial fill in about 10 min. For perspective, the maximum possible power that can be delivered by the 200A 230 VAC service of a modern home is 44 kW. Fast charging cannot be done at a residence.

Recent battery technology developments to increase the charge acceptance rate of EV batteries promise even greater battery charge rates and shorter charge times.

Battery charge rates (called C-rate, measured in RMS amps) are usually stated as a fraction or multiple of the battery's amp-hour capacity.

$$\text{Normalized charge rate } C = \text{charge rate}\left(\text{amps}\right) / \text{battery capacity}\left(\text{amp} - \text{hours}\right)$$

For example, a nominally 400 Volt, 65 kWh battery would have an amp-hour rating of 163 Ah. Charging the battery at 163 amps or 65 kW would be a C-rate of C1.

Almost all newer electric vehicles include liquid battery cooling during charging to mitigate the waste heat generated at high currents. Nevertheless, contrary to assurances from some EV makers or marketers, high-rate charging is always at least slightly worse for the battery than slower charging, and the heat generated during charging is wasted energy, that shows up as reduced range or lower MPGe. Temperature management during fast charging is paramount to avoid damage to the battery or fires. Liquid-cooled lithium batteries that are maintained in a narrow temperature range of around 30 °C can usually accept charge rates as high as C3 if the charge current is actively managed based on the temperature, the state of charge (SoC), and internal parameters such as battery impedance. Figure 14.80 shows the charge power profile for the 800V battery in a Tesla Cybertruck while being charged at a 250 kW Tesla Supercharger. Note that the 250kW charger is only providing that power level for approximately the first 12 minutes of charging.

Battery aging mechanisms are described in citation [177]. The state of the art in battery design is exemplified by Fig. 14.81 that examines the impact of fast charging on battery life. A Tesla Model 3 LMNC battery charged 90% of the time at "high rate" chargers is compared with high-rate charging only 10% of the time. The plot is intended to demonstrate in this survey that, averaged over 6300 vehicles, the difference in battery degradation as measured by user range was nearly inconsequential.

Another tradeoff encountered in the optimization of the battery is between its acceptable charging C-rate and its self-discharge rate—the amount of battery capacity lost when the battery is charged but unused. A typical battery self-discharge rate for lithium-ion EV batteries is 2–3% of capacity per month. Batteries designed to accept high charge and discharge rates usually have lower internal resistances, which leads to higher internal discharge rates, typically 3–5% per month [178].

For high-rate charging, the waste heat generated is higher than with slow charging, so the charging losses are greater. This can be explained simplistically by considering the simple electrical battery model of a lossless voltage source in series with the battery internal resistance. Power loss in the form of heat generated in the

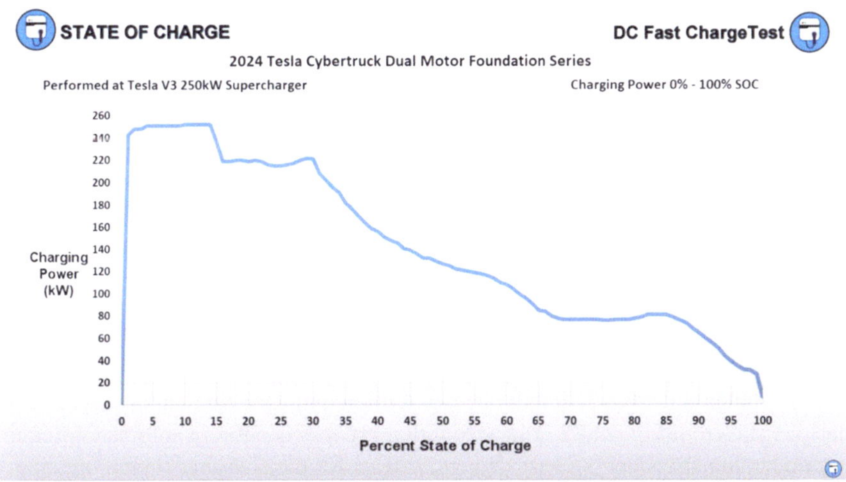

Fig. 14.80 Measured charge profile of a Tesla Cybertruck battery. Data and graphic from Tom Moloughney of InsideEVs, 30 May 2024. Online at https://insideevs.com/news/721583/tesla-cybertruck-v3-supercharger-test/. Academic fair use

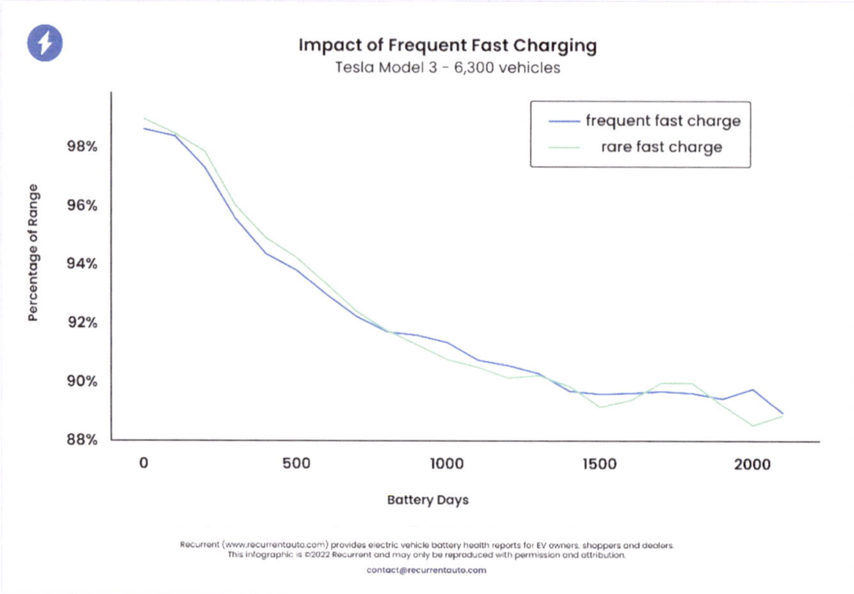

Fig. 14.81 Impact of frequent fast charging with infrequent fast charging on Tesla Model 3 battery. Data and graphic from Recurrent Automotive, 2022. https://cdn.motor1.com/images/custom/64ebcd668e5ecaab5f3b15cb-tesla-model-3-3.jpg. Academic fair use

resistance increases as the square of the current. Doubling the current reduces the charging time by half, but produces four times the wasted heat energy during charging.

The higher cost of the equipment and infrastructure required for fast charging justifies a higher cost for the energy delivered. In California (Oct 2023), the rate is $0.60/kWh for Tesla Supercharger stations, while residential SC-1 base rates are $0.12–$0.14 per kWh. It may be interesting to note that for an EV with an energy use rate of 3 miles per kWh, $0.60/kWh is cost-equivalent to an ICV that gets 25 mpg at a $5/gallon of gasoline. In California, EV charging network providers are supposed to self-report aggregate energy use to the California Air Resources Board, and (optionally) the National Renewable Energy Laboratory (NREL). But public EV charging rates are not regulated by any state or federal agency, which may explain why fast-charge rates have risen to the level that is equivalent to or may even exceed the per-mile gasoline cost of a not-particularly-efficient ICV.

The billing/cost recovery mechanisms for public EV chargers are actually no more complex than those used to pay for gasoline with a credit card. A typical scenario is shown in Fig. 14.82. The accounting process is entirely automated, requiring no human attendant—a cost savings, at least for the charging station operator.

Since government incentives support the installation costs but (usually) not the maintenance costs of privately-owned public chargers, these have a less-than-satisfactory record for reliability. They are often treated as single-cost investment/

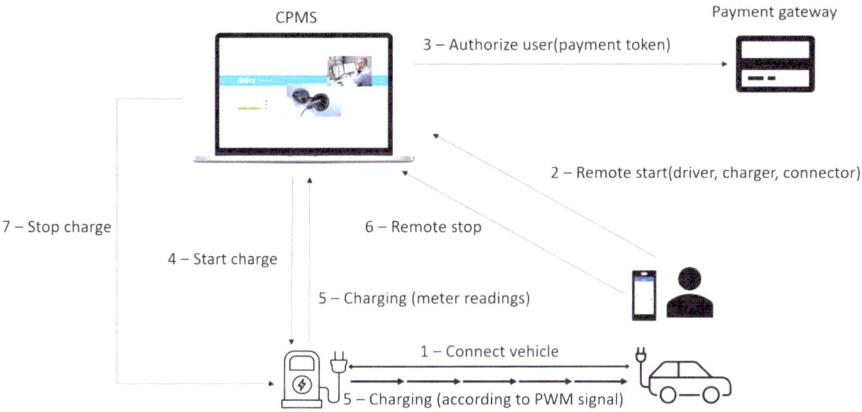

Fig. 14.82 Financial data paths for public EV chargers. Graphic from Jeff St. John, "What's behind the epidemic of unreliable EV chargers?" 12 December 2023. https://www.canarymedia.com/articles/ev-charging/whats-behind-the-epidemic-of-unreliable-ev-chargers Academic fair use

subsidy harvesting opportunities, lacking a commitment to continuous uptime despite (unmonitored) grant requirements for 97% uptime in California. Probably the most common complaints heard on general EV owner forums are frustrations with "out-of-service" chargers that EV drivers were relying upon to complete a trip. There is apparently also a problem with how operators report the reliability of chargers. For example Tesla, in its 2022 Impact Report, proudly reported that its supercharging network maintained near-perfect reliability in 2022. As reported by Inside EVs [179],

> The company says that the average uptime of Supercharger sites last year amounted to 99.95 percent (down marginally from 99.96 percent in 2021), but was higher than in 2018-2020. That sounds awesome and Tesla entitled this part of its presentation 'Chargers that just work.'
>
> However, the key is the methodology. Tesla defines the uptime of Supercharger sites as the average percentage of sites globally that had at least 50 percent daily capacity functional averaged over the year: "Uptime of Supercharger sites reflects the average percentage of sites globally that had at least 50% of their daily capacity functional for the year."

If the same metric were used for other critical consumer services such as cellular communications, an average as high as 49% of the network could be nonoperational at any given time while still earning a 100% uptime rating.

Provisions in the US 2022 IRA include specific requirements for the reliability of charging network facilities that received IRA funding, although it still lacks provisions for verification or enforcement [180].

Another (completely unnecessary) issue that makes life difficult for EV owners is the proprietary control of battery charging connectors and communications protocols, despite identical functions and the same power delivery. The debate over which type of EV charging plug will become the eventual de facto standard has limited access in the US and International fast charger markets. Multiple EV charging standards have been adopted worldwide by different EV manufacturers since the 1990s. But the major competition in the USA is between Tesla's proprietary North American Charging Standard (NACS) and the SAE CCS (Combined Charging System) standard used by most EVs sold in the USA and the EU. Tesla is capitalizing on their widespread charger deployments, the largest in the USA, in an effort to force all manufacturers to license their connector standard. As of September 2023, it appears that Tesla has won this debate via its financial clout and has made concessions allowing some of its chargers to be accessible to non-Tesla vehicles (using an adaptor and the Tesla cell phone app only), and by GM and Ford's decision to change to the Tesla interface—a move that may improve market share in the USA but certainly not in the rest of the world. Tesla's position also diminishes the authority of the SAE and other consensus-based professional automotive standard setting organizations, a conflict-generating precedent considering that tens of thousands of automotive components comply with SAE standards worldwide, to the benefit of both manufacturers and vehicle owners. The standardization of EV charging connectors and protocols is one of the objectives of the 2023 US IRA [181]. Technically, there is no advantage or disadvantage to either the SAE CCS or the Tesla NACS. It is just an issue of market control.

Daytime Charging Using Destination Solar Charging Facilities

Any mechanism by which EVs may be charged during daylight hours will benefit grid stability by matching solar availability with concurrent power demand. Midday charging at the workplace or other daytime destinations is one such way to transfer the high demand for EV charging to the time of excess electric capacity. Even better, if the source of the electric power is collocated solar PV, the impact on the grid is even less. Destination Solar Charging Canopies (or structures) are an ideal solution, providing an *hour-by-hour* match between renewable energy availability and demand—a term recently appearing in (US) incentive requirements that were previously based only on counting net kWh generation without regard to usability.

Unfortunately, many (most) of the solar canopies funded by incentives do not include EV charge ports; they are just outdoor grid-tie solar arrays. Those that do, however, can provide at least a fraction of the EV charging energy directly from a PV array mounted on a shade structure. However, the grid stability benefits of these structures should not be overstated, even for the minority that do include EV chargers. For example, for arrays such as shown in Fig. 14.83 that are integrated with vehicle charging, proud system owners like to promote the idea that cars are being entirely charged by the solar energy array. But the surface area of the array cannot even come close to providing the power required for continuous operation of even one of the vehicle chargers, especially if any of them are high-rate (e.g., 120+ kW) DC chargers. But some benefit is better than none. This might be more accurately described as a public charging station that happens to have a sun shade that is a grid-tie PV array. Regardless of the small solar capacity compared with the charging

Fig. 14.83 Lumos Solar Canopy EV Charger installed at workplace. Image from https://lumossolar.com/2015714nrg-evgo-features-lsx-modules-at-nrg-stadium/img-jpg-41/. With permission

load, *any transfer of EV charging load from non-sun hours to daytime* will help to reduce the differential load on the grid between night and day. This is one of the few situations in which EVs and solar PV are in synch.

Battery Exchange as an Alternative to In-Vehicle Charging

An Old Mechanical Solution to a New Electrical Problem

The limited range and long recharge time of battery EVs have limited their use for over 100 years. The alternative of simply exchanging the battery rather than recharging it in the vehicle has been around as long as EVs. Possibly the first successful solution to the continuous operation of electric vehicles was the exchange of a discharged vehicle battery for a fully charged one. In fleet applications with central refueling facilities, battery exchange (swapping) has been the lifeline that has kept specialized EVs in operation long before the development of advanced batteries. Battery exchange has been used for vehicles that must be operated indoors or in other enclosed spaces such as warehouses or underground mines [182].

Some early examples include electric taxis (aka hacks) in Paris as shown in Fig. 14.84 from Scientific American, 1899, or electric taxis in New York shown in Fig. 14.85 from Scientific American, 1896. Six feasible battery exchange configurations are illustrated in Fig. 14.86. A review of automated and manual rapid battery exchange technologies can be found in citation [183]

Another of several variations of battery exhange during the first renaissance of electric vehicles is shown in Fig. 14.87, the semi-automated exchange of a large battery pack from a trailer towed behind an electric bus. Twenty electrified type SL-E M.A.N. transit buses were placed in passenger service in various cities in West Germany 1974 through 1981. Each battery provided a range of 40km. Several battery exchanges were performed during each 14-h service shift, with each exchange reported to take 5–8 min to complete. Collectively, the 20 buses were operated successfully at a total distance greater than 4.5 million km [184].

Figure 14.88 shows a 1991 design study for an exchangeable battery urban bus system commissioned by Pacific Gas and Electric Co, and the California Energy Commission. Several modular battery packs are charged in the below-pavement exchange apparatus. Discharged packs from the bus are automatically replaced with fully-charged packs by the exchanged apparatus. An above-ground mock-up of the

Fig. 14.84 One of over 1000 Electric "Hacks" Serviced at the Rue Cardinet Battery Exchange and Charging station, Paris, 1899. From Scientific American 1899, with permission

Fig. 14.85 New York City Electric Taxi and battery Exchange Station, 1896. From Scientific American, 1899, with permission

rapid battery exchange (or Interchange) aka RBI or RBX system was constructed and demonstrated in 1992 and became known as the "Coke® Machine," because of the way it serially inventoried multiple batteries while charging them, dispensing "full" batteries and accepted "empties" that are returned to the charging queue or sorted out as defective. The batteries are coded with active RFID tags containing their individual history, used by the exchange apparatus to help optimize individual charging and automatically identify end-of-life batteries which are diverted to a recycling queue.

Automated battery exchange is only sensible for recharging/repowering identical vehicles equipped with receptacles to allow fully automated battery removal and insertion by a specific apparatus. Attempts by multiple startup companies in the 2000s to commercialize battery exchange for the general EV population all eventually failed despite absorbing large pools of investor and government funding [185].[21] But it remains a nearly ideal solution for applications in which one or more identical electric buses or specialized vehicles travel in a fixed continuous loop, repowered by

[21] Failed battery exchange startup companies 1995-2020 have included *A Better Place* (Israel/China), Nio (China) and Ample (USA).

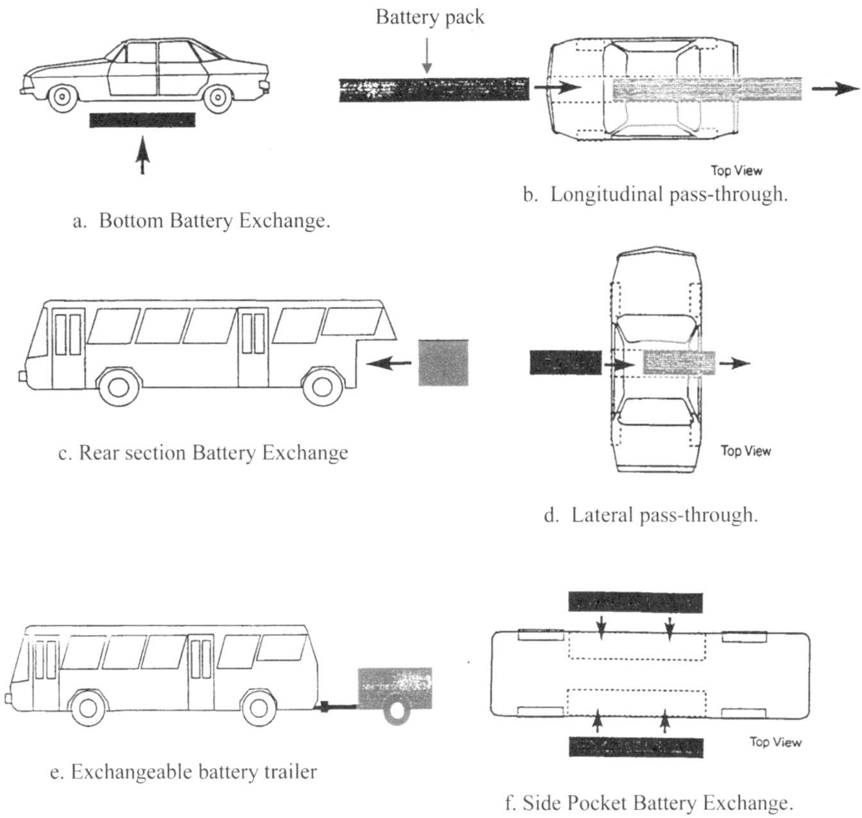

Fig. 14.86 Feasible battery exchange configurations, 1991: (**a**) Bottom battery exchange. (**b**) Longitudinal pass-through. (**c**) Rear section Battery Exchange. (**d**) Lateral pass-through. (**e**) Exchangeable battery trailer. (**f**) Side Pocket Battery Exchange

Fig. 14.87 1974 battery exchange for urban buses: M.A.N. Battery Exchange Trailer. Photo from Collie, M.J. ed. Electric and Hybrid Vehicles. Appendix C - Batteries for Electric and Hybrid Vehicles. Noyes Data Corp., Park Ridge, New Jersey, 1979, pp. 369-71. Original photo public domain

Fig. 14.88 Rapid Battery Interchange (RBI) for transit buses, "Coke machine" Concept. Developed for the California Energy Commission and Pacific Gas and Electric Co, 1991

Fig. 14.89 June 14, 2014: With much fanfare, Tesla CEO Elon Musk demonstrates the "revolutionary" Tesla 95-second automated battery exchange concept using a modified Model S. Screen capture from YouTube video. Unrestricted use

a central automated exchange unit, possibly at a regular bus stop. Examples include airport shuttles to remote parking areas or rental car lots, passenger transport in car-free downtown mall areas, and conventional or driverless taxis with limited service areas. Fun Fact: A futuristic depiction of RBX for a fictional flying taxi appeared in the 1997 Bruce Willis/Milla Jovovich action film "The Fifth Element." [186].

Figure 14.89 shows a still frame from a YouTube video showing Tesla's 2014 automated battery exchange apparatus for a Tesla Model S that demonstrated the capability to exchange the battery pack from beneath the car in 90 seconds. Figure 14.90 shows a 2016 photo of a modified 1992 GM/Conceptor Electric G-Van and automated exchange apparatus that is capable of exhanging a 216V battery pack from underneath a shuttle van in 63 seconds without the driver having to leave vehicle [187, 188]. The apparatus was constructed by a volunteer team of electrical engineering students at Cal Poly San Luis Obispo in 2014. The intended application was a 24-7 safe campus shuttle to remote campus parking lots and dorms, with the objective of performing the exchange in less time than required for passenger disembarking and boarding, giving the perception that the vehicle never needs to stop to recharge. The above-ground apparatus shown in the photo is installed non-intrusively in the pavement at the bus stop, allowing the continued use of the travel lane by other vehicles. Unfortunately, the system was not deployed by the university.

Fig. 14.90 California Polytechnic RBX (Rapid Battery Exchange) 63-second public automated exchange demonstration, 12 April 2016 and 14 April 2022. Photo: author

Fig. 14.91 Ample battery exchange station, 2022. https://ample.com/ Unrestricted use

Battery exchange as an alternative to high-rate battery charging has been one of those ideas (like hydrogen) that gets rediscovered once every 15 years or so. It routinely fails to proliferate because the assumed use case is individual EV owners that rely on public or subscription "battery swapping stations," for example, the Israeli-Chinese joint venture *Better Place* (2007) [189] or the international company Ample (2022) [190] shown in Fig. 14.91. The infrastructure requirements, ownership economics, and the impossibility of standardizing battery packs between multiple manufacturers make this approach impractical. But this is only because it is an inappropriate use case. None of these limitations apply if identical vehicles operating out of a common service location drive closed loops, similar to the battery-exchanging electric taxi services operating in New York, Paris, Boston, and Baltimore in 1899, 1900, 1906, and 1907, respectively [191]. Nevertheless, entrepreneurs and investors seem to be repeatedly captivated by the possibility of the widespread adoption of a proprietary battery and exchange standard that would require the consensus of many vehicle manufacturers. Arguments based on this assumption continue, see citations [192] or [193].

A more recent epiphany is the compatibility of automated battery exchange with another hopeful scheme, vehicle-to-grid (V2G) energy storage, the "uncharging" of EV batteries to help meet peak grid demands. This idea has periodically gained traction with utilities attempting to implement grid energy storage at no cost, but has been a consistent non-starter with EV owners. V2G will be discussed in greater detail later in this chapter. The salient point here is that V2G could be easily implemented at a battery exchange facility that maintains an inventory of multiple batteries in the exchange apparatus. The batteries retained in the exchange apparatus could be charged at any time, not just when a vehicle is out of use. This allows off-peak power to be exclusively used to charge the battery packs, as well as the option to "uncharge" these stored batteries during peak grid demand. This separation of battery charging/uncharging from vehicles avoids the obvious concerns of EV owners required by utility V2G policies that could uncharge their cars during 5pm–9pm peak load periods, possibly without their knowledge.

Intermittent Generators and EVs

The dependency of future electric energy production on solar and wind generation creates challenges for all electric energy users, including electric transportation. Electricity is an energy carrier that is simultaneously the most flexible in its uses, the most difficult to store, and a common denominator between the largest number of actual energy sources.

In the utility industry, two primary classes of power generation have been traditionally recognized: *central* and *distributed.* Central refers to large hubs of power generation such as fossil fuel or nuclear power plants, although large-scale hydroelectric is usually included in this category if it is not limited by capacity or weather. Utility-operated solar or wind facilities can also be considered central resources simply because they are large and centrally located to serve or supplement power delivered to large customer bases. *Distributed generators* (DGs) are everything else: home solar, local wind, geothermal, small-scale hydro, local diesel or gas turbine generators for power backup, or anything that is neither large in capacity nor under the direct control of the utility. This distinction was of primary relevance until the advent of local renewable energy sources, but in the present and future mix of large and small scale and renewable and non-renewable resources, is relevant only to the utility's ability to operate and control the facility.

Modern distinctions fall less along the lines of size or operating responsibility, and more on availability and predictability. With respect to renewable sources deployed in the last decade, the most important distinction is availability when the energy is actually needed. Traditional power plants can provide power continuously, i.e., it is *non-intermittent,* or power when needed (aka *dispatchable*). These sources can be relied on to provide "base" generation capacity. It is usually the largest in scale and lowest in cost for utilities to operate, but not the most environmentally sustainable, dominated by fossil fuels. Solar and wind are environmentally sustainable but not always available and are therefore *intermittent (aka non-dispatchable)* generators. Within this latter class are two levels of predictability with some overlap: *Stochastic* generators, such as solar and wind with outputs that are subject to unpredictable availability depending upon the instantaneous level of availability, e.g. solar or wind. This class represents the majority of "renewable" or "carbon-free" energy sources. *Deterministic* generators differ in that although they are intermittent, the times or conditions of their availability are predictable. Solar can *usually* be counted on to produce power during mid-day. Wind may have seasonal dependencies, but can usually be expected to provide power predictably. All forms of grid energy storage including *Pumped Storage Hydroelectric* (PSH) and Battery Energy Storage Systems (BESS) are dispatchable but have finite capacities, and are therefore best treated as deterministic intermittent generators.

When the generation and distribution facilities of the utility grid were originally built, intermittency was rarely a consideration: the utility-operated generation facilities were almost always dispatchable and load planning was cyclical and predictable. This changed fundamentally when grid-connected distributed intermittent generators started providing significant energy inputs to the grid in the early 2000s. Solar or wind generation is bounded by daylight hours and weather, respectively. How common it is to hear simplistic arguments for solar and wind energy that assume that their annually or daily averaged MW capacity is fungible with base utility generation (e.g., fossil fuel, hydroelectric, or nuclear power plants). This incompatibility is further exacerbated by the reality that energy demand is highly variable and often unpredicitble. And now EV charging is part of this intermittent demand. This mattered little twenty years ago, when solar and wind were trivial energy contributors and EV charging was nonexistent. Load fluctuations were managed by utilities using traditional peak-load-leveling provisions such as PSH and fast-start natural gas *peaking* (or *peaker*) plants.

But solar and wind are no longer trivial energy contributors. They are no longer playthings driven by altruistic intentions (or political careers), but by market economics enhanced by government incentives at a scale only dreamed of in the 1970's. Some jurisdictions (e.g., California, USA) have gone so far as to require that all new construction includes solar PV systems, or the provision for easy addition. And that at least half of all-electric power generation be from renewable sources in less than ten years. Large incentives at the state and federal levels accelerate the deployment of solar generation, electric vehicles, and elimination of natural gas for heating and cooking. Climate pledges are easy for politicians to make knowing that they will not be in office when the promises come due. If near-term carbon free future goals are to have even a chance of becoming reality, we have no choice but to rely more and more on intermittent generators. But for these sources to be usable, they must be matched by a combination of demand discipline (no you cannot charge your EV or run your data center during peak demand hours) and/or by local and utility-scale grid electric energy storage at a scale beyond the imagination of most advocates.

In the large interconnected grids of the contiguous 48 US states, there is a reasonable degree of robustness to intermittent generation and the growing demand for electric vehicles, due to the large size of the unified grid. But in isolated locations such as Hawaii that are already saturated with solar capacity, the unexpected problem of *getting rid of excess energy* during peak sun hours has created a new set of challenges. Similar solar or wind oversupply situations are now driving current political

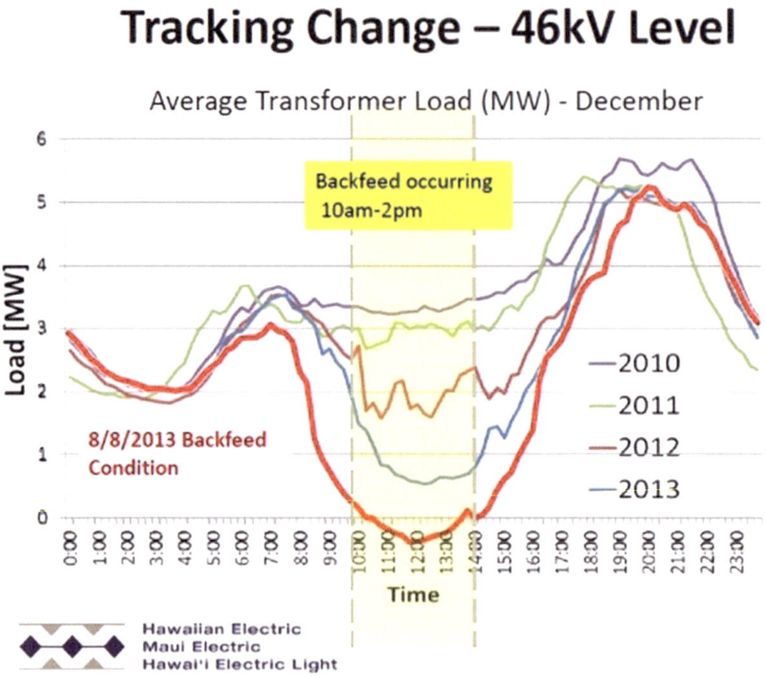

Fig. 14.92 The utility electricity demand "Duck Curve" for the Hawaiian Islands 2010-2013, showing back-feed condition mid-day due to excess solar PV generation. https://cdn.vox-cdn.com/thumbor/ExeZu3oAixQOqWApWphYoazSZBg=/0x0:544x408/1200x0/filters:focal(0x0:544x408)/cdn.vox-cdn.com/uploads/chorus_asset/file/6025225/heco-nessie-curve.jpg. Accessed 7 June, 2021. Public domain

decisions in Germany and France, which initially assumed that their grids could always absorb the intermittent power, but are now dealing with major load/generation imbalances. Radical solutions such as CEC Rule 21 in California or HECO Rule 14H in Hawaii include provisions that require all solar inverters be "advanced" or "smart" inverters that have a communications feature that allows them to be shut off by the utility with zero notice during times of overcapacity. (Fig. 14.92)

We have long been counting on solar and wind to offset the increased electric energy demand caused by electric vehicles. If EV charging occurred exclusively during peak sun or wind hours, or specifically only during periods in which the grid can support them, they could be a great benefit in helping to stabilize the grid. But there humans in the loop here. For most EV owners, this restriction is not an option because the cars are driven during the day or parked in corporate or campus parking lots with a few charge ports for hundreds of vehicles (Fig. 14.93). Popular excuses such as "EVs are only used for an hour or so each day, so there is plenty of opportunity for daytime charging" are misleading unless the vehicle has reliable access to charging at any time during the day, which is rarely the case.

Fig. 14.93 Mitsubishi Destination Solar Charging Canopy. 2023. From https://i1.wp.com/green-diary.com/wp-content/uploads/2012/08/mitsubishi_solar_powered_ev_charging_station_image_title_viehr.jpg?resize=600%2C378&ssl=1. With permission

As far as renewable transportation solutions go, direct solar charging from a dedicated PV array is the only truly zero-emission proposition: *sun to EV battery*. But this restricts vehicle use and leaves charging opportunities to the whims of solar insolation.

Utility-scale or local battery energy storage is costly. For the charging of even a single EV, a bank of batteries of nearly the same kWh capacity as the expected EV battery is required. The EV charging power demand being much larger than the peak residential load, the luxury of being able to grid-charge an EV whenever needed is too much to ask of the current infrastructure.

The cost of such a home solar +storage installation would be astronomical. Until (if ever) local or grid energy storage matches EV charging demand, it will depend on the charging habits of vehicle owners [194].

With the development of hybrid (aka all-in-one) solar inverters in approximately 2018, most solar PV equipment manufacturers and installers now offer some option for DC-integrated or AC-coupled (add-on) battery storage, usually lithium iron phosphate (LiPO4) batteries which have lower specific energy density, but lower cost. The importance of this relatively new practice for reducing solar grid destabilization cannot be underestimated. If every home with a solar PV systems and an electric vehicle included enough battery storage capacity to supply the peak period needs of the residence AND whatever portion of the EV charging energy is dispensed during peak load hours, there would no longer be a problem. Is this realistic? Probably not. But it puts the extent of the problem into perspective. Compared with the cost of infrastructure improvements needed to accommodate misalignment of generation with demand, it might be the more cost-effective remedy.

Battery storage remains an expensive add-on to most home solar installations. And with recent world political events (2025) it may not get any cheaper. As a point of reference, in 2024 (California) a 13.5 kWh Tesla Powerwall cost approximately $11,000 (installed, after 30% tax credit) [195]. That's $815 kWh of capacity; the $LiFePO_4$ batteries alone cost $139 kWh) [196]. If, hypothetically, 100 kWh capacity (8 Powerwalls) was installed solely to allow the owner of an EV the freedom to fully charge their Tesla S P100 whenever they wanted without any grid impact (or off-grid), it would cost them $88,000, which would likely exceed the cost (after tax credits) of the car.

How Much Grid Storage Is Needed to Accommodate EV Charging Demands?

Utility-scale grid energy storage is absolutely necessary to solve the mismatch between periods of excessive electric power generation and periods of excessive power demand. Worldwide, the dominant means for electric energy storage is PSH, which involves using electric power to pump water uphill to a reservoir, either natural or human-made, and then later using the stored water to drive hydroelectric turbines to generate electricity. Round-trip efficiencies of such facilities are the subject of multiple published analyses with widely conflicting conclusions,[22] but since the typical efficiency, including the motor and hydraulic losses of the pumping process, is usually no better than 90%, and the same for the turbine generation process, an RTE of 80% is probably a realistic, possibly slightly optimistic, assumption.

The idea of local electric energy storage is certainly not new. In an 1899 newspaper interview of Thomas Edison, he stated:

> In 1879 I took up that question, and devised a system of placing storage batteries in houses connected to mains and charging them in the daytime, to be discharged in the evening and night to run incandescent lamps. I had the thing patented in 1879 … [197]

In the summer of 2016, prior to the large-scale production of electric vehicles, the State of California experienced a major blackout. In the post-incident analysis, the CPUC concluded that California was shy of 53GWh of generation capacity. They also concluded that if EVs proliferate as expected, California alone will need *at least 100 GWh of distributed energy storage online by 2035* [198]. And this does not include the even greater cost of the ungraded electric distribution infrastructure to accept power from solar and wind, and dispensing the power for charging electric vehicles.

For infrastructure planning purposes in the USA, we can estimate (but not predict) the amount of grid energy storage needed to support the future incremental energy demands of electric vehicles, generated by the expected mix of deterministic and stochastic sources. Many worst-case scenarios for the unavailability of intermittent renewable power have been suggested, to help plan the magnitude of storage required, both utility scale and local, possibly under utility control to serve as *virtual powerplants* (VPPs). As with all attempts to predict future trends and requirements, the results are dependent upon the actual solar/wind generation percentage and the power demands of the population of electric vehicles. But public

[22] PSH round-trip efficiency (RTE) has been reported to be as high as 98.8% in industry publications. But this is doubtful considering the efficiency of motors and generator in general are rarely this high.

expectations of reliable power must be based on *the most pessimistic* combinations of events: minimum solar/wind and greatest demand, both likely to become worse and less predictable in time as a result of climate change.

A 2021 study [199] by the National Renewable Energy Laboratory (NREL) estimated that between 6 and 10 terawatt-hours (TWh) of energy storage would be needed to accommodate 94% of the electricity from solar and wind in the USA by 2050. Conservative (but somewhat outdated) estimates [200] suggest that by 2050 wind and solar are expected to grow 4–5 times faster than every other source of electric power.

According to NREL, which has long been a cheerleader for solar and wind energy, the additional grid energy storage requirement by 2050 will be, at a minimum, 85 times the amount of energy storage that is currently deployed on the grid, most of it currently in the form of PSH. Yet the prospects for significantly increased PSH energy storage capacity are poor, since even when additional geographic locations are identified, the associated land-use and ecology destruction issues are almost always deal-breakers for the public since these facilities must usually be located in environmentally sensitive locations. As of 2024, 3,204 US patents and 58,000 Chinese patents have been filed on (sometimes far-fetched) schemes for grid-scale electric energy storage[23]. None have yet been considered viable alternatives to either PSH or battery storage, with the possible exception (in my opinion) of local thermal storage. And the prospect of additional peaker plants to handle periods of high demand is inconsistent with environmental sustainability objectives, since they are almost exclusively fueled by natural gas.

Fully meeting this storage requirement is an ambitious, maybe even impossible goal. The above-referenced NREL study estimated a need for 6 TWh of energy storage just to cover the predicted 4 PM–9 PM peak daily demand period for a single day, assuming that solar is generating at peak capacity during the day and that wind generates at nearly maximum capacity 24-7. In the alternative NREL scenario that was less presumptuous, the study predicted that 10 TWh of energy storage would provide enough storage to cover the predicted peak demand for electricity for up to 2 days, assuming that solar and wind are generating at their average capacity.

Is this predicted requirement realistic? Is it financially realistic and achievable in the required time frame?

The annual average US grid electric power demand in 2022 was 4,050 TWh [201]. (Worldwide, this number was 25,530 TWh.) From the same source, US electricity demand is expected to increase to 5,178 TWh by 2050 but it is unclear (and unlikely) if this prediction includes the misalignment between periods that EVs are driven and EVs are charged—usually after sunset. On a daily average basis, an annual demand of 5,178 TWh is equivalent to an average daily electricity demand of 14.2 TWh. Using the NREL 94% solar/wind generation figure for 2050, this would mean that 13.1 TWh would be intermittent on a daily basis, which falls short but is not excessively greater than the NREL-estimated 10 TWh storage requirement for 2-day grid reliability.

[23] Patent search using Google Patents and USPTO online search tools

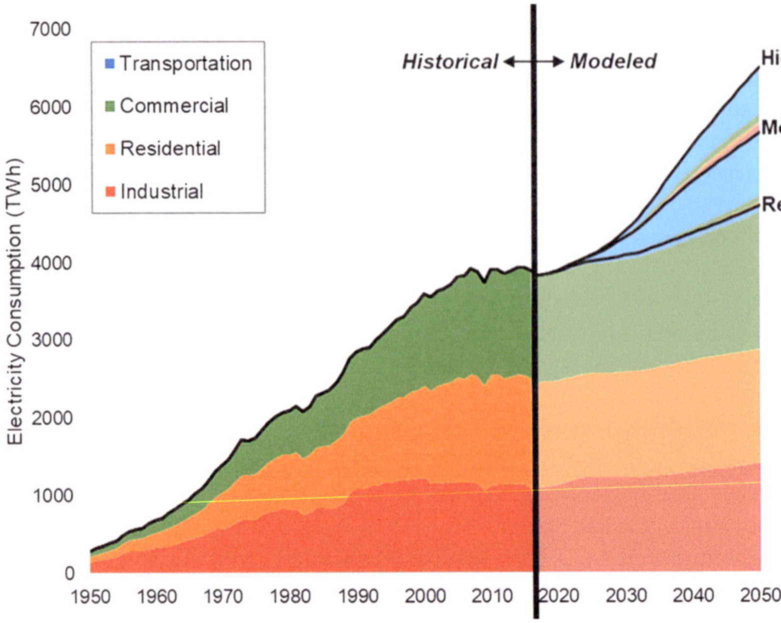

Fig. 14.94 Past data and future predictions for US electric energy consumption, overwhelmingly in the transportation sector. Graphic and data from DOE National Renewable Energy Laboratory. Public domain

Another perspective is provided by a 2018 DOE study [202] that predicted that US electric energy demand is expected to increase by 80 TWh per year by 2050, resulting in a 38% increase in total electricity use. By far the largest increase in electric power demand is due to the shift to electric fueling of vehicles, as shown in the NREL plot of Fig. 14.94 [202].

Extrapolating from 2022 electricity use data, the increased demand for electric power in the transportation sector in 2050 will alone account for 2240 TWh, with a total demand of approximately 6500 TWh. Assuming that power generation capacity in all forms will keep pace with the increased electricity demand, but that 94% of electricity in the USA will come from intermittent solar and wind in 2050, the average daily intermittency requirement would be 16.7 TWh, which makes the predicted 6 TWh energy storage requirement seem even more inadequate.

As of 2020, the claim of being the largest operating battery storage facility in the world was shared by the LS Power "1-1.5 GWh" facility in San Diego California, and the 730 MWh (which may be expanded to 1.2 GWh) BESS (Battery Electric Storage System) at Moss Landing facility on the Monterrey peninsula of California [203, 204]. The phase one facility of the Moss Landing facility was connected to California's power grid in December 2020. An adjacent battery facility operated by Tesla Inc. suffered a series of three fires during the first 3 years involving *Tesla MegaPacks*. On September 20, 2022, residents within a 4-mile radius of the facility

Fig. 14.95 Completed in 2022, the 350 MW/1400 MWh grid energy storage facility adjacent to solar PV facility at the Crimson Energy Solar Project, Blythe, California. Operated by Axium Infrastructure and Canadian Solar. https://www.rawpixel.com/image/9648646/the-crimson-energy-storage-project-solar-power Public Domain. View CC0 Licenseracks provided by LG Energy Solution located in former turbine halls at Moss Landing Energy Storage Facility, California. Operational since 2021. Image: LG Energy Solution. https://www.energy-storage.news/moss-landing-worlds-biggest-battery-storage-project-is-now-3gwh-capacity/ Permission pending

area were told to shelter in place. The Moss Landing battery complex was subject to public outcry following the incident but remains in operation as of 2024.

Figure 14.95 shows the battery storage containers and inverters of what is claimed to be the largest *single-phase power* battery energy storage facility, the 350 MW / 1400 MWh plant adjacent to a solar PV facility at the Crimson Energy Solar Project in Blythe, California.

Many other large-scale battery electric storage facilities are currently under construction, the majority in China. One example scheduled for completion in 2025 is a 1GWh BESS under construction in Manchester, England, which is also claimed to be the world's largest [205]. Other large-scale energy storage options that have actually been placed in service, but not necessarily for grid electric energy storage, include thermal (high temperature), thermal (ice), compressed air, and flywheel storage. A periodically updated list of all grid energy storage facilities other than PSH, either in operation or under construction, can be found in citation [206].

In the previously cited 2022 NREL report, the cost of lithium batteries for use in BESS facilities was assumed to continue to decline from its 2020 rate of \$350 USD per kWh to a mean estimated cost of \$150 USD per kWh over the 27-year period from 2023 to 2050. Considering the increasing scarcity of critical battery materials and recent political and trade issues between the USA and the People's Republic of China, these goals may be optimistic. 80% of the world's lithium batteries are produced in the PRC.

An extensive 2020 DOE study of BESS costs [207] covered the installed costs of several candidate battery technologies. The most common at this time, and probably for at least the next decade, is lithium iron (ferrous) phosphate (LFP). The *installed* 2020 cost for this technology was estimated in the study to be \$356/kWh. With this guideline, the fully built-out cost of the eventual 1.2 GWh facility would be \$427 million USD. This approximately agrees with the published cost of "more than \$400 million" [208] approved by the CPUC and Pacific Gas and Electric Company, the operator of the facility.

What would this imply for the large number of BESS facilities that will be required to meet the near-future electric energy storage demands of the USA? Using the (probably low) NREL estimated requirement of 10 TWh storage capacity needed by 2050, the cost to ratepayers or taxpayers (or both) in 2020 US dollars was estimated to be \$3.6 trillion. For perspective, the total Federal tax revenue in 2021 was only \$2.0 trillion [209].

Vehicle-to-Grid: Electric Cars Become Grid Storage Batteries

Another proposed solution for grid energy storage is based on the observation that with growing numbers of electric vehicles connected to home or public chargers, there is already a huge distributed battery storage capacity intermittently connected to the grid. If utilized for supplemental energy during peak demand periods, vehicle batteries could be "uncharged" under utility control during peak load periods. This idea actually dates back to the Golden Age of electric vehicles, when one of the selling points of electric vehicles was the ability to power household loads, at least for homes on DC (Edison) distribution grids. The idea resurfaced in its present form in the early 2000s and became a significant research funding sink in the USA, despite public reservations that were confirmed in every institutional trial at that time [210].

The idea makes sense from a technical point of view since in 2024, BEVs have battery capacities between 60 kWh and over 200 kWh, enough storage roughly equivalent to the daily average use of between 3 and 10 homes. An attractive proposition on paper, but in practice, there have been only token demonstration projects since, if adopted as suggested by its proponents, this would require private EV owners to allow their EV batteries to be discharged when needed by the grid, notably at peak demand times between 16:00 and 21:00 (4:00 and 9:00 PM). Even if reimbursed at a premium for the electric energy supplied by the vehicles, the timing of the energy exchange remains terminally problematic. People generally use their cars during the day when the grid is flooded with excess power from solar generation, and they home-charge their cars at night when electrical demand is generally much low. Peak demand occurs in the early evening, at a time when drivers return to homes or commercial bases with discharged batteries—the time of greatest need for supplemental power to the grid is also the time when most EV batteries are least capable of providing power to the grid. This mostly defeats the premise of V2G as a grid energy storage solution (Fig. 14.96).

There are a few specialized cases in which V2G could and has worked: Institutional facilities with vehicle fleet operations, or situations in which private

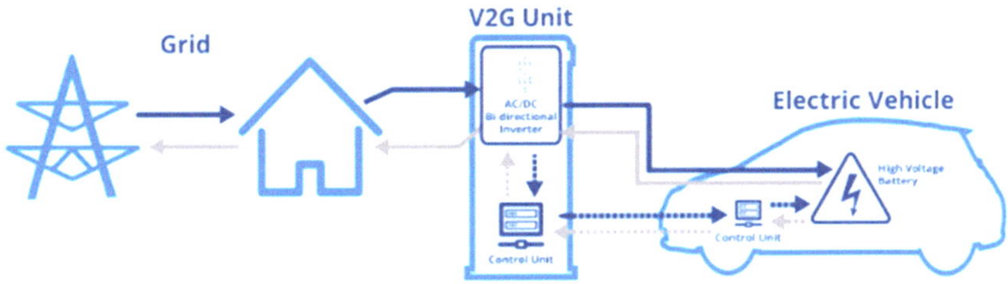

Source: fleetcarma.com

Fig. 14.96 Vehicle-to-grid concept, bidirectional charging. Graphic originally from Fleetcarma, division of EcoTab Inc., now defunct. https://cleantechnica.com/2022/09/13/electric-vehicles-101-v2g-charging-the-grid/ Public domain

vehicle owners have the final say about the direction of power flow through their chargers at any and all times. High power purchase rates during peak periods could also provide an incentive for EV owners that do not drive their cars on any particular day. A high reimbursement rate to battery sharers would also help to offset the accrued cost of increased battery wear, which reduces battery lifetime with every charge-discharge cycle. If high enough, EV purchases could potentially be motivated as a means to monetize V2G needs, in addition to providing transportation as a lower priority. But vehicle-owner discretion at any time adds yet another random factor to grid reliability, this time not just due to a lack of sunlight or lack of wind, but rather due to the randomness of personal choice or an unexpected need to use the vehicle for transportation.

However, one mutually beneficial arrangement involves employer incentivized cooperation of commuting employees, offering, for example, no-cost vehicle charging and preferred parking at work in exchange for the concession that their EV batteries could be at least partially discharged at home during the 16:00–21:00 peak load period [211].

Another potential but speculative exception, previously mentioned, V2G utilizing automated battery exchange systems, either public or private, in which several fully charged batteries are in storage in exchange stations at the start of the peak demand period. Discharging these batteries to support grid stability is a more attractive proposition for the operator of the battery exchange station, with multiple batteries in various stated of charge in the exchange facility. V2G continues to be rediscovered every few years but has remained more of a curiosity than a practical grid stability solution, except in specialized situations.

While not really V2G, the option to used the large battery of an EV for home backup power has become a major selling point for some electric trucks and SUVs. This has become particularly popular in California, Oregon, Washington and Texas since the horrific experience of the wildfires of 2023 and '24. After all, the 100 kW battery in an EV with a 300-400 mile range contains enough energy to power a typical American home for 5 days, based on the US average home electric power use of 20 kWh. The EV-as-a-backuppower source debuted in 2021 on the Ford F150 lightning [212] which included an onboard high-power inverter capable of supplying 230 VAC 60Hz electric power through a vehicle power outlet. It is now

Fig. 14.97 2021 Ford F150 Lightning Truck with *reverse charging* capability for emergency backup power. Image from https://commons.wikimedia.org/w/index.php?curid=134069096 Wikipedia Creative Commons, CC-BY-SA-4.0

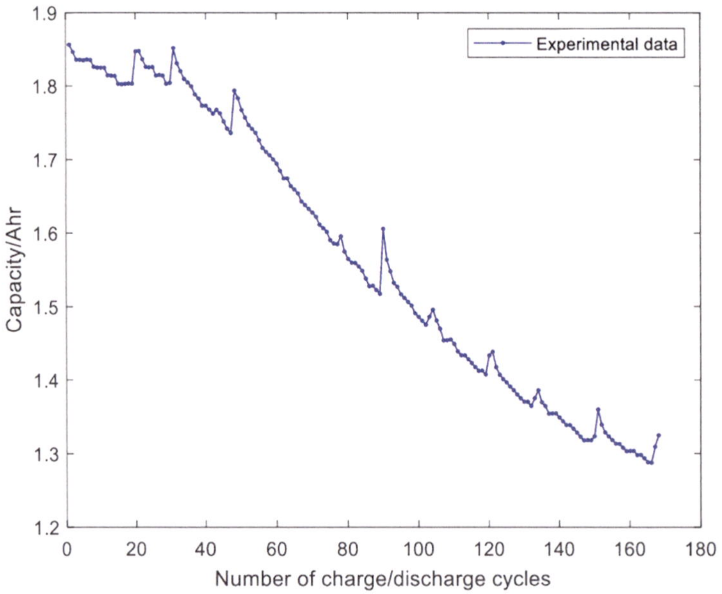

Fig. 14.98 Degradation of typical lithium battery vs cycles, from "State of Health Prognostics of Lithium Batteries", by Huixing Meng and Yan-Fu Li, CRC Press, July 2020. With permission

a popular feature of most newly introduced electric trucks and large SUVs, and is particularly popular for emergency service vehicles, such as shown in Fig. 14.97 [213].

Aside from the concerns of EV owners fearing a discharged EV at the time they need it, there is the (usually ignored) life-cycle cost of the battery to consider. Every charge/discharge cycle slightly degrades the battery, shortening its useful life, as illustrated by the plot of Fig. 14.98, which shows the degradation of the Ah capacity of a single 10 AH cell over 180 charge-discharge cycles. Expected cycle lives vary from 1000 to 2000 for NMC to as much as 10,000 (3000 average) for LFP batteries. If a battery is used daily for V2G in addition to powering the vehicle, the life could be reduced significantly. There are currently no mechanisms for compensating EV owners for "borrowing" their batteries for grid service except possibly in private demonstration projects [214].

Battery Recharging Considerations

In the USA, EV and FCV charging/refueling times are sometimes expressed in *miles per hour* (MPH), in an effort to make the speed of refueling or charging more intuitive. This is not the familiar measure of vehicle speed. It is the number of miles of range that are added to the vehicle's battery per hour of charging:

$$\text{Charge rate}\left(\text{MPH}\right)=\frac{\text{Range added}\left(\text{miles}\right)}{\text{Charging time}\left(\text{h}\right)}$$

The MPH of a vehicle+charger combination under this definition is a function of the power (kW) that is provided by the charger, limited by the maximum power (kW) that can be safely accepted by the battery, and the operating efficiency of the vehicle (miles/kWh).

The practice of classifying DC fast charging facilities in terms of their MPH is obviously also dependent upon an assumed vehicle and charging parameters (temperature, battery SOC, etc.), with different vehicles having different charge power limits and different mileage (miles/kWh) ratings.

Figure 14.99 shows the charge curve of a state-of-the-art 144V nominal voltage lithium iron phosphate (LFP) battery manufactured by the Chinese company Polinovel. Battery charging current and voltage profiles differ considerably between battery chemistries. Most segregate the charge cycle into periods of constant current, and constant or tailored voltage, while monitoring temperature and the learned charging history of the battery. Even if the battery is engineered to accept a constant current charge over most of the charge period, the battery voltage will remain nearly constant during most of the charge, but will increase rapidly at the start of the charge and when nearing 100% SOC. The charge power (W = V × I) changes accordingly. It is imperative that the current tapers down when approaching the 100% SoC state to avoid overcharging and overheating.

Regarding range estimation for BEVs: Since different BEVs have different battery types and capacities, and vehicles have variable efficiencies under different conditions, the owner must trust the accuracy and linearity of the algorithm that estimates and displays the vehicle's "range remaining". The accuracy of the range estimation has improved greatly from the optimistic estimates of early (2012–17) EVs, it has been the cause of lawsuits against EV manufacturers that have deliberately built bias into range remaining display to create the perception that the total vehicle efficiency (mileage) is higher than actual [215].

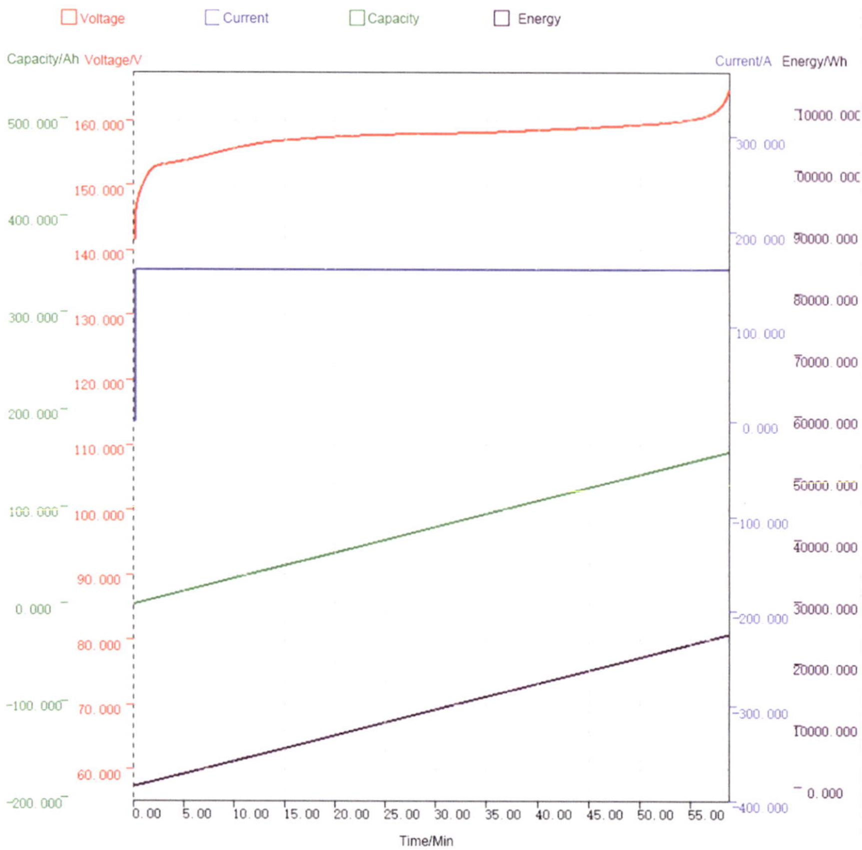

Fig. 14.99 Constant current charging data from a 144V nominal Polinovel LFP battery. Plots prepared by manufacturer for author

For most batteries, including all lithium chemistries, it is potentially harmful to discharge the battery to zero or charge to 100% of rated capacity. This means that the *rated* kWh capacity of the battery is usually somewhat greater than its useful capacity. And the amount of stored energy that constitutes "full" (100% SOC) is also variable, depending on the age and accumulated cycles of the battery, and the battery temperature.

Peukert's Formula and Vehicle Range

In 1897, German engineer Wilhelm Peukert observed that for a lead-acid Plante cell, the usable amount of energy that a battery delivered was reduced if discharged or charged at higher rates. There are mixed opinions about the accuracy or applicability of Peukert's formula to other battery chemistries, but in the years since, it has been applied with various degrees of success to most rechargeable battery chemistries. Different batteries are characterized by different experimentally determined Peukert coefficients k, which describe their sensitivity to this effect. The basic exponential relationship between discharge rate and usable energy seems to have applicability to most batteries including lithium. This dependency is rarely mentioned in battery or vehicle manufacturers [216].

The applicability of Peukert to lithium batteries has been a subject of debate in both scientific and popular media, usually examining its applicability to predict battery capacity or vehicle range in an absolute sense. But this was never the intended application of Peukert's formula. It is only applicable if applied to *the same battery* at two different rates of discharge. For example, it is helpful for translating the *amp-hour* (Ah) rating of a battery as stated by the battery manufacturer into a practical Ah rating for vehicle range estimation under actual use conditions.

It is standard practice for battery manufacturers to rate all automotive Starting Lighting and Ignition (SLI) and most deep-cycle (traction) batteries over an unrealistically optimistic 20-h continuous discharge period. In fact, the Ah rating stated on a battery nameplate is almost certainly a 20-h discharge rate unless stated otherwise. Peukert can assist with translating this unrealistic rating into a more realistic rating for the type of battery service intended. Again, this is just an experimental observation, not derived from the chemistry of the battery.

Regarding battery capacity: the energy storage in kWh is equal to the product of the battery's amp-hour rating with the (usually assumed constant) terminal voltage:

$$\text{Energy (Wh)} = \text{Amp-hours (Ah)} \times \text{terminal voltage (V)}$$

The 20-h (20 hour) discharge rate can be *manually* tested by starting with a fully charged battery at 25° C, and discharging it at a constant current rate. The constant test current i_d is an estimate based on prior experience with similar batteries or previous test iterations, so that the battery approximately reaches its cutoff voltage (for lead-acid, 1.75V for cell, 10.5V for a 12V battery) in approximately 20 hours. The actual time t_d required to reach the cutoff point is measured. If t_d is significantly longer or shorter than 20 hours, the test must be repeated with i_d increased or reduced respectively. Several iterations may be required, until t_d is sufficiently close to twenty hours, with "sufficient" subject to some interpretation. The AH rating is calculated from experimental observation:

$$Ah_{20} = i_d(\text{amps}) \times t_d(\text{hours}).$$

In practice, automated battery test protocols replace this laborious process, although not necessarily with great accuracy. A simpler procedure involves discharging the battery through a fixed resistance, which is less time-consuming. The time-integrated value of the recorded current until the battery is discharged to its cutoff voltage, and the final integrated time-current product is reported as the amp-hour capacity. If the objective is to experimentally determined the Peukert constant k for the battery, two test at different discharge rates are conducted, and these to data pairs (i_{d1}, t_{d1}) and (i_{d2}, t_{d2}) are used to calculate k as described below.

Peukert assigned the symbol *C* with units of Amp-hours (Ah) to represent energy storage E (Watt-hours) in a voltage-normalized form, independent of the battery terminal voltage V_t: Energy E (Wh) = C (Ah) × V (volts) Note that this "C" is NOT the same as the "*C-rate*" *in amps* commoly used for the normalized charge or discharge rate of a battery.

The original form of Peukert's formula is not very intuitive, but it can be rewritten in a more useful form as

$$\frac{C_2}{C_1} = \left(\frac{t_2}{t_1}\right)^{1-\frac{1}{k}} \tag{14.1}$$

where

C_1 is the Ah (Amp-hour) capacity of the battery measured as the constant current that will fully discharge (or charge) the battery in time $t_{1.}$

C_2 is the Ah capacity for discharge or charge over a different time period $t_{2.}$

Since time is used only as a ratio, the units of time are irrelevant. Fractional hours are typical.

k is the Peukert constant, unitless.

Typical values of the Peukert constant for different batteries' chemistries are listed below but is also related to the construction of the battery; note the difference in *k* between the three types of lead-acid batteries (flooded, gel, AGM):

k = 1.4 for flooded lead acid
k = 1.2 for gel lead-acid battery
k = 1.1 for absorbed glass mat (AGM)
k = 1.1–1.2 for lithium chemistries

With the Peukert formula in the form shown above, a ratiometric change in the usable battery Ah capacity is related to a change in the time it takes for the battery to fully discharge as plotted in Fig. 14.100, parametric with the Peukert constant [217].

For a given untested battery, the Peukert constant can be estimated by running two discharge tests, the first measuring C_l over interval t_l, and the second measuring C_2 over interval t_2. Then calculate *k* from:

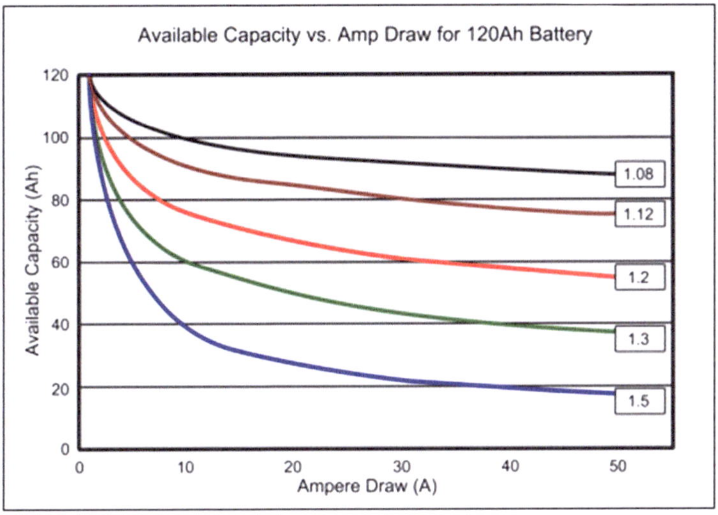

Fig. 14.100 Relationship between battery capacity C (Ah) and discharge time (h) or current i (A) parametric with the Peukert coefficient, each value of k identified by a label at the right of its curve. From A. R. Utomo, F. Husnayain, P. Priambodo "Analysis of Lead Acid battery operation based on Peukert formula" Intl Conf on Electrical Engineering and Computer Science 18 February 2014. Creative Commons CC-SA 4.0

$$k = \frac{1}{\left(1 - \frac{\ln\left(\frac{C_2}{C_1}\right)}{\ln\left(\frac{t_2}{t_1}\right)}\right)} \tag{14.2}$$

This form of Peukert is useful when a battery such as shown in Fig. 14.101 has a "nameplate Ah rating" measured at a 20-h rate.

For example, if an application will fully discharge a 100 Ah (20 h rating) battery in say, 4 hours, assuming the Peukert constant $k = 1.15$,

$$C_2 = C_1\left(\frac{t_2}{t_1}\right)^{1-\frac{1}{k}} = (100\ \text{Ah})\left(\frac{4\ \text{h}}{20\ \text{h}}\right)^{1-\frac{1}{1.15}} = 83.5\ \text{Ah}$$

With some manipulation, Peukert can also be written to examine the effect of increasing the current I on the effective storage capacity of the battery:

$$E_2 = E_1\left(\frac{I_1}{I_2}\right)^{k-1}\ \text{kWh}$$

Since $k > 1$, usable energy storage E decreases with the discharge current i. Practically speaking, vehicle *range decreases* with aggressive driving.

Fig. 14.101 Single prismatic LFP cell made by CATL, China. Photographer: Aeroid. 2001. Creative Commons Attribution-Share Alike 4.0 International license

A Practical Application

Translation of Battery Nameplate Capacity into Practical Capacity

The practical problem facing anyone designing or converting an electric vehicle: What battery capacity should be selected to ensure a target range under normal driving conditions? For this, we need to know the expected vehicle efficiency in miles/kWh (USA) or kWh/100 km (everywhere else). Passenger EVs in the USA have EPA combined average efficiency values ranging from 2.0 to 4.0 miles/kWh, an average of 3.0 miles/kWh (or approximately 20 kWh/100 km). This efficiency rating includes the losses of the motor, drivetrain, and controller electronics. So for the desired vehicle range of 300 km, one would logically specify a battery with a capacity of 20 kWh/100 km × 300 km = 60 kWh. For a battery with a nominal 300V terminal voltage, this means a 200 Ah battery.

The 60 kWh energy storage requirement is met by a 300V, 20Ah battery pack. The test conditions for determining the battery Ah rating are not known. In such a situation, it's reasonable to assume the 20-h rate:

$$T_1 = 20 \text{ h}$$
$$C_1 = 200 \text{ Ah}$$

200 Ah is expended in 20 h (10 amps average) to discharge the entire capacity of the battery.

The US Federal urban drive cycle (FTP-72) applicable to electric vehicles usually requires 22.8 min and covers 12 km. The highway drive cycle (HWFET) usually takes 12.7 minutes and covers 16.5 km. Assuming the average EV combined efficiency of 0.20 kWh/km, this means that 0.20 kWh/km × 28.5 km = *5.7 kWh or*

19 Ah @ 300V is expended over a time period of 35.5 minutes. These numbers provide a good estimate of the average power output of the battery in typical driving.

$T_2 = 35.5 \text{ min} = 0.59 \text{ h}$
$C_2 = 19 \text{ Ah}$

19 AH (reduce but do not eliminate the space between 19 and AH) is expended in 0.59 h (32 amps average) normal driving as defined by Federal emission tests.

These two test conditions can be used to calculate the Peukert coefficient k for this battery.

$$k = \frac{1}{\left(1 - \frac{\ln\left(\frac{C_2}{C_1}\right)}{\ln\left(\frac{t_2}{t_1}\right)}\right)} = \frac{1}{\left(1 - \frac{\ln\left(\frac{19.2 \text{ Ah}}{200 \text{ Ah}}\right)}{\ln\left(\frac{0.59 \text{ h}}{20 \text{ h}}\right)}\right)} = 1.02$$

What is the *actual* Ah capacity for the full battery when 19 Ah is discharged in 0.59 hours => average current = *32 amps*, rather than the manufacturer's 20 hour test rating at a much lower current: 200 Ah discharged in 20 hour = *10 amps*:

$$E_2 = E_1\left(\frac{I_1}{I_2}\right)^{k-1} = (60 \text{ kWh})\left(\frac{10.0 \text{ amps}}{32.0 \text{ amps}}\right)^{1.02-1} = 58.6 \text{ kWh}$$

The usual battery capacity under actual operating conditions is 58.6 kWh rather than the 60 kWh battery nameplate rating. Since range equilibrates to usable battery energy, the actual vehicle range is reduced by 1.4% compared to what would have been calculated using the nameplate Ah rating.

Another important factor in estimating range is the increased energy use per km as the vehicle speed increases. Motor power increases approximately with the cube of the vehicle speed due to air resistance. When Peukert is combined with the increase in energy per distance due to wind resistance, the net effect is an even greater reduction in usable vehicle range. While not a large effect, EV range is more dependent on vehicle speed than an ICV because, in addition to the cubic road load/speed relationship that affects any vehicle, the usable capacity of the battery effectively decreases as the battery output current (or power) increases.

Round-Trip Efficiency (RTE)

Battery RTE is the ratio of the electric energy that can be extracted from the battery divided by the energy amount that was put into it during charging. RTE can vary from a maximum of 96% for a new Li battery slow charged to as low as 80% for older batteries or batteries charged at very high rates [218]. But it is the dependency of RTE on the vehicle speed that we are concerned about here.

As previously discussed, charge rate can affect battery lifetime as well as charging efficiency. This dependency has been minimized but not completely eliminated in the design of newer vehicle batteries that can handle very high charge and discharge rates, with at least one manufacturer reporting no discernable dependency.

Summary of Refueling Times and Ranges: BEV, FCV, NGV, ICV

For comparison, listed here are typical times required to refuel or recharge vehicles that run on different fuels or energy sources.

Based on published data and personal observations, here are some comparative estimates for EVs, hydrogen FCVs, CNGVs, and gasoline ICVs.

BEV: 400V, 100 kWh battery (Tesla S P100D), 80% (80 kWh) charge	
Level-1 115 VAC@20A (2.3 kW)	35 hours
Level-2 Typical, 230 VAC@30A	12 hours
Level-2 Maximum NEMA limit 230 VAC@50A	8 hours
Level-3 Tesla V3 DC fast charger 625A@400V (250kW) [219].	20 minutes
Automated battery exchange	
Cal Poly RBX	63 seconds
Tesla S demonstration	90 seconds
Ample (first generation)	5.0 minutes
FCV: Automated fill of 5L compressed H_2 tank (Toyota Marai)	
35 MPa precooled, 80% fill	3–10 minutes
70 MPa precooled, 80% fill	5–20 minutes
NGV: Manual fill, 25 MPa, 8 GGE CNG tank	
Honda NGV	3–6 minutes
ICV: (gasoline or any liquid fuel): 50 L fuel tank, 0.5 L/s	
2023 Honda Civic EX	1 minute 40 seconds
Sampe Vehicle Range: BEV, FCV, NGV, ICV	
BEV: 2023 Tesla S P100D, (100 kWh battery)	576 km (358 miles)
FCV: 2023 Toyota Marai XLE with 6L, 70 MPa H_2	647 km (402 miles)
NGV: 2015 Honda Civic NGV	419 km (260 miles)
ICV: (gasoline): 2023 Honda Civic EX	716 km (445 miles)

What About Home Solar Charging?

Can I charge my EV exclusively using the output of my home solar system? This is usually one of the first questions from the owner of an EV that also has a home solar PV system.

Yes, to the extent that the solar PV system can supplement the grid AC power going to the vehicle charger from the grid. But few home solar arrays are large enough to fully charge an EV in a reasonably time.

Can it be charged entirely from your home solar system?

Yes, but it would take a very long time.

Some energy accounting …

A typical home solar array is rated at 5 kW max power. At the thirty-fifth parallel (middle of California), a well-placed array will annually average about 4 "full sun hours" per day, which means that a 5 kW nominal array can produce 20 kWh per day. And if all of this energy goes into the EV battery, that's up to about 60 miles of range per day. Not bad.

But to use this energy exclusively to charge the EV, the EV must be connected to the output of the solar inverter during all sun hours, every day. Alternatively, a battery storage system of 10 kWh capacity must be available to store the PV energy during sun hours and release it at a constant rate whenever the vehicle is available for charging.

Without battery storage, if even part of the charging is done at night, that fraction will come from the grid, and without net-zero metering (as is the case in California since 2022), the advantage of solar charging is lost for that fraction of the energy.

The larger the vehicle battery, the longer that charging will take. For example, if a 100 kWh battery in a Tesla Model S was charged exclusively from the solar PV output of a 5kW array, it would theoretically require 100 kWh/(20 kWh per day) = *5 full days to fully charge*, during which the vehicle can only be taken off the charger at night.

But from a more positive point of view… For the ~3 miles/kWh efficiency of the Tesla S, 60 miles of travel per day would only use 20 kWh of battery charge. So if you drove it only on alternate days, worked from home and charged on the off-days, the entire energy budget of the car could be met. And if you only drove 30 miles per day, such as a local commute to work, you could get away with only having the car on the charger half of the day, or having a home solar storage battery capacity of 10 kWh (one Powerwall). For those times that your daily travel exceeded 30 miles, you could feel guilty about only the fraction of the extra energy that came from the grid at its 400 g CO_2/kWh emission rate. If this scenario works for you, then it might make sense to trade in the 2200 kg Tesla and buy a 1079 kg Smart Fortwo (Smart EQ) EV that has only a 17.6 kWh battery and a higher mpge. *The Smart EQ could be fully charged every day by your home solar system* (except on cloudy days)

Why can't one just place solar PV panels all over a car, and use that electricity to power the car?

This might be possible for an extremely lightweight aerodynamic vehicle such as the solar competion vehicle shown in Fig. 14.102, but the incident sunlight (about 1kW/meter^2) could never provide sufficient power to move a practical car at a realistic speed.

EV Fuel Economy and GHG Emissions

In November 2010 the US EPA introduced the first equivalent fuel economy ratings for all-electric BEV and plug hybrid PHEV vehicles, replacing the overoptimistic and inconsistent range estimates provided by the vehicle manufacturers. These

Fig. 14.102 Sunswift Violet solar competition car built by students at the University of New South Wales. Photo: CathyeeLiao, Wikipedia Commons. CC BY-SA

ratings were based upon the same ensemble of test drive cycles that make up the "Federal Test Procedure" (FTV) and in California, the CARB Test Procedure used for gasoline and diesel ICVs, but the methods for determining the fuel use had to be different. Ratings are reported in gasoline gallons equivalent (GGe), a unit that would be more familiar to EV buyers than kWh/100 km used elsewhere in the world. As discussed below, one GGe was equilibrated to 33.7 kWh of electric energy and is intended to equal, with a small correction factor, to the energy content of one gallon of US regular gasoline. For example, an EV that can travel three miles per kWh would be EPA-rated at 3 mi/kWh × 33.7 kWh/GGe = 101 MPGe.

Since 1959, Federal law required the posting of a standardized "Monroney" window sticker (named for Senator Mike Monroney who introduced the Automobile Information Disclosure Act of 1957), an example shown in Fig. 14.103. Prior to 2008, the efficiency units and test procedures for BEVs and PHEVs were at the discretion of the manufacturer, with different manufacturers quoting efficiencies and vehicle range in different units. With the passage of the 2007 Energy Independence and Security Act (EISA), kWh/mile and MPGe efficiency units were required, and vehicle test procedures were standardized. This helped to correct the common practice of overstating EV range based upon the manufacturer's own test procedures, although misrepresentation of EV range remains common in 2023 [220].

US EPA Fuel Economy Testing of EVs and PHEVs

Regardless of whether ICE, hybrid, or battery electric, new model vehicles in the USA are tested and certified for fuel economy (and emissions for non-BEVs) using simulated driving over two types of standardized courses on a chassis dynamometer. These are referred to as the *urban and highway drive cycles* of the EPA (or

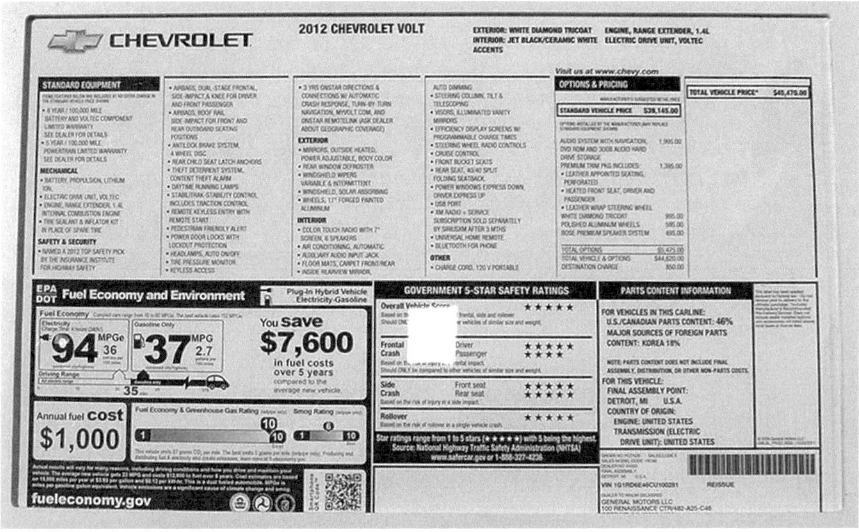

Fig. 14.103 Standardized US "Monroney" window sticker for a 2012 Chevrolet Volt Plug Hybrid electric vehicle. Sticker must state the EPA-certified mileage in MPGe, vehicle standard and optional equipment, government safety rating (5 stars max), extra-cost options, MSRP and estimated annual fuel cost based on sample. For plug hybrids, it must separately state the electric-only and gasoline-only mileage and range. Since approximately 2010, the greenhouse gas and smog (ICE operation) ratings must be stated. Public domain

CARB) Federal Test Procedure (FTP) which may be conducted in as many as five separate tests. A drive (or driving) cycle is a time-dependent trace of vehicle speed and load that represents the motion of the vehicle traveling over a simulated road course.

For electric vehicles, the total electric energy used during each test is measured by fully charging the battery before each test, and fully recharging the battery after each test. Energy use per mile is determined by measuring the total energy in the battery and dividing it by the distance that the vehicle "travels" on a dynamometer over each drive cycle.

Vehicle range is measured (somewhat controversially) by driving the car over a repeating version of each drive cycle, starting with a full battery charge, until the vehicle can no longer maintain the speed required by the test cycle—a point of some ambiguity since different points in the drive cycle are at difference loads. This protocol includes the efficiency of the onboard battery charger and the overnight RTE of the battery. Non-traction electrical loads (lights, climate control, battery cooling or heating, electric power assist steering and brakes, ADAS, accessories) are disabled during all except one of the cycles (SFTO SC03). From citation [222]:

The step-by-step procedure for a BEV is as follows:

1. Fully charge a vehicle, leave it parked overnight, and put it on the dyno the following morning.
2. The electric vehicle is then put through multiple UDDS and HWFET cycles until *the battery is completely discharged and the car can no longer drive*. The dyno measures how many miles the EV 'travels' during testing to provide a preliminary range number.
3. The car is then plugged in with the charger provided by the automaker and charged back to full. The charger is on a meter to precisely measure how much electricity the charger draws during charging—including electricity lost to inefficiency.
4. The numbers from the closed-environment test at room temperature and relatively low speeds typically overstate the amount of range and efficiency an electric vehicle would see in real-world use.

For example, the Tesla Model Y will travel 451 miles on the Multi-Cycle City/Highway Test Procedure—which is a much higher range than anyone would actually see driving the car in normal conditions. Based on sampled experience, the EPA will multiply the preliminary range and efficiency numbers from the tests by 0.7 to provide a final rating more in line with what drivers can expect from their cars. For the Model Y, that's 316 miles of range. Obviously, this factor is the subject of much debate between manufacturers and regulatory agencies.

About Range Claims

In our enthusiasm to promote EVs as the lynch pin of our climate remediation strategy, there has always been a tendency toward overstatements, usually supported by biased analyses. Range, recharge time and power output are common sources of over-optimism. These distortions can be self-defeating because they undermine the credibility of the otherwise good arguments in favor of electric propulsion.

Unrealistic range claims have long been the norm among EV manufacturers. Even Thomas Edison complained in 1883 about this, as previously mentioned in this chapter [223].

The first mass-market EV sold in the USA, the MY 2011 Nissan Leaf, became the industry litmus test for EV range specifications. Buyers who placed deposits as early as 2 years in advance of its release were surprised to find that the promised 161 km (100 miles) range was about 40% overstated and the practical maximum range was only 112 km (70 miles) maximum. It was in late 2010, just as Leaf sales started to take off, that the EPA established formal testing procedures for EV mileage and range. The well-known MPGe equivalence of *1 gallon of gasoline = 33.7 kWh* was published in November 2010, based on input from industry and consumer focus groups as much as physics. This equivalence formula as well as testing procedures for EVs and hybrid cars remained subject to debate for several years. This evolution is discussed in detail in the following section.

The Leaf and the other first-generation BEVs that followed had been under development for years, and lithium batteries were still evolving. Nissan had a reasonable excuse for the inaccuracies of range projections made 2 years before formal standards. Nevertheless, in 2012, Nissan was the target of a class action lawsuit by owners. The 100 mile advertised range was based on 100% discharge of a full battery, which would have damaged the battery. The lawsuit was settled in 2015 with Nissan compensating owners for the deficient range and replacing battery packs found to have the worst performance [224].

Other range lawsuits have followed over the years. One of the more interesting was a lawsuit filed in California in 2021 by the owner of a 2019 Hyundai Kona. In the wake of a series of battery fires in 2019–2021 Kona EVs and 2020 Ionic EVs, Hyundai's preventative solution was to reprogram the onboard charger to limit charging to no more than 90% of the advertised range. A prudent measure, but Hyundai neglected to inform owners that this was the solution being done to their cars when they brought them in under a recall order. The loss of range amounted to only about 26 miles, but as with the Leaf, this matters much more for an EV than for an ICV. 82,000 vehicles were affected worldwide. As of 2023, the case remains pending [225].

Despite eventual agreement on kWh/gasoline energy translations and the refinement of EPA testing procedures, malpractice continues. On August 3, 2023, a class action lawsuit against Tesla Inc. alleged "fraud and false advertising" for all MY2022 Tesla cars, with evidence that three of the four Tesla car models fell short of their advertised range by an average of 26% [215]. The deception in this case was a bit more sophisticated than simple overoptimistic advertisements. The lawsuit alleged that nonlinearity had been built into the range estimation algorithms that generated the "range remaining" readout for the driver. This *fuel gauge* optimistically used a higher efficiency in the upper part of the range remaining prediction, misleading drivers into mentally extrapolating a longer total range. But as the battery SoC was closer to "empty," the rate of energy use seemed to increase. In fairness, the accuracy of the range estimating software was probably no worse than the abysmal accuracy of analog gasoline gauges. But it's a much bigger deal if the driver is relying on the range specification for planning a trip or reaching the next recharging stop.

But Tesla's insensitive management of the complaints made the conflict much worse. According to a Reuters report cited in [215], in mid-2022, Tesla started routing range complaints to a special "Diversion Team" which fielded up to 2000 cases a week and closing about 750 cases a week, by simply dismissing the complaint. The outcome of the litigation is not yet known.

Range remains the specification of greatest concern for most buyers of electric cars, SUVs, and trucks. It seems that only lawyers have benefited from the range (and safety) faux pas of the EV industry.

How the Equivalence Between MPGe and Miles/kWh Was Determined

In 2022, the Miles Per Gallon Equivalent (MPGe) was calculated by the EPA using the assumption that **33.7 kWh** of electricity is equivalent to one gallon of gasoline, and for plug hybrids, summing energy provided from both electric energy stored in the battery and the gasoline used to complete each FTP drive cycle. The overall MPGe ratings, like all MPG ratings, are reported as (1) Combined city/Highway, and (2, 3) the weighted means of the three parts of the urban cycle and two parts of the highway cycle.

This can be verified approximately by calculation, based on the 2022 EPA or EEA (European Environment Agency) [226] average combustion energy (LHV) of US-specification gasoline (not E10), 41.7 MJ/kg and density of 0.76 kg/L:

$$1\ gal(US)\ gasoline \times \left(\frac{3.79\ L}{gal}\right)\left(\frac{0.76\,kg}{L}\right)\left(\frac{41.7\,MJ}{kg}\right)\left(\frac{1\,kWh}{3.6\,MJ}\right)$$

$$= 33.4\ kWh\ per\ gal(US)\ gasoline$$

For example, the 2020–2023 Chevrolet Bolt BEV uses 0.28 kWh/mile during the combined FTP cycles. The equivalent gasoline mileage reported to consumers is **118 MPGe** [227]. Checking this with the EPA equivalence factor:

$$\frac{33.7\ \text{kWh / equivalent gal}}{0.28\ \text{kWh / mile}} = 120\ \text{MPG}_e$$

Not Everyone Agrees

There is not a general consensus about the fairness of the EPA's MPGe, fuel cost, and range measurement methods. For example, when the MPGe mileage equilibration was first published in 2010, Forbes magazine [228] argued strongly that it is skewed politically in favor of EVs. A major point of disagreement was that the EPA formula calculated EV energy use starting at the point of refueling rather than at the original energy source [229–232].

Their point was that comparisons between different motive energy forms (gasoline, electricity, hydrogen, E85, etc.) should go back to a common form of energy, such as the fossil fuels used directly in an ICV, and for electricity generation at power plants. This would more fairly match the energy resource consumption as well as the actual CO_2 effectively emitted by the EV. The efficiency of the power plant and the transmission/distribution losses had been ignored. The retort is a slippery-slope argument: Why not include ALL sources of GHGs going back to extraction and refining? This is easier said than done ….

Additional energy losses for **gasoline vehicles** measured from the oil well to the point of dispensing at the service station include:

- Oil field extraction energy
- Rail transport or pipeline energy, crude oil to refinery
- Refinery energy required for gasoline production
- Gasoline distribution transport energy
- Energy to operate fuel stations
- Fuel loss due to leakage or vaporization during transport or processing

Additional energy losses from the original energy source to the charging plug of the EV.

Here we have a problem since electric energy comes from many original energy sources. For example, if we assume that all of the electricity comes from natural gas power plants, we can track energy from the gas well to the electric vehicle charging plug:

- Gas field extraction energy
- Pipeline or liquefaction and transport energy to the point of distribution
- Distribution to power plant
- Energy conversion efficiency of natural gas power plant
- Electric energy transmission, distribution and power conversion losses
- Efficiency of battery charger if fixed DC battery charger used for rapid charging
- For electricity generated using coal, include the energy for coal extraction and transport.
- For electricity generated from solar or wind. If the vehicle is charged directly from a solar or wind inverter, we can consider this to be "free" energy with no additional losses. But if solar or wind power is pushed to the electric grid and used to charge the vehicle at some other time, many losses become involved: round-trip through local secondary transformer, grid distribution losses, or RTE of local solar energy storage batteries.
- For nuclear, it is impossible to go back to the full energy content of the source uranium, so only grid distribution and conversion losses can be considered.

Comparing net efficiencies and carbon emissions is heavily based upon the assumptions, especially when Tier-3 [233] emissions beyond the control of the utility or fuel provider become involved. Consequently, restricting the vehicle efficiency measurement to just what happens in the vehicle is usually the only reasonable rating method, save for advanced climate models.

Carbon Dioxide Emissions Attributed to EVs

While electric vehicles have no exhaust emissions, the sources of the electricity used to charge them do. The operating CO_2 emissions of battery electric vehicles are indirect, that is, they are from the production of the electricity used to charge them rather than the vehicle tailpipe. So other than the manufacturing-related emissions, the answer to the question "how clean are EVs" is equivalent to "how clean is the electricity that EVs run on." Possibly all EVs in the USA, and most worldwide, are charged using utility grid electricity. As previewed earlier in this chapter, a 2023 EPA report assigned an average of 386 g/CO_2 to grid electricity nationwide, from the entire mix of energy sources, both non-renewable and renewable. In the USA and the majority of countries, the majority of electric power generation comes from fossil fuels, specifically natural gas and coal. Despite aggressive efforts to reduce this dependency using solar and wind, the intermittency of these sources and lack of grid energy storage will limit these renewables until lagging electrical and energy storage infrastructure catches up. From a 2023 report by the International Energy Agency [234], this is a particular impediment in countries with older centralized power grids and high electric energy demand:

> **The share of solar PV and wind in global electricity generation is forecast to double to 25% in 2028 in our main case.** This rapid expansion in the next five years will have implications for power systems worldwide. … grid bottlenecks will pose significant challenges and lead to increased curtailment in many countries as grid expansion cannot keep pace with accelerated installation of variable renewables.

Natural gas is currently the largest energy source for power generation, providing approximately 43% of the electricity in the USA in 2023 [235]. But, as discussed earlier, the other 16% coal generation is responsible for the largest total CO_2 emissions from electric power generation (Fig. 14.104).

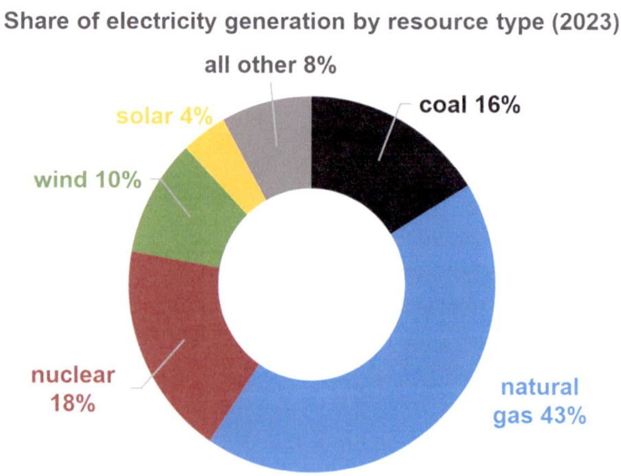

Fig. 14.104 Sources of energy used for power generation in the USA, 2023. From EIA Today in energy 22 February 2024. https://www.eia.gov/todayinenergy/detail.php?id=61444 Public domain

Much has been promised by federal and state agencies, and research or commercial enterprises about how carbon capture and storage or CCUS can be counted on (eventually) to make existing power plants "clean" by removing carbon dioxide at the power plant stack. But external investigations such as that done by the New York Times in 2023, have observed:

> None of the nation's 3,400 coal- and gas-fired power plants are currently using carbon capture technology in a significant way, raising questions about the viability of that approach. [236]

This is both good and bad news for EVs. First, it suggests that the hopeful objective of capturing 90% of carbon from power generation facilities will probably never be cost or energy effective. Second, it means that BEVs charged in the USA will continue to produce significant CO_2 at the electricity source. Overall, the substitution of new EVs for new ICVs in the near future can be counted on for no more than a 40% reduction in their comparable GHGs over a 15-year operational lifetime. EVs will help to lower but cannot eliminate carbon emissions from transportation.

Of course, this situation will improve as the overall carbon footprint of electric power generation is reduced. Such an improvement in the carbon emissions of electricity is now more necessary than ever as so many sectors of energy use shift from fossil fuels to electricity all at the same time.

It is common to find comparisons of the total GHG emissions of electric vs. IC engine vehicles published by advocates of each. Indeed, the lack of tailpipe emissions makes EVs hugely beneficial in areas prone to pollution of regulated emissions (HC, CO, NO_x, particulates), improving the health of urban residents. But it is a different situation for GHGs.

As discussed in earlier chapters, it is important to distinguish between automotive GHG emissions and EPA-regulated (toxic and carcinogenic) emissions. Regulated emissions are directly toxic or carcinogenic, with acute health effects for all life that are proximate to the point where they are emitted, i.e., freeways and dense urban areas. But GHGs are usually not directly toxic, nor even reactive. It is the *homogeneous* accumulation of GHGs in the atmosphere that causes harm by altering the thermal balance of the Earth. The global warming effects of atmospheric GHGs *are independent of the location where they are emitted.* A given quantity of carbon dioxide emitted by a diesel truck on a Los Angeles freeway or a coal power plant in Indiana (USA) *have exactly the same impact on global warming and climate change*. CO_2 emissions must be treated and remediated very differently than regulated emissions. There is no automotive emission control system that can reduce carbon emissions. This makes it critical to distinguish between a car with clean regulated emissions and a car that has low GHG emissions, whether they are released at a tailpipe or at a coal power plant in a thousand miles away.

In addition to emissions from per-km (mile) energy use, a significant part of the GHG footprint of any type of powered vehicle occurs in materials production, vehicle manufacturing, and ancillary (called Tier-3) emissions from suppliers to manufacturers. Among these are:

CO_2 Emissions Specific to EVs

The mining, manufacturing, and recycling required for lithium-based batteries, especially those that include nickel, cobalt, manganese, aluminum, or copper, are a large *additional* CO_2 generator that penalize only EVs.

CO_2 Emissions Specific to ICVs

The extraction, refining, and distribution of gasoline or diesel fuel are large additional CO_2 generators that penalize only ICVs. These factors are hard to fully and fairly quantify and are subject to wide variation due to different assumptions. The effects of these variations can profoundly influence the conclusions of comparative analyses, even in reports from highly credible sources such as the US EPA (pro-EVs) or the US Department of Agriculture (pro-biofuels).

EV Data Soup

Like all areas of energy (fossil fuel or alternative fuel), completely unbiased results are difficult to come by, sometimes even from government agencies. Climate impact comparisons between EVs and ICVs are particularly susceptible to manipulation to favor a particular conclusion. For example, consider the well-intended excerpt below from an online EPA report that has been widely cited by commercial websites.

Example 1: EPA-420-F-18-008 [237]

From Office of Transportation and Air Quality EPA-420-F-18-008, March 2018.

> **Greenhouse Gas Emissions from a Typical Passenger Vehicle: Questions and Answers.**
>
> The U.S. Environmental Protection Agency (EPA) developed this fact sheet to answer common questions about greenhouse gas emissions from passenger vehicles. This fact sheet provides emission rates and calculations consistent with EPA's regulatory work.
>
> **How much tailpipe carbon dioxide (CO_2) is emitted from driving one mile?**
>
> The average passenger vehicle emits about 404 grams of CO2 per mile. This number can vary based on two factors: the fuel economy of the vehicle and the amount of carbon in the vehicle's fuel. *Most vehicles on the road in the U.S. today are gasoline vehicles, and they average about 22.0 miles per gallon.* Every gallon of gasoline creates about 8,887 grams of CO_2 when burned. (Italics for emphasis.)

Therefore, the average vehicle when driving one mile has tailpipe CO_2 emissions of about:

$$\text{Annual } CO_2 \text{ emissions} = \frac{CO_2 \text{ per gallon}}{\text{MPG}} \times \text{miles} = \frac{8887}{22.0} \times 11{,}500$$
$$= \mathbf{4.6\ metric\ tons}$$

A typical passenger vehicle emits about 4.6 metric tons of carbon dioxide per year. This number can vary based on a vehicle's fuel, fuel economy, and the number of miles driven per year.

The reported 404 g CO_2/mile figure stated for "most vehicles on the road today" is consistent with the assumption of a 22 MPG ICV, since CO_2 emissions for gasoline are proportional to fuel consumption. The 2018 state of the art for electric vehicles is then compared with an ICV cohort consisting of "most vehicles on the road in the USA." This includes everything from a 1980 Yugo to a 9-passenger Chevy Suburban, and the muscle cars from the 60's that inspired jokes about the gas pedal and the fuel gauge being coupled together. The argument seems to be "look at how much better the world would be if every car on the road today was replaced with a new electric car." The comparison sounds reasonable, even inspiring. No doubt the 22 mpg used in the comparison was a correct *average for every passenger vehicle on the road* in the USA. But the comparison of relevance to someone contemplating the purchase of a new or used EV or ICE is between equivalent vehicles, or at least between vehicles of the same size and model years.

The last year that the USA CAFE (Corporate Average Fuel Economy) standard was 22 mpg was 1981 [238]. The median vehicle in that population was a 22 mpg 1981 Ford 150 truck with a V8 engine. The comparison can therefore be stated as between a highly efficient electric vehicle in 2018 and a 1981 gasoline full-size Ford truck. This is hardly a compelling argument for EVs.

A more appropriate comparison using the same data cited in the EPA analysis might be between two *equivalent* vehicles of the same generation, e.g., *the two highest MPGe vehicles of each type, electric vs ICE*. This would reflect the actual choice to be made by a car buyer. That would pair the 2018 world MPGe leader Tesla Model 3 that achieves 127 MPGe with the highest mileage gasoline-only car in the same year, the 2018 Toyota (non-plug) Prius that achieves an EPA-rated combined 57 MPG. Following the same simple arithmetic, the gasoline-fueled Toyota would produce annual CO_2 emissions of

$$\frac{8887 \text{ g}CO_2 \text{ per gallon}}{57 \text{ MPG}} \times 11{,}500 \text{ mi / year} = 1.8 \text{ metric tons } CO_2$$

which differs significantly from the 4.6 metric tons assumed in the EPA analysis.

Now take the EPA 127 MPGe rating of the Tesla 3 (based on the EPA's gasoline/kWh equivalence factor of 3.77 kWh/GGe), and use the 2018 DOE number from the EPA report's analysis for the carbon intensity of the US electric power grid including both renewable and non-renewable sources, 392 gCO_2 per kWh. The comparable EV emissions of the Tesla are

$$\frac{392\ \text{gCO}_2\ \text{per kWh}}{\left(\dfrac{127\ \text{MPG}_e}{33.7\ \text{kWh / gal eqivalent}}\right)\left(0.94^*\right)} \times 11{,}500\ \text{mi / year} = 1.3\ \text{metric tons CO}_2$$

[*] The 0.94 (94%) denominator factor is the 2018 US average electrical transmission efficiency between the point of generation and the point of use. The inclusion of this factor is necessary because some of the electric energy produced at the power plant is lost on the way from the generator to car's battery. This number is conservative, since it does not include local distribution losses or the 80% efficiency of a high-rate charger when it is occasionally used.

From this simple operating energy/fuel comparison for each vehicle, the EV provides a 28% CO_2 reduction (1.3 ton_e for the EV vs 1.8 ton_e for the ICV). A positive but much smaller improvement compared with the extreme conclusion of the EPA tutorial.

The carbon intensity of electricity generation is the largest factor affecting EV CO_2 emissions. But we (hopefully) can count on CO_2/kWh decreasing in the future as additional renewable electricity *with the necessary energy storage* comes online. This would continue to make EV's net cleaner than ICVs. But the average EPA-reported mileage for ICVs has also followed a steeply improving trend, and will probably continue to do so for a few more years. The 2023 CAFE requirement for passenger cars was 50.5 mpg, which increased to 52.9 mpg in 2024 and was planned to increase to 55.3 mpg in 2025 [239] (although recent political changes in the USA may dramatically reduce this target).

The simple comparisons above addressed only the operating emissions for the ICV and the EV, not the lifetime CO_2 emissions accounting for manufacturing, servicing and end of life. A more complete analysis of this and other scenarios will be presented later in the chapter. As will be seen in the following section, the non-operating emissions are much more significant for EVs than ICVs, mostly due to battery production.

Example 2: EPA Website "Electric Vehicle Myths"

A more recent (2024) published comparison can be found in another online EPA report *Electric Vehicle Myths,* intended to debunk popular criticisms of EVs [240]. Quoting the web page:

> **Myth #1:** Electric vehicles are worse for the climate than gasoline cars because of the power plant emissions.
>
> **FACT:** Electric vehicles typically have a smaller carbon footprint than gasoline cars, even when accounting for the electricity used for charging.
>
> Assumptions: EV with 300-mile range; vehicle lifetime of 173,151 miles for both EV and gas car; 30.7 mpg gas car; and U.S. average grid emissions.

In the accompanying analysis, the US overall CO_2 grid emissions, including renewables, were assumed to be *200 g/kWh*. This number varies between different agencies or reports from the same agency. The US EIA reported that in 2023 this number was 367 g/kWh [241]. The EPA on one website reported 373 g/kWh for 2023 [242]. On another EPA site, for 2022: 394 g/kWh [243]. The mean of these numbers is 378 g CO_2/kWh.

Why the difference? Usually because of the use of a selected local rather than the US average for electric power, including all sources (non-renewable and renewable), often (as in the present case) without this clarification. A 2022 report [244] by a group at MIT pointed out that the scope of the observation makes a huge difference, since states such as Oregon or Washington in the northwest USA derive their power almost entirely from zero-carbon hydroelectric, while states such as Wyoming or West Virginia derive most of their electricity from coal:

> The report shows how much these stats can swing based on a few key factors. For example, when the researchers used the average carbon intensity of America's power grid, they found that a fully electric vehicle emits about 25 percent less carbon than a comparable hybrid car. But if they ran the numbers assuming the EV would charge up in hydropower-heavy Washington State, they found it would emit 61 percent less carbon than the hybrid. When they did the math for coal-heavy West Virginia, the EV actually created more carbon emissions than the hybrid, but still less than the gasoline car.

The analysis also assumed that the average fuel economy of a comparable ICV is 30.7 mpg. This fuel economy figure is more realistic than the 22 mpg used in the previous EPA example, although it was not clear what was being considered an equivalent ICV, with 33.3 mpg reported for 2022 [245] from certification data, and according to Code of Federal Regulations *ecfr.gov* [246], the minimum domestic passenger car standard (MDPCS) in 2023 was 41.48 mpg [247].

The EPA Electric Vehicle Myths report did factor in emissions for the manufacturing and end of life for each type of vehicle, concluding that 18% of the cumulative 150 g CO_2e/mile reported for the EV, or 27 g CO_2e/kWh capacity, came from the manufacturing of the battery. Battery manufacturing data from multiple sources vary from 61 to over 400 kg [248] CO_2e/kWh of battery capacity, mostly related to the country of manufacture. As will be justified in the following section, we will use 152 kg CO_2e/kWh of battery capacity. We can check this by referencing the top-mpg_e 2022 Tesla 3 Long Range that has a 75 kWh battery that is assumed to last the entire 173,151 miles, over the 15-year lifetime of the vehicle:

$$\frac{(152\ \text{kg CO}_2\ /\ \text{kWh})(75\ \text{kWh})}{173{,}151\ \text{mi vehicle life}} = 65.8\ \text{g CO}_2\ /\ \text{mile of vehicle life}$$

Amortizing the battery manufacturing emissions over the vehicle lifetime works out to 65.8 *g CO_2e/mile* or about 44% of the reported 150 g CO_2e/mile total CO_2e, which differs from the the 18% reported. A more detailed analysis of this comparison will be presented in case studies K and L in the CO_2 emissions comparison section below.

About EV Battery Life

Another important consideration apropos to lifetime EV CO_{2e} emissions analyses is the possibility of battery replacement during the EV lifetime. The impact on the lifetime of CO_2 emission can be very significant.

The useful service life of an EV battery remains a matter of some disagreement and much speculation [244]. According to Motor Trend Magazine in 2023 [249], Tesla CEO Musk stated that

> The battery pack in the Model 3 and Model Y was designed to last 1,500 charging cycles, which translates to about 300,000 miles for Standard Range models and about 500,000 miles for Long Range versions.

Other published battery lifetime predictions vary: 8–12 years [250], 10 years [251], or 15 years [252]. This variability is understandable since there are many different variations of lithium battery chemistries. Any assessment of the lifetime emissions of an EV will depend upon the ability of the battery to last the lifetime assumed in the analysis. It is probably too soon for reliable evidence-based numbers, since even the first Tesla S models are just now reaching 12 years of age. Motor Trend Magazine addressed this question by pointing to the fact that the Tesla Model 3 RWD has a battery warranty of 8 years or 100,000 miles (equal to 6.7 years at 15,000 miles per year). The US Advanced Battery Consortium (USABC) [252] defines "dead" as 80% of the original capacity. Tesla defines the warranty service limit as 70% [253]. BYD (China) uses 60% as the end-of-service limit [254].

In 2023, the average lifetime of an automobile or light truck reported by the US Department of Transportation and Autotrader.com was 12.5 years [255, 256]. Federal law in the US requires that the battery life of an EV be warranted for 8 years or 100,000 miles, *whichever comes first* [257]. Only in the most optimistic case is it reasonable to assume that an average EV battery will last 15 years or 173,151 miles as assumed in the EPA *Myths* analysis above. This is important to the carbon footprint analysis. If the car is driven an average of 15,000 miles every year, a warranty limited to 100,000 miles would expire in 6.7 years. If the EV needs a replacement battery at any time before the end of the 15-year vehicle lifetime, its *battery manufacturing and end-of-life* carbon impacts *will double*.

This dichotomy leads to another little-discussed consequence, to the extent that anyone is concerned about it: Due to the simplicity of the drivetrain of an EV, it has the potential to last much longer than an ICV, *except for the battery*. A 12-year-old EV with a dead battery has very little residual value, and the cost of replacing the battery, between $5000 for a hybrid battery and $30,000 [258] for a long-range sedan or truck battery, will probably exceed the residual vehicle value. From a financial point of view, the logical choice is to replace the car. This suggests that the practical lifetime of an EV is the lifetime of the battery, not the longer possible lifetime due to the simpler drivetrain as often assumed. This is not the case for ICVs. The engine and drivetrain of an ICV indeed requires more maintenance, and certain components (e.g., accessory and cam drive belts) can be expected to need replacement during its lifetime, but it is unusual for the entire engine to need replacement before the vehicle's end of life.

This situation is exacerbated by the "do-not-touch" contractual requirements that accompany the sale of many EVs. The truncated lifetime of the EV and exclusive dealer-only service requirement are hidden costs to the vehicle owner, and sources of increased revenue for manufacturers and dealers. This also conflicts with the environmental benefits of retaining an asset for as long as it is serviceable. The "right to repair" movement that has led to consumer protection laws in 27 states (as of 2024) affects automobiles in general, but almost all states exempt repairs or access to the battery or high-voltage systems in the interest of safety [259].

Based on limited data on EVs that have been "retired", the trend among EV owners has been to replace the vehicle rather than the battery when it has degraded to an unacceptable degree [260]. This creates two scenarios when attempting to assess lifetime EV emissions: (1) assume that the battery will last the full vehicle life, usually 15 years, even though the evidence suggests otherwise, or (2) replace the battery at the end of its useful life, which could be as short as an 8-year warranty or twice that long. The effects on the lifetime CO_2 emissions are very significant. Partially redeeming is that when the vehicle is eventually retired during the life of its second battery, the undamaged EV battery can be repurposed at reduced capacity as a stationary energy storage battery, such as for solar installations.

Lifetime EV and ICV CO_{2e} Emission Calculations

There are many facets to environmental and energy sustainability. CO_2 and energy efficiency are ultimately but not exclusively important. There are very sophisticated simulation models, notably the National Renewable Energy Laboratory *Battery Lifetime Analysis and Simulation Tool for Vehicles (BLAST-V) model* that have been used to generate the hourly EV electrical loads given electricity generation scenarios specific to each US state, and for a generic range of vehicle types segregated by type (BEV or PHEV) and range (10 to 300 miles). The model also incorporates data on the behavioral driving tendencies of EV drivers [261].

But reasonably accurate comparisons of GHG emissions can be made based on only a few dominant factors. I put together and included with this book an Excel spreadsheet for calculation and comparison of lifetime CO_2 emissions of various vehicles. It can be easily modified by readers. It requires only data that is readily available to the public online from the vehicle manufacturers and the US EIA/EPA or EU EEA websites. I use this spreadsheet below to estimate and plot year-by-year cumulative CO_2 emissions for any vehicle, EV or ICV.

This reader may note how easily results can be driven by the choice of data, assumptions, and comparison cohorts. The data sources for all numbers used in the test cases below were obtained from 2018 to 2023 scientific publications, individually cited below.

The plotted results from nine of the ten comparisons are presented below. The spreadsheet can be downloaded free and updated as desired using current data:

https://*www.springer.com/Non-petroleumTransportation/CO2_emissions_comparison_tool*

Data for Test Cases

In all of the comparative analyses examined below, the useful life of the batteries of all electric or hybrid vehicles were assumed to be the same, 15 years, to match the practice used by the DOE, EPA, and most published university research reports. This was done only in the interest of consistency and transferability of results, aware that this exceeds auto industry vehicle data and the lifetime reflected in battery warranties. The replacement of the battery prior to the end of vehicle life can be easily accommodated in the spreadsheet analysis.

The CO_2 consequences of both the vehicle and battery manufacturing are highly variable depending on country and data source; mean values of multiple sources had to be used, many from a 2018 meta-study by the ICCT [262]. The CO_2 emissions from the extraction and refining of gasoline were derived from industry data on refinery efficiency and carbon footprint.

This analysis does not include the effect of the time value of energy/money.

In the sample case plotted below, the carbon intensity of electricity was 394 g CO_2/kWh (from all energy sources), based on 2023 EPRI (390 g CO_2/kWh) and EPA (401 g CO_2/kWh) data, except where noted.

The results of each comparison are illustrated via line plots of the *cumulative* annual CO_2 generated by each vehicle over its 15-year lifetime, starting with manufacturing and ending with recycling/salvage, if any.

Case A compares a 2021 Tesla Model 3 with a 2012 Honda Civic gasoline (weak) hybrid ICV.[24]

The expected conclusion is confirmed: While the ICV starts its life with lower cumulative CO_2e emissions due to the emissions of battery manufacturing, the EV reaches parity in 3.5 years, and produces about 40% lower CO_2 total emissions over a 15-year lifetime. Due to the initial manufacturing emissions contribution, a longer lifetime would linearly increase this margin.

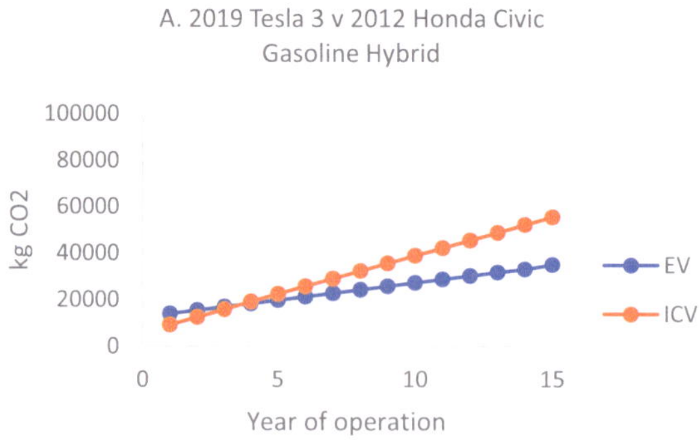

[24] Note that a non-plug hybrid is just a high-mileage ICV since all of its fuel energy comes from gasoline.

Case B compares a 127 mpge 2023 Tesla Model 3 with a 57 mpg 2023 Toyota Prius (non-plug) gasoline hybrid. (Again, a non-plug hybrid is just a high-mileage ICV.) Both represent the highest mileage vehicles of each type in 2023.

This is a more fair baseline instead of the *average of all gasoline cars on the road* used in EPA-420-F-18-008. The CO_2 emissions of each are nearly identical, with the EV slightly favorable at 15 years, but the first 5 years favoring the ICV.

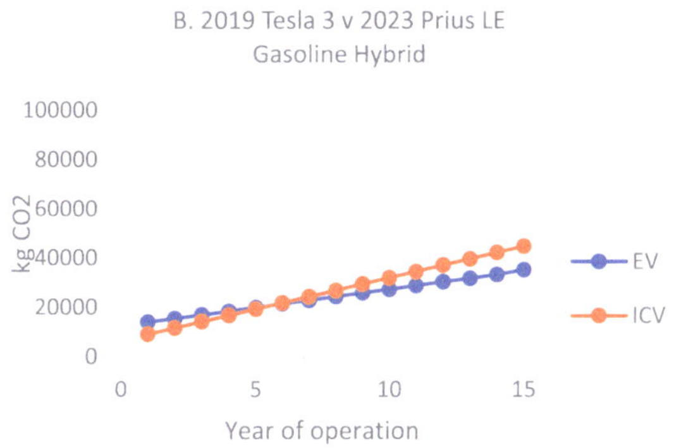

Case C is the unrealistic comparison used in EPA-420-F-18-008 [263] that paired a state-of-the-art 2018 EV with a 22 mpg ICV. Obviously, a 22 mpg ICV generates much higher CO_2 emissions than a 2018 Tesla Model 3.

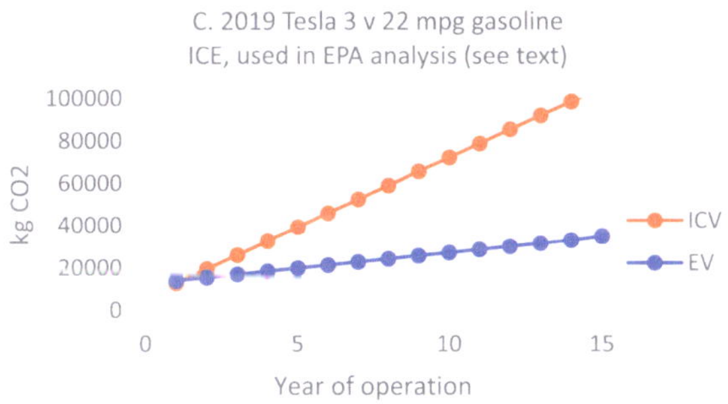

Case D was something that potential truck buyers are curious about. It compares one of the most talked-about new EVs, the 2023 Tesla Cybertruck AWD, with a 2023 gasoline truck having almost identical capabilities, a Ford F150 Regular Cab 4WD. The Cybertruck, despite its greater size and mass, achieves much lower lifetime CO_{2e} emissions than the gasoline truck, although the exotic materials used in the manufacture of the Tesla could slightly skew the results if they could have been considered.

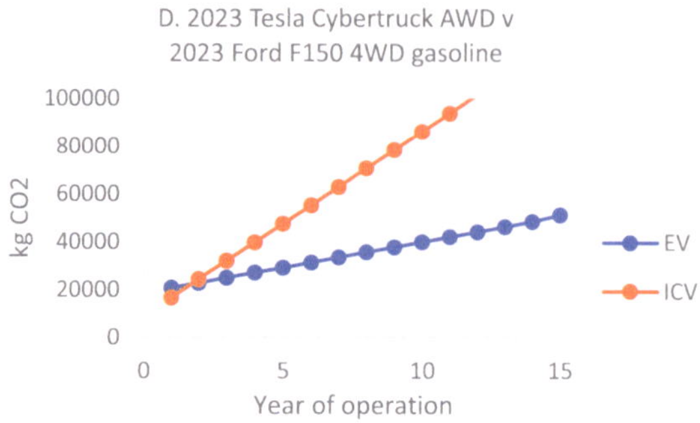

Case E compares a 2023 GM Hummer long-range electric truck with the same 2023 Ford F150 4WD gasoline truck as Case D. This is a comparison between two grossly inefficient vehicles. The lifetime CO_{2e} emissions of the 9600 lb electric Hummer are lower than the F150 at the end of life, but the crossover point is delayed until 7 years of operation. If the vehicle is owned less than 7 years from new, the Ford F-150 gasoline truck would be the more environmentally beneficial option.

Case F compares two "Premium Midsize"* luxury sedans, the 2021 Tesla S P100D with the 2021 Mercedes C220.

The choice of the Mercedes C220 luxury sedan is probably unfair to ICVs when its initial and operating CO_2 emissions are considered. Lifetime CO_2e emissions clearly favor the Tesla by more than 2:1, and the crossover point occurs in the second year of operation.

[*] Note: Tesla defines the "Premium Midsize" class in its annual impact reports to conclude that the Tesla S *produces 60% lower* lifetime CO_2 emissions compared with luxury ICVs such as the Mercedes C220. The most extreme CO_2 differences are found in this grouping, since very few ICV sedans that could be called "Premium Midsize" are known for good mileage or low CO_2.

Case G takes advantage of an increasingly rare car model offering: The 2023 Geely/Volvo X40 SUV could be purchased as either a gasoline (non-plug) hybrid or as a full BEV. Other than the powertrains, these vehicles look and drive identically.

The plots show that after 15 years, the electric version generates about half the lifetime CO_{2e} as the gasoline model. The year one emissions include manufacturing CO_2, and are nearly identical. This might be attributed to the combination of the larger-than-usual hybrid battery, and its lower-than-average 27 mpg gasoline fuel economy.

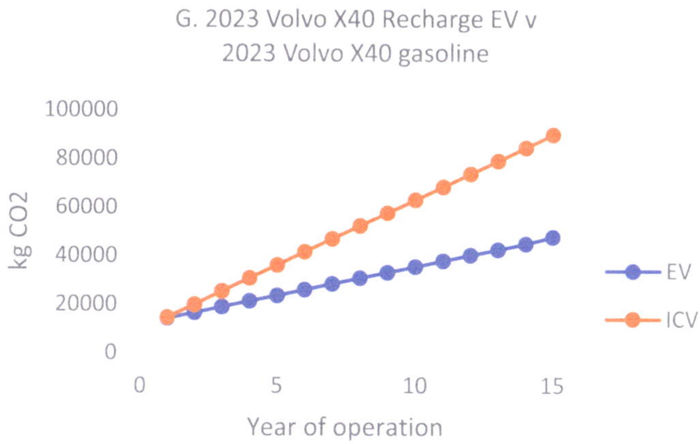

Case H compares an EV with an ICE that *runs on a fuel other than gasoline*. The Geeley/Volvo X40 EV is compared with the same X40 ICV used in Case G, except that the fuel system is modified to use M100 (methanol), as described in the Chap. 6. For this comparison, the methanol is produced from natural gas and the EV uses average grid electricity. If the methanol was produced from a renewable source, its net carbon footprint would be much lower, and its net CO_2 graph would be nearly flat after the first year.

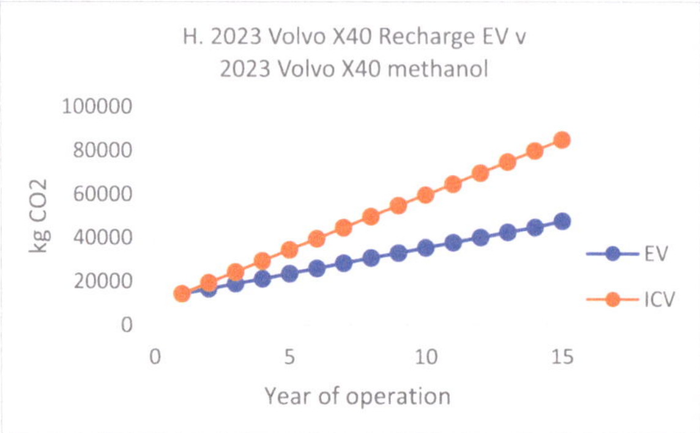

Case I compares the average efficiency of all MY 2021 EVs with the average efficiency of all MY 2021 ICVs. The vehicle mixes are similar but not certainly not 1:1 comparison sets.

The conclusions are similar to previously discussed results. In the 2021 MY, EVs on average produced about 40% lower CO_2 emissions over a 15-year lifetime, with the mean crossover point at about 2.2 years.

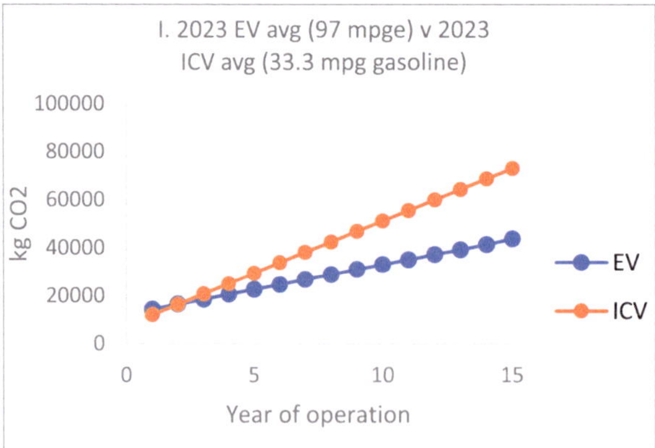

Case J was an attempt to estimate the gasoline passenger car mpg that would create CO_2 emissions per mile equivalent to the average of all electric passenger cars in 2021. It is the same as Case I except the ICV average CO_2 emissions (and subsequent mpg) were varied iteratively until the EV and ICV plots overlaid as close as possible. *The final iteration shows that a 55 mpg gasoline car would have about the same CO_2 impact as the average 2023 passenger electric vehicle.* For reference, the 2021 CAFE requirement for passenger cars was 40.1 mpg.

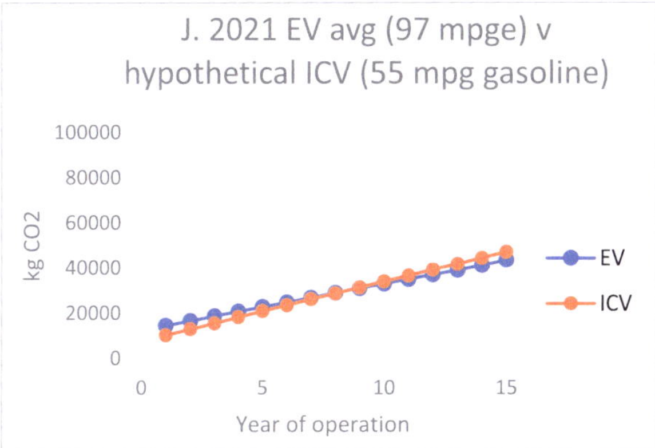

Case K is the example from the 2024 EPA online report *Electric Vehicle Myths* that used 200 g CO_2/kWh for the carbon intensity of the charging electricity and did not include distribution losses. As expected, the results conclude that at *the end of its service life, the ICV has generated 6.7 times as much CO_2 as the EV.*

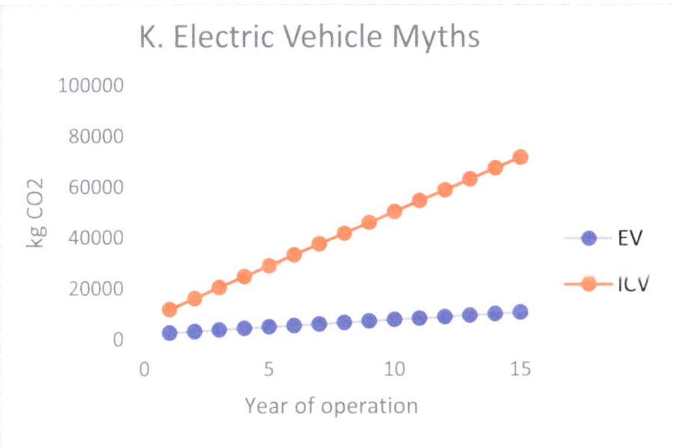

Case L is repeats case K except that 394 g CO_2/kWh is used for the grid carbon intensity, and it included 6% grid distribution and losses. At the end of its service life, *the ICV has generated 2.0 as much CO_2 as the EV, instead of 6.7 times as much reported in Case K.*

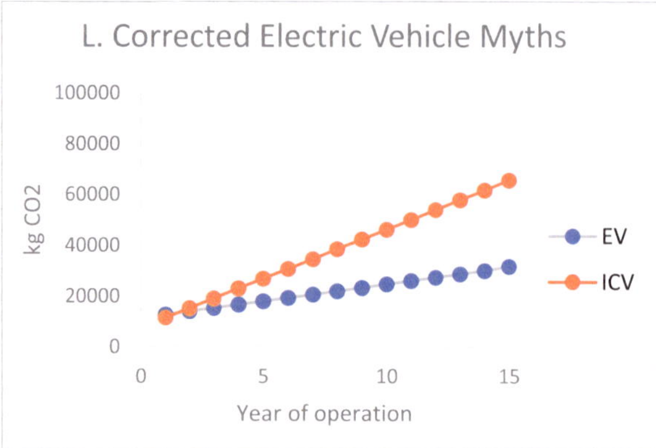

Case L vs. **Case K** demonstrate the large difference that a seemingly small difference in assumptions can make in the results.

Comments on the results of test cases

1. Electric vehicles, considered over their lifetime, are consistently less impactful to the climate than equivalent ICVs, as long as the battery is not replaced during its 15-year lifetime. Replacement of the battery at the end of an 8-year battery warranty could restart the GHG accumulation clock, as much as doubling the fraction of its life in which its emissions are inferior to an equivalent ICV or hybrid.
2. Early in the life of an EV, while the vehicle and battery manufacturing impacts still dominate, the cumulative CO_2 emissions of EVs are no better, and usually worse than equivalent ICVs. For example, Case A showed that in the first 4 years, a 2012 Honda Civic Gasoline Hybrid had a lower cumulative carbon impact than a 2019 Tesla 3 (which in that year had the highest efficiency rating of all EVs).
3. The carbon intensity of the electric power grid is the single most significant factor affecting the climate performance of EVs. *Reduction of GHG emissions from electric power generation* would almost surely be the most effective measure that could reduce the climate impact of EVs, even considering their greater start-of-life CO_2 footprint. While not a factor in the spreadsheet analysis, the time of day during which the EV is charged has a significant impact on the carbon intensity of the electric power. An EV charging during peak demand hours, say, 5–9 PM uses power from fossil fuel peaker plants that produce higher CO_2 emissions than the overall grid average. And intermittent generator such as solar or wind available only during off-peak periods contributes nothing to alleviate this problem unless an equal amount of electric energy storage receives and returns the intermittent energy to charge the EV.

Marketing of EVs

In an attempt to scratch the surface of the economic sustainability of EVs, a brief look at the current and expected near-term future of EVs is appropriate. There is much more to this topic that is outside the focus of this book. One of several comprehensive marketing analyses can be found in a 2023 McKinsey & Co. report [264]. Figure 14.105 is copied from the above-cited report, which shows a significant change in attitudes toward EVs over the course of one year, 2021–2022:

Product marketing is essential to all commerce. In the marketing of automobiles, a legacy of overreaching claims and nondisclosed limitations are often just part of what the industry refers to as "marketing strategy." Performance, fuel economy and low emissions play a role in automotive marketing, but buyers are just as concerned about non-technical attributes such as vehicle image. For over a century, automobiles have been much more than a form of transportation; they are symbols of our identity or status, and this does not change for EVs. What makes an EV a more desirable purchase than an equivalent ICV? Intended use, government subsidies, operating cost, appearance, power and the satisfaction of doing a good deed are certainly factors. But once the decision is made to go with an EV, the top of the wish list is almost always range, with recharge time a close second. The range and power wars previously discussed are simply attempts to meet these needs to achieve commercial success for products.

As the analyses of the previous section confirm, electric vehicles are consistently less harmful to the environment than gasoline or diesel vehicles. But the improvement is not as sweeping as consumers have been led to believe, and the necessary electric distribution, generation and charging infrastructure lag electric vehicle

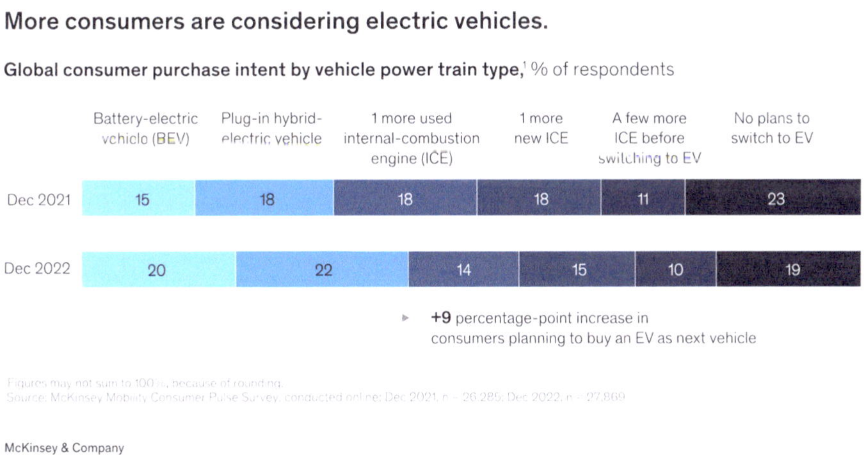

Fig. 14.105 2023 survey of consumer attitudes toward the possible purchase of an electric vehicle. From "Electric-vehicle buyers demand new experiences", July 2023, McKinsey & Company, www.mckinsey.com.

sales. These realities are slowly catching up with EVs. Starting in late 2023, an unexpected downturn of interest in electric vehicles caught the industry by surprise. Some media commentators have speculated that the "EV bubble finally burst" [265], while others see this as "just a speed bump" [266]. There seems to be growing suspicion that the promotion of EVs has creating unrealistic expectations resulting in some amount of owner backlash.

Green sells. Two random examples of recent automotive products marketed with overstated claims of dramatic efficiency or environmental benefits are shown in Figs. 14.106 and 14.107.

Specific to EVs, a frank assessment by the UK fleet financing firm *FleetEvolution in 2024* addressed the range issue and how overly aggressive promotion may have been detrimental to the proliferation of EVs [267].

Fig. 14.106 A recurring scam: magnet on a fuel line "aligns molecules of gas to improve engine efficiency." After decades of debunking, these devices still sell. Image from AliExpress, public domain

Fig. 14.107 Improved fuel economy without guilt: a 2010 Volkswagen Golf TDI "clean diesel" at the 2010 Washington Auto Show. The Volkswagen TDI diesel emissions scandal, often referred to as "Dieselgate," broke in September 2015. Photo by Mariordo Ortiz 27 January 2010. Wikipedia CC 3.0 Unported license

Table 14.3 EPA real-world mileage and GHG emissions for 2021-2023 cars and light trucks

Regu... Class	Vehicle Type	Model Year	Product... Share	Real-world MPG	Real-world MPG_City	Real-world MPG_Hwy	Real-world CO2 (g/mi)	Real-world CO2_City (g/mi)	Real-world CO2_Hwy	Weight (lbs)	Horsepower (HP)	Footprint (sq. ft.)
Car	All Car	2021	0.371	31.8	27.6	36.1	272	316	240	3631	214.6	46.9
Car	Car SUV	2021	0.114	31.0	27.5	34.2	278	314	250	3786	214.8	46.8
Car	Sedan/Wagon	2021	0.257	32.2	27.6	37.0	270	317	235	3562	214.4	46.9
Car	All Car	Prelim. 2023	-	34.9	30.8	38.8	239	274	213	3797	238.4	47.5
Car	Car SUV	Prelim. 2023	-	36.8	33.7	39.6	219	243	201	4010	251.8	47.8
Car	Sedan/Wagon	Prelim. 2023	-	33.9	29.4	38.4	250	291	220	3682	231.2	47.3

US EPA Estimated Real-World Fuel Economy, CO_2 Emissions, and Vehicle Attributes, December 2023. https://www.epa.gov/automotive-trends/explore-automotive-trends-data#SummaryData. Public domain

Other examples abound, but a convenient one to pick on is Tesla's 2022 online annual "Impact Report: *Paving the Way for a Sustainable Future*" [268] in which Tesla EVs are compared with ICVs.

As of 2024, after a decade of leadership in EV innovation and production, Tesla Inc. now shares this distinction with BYD of China. Tesla's engineering is among the best in the auto industry, but marketers will be marketers. For example, the 2022 Tesla annual Impact Report states:

> For ICE fuel consumption, we used data provided by Consumer Reports, which reports model year 2021 mid-size *premium sedans*[25] achieve ***24.3 MPG*** on average. This translates to over ***400 grams of CO_2e*** per mile once we account for emissions generated through the extraction, refining and shipment of oil.

However, according to "real-world" (sampled user) data published by the EPA [269], every 2021 gasoline passenger automobile that is comparable to the 2021 Tesla models achieved significantly more than the assumed 24.3 mpg and produced less than the stated 400 g CO_2e/mile, even accounting for gasoline production emissions. Table 14.3 from the US EPA provides "real-world" data for all MY2021 through 2023 vehicle classes, with the relevant data identified in the red box. Average real-world mileage and CO_2 emissions for *all cars* in 2021 were found to be 31.8 mpg and 272 g CO_2/mile respectively.

Great improvements have been made in just the past few years in electric and hybrid vehicles. But ICE technology has also improved significantly. So while the comparison numbers in Tesla's statement may have been accurate for the group of vehicles they selected for their comparison, it does not match the *real-world data* reported by the EPA for any vehicle class.

[25] *Mid-sized premium sedans* is an unspecified cohort of ICVs selected by Tesla, used in the 2022 Tesla *Impact Report*, not a classification defined by Consumer Reports or by Federal or State agencies.

EV Purchase Cost

The MSRPs of electric cars sold in the USA and Europe have always exceeded those of equivalent ICVs, although the margin has been shrinking due to increased competition, especially from China [270]. Table 14.4 shows the cost of the top ten best-selling EVs in the USA in January 2024.

As reported in citation [271]

> **The average price of the top ten electric vehicles in the US is about $53,758,** with an average of $48,430 for the low end trim of each model and $64,936 for the high-end trim of each model.

This price admits only the more affluent buyers, not the wider demography that the EV revolution was assumed to empower. The higher profit margins associated with expensive upscale vehicles continue to motivate EV manufacturers.

The total realized cost of ownership of an automobile includes not only the acquisition cost but also the ongoing operating costs of a vehicle. In an effort to improve this awareness, Fig. 14.108 is a copy of one of a set of "environmental postcards" printed by the Environmental Journal of Canada [272] reminding consumers that the purchase cost is not necessarily the main component of the total cost of ownership when comparing an EV to and ICV. Comparing two cars of equivalent size and capabilities, the 2022 Chevy Bolt has a calculated total cost of ownership over its lifetime that is lower than the 2022 Toyota Corolla, despite the significantly higher purchase cost of the Bolt.

Differential Cost to Purchase

Table 14.4 lists the purchase prices of the top-selling EVs in the USA in 2024. But Fig. 14.108 demonstrates that the purchase cost is actually less significant than the overall cost of ownership over the life of the vehicle.

According to the automotive valuation firm *Edmunds* in 2024 [273]:

Table 14.4 Average MSRP of the top ten best selling EVs in the USA, Q1 2024

Sales rank	Make and model	Price range	US units sold
1	**Tesla Model Y**	$43,990–$52,490	394,497
2	**Tesla Model 3**	$38,990–$45,990	220,910
3	**Chevrolet Bolt EV**	$26,599–$29,700	62,450
4	Ford Mustang Mach-E	$43,495–$65,000	40,771
5	VW ID.4	$38,995–$55,245	37,789
6	Hyundai IONIQ 5	$41,650–$53,350	33,918
7	Rivian R1S	$78,000–$99,000	24,783
8	Ford F-150 Lightning	$49,995–$91,995	24,165
9	Tesla Model X	$79,990–$94,990	23,015
10	Kia EV6	$42,600–$61,600	18,879

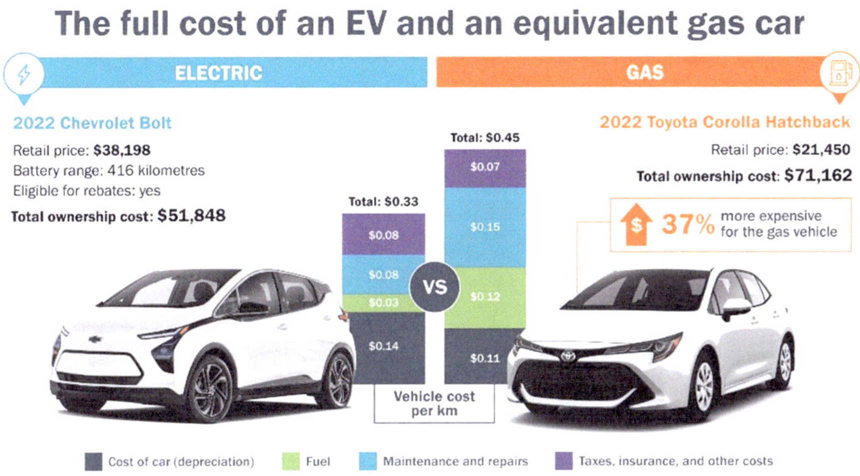

Fig. 14.108 Copy of an "Environmental postcard" by the Environmental Journal of Canada for the Canadian Transportation Agency, comparing acquisition and operating costs for two similar vehicles in 2022. https://www.canada.ca/en/environment-climatse-change/news/2023/12/canadas-electric-vehicle-availability-standard-regulated-targets-for-zero-emission-vehicles.html. Public domain

> An unpleasant truth about EVs is that they almost always have higher retail prices than their internal combustion engine (ICE) counterparts. In the first quarter of this year, there was a 42% gap in the average price of an electric car vs. a gas car.
>
> The biggest gap in MSRPs, according to Edmunds price data, was 58.5% in the popular and relatively crowded compact SUV category. The average fully electric model, with a starting MSRP of $53,048, was a wallet-draining $17,326 more than the average of $35,722 for a gas-burning compact crossover.
>
> The smallest gap was for large pickups: 18% at $76,475 for electrics versus $64,784 for ICE vehicles.

Across the board, EVs are pricier than ICVs, but the cost differential appears to be strongly related to the choice of the vehicles that are assumed to be comparable. The 2022 Chevy Bolt vs. Toyota Corolla comparison of Fig. 14.108 reports a 37% greater operating cost for the ICV compared to the EV, despite a 44% lower initial purchase cost.

The popular consensus seems to be that EVs cost 30–40% more than equivalent ICVs, usually justified by the cost of the battery, since the cost of the rest of the vehicle and powertrain is no more expensive than that of an ICV. Larger batteries (longer range) mean larger price differentials.

In the USA, until the 2022 Inflation Reduction Act (IRA) took effect in 2023, a generous Federal tax incentive of $7500 was provided for the purchase of any new EV. This blanket sum was modified under the IRA in 2023 to limit the incentive money to vehicles that met minimum US content requirements. For instance, the first half ($3750) of the credit required that at least a fraction of the car's battery minerals must be sourced from the US or from a country that has a free-trade agreement with the USA. The other $3750 required that a fraction of battery components be manufactured or assembled in these countries. A similar but even more protective subsidy program had already been implemented in China with great success in growing EV manufacturing at an unprecedented rapid pace. Starting 2025, it is uncertain what the incentive policy will eventually be, but it appears that the 7,500 tax credit is restored for EVs costing up to $80,000 USD, probably the first time in US history that taxpayer incentives were issued to buyers of luxury and high performance cars that are beyond the reach of the majority of consumers.

Differential Operating Cost

While affluent EV early opters probably did not consider operating costs an impediment to their purchase and usage decisions, as EVs have become mainstream, the total cost of ownership is an important consideration when comparing electric with non-electric purchase options.

With the exception of routine/required maintenance costs (e.g., tires, dealer-required "safety checks"), insurance, registration, and the amortized battery replacement cost, the cost of operation for an electric vehicle is almost entirely for the electricity that it consumes. This cost usually shows up on the utility's home power bill or credit/debit card statements for public charging.

In 2022, the EPA "Estimated Cost of Operation" found on the EPA Fuel Economy and Environment window sticker, example shown in Fig. 14.109 for an electric or plug hybrid vehicle. It was calculated *assuming that electricity costs $0.12 per kWh* [274], which was the national average base residential electric power cost in 2010 (or 1988 in Coastal California). It also assumes vehicle usage of 15,000 miles/year based on an assumed weekday commute distance of 57.5 miles, which is above the FHWA national average of 13,476 annual miles per year for all automobiles [275].

Electric power costs can vary significantly. The average residential E-1 base rate for electricity in California in 2024 [276] is listed by the CPUC as $0.1534/kWh. However, in Coastal California, the E-1 residential electric power rate is stated on monthly bills as [277]:

Tier 1 Usage (0–100% of baseline) $0.42676 (I)
Tier 2 Usage (101–400% of baseline) $0.53406 (I)
Tier 2 Usage continued (over 400% of baseline) $0.53406 (I)
"Smart Meter" fixed cost ($ per day even if no power used) $0.39167

Tiers 1 through 3 are time-of-use periods, with Tier 1 being low demand (night), Tier 2 being day except for Tier 3 period, which is the peak demand period usually 4–9 PM on weekdays. These times are set by the utility with the approval of the Public Utility Commission (PUC) and vary locally.

During the past 7+ years, rate increases by California's three major electric utilities have been approved with minimal debate by the California PUC, at least partially justified by the cumulative cost of hundreds (possibly thousands) of homeowner

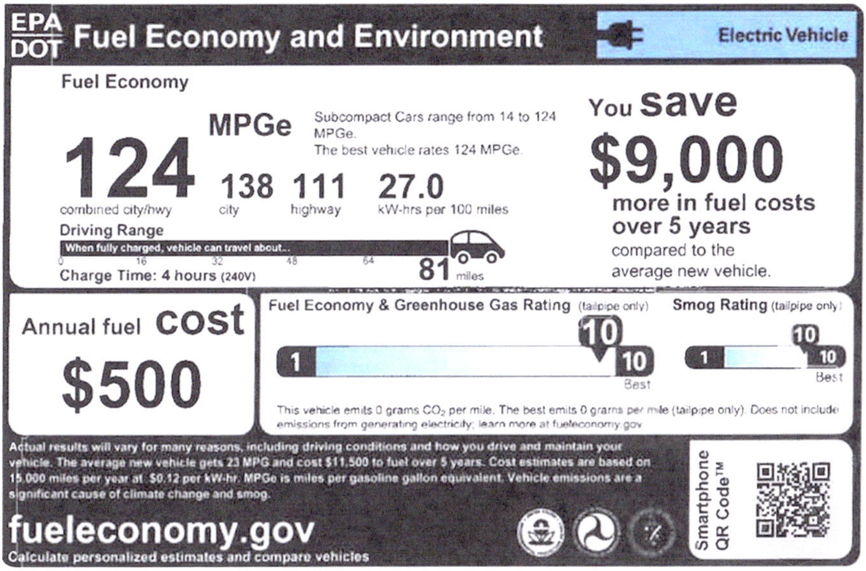

Fig. 14.109 EPA certification for 2017 Gen 2 Nissan Leaf with 30 kWh battery. Annual fuel cost was based on electric power at fixed rate of \$0.12/kWh. EPA publication. Public domain

lawsuits related to the summer wildfires in 2023 and 2024, attributed to poorly maintained transmission and distribution infrastructure. Concurrently, the electricity costs enjoyed by grid-tie solar PV households were torpedoed by ending solar *net metering* in 2023, despite the outcry of solar households. Of greatest interest to EV owners was the CPUC-approved restructuring of electricity costs for homes that charged EVs. The cost of electricity in these situations falls under two renamed rate schedules: Rate EV-A if vehicle charging power comes from the same meter as the rest of the home, or Rate EV-B, a slightly less expensive time-of-use rate that requires the homeowner to install and pay an additional monthly fee for a separate meter solely for the electricity used to charge an EV [278]. Energy units below are in kWh

Rate B			
Total energy rates (\$ per kWh)	PEAK	PART-PEAK	OFF-PEAK
Summer usage	\$0.72878	\$0.48467	\$0.37212
Winter usage	\$0.54624	\$0.41423	\$0.34250
Plus: 'Smart Meter' charge per day \$0.04928			

Assuming that the summer "part-peak" rate can be cited as a fair average, \$0.49 (including meter charge) can be used to equilibrate the per-km (per-mile) cost of a home-charged EV with the per-km cost of an equivalent ICV.

$$\left(\$0.49/\text{kWh charge}\right)/\left(5.15\,\text{km}/\text{kWh for 2023 Chevy Bolt}\right)$$
$$= \$0.095/\text{km}\left(\$0.153/\text{mile}\right)$$

Using a 2023 Honda Civic as an equivalent ICV, its gasoline fuel cost at 58 km/gal (36 miles/gal) makes the energy cost of operation of the EV equivalent to gasoline costing $5.51/gallon.

For reference, the national average cost of gasoline in the USA on May 15, 2024, was $3.607/gallon, while the average cost in Coastal California was $5.29/gallon [279].

Only a few years earlier, the "fuel" cost of an EV was considered nearly trivial compared with the amortized cost of the vehicle, especially the battery. But that is no longer the case in areas such as California that have higher electric power rates. The electric fuel cost per km or mile is now about the same or more than an ICV. If the EV is charged at a public charger, the per-kWh price is even higher. As of January 2024 power for level-II public charging costs between $0.41 and $0.50 USD [280] per kWh in California, according to data from Stable, an EV charger software and data supplier [281].

From a global perspective, the current (May 2024) world average retail cost of electricity is approximately $0.152 USD/kWh for residential users and $0.149 USD/kWh for business users [282].

The range of costs for Level 3 DC fast charging (DCFC) is reported by Stable to be between $0.51 and $0.56 nationwide, although some facilities in California are considerably higher (e.g., Tesla Superchargers). These fees are not directly regulated at this time, so they are whatever the market will bear. EV owners are a captive market.

Using the least expensive fast charger rate:

$$(\$0.51 / \text{kWh charge}) / (3.2 \text{ miles} / \text{kWh for 2023 Chevy Bolt}) = \$0.16 / \text{mile}$$

For the (2023) 36 miles/gal Honda Civic this is distance-equivalent to gasoline costing $5.76/gallon.

At the time of installation, most public EV chargers added a small premium to the actual cost of the electricity. The current and clear future trend is that operators build in a more significant margin, either as profit or to offset the usage value of the parking space at the charger. EV owners are captive to these facilities, sometimes just to get home from work. This is disturbing because the installation cost was heavily subsidized by state and federal grants and tax incentives.

Another sensitive issue among EV owners is the reliability and availability of public chargers. As discussed earlier in this chapter, it is (2024) estimated that at least 20% of all public chargers in the US are not operational. This is incongruous because provisions in the government grants and tax credits received by operators to install the chargers usually required a 95+% minimum level of availability. There is apparently no legal mechanism to enforce uptime provisions, and even if there were, local energy agencies or law enforcement have not gotten involved because it is outside their scope or jurisdiction [283].

Subsidies and Incentives for EVs

The financial situation for electric vehicles is more than a little artificial. Since 2008, the US Federal and California State governments have provided generous tax incentives and rebates to buyers of new EV. Even stronger financial incentives have been provided in China in an effort to support their already world-dominant EV and battery industry, and reduce or eliminate the need for petroleum imports.

As mentioned earlier, a Federal tax credit of $7500 USD is provided for vehicles meeting certain "country of origin" requirements. Some states such as California provide further direct and indirect incentives for EV purchases and ownership. The federal credit under the 2022 IRA has reduced the number of eligible vehicles at this time to only nine. Only five electric vehicle models qualify based on meeting minimum US content thresholds. Only one of these was a compact car, the Chevrolet Bolt, which in 2023 was discontinued and has been replaced in 2025 with a larger and more expensive car with the same name. *All others are SUVs or trucks.* Limitations are based on a buyer's household income and the vehicle cost. MSRP's greater than $80,000 USD are fully excluded, with other tiers depending upon income. A list with understandable requirements can be found at the Consumer Reports public website [284] and at Edmunds (USA) online [285].

Local governments also provide a menagerie of other credits and incentives directly to buyers, notably in Los Angeles and San Francisco California. Outside the view of the general public, taxpayer-funded incentives are also provided to the manufacturers, component suppliers, and local car dealers that sell and service the EVs. Considering the tax credits and despite the increasing cost per mile (km) of charging, the financial incentives for EV purchases have been very effective at promoting ownership, despite any deficiencies in the public charging infrastructure and the operational limitations of EVs.

But another lesser known reason has driven the high costs of EVs: higher profit margins, especially for the electric trucks and SUVs that now dominate the US market. As recent as Q1 2022, Tesla's gross margin on vehicle sales was reported to exceed 30% [286], which was previously unheard of in automotive manufacturing. Margins are currently falling as competition from China is driving lower costs internationally.

Considering the record tax incentives being provided for each new (and under the IRA, also used) EV purchase, this has led to criticism that EVs are subsidized playtoys for the rich [287], citing economic data showing that "80% of EV tax credits are claimed by people that are making at least $100,000 USD per year" [288] (the only purchasers that can afford EVs, even with the tax credits).

Also, as discussed in the Chap. 17, another non-obvious driver of costs is the elaborate and weakly regulated carbon credits (offset and regulatory credits) that have been controlling much of energy technology worldwide (especially in the USA) for over a decade. The carbon offset market is considered by some to be a surrogate international stock market, since once credits are issued, most can be freely traded and have become investment instruments. Carbon offset and regulatory credits are discussed in Chap. 17 *The artificial economy of subsidies*.

EVs as a Catalyst for Economic and Political Change

Only recently have the far-reaching impacts of the transition from petroleum to electric automobiles reached public awareness. How will the massive gasoline and diesel production and distribution industries adapt or decay? Tire shops will always be around, but what about the aftermarket parts industries that produce spark plugs, fuel injectors, and valves for IC engines, when EVs have far fewer parts to wear and replace? The labor needs of auto manufacturers will drop significantly. These are not new questions—the proliferation of electric and petroleum vehicles in 1900 lead to the collapse of the draft horse and carriage industries. Technology changes always have profound and rapid effects on society, but in the EV transition, these are felt disproportionately by lower income classes. The massive government incentives helping to drive the transition to electric vehicles are fairly accused: (1) they disproportionately benefit affluent consumers that can afford EVs and afford to live within an EV-commute distance from their places of employment, and (2) when incentives are sunsetted, the increased costs to buyers and users will drive even greater economic disparity.

As automotive transportation transitions to electric, world economic changes will follow. The trend has reshuffled the post-World War II economic world order from the USA and Western Countries toward the PRC and its economic partners. In the world automotive industry, this has been especially true for electric vehicles and alternative energy. The political ramifications are outside the scope of this book, but they are just the beginning of the collateral economic effects of the world transition away from petroleum.

As reported in Fig. 14.110, in December 2023, the PRC performed over 60% of the world's EV manufacturing, producing more electric cars and trucks than all other manufacturers in the world combined [289]. The PRC is also the largest market for EVs and leads in global exports (Fig. 14.111) [290]. And more than 90% of all EV batteries in the world, and 99+% of all LFP batteries are manufactured in China. The trend is similar in other emerging economic powerhouses in Southeast Asia, especially India and Vietnam. Supporting analyses and projections can be found in citation [291] among other world financial projections.

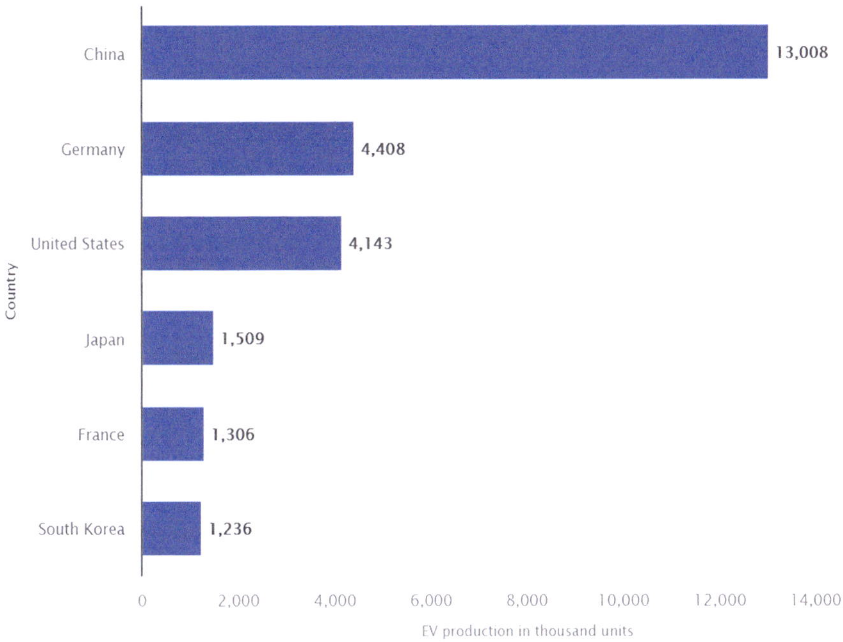

Fig. 14.110 Electric vehicle production 2018–2023, top six countries. (Data and graphic from Statistica, https://www.statista.com/statistics/270537/forecast-for-electric-car-production-in-selected-countries/ accessed 10 Sept 2023. University license

Fig. 14.111 EV production assembly line in Shenzhen, China, 2023. Creative Commons Attribution 3.0 Unported license

References

1. Zero-emission vehicle program. (California Air Resources Board), https://ww2.arb.ca.gov/our-work/programs/zero-emission-vehicle-program. Accessed 15 June 2023
2. M. Bellis, The history of electric vehicles began in 1830. (ThoughtCo History and Culture), https://www.thoughtco.com/history-of-electric-vehicles-1991603. Accessed 4 Jan 2024
3. A. Grove, J. Verne, His life and writings. (2019), https://www.thoughtco.com/jules-verne-biography-4151934. Accessed 4 Aug 2023
4. Malevus, Gustave Trouvé: the man who built the first electric car, https://malevus.com/gustave-trouve/. Accessed 1 June 2023
5. Electric Bike history. (2013), https://www.electricbike.com/e-bike-patents-from-the-1800s/. Accessed 1 Aug 2024
6. Image from *The Ayrton and Perry Electric Tricycle of 1881*. (UPS Battery Center), https://www.upsbatterycenter.com/blog/ayrton-and-perry-electric-tricycle/#prettyPhoto. Accessed 10 Sept 2021. Citation from New Atlas, publication of the German Autovision Museum, https://newatlas.com/autovision-museum-recreates-ayrton-and-perry-electric-car/20528/. Accessed 10 Sept 2021
7. M. von Jacobi, (Wikipedia), https://en.wikipedia.org/wiki/Moritz_von_Jacobi. Accessed 15 Jan 2023
8. The History of Electric Cars, Charge Cars, 29 November 2021, https://chargecars.medium.com/the-history-of-electric-cars-27f6ec0e7085, Accessed 1 November 2023
9. H. Ayrton, *The Electric Arc* (Cambridge University Press, 1902)
10. Hertha Marks Ayrton. (Britanica), https://www.britannica.com/biography/Hertha-Marks-Ayrton
11. Archives biographies: Hertha Ayrton. (The Institution of Engineering and Technology), https://www.theiet.org/membership/library-archives/the-iet-archives/biographies/hertha-ayrton/. Accessed 1 May 2013
12. Wikipedia, History of the electric vehicle, https://en.wikipedia.org/wiki/History_of_the_electric_vehicle. Accessed 1 Oct 2024
13. Wikipedia, Electric Locomotive, https://en.wikipedia.org/wiki/Electric_locomotive. Accessed 10 Apr 2024
14. E. Wakefield, *History of the Electric Car* (SAE Press, 1994)
15. B. Fultz, Addressing Edison's concern, https://www.its.caltech.edu/~matsci/btf/EdisonText.html. Accessed 1 Nov 2024
16. K. Wilson, Worth the Watt: a brief history of the electric car 1830 to present. (Car and Driver Magazine, 31 Mar 2023).
17. J. Voelcker, All-electric sports car coming next month from detroit startup? Green car reports. (Wikipedia), https://en.wikipedia.org/wiki/Detroit_Electric#cite_note-GCR032013-1. Accessed 20 May 2023
18. 1908 Baker electric model V Victoria. (Hyman Ltd., 2022), https://hymanltd.com/vehicles/5009-1908-baker-electric-model-v-victoria/.
19. Edison's alkaline battery. (Engineering and Technology History Wiki), https://ethw.org/Edison's_Alkaline_Battery. Accessed 1 May 2023
20. The new Edison storage battery. (Scientific American, 1 June 1901), https://www.scientificamerican.com/article/the-new-edison-storage-battery/. Accessed 1 May 2023
21. W. Garrison, D.M. Levinson, The transportation experience policy, planning, and deployment. (OUP USA, 2014), https://en.wikipedia.org/wiki/Exide. Accessed 10 June 2023. Image from https://commons.wikimedia.org/w/index.php?curid=29533891. Accessed Nov 2021
22. Live Science Staff, Edison's electric car battery gets a makeover. (2012), https://www.livescience.com/21227-edison-electric-car-battery.html
23. Green Car Reports, Thomas Edison's 1912 electric car gets a chance to shine, https://www.greencarreports.com/news/1049744_thomas-edisons-1912-electric-car-gets-a-chance-to-shine. Accessed 12 June 2023

24. Detroit Public Library Digital Collections, Girl charging a Baker electric automobile, National Automotive History Collection. (Detroit Public Library), https://digitalcollections.detroitpubliclibrary.org/islandora/object/islandora%3A221604
25. *The Horseless Age*, a trade publication for electric vehicles started in 1895, New York. Image of Volume 1 Issue 1 cover page from Hathi Trust Digital Archive, https://babel.hathitrust.org/cgi/pt?id=mdp.39015080361374&view=page&seq=7. Accessed 1 May 2023
26. GObankingrates, 8 decades of EVs costs when they were first released vs. now, https://finance.yahoo.com/news/8-decades-evs-costs-were-160159807.html?guccounter=1. Accessed 1 Dec 2024
27. The electric starter is patented. (Automotive History.org, 15 June 1911), https://automotive-history.org/june-15-1911-the-electric-starter-is-patented/. Accessed 1 Dec 2024
28. https://electricvehiclesnews.com/History/images/JamaiscontenteVehicle.jpg. Accessed 14 June 2023
29. The orphan chronicles: baker electric automobile company gallery. (Car and Driver Magazine, 2 July 2018), https://www.caranddriver.com/features/a21951241/the-orphan-chronicles-baker-electric-motor-vehicle-company-feature/. Accessed 31 May 2023
30. B. Brownell, The first car to ever go over 100 miles per hour was an EV (Jalopnik, 28 Feb 2024), https://jalopnik.com/the-first-car-to-ever-go-over-100-miles-per-hour-was-an-1851152622. Accessed 30 Mar 2024
31. C. Automedia, https://automedia.revsinstitute.org/web-stories/1902-when-the-worlds-fastest-car-was-electric. Accessed 10 Mar 2024
32. K.V. Shaw, America's first car to exceed 100 mph is headed to auction, it'll fetch Over $1 M. (The Drive, 10 Feb 2024), https://www.thedrive.com/news/americas-first-car-to-exceed-100-mph-is-headed-to-auction-itll-fetch-over-1m. Accessed 31 Mar 2024
33. W.I. Lewis, Top ten fastest electric cars from 0–60 mph that We've Tested. (Motor Trend, 2023), https://www.motortrend.com/features/fastest-electric-cars/. Accessed July 2024
34. Image from Hemmings Motor News, Archive, https://www.hemmings.com/stories/article/1974-81-citicar-and-commutacar. Accessed 1 Mar 2022
35. US Dept of Transportation, National Highway Traffic Safety Administration 49 CFR Part 571 Federal Motor Vehicle Safety Standards.
36. https://earthsky.org/human-world/when-will-the-age-of-aquarius-begin/. Accessed 1 Dec 2024
37. Ever-larger cars and trucks are causing a safety crisis on US streets. (The Conversation, 2 Aug 2023), https://theconversation.com/ever-larger-cars-and-trucks-are-causing-a-safety-crisis-on-us-streets-heres-how-communities-can-fight-back-206382. Accessed 4 Aug 2023
38. General Motos EV-1. (Wikipedia), http://en.m.wikipedia.org/wiki/General_Motors_EV1. Accessed 1 June 2024
39. G. Collantes, The California ZEV Mandate: policy origin and dynamics. (2006), https://www.belfercenter.org/sites/default/files/legacy/files/Gustavo%20ZEV%20Presentation.pdf
40. N. Bomoy, EVs are much heavier than gas vehicles, and that's posing a safety problem. (Axion, 28 Aug 2023), http://www.Axios.com/2023/04/28/evs-weight-safety-problems
41. C. Krome, 10 PZEV cars for lovers of low emissions. (Autobytel, 10 June 2015), https://www.autobytel.com/10-pzev-cars-for-lovers-of-low-emissions. Accessed 1 June 2024
42. Wikipedia, Paice (company), http://en.m.wikipedia.org/wiki/Paice_(company). Accessed 1 May 2023
43. Wikipedia, Energy conversion devices, https://en.wikipedia.org/wiki/Energy_Conversion_Devices. Accessed 30 Jan 2023
44. B. Winfield, Tested: 1997 General Motors EV1 proves to be the start of something big. (Car and Driver Magazine, Mar 1997), https://www.caranddriver.com/reviews/a32944084/tested-1997-general-motors-ev1-proves-to-be-the-start-of-something-big/
45. Video trailer. http://www.electric-vehiclenews.com/2011/08/old-top-gear-ev1-electric-video.html
46. US Patent 5,703,462.
47. http://altfuels.org/events/testdriv/graphics/adapter.jpg

48. Magne Charge, Wikipedia, https://en.wikipedia.org/wiki/Magne_Charge. Accessed 1 June 2023
49. AC Propulsion tZero, (Wikipedia), https://en.wikipedia.org/wiki/AC_Propulsion_tzero. Accessed 1 June 2023
50. Enviro-nut. (Slang Define), https://slangdefine.org/e/environut-2d66.html. Accessed 15 June 2023
51. Market Realist, Elon Musk bought Tesla ownership stake in the early years, has served as CEO Since, May 13, 2022. https://marketrealist.com/p/did-elon-musk-buy-tesla/. Accessed 1 May 2023
52. CleanTechnica, AC, https://cleantechnica.com/2017/06/17/ac-propulsion-t-zero-godfather-modern-evs/
53. A. Hartmans, Mr. Musk openly stated that "Tesla wouldn't exist without a little-known electric convertible called the tZero." (Business Insider, 19 Mar 2021), https://www.businessinsider.com/elon-musk-tesla-roadster-history-tzero-electric-car-2020-7?op=1. Accessed 1 Dec 2022
54. G. Berdichevsky et al., The Tesla roadster battery system. (Tesla Motors, 2006), http://large.stanford.edu/publications/power/references/docs/tesla.pdf. Accessed 1 May 2023
55. Tesla Motors, A bit about batteries, https://www.tesla.com/blog/bit-about-batteries. Accessed 15 Apr 2023
56. IMCDB (Internet Movie Cars Database), https://www.imcdb.org/vehicle_420809-Tesla-Roadster-Sport-2010.html. Accessed 1 June 2023
57. Tesla Roadster, First Generation. (Updated 19 Sept 2023), https://en.wikipedia.org/wiki/Tesla_Roadster_(first_generation)#Production. Accessed 19 Sept 2023
58. Lotus Elise model comparison, handling, power. Sands Mechanical Museum. (Lotus Elise), https://en.wikipedia.org/wiki/Lotus_Elise#Series_2. Accessed 18 Sept 2023
59. Wikipedia, Tesla roadster (first generation), https://en.wikipedia.org/wiki/Tesla_Roadster_(first_generation). Accessed 1 May 2023
60. 2010 Tesla roadster sport tested, https://www.caranddriver.com/reviews/a16579636/2010-tesla-roadster-sport-instrumented-test/
61. Weightle Immediate Certification, Lotus Elise curb weight in easy-to-read graphs, https://weightle.com/lotus-elise. Accessed 1 May 2023
62. Tesla Motors Club, (Blog Post, 12 Mar 2011), https://teslamotorsclub.com/tmc/threads/on-the-race-track.5224/. Accessed 1 May 2023
63. United States Federal Register, Federal motor vehicle safety standards: test procedures. FR 79456 **85,** 79456–79460, 49 CFR 571 (2020), https://www.federalregister.gov/documents/2020/12/10/2020-27001/federal-motor-vehicle-safety-standards-test-procedures
64. International Energy Agency, Global EV Outlook 2023: trends in charging infrastructure, https://www.iea.org/reports/global-ev-outlook-2023/trends-in-charging-infrastructure. Accessed 5 May 2024
65. F. Lambert, Tesla now has 40,000 Superchargers around the world. (Electrek, 22 Nov 2022), https://electrek.co/2022/11/22/tesla-supercharger-400000-world/. Accessed 1 Mar 2024
66. Global EV outlook, nearly one in five cars sold in 2023 was electric. (International Energy Agency, 2024), https://iea.blob.core.windows.net/assets/a9e3544b-0b12-4e15-b407-65f5c8ce1b5f/GlobalEVOutlook2024.pdf. Accessed 2 Apr 2024
67. R. Sajwar, Watch the 'Little Giant' break the land speed record for EVs. (Wonderful Engineering, 2021), https://wonderfulengineering.com/watch-the-little-giant-break-the-land-speed-record-for-evs/
68. D. Vanderwerp, Tested: 2021 Tesla model S plaid is absurdly quick but also has a few key flaws. (Car and Driver Magazine, 8 Dec 2021), https://www.caranddriver.com/reviews/a38423992/2021-tesla-model-s-plaid-by-the-numbers/. Accessed 1 Feb 2024
69. Tesla Roadster, https://www.tesla.com/roadster
70. Bugatti Rimac d.o.o., Republic of Croatia, https://www.rimac-automobili.com/nevera/. Accessed 10 June 2023

71. Rimac nevera sets world speed record for electric production cars at 258 mph. (Forbes Magazine, 15 Nov 2022), https://www.forbes.com/sites/markewing/2022/11/15/rimac-nevera-sets-world-speed-record-for-electric-production-cars-at-258-mph/?sh=28cfd81d2503; https://www.forbes.com/sites/markewing/2023/05/17/electric-rimac-nevera-sets-road-car-record-60-mph-in-174-seconds/?sh=6cbdc75b7f83.
72. Vehicle diagram: Electric vehicle (EV) drivetrain system. (Infinion), https://www.infineon.com/cms/en/applications/automotive/electric-drive-train/#!?fileId=8ac78c8c8779172a0187e052a3634ab8. Accessed 25 Nov 2023
73. R. Keim, F.J. Sprague, The man behind the motors that remade the urban landscape. (All about Circuits, 2023), https://www.allaboutcircuits.com/news/frank-j-sprague-the-man-behind-the-motors-that-remade-the-urban-landscape/. Accessed 27 Feb 2024.
74. M. Robins, The early years of electric traction, Invention, development, exploitation. J. Transport History **21**(1), 92–101 (2000). https://doi.org/10.7227/TJTH.21.1.6
75. *Electric G-Van Service Manual*. (Conceptor Industries, 1991).
76. N. Radwan-Praglowska, T. Wegiel, D. Borkowski, Modeling of axial flux permanent magnet generators. Energies **13**, 5741 (2020). https://doi.org/10.3390/en13215741 2
77. N. Zart, AC Propulsion tzero: the Godfather of Modern EVs. (CleanTechnica, 2017), https://cleantechnica.com/2017/06/17/ac-propulsion-t-zero-godfather-modern-evs/. Accessed June 2023
78. Oregon Public Broadcasting: Electric Car Drag Racing | Oregon field Guide. (YouTube), https://www.youtube.com/watch?v=UAYrsEOxqYc. Accessed 1 Nov 2024
79. The Tesla Model S Plaid Finally Tastes Defeat At The Hands Of A 1,000-Horsepower McLaren 720S! (Topspeed), https://www.topspeed.com/cars/car-news/the-tesla-model-s-plaid-finally-tastes-defeat-at-the-hands-of-a-1-000-horsepower-mclaren-720s/. Accessed 20 Dec 2023
80. P. Ridden, German Museum team recreates and rides world's first electric car. (New Atlas, 17 Nov 2011), https://newatlas.com/autovision-museum-recreates-ayrton-and-perry-electric-car/20528/
81. H. Cochrane, Electric-motor controller US Patent 547743A, 15 June 1899.
82. A system of control for electrically propelled vehicles, US Patent to W.B. Potter US715291 8 Apr 1899.
83. K.R. Pullen, A. Dhand, in *Alternative Fuels and Advanced Vehicle Technologies for Improved Environmental Performance*, (Woodhead, 2014), pp. 476–504, https://www.sciencedirect.com/topics/engineering/hydrostatic-transmission. Accessed 20 Feb 2023
84. I. Buchmann, *Batteries in a Portable World*, 4th edn. (Battery University, 2016)
85. Diagram of the first battery, which was made of alternating metal layers and saline-soaked leather pads, and proved electricity is caused by contact with dissimilar metals / Basire, sc. (US Library of Congress, 1800), https://www.loc.gov/item/92518975/
86. Gaston plante. (Britanica), https://www.britannica.com/biography/Gaston-Plante. Updated 17 May 2024. Accessed 25 May 2024
87. Stanford Ovshinsky. (MIT Lemelson Institute), https://lemelson.mit.edu/resources/stanford-ovshinsky. Accessed 15 May 2024
88. ION innovation: a true platform for solid-state. (ION Storage Systems), https://ionstoragesystems.com/technology/. Accessed 10 Oct 2023
89. K. Hede, Longer Lasting Sodium-Ion Batteries on the Horizon—an advanced sodium-ion battery design overcomes technical hurdles. (Pacific Northwest National Laboratory, US DOE, 13 July 2022), https://www.pnnl.gov/news-media/longer-lasting-sodium-ion-batteries-horizon. Accessed 10 Jan 2024
90. M. Crittenden, With ultralight lithium-sulfur batteries, electric airplanes could finally take off. (IEEE Spectrum, 19 Aug 2020), https://spectrum.ieee.org/with-ultralight-lithiumsulfur-batteries-electric-airplanes-could-finally-take-off. Accessed 31 Mar 2024
91. Automotive Li-Ion batteries: current status and future perspectives. (US Dept of Energy, OSTI report, 1 Jan 2019), p. 26, https://www.osti.gov/pages/biblio/1561559. Accessed 1 Nov 2022

92. Lithium sulfur battery. (Wikipedia), https://en.wikipedia.org/wiki/Lithium%E2%80%93sulfur_battery. Accessed 20 Apr 2024
93. J.F. Cooper, R. Krueger, *The Refuelable Zinc-Air Battery: Alternative Techniques for Zinc and Electrolyte Regeneration* (Lawrence Livermore National Laboratory, UCRL-TR-218414, 2006)
94. 5 startups leading metal-air batteries in 2024. (GreyB Services, 2024), https://www.greyb.com/blog/metal-air-batteries-startups/. Accessed 12 Apr 2024
95. High voltage vehicles: why 800-volt EVs are on the rise. (EdisOsman, Engineering.com, 2023), https://www.engineering.com/story/high-voltage-vehicles-why-800-volt-evs-are-on-the-rise. Accessed 10 Jan 2024
96. C.A. MacCarley, Electric Vehicle Safety: guidelines for safely working on electric vehicle conversion projects. (Nov 2011), https://digitalcommons.calpoly.edu/cgi/viewcontent.cgi?article=1293&context=eeng_fac
97. M.A. Raza, What is the Safe Limit of DC-Voltage for humans to touch? (Electrical Engineering, 2023), https://electricalengineeringx.com/what-is-the-safe-limit-of-dc-voltage-for-humans-to-touch/. Accessed 20 Apr 2024
98. OSHA standards, electrical. (U.S. Department of Labor, Occupational Safety and Health Administration), https://www.osha.gov/electrical/standards. Accessed 20 Apr 2024
99. Risks to emergency responders from high-voltage, Lithium-Ion battery fires addressed in safety report, 1/13/2021. (National Transportation Safety Board), https://www.ntsb.gov/news/press-releases/Pages/NR20210113.aspx. Accessed 1 June 2023
100. EV training. (National Institute for Automotive Repair Excellence), https://www.ase.com/ev. Accessed 20 Dec 2023
101. Hybrid and Electric Vehicle Repair Alliance (HEVRA). https://www.fixev.org/. Accessed 20 Dec 2023
102. Impact of battery electric vehicle and hybrid electric vehicle repair and service at independent repair shops. (IMR), https://www.automotiveresearch.com/insights/battery-electric-vehicle-hybrid-electric-vehicle-impact-auto-repair-shops. Accessed 20 Dec 2023
103. The environmental impact of Lithium batteries. (Institute for Energy Research, 12 Nov 2020), https://www.instituteforenergyresearch.org/renewable/the-environmental-impact-of-lithium-batteries/. Accessed 30 Aug 2023
104. Making mining safe and fair: artisanal cobalt extraction in the Democratic Republic of the Congo. White Paper, World Economic Forum, 91–93 route de la Capite CH-1223 Cologny/Geneva, September 2020, Switzerland. (Sept 2020)
105. D. Hall, N. Lutsey, Effects of battery manufacturing on electric vehicle life-cycle greenhouse gas emissions. (International Council on Clean Transportation, 2018), https://theicct.org/sites/default/files/publications/EV-life-cycle-GHG_ICCT-Briefing_09022018_vF.pdf. Accessed 10 Sept 2023
106. How much CO2 is emitted by manufacturing batteries? (MIT Climate Portal, updated 15 July 2022), https://climate.mit.edu/ask-mit/how-much-co2-emitted-manufacturing-batteries. Accessed 10 Sept 2023
107. Electric car battery replacement costs. (Edmunds), https://www.edmunds.com/electric-car/articles/electric-car-battery-replacement-costs.html. Accessed 15 Feb 2024
108. China's abandoned EV graveyard: thousands of cars rot in huge fields. (InsideEVs), https://insideevs.com/news/672926/china-abandoned-electric-car-graveyard-byd-geely/. Accessed 2 Apr 2024
109. D. Pratt, What happens to the old batteries in electric cars? (Consumer Reports, 23 Feb 2022), https://www.consumerreports.org/cars/hybrids-evs/what-happens-to-the-old-batteries-in-electric-cars-a1091429417/. Accessed 15 Feb 2024
110. Electric vehicle pros. (The Hybrid Shop), https://www.thehybridshop.com/battery-conditioning/. Accessed 15 Feb 2024
111. Lithium-Ion battery recycling. (US Environmental Protection Agency, 2023), https://www.epa.gov/hw/lithium-ion-battery-recycling. Accessed 4 Mar 2024

112. Z.J. Baum, R.E. Bird, X. Yu and J. Ma. *Lithium-Ion Battery Recycling - Overview of Techniques and Trends*, ACS Energy Letters, American Chemical Society, Volume 7, Issue 2, February 11, 2022, pp. 560–907
113. Battery cell comparison. (US National Aeronautics and Space Administration, Table from EPIC Technologies), https://www.epectec.com/batteries/cell-comparison.html. Accessed 31 May 2023
114. The research on characteristics of Li-NiMnCo Lithium-Ion batteries in electric vehicles. J. Energy **2020**, 3721047. https://doi.org/10.1155/2020/3721047
115. A closer look at Lithium Iron Phosphate batteries. (Tesla's New Choice of Battery, 28 Oct 2021), https://www.allaboutcircuits.com/news/a-closer-look-at-lithium-iron-phosphate-batteries-teslas-new-choice-of-battery/
116. Mark Gaige, *Electric Vehicle Fire Staged to Study Environmental, Health Ramifications.* Miller School of Medicine, University of Miami, 9 September 2024. https://news.med.miami.edu/electric-vehicle-fire-staged-to-study-environmental-health-ramifications/. Accessed 1 January 2025.
117. J. Sterman, A. Brauer, A. Nejman. EV fires are tough to fight and pose new hazards. Inside the race to get responders ready. (Fox News Baltimore, 22 July 2021), https://foxbaltimore.com/news/nation-world/ev-fires-are-hard-to-fight-and-pose-new-hazards-inside-the-race-to-get-responders-ready
118. Safety risks to emergency responders from Lithium-Ion Battery fires in electric vehicles. Safety Report NTSB/SR-20/01, PB2020-101011. (National Transportation Safety Board, 13 Nov 2020)
119. Image from Autoevolution, https://www.autoevolution.com/news/new-tesla-fires-were-caused-by-a-flood-and-a-crash-is-that-ok-170797.html. Accessed 10 June 2023
120. Federal Register, https://www.federalregister.gov/documents/2017/09/27/2017-20350/federal-motor-vehicle-safety-standards-electric-powered-vehicles-electrolyte-spillage-and-electrical
121. F-150 XLT specifications. (Ford Motor Co., 2023), https://www.ford.com/trucks/f150/models/f150-xlt/. Accessed 6 Oct 2023
122. F0150 Lighting Pro specifications. (Ford Motor Co., 2023), https://www.ford.com/trucks/f150/f150-lightning/models/f150-pro/. Accessed 6 Oct 2023
123. The 2020 EPA Automotive Trends Report, EPA-420-R-21-003. (Jan 2021), https://nepis.epa.gov/Exe/ZyPDF.cgi?Dockey=P1010U68.pdf
124. Average car weight. (Lizzie Nealon, 20 Oct 2021), https://www.bankrate.com/insurance/car/average-car-weight/. Accessed 2 Oct 2023
125. Most pickup truck owners don't actually do any truck stuff. (MotorBiscuit, 31 Oct 2023), https://www.motorbiscuit.com/most-pickup-truck-owners-dont-actually-truck-stuff/. Accessed 1 Feb 2024
126. TnT Automobile News blog, https://tntautomobilenews.wordpress.com/2021/11/25/2022-gmc-hummer-ev-range-payload-and-towing-capacity/. Accessed 10 May 2024
127. Why hummer EV sales increased by 83,000% this quarter. (The Autopian, 2 Apr 2024), https://www.theautopian.com/hummer-ev-sales-increased-by-83000-because-they-built-some-cars/. Accessed 15 May 2024
128. Electric cars from heaviest to lightest. (InsideEVs, 21 Aug 2021), https://insideevs.com/news/527966/electric-cars-from-heaviest-lightest/. Accessed 1 June 2023
129. How Green is your electric vehicle really? (10 Aug 2023), https://www.economist.com/business/2023/08/10/how-green-is-your-electric-vehicle-really. Accessed 2 Oct 2023
130. Volve XC40 specifications, https://www.volvocars.com/us/compare-specs/xc40/. Accessed 7 Oct 2023
131. Ford F350 dimensions 1988. (Carsguide), https://www.carsguide.com.au/ford/f350/car-dimensions/1988. Accessed 2 Apr 2024
132. Braking distances from test results, https://fastestlaps.com/lists/top-quickest-stoppers-60mph. Accessed 2 Nov 2023

133. Tesla Model 3 shows 60-0 mph braking weakness in CR tests, Tesla fires back with own test results. (Electrek, 2018), https://electrek.co/2018/05/21/tesla-model-3-braking-weakness-cr-tests-tesla-denies/. Accessed 6 Oct 2023
134. Consumer Information Regulations; Federal Motor Vehicle Safety Standards; Rollover Resistance. National Highway Traffic Safety Administration ruling. (US Federal Register, 12 Jan 2001), https://www.federalregister.gov/documents/2001/01/12/01-973/consumer-information-regulations-federal-motor-vehicle-safety-standards-rollover-resistance. Accessed 1 Dec 2023
135. How much damage do heavy trucks do to our roads. (Inside Science, 12 Oct 2020), https://www.insidescience.org/news/how-much-damage-do-heavy-trucks-do-our-roads. Accessed 1 May 2023
136. C. Jacobs, Nobody told EV owners how quickly they burn through tires. (The Drive, 27 Mar 2024), https://www.thedrive.com/news/nobody-told-ev-owners-how-quickly-they-burn-through-tires. Accessed 1 July 2024
137. EV tyre particulate pollution—a quiet problem. (Helixx, 3 May 2023), https://www.helixx.tech/insights/ev-tire-particulates. Accessed 1 July 2024
138. D. Zipper, Electric cars fix one pollution problem—and worsen another. (The Atlantic, 19 July 2023), https://www.theatlantic.com/technology/archive/2023/07/electric-vehicles-tires-wearing-out-particulates/674750/. Accessed 1 July 2024
139. I. van Hagen, Experts fear deteriorating parking lots could collapse under the weight of heavy electric vehicles, says report. (Business Insider, 8 Apr 2023), https://www.businessinsider.com/heavy-weight-electric-cars-could-cause-aging-parking-lots-collapse-2023-4?op=1. Accessed 5 June 2023
140. L. Day, Heavy EVs could collapse old parking structures. (The Drive, 10 Apr 2023), https://www.thedrive.com/news/heavy-evs-could-collapse-old-parking-garages-report. Accessed 5 June 2023
141. Impact Report 2022. (Tesla Motor Co.), p. 68, https://teslanorth.com/2023/04/24/tesla-2022-impact-report/
142. G. Kuntzman, CRASH TEST: electric cars are more dangerous than conventional cars, says Global Insurance Company. (StreetBlog NYC, 9 Sept 2022), https://nyc.streetsblog.org/2022/09/09/crash-test-electric-cars-are-more-dangerous-than-conventional-cars-says-global-insurance-company. Accessed 1 Oct 2024
143. Axios (Online analytics), https://www.axios.com/2023/04/28/evs-weight-safety-problems. Accessed 20 Dec 2023
144. National Academy of Engineering, Transportation Research Board, Pedestrians, Bicycles, Human Factors Committee. https://www.mytrb.org/OnlineDirectory/Committee/Details/5070
145. Motor vehicle registrations by Vehicle Type and State. (DOT Data Hub, 7 May 2024), https://datahub.transportation.gov/Roadways-and-Bridges/Motor-Vehicle-Registrations-by-vehicle-type-and-st/hwtm-7xmz/about_data. Accessed 15 May 2024
146. NADA issues analysis of 2022 auto sales and 2023 sales forecast. (National Automobile Dealers Association, 11 Jan 2023), https://www.nada.org/nada/press-releases/nada-issues-analysis-2022-auto-sales-and-2023-sales-forecast. Accessed 20 July 2023
147. Battery bloat could backfire on electric vehicle manufacturers. (Bloomberg Hyperdrive, 2024), https://www.bloomberg.com/news/newsletters/2023-08-01/battery-bloat-could-backfire-on-electric-vehicle-manufacturers?cmpid=BBD080123_hyperdrive&utm_medium=email&utm_source=newsletter&utm_term=230801&utm_campaign=hyperdrive. Accessed 1 Feb 2024
148. Biden is ignoring the Menace of Mega-EVs. (Atlantic Magazine and online newsletter, June 2023), https://www.theatlantic.com/ideas/archive/2023/06/large-electric-vehicles-road-safety-crashes/674249/. Accessed 30 June 2023
149. NTSB head warns of risks posed by heavy electric vehicles colliding with lighter cars. (From National Public Ratio, 11 Jan 2023), https://www.npr.org/2023/01/11/1148483758/ntsb-heavy-electric-vehicles-safety-risks

150. Safety Consequences of vehicle size and weight. (YouTube), https://www.youtube.com/watch?v=RfeT0NJlIDE. Accessed 25 June 2023
151. Cybertruck crash test is HORRIFYING, here's why, https://www.youtube.com/watch?v=9lI2_BDZpI4
152. A.J. Hawkins, Truck bloat is killing us, new crash data reveals. (The Verge), https://www.msn.com/en-us/autos/other/truck-bloat-is-killing-us-new-crash-data-reveals/ar-AA1jVhYN. Accessed 10 Apr 2024
153. Cybertruck: specs. (Tesla), https://www.tesla.com/cybertruck. Accessed 18 July 2024
154. N. Pahwa, Rear wheels breaking off, floppy wet noodle windshield wipers, easily humiliated by a Subaru. This is the Cybertruck. (Slate, 12 Mar 2024), https://slate.com/technology/2024/03/tesla-cybertruck-elon-musk-accident.html?sponsored=0&position=3&category=fascinating_stories&scheduled_corpus_item_id=a4ccac1a-ae6b-4c90-8b59-d7e12af208f2&url=https://slate.com/technology/2024/03/tesla-cybertruck-elon-musk-accident.html. Accessed 18 Mar 2024
155. The 2020 EPA automotive trends report. (United States Environmental Protection Agency, Jan 2021), https://nepis.epa.gov/Exe/ZyPDF.cgi?Dockey=P1010U68.pdf EPA-420-R-21-003
156. How much does the average car weigh? (Measuring Stuff, 18 Dec 2023), https://measuringstuff.com/how-much-does-the-average-car-weigh/. Accessed 10 Apr 2024
157. J. Caldwell, Edmunds guide. (Insights, online newsletter), https://www.edmunds.com/about/authors/jessica-caldwell.html
158. Why is everyone building an electric pickup truck? (Wired, 22 Oct 2020), https://www.wired.com/story/why-everyone-building-electric-pickup-truck/. Accessed 1 May 2023
159. Graphic copied from preceding citation, Data source: American Council for an Energy-Efficient Economy (ACEEE) GreenerCarsRating. (14 Sept 2023).
160. December to cap off another profitable year for dealers as consumers spend record $578 billion on new vehicles in 2023. (J.D. Power, 21 Dec 2023), https://www.jdpower.com/business/press-releases/jd-power-globaldata-forecast-december-2023. Accessed 5 Apr 2024
161. E. Shilling, Trucks and SUVs are now over 80 percent of new car sales in the U.S. (Jalopnik, 27 Jan 2023), https://jalopnik.com/trucks-and-suvs-are-now-over-80-percent-of-new-car-sale-1848427797. Accessed 4 Apr 2024
162. E-bikes aren't just faster, they're a different ride. Here are safety tips. (Los Angles Times, 16 Aug 2022), https://www.latimes.com/california/story/2022-08-16/e-bike-safety-not-just-faster-theyre-a-different-ride. Accessed 7 Oct 2023
163. Big jump in e-scooter, e-bike injuries the last four years, CPSC finds. (CBS News Money Watch, 4 Oct 2021), https://www.cbsnews.com/news/electric-scooter-electric-bike-accidents-rise-last-four-years/. Accessed 7 Oct 2023
164. Safety with E-Bikes and E-Scooters. (National Fire Protection Association), https://www.nfpa.org/Public-Education/Fire-causes-and-risks/ebikes. Accessed 7 Oct 2023
165. Swind EB-01 maker addresses Simon Cowell's back-breaking e-Bike incident. (Autoevolution, 17 Dec 2020), https://www.autoevolution.com/news/swind-eb-01-maker-addresses-simon-cowells-back-breaking-e-bike-incident-153209.html. Accessed 20 Dec 2023
166. Collisions and close calls: The e-bike boom is getting increasingly dangerous, creating a push for stricter cycling rules. (Forbes Magazine, 7 Oct 2022), https://fortune.com/2022/10/07/e--bikes-cities-accidents-danger-safety-regulation/. Accessed 7 Oct 2023
167. Overview of motor vehicle crashes in 2019, DOT HS 813 060. (National Highway Transportation Administration, Dec 2020), https://crashstats.nhtsa.dot.gov/Api/Public/ViewPublication/813060
168. Most e-scooter rider injuries happen on sidewalk, study finds. (Insurance Institute for Highway Safety, 15 Oct 2020), https://www.iihs.org/news/detail/most-e-scooter-rider-injuries-happen-on-sidewalk-study-finds. Accessed 7 Oct 2023
169. North American Electric Reliability Corporation, https://www.nerc.com/Pages/default.aspx. Accessed 12 May 2024

170. Avoiding gridlock: The impact of electric vehicles on transmission vs. distribution systems. (EnergyHub, 22 Jan 2024), https://www.energyhub.com/blog/avoiding-gridlock-the-impact-of-electric-vehicles-on-transmission-vs-distribution-systems/. Accessed 12 May 2024
171. Electrification. (Environmental and Energy Study Institute, a US non-profit political congressional advisory group), https://www.eesi.org/topics/electrification. Accessed 12 May 2024
172. K. Clement-Nyns, E. Haesen, J. Driesen, The impact of charging plug-in hybrid electric vehicles on a residential distribution grid. IEEE Trans. Power Syst. **25**(1) (2010)., https://ieeexplore.ieee.org/abstract/document/5356176?casa_token=rAeqWLf5y_EAAAAA:mxD-OsXxA0yAftw-lCa9HuxrMk4tpQoOqtZlTFMijmyDrFCkb2jMGT4a-Kg6wRlyt670SOKLsA
173. Plug-in electric load profiles. (Northwest Power and Conservation Council, 2021), https://www.nwcouncil.org/2021powerplan_plug-electric-load-profiles/. Accessed 15 May 2024
174. R. Frotscher et a., Reverse power flow impact on transformers. (IEEE PES Transformers Committee Spring 2021 Meeting, Thursday, 29 Apr 2021), https://grouper.ieee.org/groups/transformers/meetings/S2021-VirtualMtg/Documents/S21-ThuTutorial-ReversePowerFlowImpactOnTransformers.pdf. Accessed 5 May 2024
175. J. Timmer, Updating California's grid for EVs may cost up to $20 billion. (ARS Technica, 24 Apr 2024), https://arstechnica.com/cars/2024/04/updating-californias-grid-for-evs-may-cost-up-to-20-billion/. Accessed 15 May 2024
176. SDG7 world access to electricity. (International Energy Agency, 2023), https://www.iea.org/reports/sdg7-data-and-projections. Accessed 15 May 2024
177. W. Vermeer, G.R. Chandra Mouli, P. Bauer, A comprehensive review on the characteristics and modeling of Lithium-Ion battery aging. IEEE Trans. Transport. Electr. **8**(2), 2205–2232 (2022)., https://ieeexplore.ieee.org/document/9662298
178. S.M. Bhagavathy et al., Impact of charging rates on electric vehicle battery life. (Transport Findings, 23 Mar 2021), https://findingspress.org/article/21459-impact-of-charging-rates-on-electric-vehicle-battery-life. Accessed 30 May 2024
179. M. Kane, Tesla says that supercharger average uptime is near perfect *100%. (Inside EVs, 30 Apr 2023), https://insideevs.com/news/664737/tesla-superchargers-average-uptime-2022/. Accessed 1 Dec 2024
180. The inflation reduction act drives significant emissions reductions and positions america to reach our climate goals: DOE/OP-0018. (US Dept of Energy, Office of Policy, Aug 2022), https://www.energy.gov/sites/default/files/2022-08/8.18%20InflationReductionAct. Accessed 1 May 2024
181. FACT SHEET: Biden-Harris Administration announces new standards and major progress for a Made-in-America National Network of Electric Vehicle Chargers. (US Whitehouse Public Briefing), https://www.whitehouse.gov/briefing-room/statements-releases/2023/02/15/fact-sheet-biden-harris-administration-announces-new-standards-and-major-progress-for-a-made-in-america-national-network-of-electric-vehicle-chargers/. Accessed 13 Dec 2023
182. N. Winton, Battery-swapping revival could threaten electric car charging networks. (Forbes Magazine, 12 Sep 2022), https://www.greencarreports.com/news/1076202_better-place-battery-swapping-the-technology-behind-the-idea. Accessed 10 June 2023
183. C.A. MacCarley, A review of battery exchange technology for refueling of electric vehicles. (SAE Future Car Congress 2000, Society of Automotive Engineers International, 2000)
184. E.O. Dietrich, Field test experience with electric vehicles. (SAE paper No. 810489, contained in Electric and Hybrid Vehicle Progress, SAE publication P-91, 1981)
185. L. Ulrich, How is this a good idea? EV Battery Swapping > Swap this technological dead-end out for better batteries, improved superchargers and more universal EV charging standards. (IEEE Spectrum, 13 May 2021), https://spectrum.ieee.org/ev-battery-swapping-how-is-this-a-good-idea. Accessed 10 Jan 2023
186. IMDB The Fifth Element 1997 Gaumont Productions (FR), https://www.imdb.com/title/tt0119116/
187. Automated Rapid Battery Exchange instructional video. (2014), https://www.youtube.com/watch?v=yi7NDjSNVZ0. Accessed 10 June 2023

188. Automated Rapid Battery Exchange public demonstration 2016, https://www.youtube.com/watch?v=DZvteZl77do. Accessed 10 June 2023
189. Better Place. (Wikipedia), https://en.wikipedia.org/wiki/Better_Place_%28company%29. Accessed 10 June 2023
190. Let's get one million electric cars on the road. (Ample, USA), https://ample.com/. Accessed 31 May 2023
191. C.A. MacCarley, Design and safety considerations for automated battery exchange electric vehicles, FISITA World Automotive Congress, paper F2000H261, Seoul, Korea. (12–15 June 2000).
192. How 5-minute battery swaps could get more EVs on the road. (MIT Technology Review, 17 May 2023), https://www.technologyreview.com/2023/05/17/1073265/how-5-minute-battery-swaps-could-get-more-evs-on-the-road/
193. L. Ulrich, How is this a good idea? EV battery swapping. (IEEE Spectrum, 13 May 2021), https://spectrum.ieee.org/ev-battery-swapping-how-is-this-a-good-idea. Accessed 14 June 2023
194. Photovoltaics for electric vehicle charging, regulatory and policy considerations. NREL/BR-6A20-62366. (NREL Office of Energy Efficiency and Renewable Energy, Sept 2014)
195. https://www.solarreviews.com. Accessed 5 Aug 2023
196. Global EV outlook 2023—trends in batteries. (International Energy Agency), https://www.iea.org/reports/global-ev-outlook-2023/trends-in-batteries. Accessed 15 May 2024
197. Storage batteries. As reported in *Seeking Alpha, 1883 Interview With Thomas Edison On Energy Storage*. (The Electrician, 17 Feb 1883), https://seekingalpha.com/instablog/227454-john-petersen/2479201-1883-interview-with-thomas-edison-on-energy-storage. Accessed 27 July 2023
198. Phase III construction begins on expansion of world's largest lithium battery bank. (Solar Power World, Jan 2022), https://www.solarpowerworldonline.com/2022/01/phase-iii-construction-begins-on-expansion-of-worlds-largest-lithium-battery-bank/. Accessed 1 Oct 2023
199. Storage futures study. (NREL, Report and online Video, 2021), https://www.nrel.gov/analysis/storage-futures.html. Accessed 1 Oct 2023
200. Energy 2050: Insights from the ground up. (McKinsey, 2016), https://www.mckinsey.com/industries/oil-and-gas/our-insights/energy-2050-insights-from-the-ground-up
201. Electricity end use in the United States from 1975 to 2022. (Statistica Research Department), https://www.statistica.com/statistics/280704/world-power-consumption. Accessed 1 Oct 2023
202. EVs could drive 38% rise in US electricity demand, DOE lab finds. (Utilitydive), https://www.utilitydive.com/news/evs-could-drive-38-rise-in-us-electricity-demand-doe-lab-finds/527358/. Accessed 1 Oct 2023
203. Moss landing battery storage project. (NS Energy, 2021), https://www.nsenergybusiness.com/projects/moss-landing/. Accessed 1 Oct 2023
204. A. Colthorpe, Moss Landing: World's biggest battery storage project is now 3GWh capacity. (2 Aug 2023), https://www.energy-storage.news/moss-landing-worlds-biggest-battery-storage-project-is-now-3gwh-capacity/. Accessed 1 Oct 2024
205. World's largest battery energy storage park approved. (Construction Inquirer, 25 July 2023), https://www.constructionenquirer.com/2023/07/26/worlds-largest-battery-energy-storage-park-approved/. Accessed 28 July 2023
206. List of energy storage power plants. (Wikipedia, 3 May 2023), https://en.wikipedia.org/wiki/List_of_energy_storage_power_plants. Accessed 28 July 2023
207. 2020 Grid Energy Storage Technology Cost and Performance Assessment, Energy Storage Grand Challenge program. (US Dept of Energy Technical Report Publication No. DOE/PA-0204, Dec 2020).

208. Energy storage in moss landing: a smoky challenge to a new chapter. (Monterey Herald, 12 June 2022, Print article)
209. Revenues in fiscal year 2021: an infographic. (US Congressional Budget Office, 20 Sept 2022)
210. G.R. Parsons, M.K. Hidrue, W. Kempton, M.P. Gardner, University of Delaware, 2014: Willingness to pay for vehicle-to-grid (V2G) electric vehicles and their contract terms. Energy Econ. **42**(2014), 313–324
211. Case study (UK): electric vehicle-to-grid charging. (Ofgem, the Office of Gas and Electricity Markets, 6 July 2021), https://www.ofgem.gov.uk/publications/case-study-uk-electric-vehicle-grid-v2g-charging. Accessed 12 Dec 2023
212. What Is V2H? V2G? V2V? Ford's electric lightning pickup truck spreads the power around. (Forbes, 27 Dec 2021), https://www.forbes.com/sites/billroberson/2021/12/27/what-is-v2h-v2g-v2v-fords-electric-lightning-pickup-truck-spreads-the-power-around/?sh=2225039c4f3b
213. Volkswagen announces its Modularer E-Antriebs-Baukasten (MEB) platform will include vehicle-to-grid (V2G) capabilities starting in 2022. (CleanTechnica, 15 Apr 2021), https://cleantechnica.com/2021/04/15/vw-vehicle-to-grid-is-15-years-too-early-except-maybe-in-texas/
214. Critical Elements of Vehicle-to-Grid (V2G) economics. (National Renewable Energy Berkeley Lab, NREL/TP-5400-69017, Sep 2017), https://www.nrel.gov/publications
215. J. Brodkin, False advertising—Angry Tesla customers sue firm over "grossly" exaggerated EV range. (ARS Technica, 3 Aug 2024), https://arstechnica.com/tech-policy/2023/08/angry-tesla-customers-sue-firm-over-grossly-exaggerated-ev-range/. Accessed 15 May 2024
216. S. DeGeyter, Useful battery capacity varies with discharge rate as approximately predicted by Peukart's Law, explained in layman's terms. (Peukert's Law | A Nerd's Attempt to Explain Battery Capacity, 3 Mar 2022), https://www.batterystuff.com/kb/tools/peukert-s-law-a-nerds-attempt-to-explain-battery-capacity.html. Accessed 20 Mar 2024
217. I. Buchman, *Batteries in a Portable World*, 4th edn. (Battery University and CADEX, 2017)
218. Round trip efficiency of battery energy systems worldwide in 2023, by technology. (Statista Research Department, 10 Nov 2023), https://www.statista.com/statistics/1423012/efficiency-of-battery-energy-systems/. Accessed 10 Jan 2024
219. S. Ghaboussi, DC fast charging explained. (Electric Driver, 28 Jan 2024), https://electric-driver.co/articles/dc-fast-charging-explained/. Accessed 24 Mar 2024
220. J. Klender, Tesla range exaggeration lawsuit: a breakdown. (Teslarati, 3 Aug 2023), https://www.teslarati.com/tesla-range-exaggeration-lawsuit-breakdown/. Accessed 1 Dec 2023
221. Autoexpress, UK, https://www.autoexpress.co.uk/tips-advice/90816/euro-6-emissions-standards-what-do-they-mean-you; https://www.etags.com/blog/comparing-emission-standards-laws-us-eu/. Accessed 1 June 2023
222. How does the EPA calculate electric car range? (Cleantechnica), https://cleantechnica.com/2020/08/18/how-does-epa-calculate-electric-car-range/. Accessed 1 Mar 2022
223. T. Edison, Mr. Edison. (The Electrician, 17 Feb 1883), p. 329. Also reported in online essay by B. Fultz, Addressing Edison's concerns, https://www.its.caltech.edu/~matsci/btf/Edison-Text.html
224. S. Edelstein, Nissan leaf battery capacit Lawsuit: court approves settlement. (Green Car Report, 20 July 2015), https://greencarreports.com
225. C. Spicer, Hyundai electric vehicles pose fire risk, low range, alleges class action lawuit. (Top Class Actions, 10 Aug 2021)
226. Transport and mobility. (European Environment Agency), https://www.eea.europa.eu/en/topics/in-depth/transport-and-mobility. Accessed 10 June 2023
227. Compare side by side fuel economy. (US Environmental Protection Agency), https://www.fueleconomy.gov/feg/Find.do?action=sbs&id=42191. Accessed 10 June 2023
228. W. Meyer, The EPA's electric vehicle mileage fraud. (Forbes, online edition), http://www.forbes.com/sites/warrenmeyer/2010/11/24/the-epas-electric-vehicle-mileage-fraud/. Accessed Jan 2016

229. R. Schreiber, Is the EPA fudging EV MPGeFig.s? (The Truth About Cars), https://www.thetruthaboutcars.com/2012/06/is-the-epa-fudging-ev-mpge-Fig.s/. Accessed Jan 2016
230. T. Murphy, Using physics and estimation to assess energy, growth, options—MPG for Electric Cars? (Do the Math, UCSD), https://dothemath.ucsd.edu/2011/08/mpg-for-electric-cars/. Accessed Jan 2016
231. Autos, The myth of the EPA's electric vehicle mileage. (Entourage Legacy), https://www.secretentourage.com/lifestyle/autos/myth-of-the-epa-electric-vehicle-mpg/. Accessed Jan 2016
232. Miles per gallon gasoline equivalent. (Wikipedia), https://en.wikipedia.org/wiki/Miles_per_gallon_gasoline_equivalent. Accessed Jan 2016
233. Regulations for Emissions from Vehicles and Engines, Final rule for control of air pollution from motor vehicles: Tier 3 motor vehicle emission and fuel standards. (US Environmental Protection Agency, 26 Oct 2023), https://www.epa.gov/regulations-emissions-vehicles-and-engines/final-rule-control-air-pollution-motor-vehicles-tier-3. Accessed 10 May 2024
234. International Energy Agency, Renewables 2023 analysis and forecast to 2028, https://iea.blob.core.windows.net/assets/96d66a8b-d502-476b-ba94-54ffda84cf72/Renewables_2023.pdf. Accessed 25 May 2024
235. US Energy Information Agency, Use of natural gas-fired generation differs in the United States by technology and region. (22 Feb 2024), https://www.eia.gov/todayinenergy/detail.php?id=61444. Accessed 10 May 2024
236. B. Plumer, New rules for power plants could give carbon capture a boost. Here's how. (New York Times, 26 Apr 2023), https://www.nytimes.com/2023/04/26/climate/carbon-capture-power-plants.html. Accessed 15 Mar 2024
237. Greenhouse Gas Emissions from aTypical Passenger Vehicle, US EPA, Offce of Transportation and Air Quality, EPA-420-F-18-008, Mar 2018), https://nepis.epa.gov/Exe/ZyPDF.cgi/P100U8YT.PDF?Dockey=P100U8YT.PDF. Accessed 1 May 2024
238. Title 49 Subtitle B Chapter V Part 531 Fuel Economy Standards. (US Code of Federal Regulations, 8 Jan 2023), https://www.govinfo.gov/content/pkg/CFR-2023-title49-vol6/pdf/CFR-2023-title49-vol6-sec531-5.pdf
239. Vehicle Fuel efficiency CAFE Requirement by Year. (US Dept of Energy Alternative Fuels Data Center), https://afdc.energy.gov/data/mobile/10562. Accessed 15 Dec 2023
240. Electric vehicle myths. (US Environmental Protection Agency), https://www.epa.gov/greenvehicles/electric-vehicle-myths#Myth1. Accessed July 2023
241. How much carbon dioxide is produced per kilowatthour of U.S. electricity generation? (US Energy Information Agency, 2024), https://www.eia.gov/tools/faqs/faq.php?id=74&t=11
242. Measure the impact of your energy use on the environment. (US EPA), https://www.epa.gov/energy/measure-impact-your-energy-use-environment. Accessed 1 Dec 2024
243. Greenhouse gas equivalencies calculator—calculations and references. (US EPA), https://www.epa.gov/energy/greenhouse-gas-equivalencies-calculator-calculations-and-references. Accessed 1 Dec 2024
244. Are electric vehicles definitely better for the climate than gas-powered cars? (MIT Climate Patrol, Oct 2022), https://climate.mit.edu/ask-mit/are-electric-vehicles-definitely-better-climate-gas-powered-cars. Accessed 1 Dec 2024
245. FOTW #1306 4 September 3032: model year 2022 light duty vehicles sold in the US. averaged 26.4 Miles Per Gallon. Exact statement is "The average for passenger cars in MY 2022 was 33,3 mpg." (US Dept of Energy), https://energy.gov
246. US Code of Federal Regulations. Title 49 Subtitle B Chapter V Part 531, https://www.ecfr.gov/current/title-49/subtitle-B/chapter-V/part-531
247. CFR Title 49 Subtitle B Chapter V Part 531.5(d) Table 4 MY 2023. Minimum fuel economy standards for domestically manufactured passenger automobiles. https://www.ecfr.gov/current/title-49/subtitle-B/chapter-V/part-531. Accessed 20 May 2024
248. Factcheck: how electric vehicles help to tackle climate change. (Carbon Brief. FACT-CHECKS, 13 May 2019), https://www.carbonbrief.org/factcheck-how-electric-vehicles-help-to-tackle-climate-change/. Accessed 12 Dec 2023

249. A. Lau, How long does a Tesla battery last? (Motor Trend Magazine, 27 Feb 2023), https://www.motortrend.com/features/how-long-does-a-tesla-battery-last/photos/. Accessed 10 Jan 2023
250. J. Bellantuono, Unveiling the lifespan of electric car batteries: how long do they truly last? (Dynamic Manufacturing, 22 Sept 2023), https://www.dynamicmanufacturinginc.com/unveiling-the-lifespan-of-electric-car-batteries-how-long-do-they-truly-last/. Accessed 1 Dec 2024
251. Electric car battery life: fact or fiction. (Unitil, 30 Sept 2024), https://unitil.com/blog/electric-car-battery-life-fact-vs-fiction. Accessed 1 Dec 2024
252. J. Deng et al., Electric vehicles batteries: requirements and challenges. Joule **4**(3), 511–515, https://www.cell.com/joule/fulltext/S2542-4351(20)30043-X. Accessed 1 Dec 2024
253. Vehicle warranty, model 3. (Tesla), https://www.tesla.com/en_CA/support/vehicle-warranty. Accessed 1 Dec 2024
254. BYD expert: this way you can extend the lifetime of your battery. (Latam Mobility), https://latamobility.com/en/byd-expert-this-way-you-can-extend-the-lifetime-of-your-battery/. Accessed 1 Dec 2024
255. Average age of automobiles and trucks in operation in the United States. (US Dept of Transportation, Bureau of Transportation Statistics), https://www.bts.gov/content/average-age-automobiles-and-trucks-operation-united-states. Accessed 15 May 2024
256. D. Demuro, How many miles can a car last? (Autotrader online, 21 Feb 2024), https://www.autotrader.com/car-shopping/buying-car-how-long-can-you-expect-car-last-240725. Accessed 15 May 2024
257. Electric car battery life: everything you need to know, including how long they last. (Car and Driver Magazine, 10 May 2024), https://www.caranddriver.com/features/a31875141/electric-car-battery-life/. Accessed 15 May 2024
258. S. Nichols, Cost to replace and EV battery. (Green Cars, Sept 2024), https://www.greencars.com/greencars-101/cost-to-replace-an-ev-battery. Accessed 1 Dec 2024
259. M. Stone, Tesla and Rivian signed a right-to-repair pact. Repair advocates are skeptical. (Grist, 12 Oct 2023), https://grist.org/transportation/tesla-and-rivian-signed-a-right-to-repair-pact-repair-advocates-are-skeptical/. Accessed 1 Dec 2024
260. Repairing or replacing an EV battery. (EV Friendly), https://www.evfriendly.ca/consumers/repairing-or-replacing-an-ev-battery/. Accessed 1 Dec 2024
261. J. McLaren et al., Emissions Associated with electric vehicle charging: impact of electricity generation mix, charging infrastructure availability, and vehicle type. (National Renewable Energy Laboratory Technical Report NREL/TP-6A20-64852, Apr 2016), https://afdc.energy.gov/files/u/publication/ev_emissions_impact.pdf
262. Effect of battery manufacturing on electric vehicle life cycle greehouse gas emissions. (International Council on Clean Energy, Feb 2018), http://www.theicct.org
263. Greenhouse Gas Emissions from a typical passenger vehicle. (US EPA, Offce of Transportation and Air Quality, EPA-420-F-18-008, Mar 2018), https://nepis.epa.gov/Exe/ZyPDF.cgi/P100U8YT.PDF?Dockey=P100U8YT.PDF. Accessed 1 May 2024
264. T. Furcher et al., Electric-vehicle buyers demand new experiences. (McKinsey, July 2023), https://www.mckinsey.com/~/media/mckinsey/industries/automotive%20and%20assembly/our%20insights/electric%20vehicle%20buyers%20demand%20new%20experiences/electric-vehicle-buyers-demand-new-experiences.pdf?shouldIndex=false. Accessed 20 May 2024
265. The EV bubble has popped. There's no denying the numbers. (MSN reporting based on Barron's Report), https://www.msn.com/en-us/money/markets/the-ev-bubble-has-popped-there-s-no-denying-the-numbers/ar-AA1l7YlY. Accessed 24 May 2024
266. R. Rapier, Electric vehicle market hits speed bumps. (Forbes | Energy), https://www.forbes.com/sites/rrapier/2024/02/23/electric-vehicle-market-hits-speed-bumps/?sh=3f9ec478568f. Accessed 24 May 2024

267. Are manufacturers exaggerating EV range by up to 30%? (Fleet Evolution, 2024), https://www.fleetevolution.com/manufacturers-exaggerating-range/. Accessed 1 Sept 2024
268. A sustainable future is within reach. (Tesla, Impact Report, 2022), p. 28. https://www.tesla.com/ns_videos/2022-tesla-impact-report.pdf. Accessed 30 July 2023
269. The 2023 EPA Automotive Trends Report: Greenhouse Gas Emissions, Fuel Economy, and Technology since 1975. (US Environmental Protection Agency, Dec 2023), https://www.epa.gov/system/files/documents/2023-12/420r23033.pdf. Accessed 5 May 2024
270. Global EV outlook 2024: trends in the electric vehicle industry. (International Energy Agency), https://www.iea.org/reports/global-ev-outlook-2024/trends-in-the-electric-vehicle-industry. Accessed 20 May 2024
271. Find my electric electric car prices: average electric car cost in 2024, https://www.findmyelectric.com/blog/electric-car-prices/. Accessed 1 May 2024
272. C. Vitello, What's the true cost of electric versus gas vehicles? (Environmental Journal (Canada), 14 Apr 2022), https://environmentjournal.ca/whats-the-true-cost-of-electric-versus-gas-vehicles/. Accessed 1 May 2024
273. J. O'Dell, Big Gap remains in average price of electric car vs. gas car. (Edmunds Guide, 8 May 2024), https://www.edmunds.com/car-buying/average-price-electric-car-vs-gas-car.html. Accessed 10 May 2024
274. J. Elfalan, The true cost of powering an electric car. (Updated 29 Mar 2022), http://www.edmunds.com/fuel-economy/the-true-cost-of-powering-an-electric-car.html. Accessed 1 Nov 2023
275. US DOT FHWA 2023 data, as reported in *What is Average Mileage Per Year*. (Car and Driver Magazine, 24 Feb 2023)
276. California Electric Rates, 2024. https://www.electricitylocal.com/states/california/
277. PG&E E-1 residential electric power tariff schedule, https://www.pge.com/tariffs/assets/pdf/tariffbook/ELEC_SCHEDS_E-1.pdf
278. https://www.pge.com/tariffs/assets/pdf/tariffbook/ELEC_SCHEDS_EV%20(Sch).pdf
279. Gas cost calculator. (American Auto Club), https://gasprices.aaa.com/?state=CA. Accessed 15 May 2024
280. EV charging trends, Q1 2024. (Stable Insights), https://stable.auto/insights/electric-vehicle-charger-price-by-state. Accessed 10 May 2024
281. A. Neuhauser, M.R. Dickey, The costs of charging an electric vehicle in California. (Axios San Francisco, 13 Mar 2024), https://www.axios.com/local/san-francisco/2024/03/13/electric-vehicle-charging-cost. Accessed 10 May 2024
282. Global average electricity price. (ElectricRate, Updated May 2024), https://www.electricrate.com/data-center/electricity-prices-by-country/. Accessed 26 May 2024
283. J. St. John, What's behind the epidemic of unreliable EV chargers? (Canary Media, 12 Dec 2023), https://www.canarymedia.com/articles/ev-charging/whats-behind-the-epidemic-of-unreliable-ev-chargers. Accessed 10 May 2024
284. K. Barry, Electric cars and plug-in hybrids that qualify for federal tax credits: 2023 and 2024. (Consumer Reports, 8 Apr 2024), https://www.consumerreports.org/cars/hybrids-evs/electric-cars-plug-in-hybrids-that-qualify-for-tax-credits-a7820795671/. Accessed 23 July 2023
285. R. Montoya, Electric vehicle tax credits 2024: what you need to know. (Edmunds, 5 Apr 2024), https://www.edmunds.com/fuel-economy/the-ins-and-outs-of-electric-vehicle-tax-credits.html. Accessed 8 Apr 2024
286. F. Lambert, Tesla's gross margin and profits are surging along with prices, and it had to explain itself. (Electrek, 22 Apr 2022), https://electrek.co/2022/04/22/tesla-gross-margin-profits-surging-along-with-prices-explained/. Accessed 10 May 2024
287. E.J. Antoni, A.F. Esposito, Taxpayers are subsidizing rich electric-vehicle owners to the tune of billions. (New York Post, 7 Nov 2023), https://nypost.com/2023/11/07/opinion/taxpayers-are-subsidizing-rich-electric-vehicle-owners-to-the-tune-of-billions/. Accessed 3 Apr 2024
288. The EV tax credit can save you thousands, it you're rich enough. (Grist), https://grist.org/energy/the-ev-tax-credit-can-save-you-thousands-if-youre-rich-enough/

289. Global EV outlook 2023. (International Energy Agency), https://iea.blob.core.windows.net/assets/dacf14d2-eabc-498a-8263-9f97fd5dc327/GEVO2023.pdf. Accessed 5 May 2024
290. Trend in electric cars, global EV outlook 2024. (International Energy Agency), https://www.iea.org/reports/global-ev-outlook-2024/trends-in-electric-cars. Accessed 10 May 2024
291. Z. Yangarchive, How did China come to dominate the world of electric cars? (MIT Technology Review, 21 Feb 2023), https://www.technologyreview.com/2023/02/21/1068880/how-did-china-dominate-electric-cars-policy/. Accessed 20 May 2024

Hybrid Vehicles 15

Much of the technology of hybrid vehicles is covered in Chap. 14 *Electric Vehicles* since they share the same electric drive components. The focus of this chapter is limited to the technical features that distinguish hybrid–electric vehicles (HEVs) from battery-electric vehicles (BEVs) and internal combustion vehicles (ICVs).

Hybrid Versus Electric Vehicles

Hybrid implies multiple motive power sources. *Hybrid–electric vehicles* (HEVs) use both an IC engine and an electric motor to power the vehicle and are usually just referred to as *hybrids*. When modern hybrids were introduced into the world in the late 1990s, they were initially characterized as a technology transition from combustion to electric propulsion. But since the introduction of *plug hybrid–electric vehicles* (PHEVs) circa 2007, they have taken on a new role as a compromise solution for the majority of vehicle owners, providing the high efficiency and zero emissions of an electric motor, the energy recovery capability of regenerative braking, and the refueling options of both gasoline and electric charging, without the high cost and weight, reduced range, long recharge time, and additional safety concerns of BEVs.

Prior to the 1990's, almost all "hybrids" were EVs with an ICE-fueled generator onboard, making them *series hybrids*, although some are now described as BEVs with *range extender* (REX) units.

There are many ways that the combination of power sources can be coupled to the drive wheels of a vehicle and, of course, new acronyms, jargon, and classifications of these configuratioins. Here are the most popular classifications based on the drive train configuration. There are also classifications of hybrid vehicles based on the type of combustion power plant, the method of energy storage, the system

C. A. MacCarley, *Non-Petroleum Automotive Transportation*,
https://doi.org/10.1007/978-3-031-85028-8_15

voltage and the battery size, and the electric drive accessories, so that a complete description of a hybrid vehicle may involve strings of multiple classifications.

Since 1896
- Series hybrid
- Parallel hybrid

Since 1990
- Weak (or mild) hybrid
- Strong (or full) hybrid

Since 2000
- Power-split hybrid (a subset of parallel hybrids)

Since 2007
- Plug-in (or just plug) hybrid

Since 2010
- Dual power-split hybrids
- Dual power-spilt plug hybrids
- Fuel-cell/ICE series hybrids

Why Hybrids?

HEVs require two different power sources and are more complex than either ICVs or BEVs. But they provide the benefits of both: the range and refueling convenience of an ICV with the reduced emissions and high efficiency of a BEV. When introduced into the mass market in the late 1990s, they filled a gap between conventional combustion engine power and electromotive power when battery technology had not yet reached the energy density of lithium chemistries. Had better batteries been available just a few years earlier, history might have been different. The lessons of early second generation battery–electric vehicles such as the GM EV-1 were learned; BEVs were premature until better batteries were ready for automotive use. Hybrids filled the gap, but on the way, introduced unique benefits of their own. Gasoline hybrid–electric vehicles filled the need for something better during a transition period in which the problems of air pollution and fossil could no longer wait for better batteries. They were minimally disruptive to the automotive and fossil fuel status quo, since (until plug hybrids) they were entirely fueled by petroleum, requiring no additional refueling infrastructure, but shared some of the capabilities of electric vehicles, such as braking energy recovery and quiet operation at low speeds. They were an instant hit with the public, first in Japan, then the USA and Europe.

As discussed in the chapter *Electric Vehicles*, one pivotal political event can be given much of the credit for moving up the timetable for hybrid automobiles in the USA: the *Low Emission Vehicle* (LEV) Regulation of 1990, aka, *The California EV Mandate*, that created requirements for electric vehicle sales in California starting in 1998. It was the introduction of the Honda Insight (Japan in 1997, USA in 1999) that really kicked off the new generation of hybrid vehicles that continues to this day (Fig. 15.1). The Insight was advanced for its time but was considered more of a niche market product than a mainstream car. It was a gasoline-powered 'weak hybrid' with a 1.0 L 3-cylinder ICE that produced 69 HP (51 kW), combined with Honda's proprietary Integrated Motor Assist (IMA) electric drive system that added 13.4 bhp (10 kW) @ 3000 rpm. It was similar to some of the experimental vehicles build by university teams attempting to achieve ultra-high fuel economy and ultra-low low emissions. It weighed only 820 kg (1808 lbs) [1], and its aerodynamics were exceptional, with a drag coefficient of 0.25.

The real value of the IMA was not the additional power it could add to the ICE output, but its ability to recover some of the braking energy by using the motor as a generator, implementing *Regenerative Braking*. With a tiny engine, regenerative braking, and incredible aerodynamics, the 2006 Insight shown in Fig. 15.1 established the highest fuel economy ever measured by the US EPA for a production gasoline vehicle, a combined 69.2 mpg (3.4 L/100 km), a record that still stands today for gasoline-powered production cars.

There have certainly been many specialty and DIY hybrid electrics, some dating back to the golden age of EVs (1890–1910), but these were almost all series hybrids. The Insight took a different approach: the electric motor could supplement the power output of the IC engine; the first mass-market *parallel hybrid*.

While lithium battery technology was not yet ready to allow production of BEVs, the popularity of hybrids using small NiMH batteries took off, with every major automobile manufacturer in the world soon offering hybrid versions of some of their existing cars, and a few completely new hybrid models. These were all *non-plug hybrids,* which from a consumer's point of view were just high-mileage

Fig. 15.1 2006 Honda Insight, Generation 1. Last production year of the first modern gasoline hybrid sold in the USA 1999–2006. Creative Commons CC BY-SA 2.0

gasoline cars. It was obvious to all that with a larger battery, a hybrid could drive a limited distance on electricity alone and the battery could be charged directly rather than only from the gasoline engine acting as a generator. But the "plug hybrid" would have to wait almost 10 years until lithium battery technology had improved enough to safely and economically meet the "larger battery" requirement.

Between 2005 and 2011, hybrid cars were allowed access to High Occupancy Vehicle lanes on California freeways. Some municipalities provided exclusive parking for hybrids. Some states offered rebates for hybrid purchases as much as $1500 USD. But these incentives were short-lived for gasoline-only hybrids, replaced in 2007 by the *Clean Vehicle Rebate Program* (CVRP) that discontinued incentives for hybrids, but provided a purchase rebate of up to $7500 for BEVs and the *plug hybrids* that would enter the US market in 2012.

Surely, the most prominent among the new wave of hybrids was the Toyota Prius, introduced in Japan in 1997 and in the US market in 2001 shown in Fig. 15.2. It incorporated a major innovation, the *Toyota Hybrid System* (THS), later renamed the *Hybrid Synergy Drive* (HSD) at the time of its introduction into the US market. This was the first *power-split hybrid* drivetrain which allowed the car to operate in gas mode, electric mode, or both. The ICE was electronically controlled by the hybrid drive controller. The 1.3 L engine of the Gen-1 Prius would start and stop when needed to supplement the 35 kW primary drive electric motor, and a smaller 10 kW secondary drive motor that served as the control mechanism for a 3-input planetary drive transmission, similar in function to a rear axle differential.

Like most early hybrids, the Prius used a small 1.3 kWh nickel metal hydride (NMh) battery for storage of regenerative braking energy and to provide a small electric-only range that allowed electric-only low speed operation for several seconds after stopping, as well as for starting of the ICE. The Prius could silently start across an intersection before the ICE started, a disconcerting feature for some motorists and pedestrians.

The Prius was far more than an incremental improvement of previous limited-market hybrids. Toyota's choice of the name "Prius" changed the meaning of the

Fig. 15.2 2001 Toyota Prius shown in introductory print advertisement, just prior to introduction into the US market. Image from aldenjewell. Creative Commons CC BY 2.0

obscure noun "prius" into a cult identifier. Aftermarket accessories quickly appeared: the engine tachometer that Toyota chose to leave out, custom wheels, and a switch to prevent the automatic startup of the ICE so that owners could noiselessly arrive at destinations, a boon for quiet neighborhoods and cheating spouses.

The Prius represented a major advance in hybrid technology that set the standard for all hybrids that followed. Its influence on car culture was significant, eliciting the ire of drivers of other vehicles having much lower mpg ratings, including vindictive jokes, some illustrated in Fig. 15.3. Regarding the right-most meme in Fig. 15.3: Even car-culture magazines unanimously agreed that the 2000-03 Gen-1 Prius was unattractive compared with other cars from Toyota or competitors. The most similar body style in the Toyota lineup was their least expensive vehicle, the Echo [2]. Supportive bloggers and conciliatory dealers speculated that its "computer mouse" shape made the car distinctive as an eco-friendly vehicle. But technical reviews revealed that the advanced drivetrain almost surely made the car more costly to manufacture than its MSRP, suggesting a deliberate effort by Toyota to "uglify" the car to limit sales to enthusiasts only. One of many sample Prius blog posts appears in citation [3]. Toyota has never commented on the "uglification" conspiracy theory, but it would be resurrected later when the first mass market BEV, the technical masterpiece but frog-like Nissan Leaf was introduced in 2011.

By 2007, lithium-ion battery technology became affordable and reliable enough to justify its use in hybrid or battery–electric vehicles. The 2006 AC Propulsion tZero and the 2008 Tesla Roadster were the first to adopt lithium batteries. In 2012, Toyota introduced the first *plug-hybrid* version of the Prius that used a 4.4 kW lithium-ion battery, weighing only slightly more than the previous 1.3 kWh NMh battery that continued to be used in nonplug versions of the Prius. It adopted the new standard SAE J1772 EV conductive charging connector to allow it to be charged at home or from (then rare) public EV charge ports. Although only providing 11–13 miles of electric-only range, this was long enough to allow it to be used as a pure BEV for short trips, which placed it somewhere between a gasoline hybrid and a full BEV.

Fig. 15.3 Collection of random vindictive memes following the US introduction of the Toyota Prius. From Tundras.com https://www.tundras.com/threads/if-you-need-to-make-fun-of-a-prius-do-it-here.2045/

Summary of Attributes of Hybrid–Electric Vehicles

- A transition technology between ICVs and BEVs, with attributes of both.
- Avoids need for supporting charging infrastructure.
- Except for plug hybrids, a hybrid is just high-mileage gasoline vehicles.
- Plug hybrids overcomes range and recharge time objections to BEVs while still allowing some amount of electric-only range.
- Regenerative braking. Recover energy that would have been lost when braking.
- For parallel and power split hybrids only:
  - Low-speed operation on electric motor is much more efficient than IC engine.
  - Ability to stop ICE engine and run on electric motor only for short distances or while starting away from a stop.
  - Integrated control of ICE and electric motors maximizes efficiency.

Hybrid Drivetrains

The basic configurations of hybrid drivetrains are compared in Fig. 15.4. The second is a ***series hybrid***, and the third is a ***parallel hybrid.*** A ***power-split hybrid*** falls within the broader class of parallel hybrids.

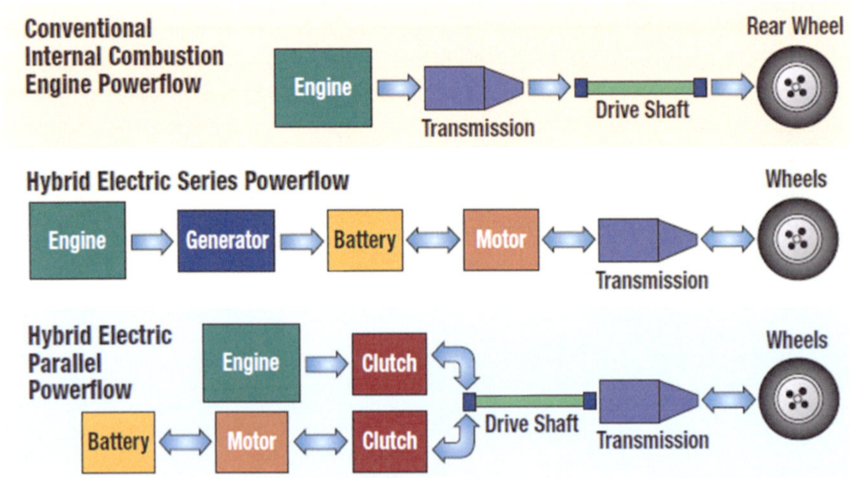

Fig. 15.4 Diagram distinguishing hybrid drivetrain configurations. From US Dept of Energy, Energy Efficiency, and Renewable Energy. Energy. http://www1.eere.energy.gov/vehiclesandfuels/pdfs/basics/jtb_electric_vehicle.pdf Public domain

Weak (or Mild) Hybrid

A weak/mild hybrid drivetrain uses an electric motor with a much lower power output than the ICE. The motor can *assist* the ICE or serve as a generator for regenerative braking. The battery is small, typically less than 1 kWh. The *Honda integrated motor assist (IMA)* hybrid drive system made the Honda Insight a weak hybrid.

Weak hybrids provide some improvement in mileage due to regenerative braking, and sometimes automatic engine start–stop control which shuts off the engine after the vehicle has been stopped more than a few seconds. But the contribution of the electric motor to moving the vehicle forward is usually small. Sometimes even trivial, as in the example of Fig. 15.5, where the GM *eAssist* system actually *reduced* the EPA mileage rating of the 2014 Chevrolet Malibu compared with the less expensive non-hybrid version [4]. But it won the car the right to be called a hybrid, incurring both PR and subsidy benefits. The addition of a hybrid assist system to an existing model was a quick fix for manufacturers seeking to avoid the engineering of an entirely new hybrid vehicle. An existing product could easily be converted to a hybrid using a simple add-on motor/generator coupled to the engine by an accessory drive belt. It would also save manufacturing cost by eliminating the need for a separate starter and/or alternator. The coveted "hybrid" badge commanded a higher MSRP, a "green" image, and could qualify the vehicle for PZEV status, enabling incentives to both the manufacturer and potential buyer.

Strong (or Full) Hybrid

The criteria for what constitutes a strong or full hybrid is subject to interpretation. But one agree-upon attribute of a strong hybrid is that the electric motor and batteries provide enough power to allow electric-only operation at low speeds, in addition to power assist when required and greater energy recovery from regenerative braking energy (Fig. 15.6).

For example, the HSD system of the MY 2004-07 Gen-2 Toyota Prius used a 57 kW gasoline engine, a 50 kW main drive motor (MG2), a 15 kW control motor (MG1), and a 1.31 kWh nickel metal hydride battery. The Gen-2 was much more successful than the Gen-1, and by 2005, the name Prius had become almost synonymous with hybrid.

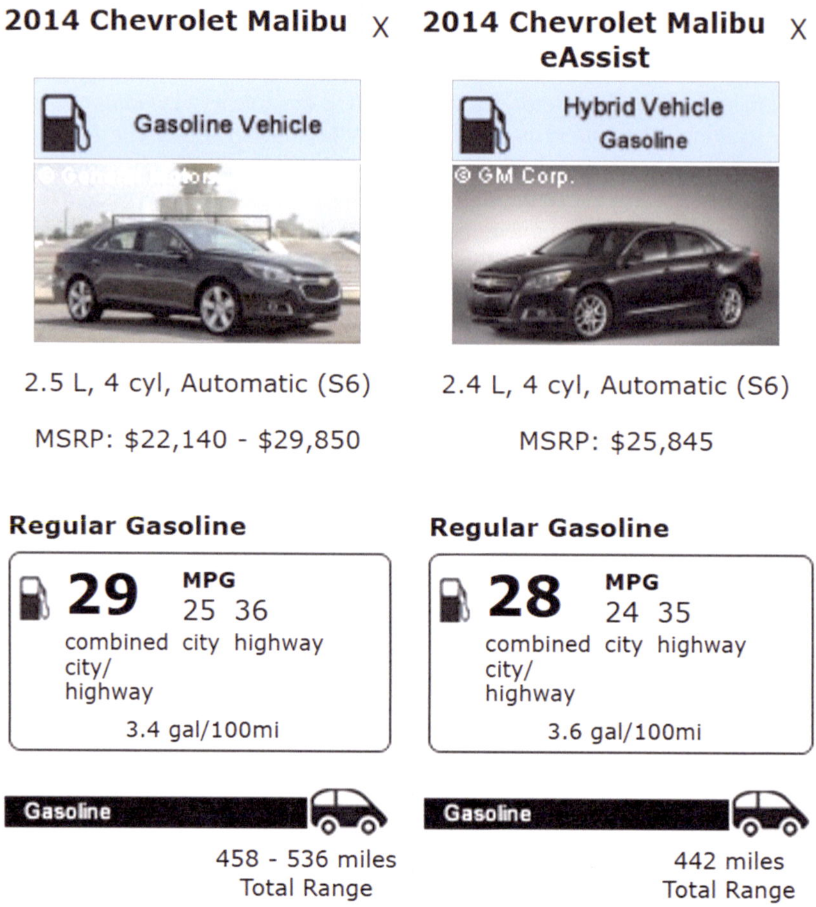

Fig. 15.5 Comparison of 2014 Chevrolet Malibu, available with and without the GM eAssist option. From US Dept of Energy FuelEconomy.gov web site. A rare case in which the hybrid's mileage was worse than the gasoline-only version of the car. The eAssist system was an engine add-one that replaced the alternator and starter, and could assist the ICE engine when needed. https://www.fueleconomy.gov/feg/Find.do?action=sbs&id=34311&id=33844 Public domain

Fig. 15.6 2006 Toyota Prius Gen-2. A very popular strong hybrid. Photo by Michael Pereckas, 2006. Creative Commons Attribution 2.0

Regenerative Braking

One of the most important benefits of both hybrids and electric vehicles is the ability to recover energy that would normally be dissipated as heat during braking, and use it to charge the battery. This is called *regenerative braking*. It can significantly improve the overall fuel economy of a hybrid vehicle in driving that requires frequent use of the brakes, e.g., stop-and-go traffic on a congested freeway. This is why, for almost all hybrids and EVs, the EPA City Cycle mileage is actually higher than the Highway Cycle mileage, the opposite ordering for ICVs. Electric drivetrains can regeneratively brake so seamlessly that most drivers of hybrids and BEVs are not even aware it is happening. Until they notice that at 100,000 miles there is almost no wear on the disk brake pads. Since regenerative braking is a feature of both EV and hybrid vehicles, additional information can be found in Chap. 14, *Electric Vehicles*.

Figure 15.7 is a high-level diagram showing the power-flow direction during acceleration and during regenerative braking in a parallel hybrid.

Nonelectric Energy Recovery

Many novel schemes have been proposed and/or attempted for mechanical recovery of braking energy. One of the last commercial attempts was a hydro-pneumatic braking energy recovery system that used an air-over-oil accumulator for energy storage. The Eaton HLA hybrid pneumatic-hydraulic braking power recovery system, shown in Figs. 15.8 and 15.9, was intended to be an add-on system for gasoline and diesel trucks and vans that were required to start and stop frequently. Garbage

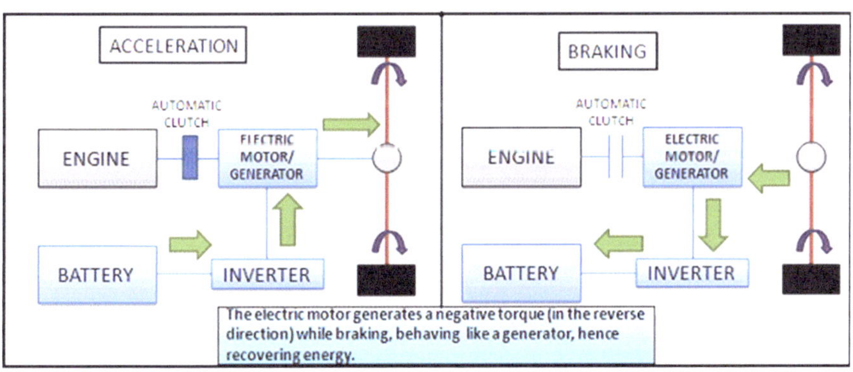

Fig. 15.7 Power flows in a parallel hybrid electric vehicle. During acceleration (forward torque) electric power flows out of the battery to power the wheels, in parallel with the ICE. During braking, the electric motor acts as a generator powered by the inertia of the vehicle, charging the battery while slowing it down. Image from Clemson University Vehicle Electronics Laboratory https://cecas.clemson.edu/cvel/auto/systems/regenerative_braking.html. With permission

Fig. 15.8 2012 Print advertisement for Ameripride custodial service promoting their use of vans equipped with Eaton Pneumatic-hydraulic braking energy recovery system. https://www.hydraulicspneumatics.com/applications/rail-truck-bus/article/21884481/ameripride-takes-delivery-of-first-hydraulic-hybrid-truck. With permission

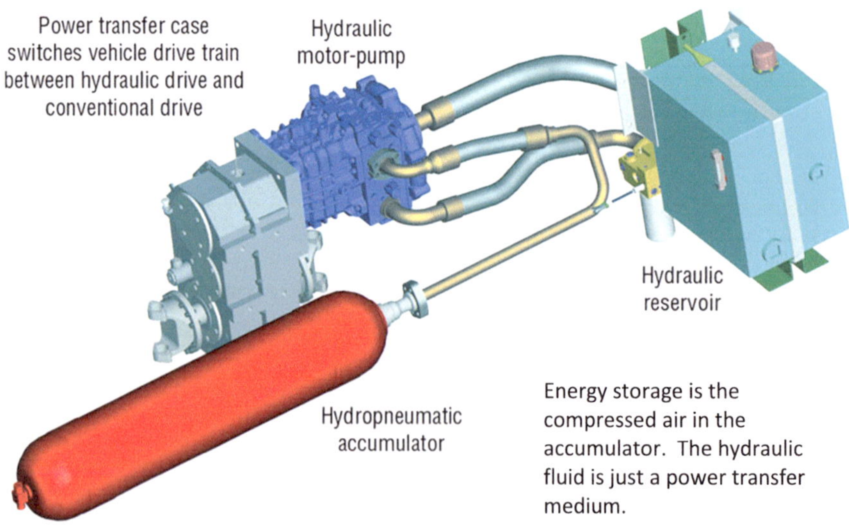

Fig. 15.9 Main components of the Eaton HLA (hydraulic launch assist) hydro-pneumatic braking energy recovery system, mid-2000s. From https://www.powermotiontech.com/news/article/21884504/eaton-dumps-hla-hybrid-hydraulic-system

trucks and delivery vans were the target market, since their operation involved frequent braking and short accelerations that were wasteful of fuel energy as well as hard on the vehicle brakes.

A dual-purpose hydraulic pump/motor was used to provide braking torque as well as forward torque. During braking, hydraulic fluid was pumped into the hydraulic accumulator—like compressing a spring. When starting forward from the stop, the pressurized hydraulic fluid would then be used to assist the ICE in accelerating the vehicle. The accumulator energy storage capacity was only sufficient to store the energy from one compete stop from a modest speed. But assuming that every stop was followed by a start, the energy exchange nearly balanced out.

Unfortunately, like many prior attempts using hydrostatic drives, this energy recovery system had very poor round trip efficiency (RTE) due to viscous fluid friction and losses from the cyclic adiabatic compression heating of air in accumulator. The system ended up being very costly to manufacture, and just one more thing for fleet operators to maintain, and was discontinued in 2013 due to cost and lack of sales. Perhaps Eaton is owed an honorable mention award for this almost-hybrid drivetrain.

Electric RegenerativeBraking

All modern hybrid or electric vehicles incorporate regenerative braking. The topic was introduced previously in Chap. 14 Electric Vehicles, but the details of its mechanical and electronic implementation are covered here.

Electric regenerative braking uses a hybrid or BEV traction motor as a generator, to recharge the vehicle battery using the otherwise wasted vehicle inertial energy during deceleration.

Electric current (therefore power) always flows from a higher to a lower voltage. But while the battery voltage is reasonably constant, the motor/generator voltage during braking can be either higher or lower than the battery. The toughest condition is slow speeds, as the vehicle is coming to a stop, since the motor/generator voltage is much lower than the battery. Figure 15.10, a MATLAB Simulink simulation diagram, illustrates the electrical situation. Regenerative braking can be explained starting with the simple macroscopic electrical model of a commutated permanent magnet (ECPM) DC motor discussed in Chap. 14, *Electric Vehicles,* and shown again in Fig. 15.11.

Here are the electrical relationships for the motor, including the effect of the motor speed N (RPM):

- Motor torque T_m (Nm) is proportional to the motor current i_m (amps):

$$T_m = K_m \times i_m \quad where \;\; K_m = \text{Motor Torque Constant (Nm/amps).}$$

- The back-emf voltage v_b (volts) increases with motor speed:

$$v_b = K_b N \quad where \;\; K_b = \text{Back-emf coefficient (volts / RPM).}$$

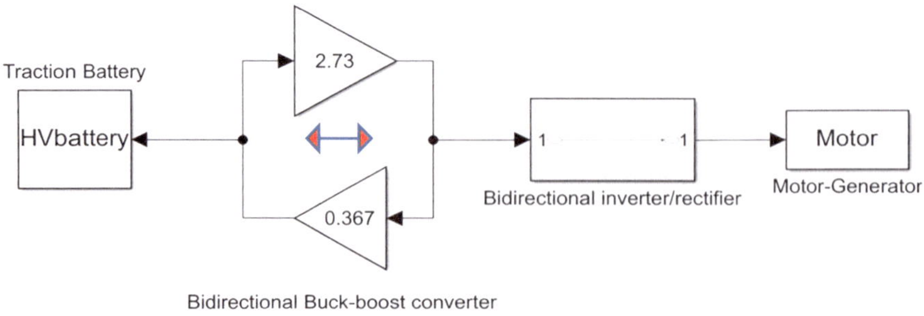

Fig. 15.10 Block diagram of a MATLAB® simulation of a BEV or hybrid power control system, showing the bidirectional power flow capability between the high-voltage battery and the motor that makes possible regenerative braking

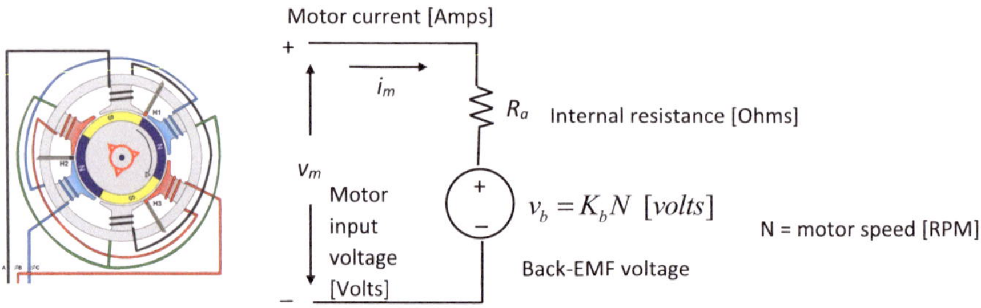

Fig. 15.11 Simplified electrical model of a commutated (electronically or brushed) permanent magnet motor of the type used in most new hybrid EVs. Graphic on left from https://electronics.stackexchange.com/questions/186330/in-order-to-understand-bldc-motors-and-their-working-principle-i-read-a-lot-on Creative commons license

- Positive current flows from the battery to the motor, from a higher to a lower voltage: $i_m = \dfrac{v_m - v_b}{R_a}$

  where v_m = motor voltage (volts), and R_a = Armature Resistance (ohms), armature could be either rotor or field depending on motor configuration.

- Combined: $T_m = \dfrac{K_m}{R_a}\left(v_m - K_b N\right)$

Motor torque T_m is therefore a function of voltage v_m and motor speed N

Electric Regenerative Braking, by the Numbers

1. When the motor input voltage v_m is greater than the back-EMF voltage v_b at some speed N, the motor current i_m is positive, and the motor produces positive torque T_m.

2. When the back-EMF voltage $v_b = K_b N$ is greater than the motor input voltage v_m at a given speed N, the current i_m flows in the opposite direction, producing negative torque T_m.
3. The transformation from motor into generator is controlled by increasing (boosting) or decreasing (bucking) the motor/generator terminal voltage v_m to a value greater or less than the motor back-EMF voltage vb.
4. Braking torque increases with the difference between the bucked/boosted voltage from the motor and the fixed battery voltage v_{bat}, since this controls the current that charges the battery while braking.
5. A circuit called a *bidirectional buck-boost converter* (BDBBC) is used between the battery and the motor input. It can change voltages up or down in either direction, and allow current to flow in either direction. The motor input voltage v_m can be increased or decreased to make it more or less than the battery voltage by adjusting the pulsed signals that control the on or off states of the power transistors in the BDBBC, shown in the circuit schematic of Fig. 15.12 below.
6. When the motor is working as a generator, it slows (brakes) the vehicle. The energy produced by the generator during braking is returned to the battery where it can be used later to power the motor in the forward direction, thus the term *regenerative*.

Figure 15.12 is an LTSpice® simulation schematic of a bidirectional buck-boost converter. The components in this case are sized for a 200 V traction battery in a compact parallel hybrid. The red arrows indicate the direction of bulk current flow in ether drive (motor) mode or regen (generator) mode.

Like most DC–DC converter circuits capable of boosting the voltage, it uses the reverse polarity high-voltage spike that occurs across an inductor (L1) upon abrupt cessation of current to cause net current to flow in the desired direction, from the battery to the motor or from the motor (acting as a generator) to the battery. The direction of current flow into or out of the battery and motor is controlled by the relative durations of the "on/off" periods of each of the four IGBTs (insulated gate bipolar transistors), which serve as switches. In this case, the 50V back-EMF voltage corresponds to

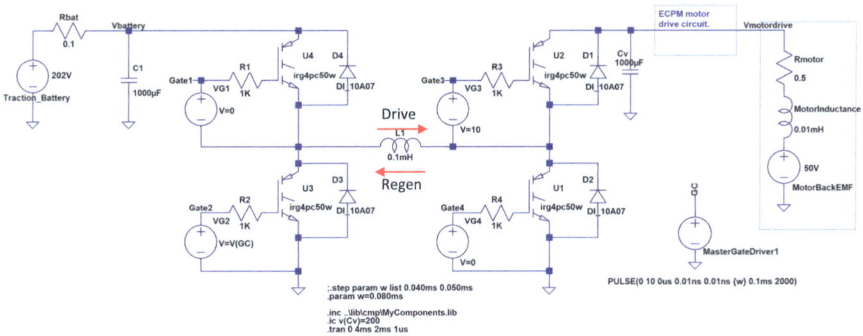

Fig. 15.12 Circuit schematic for a generic bidirectional buck-boost converter as used in a hybrid electric drivetrain. Created using LTSpice®

a vehicle speed of approximately 50 km/hr. To implement regenerative braking, the 50V back-EMF of the motor in this case is "boosted" to slightly above the 200V battery voltage, causing current to flow from the motor (acting as a generator) into the battery. The LTspice simulation file can be downloaded at this link:

LTspice model for EV bidirectional buck-boost converter in forward power and regenerative braking modes (ASC 7 kb)

Power-Split Hybrid Drive System

Most new hybrids and plug hybrids use *power-split hybrid drive* systems derived from the Toyota HSD, shown in Fig. 15.13.

Referring to Fig. 15.14, the planetary drive is located between MG1 (labeled as Generator 33 kW) and MG2 (labeled as E-motor 50 kW). The ring gear of the planetary drive is connected by HiVo chain to the final reduction and differential gears that are connected to the drive wheels.

Fig. 15.13 Gen 1 (1997) Toyota HSD transmission cutaway showing MG1 (labeled generator), MG2 (labeled E-motor) and planetary gearset between them in common housing. Noncopyrighted photo, posted on user forum https://www.yarisworld.com/forums/showthread.php?t=10424&page=6

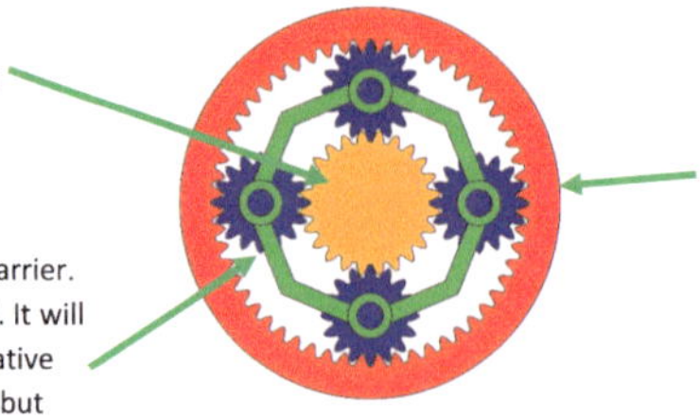

Fig. 15.14 Graphic from animated diagram of Gen 1 HSD epicyclic transmission, accessible online at https://commons.wikimedia.org/wiki/File:Epicyclic_Gearing_Stationary_Ring.gif CC0-1.0 open use dedication

The two electric motors (MG1 and MG2) and the ICE are mechanically coupled via the planetary transmission to exchange their torque contributions or loads, and provide forward or reverse net torque or maintenance of the battery state of charge during cruise conditions. The diagram shows the HSD planetary drive transmission. Note that electric motor generators MG1 and MG2 can rotate in either direction, but the ICE can only rotate in a forward direction and contribute positive torque above some minimum speed, typically 1000 RPM.

The relationship between the speeds (N, in RPM) of each of the three inputs is

$$N_{MG2} = 1.385 \times N_{ICE} - 0.385 \times N_{MG1} \left(\text{RPM}\right)$$

(N_{ICE} is always positive since the engine can't rotate backwards.)

Since MG2 is connected directly to the drive wheels

$$\text{Vehicle speed}\left(\text{mph}\right) = 0.0169 \times N_{MG2} = 0.234 \times N_{ICE} - 0.0065 \times N_{MG1}$$

When $N_{MG1} < 3.40 \times \text{ICE}$, $N_{MG2} > 0$ (vehicle in drive)
When $N_{MG1} = 3.40 \times \text{ICE}$, $N_{MG2} = 0$ (vehicle stopped)
When $N_{MG1} > 3.40 \times \text{ICE}$, $N_{MG2} < 0$ (vehicle in reverse)

The torque relationship between each of the inputs is

$$T_{MG2} = 0.72 \times T_{ICE} - 2.60 \times T_{MG1} \left(\text{Nm}\right)$$

During forward torque, the ICE, MG1, and MG2 are controlled so that $T_{MG2} > 0$

During braking, the ICE, MG1, and MG2 are controlled so that $T_{MG2} < 0$

The algorithms used to control the torque and speed of MG1, MG2, and the ICE are quite sophisticated. They maintain the traction battery in a mid-charge state so that it can accept braking current at any time, while also being ready to source current to supplement the ICE when needed.

- The back-EMF of MG1 is proportional to its speed. V_{bemf} (volts) = 0.08 × MG1 (rpm)
- MG1 functions as a motor when $V_{bemf} < V_{MG1}$
- MG1 is a generator when $V_{bemf} > V_{MG1}$
- Speed and therefore the back-EMF V_{bemf} of MG1 is maximum when the vehicle stopped
- For regenerative braking at 10 mph: MG1 = 3,863 rpm, $V_{bemf} = 309$ V
- For regenerative braking at 0 mph (MG2 = 0 rpm): ICE = 1500 rpm, MG1 = 5400 rpm, $V_{bemf} = 432$ V
- Maximum regenerative braking force is available at 0 mph.

Dual-Motor Power-Split Drive Electronics

The power electronics for the Prius differ from conventional parallel hybrid electronics in that they power two motors, MG1 and MG2, each with its own 6-switch 3-phase inverter. The mechanical power-split transmission of the HSD makes possible the use of a simpler 2-switch version of the generic 4-switch bidirectional buck-boost converter (BDBBC) described in Figs. 14.53 and 14.54 of Chap. 14 *Electric Vehicles*.

While not necessary to understand the operation of the HSD, an explanation of the electronic operation of its 2-IGBT BDBBC is presented below using LTSpice® simulation first in forward drive mode and next in regenerative braking mode.

Forward Drive Mode

Bidirectional buck-boost converter is operating in boost mode, converting 200 V battery voltage into 500 V (max) drive voltage (Figs. 15.15 and 15.16).

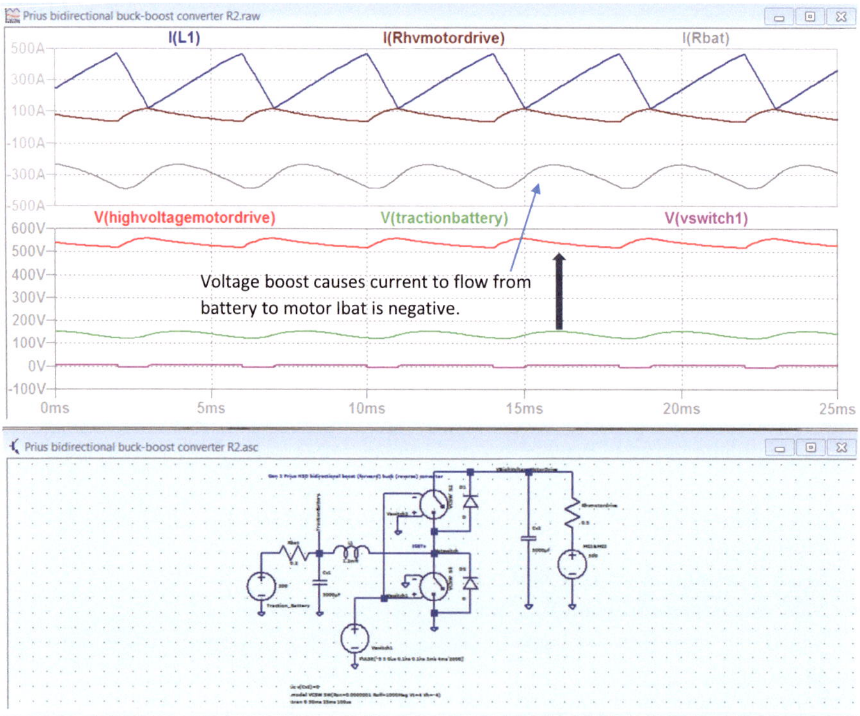

Fig. 15.15 LTSpice simulation of a buck-boost converter used for regenerative braking, showing internal voltages and currents during forward drive mode, in which current flows from the battery to the motor

Regenerative Braking Mode

The motor acts as generator driven by vehicle inertia. The bidirectional buck-boost converter operates in buck mode, reducing, in this case, the 500 V generator voltage to 200 V to charge battery.

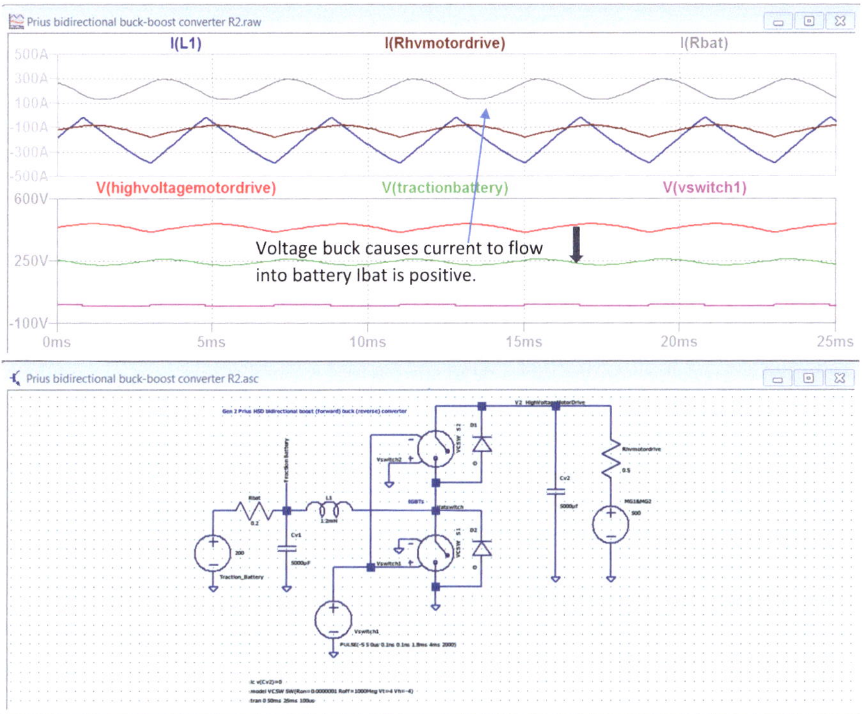

Fig. 15.16 LTSpice simulation of a buck-boost converter of regenerative braking circuit, showing internal voltages and currents during regenerative braking, in which current flows from motor (acting as generator) into battery

An Abbreviated History of Battery-ICE Hybrid Vehicles

Diesel–Electric Hybrid Locomotives

1895. Early electric locomotives that were popular in Europe used overhead catenaries (wires) for power to their electric motors. In the USA, General Electric was the main manufacturer of electric engines. But the huge cost of electrification of thousands of miles of existing tracks intended for steam locomotives lead GE to experiment with *putting the power plant on the locomotive*. GE and manufacturers in Europe used Rudolph Diesel's recently patented compression ignition IC engine to power a generator that charged onboard lead-acid batteries that provided power for the electric motors.

1914. The pairing of the diesel engine with an electric drivetrain had become an obvious alternative to dirty coal/steam and oil/steam engines on nonelectrified railways. This solved several technical obstacles to the direct use IC engines for railway locomotives. First was the challenge of using an IC engine to start a large freight or passenger train from a dead stop, which would require a massive clutch and multispeed transmission. Unlike the diesel engine, electric motors produced maximum torque at zero speed, a simple and ideal solution for trains that required high torque at zero or low speeds. It was a simple technical transition for existing electric trains designed for the use of overhead electrical catenaries since they were already powered by electric motors. The diesel engine could run a constant speed, powering a generator that continuously charged batteries, which power the electric motors: *A series hybrid.*

1914. Regulation of the electric motor power (thus the train speed) had always been a challenge for electric locomotives and continued to be a challenge for hybrid electrics. But with the added flexibility of regulating the power output of the diesel engine, generator, switchable battery taps, and field control of the electric motors, the diesel–electric hybrids were a major improvement. In 1914, a power controller was invented by Hermann Lemp, a GE electrical engineer, that used combined engine and generator control (DC field control) to great effect.

1917-18. GE produced three experimental diesel–electric locomotives using Lemp's control design.

1922. According to citation [5], the first two functional diesel–electric locomotives to be used in passenger service were built by the German company Waggonfabrik Rastatt with electric equipment from Brown, Boveri & Cie and diesel engines from Swiss Sulzer AG. In 1922, they were sold to Swiss Compagnie du Chemin de fer Régional du Val-de-Travers, where they were used in regular service up to the electrification of the line in 1944.

1923. The Kaufman Act banned steam locomotives from New York City because of severe pollution from coal or oil. The solution was to electrify high-traffic rail lines, which was justified in high density urban areas, but uneconomical in lower-traffic areas outside of cities.

1930. Hybrid diesel electric locomotives gained general acceptance, initially as switch engines.

1934. First diesel–electric hybrid passenger train, the Pioneer Zephyr, reached 112 mph, a previously unheard-of speed for rail travel. With its luxurious accommodations, it was the subject of 1935 film "Silver Streak" (Fig. 15.17). Designed and built by the Budd Company, powered by 8-cyl 8-201-A Winton diesel engine developing 600 horsepower (447 kW) at 750 rpm, which drives a generator to charge lead-acid batteries that powered multiple electric motors, with speed control equipment from General Electric. A fleet of Zephyrs remained in service 1934–1960.

Post-WWII. The *diesel–electric* powertrain arrangement is still used in almost all trains operating on nonelectrified right-of-ways, which are the majority of both freight and passenger railways in the USA.

Fig. 15.17 1935 poster for fictional film about one of the actual diesel electric Zephyr passenger trains (the Silver Streak) that entered service in 1934. Broke all railroad speed records at the time. Public domain

Fig. 15.18 Believed to be the first petrol/electric hybrid vehicle, the 1896 Armstrong Phaeton was produced in small numbers by the American Horseless Carriage Company in the USA. Photo from Hemmings Motor News, https://www.hemmings.com/stories/perhaps-the-worlds-first-hybrid-an-1896-armstrong-sells-for-483400-at-amelia-island/ with permission

Hybrid-Electric Automobiles

For 120 years, gasoline or diesel hybrids have been the solution to the range limitations of battery EVs [6]. Here is a snapshot, from the first hybrids of the nineteenth century to the present, 100 years later.

The 1896 Armstrong Phaeton

The origin of ICE/electric hybrid automobiles remains a bit murky, but according to multiple historic publications and the 1896 issue of the newly founded *Horseless Age* magazine,[1] the first practical hybrid-electric vehicle in the world was likely the 1896 Armstrong Phaeton, a vehicle invented by Harry E. Dey in Connecticut, USA in 1895 and manufactured by the Armstrong Manufacturing Company (later renamed the American Horseless Carriage Company) of Connecticut, USA. The chassis was based on a French ICE horseless carriage manufactured by the Rogers Mechanical Carriage Company. It is uncertain if the Armstrong ever entered actual production, which seems unlikely because of the cost of all the drivetrain components compared with competing horseless carriages of the time that all had simpler electric or ICE drivetrains.

It is amazing that this example of an Armstrong Phaeton survived in storage at the former factory for nearly a century and was just recently restored. This type of backstory is usually cause for skepticism among auto collectors. But its original construction had been well documented, and it sold for $483,400 at the elite Amelia Island auto auction on 29 February 2016 [7]. Figure 15.18 is a recent photo of the fully operational restored Armstrong Phaeton.

As detailed in Hemmings Motor News and other articles [8, 9], Dey had intended it to be an electric motor carriage. But following his "elevator pitch" to executives

[1] Horseless Age, 1896.

at the Armstrong Manufacturing Company in Bridgeport, Connecticut, the idea was rejected in favor of an ICE version of the carriage due to concerns over the range of a battery–electric vehicle. Dey compromised and designed it to be both a combustion and electric vehicle, inadvertantly creating the first battery-electric hybrid available for public purchase. His merged powertrain design endowed the Armstrong with a jaw-dropping array of innovative technologies, some that would not appear again until over a century later.

From information in the above citations, the Armstrong Phaeton used a 6.5-L opposed two cylinder petrol engine with a flywheel that served as the rotor of an electric motor generator (a fully-integrated ICE/motor power unit). The flywheel/motor/generator also worked as an electric starter for the ICE (15 years before the Kettering electric starter patent in 1911) [7] and dynamo (generator) to power the electric ignition system (21 years before the Kettering ignition patent in 1917) and the vehicle's incandescent (Edison-type) electric headlamps (2 years before the Columbia Electric Car in 1898). The Armstrong could travel short distances on battery–electric power alone. It is reported that the dynamo function of the flywheel provided a degree of regenerative braking. With the ICE and electric motor integrated in the Armstrong, it also qualifies as the world's first parallel hybrid. Additionally, from Hemmings Motor News, citation above,

> Dey's design also included features like solenoid-controlled intake valves to assist starting, and an innovative electro-magnetic clutch that transmitted more power to the dynamo as engine speed increased. The car's transmission featured three forward speeds and a reverse gear, and Dey's design specified that some of the gears be cut from rawhide to reduce noise.

And from Autoblog, cited above,

> Dey also gave the Armstrong a semi-automatic transmission. With three forward gears and a reverse gear, the driver swaps cogs with a selector on the steering column. When changing gears, an electric clutch automatically disengages and reengages, negating the need of a clutch pedal." "Interestingly, the Armstrong hybrid's motor was too powerful for its own good, as the torque repeatedly damaged its carriage wheels.

The 1901 Lohner–Porsche Mixte

Perhaps the best-publicized early hybrid was the 1901 Lohner–Porsche Mixte (battery electric + ICE) drive series hybrid. Designed by Ferdinand Porsche, age 26 (presumed to be in the photo of Fig. 15.19, at far right), an employee of the Jakob Lohner Co. of Vienna. Originally designed as a battery–electric vehicle with a massive lead-acid battery pack but inadequate range, so a small ICE powered generator was added that charged the batteries. Among its innovations was, in the version shown in Fig. 15.19, individual electric wheel motors for each of the four wheels. While several variations of the prototype were constructed, it was never produced commercially, although it was an important technical milestone nearly a century before the term "hybrid" existed.

The 1997–2006 Honda Insight

Shown in Fig. 15.1 and previously discussed, the 1997–2006 Honda Insight can be credited with starting the modern hybrid–electric renaissance and, to this day, still

Fig. 15.19 Lohner-Porsche ICE/electric hybrid, 1901. Image and info from Porsche AG, "Gamechanger: how Ferdinand Porsche designed the first-ever hybrid car" 27 June 2023. https://www.porsche.com/stories/innovation/gamechanger-how-ferdinand-porsche-designed-first-hybrid-car/. Photo originally from Scientific American 1901. Public Domain

Fig. 15.20 1997 Toyota Generation 1 Prius, first mass market hybrid but sold only in Japan until 2002. Note right-hand-drive for Japanese market only. A power-split strong hybrid with a 35 kW electric drive motor. https://commons.wikimedia.org/wiki/File:1997_Toyota_Prius_01.jpg. Creative Commons Attribution-Share Alike 3.0 Unported, 2.5 Generic license CC BY-SA 3.0

holds the record for the highest EPA combined mileage for a gasoline powered production automobile: 69.2 mpg (3.4 L/100 km) due to its tiny 1.0 L engine, regenerative braking, and an extremely aerodynamic body design [10]. The original Insight is now considered a cult car, with few seen on public roads due to its collector value.

The 1997–Present Toyota Prius

The Honda Insight was soon followed by the Toyota Prius that debuted in Japan in 1997, and 2001 in the USA. The first generation Prius, Japanese market model, is shown in Fig. 15.20.

Fig. 15.21 The 2005 Ford Escape was the first hybrid car from a US manufacturer. It was also the first E85 hybrid. It utilized a power-split hybrid drive partially licensed from Toyota. Photo by Aude, 27 January 2006, at Washington Auto Show. Creative Commons Attribution-Share Alike 2.5 Generic license

Early US Hybrid–Electric Attempts

In the USA, GM, Ford, and Chrysler all exhibited grant-funded hybrid prototypes at the *SAE Future Car 2000* Conference in Washington. In the conference sessions, US automakers insisted that it would not be possible to move from the current concept phase to preproduction prototype vehicles by 2004[2] as was required by their federal grants. In the wake of the 11 September 2001 terrorist attacks, government funding was diverted entirely to the hydrogen-focused FreedomCAR initiative by the Bush administration. The program eventually was criticized as a diversionary tactic intended to perpetuate status-quo auto industry and oil interests. As summarized in 2007 by Ashok Gupta, the lead energy economist at the Natural Resources Defense Council, "The FreedomCAR [program] is really about Bush's freedom to do nothing about cars today [11]."

The 2005 Ford Escape

The 2005 Ford Escape was the first hybrid-electric car from a US manufacturer. It was also the first E85 hybrid. It utilized a power-split hybrid drive system partially licensed from Toyota, which transformed the 19 city/25 hwy mpg fuel economy of the 2004 V6 pre-hybrid model into 36 city /31 hwy mpg for the 2005 hybrid with the same IC engine (Fig. 15.21).

Plug Hybrid-Electric Automobiles

A *plug hybrid electric vehicle* (PHEV) has the option to charge the battery externally rather than only from the vehicle's ICE using gasoline. It is equipped with a battery charging port and a larger battery than nonplug hybrids to provide a more acceptable electric-only range. PHEVs generally have larger batteries to increase

[2] Personal observation as an invited conference participant, April 2000.

their electric-only range, and an EV-type charge port to allow at least Level-II charging. Other than these, there is usually no appreciable difference in appearance or performance between a plug and a nonplug hybrid.

The first mass-market plug hybrid sold in the USA was the 2012 Toyota Prius Prime, shown in Fig. 15.22 that was huge success in the USA, despite its considerably higher MSRP.

Another milestone plug hybrid is shown in Fig. 15.23, the 2014-2021 BMW i8. With an EPA combined mileage of 76 mpge, top speed of 250 km/h (155 mph), and the appearance of an uber-exotic supercar, it defined a new genre for hybrids, doing everything well at the same time. It was, unfortunately, denigrated by automotive media as being underpowered compared with similar-looking exotics of that time. A 105 kW (140 HP) ECPM motor powers the front wheels, while a mid-engine turbocharged 1.5L engine producing 170 kW (228 HP) drives the rear wheels, a configuration similar to Formula-1 race cars. Gull-wing doors complete the package.

One of the new generation of plug hybrid SUVs is the 2025 Kia Carnival shown in Fig. 15.24. A 33 mpg 8-passenger family hauler with a turbocharged 1.6 L 4-cyl engine producing 180 kW paired with a 55 kW ECPM motor. Comparable to a 19 mpg Ford Expedition or 17 mpg Chevrolet Surburban.

Introduced in 2022, the *Ford Maverick hybrid* was the first compact hybrid truck from a US manufacturer. The 2024 model is shown in Fig. 15.25. While only a nonplug hybrid, it represents a positive first step toward reversing the decade-long trend toward ever-more-massive and expensive trucks and SUVs, especially electrics.

Fig. 15.22 2012–2021 Prius Prime Plug Hybrid (PHEV), USA spec. Introduced in Japan in 2009. https://commons.wikimedia.org/wiki/File:Toyota_PRIUS_PLUG-IN_HYBRID_Concept_(4058850716).jpg. Creative Commons Attribution-Share Alike 2.0 Generic license

Fig. 15.23 Another historically significant plug hybrid is the 2014 BMW i8 coupe. Arguably, the first plug hybrid in the high-end luxury-performance market. A power split strong PHEV with a 34 mile electric-only range and combined EPA fuel economy rating of 76 mpge. Photo by author at BMW Museum, Munich, August 2014

Fig. 15.24 The 2024-25 Kia Carnival plug-hybrid SUV. Photo from https://www.pexels.com/photo/black-kia-carnival-on-parking-lot-under-evening-sky-20540470/. Free stock photo

Fig. 15.25 2024 Ford Maverick 2.5 L nonplug hybrid. Assumed to be a Ford marketing photo, obtained from https://www.pexels.com/photo/ford-maverick-on-desert-17157308/ https://www.pexels.com/license/. No copyright restriction.

It is unfortunate that as of 2024, there are no plug-hybrid (PHEV) trucks of any size for sale in North America. It seems an enigma that Ford has announced a 2025 PHEV Maverick for overseas markets but has no plans to sell it in North America.

Hybrids in Racing

Hybrids need no introduction in professional racing. As listed in Table 15.1, hybrid-electric LM-1 (unlimited class) race cars have won the annual Le Mans endurance race every year since 2012: Audi (2012–14), Porsche (2015–17), Toyota (2018–22), and Ferrari (2023–24) [12]. The ability to store braking energy entering corners and release it to supplement ICE power while exiting corners provides a competitive edge over nonhybrids. The reduced fuel consumption of the Audi R18 e-tron hybrid diesel electric shown in Fig. 15.26 reduced the number of pit stops required to refuel, a factor in its multiple victories.

Table 15.1 Le Mans constructor records: starting in 2012, ICE-EV hybrids have won the 24 Heurs de Le Mans endurance race every year

2012	Audi Sport Team Joest (DEU)	Audi R18 e-tron quattro hybrid
2013	Audi Sport Team Joest (DEU)	Audi R18 e-tron quattro hybrid
2014	Audi Sport Team Joest (DEU)	Audi R18 e-tron quattro hybrid
2015	Porsche Team (DEU)	Porsche 919 Hybrid
2016	Porsche Team (DEU)	Porsche 919 Hybrid
2017	Porsche LMP Team (DEU)	Porsche 919 Hybrid
2018	Toyota Gazoo Racing (JPN)	Toyota TS050 Hybrid
2019	Toyota Gazoo Racing (JPN)	Toyota TS050 Hybrid
2020	Toyota Gazoo Racing (JPN)	Toyota TS050 Hybrid
2021	Toyota Gazoo Racing (JPN)	Toyota GR010 Hybrid
2022	Toyota Gazoo Racing (JPN)	Toyota GR010 Hybrid
2023	Ferrari – AF Corse (ITA)	Ferrari 499P hybrid
2024	Ferrari – AF Corse (ITA)	Ferrari 499P hybrid

Data from https://en.wikipedia.org/wiki/List_of_24_Hours_of_Le_Mans_records

Fig. 15.26 A technical miracle: The Audi R-18 e-Tron *diesel*-electric hybrid that used a flywheel-electric accumulator for electric energy storage, winning the 24 Heures du Mans 2012–2014. Lithium batteries replaced the flywheel electric storage system in 2016, but by then, most other teams also fielded advance hybrids that used lithium batteries. Photo by kevinmcgill, 22 June 2013. Creative Commons Attribution-Share

References

1. All production vehicle specifications from CarSpecs A-Z. https://www.carfolio.com/honda-insight-174163. Accessed 20 June 2024
2. Larry Webster, Car and Driver Magazine, February 2000. Toyota Echo. https://www.caranddriver.com/reviews/a15140671/toyota-echo-road-test/. Accessed 1 October 2024
3. Sample blog post commenting on sophistication of Gen 1 Prius drivetrain: https://www.reddit.com/r/askcarguys/comments/169yirx/why_did_it_take_toyota_so_long_to_make_the_prius

4. Michael Vaughan, CTVNews.ca. February 9, 2013 Is GM's 'mild hybrid' eAssist System Worth the Money? https://www.ctvnews.ca/autos/is-gm-s-mild-hybrid-eassist-system-worth-the-money-1.1149018. Accessed 1 Dec 2023.
5. Wikipedia. Diesel locomotive. https://en.wikipedia.org/wiki/Diesel_locomotive Accessed 10 April 2024.
6. E.H. Wakefield, *History of the Electric Automobile* (Society of Automotive Engineers, Publication No. R-122, 1994)
7. John Snyder, Autoblog 6 March 2016. World's first hybrid, built in 1896, goes up for auction. https://www.autoblog.com/news/worlds-first-hybrid-1896-armstrong-phaeton-goes-up-for-auctio Accessed 1 October 2024.
8. Stephen Edelstein, Green Car Reports, 18 March 2016. Is this 1896 horseless carriage the world's very first hybrid car? https://www.greencarreports.com/news/1102929_is-this-1896-horseless-carriage-the-worlds-very-first-hybrid-car. Accessed 1 October 2024
9. Bonhams Cars, 10 March 2016, 1896 Armstrong Phaeton Gasoline Electric Hybrid, The Amelia Island Auction, Fernandina Beach Golf Club https://cars.bonhams.com/auction/23133/lot/152/1896-armstrong-phaeton-gasoline-electric-hybrid/ Accessed 1 October 2024.
10. Wikipedia. Hybrid electric vehicle. http://en.wikipedia.org/wiki/Hybrid_electric_vehicle. Accessed 15 July 2024
11. Amanda Griscom Little for Grist.org. What can we learn from Bush's FreedomCar Plan? 26 February 2003. Quoted in Wikipedia freedom car. https://en.wikipedia.org/wiki/FreedomCAR_and_Vehicle_Technologies#cite_note-6. Accessed 15 July 2020
12. Wikipedia. Winners, 24 hours of Le Mans. https://en.wikipedia.org/wiki/List_of_24_Hours_of_Le_Mans_records. Accessed 15 July 2024

Fuels for Desperate Times 16

The transportation fuels discussed in the previous 15 chapters are all at least partially motivated by some benefit in renewability, GHG emissions, or other sustainability benefits. The fuels and technologies discussed in this epilog have none of these meritorious attributes. They are lesser-known fuels for IC engines based on nonpetroleum feedstocks that are repeatedly "rediscovered" whenever gasoline and diesel fuel become unavailable or unaffordable, such as during the politically-motivated 'gas crises' in 1973 and 1978 in the USA, photo shown in Fig. 16.1. Such scenarios have been recurring themes of dystopian films such as the 1997–2015 Mad Max film franchise [1]. None of these fuels are convenient, and some are significantly more environmentally harmful than fossil fuels.

Note that there are also motive power options involving different types of combustion engines or mechanical energy storage methods (steam, gas turbine, Stirling flywheels, rubber bands, etc.). But the focus of this chapter is on fuels that can retrofit existing ICE gasoline or diesel vehicles.

Disclaimer... Like all unconventional combustion fuels, the options described here can only legally be used in off-road vehicles, and sometimes even that is dicey. In the USA and many other countries, automotive emission standards prohibit non-OEM engine modifications of any kind. But if there were unexpectedly no available fuels or electric energy sources for transportation, this might not matter.

Homemade Alcohols

Ethyl Alcohol (aka Moonshine)

"Moonshine" is home-brewed ethanol intended for consumption, but also an excellent motor fuel. It has been used episodically as an alternative fuel for over a century, gaining broader appeal during the US prohibition years 1920–1933. It can be fermented from a wide range of cultivated food crops. This makes it a viable

C. A. MacCarley, *Non-Petroleum Automotive Transportation*,
https://doi.org/10.1007/978-3-031-85028-8_16

Fig. 16.1 "Automobiles lining up for fuel at a service station in the U.S. state of Maryland in the United States," June 15, 1979. The scene exemplifies the chaos and sense of vulnerability felt by Americans in 1973 and 1979 when gasoline became scarce as a result of restrictions by OPEC (Organization of the Oil Exporting Countries) in the Middle East. From photo archive of US Library of Congress. https://loc.gov/pictures/resource/ppmsca.03433/

"freedom fuel" in rural areas, although even in desperate times, ethanol would almost surely have a higher value as an alcoholic beverage.

Fermentation, distillation, and engine conversion to use ethanol as a fuel are discussed in Chapter 7 *Ethanol*.

Wood Alcohol

Wood alcohol is methanol, also an excellent motor fuel. With minimal engine modifications, it can power almost any IC engine. As early as 1819 wood alcohol was produced in France, although large-scale production did not begin until 1910 in the USA [2]. It can be easily made from a wide variety of readily available renewable or waste sources, from biomethane to woody biomass. Because it does not depend on food crops, methanol avoids competition with food sources that may also be scarce in desperate times.

Chapter 6 *Methanol* describes how crude methanol can be made using a simple DIY apparatus, starting with wood chips, sawdust, crop residues, or natural biomass.

Wood Gas

The use of wood to power cars during times of gasoline unavailability is well documented. Wood, cellulosic biomass, or charcoal is the energy source, but the actual fuel powering the engine is *wood gas*, a type of *syngas* generated onboard by pyrolysis and gasification. The technology is simple. Firewood is combusted with restricted air in a sealed *gasifier* to decompose it into syngas, comprised of carbon monoxide and hydrogen, with lesser amounts of methane and various organic gases depending on the biomass source. The gas is of low calorific (energy) value and also contains inert contaminants nitrogen, carbon dioxide, and water vapor, depending on the gasifier temperature, pressure and available oxygen. It is filtered, cooled, and fed to the intake manifold of the engine, replacing the gasoline that would have been aspirated into the intake air stream by a carburetor. This is essentially the first step of the process used to generate syngas for making hydrogen, methanol, or dimethyl ether from biomass, but need not be very well controlled. The syngas is burned directly as the fuel rather than being fractioned or reformed.

Wood gas vehicles were common in much of Europe during the war years of 1914–1918 and 1940–45 when gasoline was unavailable for nonmilitary use. A nascent industry evolved, with over one million automotive gasifiers produced during World War II alone [3, 4]. Interest disappeared immediately after the war, as soon as gasoline became available again.

Wood gas was rediscovered and had a small cult-like following in the USA during the energy independence fervor following the petroleum shortages of 1973 and 1979. Two of the many historical accounts of wood-powered cars appear in citations [5, 6].

The essential components of a wood gas fuel system are shown in Fig 16.2. Wood chunks or pellets are loaded into the top of the gasifier (A) and partially combusted with limited air (C, D). Hot synthesis gas from the gasifier outlet (E) is cyclone-filtered (F, G), cooled (H), final filtered (I, J, K), and fed to an air-fuel gas mixer (L) that replaces the carburetor of an engine. A wood gas vehicle is hardly convenient to operate and maintain. Before starting the engine, nearly 30 min is required for the gasifier to be loaded with wood, ignited, and reach a temperature high enough for adequate wood gas production to run the engine. During warm-up, a large amount pollutants are released and energy is lost, but since the source is

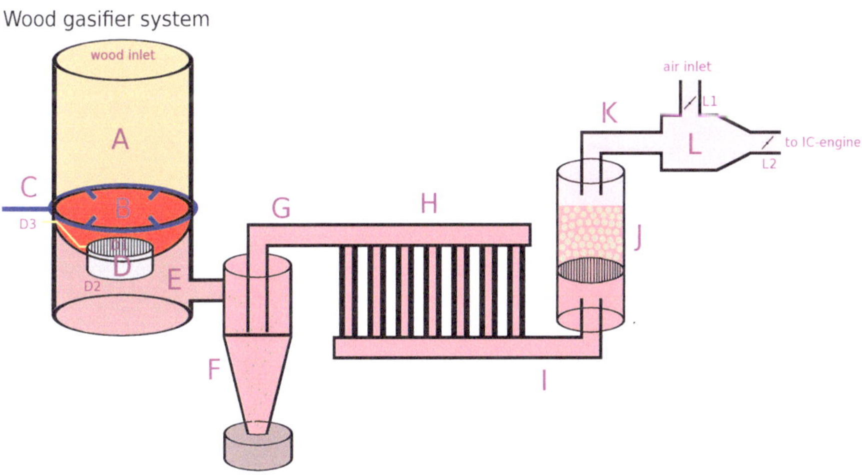

Fig. 16.2 Components of an onboard wood gasifier. Wikipedia. https://en.wikipedia.org/wiki/Wood_gas#/media/File:Planet_Mechanics_wood_gasifier.png. Wikimedia Commons KVDP, CC0 license

Fig. 16.3 1970s pickup truck converted to run on wood gas by Mother Earth News. Photo from 1981 Future Fuels Rally, a transcontinental alternative fuel competition. Used with permission

Fig. 16.4 1934–39 Adler Diplomat wood-powered limousine in Deutsches Museum, Munich, Germany. The large roof rack was used to carry extra wood chunks for the gasifier. Photo by author

assumed to be plant waste that is essentially free, it is simply accepted. Shutting it down takes another 30 min for the wood gas production to (mostly) end, during which time the remaining wood gas should be flared (but usually isn't) to avoid releasing the raw CO, HC, and $CO2$ into the air[1].

Examples of wood powered vehicles from 1981, 1939, 1990 and 1995 are shown in Figs. 16.3, 16.4, 16.5 and 16.6 respectively.

There is always a mismatch between the gas production rate and the demand of the engine, making control of the onboard gasifier challenging. Large amounts of wood gas are wasted whenever production exceeds the engine demand. There have been some improvements over the years. An example of a high-quality conversion that is less lossy was constructed by former Finnish Prime Minister Juha Sipilä: a converted 1987 Chevrolet El Camino shown in Fig. 16.6. It utilized a more efficient gasifier design and a control system that presumably provided a better match between gas production and engine demand.

[1] The 30-minute shutdown time estimate is based on personal experience. It could vary widely for a given system and fuel type.

Fig. 16.5 North Korean military truck powered by wood, 2018. (Unfortunate for the soldiers in the back of the truck.) from *North Korean transport: cars powered byb firewood*, Chi Quan, August 7, 2013. https://soha.vn/quoc-te/dac-san-giao-thong-trieu-tien-o-to-chay-bang-cui-20130708005034725.htm (Vietnam). Public domain

Fig. 16.6 Wood gas-converted 1987 Chevrolet El Camino, converted and owned by former (2015–19) Finnish Prime Minister Juha Sipilä. https://www.iltalehti.fi/autot/a/2012061101003524#gallery. Public domain

Despite the inconvenience and environmental impact, the almost universal availability of woody biomass, even in desperate times, has historically made wood gas the go-to solution to replace petroleum fuel.

The design of automotive gasifiers and selection of their feedstocks is steeped in tribal knowledge rather than rigorous science. Based on a median analysis of the composition of wood gas and average reported efficiencies for automotive gasifiers [7], the calculations in Appendix 2 predict that the CO_2 generation of a wood gas vehicle are about 50% greater than those of a gasoline car. But this is just for the wood gas combusted in the engine. The partial combustion in these single-chamber gasifiers typically consumes 25% or more of the wood to generate the reaction heat, and the additional CO_2 emission must be considered also. Any wood gas that is vented or leaked because of the mismatch between engine demand and gasifier production is environmentally harmful due to its CO, CO_2, CH_4, HC, H_2 and particulate content. The air pollutants are worse than almost any combustion fuel except coal, but probably not much different than using firewood for home heating.

Despite their emissions, wood gas vehicles can and have been classified as renewable fuel vehicles, based on the distorted arithmetic sometimes used to identify a closed carbon cycle: the original energy source was trees that sequestered carbon dioxide from the air, which is returned to the air by its combustion in the engine [8]. In fact, the *char* residue that must be periodically cleaned out of the gasifier and first-stage filter, if buried, can be counted as sequestered carbon, which would give wood fuel a theoretically negative carbon footprint. This analysis ignores the high level of other air pollutants and the wasted energy mentioned above.

Wood or other forms of cellulosic plant waste are possibly the ultimate desperation fuel, since virtually no supporting infrastructure is required, and some form of waste cellulosic biomass can be found almost anywhere except deserts. It is still in use today in North Korea, where wood-powered vehicles such as that shown in Fig. 16.5 are used due to a long-standing international oil embargo. A rough analysis of the military truck in the figure is done in Appendix 2, using assumed technical information gleaned from the web [9] for similar vintage trucks:

MPGe: 12 MJ/km (equivalent to about 6 MPG US gasoline)
LHV of wood gas is calculated in Appendix 2 to be 5.37 MJ/kg
CO_2 generated by wood gas is 0.62 kg CO_2 per kg of wood gas

The CO_2 emissions from just the engine are estimated to be 1390 g/km. However, the actual CO_2 emissions are considerably higher.

From calculations in the Chapter 4 *Engines and Fuels*, the maximum possible power output is 178 HP; considerably less than a gasoline engine of the same displacement.

Wood gas has enduring appeal for survivalists seeking self-sufficiency. Even in less desperate times, it remains a popular statement of energy independence or an earth-centered despite its environmental consequences.

Waste Vegetable Oil (WVO)

As discussed in Chap. 12, vegetable and seed oils have been used as fuels for compression ignition engines since the time of Rudolf Diesel himself, who ran his early engines on peanut oil [10]. The unprocessed plant oils used to make biodiesel fuel actually have a higher energy content than the biodiesel fuel made from them. The organic oil is processed by transesterification to transform the high-viscosity waxy plant oils into a liquid fuel with properties sufficiently similar to that of mineral diesel fuel so that they are compatible as a 20% maximum blend with D2 in the fuel injection and emission control systems of modern diesel engines. However, older diesel engines with mechanical injection systems can usually run directly on minimally processed straight vegetable oil (SVO), at least until something in the fuel injection system clogs up.

Fig. 16.7 A still frame from a YouTube video by Clay Hayes, July 2019. Public domain. https://www.youtube.com/watch?v=gKntRyE4cfE

But for many years prior to government-incentivized biodiesel, owners of pre-1990's diesel cars and trucks, and unregulated off-road equipment routinely substituted, or blended with D2 a wide assortment of cheap or free waste oils that were never intended for use as fuels: waste vehicle lube oil and the waste cooking/frying oils that previously cost restaurants and producers of fried food a substantial fee to dispose. Referred to as Waste Vegitable Oil (WVO) it is no longer a free fuel option since biodiesel producers now buy almost all available waste cooking oils for use as feedstock for biodiesel fuel (Fig. 16.7).

Converting a mechanically injected diesel vehicle to run on WVO is within the skill of mechanically inclined owners of older diesel cars and trucks. Complete DIY conversion kits used to be very common, and a few are still (2023) available online from a vendors in the USA and Australia [11, 12]. WVO is *not* a desirable fuel—it has a high risk of clogging fuel injectors and injection pumps and forming noxious deposits in combustion chambers—but it remains, in desperate times, an alternative for engines that will tolerate it, and combustible oils of some kind will probably still be available if mineral diesel or biodiesel fuel becomes scarce.

No modifications of a diesel engine's mechanical components are required. But several modifications are required for the fuel system, partially shown in Fig. 16.8:

- A separate fuel tank for the WVO, ideally containing a heating loop in the tank that uses engine coolant to heat the WVO that can solidify into lard at temperatures below 20 °C.
- A large high-efficiency cartridge filter at the fuel outlet of the tank. An electric heating element in the filter body is recommended to prevent solidification of the WVO in the filter.
- Heated fuel lines and hoses from the WVO tank to the engine compartment to keep the WVO at a sufficiently high temperature to prevent solidification in the lines.
- An electrically actuated 6-way valve to allow switching both the fuel feed and fuel return lines from the engine to either the vehicle's original diesel fuel tank or the WVO tank.

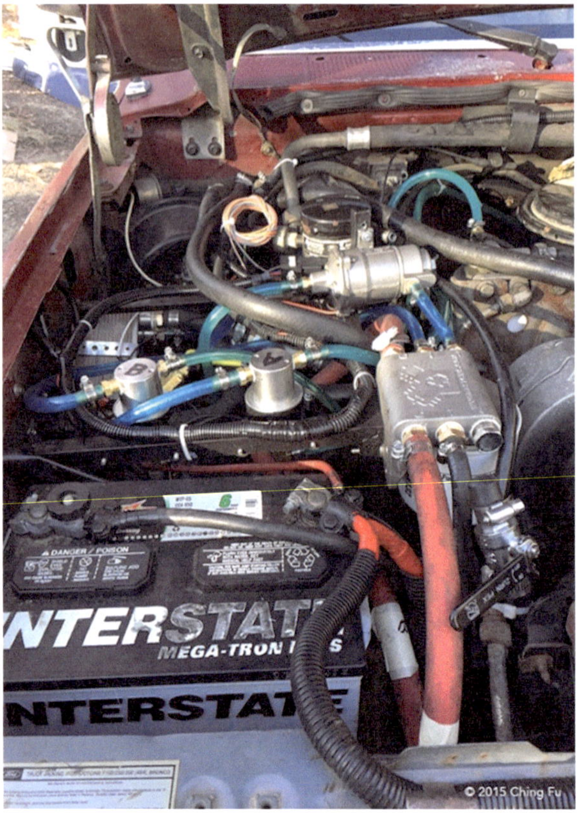

Fig. 16.8 Underhood view of DIY greasel conversion of a 1987 Ford F350 diesel truck. From https://www.livesmallridefree.com/blog/wvo-101-waste-vegetable-oil-vehicle-basics. With permission

- A switch in the vehicle cabin to actuate the fuel selector valve.
- Optionally, a temperature gauge located inside the vehicle that monitors the temperature of the WVO in the tank.

Depending on how dirty the waste oil is, it requires multiple filtrations and possibly other cleanup processes, but is otherwise a 1:1 replacement for Diesel No. 2. Filters will load up quickly, and centrifuging is recommended to remove solid food residues, saturated fats, and partially polymerized organic compounds that form in rancid oil. And if not already done by a WVO recycler (if you purchase WVO), a *dewatering* step is necessary since it usually contains a lot of water from its use in cooking. As little as 0.3% water in diesel fuel significantly reduces the combustion quality and can cause the engine to stall if it phase-separates in the fuel system. Dewatering could be as simple as leaving the raw WVO in a drum for a few days and waiting for water to settle out into a lower stratum, but complete water removal involves heating the raw WVO to near 100 deg C long enough for any water to evaporate or boil off.

This dewatering method is also beneficial for disinfecting the WVO to prevent algae formation. It is hard to believe that certain aquatic plant species and bacteria thrive in diesel fuel, but the growth of bacteria, algae and fungi in stagnant diesel fuel is a problem for *all* diesel vehicles; it is just much worse for WVO containing the remnants of food products. This happens without the awareness of the vehicle owner until the engine stalls due to a clog somewhere in the fuel path, hopefully just in the fuel filter, but possibly in the fuel injection pump and injectors. An effective diesel biocide fuel additive [13] is absolutely necessary.

In a converted diesel vehicle, the engine is started on regular diesel fuel (D2 throughout most of the USA). The fuel lines, fuel filters, and the fuel tank itself slowly warm up from the circulated engine coolant. When the engine and the fuel in the WVO tank reaches a sufficient temperature (e.g., 40 °C), the switch is activated to change the fuel connection from the D2 tank to the WVO tank. Later, a few minutes before shutting off the engine, the fuel feed is switched back to the diesel tank to purge any remaining WVO out of the fuel injection system.

Prior to about 1990 in the USA, WVO was a popular cost-saving but labor-intensive conversion for diesel engines, and was of no legal consequence because diesel cars and light trucks were exempt from automotive emission standards [14]. But with the implementation of particulate and NO_x standards for diesel vehicles in the USA in 1994, and the development of high-pressure common-rail electronic fuel injection and advanced emission controls capable of meeting the new emission restrictions, the appeal of *greasel* vehicles disappeared except for older trucks and off-road use such as older farm equipment. With California requirements for use of ULSD (Ulta-low-sulfur) diesel fuel, the prospect of even blending a fraction of a percent of WVO is out of the question. Throughout the USA, even the use of mineral diesel fuel containing more than a maximum of 20% biodiesel in a modern diesel engine will invalidate the warranty and can clog or damage the fuel injection and emission control systems. As little as 1% WVO can do the same damage. As of 2023, conversion of a diesel vehicle to WVO with the objective of saving money on fuel is no longer an option unless one has a captive supply of WVO, and the time and skills to handle it.

Aside from the problems of using WVO or SVO (straight vegetable oil) in a diesel engine, the CO_2 emissions of a diesel engine powered by SVO or WVO can be expected to be about the same or only slightly greater than mineral diesel or biodiesel, because the H:C mass ratio is usually about the same and the fuel energy content of vegetable oil lies between biodiesel and mineral diesel. Regarding regulated emissions, a rare 2021 study found that compared with mineral diesel, CO, HC, and NOx emissions were higher but the soot emissions were lower for WVO [15]. It is probably safe to assume that the air quality and climate impacts of using WVO in a diesel engine are similar enough to biodiesel and mineral diesel that the minor use of WVO is not a cause for regulatory concern, although it apparently is in California.[2] Note that detection of WVO vehicles (aka *greasels*) is easy because the exhaust smell is characteristic of the foods that were fried in the original cooking oil. The scent of Kentucky Fried Chicken, French fries or wok-fried cuisine on a highway is a dead giveaway.

[2] Personal experience. In 2015, the author was the recipient of a warning letter from the EPA after registering a small alternative fuel dispensing facility that provided free WVO and methanol for qualified student and community projects.

Biomethane (Sewage or Landfill Gas)

Sewage gas is biomethane produced by the anaerobic digestion and decomposition of animal or human waste by bacteria. Landfill gas is similar but created by the anaerobic decay of buried refuse.

Both are highly variable fuels that are equivalent to very-low-quality natural gas, but are desirable because methane vented to the atmosphere is a much more potent GHG than CO_2. Sewage gas usually contains other noxious gases such as hydrogen sulfide. Fuel handling may be hazardous and leaks quite consequential. But if safe collection collection mechanisms are in place; it can usually be obtained at zero or negative cost.

According to the Bosch automotive handbook [16], the major components of typical *sewage gas* are 46% CH_4 and 54% CO_2 by mass[3] although this varies greatly depending on the source. It is only the methane content that provides fuel value, which is less than half of the fuel mass. But the handling issues must not be underestimated.

A historic example (from personal experience):

In 1974, a 1968 Cadillac limousine was modified to run on municipal sewage gas by engineering students at the California Polytechnic University, San Luis Obispo, California. The vehicle had a 7.0 L V8 engine with a maximum speed of 4600 RPM and a (generously assumed) 30% thermal efficiency. Running on gasoline, this land yacht used an average of 10 kJ/mile of energy. The assumed fuel composition was that specified above, although it clearly also contained substantial hydrogen sulfide.

Maximum power output: **68.7 HP** (calculated in Chap. 3).

Average CO_2 emissions: **664 g/mile** (calculated in Appendix 2).

The Cal Poly team entered it in the 1974 Intercollegiate Reduced Emission Devices (RED) Rally in Los Angeles. The UCLA team (including the author) was downwind while the car was being refueled. A fuel coupling was accidentally disconnected. No one that has experienced an incident like this would recommend raw sewer gas as an alternative fuel.

Numerous biomethane vehicle conversions have been demonstrated since the 1970s, including the 2020 English Bio-Bus fleet shown in Fig. 16.9. Effluent gas from the local sewage treatment facility is captured, filtered, compressed, and used in place of natural gas to fuel a fleet of 77 Scania buses. Apparently, the project is still operational. Hopefully there have been no leaks.

Biomethane continues to be used to power agricultural equipment, including a new (2022) tractor from New Holland, purpose-built to run on livestock gas effluent [17]. The capture and combustion of agricultural or landfill methane into carbon dioxide has a net positive climate benefit by converting methane into less intensive carbon dioxide. And biomethane has become a source of supplemental revenue to farm or waste management operations that sell carbon offset credits based upon the estimated amount of methane captured from livestock operations.

[3] Robert Bosch GmbH. [16].

Fig. 16.9 Bristol City Bio-Bus, 2020. One of a city fleet of 77 biomethane-converted Scania buses. https://www.bioenergy-news.com/news/77-biomethane-powered-buses-launched-in-bristol-uk/. Public domain

Acetylene

Acetylene C_2H_2 is an unsaturated (alkyne) hydrocarbon gas that is best known for its use for oxy-acetylene welding. Burned as a free flame in air, it produces a large amount of soot due to its carbon triple bond and hydrogen deficiency. It may come as a surprise that it is actually a usable (but somewhat dangerous) fuel for SI (and potentially CI) engines.

It actually has some distinct advantages compared with some other gaseous fuels that have been seriously considered. It has a high energy storage density, whether in a welding pressure cylinder or generated onboard by the reaction of water with calcium carbide—the reaction used in miner's headlamps or early automobile and railway headlights before electrical systems and incandescent headlights were available. Another consideration is that the acetylene supply chain is independent of fossil fuels, relying almost entirely on electricity for its production, which affords a wide array of energy source options including solar or wind generated electricity.

Although largely lost to history, the use of acetylene as a motor fuel was remarkably common in the first half of the twentieth century. An article in the May 1896 issue of newly founded "Horseless Age, a monthly journal published in the interests of the motor vehicle industry" stated [18]:

> It is proposed to employ the acetylene stored in its liquefied condition for the propulsion of cars, carriages, boats, and vessels, bicycles, and also for general power purposes. In the estimation of those familiar with it, it will be found the cheapest and most positive storage of power yet known.

While an overestimation of its potential, the 1896 article proved prescient of the situations that followed in both World Wars during which petroleum fuels became scarce in Europe. Figure 16.10 is a photo of what was probably the earliest working example of an acetylene-powered vehicle, a quadricycle built by Cornelius Crastin in England in 1896. Figure 16.11 shows an example of a production acetylene-powered vehicle circa 1940 in Germany.

Commercially, acetylene is distributed in moderate-pressure steel tanks in solution with an organic solvent such as acetone, which is absorbed in porous calcium

Fig. 16.10 Cornelius Crastin and his motor-assisted quadricycle, shown here powered by acetylene from a calcium carbide gas generator, 1896. From http://www.douglas-self.com/MUSEUM/POWER/acetylene-eng/Crastin-1%201896.jpg. Image originally from Scientific American 1896, public domain

Fig. 16.11 Calcium carbide acetylene gas generator built into trunk (boot) of a passenger sedan, circa 1930, Germany. Image public domain, from museum archive. http://www.douglas-self.com/MUSEUM/POWER/acetylene-eng/excelsior2.jpg

silicate granules that fill the tank. A Type-4 140 CF acetylene tank is 39.5" high, 8.5" diameter, weighs 35 kg, and contains about 4.5 kg of acetylene when full. The static tank pressure is 15–20 bar (225–300 psi), slightly higher than an LPG tank. The fuel energy content of acetylene is 49 MJ/kg, about the same as methane and slightly higher than LPG. The energy content of one tank (4.5 kg) of acetylene is 216 MJ, or 6.2 MJ/kg including the 35-kg steel tank.

For comparison, a larger 244 CF Type-K hydrogen cylinder is 51" high, 9" in diameter and weighs 60 kg. It stores 0.4 kg of hydrogen at 138 bar (2000 psi), with a fuel energy value of 49 MJ, or 0.8 MJ/kg including the 60-kg steel tank. Acetylene in standard commercial tankage provides 7.75 greater mass energy density and over 15 times the volume energy density of compressed hydrogen.

To be fair, hydrogen onboard a state-of-the-art fuel cell vehicle (FCV) is stored at 70 MPa (10,000 psi) in a composite tank that is much lighter than steel. But even the roughly fivefold increase in the hydrogen density at 70 MPa, including a carbon fiber tank, would provide less than 2/3 the mass energy density of acetylene in a steel industrial tank. This translates to greater range as an ICE fuel.

A more likely and lighter weight storage method is on-demand generation from the reaction between calcium carbide and water. In this case, the refueling operation is simplified to loading bulk calcium carbide fuel granules or cartridges into a bin. The onboard apparatus would be nearly identical to that used to produce hydrogen gas from sodium borohydride—a *Kipp Gas Generator* [19]:

$$CaC_2 + 2H_2O \rightarrow C_2H_2 + Ca(OH)_2$$

The German Excelsior acetylene generator of Fig. 16.12 operated by this mechanism. The byproduct, calcium hydroxide $Ca(OH)_2$ aka slaked lime, has many common uses, especially as the main ingredient in mortar, plaster, and cement. To get things into the right ballpark, here is a rough energy/range calculation of the energy storage density provided by solid calcium carbide compared with compressed hydrogen.

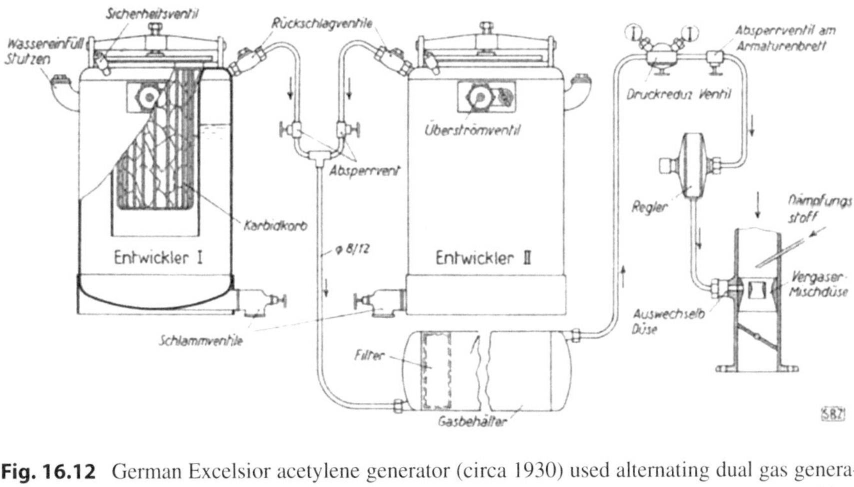

Fig. 16.12 German Excelsior acetylene generator (circa 1930) used alternating dual gas generators so that the vehicle could continue to operate during refueling. Image from museum archive, public domain, http://www.douglas-self.com/MUSEUM/POWER/acetylene-eng/acetyleneeng.htm

Fig. 16.13 Raw calcium carbide, used to make acetylene by reaction with water. Image from https://chemcess.com/calcium-carbide-properties-production-and-uses/. Used with permission

The density of calcium carbide is 2.22 kg/L. A 40 L (10.6 gallon) bin of pelletized calcium carbide would weigh about 90 kg. 1 kg of calcium carbide produces about 0.33 kg acetylene. But commercial calcium carbide is usually only 80–85% pure, so the yield should be adjusted to 0.28 kg acetylene. The 40 L of CaC_2 pellets would generate 90 kg × 0.28 kg C_2H_2/kg CaC_2 = 25 kg C_2H_2 that would have a fuel value of 1200 MJ, providing nearly 6 times the range of the acetylene tank and 9 times that allowed by 70 MPa compressed hydrogen storage tanks of similar volume.

Calcium carbide is rarely found in nature. It is manufactured by heating lime (calcium oxide) with carbon (coke, anthracite coal, char, or sequestered carbon from CCS) to 2000+ deg C in an electric arc furnace:

$$CaO + 3C \rightarrow CaC_2 + CO$$

Figure 16.13 is a photo of calcium carbide briquettes, which could be used directly in a vehicular acetylene gas generator. Speculatively, the process might also be accomplished using the high temperature at the focal point of a concentrating solar heliostat array such as the Ivanpah [20] solar thermal power plant in California. Lime is a common mineral found in nature, with many applications in industry and agriculture. It occurs naturally in small quantities, but most is made by heating very abundant calcium carbonate (limestone, chalk, bones, sedimentary rock) to 1100 °C in a kiln. CO is co-generated by this process, which is a pollutant of concern if not captured and combusted to offset the heat energy requirements of the kiln. The total energy required for making calcium carbide is almost entirely the electricity required by the arc furnace, which could potentially be renewably sourced (but like all eFuels, probably not).

A simplified analysis is worthwhile to allow comparison of the energy efficiency of a hypothetical acetylene vehicle with the two other types of vehicles that also use electric energy as their fuel: a battery electric vehicle and a hydrogen FCV that uses green (electrolytic) hydrogen. The electric energy required to produce calcium carbide from lime and carbon is reported to be 2.48–3.26 kWh/kg, and the acetylene yield from the calcium carbide 0.28 kg C_2H_2/kg CaC_2.

Process efficiencies vary worldwide, but starting with calcium carbonate (lime) and carbon (coal), and not taking advantage of the energy value of the byproduct CO to assist in kiln heating, the overall process to make calcium carbide has been reported to require between 2.48 and 4.0 (average of 3.24) kWh/kg CaC_2. With an acetylene yield from calcium carbide of 0.28 kg C_2H_2/kg CaC_2, and acetylene's LHV of 48.1 MJ/kg, this translates to an electric energy requirement of

$$\frac{3.24\,\text{kWh}_e(\text{avg})}{\text{kg CaC}_2}\frac{1\,\text{kg CaC}_2}{0.28\,\text{kg C}_2\text{H}_2}\frac{\text{kg C}_2\text{H}_2}{48.1\,\text{MJ}} = 0.24\,\text{kWh}_e\,/\,\text{MJ acetylene fuel}$$

If this fuel powers our usual reference ICE 2012 Honda Civic that uses 1.66 MJ/km, the original electricity requirement would translate to 0.40 kWh_e/km. For comparison, a typical electric vehicle uses 0.21 kWh_e/km, and a hydrogen fuel cell car with electrolytically generated hydrogen uses 0.53 kWh_e/km. This would make acetylene about 53% as energy efficient as a BEV, but 25% more efficient than an H_2 FCV in terms of the electric energy used to travel any given distance. Inverting the ratios provides a direct comparison of the CO_2 per distance for each of the three vehicle power systems: acetylene—1.9 times the CO_2 emissions per distance of a BEV, but only 25% times the CO_2 emissions assignable to an H_2 FCV. Electric distribution losses are ignored since they apply equally in all cases (Please see Chapter 14 *Electric Vehicles*).

The energy and environmental sustainability look favorable for acetylene. But now the bad news....Acetylene is an unstable, potentially explosive gas with a very low ignition temperature of 305 °C and very wide flammability limits 1.5–80 V_{fuel}/V_{air},[4] similar to hydrogen. The instability of acetylene is arguably the greatest concern. Professional welders are aware that acetylene is susceptible to shock ignition, even in the absence of air. At high pressure or temperature, when subjected to mechanical impact, acetylene in a hose or regulator can undergo explosive decomposition into hydrogen and elemental carbon. A considerable amount of heat and pressure is generated by this dissociation [21]. The resulting decomposition products (hydrogen and carbon) can then ignite when exposed to air. This is a significant impediment to the use of acetylene as a portable fuel, but it has been reported that the addition of 7% butane-propane (LPG) to the acetylene will effectively prevent such spontaneous decomposition [22].

As an SI engine fuel, the octane number of acetylene has been reported to be low, but it (probably) has never been tested since it has not been considered to be a viable fuel since the early twentieth century, before octane testing methods were standardized [23]. It is subjectively known to be susceptible to engine knock at higher compression ratios. Possibly of greater concern is that acetylene-air mixtures inducted/premixed into an ICE engine are susceptible to the same intake backfiring problem as hydrogen and methane, due to its similar low ignition energy and temperature requirements. But like hydrogen, in one experiment it was found to be amenable to the same solution: delayed fuel injection to allow precooling and purging of hot residual gases or thermal preignition sources in the cylinder prior to introduction of fuel [24].

The low ignition temperature should be expected to make acetylene an excellent compression ignition engine fuel, but direct cylinder injection of acetylene is challenging because, unlike dimethyl ether, it cannot be safely maintained in liquid form under pressure in an engine operating environment.

[4] Robert Bosch GmbH. [16].

Other considerations affecting the design of an onboard acetylene generator: The calcium carbide-water reaction is mildly exothermic, so a heat exchanger is required following the gas generator to cool the acetylene before introducing it to the engine. And as with any onboard-generated gas, there is a substantial lag between engine demand and generator production rate, which requires either a low-pressure buffer tank or a very effective predictive control algorithm.

On the positive side, the high flame speed and peak combustion temperature create the potential for increased engine efficiency due to a closer-to-ideal engine cycle. And lean mixtures have been reported to produce lower regulated emissions than gasoline or diesel fuel [23]. Acetylene is nontoxic and has a characteristic garlic-like smell that facilitates leak detection without the need for an added odorant.

There have been attempts to ameliorate the problematic combustion properties of acetylene by blending with other fuels that have opposite properties. For example, acetylene and methanol are complementary, since methanol has a very high octane number and heat of vaporization. Acetylene is reasonably soluble in methanol, and could be stored in this form. Injection of methanol as a 5% additive to aspirated low-pressure acetylene has been claimed to accomplish the desired complementary effect. US patent US6076487A was issued in 2000 for this idea [23], but it is unknown if it was ever implemented in practice.

The CO_2 emissions per km of an acetylene-fueled vehicle can be calculated in the same way as any gaseous fuel. For a stoichiometric mixture (which is not preferred due to the high combustion temperature):

$$C_2H_2 + 2.5\,O_2 \rightarrow 2\,CO_2 + H_2O$$
$$26\,g + 80\,g \rightarrow 88\,g + 18\,g$$

$$26\,g\ \text{fuel} + 4.31 \times 80\,g\ \text{air} \Rightarrow 13.3\ \text{mass AFR}$$
$$\text{LHV} = 48.1\,\text{MJ/kg}$$

For the 2012 Honda Civic used in Appendix 2 as a reference that uses 1.66 MJ/km of fuel energy, for a slightly lean AFR,

$$\left(\frac{1.66\ \text{MJ}}{\text{km}}\right)\left(\frac{\text{kg}\ C_2H_2}{48.1\ \text{MJ}}\right)\left(\frac{24\ \text{kgC}}{26\ \text{kg}\ C_2H_2}\right)\left(\frac{44\ \text{kg}CO_2}{12\ \text{kgC}}\right) = 117\,\text{g}\ CO_2\,/\,\text{km}$$

which is about the same as the 118 g CO_2/km CO_2 emissions of ethanol, and less than the 129 g CO_2/km that this vehicle emits running on E10 gasoline. This is just for the combustion of the fuel in the engine, not including whatever CO_2 may be released in the generation of the electricity used to produce calcium carbide. It should be noted that if the AFR is allowed to become richer than stoichiometric over the operational range of the engine, high levels of elemental carbon (soot) will be emitted—a highly objectionable situation.

Comparative power output in the hypothetical 1 L reference engine of Appendix 4 is 44.0 kW or 58.9 HP. This is approximately the same as the engine's output on pure methane, which is about 7.7% less than gasoline.

There do not appear to have been any acetylene engines or vehicles since the Second World War, other than research projects such as in 1976,[5] 2016 [25], and 2023 [26].

References

1. The Mad Max series of six films 1979–2015 depicted a post-apocalyptic dystopian world fighting over gasoline. https://www.imdb.com/list/ls539934352/. Accessed August 10, 2024
2. Methanex Corp. Methanol: 19th century wood alcohol emerges as an alternative fuel source for the 21st century. Historical perspective. https://www.methanex.com/sites/default/files/news/media-resources/Methanol_19th_Century_Wood_Alcohol_2017.pdf. Accessed November 1, 2024
3. Drive on Wood. https://www.driveonwood.com/library/imbert-gasifier/. Accessed July 1, 2024
4. Wikipedia. Wood gas. https://en.wikipedia.org/wiki/Wood_gas. Accessed July 1, 2024
5. K.D. Decker Wood gas vehicles: firewood in the fuel tank. Low-Tech Magazine, January 18, 2010. https://solar.lowtechmagazine.com/2010/01/wood-gas-vehicles-firewood-in-the-fuel-tank/. Accessed June 1, 2022
6. Kris De Decker, Wood gas vehicles: fire in the fuel tank Resilience (Originally published by Low-Tech Magazine). January 20, 2010. https://www.resilience.org/stories/2010-01-20/wood-gas-vehicles-firewood-fuel-tank/. Accessed July 1, 2024
7. ChemEurope.com Energy Portal Wood gas. https://www.chemeurope.com/en/encyclopedia/Wood_gas.html#References. Accessed July 1, 2024
8. Gareth Branwyn. Make. Lost knowledge: wood gas vehicles. https://makezine.com/article/science/energy/lost-knowledge-wood-gas-vehicles/. Accessed July 1, 2024
9. China Defense Blog. North Korea's wood-burning military trucks. November 8, 2013. https://china-defense.blogspot.com/2013/11/north-koreas-wood-burning-military.html. Accessed September 1, 2022
10. Wikipedia. Rudolf Diesel. https://en.wikipedia.org/wiki/Rudolf_Diesel. Accessed July 1, 2024.
11. Ford Truck Enthusiast, Greasel conversion kits. https://www.ford-trucks.com/forums/244025-greasel-conversion-kits.html. Accessed July 1, 2024
12. WVO designs, Convert your vehicle to run on vegetable oil. https://www.wvodesigns.com/products/wvo-conversion.html. Accessed October 1, 2024
13. One of many biocide additives for diesel fuel that is available on Amazon is https://www.amazon.com/FPPF-KILLEM-BIOCIDE-BOTTLE-GALLONS/dp/B009LMW96G/ref=pd_bxgy_thbs_d_sccl_2/140-6334743-5921160?pd_rd_w=qlgLW&content-id=amzn1.sym.c51e3ad7-b551-4b1a-b43c-3cf69addb649&pf_rd_p=c51e3ad7-b551-4b1a-b43c-3cf69addb649&pf_rd_r=8SJGYW89CE3WFSH0B1KQ&pd_rd_wg=3fAKu&pd_rd_r=18577c5d-6ca8-41aa-b818-c7fffe9a1721&pd_rd_i=B009LMW96G&psc=1
14. Top Gear video produced by Jeremy Clarkson, 2009. Posted on YouTube at https://www.youtube.com/watch?v=1lqCwNReU1Y. Accessed July 10, 202.
15. M. Costa, L. Marchitto, D. Piazzullo, M.V. Prati, Comparison between the energetic and environmental performance of a combined heat and power unit fueled with diesel and waste vegetable oil: an experimental and numerical study. Renew. Energy **168**, 791–805 (2021) https://www.sciencedirect.com/science/article/abs/pii/S0960148120320395. Accessed December 1, 2023
16. Robert Bosch GmbH, *Automotive Handbook*, 10th edn. (Distributed by SAE International, 2018), p. 346

[5] C. A. MacCarley and A. Ullman, UCLA [24].

17. Successful Farming, February 8, 2022. New Holland's new tractor powered by cow farts. https://www.agriculture.com/machinery/tractors/new-holland-unveils-worlds-first-methane-powered-tractor. Accessed July 1, 2024
18. Horseless Age, May 1896. From Digitized Collection of the US Library of Congress. Acetylene as a Motive Agent. 1(7):27
19. Smithsonian, National Museum of America History. *Kipp Gas Generator*. https://americanhistory.si.edu/collections/nmah_2967. Accessed July 1, 2024
20. Bechtel Corporation. World's largest solar thermal energy facility delivers clean, renewable electricity, 2023. https://www.bechtel.com/projects/ivanpah-solar-electric-generating-system/. Accessed July 5, 2024.
21. Air Products Safety Guide, *Safetygram 13. Acetylene*. Downloadable at https://www.airproducts.com › - › media › files › en › 900 › 900-13-097-us-acetylene-safetygram-13.pdf. Accessed July 1, 2024.
22. Dmitry I. Baklanov, Sergey V. Golovastov, Victor V. Golub, Roman B. Reshetnyak, Vladislav V. Volodin. *Inhibition of Spontaneous Decomposition of Acetylene by Hydrocarbon: an Experimental Investigation*, 21st ICDERS July 23–27, 2007 Poitiers, France http://www.icders.org/ICDERS2007/abstracts/ICDERS2007-0102.pdf. Accessed July 3, 2024.
23. P. Kumar, S.J. Reddy, K. Bodukuri, Internal combustion engine using acetylene as an alternative fuel. Int J Eng Res Dev **14**(5), 56–61 (2018)
24. C.A. MacCarley, A. Ulman, UCLA engines and fuels testing laboratory. Unpublished results from tests of uncommon alternative fuels, 1978. Acetylene was tested over a wide range of AFRs, in a 0.65-L two-cylinder SI test engine. Occasional intake backfiring was observed at full load that was remediated by time-delayed port fuel injection.
25. Shaik Khader Basha, P. Srinivasa Rao, K. Rajagopal, K. Ravi Kumar. Evaluation of performance & emission characteristics of an acetylene aspirated diesel engine. in *Proceedings of the World Congress on Engineering*, Vol. II, London, 2016.
26. P. Deb, A. Paul, Effect of acetylene as a low reactivity fuel on performance, combustion, exergy and emissions of an acetylene/diesel RCCI engine with variable premix ratios. J Sustain Energy Fuels **7**(18) (2023) https://pubs.rsc.org/en/content/articlelanding/2023/se/d3se00440f. Accessed July 1, 2024

The Artificial Economy of Subsidies 17

Carbon Credits, Subsidies, and Incentives

This brief chapter is about financial and political sustainability in the transportation energy transition. It is not possible to talk about how and why some technical options have flourished and some have not without looking at the motivations underlying each. The monetization of carbon dioxide emissions via carbon credits, subsidies and incentives is certainly among the most powerful of these motivations.

There are countless authentic resources in this area. This is not one of them. It is only a compendium of definitions and cause-effect relationships from other sources, intended to help make sense of the pallet of potential solutions discussed in the preceding chapters.

The prima facie objective of government subsidies and tax incentives is to encourage and partially compensate people, companies or institutions for taking on the financial risk and inconvenience of something that is deemed to be good for society. In the context of transportation, these usually incentivize the adoption of a new technology e.g., ethanol, biodiesel, hydrogen, or electric vehicles. It is a carrot rather than a stick, at the cost of everyone's carrots. World-changing social improvements have been accomplished as a result of some subsidy programs, for example, the *Work Progress Administration* (WPA) that aided in the economic recovery of the USA in the aftermath of the economic collapse in 1929. Many of the environmental and energy security gains of the last fifty years can be attributed to public collective will to invest in the greater good for the destiny of a community, state, country, or the planet.

For subsidies and incentives in any form to be *economically sustainable*, they must be temporary, to be sunsetted when they have accomplished the intended objective. But perpetual subsidies and targeted financial assistance programs now exist for everything from farming to oil exploration to Energy Star™ appliances. Entire industries, from biofuels, to green hydrogen, now exist *solely* because of government financial support or underwriting that, if withdrawn would result in

C. A. MacCarley, *Non-Petroleum Automotive Transportation*,
https://doi.org/10.1007/978-3-031-85028-8_17

their immediate collapse. A new breed of entrepreneurs and investors sometimes referred to as "subsidy harvesters" thrive on these opportunities, sometimes accomplishing good in the process, but more often, just riding the giveaway program until another such opportunity comes by. It is remarkably easy to wrap self-serving objectives in a green cloak to take advantage of the largesse of society seeking environmental or energy solutions. Political influence is usually key to success. This includes the transition of transportation from fossil fuels to new energy options, especially EVs, hydrogen and biofuels.

It is a popular adage that the more complex a law or regulation is, the more ways there are to subvert it. Tax incentives and regulatory credits are provided by fourteen US states, with California the largest, although some are provided by federal initiatives and "Earthshot" programs. Other industrialized countries, for example China, have similar programs intended to support the growth of strategic industries such as EVs, advanced batteries and solar PV.

Types of Credits

As explained in citation [1]:

- *Carbon Credits* are usually the result of federal commitments to International Climate Agreements that are tied to some measurable elimination or offset of carbon. External verification entities (both for-profit and non-profit) certify the carbon reduction mass claimed by the companies that are claiming the right to receive these credits. Carbon and offset credits are most common in forestry, agriculture, renewable fuel production, or industries that have traditionally been major polluters [2, 3].
- *Offset Credits* are specifically tied to GHG emissions (CO_2 and CH_3) since the objective is climate change mitigation. They usually extend to all regulated pollutants (HC, VOC, CO, NOx, and particulates) in addition to carbon-based or carbon-equivalent GHGs. Carbon offset credits issued by the California Air Resources Board (CARB) apply to compliance with the state's Carbon Cap-and-Trade regulations under CARB's Compliance Offset Protocols [4].
- *Regulatory credits* are provided directly by local, state, and federal government agencies without being tied to verifiable results. e.g., it is obvious that EVs have no tailpipe emissions, so there is no need for certification of the amount of carbon reduction each vehicle provides. With respect to ZEVs, *regulatory credits* are separate and in addition to federal and tax incentives provided directly to buyers of new EVs, which currently in the USA are $7,500 per vehicle, although some restrictions on domestic content and buyer income will start in 2025 under the US Inflation Reduction Act of 2023.

  As explained in citation [1]:

Regulatory credits are like subsidies provided by the State of California and 13 other US states for State regulations that are responsible for environmental pollution including: zero-gas emission, greenhouse gas, clean fuel, and carbon emission.

- All three types of credits have monetary value, can be traded in the open market, and can be a major source of revenue for entities that can claim they are responsible for reducing emissions of carbon (carbon dioxide and methane).
- *Renewable energy certificates (RECs)* apply only to electric power generation. They are similar to regulatory credits and carbon offset credits, except that their valuation is based upon 1 MWh of electricity being worth $100 USD, as long as they can claim to be produced in some renewable way, without actually monitoring the carbon dioxide or other emissions produced by a given renewable process [5]. The largest purchaser of RECs is actually the US government, to justify claims that a given facility (e.g., a military or municipal service facility) is 100% green, when in reality it could be an environmental disaster.

Good intentions

A positive perspective on carbon credits and carbon offsets is provided by Advanced Science News [6]:

> A key feature of carbon prices—lauded by economists [7]—is that they let businesses and consumers decide where and how to cut their own use of fossil fuels. Faced with higher fossil fuel prices that reflect the environmental harm caused by carbon emissions, users can decide how best to respond, curtailing emissions wherever this is cheaper than paying the higher price. Assuming they choose the most cost-effective responses, the net effect is to minimize the total cost of achieving the desired reduction in emissions.

In other words, the idea behind carbon credits is to charge a fee for the right to pollute, with these funds hopefully going to others that will forgo polluting to an equal or greater degree, or to support businesses or emerging technologies that have the potential to reduce environmental harm in some way that may be unrelated to the pollution rights purchased by the credits. So, in theory, polluters subsidize pollution reducers. It is a relatively straightforward concept that makes sense on a spreadsheet to political decision makers anxious to show that they are doing *something* to help the climate. If carefully managed, it could yield a less harmful carbon balance, but NOT a net reduction in carbon and regulated pollutant emissions when grouping together the contributions of both the buyer and seller in the transaction. Regulatory credits are even more reliant on wishful thinking, having uncertain overall benefits since the awards are done in good faith that the recipient is actually producing the claimed carbon reductions. Some enterprises derive the majority of their profit from regulatory credits that are just resold in the open market—examples below.

Issues

Numerous environmental advocacy organizations and independent analysis teams condemn carbon credits and offset credits as actually being negative for the environment. A shell game of sorts. Carbon and regulatory credits have been generated and sold in a much broader range of industries than originally envisioned: forestry—a promise to plant seedlings that twenty years from now might offset the carbon emissions that we are creating now; real estate development—a plan to devote more land to (water and chemical intensive) green space (which could be a golf course) instead of home sites; agriculture, a—planned transition to monetize existing plans to reduce tillage that also happens to generate less methane; manufacturing—of a vehicle or fuel that in the hands of a buyer will have reduced or zero tailpipe carbon dioxide emissions. This last example notably includes EV manufacturers and renewable fuel producers. "Renewable" is a subsidy-qualifying term applied to many energy production schemes that have nothing to do with closed carbon cycles.

For example, a municipal refuse incinerator in Palm County, Florida continues (2023) to be allowed renewable energy credits since it uses some of the waste heat from incineration to partially power a steam electric generation plant, despite the fact that the carbon and other forms of pollution released by the plant are worse than those of an unregulated coal-fired power plant.

Cap and Trade

Under cap and trade rules, carbon offset credits work like this:

- Cap and trade limits are set for a company by a regulatory agency, in the USA this may be the EPA or a state agency such as the California Air Resources Board. The cap is the maximum amount of carbon the company can release into the atmosphere over some time period.
- If the company generates less than this target level, they can claim the difference as a carbon credit.
- The company submits a proposal to a certification agency. The two best known in the USA are *Verra* and *Gold Standard*. The agency, for a fee, certifies or adjusts the amount of the claimed carbon offset and issues a certificate declaring one carbon credit for every metric ton of carbon avoided.
- The company then goes to a carbon exchange that connects buyers and sellers of credits. The best known in the USA are *CarbonX*, *AirCarbon* (for airlines carbon credits), *Ecosystem Marketplace*, or *LandGate* (specializing in forestry credits).
- Meanwhile, another company is exceeding their cap limit and is required by law to purchase sufficient carbon credits to cover the excess.
- The seller is matched with the buyer, and a sale is negotiated at the current market value, which fluctuates with supply and demand.

- The buyer can now exceed their carbon cap at the rate of one metric ton of carbon per credit.
- The seller is expected to follow through with their activity that has reduced their carbon output by that same amount.
- Enforcement of compliance is often weak or nonexistent, due to the complexity and the sheer magnitude of the number of carbon credits sold and bought.

Latitude for Interpretation of Rules

There is a lot of latitude for creative interpretation of activities that can generate carbon offset credits, since the justification is "If you pay me, I will forgo some amount of pollution that I could legally create within my cap, to allow you to pollute by that same amount." Also, although not quantified in the financial transaction, purchased carbon/emission credits are used in public relations to make carbon-intensive enterprises look clean, with statements such as "we reduce greenhouse gas emissions equivalent to taking x number of cars off the road," failing to note that they actually just paid for the right to continue to pollute by that much. In fact, many business entities that are hideous polluters can legally claim that they are carbon neutral by simply purchasing the required number of carbon credits.

Explained more succinctly by the international environmental protection organization Greenpeace [8]:

> Here's how it works: Let's say you run a coal-fired power station. The coal you burn and the CO_2 you emit goes into the atmosphere and heats our world. That is incontrovertible. But offsetting then encourages you to point at a forest and say, "I'm paying for those trees not to be burned so now we're even." Of course, this does nothing whatsoever to change the fact that the CO_2 from your coal plant is now warming the atmosphere—it just lets you show a balance on paper.
>
> Carbon offsetting is a license to keep polluting, and distracts us all from the real work of cutting emissions. It is where companies and governments try to meet their carbon reduction targets while still emitting carbon.

The free-market purchase of *any* of the three types of credits make it possible for entities that are responsible for the generation of GHG or regulated emissions to legally exceed international and national carbon targets while avoiding large fines. The official equilibration at the time of issuance is *$100 USD per metric ton of carbon*, but after issuance, the value per credit is free market-negotiable, subject to the agreement of the credit certifier [9]. In other words, any entity can buy credits for the right to pollute, at least up to some maximum limit under *cap-and-trade regulations*. When sold, the money paid for each credit to the seller is supposed to be used to eliminate (inset) or forgo (offset) the generation of 1 metric ton of carbon emissions. In the case of carbon offsets, sellers need only show that they will forgo some activity that could have emitted carbon, regardless of whether they had any intention of engaging in the activity at all.

The *carbon trade market* has expanded well beyond original expectations to now include *carbon cryptos, carbon NFTs, and carbon tokens*, all traded beneath the regulatory reach of governments [10]. One online real-time listing of the current market values of carbon and other regulatory credits can be found at *Live Carbon Prices Today*, a service of carboncredits.com [11], with the values as of 28 December 2023 listed in Table 17.1.

The more complex and difficult to track a process, the easier it is to manipulate perceptions and results. This is true across all the areas of energy and environmental marketing, and the subject of many financial publications.

Table. 17.1 Live Carbon Prices Today, 28 December 2023, https://carboncredits.com/carbon-prices-today/. With permission

CarbonCredits.com Live Carbon Prices	Last	Change	YTD
Compliance Markets			
European Union	€80.15	+1.28 %	+0.19 %
UK	£44.10	+5.50 %	-39.80 %
California	$28.66	-	-1.41 %
Australia (AUD)	$33.50	+0.75 %	-0.89 %
New Zealand (NZD)	$68.65	+0.22 %	-10.17 %
South Korea	$6.99	+9.07 %	-43.63 %
China	$11.16	-0.63 %	+37.27 %
Voluntary Markets			
Aviation Industry Offset	$0.63	-	-83.59 %
Nature Based Offset	$0.92	-11.54 %	-80.00 %
Tech Based Offset	$0.59	-3.28 %	-48.25 %

Carbon Credits and the New EV Industry

As reported by *carboncredits.com*, an independent energy industry watchdog [12],

> Tesla Inc. has been criticized for being dependent on the sale of its carbon credits. These credit sales have been a major driver of Tesla's profits over the years, providing $1.78 billion revenue in 2022 alone. Their largest carbon credit customers were General Motors, Stellantis (Fiat Chrysler), and Volkswagen, the largest quantity believed to offset their huge emissions credit shortage in China. Tesla has also earned additional billions for credits provided by various local regulatory sources such as California's ZEV program. These credits are then sold on the open market to polluters.

As reported in CNN Business, Tesla operated for 17 years without posting a full-year profit, surviving on government subsidies, loans and especially the sale of regulatory credits to stay afloat. Tesla reported its first profitable full year in 2020 [13].

As reported by CNBC in 2021 [14]:

> Because Tesla only sells electric cars which come under the ZEV category, the company always has excess regulatory credits and can effectively sell them at a 100% profit.
>
> Tesla raked in $518 million in revenue from sales of regulatory credits in the first quarter of the year, helping the U.S. electric vehicle maker post another quarter of profit.

Carbon credit sales have been key to the profitability of most EV manufacturers in the USA and Europe, and they provide an investment incentive for startup companies in the carbon remediation space. But this financial crutch has also propped up startup EV firms anticipating revenue from various credits and subsidies such as Lordstown Motors, Nikola, Faraday Future and Canoo that ultimately contributed nothing except losses for investors and taxpayers.

Meanwhile, from the financial website *Seeking Alpha* [15]:

> Mainland China emerges as the market with the most vibrant opportunity for regulatory credit trading in the next decade owing to the structure of its regulatory programs.

Growing concern about the extent of fraud in the carbon and regulatory credit market even extends to the companies that provide carbon credit services. As reported in Vox *Future Perfect*, August 3, 2023 [16],

> Recent revelations have cast doubt on these schemes. In January, a high-profile investigation by the *Guardian*, German newspaper *Die Zeit*, and journalism nonprofit *SourceMaterial* asserted that over 90 percent of rainforest carbon credits issued by Verra, the world's leading carbon credit certifier, claimed reductions in deforestation that didn't actually exist. As a result, they said, the credits were "worthless," provoking painstaking rebuttals from the industry.

According to the law firm Jones Day LLP [17]:

> June 20, 2023, the United States Commodity Futures Trading Commission ("CFTC") Whistleblower Office, a unit of the agency's Division of Enforcement ("DOE"), issued a rare alert, soliciting tips about potential fraud and manipulation in the carbon markets.
>
> Less than two weeks later, the CFTC announced the formation of a new DOE Environmental Fraud Task Force ('EFTF') to 'combat environmental fraud and misconduct in derivatives and relevant spot markets.' The EFTF will be comprised of DOE attorneys and investigators who will prosecute cases, serve as subject-matter experts, and coordinate efforts with the CFTC's other divisions and offices.

Greenpeace [8] aggressively opposes carbon credit offsets, referring to them as "truly a scammer's dream scheme."

There has been little disclosure in media about the extent to which transportation industries, especially related to EVs, utilize revenue from carbon credits sales. The carbon offset potential of EVs and carbon reduction projects are real, albeit overstated. And carbon offset credits continue to be claimed by on-paper-only or questionable renewable energy businesses [18]. The overall climate change reductions attributed to carbon credit commodity trading remain unclear.

References

1. PratikshaBajikar,WhatareregulatorycreditsandhowTeslamadeitabusiness?May27,2021https://startuptalky.com/tesla-regulatory-credits-revenue-model/#What_are_Regulatory_Credits?
2. The ultimate guide to understanding carbon credits: carbon markets 101. CarbonCredits.com, https://carboncredits.com/the-ultimate-guide-to-understanding-carbon-credits/. Accessed 15 December 2023
3. California Air Resource Board, *Complete list of Carbon Offset Projects*, 2023.
4. Madeline Hume, Morningstar Market Insights, 11 August, 2023. Carbon offsets vs. carbon credits: what's the difference? https://www.morningstar.ca/ca/news/238122/carbon-offsets-vs-carbon-credits-whats-the-difference.aspx. Accessed 10 November 2023
5. National Renewable Energy Laboratory, Renewable Electricity: How do you know you are using it? NREL/FS-6A20-64558, August 2015 https://www.nrel.gov/docs/fy15osti/64558.pdf Accessed 31 Dec 2023
6. Carbon pricing is not at odds with environmental justice. Advanced Science News web site, https://www.advancedsciencenews.com/carbon-pricing-is-not-at-odds-with-environmental-justice/. Accessed 21 December 2023
7. Andrea Baranzini, Jeroen C. J. M. van den Bergh, Stefano Carattini, Richard B. Howarth, Emilio Padilla, Jordi Roca. *Carbon Pricing in Climate Policy: Seven Reasons, Complementary Instruments, and Political Economy Considerations.* Published online at *Wires* Climate Change, by the Royal Geographic Society. https://wires.onlinelibrary.wiley.com/doi/10.1002/wcc.462. Accessed 21 December 2023.
8. Chris Greenberg, Greenpeace International. Carbon offsets are a scam, 10 November 2021. https://www.greenpeace.org/international/story/50689/carbon-offsets-net-zero-greenwashing-scam/. Accessed 21 December 2023
9. Carbon Markets Are in Limbo. That's not stopping this firm. New York Times, 16 December 2023. Also online at https://www.nytimes.com/2023/12/16/climate/carbon-markets-global-warming.html.

10. Carbon Cryptos, Klimadao, Carbon NFTs and Carbon Tokens. https://carboncredits.com/carbon-crypto-guide-2023-klimadao-carbon-nfts-and-carbon-tokens/ accessed 28 December 2023.
11. Carbon Prices Today., https://carboncredits.com/carbon-prices-today/. Accessed 28 December 2023
12. Carbon Credits. https://carboncredits.com/tesla-regulatory-carbon-credit-sales-jumps-116/. Accessed 4 Sept 2023
13. CNN Business, 13 May 2020. *After years without a profit, Tesla is now poised to be a powerhouse.* https://edition.cnn.com/2020/05/13/business/tesla-profit-outlook/index.html Accessed 20 March 2025
14. CNBC Tech (Online). What 'regulatory credits' are—and why they're so important to Tesla. May 18, 2021. https://www.cnbc.com/2021/05/18/tesla-electric-vehicle-regulatory-credits-explained.html Accessed 1 Dec 2023.
15. Seeking Alpha (online) Feb. 24, 2022. *Selling Credits Is A Lucrative Revenue Stream For EV Makers.* Accessed 28 December 2023.
16. Angus Chapman, Desné Masie, Vox (online) 3 August 2023, Are carbon offsets all they're cracked up to be? We tracked one from Kenya to England to find out. Accessed 24 December 2023.
17. Jones Day, New CFTC Enforcement on Carbon Markets Misconduct, August 2023. https://www.jonesday.com/en/insights/2023/08/new-cftc-enforcement-on-carbon-markets-misconduct. Accessed 20 December 2023
18. US Department of Justice, Eastern District of Virginia, Press Release 1 December 2020. Man Pleads Guilty to Defrauding Investors in Carbon Credit Scheme. Also online at https://www.justice.gov/usao-edva/pr/man-pleads-guilty-defrauding-investors-carbon-credit-scheme

Challenges and Opportunities 18

Driven by climate change and energy security concerns, automotive technology is changing fast. Along with these changes come opportunities for new ideas, new scientific advances, new areas of commerce, and a new wave of innovation, but also risks of manipulation of well-intended efforts to serve the greater good. Navigating the evolving transportation energy landscape is a challenge different from most any that we have previously encountered.

Cars are not going away. The convenience and independence provided by personal transportation is an addiction for which there is no recovery. While older cities inherited public transportation infrastructure and traditions of using it, some cities such as Los Angeles in the USA were built around the transportation arteries designed for use by personal automobiles and trucks. It is not possible to replace them. The challenge now is to make automotive transportation less consequential to global warming and climate change.

Internal combustion vehicles can be expected to continue as the preferred option for specific applications. Fortunately, we have many new technology options, and surely more to come in the future. The notion that internal combustion engines are a thing of the past is rooted in the public assumption that IC engines can only run on petroleum products. Fossil fuels have given us a great ride for a century, a major factor in our standards of living. But the cumulative environmental consequences are now clear. It's payback time for Mother Earth. Fortunately there are many non-petroleum fuel options for automobiles, most of which have less severe environmental impacts than gasoline or diesel fuel. At the top of the list are battery electric vehicles. But the list is longer than only that option.

C. A. MacCarley, *Non-Petroleum Automotive Transportation*,
https://doi.org/10.1007/978-3-031-85028-8_18

In evaluating automotive energy options, the well-to-wheel (full cycle) efficiency and GHG emissions are good places to start but not end. Practical limitations and advantages are equally important. While it is certainly arguable, a diversity of energy sources and carriers is probably more sensible than an "all-electric" path. For example, Toyota, the leader in hybrids, plug hybrids, and hydrogen fuel cells, in June 2023 announced that they will commercialize hydrogen ICEs in addition to hydrogen FCVs. The justification is a practical one: hydrogen ICEs are typically 40% efficient, not much different than the complex drive system of an FCV and less costly.

Large vehicles for freight transport or mass transit will be a longer reach for electrification since in these applications, range is everything. The required battery mass and recharge time can be a deal-breaker. Unless of course, a radically better battery is forthcoming, which is certainly not out of the question; recent developments in chemistries and electrolyte materials may change everything. Meanwhile, low net carbon combustion fuels such as methanol, dimethyl ether, natural gas, or in some cases, hydrogen, could cover the transition.

The wholesale transfer of automotive energy from petroleum to the electric grid will have profound ramifications. The huge increase and more random schedule of electrical load on the utility grid, combined with the intermittency of solar and wind is not an evolutionary challenge - it is a completely new situation that requires accommodation of *intermittent* resources and expanding *dispatchable* energy resources: geothermal, hydroelectric, geothermal, nuclear, and energy sources not yet known. The electric energy storage capacity to overcome the temporal supply-demand mismatch is not even close to reality, trailing the proliferation of grid-tied solar and wind which are our current greatest hope for clean electric energy.

Energy resilience and security become much more critical as political uncertainties grow.

Here are a few, mostly obvious, suggestions for engineering R&D and business opportunities with the chance for positive energy and environmental sustainability outcomes, with other ideas to be found at other credible resources, such as [1].

- Electric Vehicles

  Improvements are urgently needed in electric power generation, storage and distribution. University preparation in power engineering is growing faster than any other subfield.

  A huge increase in grid and local electric energy storage is needed in order to realize climate change benefits from solar and wind generation.

  Expanded EV charging infrastructure and charging technologies, to an extent much greater than the existing gasoline distribution infrastructure. Improve the reliability, cost, and management of EV charging. Solutions to accommodate ultra-high charge rates within the capacity of the electric utility grid.

  Smaller, lighter and less expensive EVs, not more high-end massive trucks, SUVs, and hypercars.

  Improvement of batteries; both new chemistries and incremental engineering. Higher energy densities to reduce the weight consequences of longer

vehicle range. Faster charge rates without large energy losses. Intrinsically stable batteries, and the possible use of less flammable materials.
Safer and less resource-intensive tools for first responders to EV battery fires.
Alternatives to high rate charging for reducing charging time. e.g., automated battery exchange for fleet electric vehicles such as buses and local delivery trucks.

- Hydrogen
  Hydrogen ICE/FCV hybrid vehicles.
  Stimulated geologic hydrogen (orange hydrogen) technology and deployments.
  Alternatives to compressed or liquid hydrogen storage: advanced interstitial and chemical hydrides and onboard fuel technologies.
  Advancement of direct solar-to-hydrogen solar modules.
  Focus on appropriate applications of hydrogen, not sweeping goals of a hydrogen economy.
- Methanol
  Methanol distribution infrastructure. ICE/DMFC hybrids.
  Transition of methanol production from natural gas to biomass.
  Small onboard methanol reformer solely to provide hydrogen to improve cold starting.
  Self-contained transportable biomass to methanol production equipment.
  Infrastructure for methanol as both a combustion fuel and a fuel cell reactant.
  Repurposing hydrogen refueling stations as multifuel dispensing hubs: EV charging, methanol, ethanol, dimethyl ether, and CNG dispensing.
  Improved direct methanol fuel cells.
  Methanol ICE/electric hybrid vehicles.
- Ethanol
  Cellulosic ethanol - efficient production of ethanol from biomass.
  Process improvements and new feedstocks for fermented ethanol.
- Dimethyl Ether
  Development of renewable dimethyl ether (DME) fueling infrastructure.
  Engineering of fuel injection systems for DME, and CI engines more compatible with DME.
- Synthetic Aviation Fuels and Compatible Engines
  Aircraft fuel and engine technologies optimized for synthetic fuels.
  Possible application of DME as a SAF.

Reference

1. US Dept. of Energy Wind Energy Technologies Office. *Opportunities and challenges on the path to 100% clean energy* 26 September 2022. https://www.energy.gov/eere/wind/articles/opportunities-and-challenges-path-100-clean-energy. Accessed 1 November 2024

Appendix A: US Legislation Affecting Transportation Energy[1]

1975	Energy Policy and Conservation Act	• Created Strategic Petroleum Reserve • Established automobile fuel economy standards • Banned crude oil exports (repealed 2015) • Consolidated Atomic Energy Commission (AEC) into the Nuclear Regulatory Commission (NRC) and Energy Research and Development Administration (ERDA)
1977	Department of Energy Organization Act	• Created Department of Energy (DOE), absorbing ERDA
1978	National Energy Policy Act of 1978: • National Energy Conservation Policy Act • Power Plant and Industrial Fuel Use Act • Public Utility Regulatory Policies Act • Energy Tax Act • Natural Gas Policy Act	• Gas Guzzler Tax on automobiles with high fuel use • Tax credits for alcohol fuels • Deregulation of gas wellhead prices

[1] Information from:

- Wikipedia https://en.wikipedia.org/wiki/List_of_United_States_energy_acts
- US Dept of Transportation https://www.transportation.gov/priorities/climate-and-sustainability/dot-funding-programs-and-climate-change
- US Dept of Transportation FHWA https://www.transportation.gov/sites/dot.gov/files/2022-11/COP%20Fact%20Sheet_new%2011_17_22FINAL.pdf
- US Environmental Protection Agency https://www.epa.gov/inflation-reduction-act/investing-america-climate-action-funding-resource-guide.

C. A. MacCarley, *Non-Petroleum Automotive Transportation*,
https://doi.org/10.1007/978-3-031-85028-8

1980	Energy Security Act, incorporating (see footnote 1): • U.S. Synthetic Fuels Corporation Act • Biomass Energy and Alcohol Fuels Act • Renewable Energy Resources Act • Solar Energy and Energy Conservation Act • Geothermal Energy Act • Ocean Thermal Energy Act	• Dissolved ERDA (Energy Research and Development Administration) • Created Synthetic Fuels Corp. • Authorized AFUP (Alternative Fuel Utilization Program) flex-fuel vehicles • Provided loan guarantees for biofuels and alcohol fuels • Revised the Defense Production Act of 1950 to include energy supply
1992	Energy Policy Act	• Required alternative fuel vehicles in some government fleets
2005	Energy Policy Act	• Boosted ethanol production by requiring the use of 7.5 billion gallons of renewable fuel
2007	Energy Independence and Security Act, incorporating:	• Increased CAFE standards • Closed loophole allowing lax standards for light-duty trucks • Increased funding for biofuel development
2008	Public Law 110-343 (see footnote 1), incorporating: • Food, Conservation, and Energy Act • Strategic Petroleum Reserve Fill Suspension and Consumer Protection Act • Energy Improvement and Extension Act of 2008	• Increased funding for biofuels • Created tax credits for electric vehicles • Suspended oil contributions to Federal Petroleum Reserve • Extended and modified renewable energy tax incentives
2009	American Recovery and Reinvestment Act of 2009	• Provided funding for electricity smart grid. • Created new renewable energy tax cuts • Funded carbon capture and storage, electric vehicle battery manufacturing, and biofuels • Increased federal subsidy for corn ethanol to 51 cents per gallon
2015	Consolidated Appropriations Act, 2016	• Extended renewable energy and biofuel tax incentives • Lifted crude oil export ban
2020	Consolidated Appropriations Act, 2021, incorporating Energy Act of 2020	• Funded development of carbon capture and storage and fossil fuel research • Promoted grid battery storage
2021	Infrastructure Investment and Jobs Act aka Bipartisan Infrastructure Law	• Established the Carbon Reduction Program administered by the FHWA: $6.4 billion for states to develop carbon reduction strategies • Expanded funding for DOE research and development of carbon capture and storage, hydrogen technology, battery recycling, and grid battery storage

2022	Inflation Reduction Act (IRA)	• Extended carbon reduction and offset credits • Modified tax credits for the purchase of electric vehicles to require US manufacturing content and cost limits • Large funding for hydrogen infrastructure • Increased funding for electric grid, and energy storage

Appendix B: Calculation of CO_2 Emissions

Reference vehicle for all calculations: 2012–2015 Honda Civic Hybrid (non-plug) with EPA fuel economy rating: 44 mpg. Assume the same thermal efficiency for every fuel, i.e., ***the same energy per mile***.[2]

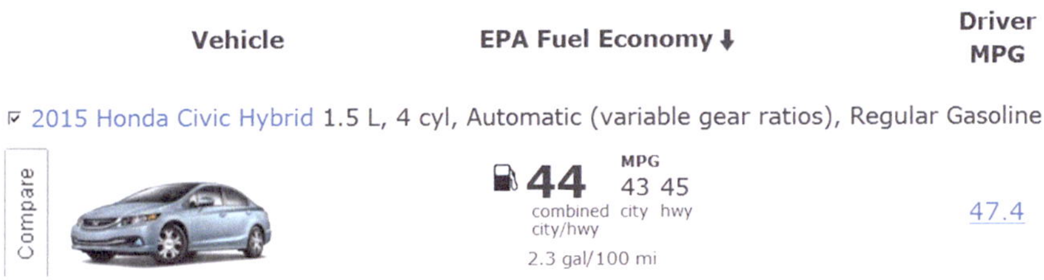

How much energy per mile is equivalent to 44 MPG gasoline?

$$44 \text{ miles / gal} \cdot \frac{1 \text{ gal}}{3.785 \text{ L}} \cdot \frac{1 \text{ L}}{0.748 \text{ kg}} \cdot \frac{1 \text{ kg}}{41.5 \text{ MJ}} = 0.375 \text{ miles / MJ}$$

or **2.67 MJ / mile or 1.66 MJ / km**

Gasoline

Gasoline without ethanol: 2020 EIA: 120,388 MBTU/gal (reported in US units on website)

$$\frac{1\text{gallon}(\text{US})}{44\text{miles}} \cdot \frac{120{,}388 \text{ MBTU}}{1 \text{ gallon } (\text{US})} \cdot \frac{168 \text{ lbs } CO_2}{10^6 \text{ BTU}} \cdot \frac{454 \text{ g}}{1 \text{ lb}}$$

$$= \mathbf{209 \ gCO_2 / mile \ or \ 129 \ g / km}$$

[2] This is a conservative assumption that gives an advantage to gasoline. Every practical alternative fuel (exceptions identified in Chap. 15, *Fuels for desperate times*) can provide higher thermal efficiency than gasoline if the compression ratio and ignition timing of the engine are optimized for that fuel.

C. A. MacCarley, *Non-Petroleum Automotive Transportation*,
https://doi.org/10.1007/978-3-031-85028-8

E10 Gasoline

Gasoline with 10% ethanol: 2020 EPA: 8877 CO_2/gal gasoline[3]

$$\frac{8877\ \text{g CO}_2\ /\ \text{gallon}}{44\ \text{miles / gallon}} = \mathbf{202\ gCO_2\ /\ mile\ or\ 125\ gCO_2\ /\ km}$$

Ethanol

Pure ethanol, E100

Density: **0.79** kg/L, energy content (LHV) = **26.8** MJ/kg

$$\frac{2.67\ \text{MJ / mile}}{26.8\ \text{MJ / kg ethanol}} = 100\ \text{g ethanol / mile}$$

Stoichiometric combustion:

$$C_2H_5OH + 3.5O_2 \rightarrow 2CO_2 + 3H_2O$$

$$12\,x\,2 + 5 + 6 + 1 = 46\,\text{g ethanol} \rightarrow 2 \times (12 + 32) = 88\,\text{g CO}_2$$

$$100.0\ \text{g}\,C_2H_5OH\ /\ \text{mile} \cdot \left(\frac{88\ \text{g CO}_2}{46\ \text{g}C_2H_5OH}\right) = \mathbf{191\ g\ CO_2\ /\ mile = 118\ gCO_2\ /\ km}$$

E85 (85% Ethanol, 15% Gasoline by Volume)

Energy per mile requirement: 2.67 MJ/mile or (1.66 MJ/km, 44 mpg)

Find the energy content of each volume fraction of 1 L of E85 having a volume ratio of 85:15.

Gasoline Fraction

Net calorific value: 41.6 kJ/g (Bosch)

Density: 750 g/L (Bosch)

Volumetric energy content:

$$\frac{41.6\ \text{kJ}}{\text{g gasoline}}\,\frac{750\ \text{g}}{\text{L gasoline}} = 31.2\ \text{MJ / L gasoline}$$

GHG emissions (EPA data): 8877 g CO_2/gallon gasoline

[3]Properties of E10 gasoline from EPA. http://www.epa.gov/oms/climate/documents/420f11041.pdf or http://www.etieco.com/content-files/EPA%20emissions%20calc%20420f05001.pdf . Accessed 10 Sept 2021.

$$\frac{8877\ \text{g CO}_2}{\text{gal}}\ \frac{\text{gal}}{3.7854\ \text{L}} = 2345\ \text{g CO}_2\ /\ \text{L gasoline}$$

Ethanol Fraction

Net calorific value: 26.8 kJ/g

Density: 790 g/L

Volumetric energy content:

$$\frac{26.8\ \text{kJ}}{\text{g ethanol}}\ \frac{790\ \text{g}}{\text{L ethanol}} = 21.2\ \text{MJ / Lethanol.}$$

One mole (46 g) or C_2H_5OH burns to produce two moles (88 g) of CO_2:

$$\frac{88\ \text{g CO}_2}{46\ \text{gethanol}}\ \frac{790\ \text{g}}{\text{L gethanol}} = 1511\ \text{g CO}_2\ /\ \text{L gasoline}$$

Both:

1 L of E85 contains 0.15 L of gasoline and 0.85 L of ethanol.

Energy content of 1 L of E85 is the sum of ethanol fraction energy and gasoline fraction energy:

$$0.15\ \text{Lgasoline}\frac{31.2\ \text{MJ}}{\text{L gasoline}} + 0.85\ \text{Lethanol}\frac{21.2\ \text{MJ}}{\text{L ethanol}} = 22.7\ \text{MJ / L E85}$$

CO_2 emissions from 1 L of E85:

$$0.15\ \text{Lgasoline}\frac{2345\ \text{g CO}_2}{\text{L gasoline}} + 0.85$$

$$\text{L ethanol}\frac{1511\ \text{g CO}_2}{\text{L ethanol}} = 1636\ \text{g CO}_2\ /\ \text{L E85}$$

CO_2 emissions per mile, using E85:

$$\frac{2.67\ \text{MJ}}{\text{mile}}\ \frac{\text{L E85}}{22.7\ \text{MJ}}\ \frac{1636\ \text{g CO}_2}{\text{L E85}} = \mathbf{192\ g\ CO_2\ /\ mile\ or\ 119\ g\ CO_2\ /\ L}$$

Methanol

Pure methanol, M100

Must calculate from fuel properties

Density: 0.79 kg/L, Energy content (LHV) = 19.7 MJ/kg:

$$\frac{2.67\ \text{MJ / mile}}{19.7\ \text{MJ / kg methanol}} = 136\,\text{g methanol / mile}$$

Stoichiometric combustion:

$$CH_3OH + 1.5O_2 \rightarrow CO_2 + 2H_2O$$

$$12 + 3 + 16 + 1 = 32 \text{ g methanol} \rightarrow 12 + 32 = 44 \text{ g}CO_2$$

$$136 \text{ g methanol} \cdot \left(\frac{44 \text{ g}CO_2}{32 \text{ g}CH_3OH} \right) = \mathbf{187 \text{ g}CO_2 / mile \text{ or } 116 \text{ g } CO_2 / km}$$

Natural Gas (H, 98% Methane)

This example uses a 2012 Honda Civic, but a different model than the 44 mpg hybrid used in previous examples. This is because Honda made a specific natural gas model in that year, the *Honda Civic GX/NGV* that has an EPA gasoline equivalent efficiency of 31 mi/gge = 50 km/GGe.

For simplicity, assume that natural gas is entirely methane. The combustion energy of CH_4 = 50.0 kJ/g, and 1 gge = 121 MJ:

$$\text{The amount of energy used to travel 1 km} = \left(\frac{1 \text{ GGe}}{50 \text{ km}} \right) \left(\frac{121 \text{ MJ}}{1 \text{ GGe}} \right) = 2.42 \text{ MJ / km}$$

The amount of methane that contains 2.42 MJ is

$$(2.42 \text{ MJ}) \left(\frac{1 \text{ kg}}{50 \text{ MJ}} \right) = 0.0484 \text{ kg} = 48.4 \text{ g methanol}$$

The amount of carbon dioxide produced by the combustion of 1 g of methanol is found in the combustion equation:

$$CH_4 + 2O_2 \rightarrow CO_2 + 2H_2O$$

$$\mathbf{16\,g\,CH_4} + 64\,g\,O_2 \rightarrow \mathbf{44\,g\,CO_2} + 36\,g\,H_2O$$

$$\frac{44 \text{ g}CO_2}{16 \text{ g}CH_4} = 2.75 \text{ g}CO_2 / \text{g}CH_4$$

The amount of CO_2 emitted per km for the Honda GX is

$$(48.4 \text{ g}CH_4)(2.75 \text{ g}CO_2 / \text{g}CH_4) = 133 \text{ g}CO_2 / \text{km}$$

If adjusted for 31 mpge Honda CX / NGV compared with 44 mpg for reference vehicle: **94 gCO_2 / km or 151 gCO_2 / mile**

Note: For all fuels considered above, CO_2 emissions are significantly lower than gasoline at 129 g CO_2/km or 209 g CO_2/mile.

Wood Gas

Assumed volume percentage composition of wood gas from [1]

Gas	Vol %	Mass %	g/mole WG	g C in 24.6 g WG	g CO_2 from 24.8 g WG	g CO_2 from 1 g WG	kJ/g	kJ in 24.8 g WG	kJ/g WG
N_2	51	57.6	14.28	0	0	0	0	0	0
CO	27	30.5	7.56	3.24	11.8	0.48	10.0	75.6	3.05
H_2	14	1.1	0.28	0	0	0	120.0	33.6	1.35
CO_2	5	8.9	2.20	0.60	2.20	0.09	0	0	0
CH_4	3	1.9	0.48	0.36	1.32	0.05	50.0	24.0	0.97
Total	100	100	24.8 g/mole WG	4.68 g C	15.32 g CO_2	**0.62**		133.2	**5.37**

WG wood gas

2012 Honda Civic (44 mpg), hypothetically converted to run on wood gas
From sums at bottom of the table above:

$$\left(\frac{1.66\text{ MJ wood gas}}{\text{km}}\right)\left(\frac{\text{kg wood gas}}{5.37\text{ MJ}}\right)\left(\frac{0.62\text{ kg CO}_2}{\text{kg wood gas}}\right)=\mathbf{190\ gCO_2\,/\,km}$$

$$\mathbf{or\ 306\ g\,CO_2\,/\,mile}$$

Korean wood gas military truck (example from Chap. 16)
Assume the truck uses 12 MJ/km avg:

$$\left(\frac{12\text{ MJ woodgas}}{\text{km}}\right)\left(\frac{\text{kg woodgas}}{5.37\text{ MJ}}\right)\left(\frac{0.62\text{ kgCO}_2}{\text{kg woodgas}}\right)=\mathbf{1390\ gCO_2\,/\,km}$$

Note that for all wood gas examples, the calculation is only for the combustion of the wood gas that reaches the engine from the gasifier. Does not include emissions from the gasifier itself or the wood gas released to the air when vehicle demand is less than gasifier output.

Biogas

Municipal sewage gas, mass percentages: 46% CH_4 + 54% CO_2 + trace hydrogen sulfide [2]

Net fuel energy: 27.2 MJ/kg:

$$\left(\frac{\left(\frac{2.67\ \text{MJ}}{\text{mile}}\right)}{\frac{27.2\ \text{MJ}}{\text{kg}}\left(\frac{1\ \text{kg}}{1000\ \text{g}}\right)}\right) = 98.1\ \frac{\text{g sewargas}}{\text{mile}}$$

$$98.1\ \frac{\text{g sewar gas}}{\text{mile}}\left(\frac{0.46\ \text{g CH}_4}{\text{g sewargas}}\right) = 45.1\ \text{g CH}_4\ /\,\text{mile}$$

$$45.1\ \text{g CH}_4\left(\frac{44\ \text{gCO}_2}{16\ \text{gCH}_4}\right) = 124\ \text{g CO}_2\ /\,\text{mile}$$

But we also need to account for the CO_2 in the sewer gas itself that passes right through the engine:

$$98.1\ \frac{\text{g sewargas}}{\text{mile}}\left(\frac{0.54\ \text{g CO}_2}{\text{g sewargas}}\right) = 53.0\ \text{g CO}_2\ /\,\text{mile}$$

$$124(\text{combustion}) + 53(\text{CO}_2\ \text{in fuel}) = \mathbf{177\,g\,CO_2\ /\ mile\ or\ 110\,g\,CO_2\ /\ km}$$

Electric Vehicle (charged from the US utility grid)

2.94 Miles/kWh (Emission from Electricity Generation Only)

2015–2018 Nissan Leaf electric car:

CO_2 from electric generation: 390 g/kWh CO_2 (US grid avg 2020, renewable and non-renewable)

Vehicle efficiency: 2.94 miles/kWh (34 kWh/100 miles):

$$\frac{34\ \text{kWh}}{100\ \text{miles}} \cdot \frac{390\ \text{g CO}_2}{\text{kWh}} \cdot \frac{1}{0.78\ \text{distribution efficiency}} = \mathbf{170\ g\ CO_2\ /\ mi}$$

$$\mathbf{or\ 105\ gCO_2\ /\ km\ \left(18.6\%\ lower\,than\,gasoline\right)}.$$

Note: Electric distribution efficiency accounts for the transmission and local distribution electric losses between the generation facility and the electric vehicle being charged (2.2–13.3%, mean value 7.8%).[4]

[4] Distribution losses for US electric power grid, 2015 average, from http://insideenergy.org/2015/11/06/lost-in-transmission-how-much-electricity-disappears-between-a-power-plant-and-your-plug/.

Appendix C: Calculation of the stoichiometric AFR

Gasoline

In practice, we just look up the mass AFR for this year's average regular or premium gasoline, with or without ethanol, as published on one of the reference websites, e.g., US EPA [3], or in the Bosch Automotive Handbook [4]. If not stated otherwise, AFR will always mean *mass* AFR, the ratio of the mass of the air over the mass of the fuel inducted with the air into each cylinder of the engine.

To calculate the stoichiometric mass AFR we can use the pseudo-molecule that represents gasoline derived in Appendix A: $C_{7.5}H_{15.2}$Combustion consumes only the oxygen in the air. The nitrogen and argon in air flow through the combustion chamber unburned. They do not participate in the combustion process but they take up volume and mass in the intake fuel-air charge and in the exhaust:

Unlike gasoline, most alternative fuels are pure fuels that have a single chemical formula, for example, methane CH_4, ethanol C_2H_5OH, or methanol CH_3OH. The stoichiometric AFR for these fuels can be calculated with only knowledge of the formula. First balance the combustion equation ignoring the non-oxygen components of air. Then find the amount of air that contains the mass of oxygen using the correct mass fraction of oxygen in the air:

$$\underbrace{C_{7.5}H_{15.2}}_{Moles\ of\ Gasoline} + \underbrace{11.3\left(O_2+\left(\frac{.78}{.21}N_2+\frac{.01}{.21}Ar\right)\right)}_{Moles\ of\ air} \rightarrow \underbrace{7.5CO_2+7.6H_2O+11.3\left(\frac{.78}{.21}N_2+\frac{.01}{.21}Ar\right)+\text{trace HC, CO, NOx}}_{Exhaust\ Moles}$$

By volume, oxygen is 21% percent of air.

$$\frac{(O_2)}{\left(O_2+\left(\frac{.78}{.21}N_2+\frac{.01}{.21}Ar\right)\right)}=\frac{(32)}{\left(32+\left(\frac{.78}{.21}\times 28+\frac{.01}{.21}\times 40\right)\right)}=\frac{32}{138}=0.232$$

By mass, oxygen is 23.2 % of air, or the air/oxygen ratio is 4.31

So the mass AFR for gasoline is

$$\frac{\text{Air mass}}{\text{Fuel mass}}=\frac{11.3\times 32\,g\,oxygen\times 4.31}{7.5\times 12\,g\,carbon+15.2\times 1\,g\,hydrogen}=\frac{1{,}558\,g\,air}{105.2\,g\,fuel}=14.8\,g\,air/g\,fuel\quad AFR_{mass}$$

C. A. MacCarley, *Non-Petroleum Automotive Transportation*,
https://doi.org/10.1007/978-3-031-85028-8

Ethanol

$$C_2H_5OH + 3O_2 \rightarrow 2CO_2 + 3H_2O$$

46 g ethanol combines with 96 g oxygen that is contained in $(96\,\mathrm{g}\;O_2)\left(\dfrac{4.31\,\mathrm{g\;air}}{1\,\mathrm{g}\;O_2}\right) = 414\,\mathrm{g\;air}$

So the mass AFR for pure ethanol is $\dfrac{414\,\mathrm{g\;air}}{46\,\mathrm{g\;fuel}} = 9.00$

Methanol

$$CH_3OH + 1.5O_2 \rightarrow CO_2 + 2H_2O$$

32 g methanol combines with 48 g oxygen that is contained in $(48\,\mathrm{g}\;O_2)\left(\dfrac{1\,\mathrm{g\;air}}{0.232\,\mathrm{g}\;O_2}\right) = 207\,\mathrm{g\;air}$

So the mass AFR for pure methanol is $\dfrac{207\,\mathrm{g\;air}}{32\,\mathrm{g\;fuel}} = 6.47$

Blends of liquid fuels

Typical Blends of petroleum + pure fuels
E10 = 10% ethanol + 90% gasoline by volume (regular gasoline in the USA)
E85 = 85% ethanol + 15% gasoline by volume (post-2000 flex fuel)
M85 = 85% methanol + 15% gasoline by volume (AFUP flex fuel 1990–2000)
A20 = 10% methanol + 90% E10 (budget racing fuel)

E10

1. First find the *mass* of each fuel component in one liter (L) of E10. E10 is 10% ethanol and 90% gasoline *by volume* specified as volume. Convert to mass fractions. Gasoline density is 748 g/L, and ethanol density is 790 g/L for E100.

   In 1 L of E10 there is

$$0.90\ \mathrm{L gasoline}\left(\frac{748\ \mathrm{g}}{\mathrm{L\ \ gasoline}}\ \mathrm{density\,of\,gasoline}\right) = 673\ \ \mathrm{g\,of\,gasoline}$$

in 1 L of E10

$$0.10\ \mathrm{L ethanol}\left(\frac{790\ \mathrm{g}}{\mathrm{L\ \ ethanol}}\,\mathrm{density\,of\,ethanol}\right) = 79.0\ \ \mathrm{g\,of\,ethanol}$$

in 1 L of E10

2. Add the masses of each component to get the total mass of 1 L of E10

$$673\,\text{g gasoline} + 79\,\text{g of ethanol} = 752\,\text{g}/\text{L overall density of E10}$$

3. Divide by the mass in grams of 1 L of E10 to get the mass of each component *in 1 g* of E10:

$$(1\ \text{g E10})\left(\frac{673\ \text{g}}{752\ \text{g E10}}\right) = 0.895\ \text{g gasoline in 1 g of E10}$$

and

$$(1\ \text{g E10})\left(\frac{79\ \text{g}}{752\ \text{g E10}}\right) = 0.105\ \text{g ethanol in 1 g of E10}$$

4. From example above, stoichiometric AFR for ethanol is = **9.0:1 ethanol mass AFR**.
5. Also from the example above, the stoichiometric AFR for gasoline is **14.8:1 gasoline mass AFR**.
6. Use the mass AFR of each to find the amount of air used by each in stoichiometric combustion.
   For the gasoline fraction: $(0.895\ \text{g gasoline})(14.8\ \text{g air/g gasoline}) = 13.2\ \text{g air}$
   For the ethanol fraction: $(0.105\ \text{g ethanol})(9.0\ \text{g air/g ethanol}) = 0.945\ \text{g air}$
7. Add these masses of air to get the *total mass of air* that combines with 1.0 g of E10.

$$13.2\,\text{g} + 0.945\,\text{g} = 14.1\,\text{g air}$$

8. Divide the total mass of air by 1.0 g of E10 (the mass of fuel) to get the AFR:

$$\frac{14.1\ \text{g air}}{1.0\ \text{g E10}} = \mathbf{14.1\ AFR\ for\ E10}$$

Note: All gasoline sold in the USA is E10, so this is the actual AFR that your engine is operating at.

E85

E85 is 85% ethanol + 15% gasoline *by volume*.

1. Convert to *mass* percentages.
   In 1 L of E85 there is

$$0.15\ \text{L gasoline}\frac{748\ \text{g}}{\text{L gasoline}} = 112\ \text{g E0 gasoline} \quad (748\ \text{g}/\text{L avg})$$

$$0.85\ \text{L ethanol}\frac{790\ \text{g}}{\text{L ethanol}} = 672\ \text{g E100 ethanol} \quad (790\,g/L)$$

2. Add the masses of each component to get the density of 1 L of E85:

112 g gasoline + 672 g of ethanol = 784 g/L overall density of E85 (5% heavier than gasoline)

3. Divide each component in E85 by the mass in grams of 1 L of E85 to get the mass ratio of each component:

$$\frac{112\ \text{g gasoline}}{784\ \text{g E85}} = 0.143\ \text{gasoline / E85 mass ratio}$$

$$\frac{672\ \text{g ethanol}}{784\ \text{g E85}} = 0.857\ \text{ethanol / E85 mass ratio}$$

4. Calculate or look up the mass AFR for each component of E100 ethanol, previously calculated:
   Ethanol: 9.0:1 mass AFR gasoline: 14.7:1 mass AFR
5. Use the AFRs to find the amount of air used by each component in stoichiometric combustion.
   For the gasoline fraction: (0.143 g gasoline)(14.7 g air/g gasoline) = 2.10 g air
   For the ethanol fraction: (0.857 g ethanol)(9.0 g air/g ethanol) = 7.71 g air
6. Add these masses of air to get the total mass in grams of air that combines with 1.0 g of E85:

2.10 g + 7.71 g = 9.81 g air

7. Divide the total mass of air by 1.0 g of E85 (the mass of fuel) to get the AFR for E85:

$$\frac{9.81\ \text{g air}}{1.0\ \text{g E85}} = \mathbf{9.81\ AFR\ for\ E85}$$

Gaseous fuels

Methane (the Main Component of Natural Gas)

$$CH_4 + 2O_2 \rightarrow CO_2 + 2H_2O$$

(12 + 4 = 16) g methane burns with (2 × 32 = 64) g oxygen which is contained in

$$(64\ \text{g}O_2)\left(\frac{4.31\ \text{g air}}{\text{g } O_2}\ \text{air / oxygen ratio}\right) = 276\ \text{g air}$$

So the mass AFR for pure methane is $\frac{276\ \text{g air}}{16\ \text{gfuel}} = \mathbf{17.3}$

Hydrogen

$$H_2 + \frac{1}{2}O_2 \rightarrow H_2O$$

2 g H_2 burns with (1/2 × 32 = 16) g oxygen which is contained in

$$(16\ gO_2)\left(\frac{4.31\ g\ air}{g\ O_2}\right) = 69.0\ g\ air$$

So the mass AFR for hydrogen is $\frac{69.0\ gair}{2\ gfuel} = \mathbf{34.5}$

Acetylene

$$C_2H_2 + 2.5O_2 \rightarrow 2CO_2 + H_2O$$

26 g acetylene combines with 80 g of oxygen that is contained in 4.31 × 80 g O_2 = 345 g air:

$$\frac{345\ g\ air}{26\ gfuel} = 13.3\ massAFR$$

$$26\,g\,fuel + 345\,g\,air \Rightarrow 13.3\,mass\,AFR$$

Blends of gaseous fuels

Some typical blended gaseous fuel compositions (mass percentages) [5]:
Natural gas H (Siberia) mass %: 98% methane + 1% ethane + 1% nitrogen
Natural gas L (EU) (mass %) 83%: CH_4 + 4% C_2H_6 + 1% C_3H_8 + 2% CO_2 + 10% N_2
Natural gas (US) = (volume %): 95% methane, 2.5% ethane, 1.6% nitrogen, 0.7% carbon dioxide, 0.2% other non-combustible gases
Producer gas (wood gas, biomass syngas): 7–22% CO + 16–20% H_2 + 2.3% CH_4 + 10–15% CO_2 + 50–55% N_2 From http://www.nariphaltan.org/gasbook.pdf
Bio-methane (from anaerobic digestion of organic waste, typical):

	Digester gas	Landfill gas	Natural gas (Dutch, average during 2007)
Methane (vol %)	60–70	35–65	89
Other HC (vol %)	0	0	9.4
Hydrogen (vol %)	0	0–3	0
CO_2 (vol %)	30–40	15–50	0.67
N_2 (vol %)	~0.2	5–40	0.28
O_2 (vol %)	0	0–5	0
H_2S (ppm)	0–4000	0–100	2.9
Ammonia (ppm)	~100	~5	0
Lower Heating Value (kWh/NM3)	6.5	4.4	11.0

Source: Peterson and Wellinger, 2009

Natural Gas H

1. Composition [5]: natural gas (H) is 98% methane, 1% ethane, and 1% nitrogen *by mass*

$$1\,\text{g of fuel contains}\,0.98\,\text{g methane} + 0.01\,\text{g ethane} + 0.01\,\text{g nitrogen}.$$

2. Find what mass of ***air*** combines with each mass of fuel. Need to calculate (or look up) stoichiometric mass AFRs for each fuel:

   For the methane fraction:
   $CH_4 + 2O_2 \rightarrow CO_2 + 2H_2O$
   16 g fuel combines with 64 g oxygen.
   But oxygen is 23.2% of air by mass, so
   16 g CH_4 combines with 64/0.232 = 276 g air

   The mass AFR for methane is $\dfrac{276\ \text{gair}}{16\ \text{gmethane}} = 17.25\ \text{gair / g methane}$

   For the ethane fraction:
   $C_2H_6 + 3.5O_2 \rightarrow 2CO_2 + 3H_2O$
   30 g C_2H_6 combines with 112 g oxygen.
   But oxygen is 23.2% of air by mass, so 30 g of ethane combines with 112/0.232 = 483 g air

   The mass AFR for ethane is $\dfrac{483\ \text{g air}}{30\ \text{gethane}} = 16.1\ \text{g air / g ethane}$

   For the nitrogen (inert) fraction:
   $N_2 \rightarrow N_2$
   This just passes through from the intake to the exhaust.

3. Using these AFRs for each fuel component, find the total mass of ***air*** that combines with a given mass of fuel by summing the amount of air that combines with each component.

   1.0 g of NG(H) combines with this much air:

$$(0.98\ \text{g}\ CH_4)(17.25\ \text{AFR}) + (0.01\ \text{g}\ C_2H_6)(16.1\ \text{AFR})$$

$$= 16.9\ \text{gair} + 0.161\ \text{gair} = 17.1\ \text{g air}$$

   So the AFR for NG(H) is $\left(\dfrac{17.1\ \text{g air}}{1\ \text{gNG(H)}}\right) = \mathbf{17.1\ AFR}$

US Average Natural Gas

Average *volume* composition: 95% methane, 2.5% ethane, 1.6% nitrogen, 0.7% carbon dioxide, 0.2% other non-combustible gases.

1. Find the mass fraction of each component of the gas.
   At 22 °C (295 K) and 1.0 atm, 1 L of air contains 0.0413 moles.
   L of fuel contains 0.95 L methane, 0.025 L ethane, 0.016 L nitrogen, 0.007 L carbon dioxide, and 0.002 L of other non-combustible gases, which we will just add to the nitrogen fraction increasing it to 0.018 L.
   Mass of each gas in 1 L of fuel at 22 °C, 1 atm:
   $(0.95 \times 0.0413 \text{ moles } CH_4)(16 \text{ g/mole}) = 0.628 \text{ g } CH_4$
   $(0.025 \times 0.0413 \text{ moles } C_2H_6)(30 \text{ g/mole}) = 0.031 \text{ g } C_2H_6$
   $(0.018 \times 0.0413 \text{ moles } N_2)(28 \text{ g/mole}) = 0.021 \text{ g } N_2$
   $(0.007 \times 0.0413 \text{ moles } CO_2)(44 \text{ g/mole}) = 0.013 \text{ g } CO_2$
   Total mass of 1 L fuel = 0.693 g fuel
2. What mass of ***air*** combines with each mass of fuel? First, calculate AFR for each fuel.
   For the methane fraction:
   $CH_4 + 2O_2 \rightarrow CO_2 + 2H_2O$
   16 g fuel combines with 64 g oxygen.
   But oxygen is 23.2% of air by mass, so
   16 g methane combines with 64/0.232 = 276 g air

   So the mass AFR for methane is $\frac{276 \text{ g air}}{16 \text{ gmethane}} = 17.25 \text{ g air / gmethane}$

   For the ethane fraction:
   $C_2H_6 + 3.5O_2 \rightarrow 2CO_2 + 3H_2O$
   30 g fuel combines with 112 g oxygen.
   But oxygen is 23.2% of air by mass, so 30 g of ethane combines with 112/0.232 = 483 g air.

   So the mass AFR for ethane is $\frac{483 \text{ g air}}{30 \text{ gethane}} = 16.1 \text{ g air / g ethane}$

   For the nitrogen and carbon dioxide (inert) fractions:
   $N_2 \rightarrow N_2$ and $CO_2 \rightarrow CO_2$. They do not combine with oxygen.
   These just pass through from the intake to the exhaust. It does not tie up any of the intake air, but it must be accounted for as a component of the exhaust.
3. Summing the mass of air that combines with each component by using the AFR of each gives the total mass of ***air*** that combines with the natural gas blend:

$$(0.628 \text{ g } CH_4)(17.25 \text{ AFR}) + (0.031 \text{ g } C_2H_6)(16.1 \text{ AFR})$$

$$= 10.8 \text{ gair} + 0.50 \text{ gair} = 11.3 \text{ g air for combustion}$$

   0.693 g of fuel (including inerts) combines with 11.3 g air.

   So the AFR for NG(H) is $\left(\frac{11.3 \text{ gairforcombustion}}{0.693 \text{ gtotal mass of fuel}}\right) = \mathbf{16.3 AFR}$

Appendix D: Calculation of Power Output

Assumptions for All Analyses to Follow

- 1.0 L displacement engine.
- For gaseous fuels, the intake air volume is adjusted to accommodate the volume occupied by the fuel.
- Full throttle, 100% volumetric efficiency (cylinder pressure when the intake valve closes is the same as atmospheric pressure).
- Air temperature 22 °C.
- Air pressure 1 atm.
- Assume perfect fuel-air mixing.
- AFR mixture is stoichiometric.
- Same thermal efficiency (28%) for any fuel.
- Same engine speed at maximum power (6000 RPM).
- Liquid fuels enter the cylinder entirely in liquid form and therefore occupy a trivial volume of the intake charge which can be ignored in calculations.

Number of moles of air occupying 1 L at 1 atm and 22 °C:
From ideal gas law PV = nRT:

$$n = \frac{\text{PV}}{\text{RT}} = \frac{(1\ \text{atm})(1\ \text{L})}{\left(0.08206 \frac{\text{atm} \cdot \text{L}}{\text{K} \cdot \text{mole}}\right)(295\ \text{K})} = 0.0413\ \text{moles or } 24.2\ \text{L/mole}$$

Gasoline

1 L of air at 22 °C and 1 atm combines with what mass of gasoline, assuming that the volume of the liquid gasoline entering the cylinder is negligible compared with the air volume?

Need: air/gasoline AFR, mass density of air, mass density of fuel.

C. A. MacCarley, *Non-Petroleum Automotive Transportation*,
https://doi.org/10.1007/978-3-031-85028-8

From above, 1 L of air at 22 °C and 1 atm contains 0.0413 moles of air which has a mass of

$$0.0413 \text{ moles air} \frac{29 \text{ g}}{\text{mole air}} = 1.20 \text{ g air}$$

For 1.20 g air and AFR = 14.8 a stoichiometric mass of fuel is

$$\frac{1.20 \text{ g air}}{14.8 \text{ air / fuel}} = 0.081 \text{ gfuel}$$

Energy content of fuel in 1 L intake charge (LHV for gasoline = 41.5)
$0.081 \text{ gfuel}(41.5 \text{ kJ/g}) = 3.36 \text{ kJ for each L of air inducted.}$
At 6000 RPM, what is the rate of airflow into the engine?

$$6000 \text{ RPM}\left(\frac{1 \text{ min}}{60 \text{ s}}\right)\left(\frac{1 \text{ intake}}{2 \text{ revolutions}}\right)\left(\frac{1.0 \text{ L air}}{1 \text{ intake}}\right) = 50 \text{ L / s air inducted}$$

by the engine

Maximum power (engine thermal efficiency = 28%):

$$50 \text{ L air / s}\left(\frac{3.36 \text{ kJfuel}}{1 \text{ L air}}\right)\left(\frac{1 \text{ kW}}{1\frac{\text{kJ}}{\text{s}}}\right)(0.28 \text{ thermal efficiency}) = \mathbf{47.0 \text{ kw} = 63.0 \text{ HP}}$$

Ethanol

Same number of moles of air in one liter of intake air: 0.0413 moles
Stoichiometric AFR = 9.0:1 from Appendix B.
How much fuel combines with this amount of air?

$$\frac{(0.0413 \text{ moles air})(29.0 \text{ gair / mole air})}{(9.0 \text{ gair / g ethanol})} = 0.133 \text{ gethanol}$$

At 26.8 kJ/g, 0.133 g ethanol releases 3.56 kJ of energy for every liter of air inducted.

From above, intake air in L/s at 6000 RPM, wide open throttle, 1 L displacement ⇒ 50 L air/s:

$$50 \text{ Lair / s}\left(\frac{1 \text{ kW}}{1\frac{\text{kJ}}{\text{s}}}\right)\left(\frac{3.56 \text{ kJfuel}}{1 \text{ Lair}}\right)(0.28 \text{ thermal efficiency}) = \mathbf{49.9 \text{ kW} = 66.8 \text{ HP}}$$

(4.7% greater than gasoline)

Methanol

Same number of moles of air in one liter: 0.0413 moles

Stoichiometric mass AFR for methanol = 6.47 from Appendix C

$$\frac{(0.0413 \text{ moles air})(29.0 \text{ g air / moleair})}{(6.47 \text{ g air / gmethanol})} = 0.185 \text{ gmethanol}$$

LHV of methanol is 19.7 kJ/g. 0.185 g methanol releases **3.64 kJ** of energy every 2 rotations of the engine. At 6000 RPM, WOT, 0.28 thermal efficiency:

$$50 \text{ L air / s}\left(\frac{1 \text{ kW}}{1\frac{\text{kJ}}{\text{s}}}\right)\left(\frac{3.64 \text{ kJfuel}}{1 \text{ L air}}\right)(0.28 \text{ thermal efficiency}) = \mathbf{51.0 \text{ kW} = 68.3 \text{ HP}}$$

(7.1% greater than gasoline)

Gaseous Fuels

As previously calculated, gaseous fuels are inducted already in the vapor phase so they displace a significant amount of the air in the intake mixture. Power output is reduced in the same proportion that the air is reduced.

Methane

Combustion: $CH_4 + 2O_2 \rightarrow CO_2 + 2H_2O$

One mole of methane combines with 2 moles of oxygen or 9.52 moles of air. This means the stoichiometric ***volume ratio*** between air and methane is **9.52:1** $\mathbf{AFR_{volumetric}}$

The fuel occupies $\frac{1 \text{ mole} CH_4}{1 \text{ mole } CH_4 + 9.52 \text{ mole air}} = 0.095$ of the volume of this total intake charge, **displacing 9.5% of the air** that the intake charge would normally contain.

This results in a power reduction proportional to the air reduction (and therefore a fuel reduction) since the amount of fuel is proportional to the mass of air inducted during each intake stroke.

What is the power output including this effective reduction in displacement?

At 6000 RPM a 1.0 L engine inducts 50 L/s total intake mixture, which contains

$$(50 \text{ Lair} + \text{fuel / s})\left(0.095\frac{\text{L methane}}{\text{L of air} + \text{methane}}\right)\left(\frac{16 \text{ g methane / mole}}{24.2 \text{ L methane / mole}}\right)$$
$$= 3.14 \text{ gmethane / s}$$

This produces a power output of

$$(3.14 \text{ g methane/s})\left(\frac{50.0 \text{ kJ}}{\text{g methane}}\right)(0.28 \text{ thermal efficiency})\left(\frac{1 \text{ kW}}{1\frac{\text{kJ}}{\text{s}}}\right)$$

$$= \mathbf{44.0\ kW} = \mathbf{58.9\ HP}$$

(7.7% less power than gasoline)

Natural Gas H

Gas composition by mass (98% methane + 1% ethane + 1% nitrogen):

1 mole of an ideal gas occupies 24.2 L at 22 °C and 1 atm.

1 g of air occupies 29 g/mole/24.2 L/mole = **1.20 g air / L** at 22 °C and 1 atm

One g of fuel can be broken down into

0.98 g CH_4: $\left(\frac{0.98 \text{ g } CH_4}{16 \text{ g/mole}}\right)\left(\frac{24.2 \text{ L}}{\text{mole}}\right) = 1.48 \text{ L } CH_4$ methane energy contribution

which contributes $(0.98 \text{ g } CH_4)\left(\frac{50.0 \text{ kJ}}{\text{g}}\right) = \mathbf{49\ kJ}$

plus0.01 g C_2H_6: $\left(\frac{0.01 \text{ g} C_2H_6}{30 \text{ g/mole}}\right)\left(\frac{24.2 \text{ L}}{\text{mole}}\right) = 0.00806 \text{ L} C_2H_6$ ethane energy contribution

which contributes to $(0.01 \text{ g} C_2H_6)\left(\frac{47.5 \text{ kJ}}{\text{g}}\right) = \mathbf{0.475\ kJ}$

and0.01 g N_2: $\left(\frac{0.01 \text{ g} N_2}{28 \text{ g/mole}}\right)\left(\frac{24.2 \text{ L}}{\text{mole}}\right) = 0.00864 \text{ L} N_2$ nitrogen contributes **0 kJ**

1 g of NGH contains 49.5 kJ and occupies 1.48 + 0.00806 + 0.00864 = 1.50 L/g or 0.667 g/L of NGH and volume energy density of 49.5 kJ/1.50 L = 33.0 kJ/L.

The stoichiometric mass AFR for NG(H) was calculated in Appendix C to be 17.1:1:

1 g of NGH combines with 17.1 g air which occupies 14.3 L air.

Energy content in kJ of 1 L of inducted air + fuel mixture:

$$\frac{\text{energy in 1 g fuel}}{\text{volume of air + fuel}} = \frac{49 \text{ kJ from } CH_4 + 0.48 \text{ kJ from } C_2H_6}{14.3 \text{ Lair} + (1.48 \text{ L}CH_4 + 0.0081 \text{ L}C_2H_6 + 0.0086 \text{ L}N_2)}$$

$$= \frac{49.5 \text{ kJ}}{15.8 \text{ L}} = 3.13 \text{ kJ/L}$$

At 6000 RPM, 50.0 L/s of air + fuel mixture flows through the engine, which at 28% thermal efficiency produces power:

$$(50.0 \text{ L/s})(3.13 \text{ kJ/L})(0.28 \text{ thermal efficiency}) = \mathbf{43.8 \ kW \ or \ 58.7 \ HP}$$

(Trivially less than pure methane, 7.8% less power than gasoline)

Acetylene

$C_2H_2 + 2.5\ O_2 \rightarrow 2\ CO_2 + H_2O$

26 g acetylene combines with 80 g of oxygen that is contained in 4.31 × 80 g O_2 = 345 g air:

$$\frac{345\text{g air}}{26 \text{ gfuel}} = 13.3 \text{ massA FR}$$

Or

$$\left(\frac{345 \text{ g air}}{26 \text{ gfuel}}\right)\left(\frac{\frac{1 \text{ mole}}{29 \text{ g air}}}{\frac{1 \text{ mole}}{26 \text{ gfuel}}}\right) = 11.89 \text{ volumeAFR}$$

The fuel occupies

$$\frac{1 \text{ mole fuel}}{11.9 \text{ moleair} + 1 \text{ mole fuel}} = 0.0775 \text{ or } 7.75\% \text{ of the intake charge volume.}$$

At 6000 RPM a 1.0 L engine inducts 50 L/s total intake mixture, which contains

$$(50 \text{ Lair} + \text{fuel/s})\left(0.0775 \frac{\text{L fuel}}{\text{L of air} + \text{methane}}\right)\left(\frac{26 \text{ gacetylene/mole}}{24.2 \text{ Lacetylene/mole}}\right)$$

$$= 4.16 \text{ gacetylene/s}$$

$$(4.16 \text{ gacetylene/s})\left(\frac{48.1 \text{ kJ}}{\text{g}}\right)(0.28 \text{ efficiency}) = \mathbf{56 \ kW = 75.0 \ HP}$$

This produces a power output of

$$(3.14 \text{ g methane/s})\left(\frac{50.0 \text{ kJ}}{\text{g methane}}\right)(0.28 \text{ thermal efficiency})\left(\frac{1 \text{ kW}}{1\frac{\text{kJ}}{\text{s}}}\right)$$

$$= \mathbf{44.0kW = 58.9HP}$$

(Identical to methane, 7.7% less power than gasoline)

Syngas: Hypothetical Volumetric Mixture of 50% Hydrogen and 50% Carbon Monoxide

1 L of an ideal gas at NTP occupies 24.2 L.

One L of wood gas can be broken down into

$$\text{Hydrogen : Energy contribution} = (0.5\ \text{H}_2\ \text{fraction})\left(\frac{2\ \text{g}}{\text{mole}}\right)\left(\frac{120\ \text{kJ}}{1\ \text{g}\text{H}_2}\right)\left(\frac{1\ \text{mole}}{24.2\ \text{L}}\right)$$

$$= 4.96\ \text{kJ/L}$$

$$\text{Carbon monoxide energy contribution} = (0.5\ \text{CO fraction})\left(\frac{28\ \text{g}}{\text{mole}}\right)$$

$$\left(\frac{10.05\ \text{kJ}}{1\ \text{g}}\right)\left(\frac{1\ \text{mole}}{24.2\ \text{L}}\right) = 5.81\ \text{kJ}$$

1 L of wood gas contains 4.96 + 5.81 = 10.77 kJ

Volume AFR for H_2:

$$2\text{H}_2 + \text{O}_2 \rightarrow 2\text{H}_2\text{O}$$

$$\frac{1\ \text{mole O}_2}{2\ \text{moles}\text{H}_2} \times \frac{1\ \text{mole air}}{0.21\ \text{mole O}_2} = 2.38\ \text{volume AFR for}\text{H}_2$$

Volume AFR for CO:

$$\text{CO} + 0.5\text{O}_2 \rightarrow \text{CO}_2$$

$$\frac{0.5\ \text{mole}\text{O}_2}{1\ \text{mole CO}} \times \frac{1\ \text{mole air}}{0.21\ \text{mole}\text{O}_2} = 2.38\ \text{volume AFR for}\text{CO}$$

Since both volume AFRs are the same, the overall volume AFR of wood gas is 2.38 L air/L of syngas.

Energy release during each power stroke is

$$(1\ \text{L}(\text{air} + \text{woodgas}))\left(\frac{1\ \text{Lwoodgas}}{2.38\ \text{Lair} + 1\ \text{Lwoodgas}}\right) = 0.30\ \text{Lwoodgas}$$

Power output is

$$(0.30\ \text{Lwoodgas/s})\left(\frac{10.77\ \text{kJ}}{\text{Lwoodgas}}\right)\left(\frac{1\ \text{intake}}{2\ \text{revs}}\right)\left(\frac{4000\ \text{revs}}{\text{min}}\right)\left(\frac{1\ \text{min}}{60\ \text{s}}\right)$$

$$(0.28\text{efficiency}) = \mathbf{30.2kW} = \mathbf{40.4HP}$$

(36% less than gasoline)

Appendix E. Ignition System Timing

Purpose of Ignition Timing

In the Ideal Otto Cycle analysis, we assumed that combustion occurred instantaneously, so that ignition occurred exactly at TDC. But combustion takes time and occurs over many degrees of rotation of the engine. ***Ignition timing*** *is when to fire the spark plug that starts the combustion process.*

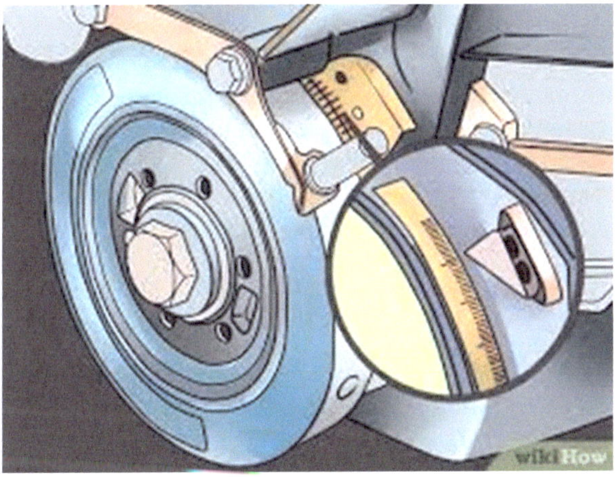

Graphic from WikiHow https://www.wikihow.com/Adjust-Timing

- The objective is usually maximum efficiency and power, without generating excessive NO_x, at any RPM and throttle setting (that controls intake charge pressure).
- For gasoline, optimum timing typically positions the combustion **pressure peak** at 10–15 degrees After Top Dead Center (ATDC), which requires that the point of ignition be 5–35 degrees in advance of TDC (called Before TDC or BTDC.
- This maximizes torque, measured as Indicated Mean Effective Pressure or IMEP.
- This timing setting is referred to as Mean Best Torque (MBT) timing.

C. A. MacCarley, *Non-Petroleum Automotive Transportation*,
https://doi.org/10.1007/978-3-031-85028-8

- Optimal ignition time varies significantly with the engine, related to the combustion chamber shape and size, and the fuel and A/F ratio.
- Optimal ignition timing also varies significantly with different fuels, with some fuels allowing less **ignition advance** because they have higher rates of combustion (shorter combustion times).

A historic example of a bad attempt at emission reduction

In the 1970s before electronic emission controls, it was common practice for a manufacturer to deliberately retard ignition timing from the optimum as a means to reduce NOx emissions to within EPA limits, which were very lax. This worked by delaying the pressure peak, reducing the time that the air spent at high pressure and temperature in the cylinder which lead to NOx formation. The consequence was reduced efficiency, power, and even increased HC emissions. This "pollution control measure" was easily defeated by resetting the ignition timing. But this and other zero-cost engineering responses to the new clean-air regulations (starting in 1968) lead to the widespread belief by the public that pollution controls were synonymous with reduced power and fuel economy. Although the practice was mostly corrected in the 1980s following feedback fuel control and three-way catalytic converters, it took the passing of an entire generation for the public to let go of this bias against cleaner cars, although it still persists today. And even today, motorheads of all generations still vilify environmentalists for the demise of the great (and hideously polluting) V8 "muscle cars" of the 1960s and early 1970s.

Mean Best Torque (MBT) timing is the ignition advance setting at which engine Torque is highest at a given speed and manifold air pressure (**MAP**).

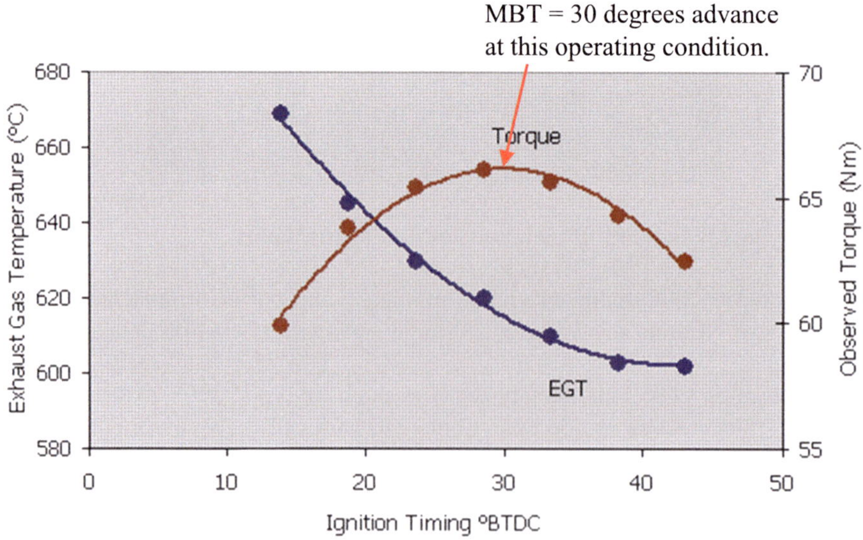

MBT ignition timing is a function of:

1. **Engine speed** (RPM)—Higher speed allows less time for completion of combustion. More ignition advance.
2. **Air pressure** (vacuum)—Higher vacuum, lower flame speed, more ignition advance.
3. **The type of fuel—**Different fuels have different flame speeds. Faster flame speed, less ignition advance. A stoichiometric hydrogen-air mixture has an extremely fast flame speed. Ammonia has a very slow flame speed.
4. **The AF ratio—**Too lean or too rich reduces flame speed. More ignition advance.

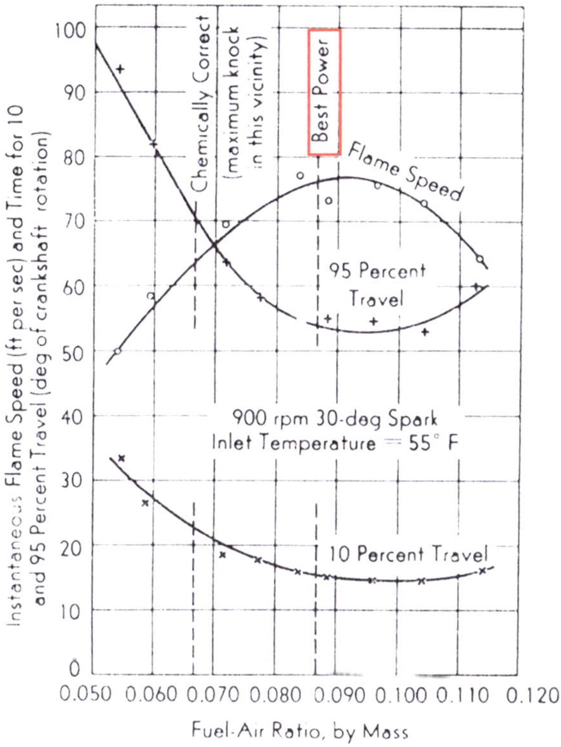

The highest torque occurs near the point of maximum flame speed, at a slightly *richer* AFR than stoichiometric. From Robert Bosch GmbH, Automotive Handbook, 9th edition

Below is a 3-D graphical rendering of an ignition timing map. X axis is RPM, Y axis is manifold vacuum, Z axis is ignition advance in degrees of crankshaft rotation, from peachparts.com :

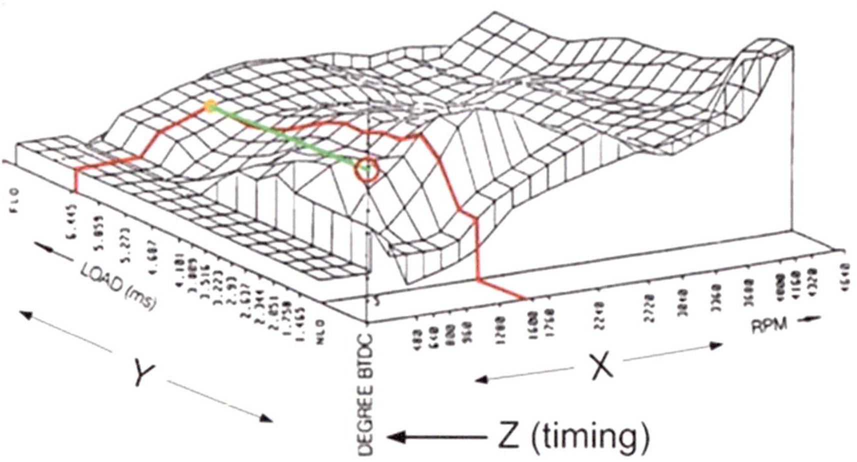

Timing Effects on Emissions and Efficiency

Effect of ignition timing on HC, CO, and NO_x emissions. See Fig. E.1.

Ignition Systems

Creates the spark that initiates the combustion process in each cylinder at exactly the correct crankshaft rotational position.

Inductive (Kettering) Ignition Systems

Uses the abrupt cutoff of electrical current in an inductor to produce a high-voltage spike (Figs. E.2, E.3).

How does an ignition circuit create a high voltage spark? Voltage across an inductor is proportional to the rate of change of current: current $i = L\frac{di}{dt}$. When the contact points are opened (or the power transistor switch is turned off), the current flowing through an inductor is abruptly stopped. A negative high voltage spike occurs since at the moment $\frac{di}{dt}$ is nearly infinite. The high voltage spike in the primary winding of the ignition coil (a transformer) is stepped 100:1 producing a very high voltage spike on the secondary winding that is delivered to a spark plug via the distributor. Typical spark energy is 100 mJ at 10,000–30,000 Volts. Figure E.4 shows a simulation of an inductive ignition circuit.

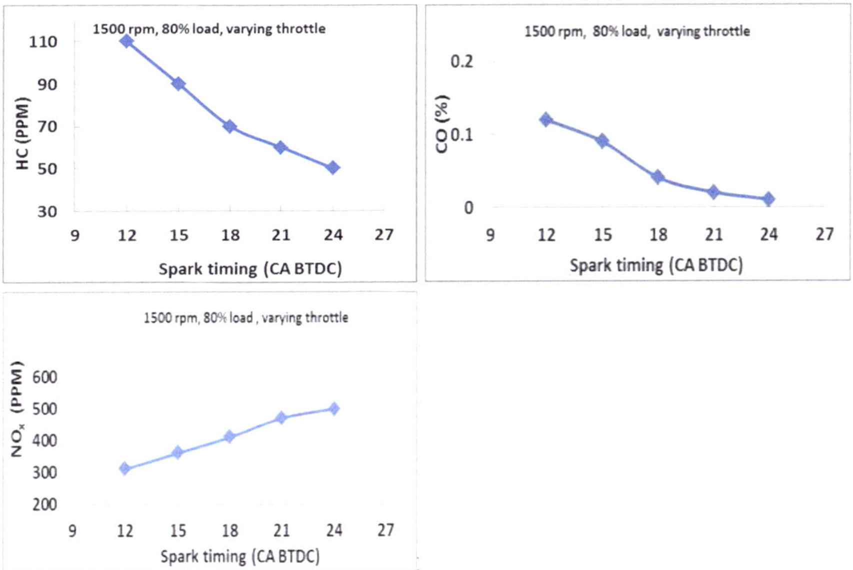

Fig. E.1 MBT (Mean Best Torque) time for these plots was 18 deg BTDC. More advance than MBT timing decreases HC and CO, but increases NO_x. From: http://www.ijritcc.org/download/conferences/ICEMTE_2017/Track_3_(MECH)/1490082908_21-03-2017.pdf

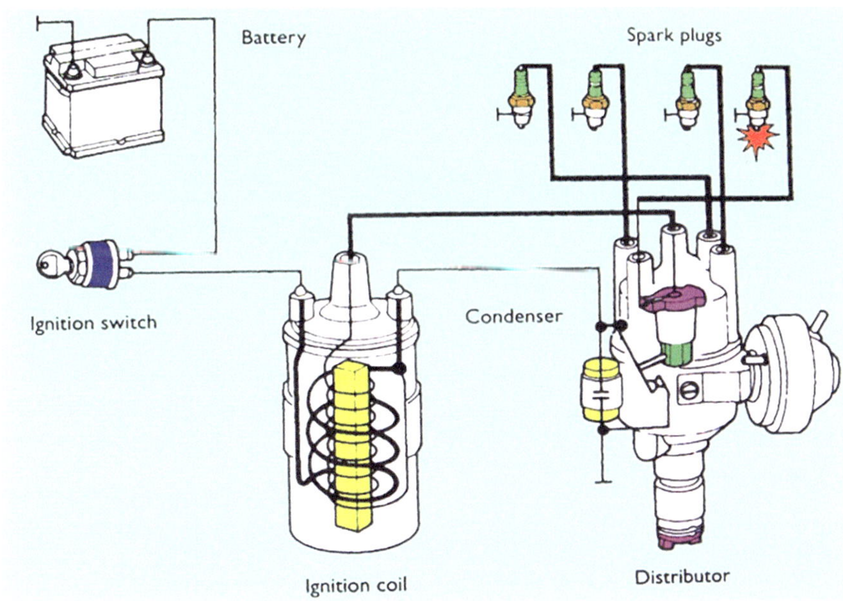

Fig. E.2 An older distributor ignition system. One ignition coil serves all six cylinders, with spark directed to the correct spark plug by the distributor that rotates at the camshaft speed

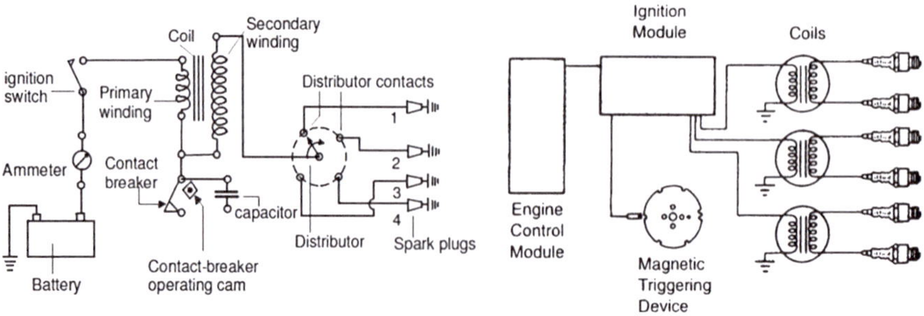

Fig. E.3 Diagram of an inductive (Kettering) ignition system, implemented with a distributor and contact breaker points on the left, and implemented with electronic breaker and individual coils for every pair of cylinders on the right. In either implementation, a spark is generated by opening the contact breaker points, which abruptly cuts off current in the primary winding of the ignition coil. An inductive voltage spike occurs that is stepped up by the coil to cause a high voltage spark at the secondary output of the coil, connected to a spark plug. Diagrams from James Halderman and Tony Martin. Hybrid and Alternative Fuel Vehicles, 2nd ed. Prentice Hall, Pearson Education, 2011

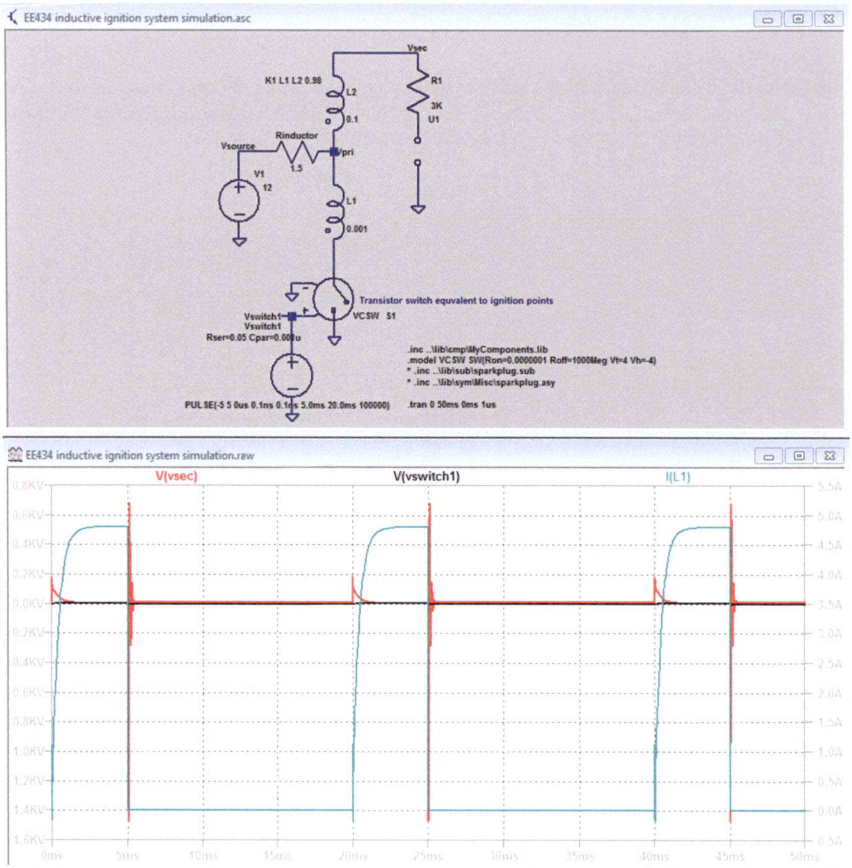

Fig. E.4 LT Spice(R) circuit simulation of inductive ignition system. Blue trace is the current through the primary winding of the ignition coil. Red trace is the primary winding voltage that is stepped up by the coil to as high as 40,000 V if there is no spark plug attached

Capacitive Discharge Ignition (CDI)

The other type of ignition system is much less common than inductive systems. It uses a capacitive current spike rather than an inductive voltage spike to generate the ignition spark.

A large capacitor is charged to 400–500 volts. To trigger an ignition spark, the capacitor is discharged through the primary side of the ignition coil by a large switching transistor turning ON. The abrupt connection causes a rapid voltage rise in the voltage at the secondary side of the coil, due to the 1:100 primary-to-secondary ratio of coil turns. Spark voltages as high as 50,000 V are possible. The very high voltage is preferred for supercharged engines, since the dielectric constant of air increases with pressure. At high boost pressures and/or high compression ratios, the output voltage of an inductive ignition might not be high enough to form an arc. *CD systems are regularly used for methanol-powered racing engines since they often use high supercharge boost pressures and high compression ratios.*

Also, unlike an inductive ignition, the energy is not limited by the tradeoff caused by the inductance of the ignition coil—higher inductance stores more energy, but it requires more time to "recharge" in the dwell time between ignition events. This limits the maximum possible engine speed since the dwell time gets shorter as the engine speed increases. This is more of a problem for distributor-type ignition systems, e.g., for an older V-8 engine, 4 ignition events must occur for every rotation of the crankshaft. In modern single coil per cylinder systems, each spark is separated by the entire 720 degrees of engine rotation in each complete engine cycle.

The main limitation of a CD ignition system, other than higher cost, is the characteristic of the circuit that the spark has a very brief duration, which can be a problem at light loads or idle when the cylinder pressure is lower and the mixture is lean requiring longer spark "burn times" to initiate the combustion flame front (Fig. E.5). The aftermarket ignition modules such as the ones in Fig. E.6 are programmed to produce a train of multiple sparks at low engine speeds. This makes up for the short spark duration by firing several sparks in rapid succession, separated by 1 ms or less.

In modern cars, ignition timing advance is calculated by the engine control computer based on the engine speed, manifold vacuum, and other engine parameters. It is programmable for different fuels and different engine configurations (Fig. E.7).

Timing control is just one of the functions of an integrated Electronic (Engine) Control Module (ECM) or Engine Management Computer. Previously done by a separate control module (Fig. E.8).

In a modern ignition system, separate ignition coils for each cylinder are used, avoiding the need for a distributor and ignition leads, a high-maintenance item (Fig. E.9).

Another input to the ignition controller: the *knock sensor*. Basically a specialized microphone that is attached to the cylinder head via the head bolts. Listens for the distinct vibrations of engine knock. If detected, the ECU instantly retards the ignition timing until the knock goes away. Sometimes causes a *check engine* light to warn the driver that something is wrong; either insufficient fuel octane rating, ignition timing, or a mechanical engine problem (Fig. E.10).

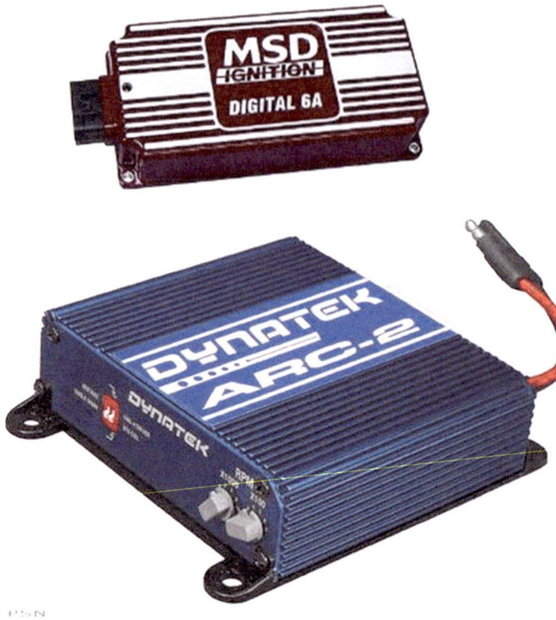

Fig. E.5 Two aftermarket CD ignition systems for racing applications. For the upper module, the company MSD (Multi-spark-discharge) was founded in the 1970s based on its patent of the same name. The ARC-2 system is specifically for four-cylinder motorcycles. Upper photo public domain from advertisement. Lower photo by author

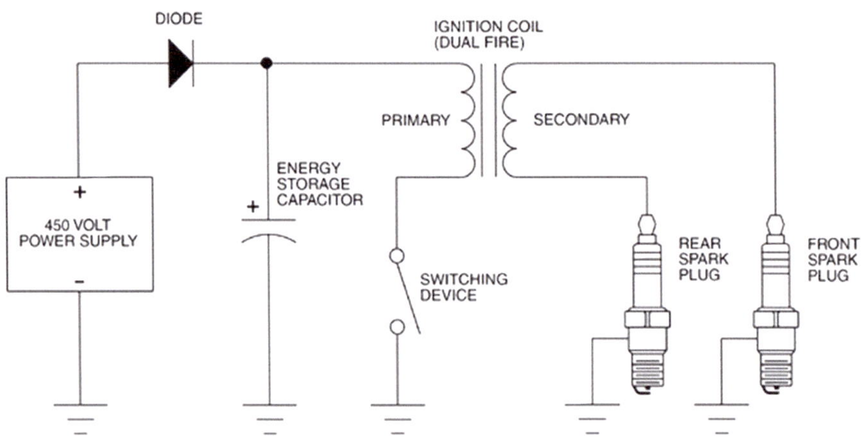

Fig. E.6 Schematic of a capacitive discharge (CD) ignition system. The application shown is a motorcycle, in which one coil is used for two cylinders

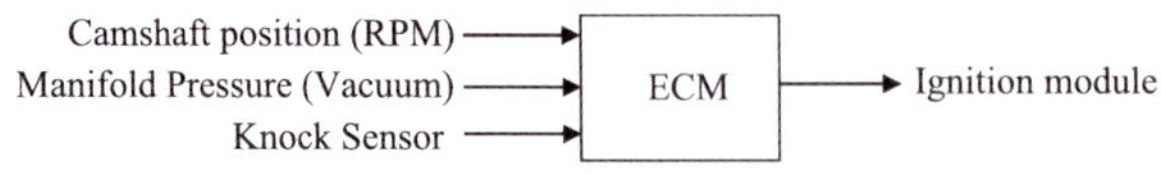

Fig. E.7 Basic diagram of a microprocessor ignition control module and engine sensors required for ignition timing

Fig. E.8 Typical engine control module (ECM) that handles both fuel and ignition control. Photo unattributed

Fig. E.9 Four-coil assembly for an inline 4-cylinder engine. Photo found online, unattributed

Fig. E.10 Knock sensor. From Robert Bosch data sheet. https://www.bosch.co.jp/tms2015/en/products/pdf/2WP_ProductDataSheet_Knock_Sensor_EN_highres_20151023.pdf

Appendix F: The Ideal Otto Cycle

The Otto[5] Cycle—a simplified model of what happens during each of the four strokes of an engine. Allows prediction of engine efficiency and power output

Also Known as a Constant-Volume (Isochoric) Process

Combustion is assumed to occur instantaneously at top dead center, so the volume in the cylinder at that moment is constant. This is not actually the case, but it is a helpful simplification that allows us to calculate the combustion temperature rise and energy release during the expansion stroke. See Fig. F.1.

The complete trip from point 1 back to point 1 occurs over **two** rotations of the crankshaft. We can calculate the theoretical work[6] (energy) or mechanical power (horsepower or kW) produced by the Otto Cycle model as follows.

The network from the ideal 4-stroke Otto cycle is the work produced in the combustion stroke less the work invested during the compression stroke. *The intake and exhaust strokes are ignored.*[7]

$$\text{Network output}\left(\text{mechanical energy}\right) = \text{expansion work}\left(4 \rightarrow 5\right) - \text{compression work}\left(2 \rightarrow 3\right).$$

[5] Nikolaus Otto, 1832–1891 https://en.wikipedia.org/wiki/Nikolaus_Otto.

[6] *Work* = Force applied over a distance. For variable force, $work = \int force(x)dx$.

[7] Energy is defined as the capacity to do *work*.

C. A. MacCarley, *Non-Petroleum Automotive Transportation*,
https://doi.org/10.1007/978-3-031-85028-8

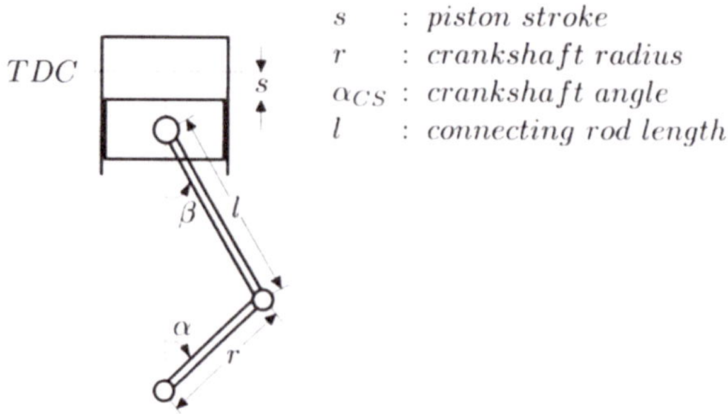

From: Kiencke, U. and Nielsen, L. "Automotive Control Systems", Springer Verlag and SAE International, 2000

Power (rate of energy flow in kW) = (network per 4 cycles in kJ × Engine Speed in RPM)/120.

Work has units of energy: Joules, Watt-hours or Horsepower-hours

Power has a unit of energy per unit time: Joules per second = Watts, horsepower

Engine power output is universally rated in with kilowatts (kW) or horsepower (HP)

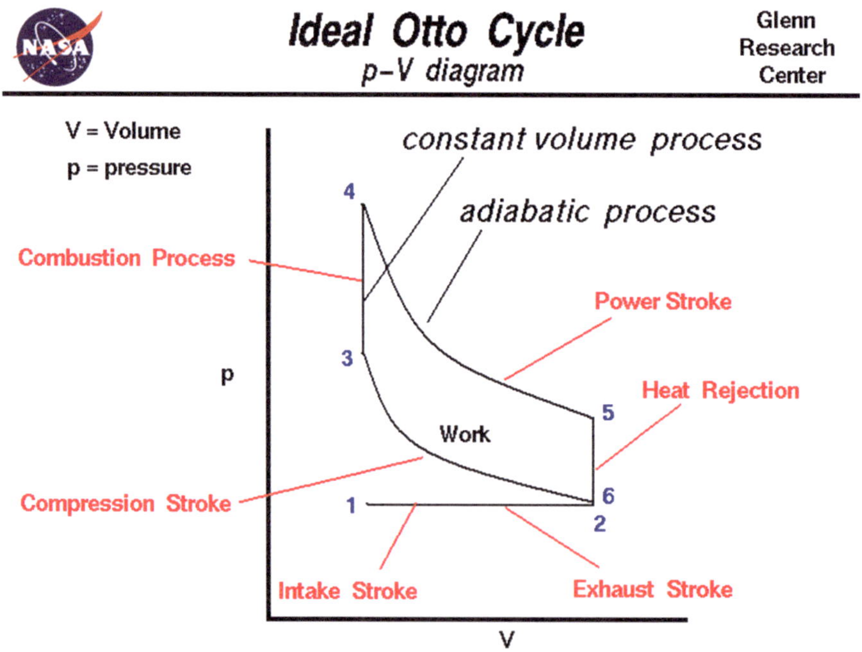

Fig. F.1 Ideal constant volume (Otto cycle) engine model. Graphic from NASA, Diagram below from: http://www.grc.nasa.gov/WWW/K-12/airplane/otto.html. Public domain

US horsepower = 550 foot-pounds per second = 0.745 kW
Metric horsepower = 75 meter-Newtons = 0.735 kW = 0.987 × 1 US horsepower

We make ***a lot of idealizing assumptions*** in the interest of computational simplicity:

1. Ideal valves: Intake and exhaust valves open and close instantaneously.
2. 100% Volumetric efficiency: The air pressure in the cylinder while the intake valve is open is the same as atmospheric air pressure.
3. No exhaust back-pressure: the pressure in the cylinder while the exhaust valve is open is the same as atmospheric air pressure.
4. Instantaneous combustion: The combustion occurs in zero time at exactly TDC.
5. Ideal gases: Both the intake air/fuel and the exhaust products are ideal gases.
6. Closed system: The compression and expansion processes occur so quickly that no heat is lost to the environment: it is an ***isentropic*** or closed system.
7. Constant heat capacities: The ***heat capacity at constant volume*** for the intake of air and fuel is constant, even though the pressure and temperature are changing.

From: **Table_of_specific_heat_capacities** $c_{v,\text{air}} = 20.85 \dfrac{\text{J}}{\text{mole} \cdot \text{K}}$

8. Constant ratio of specific heats: The **ratio of specific heats** $\gamma = \dfrac{c_p}{c_v} = 1.4$ during compression, and γ = 1.2 during the much higher temperatures during expansion, only slightly affected by temperature.[8]

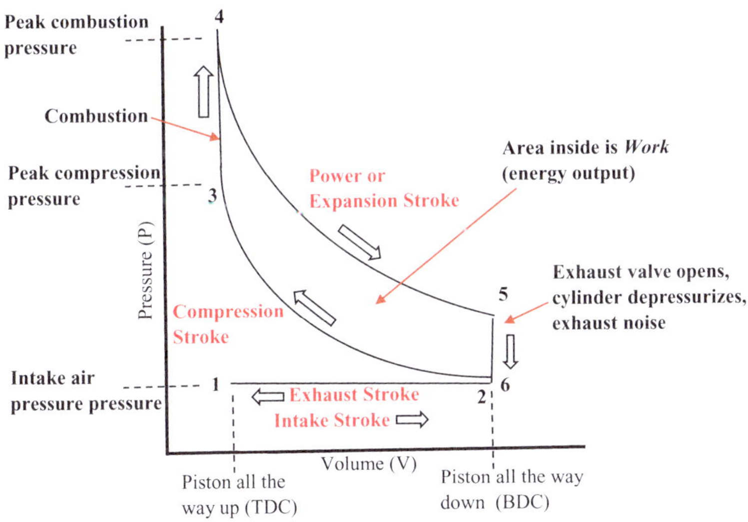

[8] http://en.wikipedia.org/wiki/Heat_capacity_ratio.

Compression stroke work:

$$w_{\text{compression}} = n_{\text{intake}} c_{v,\text{air}} \left(T_3 - T_2\right)$$

(The heat capacity equation)

Expansion stroke work:

$$w_{\text{expansion}} = n_{\text{exhaust}} c_{v,\text{exhaust}} \left(T_4 - T_5\right)$$

Then the total energy output per complete cycle of the engine is just the difference:

$$w_{\text{net}} = w_{\text{expansion}} - w_{\text{compression}}$$

Work from the expansion stroke less the invested in the compression stroke.

To find $w_{\text{expansion}}$ or $w_{\text{compression}}$ we would usually find the force on the piston over its travel during compression or expansion. Then we integrate over the piston travel distance to find the total work. But there is an **easier way** that avoids integration, using only the change in system energy at each state (numbered point) in the cycle diagram.

Notes:

- Assumptions 6–8 are mild and do not make the model unrealistic.
- Assumptions 1–4 defy the realities of mechanical or chemical systems and make the model overly idealized.

The 9-step procedure to find the **power or efficiency** of the Otto cycle for any engine or fuel.

1. Determine the **compression work** $w_{\text{compression}}$. First we find the temperature rise due to rapid (adiabatic) compression during the compression stroke, as the piston goes from the bottom of the cylinder to the top of the cylinder.

We use a variation of the *Ideal Gas Law,* to find the final (fully compressed) temperature T_3 of the intake air/fuel:

$$T_3 = \left(\frac{V_{\text{TDC}}}{V_{\text{disp}} + V_{\text{TDC}}}\right)^{1-\gamma} T_2 = \varepsilon^{\gamma-1} T_2 \text{ Kelvins}$$

where T_2 = ambient air temperature = 295 K (22 °C)T_5 = temperature of air and fuel mixture at top of compression stroke

$$\varepsilon = \frac{V_{\text{disp}} + V_{\text{TDC}}}{V_{\text{TDC}}} = \text{compression ratio, from the geometry of the engine}$$

$$\gamma = \frac{c_p}{c_v} = 1.4$$ during compression—we assume that the intake charge is almost entirely air.

2. We need to know the **mass of air and fuel** that were subjected to this compression—the displacement of the cylinder.

The number of moles of air and fuel vapor in the cylinder *just prior to the start of compression* is the displacement of the cylinder V_{disp} times the number of moles per unit volume at the ambient air temperature. The air/fuel mixture is treated as an ideal gas:

$$n_{air} = \frac{pV}{RT} = \frac{p(\text{atm}) \times V(L)}{R\left(\text{Gas constant} \frac{L \cdot \text{atm}}{K \cdot \text{mole}}\right) \times T(K)} \quad \text{moles of air and vaporized}^{*} \text{ fuel.}$$

Notice that the number of moles of vaporized fuel is much less than the number of moles of air, so that n_{fuel} is often ignored in this calculation. Most simple analyses just use n_{air} for the total intake mixture.

*For purposes of determining pressure rise due to compression, we treat the entire intake charge as an ideal gas.

How to find the number of moles of fuel n_{fuel} ? Find the number of moles of fuel for a stoichiometric mixture with n_{air} But AF ratios are usually reported as mass ratios, so we need to translate the mass ratio to a molar ratio.

Air weighs about 29.0 g/mole, so

$$(n_{air} \text{ moles air})(\rho \text{ g / mole of air}) = m_{air} \text{ g air}$$

Look up or calculate the Air/Fuel mass ratio for the fuel you are using. For example, for a stoichiometric mixture of air and gasoline, we find the mass air/fuel ratio (AFR) = 14.7 from the Bosch Handbook:

$$m_{fuel}(g) = \frac{1}{\text{AFR } (\text{g air / g fuel})}(m_{air} \text{ g air})$$

Then look up or calculate the molar density of the fuel. For example, average gasoline $C_{7.68}H_{12.9}$ weighs 105 g/mole. So the number of moles of fuel is

$$\frac{m_{fuel}(\text{g fuel})}{\rho(\text{g / mole})} = n_{fuel} \text{ moles fuel}$$

Finally, the total number of moles of air and liquid fuel entering the cylinder is

$$n_{intake} = n_{air} \text{ moles air} + n_{fuel} \text{ moles fuel} = n_{intake}$$

$$= \text{total moles of air + fuel inducted during the intake stroke}$$

In this simplified analysis, we ignore the fuel moles assuming it is small and in incompressible liquid form. We assume that the entire intake charge is just air. i.e., $n_{intake} = n_{air}$. For modern direct-cylinder injected (DCI) engines, this is actually correct.

But, n_{fuel} is still worth calculating here since we will need to know m_{fuel} later when we calculate the energy released during combustion.

3. Then using the heat capacity equation, the **compression work** in Joules is

$$w_{\text{compression}}(J) = (n_{\text{intake}}\ \text{moles})(c_{v,\text{air}}\ \text{J/mole}-K)\ (T_3 - T_2\ \text{temp rise in Kelvins})$$

where $c_{v,\text{air}} = 20.85 \dfrac{\text{J}}{\text{mole}\cdot\text{K}}$ for air

4. **Temperature after combustion**: At maximum compression when the piston reaches TDC, combustion is assumed to instantaneously convert all the fuel and oxygen into CO_2 and H_2O vapor, with the release of heat $Q_{\text{combustion}}$ which is the **net calorific value** (aka fuel energy density, lower heating value, LHV) of the fuel times the mass of the fuel in the air/fuel charge.

Properties of liquid fuels and hydrocarbons LHV

Substance	Density	Main constituents	Boiling temperature	Specific heat of evaporation	Net calorific value
	kg/l	% by weight	°C	kJ/kg	MJ/kg
Gasoline,	.748				avg gasoline = 41.5 MJ/kg
Regular	0.720...0.775	86 C, 14 H	25...210	380...500	41.2...41.9
Premium	0.720...0.775	86 C, 14 H	25...210	–	40.1...41.6
Aviation fuel	0.720	85 C, 15 H	40...180	–	43.5
Kerosene	0.77...0.83	87 C, 13 H	170...260	–	43 42.07
Diesel fuel	0.820...0.845	86 C, 14 H	180...360	≈ 250	42.9...43.1

Fuel Energy (Net Calorific Value) from Bosch Automotive Handbook, 10th ed. 2018.

With this instantaneous addition of heat, we use the *Heat Capacity Equation to* find the final temperature T_4 of the nitrogen, carbon dioxide, and water vapor after combustion:

$$T_4 = \frac{(Q_{\text{combustion}}\ J)}{(n\ \text{moles})\left(c_v \dfrac{J}{\text{mole}-K}\right)} + T_3\ \text{Kelvins}$$

For regular gasoline, from Bosch 10th ed., Q = 41.5 MJ/kg

5. **Temperature after compression:** We want to calculate the work $w_{\text{expansion}}$ done by the piston during the expansion stroke. First, we find the temperature T_5 of the air/fuel in the cylinder after it has fully expanded from TDC to BDC:

$$T_5 = \left(\frac{V_{\text{disp}} + V_{\text{TDC}}}{V_{\text{TDC}}}\right)^{1-\gamma} T_4 = \varepsilon^{1-\gamma} T_4$$

6. Then we use the temperature rise to find the amount of **work done during compression**:

$$w_{\text{expansion}}(J) = n_{\text{exhaust}}(\text{moles})\ c_{v,\text{exhaust}}\ (\text{J/mole}-K)\ (T_4 - T_5)(\text{Kelvins})$$

Note: The gases in the cylinder have changed from before combustion (nitrogen, oxygen, and fuel) to after combustion (nitrogen, carbon dioxide, and water vapor). The total number of moles and heat capacity n_{exhaust} (moles) and $c_{v,\text{ exhaust}}$ (J/mole − K) after combustion are slightly different than before the combustion $n_{\text{air}}(moles)$ and $c_{v,\text{ air}}$ (J/mole − K) . These will be treated more rigorously in the example below, but for approximation purposes, they will be assumed to be the same here.

7. Find **network** over all four strokes during two rotations of the crankshaft:

$$w_{\text{net}} = w_{\text{expansion}} - w_{\text{compression}}$$

8. **Power** is work per unit time. To find engine power output, multiply this by the number of times it happens every second:

$$\text{Power}(\text{kW}) = (w_{\text{net}}\text{ Joules})\left(\frac{1\,\text{power stroke}}{2\,\text{engine rotations}}\right)(RPM\,\text{revs / min})\left(\frac{\text{min}}{60\text{s}}\right)\left(\frac{1\text{kW}}{1000\,\text{W}}\right)$$

If preferred, multiply the power in kW by 1.34 to get HP.

However, the engine is not ideal. The net usable power output is the theoretical power output reduced by a mechanical efficiency factor due to non-ideal valve timing, frictional, thermal, vibration, and acoustic losses, so it is usually multiplied by a mechanical efficiency factor that is highly dependent on the design of the engine. This is highly variable over all different engines, from stationary irrigation engines to high-performance racing engines, but is typically in the range of 60–80% (due to 20–40% non-ideal losses). The usable output is traditionally referred to as the net *Brake Power* output:

$$w_{\text{net}}(\text{Brake Power}) = \text{ideal Power}(\text{kW}) \times \eta_{\text{mechanical}}\ (kW)$$

9. The overall ability of the engine to convert the chemical energy of the fuel into rotational output energy is the **thermal efficiency** η_T found by dividing the brake/net energy output w_{net} by the fuel energy input $Q_{\text{combustion}}$:

$$\eta_T(\text{efficiency}) = \left(\frac{w_{\text{net}}\text{ Joules}}{Q_{\text{combustion}}\text{ Joules}}\right)$$

A Complete Example—A 1.0 L Single-Cylinder Gasoline Engine

1. Adiabatic heating during the *compression stroke*: Find T_3, the temperature of the compressed air-fuel mixture at the end of the compression stroke (TDC). This does not depend on the composition of the mixture being compressed as long as we can reasonably assume ideal gases and adiabatic compression. From the Ideal Gas Law,

$$T_3 = \left(\frac{V_{\text{TDC}}}{V_{\text{disp}} + V_{\text{TDC}}} \right)^{1-\gamma} \times T_2 = \varepsilon^{\gamma-1} T_2 \text{ in K}$$

where $\varepsilon = \dfrac{V_{\text{disp}} + V_{\text{TDC}}}{V_{\text{TDC}}} = \text{compression ratio}$

and

$$\text{T}_2 = \text{ambient air temperature} = 295 \text{ K}$$

For a cylinder with 1 L displacement and a 10:1 compression ratio ε: $V_{\text{disp}} = 1$ L and $V_{\text{TDC}} = 0.111$ L

$$T_3 = \left(\frac{V_{\text{TDC}}}{V_{\text{disp}} + V_{\text{TDC}}} \right)^{1-\gamma} T_2 = \varepsilon^{\gamma-1} T_0$$

$$T_3 = 10^{1.4-1} (295°\text{K}) = 741 \text{ K} (\text{or } 468 \text{ °C})$$

Note 1: This ignores the cooling effect of fuel vaporization during the compression stroke and in the induction system after carburetion or injection. In reality, vaporization prior to intake valve closure increases the density of the initial air-fuel mixture by decreasing the temperature (an intercooling effect), while vaporization during compression absorbs heat decreasing T_{initial} at the start of combustion.

Note 2: The compression process and temperature distribution in the cylinder are not uniform, which is good since the auto-ignition temperature of regular gasoline is only 573 K.**2. What number of moles of intake air and fuel** in the volume $V_{\text{disp}} = 1$ L at ambient temperature and pressure? For air:

$$n_{\text{air}} = \frac{pV}{RT} = \frac{(1 \text{ atm})(1 \text{ L})}{\left(0.08206 \dfrac{\text{L} \cdot \text{atm}}{\text{K} \cdot \text{mole}}\right)(295 \text{ K})} = 41.3 \times 10^{-3} \text{ moles of air}$$

If we assume that the gasoline is entirely in liquid form when it is inducted, we should add the moles of fuel n_{fuel} in the intake mixture.

Air weighs about 29.0 g/mole, so

$$(41.3 \cdot 10^{-3} \text{ moles air})(29.0\text{g} / \text{mole}) = 1.20\text{g air}$$

For a stoichiometric mixture, the mass air/fuel ratio (AFR) = 14.7

$$m_{\text{fuel}}(g) = \frac{\text{g fuel}}{14.7 \text{ g air}} (1.20\text{g air}) = \mathbf{81.6 \cdot 10^{-3} \text{g fuel}}$$

Average gasoline $C_{7.68}H_{12.9}$ weighs 105 g/mole. So the number of moles of fuel is

$$\frac{\mathbf{81.6 \cdot 10^{-3}\,g\,fuel}}{\mathbf{105\ \ g / mole}} = \mathbf{0.777 \cdot 10^{-3}\ \ moles\,fuel}$$

$$n_{\text{intake}} = n_{\text{air}} + n_{\text{fuel}} = 41.3 \cdot 10^{-3}\ \text{molesair} + 0.777 \cdot 10^{-3}\ \text{moles fuel} = \mathbf{42.1 \cdot 10^{-3}\ \ moles}$$

3. Compression work

$$w_{compression} = n_{\text{intake}} c_{v,air}\left(T_3 - T_2\right) = 0.0421\ \text{moles} \cdot \left(20.85 \frac{\text{J}}{\text{mole} \cdot \text{K}}\right)\left(741K - 295K\right) = 392\ \text{J}$$

Temp after compression (T_3) — Ambient air temp (T_2) — C_v for air (20.85)

4. Combustion heat release

Now we seek T_4, the temperature that results from the assumed instantaneous combustion of the fuel at the start of the power or expansion stroke. This is the leg of the PV diagram between points 3 and 4 in the diagram.

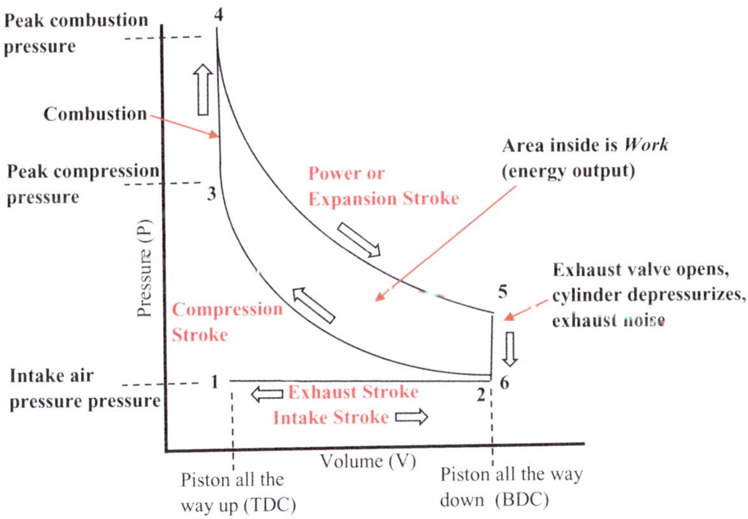

We assume that all of the heat from the combustion is added to the compressed fuel-air volume at TDC, converting it completely into exhaust products.

$$Q_{\text{combustion}} = Q_{\text{fuel}}$$

To find T_4 we need to know Q_{fuel}. This is the fuel energy (net calorific value, or lower heating value), inducted during the intake stroke. Assuming a stoichiometric AFR, the mass of fuel (gasoline) for inducted into the cylinder of the 1 L engine was previously calculated:

$$m_{\text{fuel}} = \mathbf{81.6 \cdot 10^{-3} g} \ (\text{gasoline without ethanol,or E0})$$

How much energy is this? From Bosch's book, an average value for gasoline: 41.5 MJ/kg

$$Q_{\text{combustion}} = Q_{\text{fuel}} = (41.5 \ \text{kJ/g})(81.6 \times 10^{-3} \ \text{g fuel})$$

$$= \mathbf{3.39 kJ} \ \text{released in the combustion.}$$

5. We can now find the **combustion temperature rise** due to the combustion by rearranging the previous heat capacity equation:

$$T_4 = \frac{Q_{\text{combustion}}}{nc_v} + T_3$$

where

$Q_{\text{combustion}}$ is the heat released in the combustion, in kJ.

c_v is the molar heat capacity of the combustion products, in $\frac{\text{J}}{\text{mole} \cdot \text{K}}$.

n is the number of moles of combustion end-products.

T_3 is the temperature of the air and fuel vapor before combustion, in K (Kelvins).

T_4 is the temperature of the exhaust gases immediately after combustion, in K (Kelvins).

In the combustion process, the intake oxygen and fuel are converted into exhaust carbon dioxide and water vapor.

In approximations, the number of moles of the exhaust is often assumed to be the same as the number of moles of the intake.

But these are actually different. For more accurate calculations, we should calculate

1. The different number of moles of exhaust compared with air + fuel
2. The new c_v which for the exhaust gases which are different than the air/fuel mixture

Here is how:

The number of moles n and the specific heat c_v in the heat capacity equation above refer to the **exhaust gas (nitrogen, water, and carbon dioxide)** in the cylinder after combustion rather than the air and fuel before the combustion.

The total **mass is preserved** before and after combustion, but the number of moles of the exhaust (water vapor, carbon dioxide, and nitrogen) is different than the number of moles of intake charge (oxygen, nitrogen, and fuel):

n_{exhaust} is not the same as n_{intake}

Also, since the substances are different, $c_{v,\text{ exhaust}}$ is not the same as $c_{v,\text{ intake}}$.

To be accurate, we need to know n_{exhaust} and $c_{v,\text{ exhaust}}$ for the exhaust products.

Combustion reaction: $C_{7.68}H_{12.9} + 10.9O_2 \rightarrow 7.68\ CO_2 + 6.45H_2O$ (stoichiometric, EPA gasoline).

But we also must include the non-reacting gases in the air, almost entirely nitrogen. Then the *complete* air-fuel combustion equation is[9]

$$C_{7.68}H_{12.9} + 10.9\left(O_2 + \frac{79\%}{21\%}N_2\right) \rightarrow 7.68\ CO_2 + 6.45\ H_2O + 41.0\ N_2$$

or

$$C_{7.68}H_{12.9} + 10.9O_2 + 41.0\ N_2 \rightarrow 7.68\ CO_2 + 6.45\ H_2O + 41.0\ N_2$$

$$51.9\,\text{moles air} \rightarrow 55.1\,\text{moles exhaust}$$

For 1 L intake volume containing 41.3×10^{-3} moles of air, we get[10]

$$\left(41.3 \times 10^{-3}\,\text{moles air}\right)\left(\frac{55.1}{51.9}\right) = \mathbf{43.8 \times 10^{-3}\,moles\ exhaust} = n_{\text{exhaust}}$$

Now we find the heat capacity $c_{v,\,\text{exhaust}}$ for the exhaust gas.

In the exhaust, the different gases have different heat capacities:

Water vapor: $c_{v,H_2O} = 28.03\dfrac{J}{\text{mole} \cdot K}$

Carbon dioxide: $c_{v,CO_2} = 28.46\dfrac{J}{\text{mole} \cdot K}$ (heat capacities for CO_2 and H_2O are different than air)

Nitrogen: $c_{v,N2} = 20.80\dfrac{J}{\text{mole} \cdot K}$ (nearly same as air)

We find a weighted average $c_{v,\text{exhaust}} = \left(\dfrac{n_{CO_2}c_{v,CO_2} + n_{H_2O}c_{v,H_2O} + n_{N_2}c_{v,N_2}}{n_{CO_2} + n_{H_2O} + n_{N_2}}\right)$ for this mix of gases:

$$c_{v,\text{exhaust}} = \left(\frac{\left(6.11 \cdot 10^{-3}\ \text{moles}\ CO_2\right)\left(28.46\dfrac{J}{\text{mole} \cdot K}\right) + \left(5.13 \cdot 10^{-3}\ \text{moles}\ H_2O\right)\left(28.03\dfrac{J}{\text{mole} \cdot K}\right) + \left(32.6 \cdot 10^{-3}\ \text{moles}\ N_2\right)\left(20.80\dfrac{J}{\text{mole} \cdot K}\right)}{6.11 \cdot 10^{-3}\ \text{moles}\ CO_2 + 5.13 \cdot 10^{-3}\ \text{moles}\ H_2O + 32.6 \cdot 10^{-3}\ \text{moles}\ N_2}\right)$$

[9] Air consists of 21% O_2, 78% N_2 and 1% Argon *by volume or molar ratio*. We ignore the 0.04% CO_2.

[10] Compare this with 41.3×10^{-3} moles of intake air + fuel. The mass is the same, but there are slightly more moles of exhaust products than moles of intake air mixture.

where n_{CO_2} and n_{H_2O} and n_{N_2} are the respective number of moles of each exhaust product, calculated above.

$c_{v,\text{exhaust}} = 22.7\frac{\text{J}}{\text{mole}\cdot\text{K}}$ average heat capacity at constant volume for exhaust gases.[11]

Now, **at last** we can find the ***theoretical*** peak temperature after combustion:

$$T_4 = \frac{Q_{\text{combustion}}}{nc_v} + T_3 = \left(\frac{3.39\ \text{kJ}}{\left(43.8\cdot 10^{-3}\ \text{moles exhaust}\right)\left(22.7\frac{\text{J}}{\text{mole}\cdot\text{K}}\right)}\right) + 741°\text{K}$$

$$= 4151\ \text{K}\left(\text{really high}^*\right)$$

Note 1: Actual combustion is not instantaneous but occurs over many degrees of crankshaft rotation.

Note 2: We have ignored heat loss to the cylinder walls and other nonidealities that will significantly reduce this temperature.

6. Expansion Work

The only useful work is generated during the expansion (or power) stroke between points 4 and 5 on the diagram. The hot combustion gases expand while exerting pressure on the piston, pushing it down.

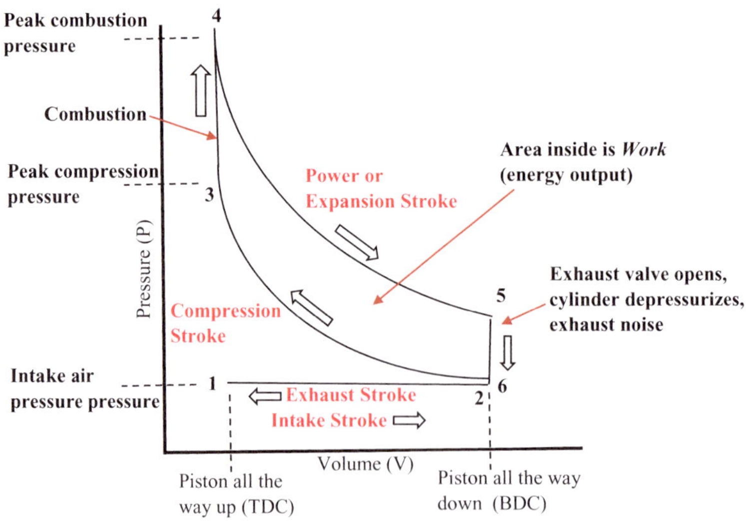

[11] Heat capacity of air is 20.85(J/mole · K). Since the difference is small, approximate that $c_{v,\text{exhaust}} = c_{v,\text{air}}$.

We again use the ideal gas law to find the final temperature T_5 just before the exhaust valve opens:

$$T_5 = \left(\frac{V_{\text{disp}} + V_{\text{TDC}}}{V_{\text{TDC}}} \right)^{1-\gamma} T_4 = \varepsilon^{1-\gamma} T_4$$

At these high temperatures, the ratio of specific heats is a bit lower, usually about $\gamma \approx 1.20$.

For our example,

$$T_5 = 10^{(1-1.2)} \cdot 4151 = \mathbf{2619\ \ K}$$

We can now calculate the work output from the expansion stroke from the difference in these temperatures.

$$w_{\text{expansion}} = n_{\text{exhaust}} c_{v,exhaust} \left(T_4 - T_5 \right) = (43.8 \cdot 10^{-3}\ moles\ exhaust) \left(22.7 \frac{\text{J}}{\text{mole} \cdot \text{K}} \right) \left(4151 - 2619^{\circ} K \right) = 1523\ \text{J}$$

High temperature at start of expansion (T_4); Lower temperature at end of expansion (T_5)

7. Net energy output per power stroke

The network from one complete Otto Cycle (four strokes, two rotations of the crankshaft) is $w_{\text{net}} = w_{\text{expansion}} - w_{\text{compression}} = 1523\ \ \text{J} - 392\ \ \text{J} = \mathbf{1131\ J}$8. Power Output (Flow rate of mechanical energy from the engine)

Expansion work occurs once every *two* rotations of the engine.

If we assume that the maximum power output of this engine occurs at 6000 RPM, then with our calculated amount of work produced every other rotation (because it's a 4-stroke engine), the Otto model power output for our 1 L engine would be

$$\text{power} = \left(\frac{1131\ \text{J}}{2\ \text{revs}} \right) \left(\frac{6000\ \text{revs}}{\text{min}} \right) \left(\frac{1\ \text{min}}{60\ \text{s}} \right) = 56.6 \cdot 10^3 \frac{\text{J}}{\text{s}} = \mathbf{56.6\ \ kW}$$

$$\text{or in US units} \quad (56.6\ \text{kW}) \left(\frac{HP}{0.746\ \text{kW}} \right) = \mathbf{75.8\ \ HP(US)}$$

Check the reasonableness of 75 HP/L output by comparing it with two actual automotive engines.

Large 70's V8:

1972 Ford 5.8 L V8: 153 HP. **26.4 HP/L** @ 5500 RPM (much less than calculated)[12]

1.0 L 4-cyl motorcycle engine:

[12] Horsepower rating of Ford 5.7 L Windsor V8 at http://www.f150hub.com/specs/351w.html.

2001 Suzuki GSXR 1000: **160 HP/L** @ 9500 RPM (much greater than calculated)[13]

9. Thermal Efficiency

The ability of the engine to convert the chemical energy of the fuel into mechanical work is the *thermal efficiency* η_T. In this ideal example,

$$\frac{w_{\text{net}}}{Q_{\text{fuel}}} = \frac{1.131\ \text{kJ}}{3.39\ \text{kJ fuel energy}} = 0.334 = 33.4\%$$

As discussed previously, engine-specific mechanical losses will usually reduce both the calculated power output and efficiency by the mechanical efficiency factor $\eta_{\text{mechanical}}$, which typically is between 80% and 90% for modern engines. So the actual *brake efficiency* of this hypothetical 1 L engine might be

$$\eta_{\text{brake}} = \eta_{t,\text{theoretical}} = 33.4\% \times 0.90 = 30.1\%$$

This is the number that when multiplied by the vehicle drivetrain, air resistance, and road losses, determines the gas mileage of the vehicle. It is a normalized value that is independent of the fuel used, since we have assumed the ideal instantaneous heat addition of a given quantity of fuel energy between the compression stroke and the expansion stroke. In practice, however, the efficiency is affected by the choice of fuel, entirely due to the factors not considered in the ideal Otto Cycle calculation.

But there is one engine-specific theoretical factor that is considered in the Otto Cycle calculation, that very significantly affects both power output and efficiency…

Effect of Compression Ratio on Power and Efficiency

For the Otto Cycle, subject to (many) simplifying assumptions, thermal efficiency can be simply related to the compression ratioε:

$$\eta_T = 1 - \frac{1}{\varepsilon^{\gamma-1}}$$

This relationship is not accurate for calculating efficiency in an absolute sense as a function of the compression ratio, since it includes many simplifications in the most simplified version of the Otto cycle calculations. But it is an excellent simple formula for predicting the ***relative*** change in efficiency or power for a given engine and fuel if changes are made in the engine compression ratio ε.

[13] Horsepower output of Suzuki 1.0 L motorcycle engine at https://en.wikipedia.org/wiki/Suzuki_GSX-R1000.

For our example, $\varepsilon = 10:1$ CR and using $\gamma = 1.2$

$$\eta_T = 1 - \frac{1}{10^{1.2-1}} = 0.369 = 36.9\% \ \left(\text{our previously calculated value was } 33.4\%\right)$$

But consider why racing engine builder almost always increase the engine and use racing fuels with a higher octane than gasoline.

Suppose the 8:1 stock CR of an engine is increased to 11:1, by replacing the pistons or milling the cylinder head:

$$\frac{\eta_{T,11}}{\eta_{T,8}} = \frac{1 - \frac{1}{11^{1.2-1}}}{1 - \frac{1}{8^{1.2-1}}} = \frac{0.381}{0.340} = 1.12 = 12\% \quad \text{relative increase in power}$$

for the same amount of fuel.

So if the power output of the 8:1 engine was 300 HP, you could expect an increase to 336 HP for the same fuel quantity.[14] Most alternative SI engine fuels allow higher CRs than gasoline.

Real Engine Cycles

Typical SI engines only approximately follow the Otto cycle, and generally never achieve Otto cycle efficiency, because the model ignores:

- The finite burn time of the fuel-air charge—not an instantaneous isochoric process
- Realistic valve opening/closing times
- The volume occupied by the fuel in the intake charge
- Pumping or throttling losses (vacuum load at part throttle) during intake
- Exhaust back pressure (or vacuum if tuned resonance can be achieved)
- Heat loss through the cylinder walls to the coolant and air
- Gas dynamics and non-uniform pressure/density distribution for intake charge
- The "supercharge" effect of a fuel vaporization
- Mechanical friction losses

A modified PV diagram to better model non-instantaneous combustion:

[14]This change in power output occurred when in 1971 Ford introduced the Boss 351, a 11:1 CR version of the previous 8:1 CR "High Output Cleveland" 351 engine.

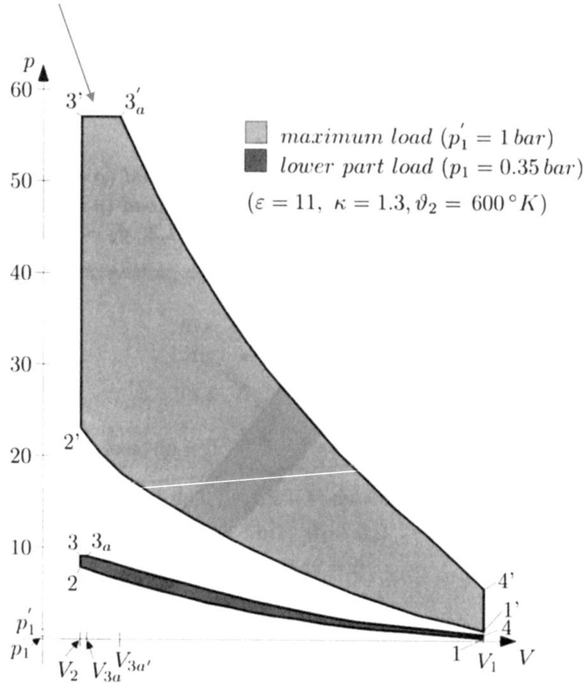

Diagram from Bosch Automotive Handbook, 10th ed. 2018. Two cases are shown: one at idle (bottom) and one at full load (top), with the combustion peak modified (flat top) to be partially a constant pressure cycle. But it still assumes instantaneous combustion at TDC and ideal valve opening and closing.

Effects of Non-ideal Combustion on the Ideal Otto Cycle

Pressure rise during combustion does ***not*** occur instantaneously at TDC (NOT really an isochoric process). What really happens?

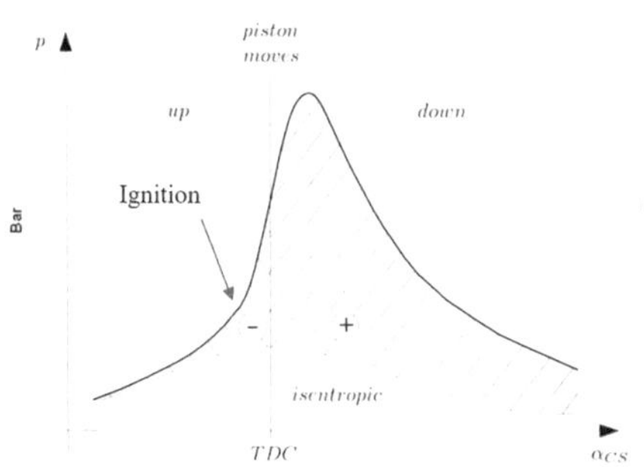

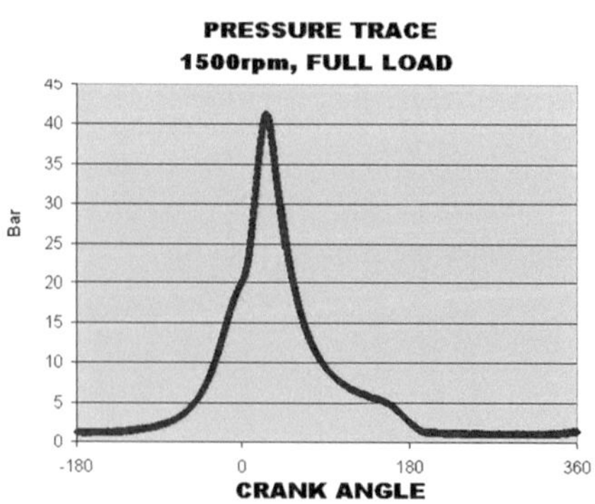

What is the relationship between cylinder pressure and engine torque?

$$HP = \frac{\text{Torque}(\text{lb}-\text{ft}) \cdot \text{RPM}}{5252}$$

A more realistic PV diagram from engine simulation which does not assume ideal valves or 100% volumetric efficiency (from Bosch Text)

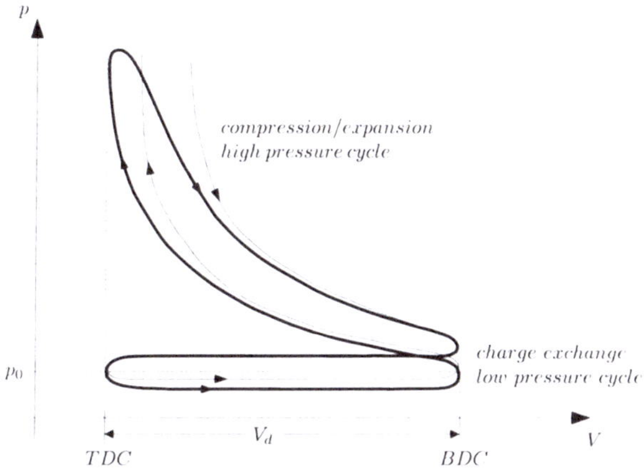

Actual Cylinder PV Pressure Traces

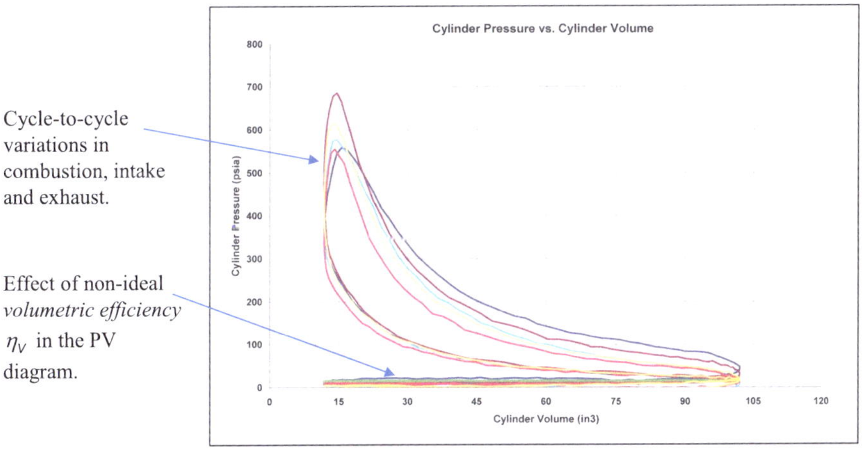

References for Appendices

1. Wood gas (Chem Europe Energy Portal), https://www.chemeurope.com/en/encyclopedia/Wood_gas.html. Accessed 1 July 2024.
2. R. Bosch, *Automotive Handbook*, 10th edn. (SAE International, 2018)
3. Gasoline standards. (US Environmental Protection Agency), https://www.epa.gov/gasoline-standards Accessed 6 July 2024.
4. R. Bosch, Properties of liquid fuels and hydrocarbons, in *Automotive Handbook*, 10th edn., (SAE International, 2018), p. 345. https://www.sae.org/publications/books/content/bosch10/
5. R. Bosch, *Properties of gaseous fuels*, in *Automotive Handbook*, 10th edn., (SAE International, 2018), p. 346

C. A. MacCarley, *Non-Petroleum Automotive Transportation*,
https://doi.org/10.1007/978-3-031-85028-8

Index

C. A. MacCarley, *Non-Petroleum Automotive Transportation*,
https://doi.org/10.1007/978-3-031-85028-8

MIX
Papier aus verantwortungsvollen Quellen
Paper from responsible sources
FSC® C105338

If you have any concerns about our products,
you can contact us on
ProductSafety@springernature.com

In case Publisher is established outside the EU,
the EU authorized representative is:
Springer Nature Customer Service Center GmbH
Europaplatz 3, 69115 Heidelberg, Germany

Printed by Libri Plureos GmbH
in Hamburg, Germany